Sara Hughes · Eric K. Chu
Susan G. Mason
Editors

Climate Change in Cities

Innovations in Multi-Level Governance

Springer

Editors
Sara Hughes
Department of Political Science
University of Toronto
Toronto, ON
Canada

Susan G. Mason
School of Public Service
Boise State University
Boise, ID
USA

Eric K. Chu
School of Geography, Earth and Environmental Sciences
University of Birmingham
Edgbaston, Birmingham
UK

ISSN 2365-757X ISSN 2365-7588 (electronic)
The Urban Book Series
ISBN 978-3-319-65002-9 ISBN 978-3-319-65003-6 (eBook)
https://doi.org/10.1007/978-3-319-65003-6

Library of Congress Control Number: 2017948621

Printed on acid-free paper

This Springer imprint is published by Springer Nature
The registered company is Springer International Publishing AG
The registered company address is: Gewerbestrasse 11, 6330 Cham, Switzerland

Aims and Scope

The Urban Book Series is a resource for urban studies and geography research worldwide. It provides a unique and innovative resource for the latest developments in the field, nurturing a comprehensive and encompassing publication venue for urban studies, urban geography, planning and regional development.

The series publishes peer-reviewed volumes related to urbanization, sustainability, urban environments, sustainable urbanism, governance, globalization, urban and sustainable development, spatial and area studies, urban management, urban infrastructure, urban dynamics, green cities and urban landscapes. It also invites research which documents urbanization processes and urban dynamics on a national, regional and local level, welcoming case studies, as well as comparative and applied research.

The series will appeal to urbanists, geographers, planners, engineers, architects, policy makers, and to all of those interested in a wide-ranging overview of contemporary urban studies and innovations in the field. It accepts monographs, edited volumes and textbooks.

More information about this series at http://www.springer.com/series/14773

The Urban Book Series

Acknowledgements

This book is the result of a set of panels first organized for the 2016 Urban Affairs Association meeting in San Diego, California. The editors would like to thank the organizers of the UAA meeting and the editors at Springer, especially Juliana Pitanguy, for their interest in and support for these panels and the ideas they sparked. The chapter authors would also like to acknowledge those who supported the research that went into their contribution.

George Homsy would like to acknowledge support from the USDA Agriculture and Food Research Initiative with a Foundational Agricultural Economics and Rural Development Grant (#2014-68006-21834).

Ann Dale and coauthors would like to acknowledge that their project, MC^3, is a tri-university collaboration between Royal Roads University, the University of British Columbia, and Simon Fraser University. In addition to the authors of their chapter, the research team includes Stephen Sheppard, Leslie King, Meg Holden, Mark Roseland, and doctoral students Rob Newell, University of Victoria and Alastair Moore, Manchester University. They also gratefully acknowledge funding from the SSHRC Insight Program, which allowed them to continue their research as a longitudinal study re-investigating the original case studies. In the second phase, the authors are exploring the nature and extent of development path change and indicators for transformative change.

Umamaheshwaran Rajasekar and coauthors would like to acknowledge support from the Rockefeller Foundation under their Asian Cities Climate Change Network (ACCCRN) program in India (#2014 CAC 308), as well as Shweta Sharma for providing inputs to the historical climate variability analysis and Lalit Dashora for providing inputs to the smart city interventions.

Jörg Kemmerzell would like to acknowledge support from the German Research Foundation and research team members Arthur Benz, Michèle Knodt, and Anne Tews.

Anna Brown would like to acknowledge project funding from the Rockefeller Foundation.

Julia Bausch and coauthors would like to acknowledge support from the U.S. National Science Foundation under Grant No. 1414052, CNH: The Dynamics of Multi-Scalar Adaptation in Megacities (PI H. Eakin); and a Food and Agriculture Sustainability Research Grant from The Neely Charitable Foundation. All opinions are those of the authors, not the sponsoring agencies. The authors thank the research participants for generously sharing their time, expertise, and perspectives, and Kathryn Kyle for kindly editing the manuscript.

Mitchell Cook and Eric Chu would like to acknowledge the generous field support provided by members of the Surat Municipal Corporation, the Surat Urban Development Authority, the Southern Gujarat Chamber of Commerce and Industry, the Surat Climate Change Trust, TARU Leading Edge, and the Rockefeller Foundation. Also, this research was made possible by financial support from the Department of Urban Studies and Planning, the School of Architecture and Planning, and the Center for International Studies at MIT, the Indian Council for Research on International Economic Relations, and the U.S. Institute for International Education's National Security Education Program.

Ryan Bellinson presented an initial draft of this chapter at the workshop entitled "Contemporary Urban Governance for Mitigation and Adaptation: Mapping, Exploring and Interrogating," which was funded by the European Union COST Action: IS1309—Innovations in Climate Governance: Sources, Patterns and Effects (INOGOV). Comments on this chapter were provided by Jeroen van der Heijden, Sirkku Juhola, Daniel Gabaldón Estevan, Nanja Nagorny, and Eric Chu. The author would like to thank local government officials from Berkeley and Rotterdam for fieldwork support.

Aniessa Sari and Nyoman Prayoga would like to acknowledge the Rockefeller Foundation in funding the Asian Cities Climate Change Resilience Network Program in Indonesia, which allowed opportunities to continuously learn and has influenced the way we do our work in the field. The authors also thank Jennifer Bielman and Malia Robello for giving helpful inputs in completing this paper, to the local partners in Semarang City, and the University of Diponegoro and the BINTARI Foundation in documenting lessons in the project case study.

Contents

Editors and Contributors

About the Editors

Sara Hughes is an Assistant Professor of Political Science at the University of Toronto and the University of Toronto-Mississauga. Her research focuses on urban politics, institutions, and policymaking for water resources, climate change mitigation and adaptation, waste management, and environmental justice.

Eric K. Chu is a Lecturer in Planning and Human Geography in the School of Geography, Earth and Environmental Sciences at the University of Birmingham in the UK. His research is on the politics of climate change governance in cities, with particular emphasis on the globally comparative perspectives of socio-spatial change, development planning, policy change, and local environmental justice.

Susan G. Mason is Emeritus Faculty in the School of Public Service at Boise State University. She was previously Director of Strategic Initiatives for the School of Public Service and Professor and Founding Director of the Department and Master and Certificate programs in Community and Regional Planning at Boise State University. Her research focuses on regional governance, urban and community development, and sustainability.

Contributors

Shujat Ali Ministry of Finance, Islamabad, Pakistan

Julia C. Bausch Arizona State University, Tempe, AZ, USA

Ryan G. Bellinson Department of Urban Studies and Planning, University of Sheffield, Sheffield, United Kingdom

Gopalkrishna Bhat TARU Leading Edge, New Delhi, India

Michael R. Boswell Department of City & Regional Planning, California Polytechnic State University (Cal Poly), San Luis Obispo, CA, USA

Eve Bourgeois University of Toronto, Toronto, Canada

Anna Brown Aequita Consulting, Freeport, ME, USA

Sarah Burch Department of Geography and Environmental Management, University of Waterloo, Ontario, Canada

Soumita Chakraborty TARU Leading Edge, New Delhi, India

Eric K. Chu School of Geography, Earth and Environmental Sciences, University of Birmingham, Birmingham, United Kingdom

Mitchell J. Cook Department of Urban Studies and Planning, Massachusetts Institute of Technology, Cambridge, MA, USA

Ann Dale School of Environment & Sustainability, Royal Roads University, Victoria, Canada

Hallie C. Eakin Arizona State University, Tempe, AZ, USA

Lars A. Engberg Danish Building Research Institute, Aalborg University Copenhagen, Copenhagen, Denmark

George C. Homsy Department of Public Administration, Binghamton University, Binghamton, NY, USA

Sara Hughes University of Toronto, Toronto, Canada

Muhammad Naveed Iftikhar University of Delaware, Newark, DE, USA

Jörg Kemmerzell Institut für Politikwissenschaft, Technische Universität Darmstadt, Darmstadt, Germany

Amy M. Lerner Universidad Nacional Autónoma de México, Mexico City, Mexico

Susan G. Mason School of Public Service, Boise State University, Boise, ID, USA

Scott Morton Ninomiya University of Waterloo, Ontario, Canada

Jacqueline Peterson University of Toronto, Toronto, Canada

Nyoman Prayoga Mercy Corps Indonesia, Jakarta, Indonesia

Umamaheshwaran Rajasekar TARU Leading Edge, New Delhi, India

John Robinson Munk School of Global Affairs and School of the Environment, University of Toronto, Toronto, Canada

Aniessa Delima Sari Mercy Corps Indonesia, Jakarta, Indonesia

Andrea Sarzynski University of Delaware, Newark, DE, USA

Chris Strashok School of Environment & Sustainability, Royal Roads University, Victoria, Canada

Chapter 1
Introduction

Sara Hughes, Eric K. Chu and Susan G. Mason

Abstract Cities are increasingly shaping the trajectory and impacts of climate change. While local actors play a central role in designing the institutions, infrastructures, and behaviors that drive decarbonization and adaptation to changing climatic conditions, their options and incentives are inextricably enmeshed within broader political and economic processes. Resolving these tensions and contradictions is likely to require innovative approaches to governing climate change in the city: new interactions, new political actors, new ways of coordinating and mobilizing resources, and new frameworks and technical capacities for decision-making. This book presents pioneering work on the range of innovative practices, experiments, and ideas that are becoming an integral part of urban climate change governance in the twenty-first century. Theoretically, the book builds on a nearly two-decade history of scholarship identifying the emergence of new urban actors, spaces, and political dynamics in response to climate change. Empirically, the chapters investigate new governance arrangements from around the world and leverage the insights they provide for both theory and practice. The book is organized around four guiding questions: 1) how do multilevel governance arrangements relate to innovation for urban climate change governance? 2) where is the greatest need for innovation? 3) where is innovation difficult or stifled? 4) how can innovation be fostered and encouraged in a multilevel governance context?

Keyword Cities and climate change · Mitigation · Adaptation · Urban governance

S. Hughes (✉)
University of Toronto, Toronto, Canada
e-mail: sara.hughes@utoronto.ca

E.K. Chu
School of Geography, Earth and Environmental Sciences,
University of Birmingham, Edgbaston, Birmingham, UK
e-mail: e.chu@bham.ac.uk

S.G. Mason
School of Public Service, Boise State University, Boise, USA
e-mail: susanmason@boisestate.edu

S. Hughes et al. (eds.), *Climate Change in Cities*, The Urban Book Series,
https://doi.org/10.1007/978-3-319-65003-6_1

1.1 Climate Change in Cities: Innovations in Multilevel Governance

Cities play a critical role in shaping the trajectory and impacts of global climate change. In the most recent Fifth Assessment Report of the Intergovernmental Panel on Climate Change (IPCC), cities were recognized as drivers of transformative action, especially in terms of their ability to address the sectoral, demographic, spatial, and ecological challenges of climate change and extreme disaster risks (Bartlett and Satterthwaite 2016; IPCC 2014; Revi et al. 2014; Ziervogel et al. 2016). Furthermore, as global policymakers are advocating for the concept of "climate change resilience" as an amalgamation of co-benefits between mitigation, adaptation, and sustainable development (Adger et al. 2011), cities are gaining political salience within current global agreements such as the Paris Agreement (2015) of the United Nations Framework Convention on Climate Change (UNFCCC), the Sustainable Development Goals (SDGs), and the New Urban Agenda (2016) of Habitat III (Paris Agreement 2015; Parnell 2016). All of these global frameworks take note of the centrality of urban communities and decision-makers in responding to climate change; however, how to achieve the changes that are necessary in cities in an equitable, accountable, and inclusive way is still unclear.

This book is motivated by two intersecting challenges underlying the task of responding to climate change in cities. First is the multilevel context in which urban governance takes place. The decisions and outcomes observed at the urban scale are often the product of a multitude of actors and forces within the city and at regional, national, and international scales. Governing cities is a collective endeavor, and climate solutions are especially likely to straddle political and jurisdictional boundaries. Urban climate change solutions are necessarily embedded in and produced by multilevel governance contexts. Second is the need for innovative approaches to governing the city, especially in the context of climate change uncertainty, resource, and capacity constraints, as well as increasing urbanization rates at a global scale (Bartlett and Satterthwaite 2016; Rosenzweig et al. 2010). Complex urban governance systems produce and demand innovative political, institutional, economic, spatial, and social approaches to governing the drivers and consequences of climate change in cities. Addressing climate change in a meaningful way therefore requires the development of new patterns and processes of engagement, finance, and collaboration.

The chapters in this volume provide theoretical and empirical insights into the relationship between innovative approaches to urban climate change governance and the multilevel context in which they are embedded. The authors draw on empirical case studies and insights from across the global North and South, and focus explicitly on those innovations that produce new relationships between levels of government, between government and citizens, and between governments, the private sector, and transnational and civil society actors. This comparative, empirical approach highlights the context-dependent dynamics that shape mitigation and adaptation innovation in cities, despite the burgeoning number of global and national climate change directives.

1.2 Scope of the Book

While climate change was initially framed as requiring a coordinated and global response, international policy action has been slow in coming. Instead, a plethora of strategies, frameworks, policies, and funding mechanisms at sub-national levels have dominated the landscape. Primary among these have been actions taken by cities—independent of national directions—to respond to climate change. Many have joined together to form city networks to facilitate joint action, knowledge sharing, and peer accountability, such as in the form of C40 Cities Climate Leadership Group and ICLEI's Cities for Climate Protection program (Betsill and Bulkeley 2004; Hakelberg 2014). Cities have also become the target for donors and nongovernmental organizations with an interest in climate change, such as the World Bank's Cities and Climate Change program and the Rockefeller Foundation's 100 Resilient Cities program. Many cities have also elected to organize regionally or domestically, such as in the case of the Southeast Florida Regional Climate Compact in the US, the Regional Adaptation Collaboratives Program in Canada, the UK Climate Change Impact Programme (UKCIP), the Asian Cities Climate Change Resilience Network (ACCCRN), and many others (Bauer and Steurer 2014; Betsill and Bulkeley 2007; Chu and Schenk 2017). Many of these actions and partnerships recognize the intrinsic value of urban knowledge and capacity, but also highlight the unique challenges of governing climate change at the urban level given the presence of complex political interests, economic priorities, and multi-scalar dynamics.

The emergence of climate change as a distinct global public policy issue corresponded with a resurgence of cities as a unit of analysis in the fields of politics, economics, and development. Across European and North American cities, economic restructuring in the post-World War II era of neoliberalization—with the growing influence of globalized trade, investment, and speculation—promoted private capital as the primary driving force behind municipal politics, planning action, and spatial development (Friedmann 1986; Harvey 1989; Lefebvre 1991). In the global South, trends in democratization meant that many cities were increasingly beneficiaries of devolved budgetary, legislative, and infrastructural powers, but were hamstrung by capacity and governance deficits that were symptomatic of the post-colonial condition (Robinson, 2011; Roy, 2011; Watson, 2009). Despite these divergent experiences, both point to contemporary cities being sites of capital accumulation, contentious politics, and the spatial manifestations of the two (Brenner and Theodore 2005). Climate change discourses therefore arose from this backdrop of concentrated power amongst small numbers of urban economic elites, structural biases towards decentralized network governance approaches, and a persistent "hollowing out" of public sector planning and decision-making authorities (Chu et al. 2016; Himley 2008; Swyngedouw 2004).

While the importance of cities as both socio-political entities and nodes of spatial-material flows is well established, the work of realizing the potential of cities to make a meaningful contribution to climate change mitigation and adaptation is just beginning. This book aims to contribute to the effort to better understand climate change responses in cities, what is missing, what it takes to realize the potential of cities, and the factors and processes at play that are shaping the outcomes we see. A primary aim of this book is to explore the extent to which cities are able to harness and facilitate the innovative and creative potential of multilevel pathways of resource support, capacity development, and authority. By drawing on theories in the fields of public policy, urban planning and administration, governance, and environment and society, the book unpacks the complex governance structures and processes through which climate change innovations arise (or not) in cities across the global North and South.

1.3 Why Multilevel Governance?

A central feature of climate change action in cities is the political and jurisdictional complexity that shapes urban decision-making and its outcomes. Efforts to facilitate transformative change in cities must confront the multiple levels and scales at which urban processes are organized. Spatially, many cities are fragmented, with political boundaries dividing what are otherwise contiguous urban regions. Cities have unique ecologies—ecosystems, built environments, and human communities—that are not clearly bounded and often spatially mismatched with one another (Albrechts 2004; Bai et al. 2010). Coordinating climate change actions across diverse landscapes and populations is challenging due to the geographical specificities of climate risks and impacts, which are determined by particular socio-cultural contexts, political or legal jurisdictions, and ecological conditions (Adger et al. 2013). The interconnectedness of different infrastructure networks across space is compounded by the fragmentation of governance scales and jurisdictions, resulting in numerous agencies and authorities with distinct yet highly interconnected roles and responsibilities.

Cities are typically embedded within wider governance regimes, with responsibilities divided across different levels of government; so many climate change actions require collaboration across jurisdictional boundaries. For example, changing urban mobility behaviors by incentivizing public transportation usage or transit-oriented development is critical for reducing greenhouse gas emissions, but such actions rely on coordinated policies and planning across regions, as transportation networks and urban agglomerations transcend political boundaries (Bollinger et al. 2014). Moreover, the patterns in which communities spread across space—driven by housing availability, zoning and land use planning, and the provision of employment opportunities—impacts travel demands, motor vehicle dependency, and greenhouse gas emissions levels (Dulal et al. 2011).

In other cases, the trans-boundary nature of infrastructure networks influences the ability of cities to coordinate climate adaptation and risk management actions across space (Davoudi et al. 2009; Zimmerman and Faris 2010). As a result, since cities cannot tackle these issues as standalone stressors on specific locations (Hallegatte 2009), they must instead approach climate change as portfolios of systemic risks on infrastructure networks that stretch across political boundaries. Beyond trans-boundary infrastructure, addressing other issues—including land use, biodiversity conservation, and solid waste management networks—is key to facilitating climate action, yet similarly hobbled by fragmented governance (Grimm et al. 2008).

The effectiveness of many climate change programs also hinges on the ability to coordinate across political jurisdictions due to the presence of trans-boundary risks —such as sea level rise and storm surges—that span ecosystems and infrastructure networks (Bollinger et al. 2014; Corfee-Morlot et al. 2010). The complexities around scale and space have profound implications for who actually has control over relevant mitigation or adaptation decisions, which in turn informs the effectiveness and legitimacy of policy-making and planning processes (Moser 2006). To design climate change interventions that are appropriately sized and scaled, cities must be able to bridge the trans-boundary and multi-scalar nature of climate change actions. The positive or negative perceptions of the other actors involved (Downs 1967; Wright 1988) and the pursuit of similar or divergent goals (Pressman 1975; Wright 1988) will in turn affect outcomes across boundaries. Furthermore, social institutions in the form of neighborhood organizations or more complex institutions such as multi-national organizations can resolve uncertainty and stabilize anticipate actions and outcomes (Knight 1992), fostering acceptable outcomes in trans-boundary and multi-scalar actions. Polycentric systems can similarly be beneficial for dealing with climate change as they have the potential to generate benefits at multiple scales (Ostrom 2010).

Scholars of multilevel governance argue that this "stretching" of planning and policy-making authority happens horizontally—i.e., across jurisdictional boundaries in space—*and* vertically between local, regional, national, and global levels of government (Bulkeley and Betsill 2005; Hooghe and Marks 2003; Sellers 2002). In an era of urban political and economic restructuring, control over many urban planning and policy-making responsibilities is increasingly devolved to non-state, network, or extra-local actors and forces (Harvey 1989; McFarlane 2009). For example, as already noted earlier, transnational networks provide necessary capacity and resource support for urban climate change actions (Fünfgeld 2015; Hakelberg 2014). The private sector also plays a variety of key roles (Mees et al. 2012; Tompkins and Eakin 2012). For instance, water and electricity systems are often privately owned or managed, yet are both integral to communities and intertwined with other infrastructure systems. Recent priorities around urban climate change resilience have favored economically important sectors, such as in the case of protecting financial institutions in central business districts from extreme risks rather than providing much needed risk reduction services to low-income vulnerable neighborhoods (Anguelovski et al. 2016; Shi et al. 2016). Recent scholarship on "splintering urbanism", notes that the privatized enclosure and increasing specialization of infrastructure sectors are exacerbating governance fragmentation and shortfalls in public service provision (Graham and Marvin 2001). These examples highlight how urban climate change action—especially when placed within political contexts with heightened ideological, economic, and political differences—may produce opportunities for elite capture, policy confusion, corruption, or inequitable outcomes.

With these governance challenges in mind, policy and planning responses to climate change in cities across the global North and South are often fraught with tension and contradictions (Hughes 2017; Rosenzweig et al. 2010). While on one hand local actors play a central role in designing institutions, infrastructures, and behaviors that drive decarbonization and adaptation to changing climatic conditions, their options and incentives are inextricably enmeshed within broader political, economic, and ecological processes (Bulkeley 2010). For example, some have argued that directives for rapid decarbonization as stipulated by the Paris Agreement (2015) may lead to stranded infrastructure and capital assets as many urban and regional production systems are already locked into the fossil fuel economy (Ansar et al. 2013; Gupta et al. 2017; Richels et al. 2009).

To facilitate climate change governance in cities, local authorities must share the political arena with actors ranging from private actors to transnational networks. Within this cacophony of competing interests, cities must find ways to appropriately balance the scope—in terms of both space and scale—of any climate change mitigation and adaptation action. A more comprehensive understanding is therefore needed of the innovative approaches being used to navigate the complex networks and relationships that constitute contemporary multilevel urban climate change governance.

1.4 Why Innovations?

To resolve the tensions and contradictions that arise in urban multilevel governance, many have argued that innovative approaches will be required to govern climate change mitigation and adaptation in cities. Innovation is "the intentional and proactive process that involves the generation and practical adoption and spread of new and creative ideas which aim to produce a qualitative change in a specific context" (Sørensen and Torfing 2011). Innovations, in the context of urban climate change governance, refer to new interactions, new political actors, new ways of coordinating and mobilizing resources, and new frameworks and technical capacities for decision-making. In most cases, innovation requires collaborative interactions between different public and private actors because municipal resources—such as financial resources, political capacity, and knowledge supply—are finite and often insufficient to meet the cross-sectoral demands of mitigation and adaptation (Anguelovski and Carmin 2011; Bulkeley and Betsill 2013). Innovations can be shaped by the multilevel governance context in which they are conceived or implemented (such as new financing tools or building design) or embodied in a novel governance arrangement itself (such as new mechanisms for citizen engagement or intergovernmental collaboration).

Though there is a long history of innovation scholarship in the domains of national economic and industrial policy (Freeman 1991; Nelson 1993), there is relatively little on innovation at the urban scale, especially beyond the fields of economic geography (see Walker and Storper 1989, for example), cultural studies (see Evans 2009, for example), and strategic urban planning (see MacCullum et al. 2009, for example). Recent work on climate change has begun to theorize the genesis and implications of urban innovations, particular through the lens of experimentation or urban "living labs" (Bulkeley, Castán Broto, and Edwards 2015; Evans and Karvonen 2014; Chu 2016). Cities often face challenges with initiating and sustaining innovation due to lack of political support, financial resources, and human capacity (Anguelovski et al. 2014; Gordon 2013). The availability of these resources, and the likelihood of urban experiments to scale up or out will be shaped by the multilevel governance context in which they are embedded. Changes in multilevel governance arrangements can drive innovation by uncoupling cities from their traditional planning models, experimenting with alternative decision-making pathways, and helping to overcome bureaucratic constraints within city governments (Gordon 2013; Okereke et al. 2009; Sørensen and Torfing 2011). Multilevel governance can also contribute to the cycle of innovation in cities by fostering cross-sectoral and inter-municipal collaboration (Bulkeley and Betsill 2013; Massey et al. 2014) and leading by example such as in the case of states and regional agencies (Bedsworth and Hanak 2010).

Innovations in multilevel governance arrangements themselves also have the potential to address structural constraints in cities by facilitating decision-making and knowledge dissemination pathways that bridge the siloed nature of urban policy and planning, assist with accessing additional resources and capacity, and gain stakeholder support from a wider variety of urban actors (Giest and Howlett 2013; Jordan and Huitema 2014; Measham et al. 2011; Moore and Hartley 2008). Such innovations are catalyzed in a number of ways. In some instances, regional or strategic initiatives will be driven by a top-down approach, while others will be from a local level or bottom-up (Baker 2001). Innovation can be driven by crises—such as a resource shortage—directed by citizen involvement, or a product of perceived mutually beneficial outcomes.

However, there is much to be learned about the nature of governance innovations and the outcomes they produce. Some argue for centralized innovation, or for federal or nation-state level policy to provide a climate plan to acquire cooperation across multiple scales (Nelson et al. 2015). Others suggest an ongoing decision support capacity at the regional level (Dilling and Berggren 2015) or climate change alliances that can reshape the landscape crating new governance regimes (Moloney and Fünfgeld 2015). Better guidance on identifying vulnerabilities to climate change and approaches to enhancing future development investments for resiliency across countries in the global South could be an important type of innovation (Carmin et al. 2013; Furlow et al. 2011). Understanding how innovations at multiple levels of governance interact with capacity (Low et al. 2003), or the mechanisms of social and societal learning could also prove fruitful to providing flexible tools to help address climate change in cities (Pahl-Wostl 2009).

With these recent theoretical and policy advancements in mind, this book aims to take an empirical deep dive into whether and how cities utilize or facilitate innovation in multilevel urban governance systems.

1.5 Guiding Questions

The interactions between multilevel governance and innovation have the potential to significantly shape urban climate responses, and their investigation is a crucial component of any effort to support cities in their efforts. The chapters in this book explore four central questions:

1) *How do multilevel governance arrangements relate to innovation for urban climate change governance?* The chapters in this book examine the multiple intersections and outcomes of multilevel governance arrangements and urban climate change innovation. The case studies and theoretical developments help to establish baseline knowledge of the relationships and the dimensions along which we might characterize it. The chapters explore the ways in which multilevel governance arrangements influence innovation in urban climate response, and the extent to which new governance approaches can in and of themselves be considered important innovations. Using a variety of methods and theoretical approaches, the chapter authors explore the many intersections of multilevel governance and innovation.
2) *Where is the greatest need for innovation?* Urban climate change innovations in multilevel governance are still in their infancy. As we gain more experience with new approaches, it is important that we take stock of the kinds of innovations that are most needed and the scale of change that is required in different dimensions. Given the scope of climate change drivers and solutions, there may be no one "greatest" area or need for innovation but rather a collective effort at each level for action. In the end, the most significant innovations are those that result in substantial reductions in greenhouse gas emissions or climate risks and impacts, as well as concerted efforts that galvanize the trust of actors across scale to commit to long-term outcomes while realizing short-term benefits.
3) *Where is innovation difficult or stifled?* This book starts from the assumption that innovation is a necessary ingredient for effective and equitable urban climate change responses. The chapters help to empirically trace the dynamics of change and improve our understanding of the role that multilevel governance arrangements play in inhibiting or preventing greater innovation. They help to lay a foundation for theorizing the conditions within which innovations do or do not transpire, their respective implications for governance change, as well as their outcomes in terms of equity and inclusiveness. The chapters help to reveal common barriers to achieving the kinds of innovations that are necessary for meaningful climate change responses in cities.
4) *How can innovation be fostered and encouraged in a multilevel governance context?* There is demand from practitioners and advocates for greater insight into the strategies and institutions that are able to best foster the innovative responses to climate change that are needed despite—or perhaps even as a result of—the multilevel context in which they are necessarily embedded. The chapters in this volume take up this challenge and provide key insights into where innovation has been successful and ways that further progress can be made going forward.

1.6 The Chapters

This book is divided into four main parts. Part I contains four chapters on intergovernmental governance innovations. Chapter 2 by Homsy argues that although there is a strong tradition of local discretion in the United States, the complexity of climate change action means smaller municipalities are more likely to engage in climate change action when their states are also acting. Chapter 3 by Kemmerzell examines European innovations in governance in the major cities of Germany. The author examines whether multilevel governance structures—specifically the Covenant of Mayors—have an impact on local climate policy. The findings reveal that both hierarchical and lateral activities are having an impact on climate policy and the Covenant of Mayors is not a driving force but rather an ancillary factor motivating action. Chapter 4 by Boswell and Mason examines the impact of the Sustainable Communities Strategy (SCS) on coordinating city and regional climate change actions in California. The authors find mid-sized to larger cities are more likely to coordinate their policies with the region than smaller cities, but that ultimately it may be too early to tell if the SCS is a driving factor in the intergovernmental innovations. Chapter 5 by Bourgeois and Hughes examines how the degree of centralization of decision-making in metropolitan Montreal creates trade-offs for climate change policy innovation and democratic decision-making. Looking specifically at the waste management sector, the authors find the more centralized the decision-making, the less autonomy and engagement is elicited from citizen participation.

Part II has four chapters devoted to innovations in citizen engagement. Chapter 6 by Sarzynski examines the nationally recognized policy and planning efforts of the City of Baltimore, Maryland, which has innovated by combining disaster preparedness with a climate change adaptation plan. The Baltimore case reveals both the resiliency of staff when working on climate adaptation planning and the challenges of obtaining community-wide ownership for the action in the plan. Chapter 7 by Sari and Prayoga describes two projects in Semarang, Indonesia, that highlight new mechanisms for improving communication pathways within cities for local public health and environmental needs. These mechanisms include the use of digital technology coupled with citizen engagement. In both projects, citizen engagement is needed to guarantee success and local knowledge, networks, and community motivation are important. Citizen awareness of governance structures is a key ingredient for success. Chapter 8 by Engberg focuses on the Danish experience, and the national government's efforts to manage large scale project needs. The author looks specifically at the case of water management and the local level impacts of collaboration with citizens in Copenhagen. Chapter 9 by Iftikhar, Ali, and Sarzynski also focuses on water using the case of Bhalwal, Pakistan. Here, the community-government partnership-based initiative succeeded in a situation where the performance of the traditional government-managed scheme was weak. The authors analyze the reason for the initiative's success and the challenges of duplicating the clean water project for other types of services or locations in Pakistan or similar urbanizing nations.

Part III focuses on innovations and city networks. Chapter 10 by Bellinson examines how transnational municipal networks influence local government climate policy processes and promote urban adaptation actions using the cases of Rotterdam, the Netherlands, and Berkeley, California, USA. The author finds communication, coalition building, and confronting conflicts to be key ingredients for adopting adaptive innovations. In Chap. 11, Rajasekar, Charkraborty, and Bhat focus on the case of India, which is actively trying to incorporate urban climate change into emerging "smart city" initiatives across the national and local government levels, but is faced with many obstacles. In Chap. 12, Brown unpacks the Asian Cities Climate Change Resilience Network (ACCCRN) successes in helping cities across South and Southeast Asia to be more resilient to climate shocks and stressors. The chapter provides insight into the way facilitating organizations can help cities navigate a path to processes that work. In Chap. 13, Cook and Chu examine the case of Surat, India, and the way fiscal constraints can be a barrier to climate change action, as well as how the city is developing new pathways for steering funding for climate change adaptation.

Part IV outlines the drivers and obstacles of multilevel innovations. Chapter 14 by Peterson reviews the relationship between municipal finance and the broader multilevel frameworks that govern climate change decision-making in cities. Focusing primarily on the United States, Peterson identifies some of the opportunities cities have to pursue innovations in municipal finance to support their climate policy objectives. Chapter 15 by Bausch, Eakin, and Lerner details how Mexico City has innovated to incorporate peri-urban agriculture into their climate change policies. The authors provide recommendations on how to successfully articulate peri-urban agriculture into the larger climate change dynamic. They also provide strategies for mediating trade-offs in order to obtain the desired program results. Chapter 16 by Ninomiya and Burch examines the case of Waterloo, Canada, illustrating how vital experimentation is for innovation in ideas prior to scaling up. The authors highlight the vital role that a forum can play in providing new actors and participatory process with a starting place and an opportunity to continue innovations in local energy systems. Chapter 17 by Dale and colleagues reflect on experiences working with "climate innovators" across British Columbia, Canada, and show how engaging local communities is essential for facilitating climate innovations in the absence of national-level directives.

The final chapter in the book synthesizes the insights provided by the rich diversity of cases and perspectives brought together in this volume. We refer back to our four guiding questions, and identify three important research needs in this area going forward: the institutional foundations for urban innovations, unbounding the urban in climate governance, and resisting the post-politics of climate innovations.

References

Adger, W. N., Barnett, J., Brown, K., Marshall, N., & O'Brien, K. (2013). Cultural dimensions of climate change impacts and adaptation. *Nature Climate Change, 3*(2), 112–117. doi:10.1038/nclimate1666.

Adger, W. N., Brown, K., Nelson, D. R., Berkes, F., Eakin, H., Folke, C., et al. (2011). Resilience implications of policy responses to climate change. *Wiley Interdisciplinary Reviews: Climate Change, 2*(5), 757–766. doi:10.1002/wcc.133.

Albrechts, L. (2004). Strategic (Spatial) planning reexamined. *Environment and Planning B: Planning and Design, 31*(5), 743–758. doi:10.1068/b3065.

Anguelovski, I., & Carmin, J. (2011). Something borrowed, everything new: Innovation and institutionalization in urban climate governance. *Current Opinion in Environmental Sustainability, 3*(3), 169–175. doi:10.1016/j.cosust.2010.12.017.

Anguelovski, I., Chu, E., & Carmin, J. (2014). Variations in approaches to urban climate adaptation: Experiences and experimentation from the global South. *Global Environmental Change, 27,* 156–167. doi:10.1016/j.gloenvcha.2014.05.010.

Anguelovski, I., Shi, L., Chu, E., Gallagher, D., Goh, K., Lamb, Z., et al. (2016). Equity impacts of urban land use planning for climate adaptation: Critical perspectives from the global north and south. *Journal of Planning Education and Research, 36*(3), 333–348. doi:10.1177/0739456X16645166.

Ansar, A., Caldecott, B., & Tilbury, J. (2013). *Stranded assets and the fossil fuel divestment campaign: What does divestment mean for the valuation of fossil fuel assets.* Oxford, UK: Stranded Assets Programme, SSEE, University of Oxford.

Bai, X., McAllister, R. R., Beaty, R. M., & Taylor, B. (2010). Urban policy and governance in a global environment: Complex systems, scale mismatches and public participation. *Current Opinion in Environmental Sustainability, 2*(3), 129–135. doi:10.1016/j.cosust.2010.05.008.

Baker, M. (2001). Some reflections on strategic planning processes in three urban regions. *Planning Theory & Practice, 2*(2), 230–234. doi:10.1080/14649350120068849.

Bartlett, S., & Satterthwaite, D. (Eds.). (2016). *Cities on a finite planet: Towards transformative responses to climate change*. Routledge.

Bauer, A., & Steurer, R. (2014). Innovation in climate adaptation policy: Are regional partnerships catalysts or talking shops? *Environmental Politics, 23*(5), 818–838. doi:10.1080/09644016.2014.924196.

Bedsworth, L. W., & Hanak, E. (2010). Adaptation to climate change: A review of challenges and tradeoffs in six areas. *Journal of the American Planning Association, 76*(4), 477–495. doi:10.1080/01944363.2010.502047.

Betsill, M. M., & Bulkeley, H. (2004). Transnational networks and global environmental governance: The cities for climate protection program. *International Studies Quarterly, 48*(2), 471–493. doi:10.1111/j.0020-8833.2004.00310.x.

Betsill, M. M., & Bulkeley, H. (2007). Looking back and thinking ahead: A decade of cities and climate change research. *Local Environment, 12*(5), 447–456. doi:10.1080/13549830701659683.

Bollinger, L. A., Bogmans, C. W. J., Chappin, E. J. L., Dijkema, G. P. J., Huibregtse, J. N., Maas, N., et al. (2014). Climate adaptation of interconnected infrastructures: A framework for supporting governance. *Regional Environmental Change, 14*(3), 919–931. doi:10.1007/s10113-013-0428-4.

Brenner, N., & Theodore, N. (2005). Neoliberalism and the urban condition. *City, 9*(1), 101–107. doi:10.1080/13604810500092106.

Bulkeley, H. (2010). Cities and the governing of climate change. *Annual Review of Environment and Resources, 35*(1), 229–253. doi:10.1146/annurev-environ-072809-101747.

Bulkeley, H., & Betsill, M. (2005). Rethinking sustainable cities: Multilevel governance and the "Urban" politics of climate change. *Environmental Politics, 14*(1), 42–63. doi:10.1080/0964401042000310178.

Bulkeley, H., & Betsill, M. M. (2013). Revisiting the urban politics of climate change. *Environmental Politics, 22*(1), 136–154. doi:10.1080/09644016.2013.755797.

Bulkeley, H., Castán Broto, V., & Edwards, G. A. S. (2015). *An urban politics of climate change: Experimentation and the governing of Socio-technical transitions*. New York and London: Routledge.

Carmin, J., Dodman, D., & Chu, E. (2013). *Urban Climate Adaptation and Leadership: From Conceptual Understanding to Practical Action* (No. 2013/26). Paris, France: Organisation for Economic Co-operation and Development (OECD). doi:10.1787/5k3ttg88w8hh-en.

Chu, E., Anguelovski, I., & Carmin, J. (2016). Inclusive approaches to urban climate adaptation planning and implementation in the Global South. *Climate Policy, 16*(3), 372–392. doi:10.1080/14693062.2015.1019822.

Chu, E. K. (2016). The governance of climate change adaptation through urban policy experiments. *Environmental Policy and Governance, 26*(6), 439–451. doi:10.1002/eet.1727.

Chu, E., & Schenk, T. (2017). Communicating about climate change with urban populations and decision-makers. In M. C. Nisbet (Ed.), *Oxford research encyclopedia of climate science* (Vol. 1). Oxford University Press: Oxford and New York. doi:10.1093/acrefore/9780190228620.013.413.

Corfee-Morlot, J., Cochran, I., Hallegatte, S., & Teasdale, P.-J. (2010). Multilevel risk governance and urban adaptation policy. *Climatic Change, 104*(1), 169–197. doi:10.1007/s10584-010-9980-9.

Davoudi, S., Crawford, J., & Mehmood, A. (Eds.). (2009). *Planning for climate change: Strategies for mitigation and adaptation for spatial planners*. London and Sterling, VA: Earthscan.

Dilling, L., & Berggren, J. (2015). What do stakeholders need to manage for climate change and variability? A document-based analysis from three mountain states in the Western USA. *Regional Environmental Change, 15*(4), 657–667. doi:10.1007/s10113-014-0668-y.

Downs, A. (1967). *Inside bureaucracy*. Boston, MA: Little, Brown and Company.

Dulal, H. B., Brodnig, G., & Onoriose, C. G. (2011). Climate change mitigation in the transport sector through urban planning: A review. *Habitat International, 35*(3), 494–500. doi:10.1016/j.habitatint.2011.02.001.

Evans, G. (2009). Creative cities, creative spaces and urban policy. *Urban Studies, 46*(5–6), 1003–1040. doi:10.1177/0042098009103853.

Evans, J., & Karvonen, A. (2014). "Give Me a Laboratory and I Will Lower Your Carbon Footprint!"—Urban laboratories and the Governance of Low-carbon futures. *International Journal of Urban and Regional Research, 38*(2), 413–430. doi:10.1111/1468-2427.12077.

Freeman, C. (1991). Networks of innovators: A synthesis of research issues. *Research Policy, 20* (5), 499–514. doi:10.1016/0048-7333(91)90072-X.

Friedmann, J. (1986). The world city hypothesis. *Development and Change, 17*(1), 69–83. doi:10.1111/j.1467-7660.1986.tb00231.x.

Fünfgeld, H. (2015). Facilitating local climate change adaptation through transnational municipal networks. *Current Opinion in Environmental Sustainability, 12,* 67–73. doi:10.1016/j.cosust.2014.10.011.

Furlow, J., Smith, J. B., Anderson, G., Breed, W., & Padgham, J. (2011). Building resilience to climate change through development assistance: USAID's climate adaptation program. *Climatic Change, 108*(3), 411–421. doi:10.1007/s10584-011-0127-4.

Giest, S., & Howlett, M. (2013). Comparative climate change governance: Lessons from european transnational municipal network management efforts. *Environmental Policy and Governance, 23*(6), 341–353. doi:10.1002/eet.1628.

Gordon, D. J. (2013). Between local innovation and global impact: Cities, networks, and the governance of climate change. *Canadian Foreign Policy Journal, 19*(3), 288–307. doi:10.1080/11926422.2013.844186.

Graham, S., & Marvin, S. (2001). *Splintering urbanism: Networked infrastructures, technological mobilities and the urban condition*. London and New York: Routledge.

Grimm, N. B., Faeth, S. H., Golubiewski, N. E., Redman, C. L., Wu, J., Bai, X., et al. (2008). Global change and the ecology of cities. *Science, 319*(5864), 756–760. doi:10.1126/science.1150195.

Gupta, J., Chu, E., Bos, K., & Kuijten, T. (2017). The Geo-ecological risks of oil investments by China and the global south: The right to development revisited. In M. P. Amineh & G. Yang (Eds.), *Geopolitical economy of energy and environment: China and the European Union* (pp. 271–304). Leiden and Boston: Brill Publishing.

Hakelberg, L. (2014). Governance by diffusion: Transnational municipal networks and the spread of local climate strategies in europe. *Global Environmental Politics, 14*(1), 107–129. doi:10.1162/GLEP_a_00216.

Hallegatte, S. (2009). Strategies to adapt to an uncertain climate change. *Global Environmental Change, 19*(2), 240–247. doi:10.1016/j.gloenvcha.2008.12.003.

Harvey, D. (1989). From managerialism to entrepreneurialism: The transformation in urban governance in late capitalism. *Geografiska Annaler, 71*(1), 3–17.

Himley, M. (2008). Geographies of environmental governance: The nexus of nature and neoliberalism. *Geography Compass, 2*(2), 433–451. doi:10.1111/j.1749-8198.2008.00094.x.

Hooghe, L., & Marks, G. (2003). Unraveling the central state, but how? Types of Multi-level governance. *American Political Science Review, 97*(2), 233–243. doi:10.1017/S0003055403000649.

Hughes, S. (2017). The politics of urban climate change policy. *Urban Affairs Review, 53*(2), 362–380. doi:10.1177/1078087416649756.

IPCC. (2014). Climate change 2014: Impacts, adaptation, and vulnerability. contribution of working group II to the fifth assessment report of the intergovernmental panel on climate change.In C. B. Field, V. R. Barros, D. J. Dokken, K. J. Mach, M. D. Mastrandrea, T. E. Bilir, et al., (Eds.), Cambridge and New York: Cambridge University Press.

Jordan, A., & Huitema, D. (2014). Innovations in climate policy: The politics of invention, diffusion, and evaluation. *Environmental Politics, 23*(5), 715–734. doi:10.1080/09644016.2014.923614.

Knight, J. (1992). *Institutions and social conflict*. Cambridge: Cambridge University Press.

Lefebvre, H. (1991). *The production of space*. Oxford, UK and Malden MA, USA: Wiley-Blackwell.

Low, B., Ostrom, E., Simon, C., & Wilson, J. (2003). Redundancy and diversity: Do they influence optimal management? In F. Berkes, J. Colding, & C. Folke (Eds.), *Navigating Social-Ecological Systems: Building Resilience for Complexity and Change* (pp. 83–114). London: Cambridge University Press.

MacCullum, D., Moulaert, F., Hillier, J., & Haddock, S. V. (Eds.). (2009). *Social innovation and territorial development*. London: Routledge.

Massey, E., Biesbroek, R., Huitema, D., & Jordan, A. (2014). Climate policy innovation: The adoption and diffusion of adaptation policies across Europe. *Global Environmental Change, 29,* 434–443. doi:10.1016/j.gloenvcha.2014.09.002.

McFarlane, C. (2009). Translocal assemblages: Space, power and social movements. *Geoforum, 40*(4), 561–567. doi:10.1016/j.geoforum.2009.05.003.

Measham, T. G., Preston, B. L., Smith, T. F., Brooke, C., Gorddard, R., Withycombe, G., et al. (2011). Adapting to climate change through local municipal planning: Barriers and challenges. *Mitigation and Adaptation Strategies for Global Change, 16*(8), 889–909. doi:10.1007/s11027-011-9301-2.

Mees, H. L. P., Driessen, P. P. J., & Runhaar, Ha C. (2012). Exploring the scope of public and private responsibilities for climate adaptation. *Journal of Environmental Policy & Planning, 14*(3), 305–330. doi:10.1080/1523908X.2012.707407.

Moloney, S., & Fünfgeld, H. (2015). Emergent processes of adaptive capacity building: Local government climate change alliances and networks in Melbourne. *Urban Climate, 14,* 30–40. doi:10.1016/j.uclim.2015.06.009.

Moore, M., & Hartley, J. (2008). Innovations in governance. *Public Management Review, 10*(1), 3–20. doi:10.1080/14719030701763161.

Moser, S. C. (2006). Talk of the city: Engaging urbanites on climate change. *Environmental Research Letters, 1*(1), 14006. doi:10.1088/1748-9326/1/1/014006.

Nelson, H. T., Rose, A., Wei, D., Peterson, T., & Wennberg, J. (2015). Intergovernmental climate change mitigation policies: Theory and outcomes. *Journal of Public Policy, 35*(1), 97–136. doi:10.1017/S0143814X14000117.

Nelson, R. R. (Ed.). (1993). *National innovation systems: A comparative analysis*. Oxford and New York: Oxford University Press.

Okereke, C., Bulkeley, H., & Schroeder, H. (2009). Conceptualizing climate governance beyond the international regime. *Global Environmental Politics, 9*(1), 58–78. doi:10.1162/glep.2009.9.1.58.

Ostrom, E. (2010). Polycentric systems for coping with collective action and global environmental change. *Global Environmental Change, 20*(4), 550–557. doi:10.1016/j.gloenvcha.2010.07.004.

Pahl-Wostl, C. (2009). A conceptual framework for analysing adaptive capacity and multi-level learning processes in resource governance regimes. *Global Environmental Change, 19*(3), 354–365. doi:10.1016/j.gloenvcha.2009.06.001.

Paris Agreement. (2015). *The Paris Agreement under the United Nations Framework Convention on Climate Change (UNFCCC)*. United Nations.

Parnell, S. (2016). Defining a global urban development agenda. *World Development, 78*, 529–540. doi:10.1016/j.worlddev.2015.10.028.

Pressman, J. L. (1975). Federal programs and city politics: *The dynamics of the aid process in Oakland*. Berkeley, CA: University of California Press.

Revi, A., Satterthwaite, D., Aragon-Durand, F., Corfee-Morlot, J., Kiunsi, R. B. R., Pelling, M., et al. (2014). Towards transformative adaptation in cities: The IPCC's fifth assessment. *Environment and Urbanization, 26*(1), 11–28. doi:10.1177/0956247814523539.

Richels, R. G., Blanford, G. J., & Rutherford, T. F. (2009). International climate policy: A "second best" solution for a "second best" world? *Climatic Change, 97*(1–2), 289–296. doi:10.1007/s10584-009-9730-z.

Robinson, J. (2011). Cities in a world of cities: The comparative gesture. *International Journal of Urban and Regional Research, 35*(1), 1–23. doi:10.1111/j.1468-2427.2010.00982.x.

Rosenzweig, C., Solecki, W., Hammer, S. A., & Mehrotra, S. (2010). Cities lead the way in climate–change action. *Nature, 467*(7318), 909–911. doi:10.1038/467909a.

Roy, A. (2011). Urbanisms, worlding practices and the theory of planning. *Planning Theory, 10*(1), 6–15. doi:10.1177/1473095210386065.

Sellers, J. M. (2002). The Nation-State and urban governance: Toward multilevel analysis. *Urban Affairs Review, 37*(5), 611–641. doi:10.1177/107808740203700501.

Shi, L., Chu, E., Anguelovski, I., Aylett, A., Debats, J., Goh, K., et al. (2016). Roadmap towards justice in urban climate adaptation research. *Nature Climate Change, 6*(2), 131–137. doi:10.1038/nclimate2841.

Sørensen, E., & Torfing, J. (2011). Enhancing collaborative innovation in the public sector. *Administration & Society, 43*(8), 842–868. doi:10.1177/0095399711418768.

Swyngedouw, E. (2004). Globalisation or "glocalisation"? Networks, territories and rescaling. *Cambridge Review of International Affairs, 17*(1), 25–48. doi:10.1080/0955757042000203632.

Tompkins, E. L., & Eakin, H. (2012). Managing private and public adaptation to climate change. *Global Environmental Change, 22*(1), 3–11. doi:10.1016/j.gloenvcha.2011.09.010.

Walker, R., & Storper, M. (1989). *The Capitalist Imperative: Territory*. Wiley-Blackwell: Technology and Industrial Growth.

Watson, V. (2009). Seeing from the South: Refocusing urban planning on the globe's central urban issues. *Urban Studies, 46*(11), 2259–2275. doi:10.1177/0042098009342598.

Wright, D. S. (1988). *Understanding intergovernmental relations* (3rd ed.). Pacific Grove, CA: Brooks/Cole Publishing Company.

Ziervogel, G., Cowen, A., & Ziniades, J. (2016). Moving from adaptive to transformative capacity: Building foundations for inclusive, thriving, and regenerative urban settlements. *Sustainability, 8*(9), 955. doi:10.3390/su8090955.

Zimmerman, R., & Faris, C. (2010). Chapter 4: Infrastructure impacts and adaptation challenges. *Annals of the New York Academy of Sciences,* 1196, 63–85. doi:10.1111/j.1749-6632.2009.05318.x.

Author Biographies

Sara Hughes is an Assistant Professor of Political Science at the University of Toronto and the University of Toronto-Mississauga. She is also a Senior Fellow at the Global Cities Institute and Bissell-Heyd Fellow at the Center for the Study of the United States at the University of Toronto. Her research focuses on the drivers of urban environmental politics and policy-making surrounding water resources, climate change mitigation and adaptation, waste management, and environmental justice.

Eric K. Chu is a Lecturer in Planning and Human Geography in the School of Geography, Earth and Environmental Sciences at the University of Birmingham in the UK. His research is on the politics of climate change governance in cities, with particular emphasis on the globally comparative perspectives of socio-spatial change, development planning, policy reform, and local environmental justice.

Susan G. Mason is Emeritus Faculty in the School of Public Service at Boise State University. She was previously Director of Strategic Initiatives for the School of Public Service and Professor and Founding Director of the Department and Master and Certificate programs in Community and Regional Planning at Boise State University. Her research focuses on regional governance, urban and community development, and sustainability.

Part I
Intergovernmental Governance Innovations

Chapter 2
Size, Sustainability, and Urban Climate Planning in a Multilevel Governance Framework

George C. Homsy

Abstract In the United States, the absence of federal leadership on climate change and a strong tradition of localism has created a system in which many greenhouse gas reduction efforts fall to the discretion of municipalities. This often leads to uncoordinated action across jurisdictional boundaries. Despite the widespread notion that cities can lead on climate policy from the bottom-up, I find, using a logistic regression analysis of data from 1837 municipalities, that local governments are more likely to enact climate change policies in an environment where higher levels of government have acted rather than in a decentralized one. Smaller municipalities, in particular, have increased odds of action when their states act. Using existing regional, state-based initiatives, I present options for a coordination and capacity building framework.

Keywords Local government · Climate change · Urban policymaking · Regionalism

2.1 Introduction

For much of its history, the United States has had trouble dealing with challenges of the commons or common-pool resources, especially pollution and natural resource protection. The rapid industrialization of the United States following the Second World War came with horrendous water and air pollution; rivers caught fire and deadly smog suffocated regions with pollution flowing easily across jurisdictional borders. In 1948, thick air pollution originating in Donora, Pennsylvania's zinc industry killed 13 people and sickened thousands in that city and downwind in the neighboring city of Webster (Snyder 1994). Municipalities pumped wastewater into the rivers from which downstream neighbors pulled their drinking water (Holloway et al. 2014). Local leaders were unwilling to shoulder cleanup costs or impose them

G.C. Homsy (✉)
Department of Public Administration, Binghamton University, 13902 Binghamton, NY, USA
e-mail: ghomsy@binghamton.edu

S. Hughes et al. (eds.), *Climate Change in Cities*, The Urban Book Series,
https://doi.org/10.1007/978-3-319-65003-6_2

on industries that threatened to close factories and cut jobs (Andreen 2003). Starting in 1970, environmental protection over some issues in the United States was nationalized through the passage of more than a dozen new federal statues (Andrews 2006). The top-down imposition of command and control regulation cleaned much of the worst air and water pollution in the United States (Fiorino 2006).

Today, climate change represents "the ultimate commons problem" (Stavins 2010) (see Chap. 6: Sarzynski). One approach to governing climate change involves the top-down imposition on local governments of rules by a central authority, such as a national government. This has made the U.S. a leader in environmental cleanup (Fiorino 2006). However, this approach does not work well in dealing with complex problems (Kettl 2002). Centralized, expert-driven solutions usually view all problems as if they are technical puzzles (Fiorino 2006) to be broken down like a machine and fixed piece-by-piece (Innes and Booher 2001). This approach is easy to administer, but does not reflect the complexity of the real world (Agranoff and McGuire 2003). Central mandates are traditionally command and control regulations, which provide little flexibility for adjusting to local conditions (Mazmanian and Kraft 1999). And, in the case of climate change, national-level climate policies in the United States are weak to non-existent. In 2015, President Barack Obama's Clean Power Plan, which imposed federal regulations on the coal-fired power plants, also issued CO_2 emission goals for states. However, implementation has been delayed by lawsuits and the administration of President Donald Trump has begun to dismantle the plan.

The second approach to environmental protection focuses on local solutions without intervention from a central authority. A number of municipalities have undertaken greenhouse gas reduction efforts and the communities at this level of government have the potential to be important actors in the greenhouse gas reduction effort (Kousky and Schneider 2003; Gore and Robinson 2009). Developed as a theory of small-scale, common-pool resource management, this decentralized approach has been applied to climate issues (Ostrom 2010) and emphasizes local solutions to fit local problems (Nagendra and Ostrom 2012). It can ensure redundancy of potential solutions and increase accountability (Sovacool 2011). The mayor of one small city claimed that the most important climate change action would happen at the municipal level: "We will save the world one plan at a time, one initiative at a time, one strategy at a time… Make no mistake, we will save the world" (Homsy and Warner 2015). Despite such boasts and the extensive policy action in some big cities, adoption of municipality-based climate action plans and general sustainability policy actions remains low overall (Svara 2011; Homsy and Warner 2012). Further, a decentralized governance approach can result in negative externalities, spillovers, regional inequity, and capacity constraints (Howell Moroney 2008; Pastor et al. 2009; Feiock 2013).

An emerging literature argues for multilevel governance as the framework for environmental sustainability including climate change (Bulkeley 2010; Homsy and Warner 2013; Balme and Qi 2014). While cities may initiate environmental protection, they must coordinate with each other, with regional and national governments, and with other non-state actors to be successful (Bulkeley and Betsill 2005). U.S. cities operate within complex governance systems and variations in local government outcomes may result from external factors, such as connections with non-governmental organizations and a central authority (Andersson and Ostrom 2008).

In this chapter, I use a survey of U.S. municipalities to test the hypothesis that jurisdictions will be more likely to adopt climate change policies in a multilevel governance environment that is encouraging of such action as opposed to a decentralized framework in which local governments must act on their own. In addition, I examine the role that a municipality's population size plays with the second hypothesis that smaller places will benefit more from a multilevel environment than bigger cities. The federated nature of the 50 American states provides a good laboratory for testing the importance of multilevel action versus a more polycentric one since each state has different regulations, policies, and incentives governing the policy options available to municipalities. Although no governance level (states, region, or federal) mandates local climate change mitigation by municipalities, some states have climate change plans and supportive policy and programmatic frameworks that may enhance emissions reduction efforts; others do not. In addition, groups of states are organized into regional initiatives specifically seeking reductions in greenhouse gas emissions; these organizations present another level of governance with which municipalities might interact.

I find municipalities take cues within a complex multilevel environment from a variety of internal and external drivers. Unlike much literature that argues decentralized municipal action on the environment is most effective, the data here show that municipalities in states supportive of climate change action are more likely to act, even without legal requirements to do so. The research also identifies internal factors that push climate mitigation. The combination of top-down and bottom-up factors indicate the advantage of multilevel governance in climate change mitigation.

The chapter also fills a hole in the literature regarding climate change actions across localities of different sizes. Sustainability and climate change research focuses on big cities, which are consistently described as leaders. However, most Americans live in small, often suburban municipalities. These smaller places have different relationships with state governments, different access to technical and fiscal capacity, and different political environments. I find that state influence and internal politics do act differently in communities of different population size and metropolitan status. Both sets of findings can reframe our understanding of local policy-making as it relates to regional and global commons issues. These understandings have important implications for research as well as policy. In the concluding section, I offer another, possibly more cohesive, policy approach that could bring comprehensive multilevel governance to local action in the United States.

2.2 American Cities and Climate Action

In the United States, the federal government has paid little attention to the climate policies of local governments. National climate change policy focuses on industrial sectors, especially the reduction of emissions from coal-fired power plants and the increase in motor vehicle fuel efficiency (Crane and Landis 2010). In 2009, the federal government announced the Partnership for Sustainable Communities, a multi-agency effort to craft a national vision for local sustainability, which included greenhouse gas emissions reduction. Although the Partnership supported scores of community and regional projects, particularly to improve coordination between land use and transportation planning, the program suffered funding losses and failed to develop the measures and tools necessary to establish concrete standards (Birch and Lynch 2012). The 2015 Clean Power Plan set goals for state emissions reductions, but 2 years later the plan was still held up in courts and in the spring of 2017, the administration of President Donald Trump began to withdraw from the plan's regulations and international commitments (Davenport and Rubin 2017).

Given the failures at the national and international levels, some have targeted municipalities as the appropriate scale for government action on climate change for four reasons. First, in an increasingly urbanized world, cities emit a significant portion of the greenhouse gases (Bulkeley 2010). Second, American municipalities can choose to construct energy efficient government buildings or retro-fit older ones; they can install more efficient street and traffic lights; and some experiment with alternative energy generation (Svara et al. 2011). Such strategies not only save power, but also model policies to the private sector. Third, local governments can impact (through incentives, regulations, or other policies) non-public activities within their borders by, for example, imposing green building requirements on private projects (Salkin 2009). Municipalities, through their land use regulations, can also require denser, more efficient developments or provide transportation alternatives to the private automobile (Jepson 2004). Others, through their municipal electric utilities, have the ability to induce energy sustainability in both the public and private sectors (Homsy 2016). Fourth, cities will be first responders to potential climate-caused disasters (FEMA 2008).

Most research on climate change action in US cities examines large, urban centers or cities that are pioneers (e.g., Berry and Portney 2013; Bulkeley and Betsill 2003; Kousky and Schneider 2003; Tang et al. 2010) However, in the United States, just over half of the population lives in smaller municipalities (fewer than 25,000 people). Only a quarter of Americans live in the fewer than 300 cities with more than 100,000 residents. Auto-centric American suburbs are the least carbon efficient (Glaeser and Kahn 2010) and rural commuting comprises a large and growing portion of total miles driven (Renkow and Hoover 2000).

The majority of municipalities do nothing about climate change. A 2010 national survey of American cities (Svara 2011) found that only 12% of responding municipalities have created a baseline of emissions or set reduction targets of some sort; 22% of respondents sought to reduce energy use in transportation fleets and outdoor light fixtures; only 5% offered energy audits to private businesses. Smaller communities lag larger ones in the adoption of general sustainability policies (Homsy and Warner 2012). The reasons why local governments, especially smaller ones, choose to act on climate change remains a significant gap in academic and practitioner understandings.

2.3 Local Governance of the Global Commons

As more municipalities initiate efforts to mitigate climate change, there is a debate over whether local governments can act on their own or not. Although no states mandate local government action on climate change, the states do have differing levels of commitment to the environmental challenge which can be conducive or not for policy action. In addition, numerous states have joined multistate initiatives that focus on climate change mitigation. In this section, we discuss the theoretical foundations for two conflicting frameworks of local action on commons issues: decentralization and multilevel governance.

2.3.1 Decentralized Governance

Decentralized governance is a public choice model in which the competition for residents and businesses drives the provision of public goods. It arose as metropolitan-level polycentrism in the 1960s, when Ostrom, Tiebout, and Warren (1961) demonstrated that some public services, such as policing and education, seemed to be best provided at the local level. They maintained that intermunicipal competition and local government's close connection to constituents result in cost-effective outcomes, local innovations, and a diversity of options. The actors in a decentralized system better understand local needs and thus better provide for local public goods than a higher authority (McGinnis 1999).

Elinor Ostrom (2009) hypothesized that this polycentric manner of public goods provision at the local level offers a model for the governance of the global commons. She contended that a variety of public and private actors (including municipalities, utilities, households, firms, nations, etc.) will be driven by competition and local advantages to create independent solutions to greenhouse gas reductions. Such a competitive approach to resource allocation envisions municipalities using strategies best suited to the local environment, citizenry, and other particulars of circumstance. Diffuse local action unburdens the dysfunctional international climate negotiation agenda by having priorities taken up by lower levels of government (Rayner 2010).

Benefits of local independent action include: more experimentation and innovation, local tailoring of action to fit circumstances, political testing of policies, and local experience in enforcement (see Chap. 16: Ninomiya and Burch). However, municipally driven initiatives also have the potential to cause an economically ineffective patchwork of regulations, duplicative enforcement efforts, cross-boundary mismatches between pollution sources and effects, shuffling of high-carbon activities to weaker regulatory areas, and confusion over responsibility between levels of government (Lutsey and Sperling 2008). Local stakeholders acting on their own can grow frustrated with the lack of coordination and express the desire for a holistic approach to greenhouse gas mitigation (Greenwood 2012).

Some empirical analysis seems to indicate that municipalities can act on their own with regards to local climate action (Pitt 2010; Krause 2011a, b). For example, communities are more likely to act when climate change mitigation is linked to a policy already on the local government agenda (Betsill 2001). Such an approach reframes global problems as more local concerns (Metz and Below 2009) and ones on which local governments have the authority to act (see Chap. 12: Brown). These local co-benefits emerge in various forms, such as: reduction in energy costs (Kousky and Schneider 2003; Svara et al. 2011), increased public health (Bloomberg and Aggarwala 2008), or sustainable economic development and local job production (Jochem and Madlener 2003).

2.3.2 *Multilevel Governance*

The multilevel governance framework emerged as a way to analyze and organize the new European Union's relationship to its member states (Bulkeley and Betsill 2003). This approach engages multiple tiers of government in a communicative process that requires the co-production of knowledge and policy up and down levels of authority (Corburn 2009; Homsy and Warner 2013). It requires a respect for local knowledge in the creation of place-specific policies. At the same time, it recognizes the role of a central authority, which has technical expertise as well as the ability to coordinate local governments and induce compliance through incentives or regulations (Homsy et al. 2016). Hooghe and Marks (2003) describe two types of governments within a multilevel framework: one is geographically bound while the other focuses on managing common-pool resources across jurisdictions.

Unlike in Europe, the United States federal government rarely participates with local governments on climate issues, leading to uncoordinated efforts, differing goals, and inconsistent time horizons (Selin and VanDeveer 2009). In 2010, 35 states either had completed or were in the process of developing climate action plans (Center for Climate and Energy Solutions 2011). No states have mandated local government action. The closest is a 2008 California law that requires urban regions to meet greenhouse gas reduction goals by coordinating land use and transportation policy (Barbour and Deakin 2012). While an increasing number of states and regions continue to enact policies on climate change (Rabe 2009), most efforts focus on industry sectors, not local governments (Selin and VanDeveer 2009).

A multilevel framework is not completely foreign to American governance structures and has led to some environmental successes. The federal government in the United States has experimented with cooperative federalism, in which local and state governments participate in the implementation of federal standards (Fischman 2005). The U.S. Environmental Protection Agency engaged in a more co-productive relationship in an effort to clean the polluted Rouge River Watershed. By threatening to impose expensive top-down regulations, the federal government successfully built a coalition of local governments and private actors to cut water pollution, reduce the danger of toxic chemicals, and improve the habitat in the almost 1,200 square kilometer watershed (Homsy et al. 2016).

While there have been some implicit (e.g., auto fuel efficiency standards) and explicit (e.g., 2015 Clean Power Plan) federal policies that have reduced greenhouse gas emissions, subnational governments remain the major drivers of action in the United States (Karapin 2016). The national government is absent due to the inability of most officials and citizens to see climate change impacts; the polarized state of the U.S. party system and the general ideology of limited government intervention; the lack of national authority over many issues; and the lack of strong international institutions (Hale 2010).

Some state governments in the United States have formed state-to-state horizontal networks focused on environmental issues with varying degrees of success. Water quality in the Great Lakes was dramatically improved through the creation in the 1950s of the Great Lakes Commission, which advised on and advocated for clean water for clean water, and the Council of the Great Lakes Governors, which provided a forum for information flows among state leaders (Rabe 1999). In 2001, the Conference of New England Governors and Eastern Canadian Premiers created a climate action plan with aggressive greenhouse gas emissions targets while the Western Governors Association established clean energy goals aimed at new technology development (Rabe 2009). Such networks could lead to greater emissions reductions than single state efforts due to greater geography and population encompassed, potential for uniformity of regulation, ability to capitalize on shared resources and economies, and development of a shared regional vision (Engel 2005). However, these networks remain state-to-state affairs with plans encompassing only the state level of operations and rarely engaging or organizing local governments.

2.4 Research Method

This chapter seeks to empirically investigate the debate around the ability of local governments to act on climate change on their own versus the need for higher level government support. My first hypothesis is that municipalities are more likely to adopt climate change policies if they are within supportive states and regions. I also examine a second hypothesis that smaller municipalities will benefit more from such support than bigger cities.

This project takes advantage of a comprehensive survey of sustainability policy-making by U.S. municipalities. The 2010 Sustainability Survey, conducted by International City/County Management Association,[1] asked county and municipal leaders about their adoption of policies and programs in areas such as climate change, water quality protection and provision, building construction, and land use. Surveys were mailed to a sample of municipalities with populations of more than 2500 people and fewer than 1,000,000 people. Within these parameters, managers of 7257 local governments received surveys and 1874 responded (25.8% response rate). Complete demographic, fiscal, and governance data was gathered for 1837 municipalities, which represents the final number of local governments in the sample.

2.4.1 Dependent Variable

The dependent variable measures whether or not a community is a *climate change actor*. The variable is based on six climate change actions that a municipality might undertake. For each community, this dichotomous variable had a value of one (1) if officials indicated on the survey that their jurisdiction created any one or more of the following:

- A baseline of greenhouse gas emissions produced by the local government;
- A baseline of greenhouse gas emissions produced by the community;
- Greenhouse gas reduction targets for local government operations;
- Greenhouse gas reduction targets for businesses;
- Greenhouse gas reduction targets for multi-family residences; or
- Greenhouse gas reduction targets for single-family residences.

Establishing a baseline of emissions for either the local government or the community is a major undertaking for a municipality and indicates a commitment to climate change action. Adoption of the various targets can be symbolic, but also indicates an official intention by local leaders to address greenhouse gas emissions. Table 2.1 shows the distribution of municipalities considered climate change action communities by population size. The adoption of policies is more prevalent among larger municipalities.

---

[1]The survey was conducted in collaboration with researchers at Arizona State University's Center for Urban Innovation and ASU's Global Institute of Sustainability. A descriptive summary of the results can be found in Svara (2011).

Table 2.1 Distribution of climate change actor communities by population size

Population size	Percent climate change actors	Number of *climate change actors*	Total number of communities in sample
2500 to 9999	8.1	69	848
10,000 to 24,999	15.2	77	504
25,000 to 99,999	34.7	135	389
100,000 to 499,999	49.5	45	91
500,000 to 999,999	60.0	3	5
Total	17.9	329	1837

2.4.2 *Independent Variables*

The independent variables and the sources of the data are described in Table 2.2 and are grouped into following subject areas.

Multilevel variables. Two dichotomous variables examine the potential link between a multilevel governance framework and local government climate action. The first measures whether a municipality's state has a climate action plan, which was true for 327 municipalities in the sample. The second indicates whether the state is a member of a regional climate change initiative. In 2010, there were four regional initiatives (Western Climate Initiative, Regional Greenhouse Gas Initiative, Midwest Greenhouse Gas Reduction Accord, and the Transportation and Climate Initiative) that covered 14 states. The data includes 526 municipalities that were in states within one of the regional initiatives.

The following variables examine the internal drivers of climate action within communities.

Local politics variables. The first variable in this category indicates whether or not a community has a council-manager form of government, which research shows enact more innovative policies (Nelson and Svara 2012) including around issues of sustainability (Svara 2011). Second, I measure political attitudes, which can impact local sustainability policy in general and climate change in particular (Krause 2011a; Slavin 2011; Barbour and Deakin 2012). I employ an index of New Political Culture, which uses demographic information to predict local adoption of progressive policies. My index is based upon one developed by Saha (2009) and is built from the standardized values of the percentages in a jurisdiction of non-family households; unmarried households; people working in professional, scientific, technical, or educational jobs; residents between ages 18 and 44 years; women in workforce; and those who hold a bachelor's degree or higher.

Table 2.2 Summary statistics for variables

	Municipalities (n = 1837)			
Variables	**Mean** (or percent 'yes' for 1/0 variables)	**St. Dev.**	**Minimum**	**Maximum**
Dependent variable				
Climate change actor[a] (1 = yes)	17.8	NA	0	1
Independent variables				
External factors				
State climate plan[b] (1 = yes)	64.1	NA	0	1
State participation in regional climate initiative (1 = yes)[b]	28.6	NA	0	1
Internal factors				
Local politics				
City manager government (1 = yes)	62.0	NA	0	1
Progressive political culture index	1.98	11.7	−5.97	10.5
Dominant economic players				
Employment change 99–09[f] (%)	4.1	19.6	−57.5	178.9
Agricultural employment[c] (%)	2.7	3.8	0	27.4
Manufact. employment[c] (%)	12.5	6.1	0.7	67.1
Local capacity				
Local govt. rev. per capita[d] ($1000 s)	984	949	0	18,280
Educ. att. (bachelor plus)[c] (%)	28.6	16.1	2.4	86.8
Per capita income[c] ($)	27,883	12,770	6399	124,327
Sociodemographic controls				
Central cities (1 = yes)	9.3	NA	0	1
Suburban municipalities (1 = yes)	59.2	NA	0	1
Rural communities (1 = yes)	31.5	NA	0	1
Population[e]	27,876	54,461	1997	741,206
Pop. change 2000–2010[e] (%)	13.8	31.8	−36.6	510.8

[a]Derived from ICMA Sustainability Survey, 2010
[b]Center for Climate and Energy Solutions, (2011)
[c]American Community Survey, 2005–2009
[d]Census of Local Governments, 2002
[e]U.S. Census, 2010
[f]County Business Patterns, 1999–2009

Dominant economic players. Environmental protection is often seen as in conflict with economic development (Campbell 1996). However, three studies focusing on climate change show no correlation between the presence of manufacturing and climate policy action or general environmental sustainability (Krause 2011a; Sharp et al. 2011; Homsy and Warner 2015), though other studies indicate that local manufacturing decreases the chances that a community would act (Bulkeley and Betsill 2003; Gustavsson et al. 2009; Krause 2011b). Three variables test this conflict in the current study. The first is the 1999 to 2009 change in the number of jobs within the municipality's county, which represents general economic development. The other two variables are the percentage of people employed in agriculture/extractive operations and in manufacturing.

Local capacity variables. Local capacity examines the ability of a municipality to carry out policies, including climate change planning. Local government revenue per capita measures the ability of a community to raise funds through taxes and fees and thus fund policy-making and programming. Educational attainment (percentage of the population with a bachelor's degree or more) is a measure of the potential for community members to provide volunteer expertise. Finally, per capita income has been shown to correlate with general sustainability policies (Lubell et al. 2009) and climate change action in particular (Zahran et al. 2008; Sharp et al. 2011).

2.5 Analysis of Results

The research hypotheses were tested using a series of six logistic models. Since the local governments are clustered within states, I used a hierarchical approach to control for the influence of states beyond the state climate variable tested. The first three models include just the presence of a state climate action plan across communities differentiated by population size: municipalities with populations of more than 25,000 people (n = 485); smaller communities between 2500 and 25,000 in size (n = 1352); and the entire sample (n = 1837). Models four, five, and six include the state's participation in a regional initiative as an additional factor, again across the different-sized local governments. If municipalities operate in a purely polycentric manner, then the influence of the state and regional initiative will be insignificant. If top-down factors push climate change action, then internal drivers will be small or insignificant. The results of the logistic regression models are presented as odds ratios in Table 2.3.

The results support the main hypothesis that multiple factors—internal and external—positively correlate with the increased odds that a municipality undertakes climate change planning. The first external factor, the presence of a state climate plan, increases the odds of local action from 1.867 times to more than 2.431 times in five of the six models—even though no state plan directly requires action by municipalities. The exception is model five, focused on bigger cities, in which the presence of a state plan is insignificant, but the other external variable, regional

Table 2.3 Results of multilevel logistic regression

	State climate plan			State climate plan and regional initiative		
Model	**1**	**2**	**3**	**4**	**5**	**6**
	All	Large	Small	All	Large	Small
	(*n* = 1837)	(*n* = 485)	(*n* = 1352)	(*n* = 1837)	(*n* = 485)	(*n* = 1352)
Multilevel factors	*Results presented as odds ratios*					
State climate plan	**2.335	*2.132	**2.431	*1.867	1.508	*2.173
Regional initiative				*1.738	**3.546	1.271
Internal factors						
Local politics						
City manager government	1.120	1.332	1.074	1.151	1.534	1.081
Progress. political culture index	**1.252	**1.388	**1.208	**1.249	**1.390	**1.205
Economic dependence						
Employment change 99–09	0.998	0.997	0.998	0.998	0.995	0.998
Agricultural employment	1.006	1.062	0.994	1.013	1.071	0.996
Manufacturing employ.	*0.965	0.987	*0.944	0.969	0.996	*0.947
Local capacity						
Local govt. rev. per capita (1000 s)	**1.257	**1.638	*1.202	**1.241	*1.402	*1.508
Educ. att. (bachelor plus)	*1.020	0.989	*1.031	**1.020	0.986	**1.031
Per capita income (1000 s)	0.990	1.042	0.977	0.989	1.045	0.976
Control variables						
Central cities	Reference			Reference		
Suburban municipalities	0.641	0.559	0.865	0.631	*0.511	1.822
Rural communities	*0.552	0.606	0.465	0.561	0.648	1.456
2010 Population (logged)	**1.629	*1.564	*1.503	**1.662	**1.691	*1.547
Pop. change 2000–2010	0.997	0.999	0.993	0.998	1.001	0.994

** indicates significance at the 0.05 level ** indicates significance at the 0.01 level* initiative, correlated to increased odds of local climate change planning by 3.546 times.[2] At the same time, internal drivers are also significant and sizable.

The significance of multilevel drivers contradicts the results of some previous research, which finds that municipalities act independently on this issue and that states play no role in local climate action (Pitt 2010; Krause 2011a, b). Two factors might account for this divergence in findings. First, the dataset in the current study is larger and broader. In her two studies, Krause only examines places with populations greater than 25,000 and 50,000 respectively. The second factor is the difference in the construction of the dependent variables. In one study, Krause (2011a) uses the U.S. Mayors Climate Protection Agreement as a dependent variable, which requires neither the resource investment of a greenhouse gas inventory process nor the political capital needed to adopt emissions reductions targets.

The second Krause study (2011b) and Pitt (2010) employ as the dependent variable an additive index of policies which could reduce greenhouse gas emissions. However, many of the included policies (e.g., tree ordinances, recycling, bike lanes, public transit incentives, and encouragement of mixed-use/pedestrian-oriented development) need not have been undertaken with the intention of reducing emissions. Do a community's efforts to encourage public transit, for example, derive from a desire to fight climate change or reduce congestion or provide transportation to low-income residents? To a practitioner, these differences are not important because the multiple facets broaden the pool of policy supporters. However, researchers seeking to study climate change need to make such distinctions; otherwise we are simply testing smart growth or general sustainability (see Chap. 4: Boswell and Mason). The dependent variable in the current study is targeted to a community's actions (emissions baselines) and intentions (adopted goals and targets) and represents specific climate change policy commitments. It offers a clear measure of policy intention and such precision is important if we are to understand what drives climate change policy at the local level.

The secondary hypothesis, that smaller municipalities would benefit more in a multilevel environment than larger cities, is also largely supported. The impact of a state climate plan is greater among small municipalities (models three and six) than larger ones (models two and five, where it is insignificant). Participation in a regional climate change initiative produces no benefit to smaller places while it increases the odds of climate change planning in larger places. In bigger cities, fiscal and technical capacity pose less of a challenge; this frees them up to more fully engage in the discourse and positive environment created by a state that has taken the extra step of joining a regional initiative. Smaller places, however, may remain tied to their states on which they rely for fiscal and technical capacity as well as political cover. This is an important difference between larger and smaller municipalities. Population change and density are not significant in any model.

---

[2]The models were also run without the presence of the multilevel variables and the results for the internal factors changed little in the six models.

One of the internal variables, percentage of manufacturing employment, also indicates an important difference between larger places, where it is not significant, and smaller places, where it has a negative correlation to the odds of climate change planning. In smaller communities, the power of such dominant economic players could work against greenhouse gas reduction by local governments; in these small places, factory management and large numbers of employees would hold the most sway. Manufacturing interests have less power in bigger cities with more diverse voter and tax bases. The other two economic variables, employment change between 1999 and 2009 and level of agricultural employment, are not significant.

Two variables test form of government and political progressiveness—important internal factors. The form of government variable (presence of a council manager) is not significant, which is opposite of what was expected given the innovative nature of managers. Despite the rhetoric, climate change planning is still a pioneering action (Tang et al. 2010), perhaps still so new that the typical advantages of city manager forms of government do not apply. Political progressiveness, another internal driver and tested in the form of the Progressive Political Culture Index, was significant across all six models.

Educational attainment, a measure of local capacity to act, is most important in smaller places. Larger places may have staff and resources to drive climate mitigation policies, but in smaller places, capacity may have to come from the populace. Local government revenue per capita, which describes a local government's ability to act on its own, was significant across all six models.

2.6 Creating a Multilevel Governance Framework for Climate Action

My analysis of a broad municipal dataset indicates that a multilevel governance framework facilitates more climate change planning by local governments than a decentralized approach. Without some leadership by state governments, larger cities will pioneer local climate change action, but the vast majority of cities will do nothing independently. Unlike in Europe where some national governments and the European Union took up the cause of climate change (in word if not in deed), the U. S. lacks an overarching climate framework for municipalities. The question for practitioners and policy makers is identifying the programs that will provide supportive environments at the state or regional level. Municipalities across metropolitan regions have coordinated to achieve affordable housing, economic development, open space conservation, and watershed protection goals (Wheeler 2002) though there is little cooperation around issues of climate change.

Here I want to introduce a potential approach to fostering the kind of governance environment that could induce more local government action. This new kind of regionalism builds on a supra-state structure already existing in the regional climate change initiatives around the U.S., such as the Regional Greenhouse Gas Initiative or the Western Climate Initiative. These state-to-state projects have little direct interaction with municipalities, but, as our findings demonstrate, they do create a supportive environment for climate mitigation in larger cities. As organizations of states, they have the authority to require or incentivize greenhouse gas reductions in all municipalities in the multistate region.

The strength of these multistate regions over single state or metropolitan regional governance is their broader geographic scope, which can more effectively eliminate free-rider problems and reduce leakage that pushes polluting industry to states or municipalities with less stringent regulations. States and municipalities may simply be too small geographically and economically to be effective. In the proposed multilevel structure, central knowledge could be gathered and targeted to specific regional initiatives. Innovative policies developed by local governments could be more relevant to other members of the region. For example, municipalities in the northeast could band together around reducing their natural gas emissions, while those in coal-producing states can develop shared outcomes for their challenges.

The recognized ineffectiveness of voluntary networks (Kern and Bulkeley 2009) might stem from their national or international scope; they are a coalition of communities with interests that are simply too different. Multistate regions might be more effective at incentivizing or requiring action because the states (and their municipalities) within these new geographically based regional networks will more likely share economic goals and political constituencies. For example, the cap and trade program run by Regional Greenhouse Gas Initiative (RGGI) in the northeastern United States, which also provides technical capacity to states, realized a net positive economic impact of $1.6 billion (Hibbard et al. 2011). Such successes make membership and action enticing, especially when states share geography, weather, negative impacts of a changing climate, and economic situation. In the case of RGGI, for example, none of the current nine-member state governments produce coal within their borders. If RGGI tried to add coal-producing Pennsylvania to the mix, it is likely the network would become unstable and policy innovations would prove to be less common to all members. In some ways, the new boundaries would create regions about the size of European nations, each with a common heritage, similar climate, and comparable economic situation.

2.7 Conclusion

Much has been written describing the contents and effectiveness of climate change planning by local governments. Less well investigated is the motivation for local action on such a global commons problem. My analysis of 1837 municipalities indicates that both internal and external factors drive climate change action in those places that do act. Since most communities simply have not adopted climate policies on their own as expected by proponents of decentralized theories of urban policy-making, policy-makers need to create a supportive multilevel environment that recognizes the importance of top-down goal setting and sanctioning power with bottom-up knowledge and buy-in (Homsy and Warner 2013).

Despite the hype, municipal level climate action planning remains disappointingly low. The new US administration of President Trump has called climate change a hoax and, therefore, will likely provide no new—and probably dismantle existing—federal efforts (Davenport 2017). Planners and other policy-makers must realize that the hope for a locally driven, bottom-up approach to climate change will remain limited to pioneering municipalities, even under the best of circumstances. In the absence of federal oversight, state and municipal leaders might build on existing regional networks that group "like" states together and create a multilevel structure within which, this analysis indicates, local action is more likely to thrive.

References

Agranoff, R., & McGuire, M. (2003). *Collaborative public management: new strategies for local governments*. Washington, DC: Georgetown University Press.

Andersson, K. P., & Ostrom, E. (2008). Analyzing decentralized resource regimes from a polycentric perspective. *Policy Sciences, 41*(1), 71–93. doi:10.1007/s11077-007-9055-6.

Andreen, W. L. (2003). The Evolution of Water Pollution Control in the United States—State, Local, and Federal Efforts, 1789–1972: Part 1. *Stanford Environmental Law Journal, 22,* 145–200.

Andrews, R. N. L. (2006). *Managing the Environment, Managing Ourselves: A History of American Environmental Policy*. New Haven, CT: Yale University Press.

Balme, R., & Qi, Y. (2014, June 5). Multi-Level Governance and the Environment: Intergovernmental Relations and Innovation in Environmental Policy. *Environmental Policy & Governance*, pp. 147–154.

Barbour, E., & Deakin, E. A. (2012). Smart Growth Planning for Climate Protection. *Journal of the American Planning Association, 78*(1), 70–86. doi:10.1080/01944363.2011.645272.

Berry, J. M., & Portney, K. E. (2013). Sustainability and Interest Group Participation in City Politics. *Sustainability, 5*(5), 2077–2097. doi:10.3390/su5052077.

Betsill, M. (2001). Mitigating climate change in US cities: Opportunities and obstacles. *Local Environment, 6*(4), 393–406.

Birch, E. L., & Lynch, A. (2012). Measuring U.S. Sustainable Urban Development. In L. Starke (Ed.), *State of the World 2012* (pp. 77–86). Island Press/Center for Resource Economics. http://www.springerlink.com.proxy.library.cornell.edu/content/r439228522h85m79/abstract/. Accessed 1 July 2012.

Bloomberg, M. R., & Aggarwala, R. T. (2008). Think Locally, Act Globally: How Curbing Global Warming Emissions Can Improve Local Public Health. *American Journal of Preventive Medicine, 35*(5), 414–423. doi: 16/j.amepre.2008.08.029.

Bulkeley, H. (2010). Cities and the Governing of Climate Change. *Annual Review of Environment and Resources, 35*(1), 229–253. doi:10.1146/annurev-environ-072809-101747.

Bulkeley, H., & Betsill, M. (2003). *Cities and climate change: urban sustainability and global environmental governance*. London: Routledge.

Bulkeley, H., & Betsill, M. (2005). Rethinking Sustainable Cities: Multilevel Governance and the "Urban" Politics of Climate Change. *Environmental Politics, 14*(1), 42–63. doi:10.1080/0964401042000310178.

Campbell, S. (1996). Green Cities, Growing Cities, Just Cities?: Urban Planning and the Contradictions of Sustainable Development. *Journal of the American Planning Association, 62*(3), 296–312. doi:10.1080/01944369608975696.

Center for Climate and Energy Solutions. (2011). U.S. Cities and States. https://www.c2es.org/us-states-regions. Accessed 25 May 2012.

Corburn, J. (2009). Cities, Climate Change and Urban Heat Island Mitigation: Localising Global Environmental Science. *Urban Studies, 46*(2), 413–427. doi:10.1177/0042098008099361.

Crane, R., & Landis, J. (2010). Introduction to the Special Issue. *Journal of the American Planning Association, 76*(4), 389–401. doi:10.1080/01944363.2010.512036.

Davenport, C. (2017, January 21). Climate Change References Are Purged From Website. *The New York Times*, p. A20. New York.

Davenport, C., & Rubin, A. J. (2017, March 28). Trump Signs Executive Order Unwinding Obama Climate Policies. *The New York Times*. https://www.nytimes.com/2017/03/28/climate/trump-executive-order-climate-change.html. Accessed 29 March 2017.

Engel, K. H. (2005). Mitigating Global Climate Change in the United States: A Regional Approach. *NYU Environmental Law Journal, 14,* 55–86.

Feiock, R. C. (2013). The Institutional Collective Action Framework. *Policy Studies Journal, 41*(3), 397–425. doi:10.1111/psj.12023.

FEMA. (2008). *National Response Framework*. Washington, DC: Department of Homeland Security.

Fiorino, D. J. (2006). *The New Environmental Regulation*. Cambridge, MA: The MIT Press.

Fischman, R. L. (2005). Cooperative Federalism and Natural Resources Law. *New York University Environmental Law Journal, 14,* 179.

Glaeser, E. L., & Kahn, M. E. (2010). The greenness of cities: carbon dioxide emissions and urban development. *Journal of Urban Economics, 67*(3), 404–418.

Gore, C., & Robinson, P. (2009). Local Government Response to Climate Change: Our Last, Best Hope? In H. Selin & S. D. VanDeveer (Eds.), *Changing Climates in North American Politics: Institutions, Policymaking, and Multilevel Governance* (pp. 137–158). Cambridge, MA: The MIT Press.

Greenwood, D. (2012). The challenge of policy coordination for sustainable sociotechnical transitions: the case of the zero-carbon homes agenda in England. *Environment and Planning C: Government and Policy, 30*(1), 162–179. doi:10.1068/c1146.

Gustavsson, E., Elander, I., & Lundmark, M. (2009). Multilevel governance, networking cities, and the geography of climate-change mitigation: two Swedish examples. *Environment and Planning C: Government and Policy, 27*(1), 59–74. doi:10.1068/c07109j.

Hale, S. (2010). The new politics of climate change: why we are failing and how we will succeed. *Environmental Politics, 19*(2), 255–275. doi:10.1080/09644010903576900.

Hibbard, P. J., Tierney, S. F., Okie, A. M., & Darling, P. G. (2011). *The economic impacts of the regional greenhouse gas initiative on ten Northeast and Mid-Atlantic states*. Boston, MA: Analysis Group.

Holloway, C. F., Strickland, C. H., Gerrard, M. B., & Firger, D. M. (2014). Solving the CSO Conundrum: Green Infrastructure and the Unfulfilled Promise of Federal-Municipal Cooperation. *Harvard Environmental Law Review, 38,* 335–369.

Homsy, G. C. (2016). Powering sustainability: Municipal utilities and local government policymaking. *Environment and Planning C: Government and Policy, 34*(6), 1076–1094. doi:10.1177/0263774X15596530.

Homsy, G. C., Liu, Z., & Warner, M. E. (2016, December 28). *Finding the Center in Multilevel Governance: Comparing U.S. and China Environmental Cleanup Efforts*. Unpublished, Binghamton, NY.

Homsy, G. C., & Warner, M. E. (2012). Off the Beaten Path: Sustainability Activities in Small Towns and Rural Municipalities. In Municipal Year (Ed.), *Book 2012* (pp. 53–61). Washington, DC: ICMA Press.

Homsy, G. C., & Warner, M. E. (2013). Climate Change and the Co-Production of Knowledge and Policy in Rural USA Communities. *Sociologia Ruralis, 53*(3), 291–310. doi:10.1111/soru.12013.

Homsy, G. C., & Warner, M. E. (2015). Cities and Sustainability Polycentric Action and Multilevel Governance. *Urban Affairs Review, 51*(1), 46–73. doi:10.1177/1078087414530545.

Hooghe, L., & Marks, G. (2003). Unraveling the Central State, but How? Types of Multi-Level Governance. *The American Political Science Review, 97*(2), 233–243.

Howell-Moroney, M. (2008). The Tiebout Hypothesis 50 Years Later: Lessons and Lingering Challenges for Metropolitan Governance in the 21st Century. *Public Administration Review, 68*(1), 97–109. doi:10.1111/j.1540-6210.2007.00840.x.

Innes, J. E., & Booher, D. E. (2001). Indicators for sustainable communities: a strategy building on complexity theory and distributed intelligence. *Planning theory & practice, 1*(2), 173–186.

Jepson, E. J. (2004). The Adoption of Sustainable Development Policies and Techniques in U.S. Cities: How Wide, How Deep, and What Role for Planners? *Journal of Planning Education and Research, 23*(3), 229–241. doi:10.1177/0739456X03258638.

Jochem, E., & Madlener, R. (2003). The forgotten benefits of climate change mitigation: Innovation, technological leapfrogging, employment, and sustainable development. In *Workshop on the Benefits of Climate Policy: Improving Information for Policy Makers*. Paris: Organization for Economic Co-operation and Development. http://www.oecd.org/env/cc/19524534.pdf.

Karapin, R. (2016). *Political Opportunities for Climate Policy: California, New York, and the Federal Government*. New York: Cambridge University Press.

Kern, K., & Bulkeley, H. (2009). Cities, Europeanization and multi-level governance: Governing climate change through transnational municipal networks. *JCMS: Journal of Common Market Studies, 47*(2), 309–332. https://doi.org/10.1111/j.1468-5965.2009.00806.x.

Kettl, D. F. (2002). *The transformation of governance: public administration for twenty-first century America*. Baltimore, MD: John Hopkins University Press.

Kousky, C., & Schneider, S. H. (2003). Global climate policy: will cities lead the way? *Climate Policy, 3*(4), 359–372.

Krause, R. M. (2011a). Policy innovation, intergovernmental relations, and the adoption of climate protection initiatives by U.S. cities. *Journal of Urban Affairs, 33*(1), 45–60. doi:10.1111/j.1467-9906.2010.00510.x.

Krause, R. M. (2011b). An assessment of the greenhouse gas reducing activities being implemented in US cities. *Local Environment, 16*(2), 193–211. doi:10.1080/13549839.2011.562491.

Lubell, M., Feiock, R. C., & Handy, S. (2009). City Adoption of Environmentally Sustainable Policies in California's Central Valley. *Journal of the American Planning Association, 75*(3), 293–308. doi:10.1080/01944360902952295.

Lutsey, N., & Sperling, D. (2008). America's bottom-up climate change mitigation policy. *Energy Policy, 36*(2), 673–685. doi:10.1016/j.enpol.2007.10.018.

Mazmanian, D. A., & Kraft, M. E. (1999). The Three Epochs of the Environmental Movement. In D. A. Mazmanian & M. E. Kraft (Eds.), *Toward Sustainable Communities: Transition and Transformations in Environmental Policy* (pp. 3–41). Cambridge, Massachusetts: The MIT Press.

McGinnis, M. (1999). Introduction. In Michael McGinnis (Ed.), *Polycentric Governance and Development* (pp. 1–28). Ann Arbor, MI: The University of Michigan Press.

Metz, D., & Below, C. (2009). *Local land use and climate change policy: summary report from focus groups and interviews with local officials in the Intermountain West* (Working Paper No. WP09DM2). Cambridge, MA: Lincoln Institute of Land Policy.

Nagendra, H., & Ostrom, E. (2012). Polycentric governance of multifunctional forested landscapes. *International Journal of the Commons, 6*(2), 104–133.

Nelson, K. L., & Svara, J. H. (2012). Form of Government Still Matters Fostering Innovation in U. S. Municipal Governments. *The. American Review of Public Administration, 42*(3), 257–281. doi:10.1177/0275074011399898.

Ostrom, E. (2009). *A polycentric approach for coping with climate change* (Policy Research Working Paper No. 5095). Washington, DC: World Bank.

Ostrom, E. (2010). Polycentric systems for coping with collective action and global environmental change. *Global Environmental Change, 20*(4), 550–557. doi:16/j.gloenvcha.2010.07.004.

Ostrom, V., Tiebout, C. M., & Warren, R. (1961). The Organization of Government in Metropolitan Areas: A Theoretical Inquiry. *The American Political Science Review, 55*(4), 831–842. doi:10.2307/1952530.

Pastor, M., Lester, T. W., & Scoggins, J. (2009). Why Regions? Why Now? Who Cares? *Journal of Urban Affairs, 31*(3), 269–296.

Pitt, D. (2010). The impact of internal and external characteristics on the adoption of climate mitigation policies by US municipalities. *Environment and Planning C: Government and Policy, 28*(5), 851–871.

Rabe, B. G. (1999). Sustainability in a Regional Context: The Case of the Great Lakes Basin. In D. A. Mazmanian & M. E. Kraft (Eds.), *Toward Sustainable Communities: Transition and Transformations in Environmental Policy* (pp. 247–281). Cambridge, Massachusetts: The MIT Press.

Rabe, H. (2009). Second-Generation Climate Policies in the States: Poliferation, Diffusion, and Regionalization. In H. Selin & S. D. VanDeveer (Eds.), *Changing Climates in North American Politics: Institutions, Policymaking, and Multilevel Governance* (pp. 67–85). Cambridge, MA: The MIT Press.

Rayner, S. (2010). How to Eat an Elephant: A Bottom-up Approach to Climate Policy. *Climate Policy, 10*(6), 615–621.

Renkow, M., & Hoover, D. (2000). Commuting, Migration, and Rural-Urban Population Dynamics. *Journal of Regional Science, 40*(2), 261–287. doi:10.1111/0022-4146.00174.

Saha, D. (2009). Factors Influencing Local Government Sustainability Efforts. *State and Local Government Review, 41*(1), 39–48. doi:10.1177/0160323X0904100105.

Salkin, P. E. (2009). Sustainability and Land Use Planning: Greening State and Local Land Use Plans and Regulations to Address Climate Change Challenges and Preserve Resources for Future Generations. *William & Mary Environmental Law and Policy Review, 34,* 1–50.

Selin, H., & Van Deveer, S. D. (2009). North American Climate Governance: Policymaking and Institutions in the Multilevel Greenhouse. In H. Selin & S. D. VanDeveer (Eds.), *Changing Climates in North American Politics: Institutions, Policymaking, and Multilevel Governance* (pp. 305–325). Cambridge, MA: The MIT Press.

Sharp, E. B., Daley, D. M., & Lynch, M. S. (2011). Understanding Local Adoption and Implementation of Climate Change Mitigation Policy. *Urban Affairs Review, 47*(3), 433–457. doi:10.1177/1078087410392348.

Slavin, M. I. (2011). Where Sustainability Stands Now: Contemporary Trends and Future Prospects. In M. I. Slavin (Ed.), *Sustainability in America's Cities* (pp. 231–246). Washington, DC: Island Press/Center for Resource Economics. Accessed 22 March 2012.

Snyder, L. P. (1994). "The Death-Dealing Smog over Donora, Pennsylvania": Industrial Air Pollution, Public Health Policy, and the Politics of Expertise, 1948-1949. *Environmental History Review, 18*(1), 117–139. doi:10.2307/3984747.

Sovacool, B. K. (2011). An international comparison of four polycentric approaches to climate and energy governance. *Energy Policy, 39*(6), 3832–3844. doi:10.1016/j.enpol.2011.04.014.

Stavins, R. N. (2010). The Problem of the Commons: Still Unsettled After 100 Years. *National Bureau of Economic Research Working Paper Series, No. 16403.* http://www.nber.org/papers/w16403. Accessed 13 September 2011.

Svara, J. H. (2011). The Early Stage of Local Government Action to Promote Sustainability. In *The Municipal Year Book 2011* (pp. 43–60). Washington, DC: ICMA Press.

Svara, J. H., Read, A., & Moulder, E. (2011). *Breaking New Ground: Promoting Environmental and Energy Programs in Local Government.* Washington, DC: IBM Center for the Business of Government.

Tang, Z., Brody, S. D., Quinn, C., Chang, L., & Wei, T. (2010). Moving from agenda to action: evaluating local climate change action plans. *Journal of Environmental Planning and Management, 53*(1), 41–62. doi:10.1080/09640560903399772.

Wheeler, S. M. (2002). The new regionalism - Key characteristics of an emerging movement. *Journal of the American Planning Association, 68*(3), 267–278.
Zahran, S., Grover, H., Brody, S. D., & Vedlitz, A. (2008). Risk, Stress, and Capacity Explaining Metropolitan Commitment to Climate Protection. *Urban Affairs Review, 43*(4), 447–474. doi:10.1177/1078087407304688.

Author Biography

George C. Homsy is an assistant professor in the Department of Public Administration at Binghamton University (Binghamton, New York, USA), and researches the factors that shape sustainability policies at the municipal level, especially in small- to medium-sized cities and towns. Homsy also explores the nexus of heritage and sustainability at the neighborhood level as well as the ways that cities plan programs and the built environment to foster lively and healthy communities for citizens of all ages. Homsy frequently links his research to practice through collaborations with professional organizations, such as the International City/County Management Association and the American Planning Association. Previously, Homsy was a planning consultant helping small- and medium-sized municipalities in the northeastern United States create environmentally and economically sustainable communities.

Chapter 3
Innovations in European Climate Governance and Their Impact on Local Climate Policy: An Analysis of German Major Cities

Jörg Kemmerzell

Abstract Cities and urban regions are not only the main originators of global climate change but particularly places where innovations towards a low carbon transition take place. Notably in the European context, local climate change policies are embedded in a dense structure of multilevel governance arrangements. On the level of the European Union, various instruments address climate change. With regard to climate change mitigation, policies usually are part of general long-term strategies, like the *Climate and Energy Package* of 2008 and its successor, the *Climate and Energy Framework* of 2014. This chapter concentrates on two facets of multilevel governance: first, the involvement of cities in vertical climate governance and second, their participation in horizontal structures of climate governance. The European Commission became an actor in this arena not only by setting up funding schemes but also by the founding of the *Covenant of Mayors*. The main research question of the chapter is: Does the inclusion of cities in multilevel governance structures and particularly the Covenant of Mayors have an impact on local climate policy? While vertical and horizontal activities show both a significant influence on the scope of local climate policy, membership in the *Covenant* seems to be rather an add-on for cities that are already active in climate change mitigation.

Keywords European climate governance · Local climate policy · Covenant of mayors · Multilevel governance

3.1 Introduction

It has become common knowledge that cities and urban regions are not only the main originators of global climate change but particularly places where innovations towards a 'low carbon transition' take place (Bulkeley et al. 2011). Notably in the

J. Kemmerzell (✉)
Institut Für Politikwissenschaft, Technische Universität Darmstadt, Dolivostraße 15, 64293 Darmstadt, Germany
e-mail: kemmerzell@pg.tu-darmstadt.de

S. Hughes et al. (eds.), *Climate Change in Cities*, The Urban Book Series,
https://doi.org/10.1007/978-3-319-65003-6_3

European context, local climate change policies are embedded in a dense structure of multilevel governance arrangements.

On the level of the European Union, various instruments address climate change. With regard to climate change mitigation, policies usually are part of general long-term strategies, like the *Climate and Energy Package* of 2008 and its successor, the *Climate and Energy Framework* of 2014. Within those general frameworks, the EU issues laws, which belong to conventional *regulatory policymaking*, and more innovative initiatives, which include *incentive-based* and *soft governance* instruments. Examples of non-regulatory instruments are programs providing funding opportunities for different actors, such as municipalities. As a unique innovation solely targeting cities, the European Commission launched the *Covenant of Mayors* in 2008. The covenant intends to combine a *vertical* mode of governance, attracting cities to contribute to the EU climate goals, with a *horizontal* layer related to network building, mutual learning and the spread of best practices among the members.

While multilevel governance *in Europe* cannot be properly understood without paying attention to the intermediate levels (national and subnational), this article concentrates on two facets of multilevel governance: first, the involvement of cities in vertical climate governance and second, their participation in horizontal structures of climate governance, particularly transnational city networks. Thus, the main research question of the chapter is: *Does the inclusion of cities in multilevel governance structures have an impact on local climate policy?* It concentrates on the climate protection domain of climate policy. Even though considerable work on the general functions of multilevel climate governance (e.g., Andonova et al. 2009; Kern and Bulkeley 2009; Bouteligier 2013) and, to a lesser extent, case studies on its impact on the local level (e.g., Benz et al. 2015) have been conducted, a broad account is still missing (Busch 2015; as an exception, see Lee and Koski 2015 on U.S. municipalities). As a first step to fill this gap, this chapter presents data from a survey of German major cities.

The next section provides a general overview of European climate governance and examines the integration of the local level into the European sphere. Subsequently, the chapter discusses possible impacts of multilevel governance on local climate policy and presents data drawn from a questionnaire-based survey on climate policy in German major cities (population of more than 100,000 inhabitants).[1] It gives insight into the scope and institutionalization of local climate policy and the involvement of municipalities into vertical (European) and horizontal multilevel governance structures. The fourth section then assesses the impact of

[1]Data presented in this chapter is drawn from joint research of the author and his colleagues Anne Tews, Michèle Knodt, and Arthur Benz. The research (*Überlokales Handeln und lokale Innovationen im Klimawandel. Eine vergleichende Analyse deutscher Großstädte/ Trans-local action and local innovations in the context of climate change. A comparative analysis of German major cities*) has been funded by the *Deutsche Forschungsgemeinschaft/ German Research Foundation DFG.*

multilevel governance and draws inference from selected findings. In the conclusion, I will summarize the findings and suggest potential areas for further analysis.

3.2 Climate Governance in the European Multilevel System and the Local Level

European cities are part of a multilevel structure and affected by policy decisions made at different levels. Climate governance in the EU started to a significant extent in the late 1980s, when the Commission acknowledged the 'greenhouse effect' in a *Communication* to the European Council (European Commission 1988). After describing the main features of the greenhouse effect, the Commission emphasized policy options to deal with the problem. These options concentrate on a call for strengthening common research activities and sketch possible fields where policy interventions may take place. Despite the preliminary nature of the document it is striking that those domains were emphasized where the EU became an active policy maker during the 1990s and 2000s, especially energy efficiency and fuels. However, beyond a rough differentiation between policy measures, which have to be taken on the international or global level, and those that should be taken at the European or national level, the *Communication* pays no attention to the proper scale of the proposed policies.

After the adoption of the *United Nations Framework Convention on Climate Change* in 1992, the European Commission essentially concentrated on representing the internally diverse EU as a unified actor in external relations. This became important as the EU signed the 1997 Kyoto Protocol as a single bloc (Jordan et al. 2012). Simultaneously, sustainability became an issue within the domain of European energy policy. In its 1995 *White Paper* 'An Energy Policy for the European Union', the Commission not only stressed environmental concerns of energy policy but also addressed the importance of regions and the local level. The Commission stated that 'the Community needs to encourage energy consumers through their local authorities to develop a more active role' and detected local communities and authorities as important sites to strengthen energy efficiency and promote *renewables* (European Commission 1995). In the following decade, the EU set up *framework programs*, providing opportunities for cities to receive financial subsidies and to get in touch with peers interested in the same issues. With the *White Paper,* the European institutions recognized cities as relevant actors and potential partners in the vertical dimension of European multilevel governance. In the field of climate policy, a process coined *multilevel reinforcement* (Schreurs and Tiberghien 2007) unfolded. Multilevel reinforcement is defined as a 'dynamic process … among different political poles within a context of decentralized governance' (ibid: 22) with actors working in different sectors at different levels in a polycentric fashion. Among cities, during the 1990s a horizontal sphere of joint action emerged, embodied by a variety of *transnational municipal networks*

(Kern and Bulkeley 2009; Busch 2015), such as the *Climate Alliance* (founded in 1990), *Energy Cities* (founded in 1994) and, on a global level, *CCP* (Cities for Climate Protection, founded as initiative within the ICLEI network in 1990). Those networks serve both as sites for information exchange and mutual learning and as bodies of interest representation at the European level (Kern 2014: 115).

The development of an integrated European climate governance culminated in the *Climate and Energy Package* adopted in 2008 by the European Council, which covered various laws (e.g., the *Directive 2009/28/EC* on the promotion of the use of energy from renewable sources), decisions and guidelines (European Commission 2008a; European Council 2009). The package sets three central targets: a 20% cut in greenhouse gas emissions (from 1990 levels), an increase of the share of renewable energy by 20% and a 20% improvement in energy efficiency by 2020. Of particular interest for cities is the *Directive 2010/31/EC* on the energy performance of buildings that applies high standards for the energy efficiency of public owned real estate. While the package largely relied on traditional regulatory instruments, it also strengthened financing and soft modes of governance. In the course of the climate and energy package, the EU explicitly addressed subnational and local entities as relevant sites of action towards a low carbon transition. The former European Commissioner on Energy Andris Piebalgs (European Commission 2008b) emphasized

> Cities are becoming the places to deliver new ideas and innovative projects against global warming. The approach to tackle the climate crisis challenge can only be holistic, integrated, long-term and, most of all, based on the participation of citizens. This complex picture is best managed at local level. Cities must therefore become leading actors for implementing sustainable energy policies, and must be supported in their effort.

The Commission gained opportunities to include local authorities in its program *Intelligent Energy Europe* and extended funding with the *Horizon 2020* framework that gives cities prospects to apply for funding of research and planning activities, the employment and training of staff or the cooperation with partners.

In addition to the climate and energy package, the Commission launched the *Covenant of Mayors* (CoM). The CoM indicates a strategic approach to the local level differing significantly from the older voluntary networks (Bulkeley and Betsill 2013; Heyvaert 2013). The CoM goes beyond the voluntary network type for two reasons: first, it provides a basis for the aggregation of interests of cities towards the European level; second, membership in this specific *network* embodies stronger forms of commitment to the network goals. In the case of the CoM these are the *20-20-20 goals* described above. Even though the *Climate and Energy Package* included binding national targets on the expansion of renewables and national emission reduction targets, the EU lacks proper instruments to enforce national commitment to such kind of long-term strategies. Therefore, it tried to mobilize subnational political authorities. The core instruments of the CoM are *Baseline Inventories* of GHG emissions and *Sustainable Energy Action Plans* (SEAP). These plans must be submitted to the European Commission within 2 years following the membership signature of a municipality. In the follow-up of the submission of a SEAP the members have to monitor their achievements every two years in *progress reports*. All three steps—baseline inventory, SEAP, and progress reports—are subject to standards defined by the *Covenant Methodology* and are reviewed by the Joint Research Centre (JRC) of the European Commission.

In 2014, the European Council adopted a new *Climate and Energy Framework*, which adjusted the climate and energy goals by 2030 (European Commission 2014). While the new framework stepped up the goals (a 40% cut in GHG emissions, 27% increase of the share of renewables, and 30% improvement of energy efficiency), compared to the *package* of 2007 it falls short in terms of institutionalization. This is mainly due to the fact that, despite the project of an *Energy Union*, the new framework sets no nationally binding targets and commitments. Observers explain this institutional shortcoming with the EU enlargements of the 2000s, which bred more conflicts within climate and energy policy (Knodt 2016) and with the economic crisis in the aftermath of 2007 (Geden and Fischer 2014).

The new *framework* pays no particular attention to the local level. However, in 2014 the CoM saw a relaunch as *Covenant of Mayors for Climate & Energy*, which adopted the *2030 goals* and broadened its focus. The initiative integrated climate change adaptation (*Mayors Adapt*) and the provision of secure and affordable energy as network purposes (Bendlin 2016: 103). This accompanies the focus of the *Energy Union* on a norm triangle of European Energy policy—sustainability, competitiveness, and security of energy supply. In overall numbers, the CoM seems to be a success story. As of December 31, 2016, the organization reports 7190 signatories and 5674 SEAP submissions; 4433 plans were accepted, 130 rejected and 1105 under review by the JRC. By December 2016, the members have provided 1263 progress reports. The CoM represents roughly 200 million people in 42 countries, including members from non-EU countries.[2] However, membership is unevenly distributed between different member states. While Italy and Spain represent the bulk of signatories, the statistics for countries from Northern and Western Europe, which are traditional forerunners of climate protection, are remarkably lower. This could be explained by the fact that many South-European municipalities suffer from flawed or absent climate policies of their national governments and resort to the European level. On the other hand, municipalities from Germany, the Netherlands, or Denmark have more extensive support structures provided by national or regional governments and thus may have weaker incentives to subscribe to the CoM. By December 31, 2016, 63 German municipalities signed the CoM. This raises the question of whether a CoM membership of German cities contributes to the format of climate governance or remains rather an add-on. Before turning to this question in Sects. 3 and 4, the following provides a brief overview of the principal functions and benefits of the CoM.

Without elaborating on these benefits in detail, we can assert that the benefits of the CoM for the EU Commission are relatively clear-cut. The Commission gains, through a soft *bypassing* of the member states, an additional advantage for supporting its goals and it increases the credibility of its leadership claims.

---

[2]See http://covenantofmayors.eu/about/covenant-of-mayors_en.html as of December 31, 2016.

For municipalities the perceived benefits depend on the *maturity* of its climate policy. Lena Bendlin distinguishes 'internal' and 'external' benefits in an early and a mature stage of policy (Bendlin 2016: 112). In an early stage of climate policy development, local actors utilize the CoM membership internally for the establishment of authoritative constraints for action, the setup of benchmarks and the introduction of a methodology for climate and energy planning, while at a later stage of policy development the membership creates opportunities to lock-in already established trajectories. External benefits in an early stage are the facilitation of external funding, an enhanced visibility of local leaders, and aspects of 'place-branding' (Busch 2015: 224). At a later stage, benefits shift to rather technical aspects like an audit for existing policies or a sustained advocacy platform. Additionally, the instruments of the CoM might disseminate among non-members. A concrete example of multilevel reinforcement took place in France, where the national policy on local climate and energy plans informally led to a broad adoption of the *Covenant Methodology* (Bendlin 2016: 113).

An appropriate understanding of climate governance in the European multilevel system shall include the intermediate, i.e., the national and regional, level. In Germany, at least two instruments are of particular importance. Already by 1997, the federal government provided guidelines on the adoption of local climate protection (*Leitfaden Kommunaler Klimaschutz*). These guidelines were backed by the establishment of a federal service agency but concentrated on the spread of information in the different areas of the issue (e.g., the development of local climate plans, exchange of ideas and mutual learning, recommendations for the relevant sectors). Comprehensive programs organized by the German states (*Länder*) occasionally amended these guidelines (Kern 2014: 119). With the establishment of the *National Climate Initiative* in 2008, the federal government extended the merely consultative approach with a funding instrument, the directive for local climate change mitigation (*Kommunalrichtlinie*). Under this directive that has been updated and amended several times, local authorities can apply for co-funding of local climate protection plans, receive financial support for the implementation of mitigation measures in buildings or the infrastructure, and receive 40% refunding for the employment of municipal climate managers (Schaefter et al. 2013). It is worth noting that the *Klimaschutzrichtlinie* has been established in the same year as the CoM. In the next section, I will discuss whether the figures for the adoption of climate protection plans and the foundation of administrative units related to climate protection coincide with the establishment of these two instruments.

3.3 Multilevel Governance and Local Climate Policy

The following section describes the involvement of cities into the structures of European multilevel governance. It starts with brief theoretical assumptions on the internal impacts of *trans-local activities of cities*, followed by the presentation of a database on climate protection policies in 71 German major cities. The section ends with the presentation of descriptive findings on the multilevel activities of the cities.

3.3.1 Theorizing the Impact of Multilevel Governance

Both the vertical and the horizontal dimension of European multilevel governance exceed the boundaries of action of local authorities. In some cases, like the *Covenant of Mayors*, we find vertical and horizontal action systematically combined. Even though the subsequent section concentrates on internal impacts, the external functions of trans-local activities (from exerting influence on agenda setting and decision-making of upper governmental levels to aspects of place-branding and showcasing) should not be neglected. A first impact of multilevel governance on the local level is *policy learning*. Learning takes place both among individual actors and organizations.[3] On an individual level, local administrators have the opportunity to learn within epistemic communities, which are established by working groups of city networks or within projects that bring different partners together. Organizations learn from best practices, if cities establish contacts with others that provide policy models on certain issues. However, interview data from qualitative in-depth interviews in three German cities indicate that context-specific constraints limit the opportunities to simply 'imitate' or 'emulate' best practices, restricting learning rather to the mode of inspiration (Benz et al. 2015).

Another function of multilevel governance is achieving *fiscal co-benefits* through the extraction of external resources. The acquisition of subsidies may persuade the local council to engage in climate protection policies or to extend the scope of measures. The practice of *co-financing* is usually part of European or federal funding schemes. Therefore, it reduces the setup-costs of climate protection policies and enables the implementation of important but non-obligatory tasks, which would otherwise be left undone.[4]

Closely connected to the issue of co-benefits is the *safeguarding* function of multilevel governance arrangements. By joining the CoM for instance, a city commits itself to pursue some shared goals and to meet common standards of action and monitoring. Such commitment can strengthen the professional position of the administration vis-à-vis the local council and the political leaders. While the fiscal co-benefits may play a decisive role at an early stage of a policy, safeguarding sustains already established policies.

An obvious function is *compliance* to rulings of higher jurisdictions, which is part of hierarchical/vertical governance. Regulatory policymaking makes a great deal of European policymaking. Its impact on municipalities is however rather indirect, aside from regulations like the EU Directive on buildings (s. a.) which have immediate consequences.

---

[3]See Kemmerzell (2016) for a more comprehensive account of the *mechanisms* of trans-local action.

[4]It is important to underline that climate protection is a widely non-obligatory municipal task in Germany lacking hierarchical elements.

3.3.2 The Database on Climate Protection in German Major Cities

To answer the question of whether *the inclusion of cities in multilevel governance structures has an impact on local climate policy* a database on climate protection policies in German major cities was assembled in 2016. Because of an uneven distribution of publicly available information and data, it became appropriate to conduct a survey of municipal senior officials on the adoption of climate policies. Questionnaires were sent to the mayor's office with a request to forward the survey to the administrative unit in charge of climate protection. The questionnaire includes 14 questions (some containing sub-level questions) related to policies, organizational issues and particularly the involvement of cities into structures of multilevel governance. One question covers a checklist of 33 policy instruments, another one asks for the integration of climate protection within the administration (see below). These 33 policy instruments are the basis of a weighted index of local climate protection.

As seen in Table 3.1 the instruments fall into five different categories. The first category involves instruments for the implementation and evaluation of local climate protection strategies. They serve as a necessary requirement for GHG reductions, while they do not contribute to these reductions by themselves. Instruments belonging to 'self-governing' (Bulkeley and Kern 2006) of a municipality represent the second category, which covers energy consumption within the local government's sphere of authority. The third category captures instruments targeting community-wide emissions, urban planning and private real estate, while the fourth category is related to urban traffic and transport. The fifth category includes instruments related to 'soft modes' of governance like consulting and enabling. The indexation of local climate protection bases upon an equal incorporation of each category. A division of realized instruments by available instruments assigned the values of each category. The five values ranging between zero and one were added and divided by five.

The index scores range from 0.19 to 0.96 with a median score of 0.63 and an arithmetic mean of 0.66. Cities that implement a high share of measures and are active in many categories score high on the index and vice versa. Figure 3.1 presents the distribution of cities within the sample. Therefore, I have rounded the index scores to the second decimal place.

Selection criterion for inclusion was the status as 'major city', i.e., a population of more than 100,000 inhabitants. According to census data from 2015, 79 cities recently exceed this number. The number of 85 can be traced back to the decision to include all cities that met the major city status during a time span of 20 years (between 1993 and 2013), which features the period where climate protection became a political issue at all. The return was 71, equating a response rate of 84 per

Table 3.1 Local climate protection instruments

Implementation and evaluation	
(1) GHG inventory	60
(2) Regularly monitoring of energy consumption	57
Self-governing	
(3) Clean energy production in municipal buildings	63
(4) Acquirement of energy efficient vehicles	59
(5) Cogeneration of heat and power (CHP) within municipal buildings	53
(6) Acquirement of energy efficient office equipment	55
(7) New Buildings and refurbishment of municipal real estate exceeding the standards of 'ENEV'[a]	41
(8) Funding of municipal energy savings through 'internal contracting'	30
(9) Funding of municipal energy savings through 'external contracting'	12
(10) Extension of the share of renewables within municipal power supply	52
(11) Divestment	2
Planning and private real estate	
(12) Model districts for climate protection	34
(13) Climate protection requirements in land-use-plans	43
(14) Compulsory connection to district- or local-heating systems	32
(15) Conservation of green and planted areas	51
(16) Register for solar energy use	60
(17) Incentives for low- and/or zero-energy buildings	21
(18) Incentives for climate friendly building equipment (solar panels, CHP)	29
(19) Incentives for energetic refurbishment exceeding 'ENEV' standards	28
Traffic and transport	
(20) Efficient street lights	63
(21) Incentives for strengthening the use of public transport	60
(22) Acquirement of low-emission public transport vehicles	53
(23) Cycle friendly provisions (cycle lanes, cycle parking lots)	67
(24) Mobility management and consulting	44
(25) Promotion of electro mobility	41
Consultation and enabling	
(26) Enabling energy suppliers to invest in renewables	19
(27) Enabling and promotion of energy cooperatives	30
(28) Incentives for households to purchase energy efficient home appliances	14
(29) Local fund supporting private climate protection projects	17
(30) Public relations concerning climate protection	65
(31) Raising awareness concerning climate and resource friendly behavior	62
(32) Consulting for investors and companies	44
(33) Energy consulting for households	54

The numbers in the right column indicate the cumulated checks
[a]ENEV: the 'Energieeinsparverordnung'/'Energy Saving Ordinance' is a German federal regulation, which defines minimum standards for the energy efficiency of buildings

cent. While distribution of response is not exactly equal among different population clusters, statistical tests found no systematic distortion within the sample of 71 cities. Table 3.2 reports descriptive findings from the basic population of 85 cities and the sub-set of 71 cities.

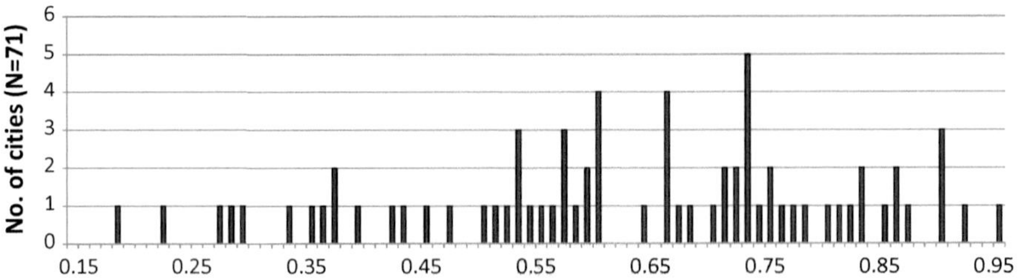

Fig. 3.1 Index scores of implemented climate protection measures

Table 3.2 Comparison of all German major cities and survey respondents

	All cities (N = 85)	Responding cities (N = 71)	Response Rate (%)
Population (2013) above 1,000,000	4	4	100
Population 500,000–1,000,000	9	9	100
Population 200,000–500,000	26	21	81
Population below 200,000	46	37	80
Average population	304,980	331,267	
Location in Eastern Germany	12	10	83
Location in Western Germany	73	61	84
Institutional Status: City State	3	3	100
County independence	68	56	82
County dependence	11	9	82
Special status (metropolitan region)	3	3	100
Median income p.c. (€, 2012)	3079	3090	
Tertiary sector employment (%, 2012)	79.4	79.7	
Municipal debt p.c. (€, 2014)	3900	4198	
Tax revenue capacity p.c. (€, 2014)	1212	1243	

3.3.3 Descriptive Findings

The subsequent section concentrates on multilevel governance. It presents findings on the application and participation in EU- and federal-funded projects and programs and on the membership in transnational city networks and the *Covenant of Mayors*. First, it gives a sense for the programmatic dimension of climate protection (showing the year of first adoption of a climate protection plan) and the administrative integration of climate protection issues within the local administration.

The vast majority of the cities (67 out of 71) have already adopted an integrated climate protection concept/program by council decision. Three cities reported that they were in the process of drafting such a program and just one city has neither adopted nor prepared a program. We asked further to indicate the year when a climate program has been adopted for the first time. It is obvious that before the turn of the millennium local climate protection was restricted to a minority of pioneering cities, while the majority refrained from institutionalizing this issue. By the year 2000, 17 cities in Germany had enacted a climate protection plan, a trend starting around the *Rio Conference* in 1992. The early 2000s saw just a moderate increase in adoptions, resulting in 25 cities having a climate protection concept in 2008. The majority of cities adopted a program after the year 2008, which indicates a strong influence of the external incentives discussed above. In 2008, both the *Covenant of Mayors* and the *federal directive* on local climate change mitigation were established (Fig. 3.2).

In addition to programmatic innovation, an important feature is organizational change within the municipal administration. Since climate change represents a *cross-cutting* issue that affects different departments, the establishment of cross-departmental units on climate issues indicates organizational innovation. A cross-departmental unit on climate issues coordinates climate- and energy-related issues citywide. Forty-nine cities have established a unit so far, the first having been founded in 1998. Similar to the adoption of climate protection concepts, we detect a rapid increase in the late 2000s. This likewise indicates an institutional response to the external incentives, in this case probably to the *federal directive* (Fig. 3.3).

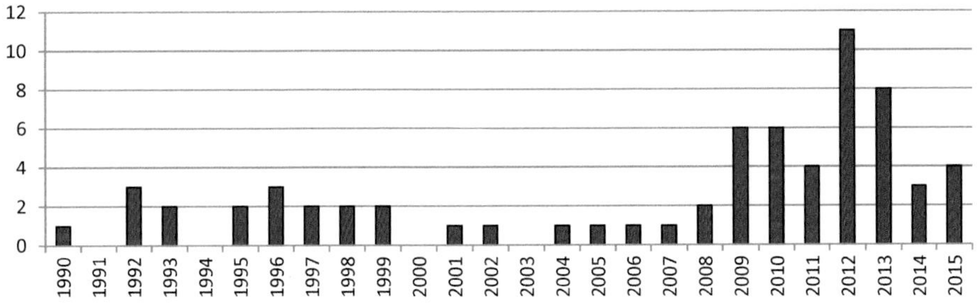

Fig. 3.2 Adoption of climate protection concepts per year

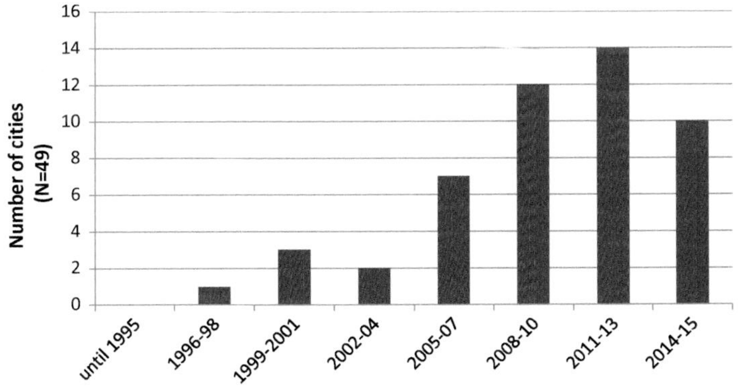

Fig. 3.3 Year of establishment of a cross-departmental unit on climate change issues

In the vertical dimension of multilevel governance, the application for and the acquisition of external funding plays a crucial role besides the implementation of regulative standards. While all cities are subject to European and national laws to the same extent, they differ, of course, in their implementation strategies: participation in funded projects is voluntary. As already mentioned, the EU as a programmatic actor in climate and energy policy provides several opportunities to participate in joint projects. On the German federal and subnational level, similar programs exist. The survey shows a significant difference between participation in national and European programs. While most cities (66) participated in at least one domestic program, 32 cities were engaged in EU-funded projects. A closer look reveals that cities engage at the European level when they are already highly active in the national arena. Thus, it would appear that involvement in EU projects represent more a complementary avenue than an alternative or compensatory one. The findings also suggest that European projects are particularly a field of activity for cities which are already engaged in climate policy. At the same time, this might dampen expectations about the reach of those projects and point to the limits of the dissemination of best practice (Figs. 3.4 and 3.5).

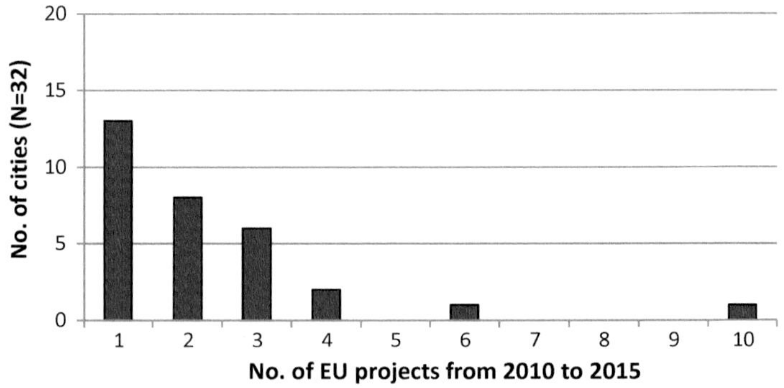

Fig. 3.4 Participation in EU funded programs

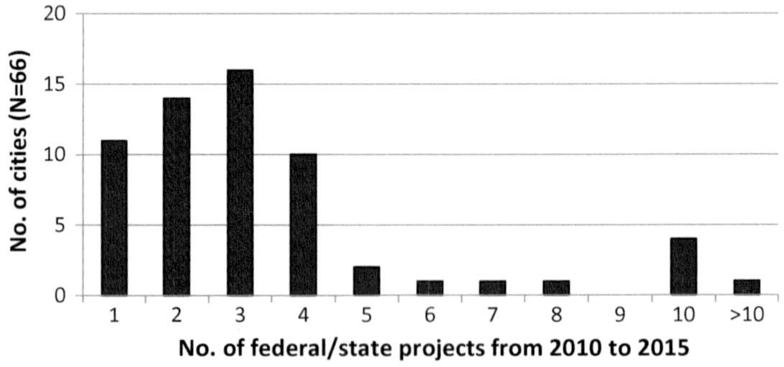

Fig. 3.5 Participation in German federal or state funded programs

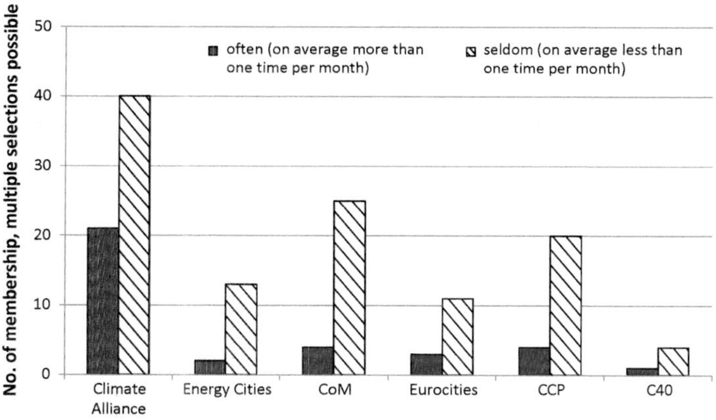

Fig. 3.6 Regular contacts with transnational municipal networks

Membership in transnational municipal networks and the regular exchange with other network members on climate issues has been regarded as an important source of innovative policy strategies. The survey collected data on the membership in the most important networks and the intensity of network activities. In addition to the networks mentioned above, we also included *Eurocities* as a non-thematic organization, which features climate and energy as a priority (*Working Group* on air quality, climate change and energy efficiency), and the *C40 Network* of self-declared urban leaders on climate change mitigation. Figure 3.6 reports data on the CoM as well, even though its scope is beyond the solely voluntary approach of the other networks. Most cities (61) are members of the Climate Alliance, which is the oldest and most common of these organizations. Twenty-nine cities in the sample are members in the CoM.[5] A rather passive style of membership appears to be dominating, since only five cities show a higher frequency of contacts. With regard to the functions and benefits assigned to the CoM, it is crucial to ask for the relationship between accession to the covenant and the adoption of climate protection concepts. An exploration of this relationship likewise informs about the function of membership, be it more policy inducing or safeguarding (s. a.). A majority of cities (38) participate in one or two networks. While ten cities abstain entirely from such activities, a group of 16 cities is engaged in at least four networks.

[5]Twenty-nine cities are full members of the CoM, one city joined *Mayors Adapt* only. In the whole population of 85 major cities, 33 are currently associated with the CoM.

3.4 Local Climate Policy and the Impact of Multilevel Governance

A broad literature on the general functions of multilevel governance arrangements considers the operation of transnational networks (Kern and Bulkeley 2009; Sippel 2011; Lee 2014; Busch 2015), accounts for explanations for joining those networks (Sharp et al. 2011), and debates repercussions of the involvement in multilevel governance structures on the adoption of local climate protection concepts (Hakelberg 2014). However, the impact of such activities on the actual reduction of GHG emissions remains in question (Bulkeley and Newell 2010; Bendlin 2016), in part because of the pitfalls of local emissions monitoring (Ibrahim et al. 2012). Of course, the subsequent analysis cannot provide an answer to this intricate issue, but it will draw some inferences on related questions.

3.4.1 *Who Joins the Covenant of Mayors?*

First, I will analyze the local functions of membership in the CoM. As noted above, we can distinguish *early* and *later stage* impacts of a covenant membership. The adoption of a climate protection concept after accession to the CoM would indicate a typical early stage impact. If cities that already have a climate protection concept join the CoM, we may assume that later stage impacts dominate. A cross-tabulation of CoM membership and the year of adoption of a climate protection concept (distinguishing between adoption up to 2008 and as of 2009) indicates that 15 out of 29 members had a concept before signing the covenant, while 14 covenant members adopted one since 2009. Ten out of 42 non-members had a climate protection concept by 2008, while 32 adopted a concept after 2009 or still have none. A simple categorical Chi-square test rejects the assumption of independence. That confirms expectations that under the conditions of German local climate policy an institution like the CoM becomes a tool for pioneering cities rather than an attempt to spread climate protection policy. A closer look on CoM membership in relation to longer established network organizations support this finding. Only three of the 29 CoM members started their 'career' in network activity after gaining membership in the CoM and 15 cities joined at least two other networks at the time of signing the *covenant*.

3.4.2 *Multilevel Impact on Local Climate Policy*

In the remainder of the section, I shall investigate the influence of European multilevel governance on the scope of local climate policy. The index presented in Sect. 3.3.2 serves as the dependent variable to test the influence of involvement in vertical structures of multilevel governance as well as in horizontal structures. For the purpose of this chapter, I abstain from running a full-fledged statistical analysis and concentrate on selected tests in both dimensions.

In the vertical dimension, the participation in EU-funded programs serves as a categorical independent variable. If we compare the two groups of participants and non-participants, we find a significant deviation. While municipalities that participate in funded projects display an average index score of 0.71, municipalities that abstain display an average index score of 0.58. The *p-value* of a *t-test* (0.004) indicates a statistically significant difference between both groups (Table 3.3).

In the horizontal dimension, I turn to two different questions. First, I examine the relationship between the level of network activity and the scope of climate protection policy. The reported cumulated network membership serves as independent variable while the index score remains the dependent variable. The independent variable varies between *zero* (no network membership) and *six* (membership in all requested organizations). A univariate regression analysis indicates a highly significant relationship between the two variables and an *adjusted R-square* of 0.298.

In a second step, I test whether membership/non-membership in the *Covenant of Mayors*, as the most generalized tool of European multilevel governance of municipalities, indicates significant differences of implemented policies. A comparison of the two groups with regard to the scope of climate protection policy shows average mean values of the index of 0.75 for CoM members and 0.55 for non-members. The t-statistics indicates a statistically significant difference between the two groups. However, if both independent variables, membership in the CoM (as a dummy) and cumulated network membership, are integrated into one model, CoM membership contributes to a lesser extent to the result. In comparison to the model that included cumulated network membership exclusively (s. a.), the *adjusted R-square* barely increases to 0.31. Within the model, membership in the CoM fails to meet the test of significance (Table 3.4).

These statistical findings support the analysis of the characteristics of covenant members. Apparently, the *Covenant of Mayors* serves as *ancilliary factor* for cities which are both active in multilevel governance structures and climate protection, even before becoming a member of the covenant.

Finally, we can investigate the relationship between the involvement in vertical and horizontal structures of multilevel governance. A cross-tabulation shows a

Table 3.3 Impact of vertical structures of multilevel governance

	Participation in EU projects	No participation	t-statistics	
N	32	39	t	−2.96
Mean	0.71	0.58	df	69
Standard deviation	0.18	0.17	sig. (two-sided)	0.004
Standard error	0.028	0.031	Mean difference	−0.125

Table 3.4 Impact of horizontal structures of multilevel governance

Dependent variable: scope of policy implementation (climate protection)	Model 1	Model 2
Const.	0.45***	0.47***
Cumulated network membership (non-standardized coeff.)	0.06***	0.04**
CoM membership (non-standardized coeff.)	–	0.08
F	30.66***	16.67***
N	71	71
R-square	0.31	0.33
Adj. R-square	0.298	0.31

Table 3.5 Cross-tabulation of vertical and horizontal multilevel governance

Cross-tabulation: vertical and horizontal activities			
Network membership, numbers of networks	Participation in EU projects	Non-participation in EU projects	Sum
0	0	3	3
1	2	8	10
2	8	11	19
3	3	10	13
4	4	5	9
5	10	1	11
6	5	1	6
			71

strong symmetry between both variables (see Table 3.5), and the Chi-square test produces a significant result. This finding supports the assumption of the interrelatedness of trans-local activities. Municipalities active in the vertical dimension display on average a higher level of horizontal activity and vice versa.

3.5 Conclusion

The chapter has shown that multilevel governance in the European Union extends to the local level. Besides vertical forms of climate governance, the horizontal level plays a growing role, because it facilitates mutual learning as well as gives municipalities opportunities to become collective actors on the European level (see Chap. 10: Bellinson). The European Commission reacted to the conjunction of both forms by the founding of the *Covenant of Mayors*. The *covenant* represents an institutional innovation, in that it gives the Commission a lever to pursue its climate and energy goals in a domain where it has no direct influence. Concurrently, it enables network building and the distribution of best practices among its members.

Subsequently, the chapter discussed possible impacts of multilevel governance on local climate policy. After presenting data on climate protection policy in German major cities, I have argued that membership in the covenant is less important for the initialization of local climate policy than for the lock-in and safeguarding of already established practices. At the time of accession, the signatories display a higher level of activity with regard to both climate protection and involvement in multilevel governance structures. The comparatively low overall membership figures (only 63 members in Germany, as opposed to Italy with more than 3000 or Spain with about 1700 members) would suggest that the opportunities for spreading best practices are rather limited. For the bulk of German members, signing the covenant seems to be rather an add-on within an already established climate protection portfolio. Vertical and horizontal activities show both a significant influence on the scope of local climate policy and are conducive to the adoption of innovations. To some extent, the data analysis supports the concern raised by Kern that in multilevel climate governance the 'split between the pioneering cities and cities which clearly lag behind […] may become even more pronounced' (Kern 2014: 125).

The chapter only could give selected impressions into the repercussions of multilevel governance on local climate governance. At least three open questions remain. First, even though the analysis shows an imbalance of membership, the theory of multilevel reinforcement points to *spillover-effects* on non-members. These should be taken into account. Second, the data set pays not attention to the *timing* of policy adoption and activities within multilevel governance arrangements. To conduct a more fine-grained analysis, process-tracing within selected cases-studies seems appropriate. Finally, the chapter did not inquire into the structural conditions (like population magnitude, wealth or political majorities) for both involvement in multilevel governance and the scope of climate policy. Isolating the effects of multilevel governance on local climate policy in a more sophisticated fashion would require an integrated analytic model that includes those structural conditions as control variables. In this respect both, a comprehensive regression analysis or a fuzzy set QCA, might be viable opportunities.

References

Andonova, L. B., Betsill, M., & Bulkeley, H. (2009). Transnational climate governance. *Global Environmental Politics, 9*(2), 52–73.

Bendlin, L. (2016). Cities' views and ownership of the Covenant of Mayors. In J. Kemmerzell, M. Knodt, & A. Tews (Eds.), *Städte und Energiepolitik im europäischen Mehrebenensystem. Zwischen Energiesicherheit, Nachhaltigkeit und Wettbewerb* (pp. 103–124). Baden-Baden: Nomos.

Benz, A., Kemmerzell, J., Knodt, M., & Tews, A. (2015). The trans-local dimension of local climate policy. Sustaining and transforming local knowledge orders through trans-local action. *Urban Research & Practice, 8*(3), 319–335. doi:10.1080/17535069.2015.1051380.

Bouteligier, S. (2013). *Cities, networks, and global environmental governance. Spaces of innovation, places of leadership*. London: Routledge.

Bulkeley, H., & Betsill, M. (2013). Revisiting the urban politics of climate change. *Environmental Politics, 22*(1), 136–154. doi:10.1080/09644016.2013.755797.

Bulkeley, H., & Kern, K. (2006). Local government and the governing of climate change in Germany and the UK. *Urban Studies, 43*(12), 2237–2259. doi:10.1080/00420980600936491.

Bulkeley, H., & Newell, P. (2010). *Governing climate change*. London, New York: Routledge.

Bulkeley, H., Castán Broto, V., Hodson, M., Marvin, S. (2011). Cities and the Low Carbon Transition. *The European Financial Review* (August–September 2011), 24–27.

Busch, H. (2015). Linked for action? An analysis of transnational municipal climate networks in Germany. *International Journal of Urban Sustainable Development, 7*(2), 213–231. doi:10.1080/19463138.2015.1057144.

European Commission. (1988). The Greenhouse effect and the community. COM(88) 656 final. Brussels: European Commission.

European Commission. (1995). White Paper. An energy policy for the European Union. COM(95) 682 final. Brussels: European Commission.

European Commission. (2008a). 20 20 by 2020. Europe's Climate Change Opportunity. COM (2008) 30 final: Brussels: European Commission.

European Commission. (2008b). Sustainable energy cities take the lead on climate change: The European commission launches the Covenant of Mayors. IP/08/103. Brussels: European Commission.

European Commission. (2014). A policy framework for climate and energy in the period from 2020 to 2030. COM(2014) 15 final. Brussels: European Commission.

European Council. (2009). *Directive 2009/28/EC of the European Parliament and of the Council of 23 April 2009 on the promotion of the use of energy from renewable sources and amending and subsequently repealing Directives 2001/77/EC and 2003/30/EC*. Brussels: European Council.

European Council. (2010). *Directive 2010/31/EU of the European Parliament and of the Council of 19 May 2010 on the energy performance of buildings*. Brussels: European Council.

Geden, O., & Fischer, S. (2014). Moving Targets. Die Verhandlungen über die Energie- und Klimapolitik-Ziele der EU nach 2020. Berlin: Stiftung Wissenschaft und Politik (SWP Studie S1 Januar 2014).

Hakelberg, L. (2014). Governance by diffusion. Transnational municipal networks and the spread of local climate strategies in Europe. *Global Environmental Change, 14*(1), 107–129. doi:10.1162/GLEP_a_00216.

Heyvaert, V. (2013). What's in a name? The covenant of mayors as transnational environmental regulation. *Review of European, Comparative & International Environmental Law, 22*(1), 78–90. doi:10.1111/reel.12009.

Ibrahim, N., Sugar, L., Hoornweg, D., & Kennedy, C. (2012). Greenhouse gas emissions from cities. Comparison of international inventory frameworks. *Local Environment, 17*, 223–241. doi:10.1080/13549839.2012.660909.

Jordan, A., van Asselt, H., Berkhout, F., Huitema, D., & Rayner, T. (2012). Understanding the Paradoxes of multilevel governing. Climate change policy in the European Union. *Global Environmental Politics, 12*(2), 43–66.

Kemmerzell, J. (2016). Überlokales Handeln in der lokalen Klimapolitik. Eine Brücke zwischen globalem Anspruch und lokaler Implementation. In M. Barbehön & S. Münch (Eds.), *Variationen des Städtischen, Variationen lokaler Politik* (pp. 245–271). Wiesbaden: Springer VS. doi:10.1007/978-3-658-13394-8_10.

Kern, K. (2014). Climate governance in the European Union multilevel system: the role of cities. In I. Weibust & J. Meadowcroft (Ed.) *Multilevel environmental governance. Managing water and climate change in Europe and North America*, (pp. 111–130). Cheltenham: Edward Elgar.

Kern, K., & Bulkeley, H. (2009). Cities, Europeanization and multi-level governance: Governing climate change through transnational municipal networks. *Journal of Common Market Studies, 47*(2), 309–332.

Knodt, M. (2016). Die Energiepolitik der Europäischen Union und Städte. In J. Kemmerzell, M. Knodt, & A. Tews (Eds.), *Städte und Energiepolitik im europäischen Mehrebenensystem. Zwischen Energiesicherheit, Nachhaltigkeit und Wettbewerb* (pp. 23–43). Baden-Baden: Nomos.

Lee, T. (2014). *Global cities and climate change: The translocal relations of environmental governance*. London, New York: Routledge.

Lee, T., & Koski, C. (2015). Multilevel governance and Urban climate change mitigation. *Environment and Planning C, 33*(6), 1501–1517. doi:10.1177/0263774X15614700.

Schaefter, S., Mohns, T., Hemmati, M., & Offenmueller, S. (2013). *Promoting local climate mitigation (scoping study)*. Berlin: Adelphi.

Schreurs, M., & Tiberghien, Y. (2007). Multi-level reinforcement: Explaining European Union leadership in climate change mitigation. *Global Environmental Politics, 7*(4), 19–46.

Sharp, E. B., Daley, D. M., & Lynch, M. S. (2011). Understanding local adoption and implementation of climate change mitigation policy. *Urban Affairs Review, 47*(3), 433–457. doi:10.1177/1078087410392348.

Sippel, M. (2011). German cities, their climate mitigation activities, and the potential of city-partnerships. *Resources, Energy, and Development, 8*(1), 37–48.

Author Biography

Jörg Kemmerzell is affiliated with the Institute of Political Science at the Technische Universität Darmstadt in Germany. He is a Senior Research Associate in the research project *System Integration and Networks for the Energy Supply*, funded by the German Ministry of Education and Research. Prior to that, he conducted two research projects on local climate policy. The contribution in this volume is based on findings of these projects. His research is on comparative politics in general, with a particular interest in energy and climate policy in multilevel settings. He is a member of the TU Darmstadt research cluster on *Future Energy Systems*.

Chapter 4
Regional Climate Planning and Local Outcomes in California

Michael R. Boswell and Susan G. Mason

Abstract In this chapter, we investigate the impact of California's landmark regional climate planning law Senate Bill (SB) 375 on local climate change planning and policy-making. SB 375 is intended to provide the framework to reduce greenhouse gas (GHG) emissions as mandated by Assembly Bill 32 by coordinating land use and transportation at the regional level through a "Sustainable Communities Strategy" or SCS. Notably, SB 375 does not carry a mandate for local jurisdictions to develop consistent strategies with the SCS. We find that, in general, mid-size to larger cities are coordinating their policies more than smaller cities. Ultimately, this may negate the negative aspect of the free-rider problem on regional outcomes. However, it still may be too early to tell the ultimate impact of SCS on local initiatives.

Keywords Climate action plans · Regional planning · Sustainable communities strategy

4.1 Introduction

Despite the recent breakthrough at the Paris Climate Conference (COP-21), international agreements alone will not be sufficient to solve the global climate crisis. Sub-national governments (e.g., cities, counties, regions, and states) must also play a part in mitigating greenhouse gas (GHG) emissions and adapting to climate change. California is a leader in taking state-directed action on climate change and

M.R. Boswell
City and Regional Planning Department, California Polytechnic State University (Cal Poly), San Luis Obispo, CA, USA
e-mail: mboswell@calpoly.edu

S.G. Mason (✉)
School of Public Service, Boise State University, Boise, ID, USA
e-mail: susanmason@boisestate.edu

S. Hughes et al. (eds.), *Climate Change in Cities*, The Urban Book Series,
https://doi.org/10.1007/978-3-319-65003-6_4

over 150 of California's cities and counties have adopted climate change policies and programs.[1] In 2008, California passed the Sustainable Communities and Climate Protection Act (Senate Bill 375), which directed regional planning and targets for reducing GHG emissions associated with land use and transportation. In this chapter, we examine the impact of this law on local climate actions to see if local plans adopt elements of the regional plans.

Several scholars have found local factors to be the primary reason cities adopt tools to address climate change in the U.S. (Krause 2011; Pitt 2010; Mason et al. 2011). Yet, until recently there were not many regional programs to follow or learn from for local municipalities. In California, the implementation of Senate Bill (SB) 375 provides a unique situation where regional planning is mandated but the localities are not compelled to follow suit, thus preserving "home rule" over land use matters. This makes California's 18 regional planning areas a unique test case to see if regional planning does in fact influence local municipalities that have no mandate themselves to engage in climate action planning or policy adoption to reduce GHG emissions, but are strongly encouraged to do so. If this is the case, then regional planning may be an effective tool to nudge communities to take action. This may also provide a new way to deal with free-riding issues that have plagued voluntary cooperation in general and climate action planning specifically (Krause 2011; Zahran et al. 2008). When considering climate change, there are concerns about the potential of free-riding by cities and citizens that can negate the positive impacts by the communities and citizens engaged in reducing GHGs. The concern is that some cities and citizens may engage in behaviors and policies that reduce GHG emissions, while others do not reduce or even produce more GHG emissions. Since it is not possible to limit the benefits of GHG reductions to only cities or citizens that reduce their emissions, others are able to benefit without contributing to lowering GHG emissions and thus undermining the true benefit for all.

4.2 Metropolitan Planning Organizations

Metropolitan Planning Organizations (MPOs) coordinate transportation planning within their boundaries and receive federal funding to help fulfill their transportation planning needs. MPO boundaries are required for urbanized areas over 50,000 people to ensure a coordinated and comprehensive transportation planning process. A policy board that is comprised of local elected officials who work on regional land use and transportation issues governs MPOs. MPOs maintain a forum for decision-making and involve the public in this decision-making process as they prepare Regional Transportations Plans (RTPs) and work on their Transportation

[1]Estimate as of March 1, 2017 from the Climate Action Planning Database maintained by Michael R. Boswell and Adrienne I. Greve at California Polytechnic State University. The database is regularly updated to include the status of all climate action plans and GHG reduction plans in California and the U.S.

Improvement Program (TIP). There are 18 federally created MPOs in California whose mission is to facilitate regional planning. In California, there are also 26 state statutorily designated Regional Transportation Planning Agencies (RPTAs) that also work on regional transportation plans. These agencies receive state rural planning assistance (RPA) funds to carry out their planning functions (California Transportation Department 2010, p. 6, California Transportation Department n.d.).

4.3 Regional Influences and Voluntary Cooperation

We know there are examples of the way inter-jurisdictional cooperation can come about with the support of higher level programs. For example, Schnieder et al. (2003) find federal programs can augment the ability of regional level actors to overcome collective action problems because they create opportunities for the development of regional institutions by providing tangible support in terms of funding, as well as with the information disseminated about successful and unsuccessful practices. The higher level support provides the foundation or opportunity to gain cooperative governance through voluntarily action. Feiock and West (1993) found cooperative governance between state and local levels with state-level policies increasing the local-level adoption of curb-side recycling. In this chapter, we ask if a similar effect might be possible for regional climate action planning on local-level planning.

There are also other influential factors such as a city's hierarchal position in the region. Central cities in California and the Pacific Northwest have demonstrated that they more often look outward and base their decisions to a greater degree on regional considerations than cities in suburbs do (Lewis 2001; Mason et al. 2011). At the same time, there is evidence that both central cities and suburban cities are prone to looking inward with regard to the localized effects of decisions in land use policy (Lewis 2001), suggesting the role of the hierarchy may be nuanced. In Krause's (2012) study, she attempts to vet whether belonging to a network such as the Cities for Climate Protection (CCP) program—run by ICLEI's CCP program—Cities for Sustainability, or the U.S. Conference of Mayors' Climate Protection Agreement (MCPA) has an effect on cities' implementation of climate, energy, and transportation policies. She finds that within these networks there are differences, where MPCA had little effect on policy implementation while being an ICLEI CCP member was a more significant factor for several GHG reducing activities implemented by cities.

Given the unique nature of the SB 375 having no mandate on localities, then the likelihood of voluntary cooperation should be a consideration. There are case studies that provide evidence for voluntary cooperation. For example, Perkmann finds voluntary cooperation in regional cooperation across borders in Europe (2003), Olberding provides evidence of voluntary regional partnerships in economic development (2002a, b), as does Lackey, Freshwater and Rupasingha's examination of cooperation in rural areas in Tennessee (2002). These cases provide a reason to believe that local-level entities may follow suit, since they have regional level examples to model. Kwon et al. (2014) find one reason that California cities take more sustainability actions than cities in the rest of the country is because of climate projection mandates such as AB 32 and SB 375. At the same time, Kwon et al. (2014) find variation among cities in terms of the cities' commitments and policy actions. One explanation for the variation in outcomes could be the fact that metropolitan areas are complex and networked communities. Innes and Rongerude (2013) find that civic networks, such as California's Collaborative Regional Initiatives (CRIs), play a special role in regions connecting individuals and agencies for more sustainable systems since they reach across dispersed organizations and systems, and connect in ways traditional government organizations do not. This, in turn, raises doubts about the influential nature of centralized government mandates. Furthermore, work by Ostrom (2010) suggests that frameworks for understanding collective action should be updated in the face of empirical evidence that small to medium scale (as opposed to global scale entities) do cooperate when faced with collective action problems. She notes that a polycentric approach where there are independent decision-making centers can, in metropolitan areas, function together to resolve conflicts such as in the case of managing water resources (Ostrom 1962). Furthermore, given the complexity of problems surrounding climate change, a polycentric approach allows for benefits at multiple scales (i.e., individual, local, regional, state) and ultimately contributes to global outcomes (Ostrom 2010).

To sum up, it appears there is evidence that voluntary cooperation and regional planning can result in effective policy action to address climate change. Yet it is not known if legislation can be a driver of coordinated behavior for planning for climate change. The impact of state mandates is unclear and an under-investigated potential point of persuasion. We examine whether a centralized approach [California's Sustainable Communities Strategies (SCS)] can influence inter-local cooperation on land use and transportation planning and spur the adoption of regionally promoted strategies. Specifically, we investigate when and why cities conform to, implement, or modify policies to meet regional climate action planning. To understand the way regional plans might elicit local cities to engage in climate change planning, we examine the policies and actions of cities within three of the 18 MPOs in California: San Diego Association of Governments (SANDAG); Sacramento Area Council of Governments (SACOG); and Southern California Association of Governments (SCAG) covering the Los Angeles metro area. We specifically examine the policies at the regional and local levels to determine which, if any, policies or policy arenas at the local level—such as energy or transportation—conform to the regional plan by the cities' own choosing. Finally, we consider what influence, if any, regional plans can have for motivating local-level climate action planning and the efficacy of the polycentric systems in this arena.

4.4 California Climate Planning

In 2005, Governor Arnold Schwarzenegger issued Executive Order S-03-05 directing state agencies to address climate change and establishing GHG emissions reduction targets. The California legislature quickly followed in 2006 by passing the Global Warming Solutions Act (GWSA or AB 32), which required the state to reduce its GHG emissions to 1990 levels by 2020. To accomplish this, the state prepared the Scoping Plan—essentially a climate action plan—that identified state-level strategies for how the targets would be met. The Scoping Plan focused on state-level action and only briefly mentioned the need for local government action. In response to a perceived gap in the Scoping Plan (with its focus on state actions), the legislature in 2008 passed SB 375 aimed at addressing land use and transportation at the regional and local level.

SB 375 requires a "regional transportation plan for regions of the state with a metropolitan planning organization to adopt a sustainable communities strategy (SCS), as part of its regional transportation plan, as specified, is designed to achieve certain goals for the reduction of GHG emissions from automobiles and light trucks in a region" (Chap. 728, §1). In addition, the California Air Resources Board (CARB), in consultation with the MPOs, established per capita GHG emissions reduction targets for each region. For the San Diego region, SANDAG, the target is 7% below 2005 by 2020 and 13% below 2005 for 2035. For the Sacramento region, SACOG, the target is 7% below 2005 by 2020 and 16% below 2005 for 2035. For the Southern California region around Los Angeles, SCAG, the target is 13% below 2005 by 2035.

SB 375 then links the SCS and achievement of reduction targets to certain types of transportation spending established in the MPO's Regional Transportation Plan. Cities and counties are not required to modify their general plans or other policies and laws to comply with the SCS. Nor does the SCS create mandates for local government. Instead the SCSs incorporate local issues and concerns and are partially dependent on the voluntary actions of local governments if they are to be implemented. For example, one policy in the SACOG SCS requests that local governments concentrate development around transit nodes. The SCS cannot mandate this but SCS initiatives could indirectly influence the choices made at the local level so there are expectations that cooperation will occur if the region is ultimately to reach its GHG reduction targets. A common critique is that SB 375 has no regulatory teeth and may fail to achieve its objectives, but clearly the legislature was respecting the well-established principle of home rule.

4.5 Methods

We used a combination of qualitative and quantitative methods to examine our primary research question: Do regional SCSs influence cities in the region to adopt climate change policies? For this chapter, we only looked at three regions:

San Diego, Sacramento, and Southern California (Los Angeles metro area). These regions adopted the first SCSs in 2011–12[2] providing sufficient time for cities to respond to them by amending and updating local plans and policies. We first reviewed the regional SCSs from the three regions to identify those policies relevant to cities. The SCSs contain policies aimed at the regional agencies exclusively, other ones that are aimed at both the regional and local level, and others still that are just at the local level. This lack of clarity around the jurisdictional level intent of the strategies proved challenging as it was often not clear for who the policy was aimed toward.

We then conducted a content analysis of local climate action plans (CAPs) and general plans for consistency with policies in the regional SCS. We looked at these types of plans because they are the primary instruments in California for establishing climate change policy.[3] We reviewed all local plans from the San Diego and Sacramento regions and took a random sample of plans from the Southern California region (due to the size); this yielded 121 plans. We recorded whether the local plans contained policy that was consistent, somewhat consistent, or not consistent (or not present) with the regional SCS. For the quantitative analysis, we only used the "consistent" responses since the general vagueness of policy made the "somewhat consistent" policies weak. Although we did not conduct a formal inter-coder reliability test we did norm the protocol before using it and we regularly compared results. In addition, we split the Southern California sample randomly and checked the mean and median scores between the coders.

We then explored the data to answer two questions. The first question is: are local plans consistent with the regional SCS? We define consistent in three ways: high consistency is having two-thirds of the policies in the regional SCS represented in the local plans; moderate consistency is between one-third and two-thirds; and, low consistency is less than one-third. The second question is: does regional climate action planning influence local planning? We conducted difference of mean tests and linear regression to determine whether the adoption of the regional SCS is correlated with a higher number of consistent policies. Some cities have local plans that pre-date and some have local plans that post-date the regional SCS, so we examined the difference in those local plans.

See Fig. 4.1 for a map of the 18 regions in California with mandated regional climate change planning.

---

[2]SCAG and SACOG updated their SCSs in 2016 when this analysis was being completed and are, therefore, considered in the study.

[3]California does not mandate that local governments prepare climate action plans, but there are numerous incentives to do so. California does mandate that all local governments prepare a general plan, though with some exceptions, the frequency that these are comprehensively updated is left to the local governments.

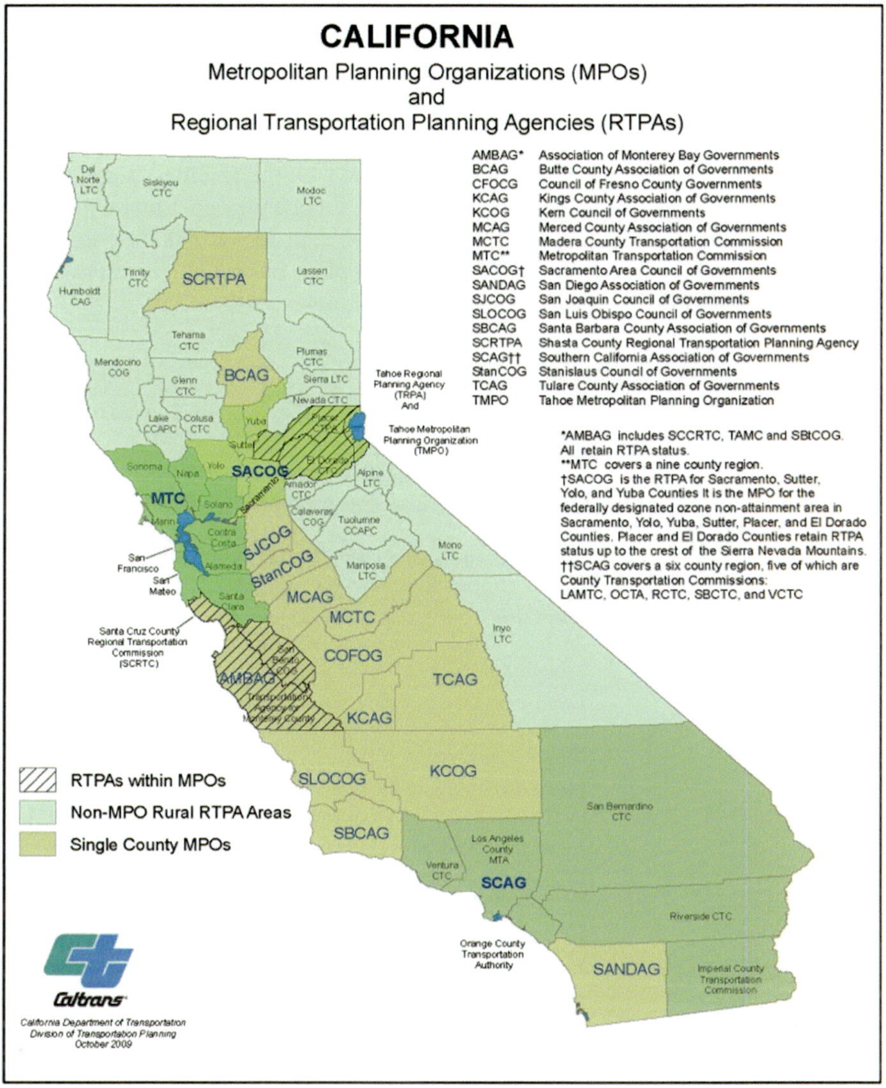

Fig. 4.1 MAP of regions in California (*Source* California Department of Transportation)

4.6 Findings

4.6.1 San Diego Region

In October 2011, SANDAG adopted their SCS for 2035. This included 31 policies with 13 of those being dependent on local government implementation. SANDAG includes one county and 18 cities. For the San Diego region, we reviewed all 18 general plans and 11 CAPs of which two were in draft format. The first question we asked was: are local plans consistent with the regional SCS? Our analysis revealed 11 of the 18 cities in the region are adopting some of the elements in the SCS and, of those, the average number adopted was 3.7 of the 13 (28%) local items in the SCS with a range of 0–10 across both the CAPs and general plans for the cities (see Table 4.1 for elements in the San Diego region SCS). Of the 18 cities, none showed a high consistency (greater than 66%), six showed a moderate consistency (between 33 and 66%), and 12 showed low consistency (less than 33%), including four with no consistency (0%). Thus we conclude that local plans have low consistency with the SANDAG MTP/SCS.

Table 4.1 SANDAG SCS items

Transportation: Identify transportation infrastructure that could be vulnerable to climate change
Transportation: Design projects to meet the needs of all potential users by following complete street development principles
Land use: Incorporate SB 375 CEQA streamlining provisions into local development review process
Land use: Coordinate at the regional level on the planning and implementation of future transportation infrastructure and habitat preserves
Land use: Design future infrastructure projects in a way that protects wildlife corridors and habitat linkages in designated habitat conservation plans
Land use: Promote the construction of sustainable housing and mixed-use projects at existing and planned transit stations
Land use: Address climate adaptation issues in the design of new projects, and when improvements are made to existing infrastructure
Energy: Assist in implementing the regional energy strategy
Energy: Support the increased use of clean, alternative fuels in vehicle fleets
Energy: Support planning and infrastructure development for alternative fueling stations and plug-in electric vehicle (EV) chargers
Energy: Work with San Diego Gas and electric and other stakeholders to mitigate the potential impacts of electric vehicles on the electric grid
Financial: Use the updated 2050 RTP smart growth concept map as a basis for allocating smart growth incentives, prioritizing transit service enhancements, and seeking additional smart growth funds
Greenhouse gas: Assist in implementing the climate action strategy

The second question we asked was: does regional climate action planning influence local planning? The plans completed before the SCS average 10% consistency and after average 39% consistency, with the most consistent city having plans post-dating the SCS. The post-SCS mean is higher, as expected, and an independent samples *t*-test shows this difference of means to be statistically significant at the 0.05 level. Thus, at this time, the SANDAG MTP/SCS *does* appear to have influenced local plans.

As an additional question, we examined the five (of 13) strategies from the SCS that had the highest level of local consistency. These five were (in order of consistency, from highest to lowest):

- Design projects to meet the needs of all potential users by following complete street development principles;
- Promote the construction of sustainable housing and mixed-use projects at existing and planned transit stations;
- Support the increased use of clean, alternative fuels in vehicle fleets;
- Support planning and infrastructure development for alternative fueling stations and plug-in electric vehicle (EV) chargers;
- Incorporate SB 375 CEQA [California Environmental Quality Act] streamlining provisions into the local development review process.

The five most consistent items are elements of the transportation, land use, and energy SCS items but not financial or GHG directly. However, the last element of SCS on GHG actually entails an additional set of items when it indicates to assist in implementing the climate action strategy unique to the San Diego region.

In the San Diego region, the explicit elements of the Climate Action Strategy are listed in Table 4.2. The policy measures of the Climate Action Strategy are intended to be a list of tools that the regional and local policy makers and the general public could use to address in updates to their regional transportation and comprehensive plans. The SCS success is ultimately reliant on the planning contained in the regional transportation plan and CAPs. There is some consistency with some of the CAPs. One of the most frequently mentioned strategies in the SANDAG cities' CAPs was leading by example. The CAPs demonstrated this through their efforts to deal with their own municipal energy use and by looking for ways to create a more energy efficient municipal fleet. Promoting measures that facilitated more green building, mixed-use development, or street connectivity were also often pursued in the CAPs which is another goal within the Climate Action Strategy. Perhaps not surprising, the local efforts were targeted toward reducing their own municipal GHG emissions based on the factors identified in the benchmarks and were often very specific actions such as increasing the use of photovoltaics. This suggests that the more targeted Climate Action Strategy items may be easier to achieve than the more general SCS action items.

Table 4.2 Major topic areas with goals in the climate action strategy—general areas goals

1. Lead by example
2. Adopt a climate action plan
Land use and transportation goals
3. Build communities for walking, bicycling and public transit
4. Minimize greenhouse gases released from vehicles
5. Support increased use of low carbon alternative fuels
6. Protect transportation infrastructure form climate change impacts, e.g., sea level rise, heat, mudslides
Clean energy and efficient building goals
7. Reduce energy use
8. Pursue energy reductions in existing residential commercial buildings
9. Promote state policy for zero net energy residential and commercial buildings

4.6.2 Sacramento Region

In April 2012, the SACOG adopted their 2035 Metropolitan Transportation Plan/Sustainable Communities Strategy (MTP/SCS). The SACOG MTP/SCS includes 31 policies and numerous supporting strategies; 15 of the strategies depend partially or completely on local government for implementation (see Table 4.3 for the 15 strategies).

The SACOG region includes 23 cities (in six counties), however, our analysis only includes 21 cities due to the availability of plans. All 21 cities have adopted general plans but only six have adopted CAPs as of 2015. Only five of the 21 cities have local plans that post-date the adoption of the MTP/SCS in April 2012, thus little local planning has occurred since the development of the MTP/SCS.

We reviewed the city general plans and CAPs for consistency with the 15 strategies in the MTP/SCS that depend on local-level implementation. The first question we asked was: are local plans consistent with the regional SCS? The analysis shows that on average the cities are consistent with only 4.2 strategies of the 15 (28%) in the MTP/SCS, with the range being from zero to ten. Of the cities, only one showed a high consistency (greater than 66%), eight showed a moderate consistency (between 33 and 66%), and 12 showed low consistency (less than 33%), including three with no consistency (0%). Thus we conclude that local plans have low consistency with the SACOG MTP/SCS.

The second question we asked was: does regional climate action planning influence local planning? The plans completed before the MTP/SCS average 27% consistency and after average 32% consistency; and an interesting observation is that the most consistent city has plans pre-dating the MTP/SCS. Although the post-MTP/SCS mean is higher as expected, an independent samples t-test does not show this difference of means to be statistically significant. Thus, at this time, the SACOG MTP/SCS does not appear to have influenced local plans.

Table 4.3 SACOG SCS items

Transportation: Develop community activity centers well-suited for high-quality transit service and complete streets
Transportation: Support preservation of the existing road system as the top funding priority
Transportation: Improve service to transit-dependent populations—disabled, low-income, senior, youth
Transportation: Support smart growth development that promotes bicycling, walking, and transit use
Transportation: Support development in Transit Priority Areas (TPAs)
Transportation: Develop a complete streets policy
Transportation: Develop transit station near economic centers and neighborhoods
Transportation: Create an interconnected system of streets, bikeways, and walkways that support a more compact development form
Land use: Grow consistent with blueprint principles
Land use: Facilitate development of housing in all price ranges
Land use: Direct greenfield developments to areas immediately adjacent to the existing urban edge
Land use: Implement the Rural-Urban Connection Strategy (RUCS)
Land use: Create a development activity tracking tool to assess growth patterns
Land use: Encourage development patterns that provide safe and efficient pedestrian and bicycle access to transit
Financial: Collect development-based fees sufficient for both local road improvements and regional-scale road, transit and/or bicycle pedestrian improvements

As an additional question, we examined the five (of 15) strategies from the MTP/SCS that had the highest level of local consistency. These five were (in order of consistency—highest to lowest):

- Facilitate development of housing in all price ranges;
- Support smart growth development that promotes bicycling, walking, and transit use;
- Direct greenfield developments to areas immediately adjacent to the existing urban edge;
- Encourage development patterns that provide safe and efficient pedestrian and bicycle access to transit;
- Create an interconnected system of streets, bikeways, and walkways that support a more compact development form.

Of note, SACOGs signature regional planning policy—its “Blueprint” plan—was identified in only two cities. The most consistent items are land use and transportation items and none are financial.

4.6.3 Southern California Region

In April 2012, the SCAG adopted their 2012–2035 Regional Transportation Plan/Sustainable Communities Strategy (RTP/SCS). The SCAG RTP/SCS includes 60 policies; 25 of the strategies depend partially or completely on local government for implementation (see Table 4.4 for the 25 strategies).

In the SCAG region, there are six counties and 191 cities. We took a stratified sample of the cities by listing the cities in order of population and dividing the cities into thirds with 63 cities in the first group and 64 cities in the second two groups. We then randomly sampled 16 cities from the first third and 17 cities from next group. We wanted to be sure to include Los Angeles in the analysis, so we selected Los Angeles from the third group and randomly sampled an additional 16 cities from the remaining 63 cities in the third sub-group. One city was dropped from the analysis due to lack of data. One author examined 25 and the other examined 24 of the 49 cities general plans and CAPs.

The first question we asked was: are local plans consistent with the regional SCS? We reviewed the city general plans and CAPs for consistency with the 25 strategies in the MTP/SCS that depend on local-level implementation. The average number of adopted items was 7.9 strategies of the 26 (30%) in the RTP/SCS, with a range of zero to 21 items. Of the 49 cities, only three showed a high consistency (greater than 66%), 15 showed a moderate consistency (between 33 and 66%), and 31 showed low consistency (less than 33%), including one with no consistency (0%). Thus, we conclude that local plans have low consistency with the SCAG RTP/SCS.

The second question we asked was: does regional climate action planning influence local planning? The plans completed before the RTP/SCS average 24% consistency and after average 33% consistency; and unlike the SACOG region, the two most consistent cities have plans post-dating the RTP/SCS. The post-RTP/SCS mean is higher, as expected, and independent samples t-test shows this difference of means to be statistically significant at the 0.10 level. Thus, at this time, the SCAG RTP/SCS *does* appear to have influenced local plans.

As an additional question, we examined the five (of 25) strategies from the MTP/SCS that had the highest level of local consistency. These five were (in order of consistency—highest to lowest):

- Identify resources that can be used for employing strategies to maintain and assist in the development of affordable housing;
- Support projects, programs, policies, and regulations to protect resources areas, such as natural habitats and farmland, from future development;
- Develop comprehensive active transportation network;
- Encourage the implementation of a complete streets policy;
- Encourage the use of range-limited battery electric and other alternative fueled vehicles.

Table 4.4 SCAG SCS items

Transportation: Provide a network of local community circulators that serve new TOD, HQTAs, and neighborhood commercial centers
Transportation: Develop first-mile/last-mile strategies on a local level to provide an incentive for making trips by transit, bicycling, walking, or neighborhood electric vehicle
Transportation: Encourage transit fare discounts and local vendor product and service discounts for residents and employees of TOD/HQTAs
Transportation: Work with transit properties to identify and remove barriers to maintaining on-time performance
Transportation: Develop comprehensive active transportation network
Transportation: Encourage the implementation of a Complete Streets policy
Transportation: Support work-based programs that encourage emission reduction strategies and incentivize active transportation commuting or ride-share modes
Transportation: Encourage the development of telecommuting programs by employers through review and revision of policies that may discourage alternative work options
Transportation: Expand the use of regional transit modes such as BRT, rail, and limited-stop service
Transportation: Encourage regional and local transit providers to develop rail interface services at Metrolink, Amtrak, and high-speed rail stations
Transportation: Prioritize transportation investments to support compact infill development
Land use: Adopt neighborhood-oriented development, suburban villages, and revitalized main streets as livability strategies in areas not served by high-quality transit
Land use: Support active and healthy community environments that encourage safe walking, bicycling, and physical activity by children
Land use: Update local zoning codes, general plans, and other regulatory policies to accelerate adoption of land use strategies included in the 2012–2035 RTP/SCS Plan
Land use: Encourage the development of complete communities, which includes a diversity of housing jobs, recreation and culture, and services
Land use: Pursue joint development opportunities to encourage the development of housing and mixed-use projects around transit
Land use: Consider developing healthy community or active design guidelines that promote physical activity and improved health

(continued)

Table 4.4 (continued)

Land use: Support projects, programs, policies, and regulations to protect resources areas, such as natural habitats and farmland, from future development
Land use: Engage in a strategic planning process for open space and park space, specifically in park-poor communities
Land use: Implement innovative strategies that enhance mobility and air quality, including those that increase the walkability of communities and accessibility to transit
Energy: Encourage the use of range-limited battery electric and other alternative fueled vehicles
Energy: Support strategies to develop infrastructure and supportive land uses to accelerate fleet conversion to electric or other near zero-emission technologies
Financial: Develop policies and prioritize funding for strategies and projects that enhance mobility and air quality
Financial: Seek partnerships with state and regional agencies to acquire funding sources for innovative planning projects
Financial: Cooperate with stakeholders to identify new funding sources and/or increased funding levels for the existing transportation network
Financial: Identify resources that can be used for employing strategies to maintain and assist in the development of affordable housing

The most consistent items cross all the elements are transportation, land use, energy and financial in the SCS. Of note, much like the city of San Diego, the city of Los Angeles touts leading by example and actually has adopted 21 of the 25 elements into their CAP and general plan. Santa Monica, another large city, had similar outcomes. The fact that these larger cities are adopting elements recommended at the regional level may point to what Lewis (2001) and Mason et al. (2011) found to be true of central cities, which is central cities tend to be more outward looking than more suburban cities.

4.6.4 Overall Regional Findings

In addition to analyzing each region independently, we combined the data to complete an overall assessment. The entire sample included 88 cities across the three regions. The first question we asked was: are local plans consistent with the regional SCSs? The analysis shows that on average the cities are consistent with 28% of the policies in the associated MTP/SCS, with the range being from 0 to 81%. Thus we conclude that local plans are generally not consistent with the MTP/SCSs. The policies that were consistent with the SCSs mostly fell into five general categories: affordable housing, active transportation, community planning (including compact form, infill development, and diversity of land uses), complete

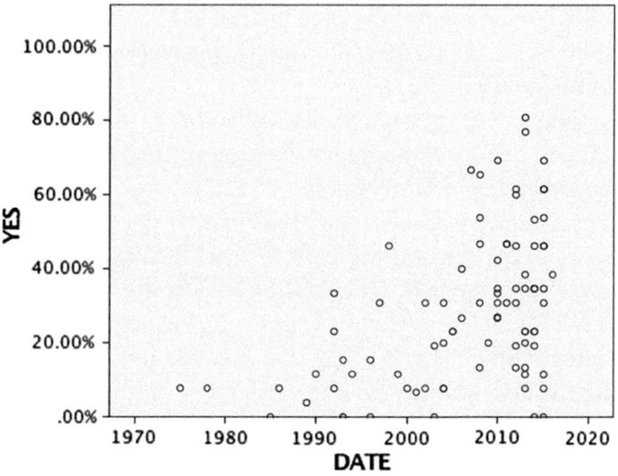

Fig. 4.2 Date of the adopted local plans and percentage of RTP/SCS policies present in those plans

streets, and alternative fuel vehicles and infrastructure. One explanation for some of the consistency across two of these categories is that housing and complete streets are mandated to be considered in the plans by state law.

The second question we asked was: does regional climate action planning influence local planning? The plans completed before the MTP/SCS average 23% consistency and after average 35% consistency. The post-MTP/SCS mean is higher as expected, and is shown to be a statistically significant difference (at the 0.01 level). But further analysis reveals a complication. A linear regression model that includes both the pre/post dichotomous variable and a time variable (age of plan) shows that time is statistically significant and pre/post is not. In other words, local plans have been generally getting better over time in terms of being more consistent with principles in the RTP/SCS and an independent effect of the RTP/SCS cannot be discerned. This is perhaps best demonstrated in Fig. 4.2, which shows the percentage consistency over time, keeping in mind that the RTP/SCSs were adopted in 2011 and 2012. This effect also appears to be present in the individual regions though we did not run a regression due to the smaller sample sizes. Thus, at this time, the RTP/SCSs do not appear to have independently influenced local plans. Perhaps the good news, though, is that local plans are improving over time. Of note, we also found the increasing population size was a statistically significant independent predictor of local plan consistency.

4.7 Discussion and Conclusions

Taken together, the regions of Sacramento, San Diego, and Southern California (Los Angeles metro area) provide little evidence to support the hypothesis that regional planning under SB 375 has affected local-level planning. It must be emphasized that it may be too early to draw firm conclusions. Only 30 of the 88 cities studied have updated their CAP or general plan since the adoption of the MTP/SCS. Given that the rate of plan updates is about two per year, change will come slowly.

Consistent with previous research, we did find that central cities are leaders in climate action planning. In looking at the relative consistency ranking in the three central cities in each region— Sacramento, Los Angeles (Southern California), and San Diego—we see that the first two of the three are clear leaders. In the Sacramento region, the central city of Sacramento had the second highest consistency with the SACOG RTP/SCS. The highest degree of consistency is the city of Davis, which is the 6th largest city, a major employment center in the region, a college town (location of the University of California, Davis), and known for progressive public policy especially on bicycle transportation. In the Southern California region, Los Angles tied for highest consistency with the SCAG RTP/SCS with the city of Santa Monica. Santa Monica arguably functions as a central city for the west side of the Southern California region. The city of San Diego in the San Diego region was not as outward looking as some of the other cities, falling somewhere in the middle of the pack of the cities adopting SCS elements in the region.

This trend is more nuanced in the San Diego region, however, where the highest levels of consistency in plans are found in the mid-sized cities of Carlsbad and San Marcos and moderate levels of consistency in the two largest cities of San Diego and Chula Vista. It may be that the progressive efforts of regions of SACOG and in SANDAG the mid-sized and larger cities in the regions that drive the goals of the SCS are being met. This, in turn, would mean some free-riding, by small cities, may not be as detrimental to the larger regional goals if mid-sized and larger cities are taking the lead. If the SCS is the benchmark of goals to be met for regional planning, they simply may not be specific enough in terms of reducing GHG emissions or manageable to achieve at the local level. For example, one SCS element says, 'Work with San Diego Gas and Electric and other stakeholders to mitigate the potential impacts of EVs on the electric grid'. This and other elements like it are so nonspecific in terms of its impact, it is hard to gauge if there would be any impact on reducing GHG emissions. Another SCS element example is the goal to "Identify transportation infrastructure that could be vulnerable to climate change" is not adopted in local plans as a tool for GHG reduction. In the case of the San Diego region, the idea of leading by example as having an "effect" on the region may be more of a result of the specific Climate Action Strategy elements dealing with GHG emissions as an element of the SCS than the overarching SCS items in total.

In the end, it appears that some localities are doing their part and following the lead of the region on several elements and actionable goals under the SCS. It remains to be seen whether individual efforts within a region can overcome or will be thwarted by municipalities that act in ways associated with free-riding by not taking action. The concern that free-riding or anything less than complete regional effort could result in a lack of net gain in a reduction or a significant reduction in

GHG emissions may need further exploration. We do have evidence that some localities are engaged and are taking very targeted actions, and those communities may be large enough to offset the concerns of smaller communities not engaging. Moreover, MPOs are using the SCSs to direct funding to local areas to incentivize climate action planning and GHG emissions reduction implementation.

The findings suggest that regions can lead by example and provide best practices or options in strategies. In some cases (such as SACOG), it does not appear the SCS is a persuasive mechanism for GHG reductions at the local level, but not in all cases there or elsewhere such as SANDAG and SCAG. The more fine-tuned and targeted initiatives at the regional level are being taken on by those engaged at the local level, and those appear more helpful to climate change mitigation efforts than more general sustainability action items. Clearly, the mission of the goals coming out of the SCS reach beyond the reduction of GHGs and extend to matters of habitat conservation and improving service to transit-dependent populations—including disabled, low-income, senior, and youth populations. As the name implies, the SCS is taking a more holistic look at what a sustainable region of communities should be (see Chap. 2: Homsy). Yet, where there is a more targeted regional strategy for GHG emission reductions as in the case of the San Diego Climate Action Strategy, the region demonstrates it might become a vehicle for helping cities to reduce their GHG emissions through their own CAPs and inventories. This suggests providing more targeted strategies within a regional plan might be a catalyst for initiatives at the city level, which in turn reaches both local and regional goals (see Chap. 7: Sari and Prayoga). This also suggests regional initiatives can be an appropriate way to help cities address climate change issues, but it may be too soon to determine the ultimate effect of the SCS on local level actions. One way to foster more targeted strategies would be for the SCSs to explicitly lay out their expectations for regional and local entities, perhaps even giving estimates needed at the local areas to meet regional targets.

References

California Department of Transportation. (2010). California regional transportation plan guidelines. Retrieved September 29, 2016 http://www.dot.ca.gov/hq/tpp/offices/orip/rtp/index_files/2010%20RTPGuidelines_Jan2011_Technical_Change.pdf.

California Department of Transportation. (N.A.). Statutes and Definitions. Retrieved September 29, 2016. http://www.dot.ca.gov/hq/tpp/offices/orip/list/agencies_files/p1-3.pdf.

Feiock, R., & West, J. (1993). Testing competing explanations for policy adoption: Municipal solid waste recycling. *Political Research Quarterly, 46*(2), 399–419.

Innes, J. E., & Rogerude, J. (2013). Civic networks for sustainable regions—Innovative practices and emergent theory. *Planning Theory & Practice, 14*(1), 75–100.

Krause, R. (2011). Policy innovation, intergovernmental relations and the adoption of climate protection initiatives by US cities. *Journal of Urban Affairs, 33*(1), 45–60.

Krause, R. (2012). An assessment of the impact that participation in local climate networks has on Cities' implementation of climate, energy, and transportation policies. *Review of policy Research, 29*(5), 585–604.

Kwon, M., Jang, H. S., & Feiock, R. (2014). Climate protection and energy sustainability policy in California cities: what have we learned? *Journal of Urban Affairs, 36*(5), 905–924.

Lackey, S. B., Freshwater, D., & Rupasingha, A. (2002). Factors influencing local government cooperation in rural areas: Evidence from the Tennessee Valley. *Economic Development Quarterly, 16*(2), 138–154.

Lewis, P. G. (2001). Looking outward or turning inward? Motivations for development decisions in California central cities and suburbs. *Urban Affairs Review, 36*(5), 696–720.

Mason, S. G., Marker, A., & Mirsky, R. (2011). Primary factors influencing green building in cities in the Pacific Northwest. *Public Works Management & Policy, 16*(2), 157–185.

Olberding, J. (2002a). Does regionalism beget regionalism? The relationship between norms and regional partnerships for economic development. *Public Administration Review, 62*(4), 432–443.

Olberding, J. (2002b). Diving into the "Third Waves" of regional governance and economic development strategies: A study of regional partnerships for economic development in U.S. Metropolitan areas. *Economic Development Quarterly, 16*(3), 251–272.

Ostrom, E. (2010). Polycentric systems for coping with collective action and global environmental change. *Global Environmental Change, 20*(4), 550–557.

Ostrom, V. (1962). The political economy of water development. *American Economic Review, 52*(2), 450–458.

Perkmann, M. (2003). Cross-border regions in Europe: Significance and drivers of regional crossborder co-operation. *European Urban and Regional Studies, 10*(2), 153–171.

Pitt, D. (2010). The impact of internal and external characteristics on the adoption of climate mitigation policies by US municipalities. *Environment and Planning C: Government and Policy, 28*(5), 851–871.

Schnieder, M., Scholz, J., Lubell, M., Mindruta, D., & Edwardsen, M. (2003). Building consensual institutions: Networks and the national estuary program. *American Journal of Political Science, 47*(1), 143–158.

Zahran, S., Himanshu, G., Brody, S. D., & Vedlitz, A. (2008). Risk, stress, and capacity: Explaining metropolitan commitment to climate protection. *Urban Affairs Review, 43*(4), 447–474.

Author Biographies

Michael R. Boswell is Professor and Department Head of the City and Regional Planning Department at California Polytechnic State University-San Luis Obispo. His research focuses on climate action planning, hazard mitigation, adaptive management and governance, local government planning and decision-making, and sustainable development.

Susan G. Mason is Emeritus Faculty in the School of Public Service at Boise State University. She was previously Director of Strategic Initiatives for the School of Public Service and Professor and Founding Director of the Department and Master and Certificate programs in Community and Regional Planning at Boise State University. Her research focuses on regional governance, urban and community development, and sustainability.

Chapter 5
Metropolitan Governance of Waste for Climate Change Mitigation: Examining the Case of Montreal

Eve Bourgeois and Sara Hughes

Abstract The increasing ambitions of municipal governments to reduce greenhouse gas emissions raises important questions about the role of municipal and regional governance structures. More or less centralized decision-making structures present unique sets of trade-offs. While centralization offers benefits of authority and standardization, it can also create challenges for public input and democratic oversight. Using the City of Montreal, this chapter examines these trade-offs for climate change policy innovation in the waste management sector. Montreal's experience highlights the trade-offs between centralization and policy innovation. Although centralization has produced positive environmental outcomes in the waste management sector through the implementation of a metropolitan area-wide compost program, the experience of Montreal reveals this comes with important trade-offs in local autonomy and engagement. While more centralized control can, in one sense, lead to more effective decision-making, this must come with measures to provide space for citizen participation and consideration of local specificities in the policy-making process.

Keywords Metropolitan governance · Urban governance · Waste management · Climate change mitigation · Montreal

5.1 Climate Change Mitigation, Municipal Waste, and Regional Governance

A growing number of cities in Canada are rethinking their waste management strategies as a way of meeting increasingly ambitious sustainability and climate change goals. Waste management practices can account for as much as 15% of a

E. Bourgeois · S. Hughes (✉)
University of Toronto, Toronto, Canada
e-mail: sara.hughes@utoronto.ca

E. Bourgeois
e-mail: eve.bourgeois@mail.utoronto.ca

S. Hughes et al. (eds.), *Climate Change in Cities*, The Urban Book Series,
https://doi.org/10.1007/978-3-319-65003-6_5

city's greenhouse gas (GHG) emissions, and many cities are facing very real limits to their ability to deposit waste in municipal landfills. Increasing residential and commercial density in cities, combined with inefficient waste pickup routes, can lead to significant air pollution from waste management services. Cities are therefore taking advantage of growing demands and opportunities for reforming the waste management sector in ways that reduce its contribution to climate change, local air pollution levels, and demand for scarce land for disposal.

Incorporating organic waste recycling is one way many cities are restructuring waste management services to meet sustainability and climate change goals. Organic waste recycling programs typically collect food waste from households and businesses and compost this waste in an industrial facility. Such programs are able to reduce demand for landfill space by diverting organic waste, and subsequently, reduce the amount of methane (a potent GHG) produced by landfills. Organic waste recycling can be particularly attractive for the growing number of cities that have set "zero waste" targets, as their aim is to soon stop sending any waste to landfills.

While the benefits of waste management reform, and particularly the introduction of organic waste recycling programs, may be increasingly attractive, the means by which such reforms can or should be introduced in a complex urban governance landscape are much less so. An important feature of the multilevel governance context for many cities is regional or metropolitan scale governments and institutions, and the role these play in shaping and delivering waste management services.

Scholars of urban governance have long debated the political, policy, and economic implications of more or less centralized metropolitan governance institutions (Savitch and Vogel 2009). Early proponents of regional or metropolitan scale institutions argued for their economies of scale and increased efficiency. Pooling resources and authorities at the metropolitan scale can facilitate economies of scale in service delivery and improve outcomes. Vincent Ostrom and others (e.g., Ostrom, Tiebuot, and Warren 1961) pushed back on this view, arguing that more decentralized institutional arrangements facilitated both efficiency and democratic accountability in service delivery by allowing municipalities to tailor services to local needs. In the 1990s, the debate began to center on the economic consequences of regional institutions, with proponents arguing that regional governance can enhance economic competitiveness in a global economy. More recently scholars have examined the political and redistributive implications of metropolitan governance (Judd and Swanstrom 2012; Swanstrom 2001).

Many have suggested that rather than pragmatic responses to service delivery needs, metropolitan governance reforms are highly contentious and are responding to emerging priorities in a quickly shifting political and economic landscape (e.g., Brenner 2002; Horak 2013). The value of metropolitan governance in this case then is in the political benefits such reforms are able to provide. However, very little research has examined the ways in which regional or metropolitan governments might influence local environmental or climate change policy outcomes. This is a critically understudied area, as we know that regional governments matter for outcomes in sectors imports for climate change mitigation such as waste management, transportation systems, and land use.

In this chapter, we examine the ways in which advocates have used metropolitan governance institutions to forward waste management reforms in Montreal, Canada, as part of a broader climate change mitigation agenda. We use the case of Montreal to better understand how metropolitan level institutions shape local climate change mitigation practices by influencing the service delivery choices of smaller units of government, and why this dynamic might vary within a metropolitan region. Montreal's experience highlights the trade-offs between centralization and policy innovation. Although centralization has produced positive environmental outcomes in the waste management sector through the implementation of a metropolitan area-wide compost program, the experience of Montreal reveals this comes with important trade-offs in local autonomy and engagement.

Montreal has four distinct types of actors involved in metropolitan government: The Montreal Metropolitan Community (*Communauté Métropolitaine de Montréal*); the Agglomeration of Montreal (*Agglomération de Montréal*); the City of Montreal and the related municipalities that de-amalgamated; and the boroughs within Montreal's city. At the higher level, the provincial government sets the waste reduction target, and municipalities have to work toward these goals. In the case of Montreal's metropolitan region, the four levels of municipal government shared responsibilities regarding waste management in order to satisfy provincial regulation. For instance, strategy development and implementation are done at the regional level—i.e., the Montreal Metropolitan Community, and the Agglomeration of Montreal—while service delivery is dealt by the City of Montreal in collaboration with its boroughs. Institutionally, Montreal is, therefore, something of an in-between case in terms of regional metropolitan government structure: decision-making about service delivery is neither entirely regional nor entirely local. It is also a case of a region that has tended to experiment with different forms of regional governance as an explicit response to emerging policy challenges, as we will outline below. We, therefore, use this case to learn more about how metropolitan governments are able to shape the service delivery choices of municipalities, particularly when they are working to persuade them to adopt more environmental or climate change oriented practices rather than requiring them, and what role institutional reform can play in helping meet urban climate change goals.

5.2 Metropolitan Governance in Montreal: Experimenting with Form

Montreal is located in the Canadian province of Quebec, which adopted in the 1960s several laws to encourage cities to amalgamate in order to pool fiscal capacities and increase administrative efficiency. The Province hoped that by sharing fiscal resources in this way, less wealthy municipalities would be able to provide better public services. This strategy had only limited success, and municipal amalgamations came back on the provincial policy agenda in the 1990s.

In December of 2000, the Quebec government, under the *Parti Québécois*, passed Bill 170, which forced the amalgamation of many municipalities (MAMOT 2010).

Bill 170 specifically targeted large urban areas, and the number of municipalities in Quebec decreased significantly (Belley et al. 2009; Collin 2002). The 28 municipalities on the island of Montreal (and the L'Île-Bizard) were amalgamated to form the new City of Montreal in order to create a "strong, big and internationally known urban core by tapping into suburban growth to redistribute property tax revenues extracted from industrial and commercial development" (Keil and Boudreau 2005). Amalgamations also took place during the same period, and with the same rationale, on the South and North shores of Montreal (and across the province). Quebec was not alone in its pursuit of reform at this time. The province of Ontario was also undergoing a metropolitan reform process, which had an even greater effect on the number of municipalities (Razin and Smith 2006). Despite a large number of amalgamations that came from Bill 170 in Quebec, there remain 1288 municipalities in the province as of 2016 (Institut de la statistique du Québec 2016).

The municipalities that were included within the amalgamated City of Montreal are referred to as boroughs. These boroughs were granted some additional powers during amalgamation as a way to soften the impacts of reform (Hamel 2009). The borough system in Montreal "is a compromise in response to the desire to preserve a place for the expression of local distinctions and, therefore, to make municipal restructuring more acceptable in the eyes of the elected officials and the administrators of the suburbs" (Collin and Robertson 2005).

The new City of Montreal was, however, very short-lived; popular opposition was strong, especially among Montreal's suburban municipalities. Opponents claimed the amalgamation threatened the local political culture of the new amalgamated cities, the integrity of their community, and hindered local democracy (Hamel 2006). As a result, municipal de-amalgamations were part of the Liberal party's strategy during their successful 2003 bid for provincial power. Following their election, the Liberal party held public consultations about how municipal de-amalgamation should proceed. Subsequently, several referenda were held across the province during summer 2004, which led 15 amalgamated municipalities on the island of Montreal to recover their independence.

However, these de-amalgamated cities did not recover the full authority they had enjoyed before the reform. In 2006 a new regional structure, the *Agglomération de Montréal* (Agglomeration of Montreal—here after, AM), was created as a third tier of government that encompasses the City of Montreal and the 15 de-amalgamated municipalities "in order to preserve an organic bond between the central city and the municipalities that chose to be reconstituted" (Belley et al. 2009). The regional government was tasked with issues such as affordable housing, fire, and police, which represent about one-third of municipal expenditures of the region (Belley et al. 2009). Other "local concerns" were left in the hands of the municipalities.[1] (Fig. 5.1)

---

[1]Or what are called "related municipalities" in the work of Belley et al. (2009).

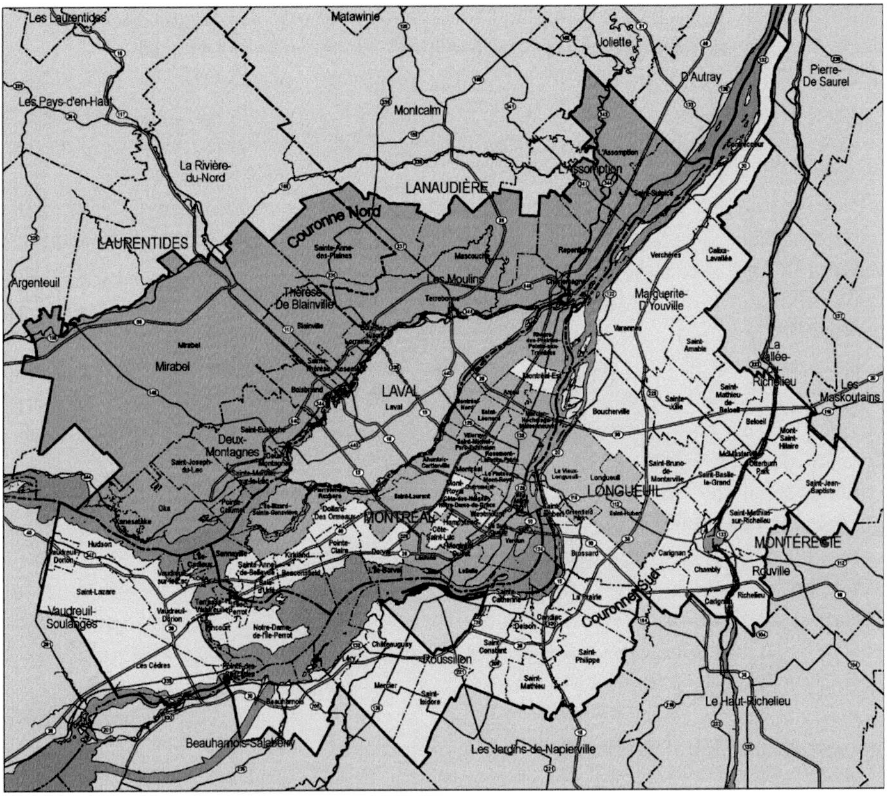

Fig. 5.1 CMM's Territory (CMM 2015)

This redistribution of powers has led some observers of urban governance in Montreal to suggest that between the boroughs and the AM, the City of Montreal lost most of its powers (Hamel 2009). This view, however, does not take into account the fact that the City of Montreal has significant influence over the AM. The seats at the AM are shared among member municipalities according to population size, and the City of Montreal holds the majority of seats (MAMOT 2016). This has created an imbalance as suburban municipalities make significant financial contributions to regional services but do not have much influence on the decision-making process.

Ultimately, coordination between the AM and its suburban regions is managed through the *Communauté Métropolitaine de Montréal* (Montreal Metropolitan Community—here after, CMM) who oversees planning and coordination of regional development, promotion of art and culture, waste management, affordable housing, and municipal infrastructure. The CMM governs 82 municipalities with 3.5 million people living in the region (almost half the population of the province of Quebec), and it corresponds broadly to the territory of the census metropolitan region defined by Statistics Canada. The decision-making process at this stage takes place through indirectly elected representation, as the CMM members are elected officials from the municipalities. While many see the CMM as a welcome improvement to a larger scale regional coordination, it has also been criticized for a lack of citizen participation and a duplication of tasks performed by municipalities (Boudreau 2013; Razin and Smith 2006).

Finally, the latest municipal reform in 2015 has taken several powers away from the boroughs and given them to the more centralized City of Montreal and the budgets of the boroughs have subsequently decreased. This shift in authority and financial resources includes several local services previously under boroughs' responsibilities including waste collection and recycling. These changes were not welcomed by the boroughs, who claim that service delivery will suffer from centralized control (Marsolais 2014). However, the city's conviction that centralized service delivery would facilitate equality in services delivery and fiscal efficiency won the day.

Because of the municipal reform, Montreal is governed through four-tier governance arrangements: the boroughs; the City of Montreal (and the related municipalities); the AM; and the CMM. The case of Montreal is, however, not an exception; such multilevel structures are in fact quite common practices in Canadian metropolitan areas (Young 2009). Despite the trends toward new regionalism that have characterized the municipal reorganization in the United States or in Europe in the 1990s, Canada has adopted coordinating structures for the municipal sector that resemble the old regionalism approach (Hamel 2006, Lafortune and Collin 2011; Razin and Smith 2006).

5.3 Governing Waste in Montreal for Climate Change Mitigation

Each of the four tiers of government in Montreal helps to shape the purpose and delivery of waste management services. The provincial government sets the guidelines and goals for waste reduction and elimination for both the municipal and the private sector (i.e., the industries, businesses, and institutions sector; and the construction, renovation, and demolition sector). Provincial legislation requires that municipal and regional governments produce a waste management plan (*Plan de gestion des matières résiduelles*) in which they detail their strategy for complying with provincial legislation. The Province will typically provide funding to municipalities to build the infrastructure needed to reach provincial goals.

In the case of the Montreal metropolitan area, strategies for meeting provincial targets are often developed by the CMM, while in non-metropolitan regions this would be taken on either by municipalities themselves or by regional county municipalities (RCM). Additionally, several municipalities in the Montreal metropolitan area have chosen to outsource waste management to the private sector.

Table 5.1 Actors and their Main Roles in Montreal's Waste Management Sector Following the 2015 Municipal Reform in Montreal

Provincial government	Establishes the waste recovery, valorization and elimination goals through legislation; funds municipalities
Communauté métropolitaine de Montréal (CMM)	Develops a waste management plan for the metropolitan region through participation of its members
Agglomeration of Montreal (AM)	Implements the waste management plan adopted by the CMM; manages waste recycling and elimination
City of Montreal & related municipalities	Gives contracts to the private sector for waste collection and transporting, or does it internally
City of Montreal's boroughs	Manage waste collection following the City's directives

Further, industrial, commercial, and institutional waste and the construction, renovation, and demolition sectors are outside of municipalities' jurisdiction. Thus, waste coming from these sources is entirely left to the private sector, which represents about two-thirds of the waste generated in the province (CMM 2016). Overall, more than 1000 organizations and businesses are working in the waste management sector on the territory of the CMM, including tens of *Éco-Centers*[2] which are under the municipal management (CMM 2015). (Table 5.1)

The Quebec government has been taking steps since the late 1990s to reduce the amount of waste sent to landfills. The first legislation was adopted in 2000 by the Quebec government with the goal to decrease the overall amount of waste produced. This legislation set a 60% target for both organic and traditional waste recycling by 2008 (Gouvernement du Québec 2000). Following the legislation, curbside collection programs for both recycling efforts were implemented across the province and municipal recycling rates increased dramatically, doubling between 2000 and 2006 (Ménard 2009). The organic waste recycling programs were the less successful of the two due to problems of odors coming from organic waste facilities. The sluggish organic waste recycling rates added an additional financial burden on facilities already facing economic challenges, as they lacked the funds to adapt the infrastructure in order to rectify the odor problem. As a result, a large portion of the organic recycling facilities in the province declared bankruptcy and shut down their activities (Chamard and Méthot N.d.). Indeed, organic waste recycling is the primary waste management challenge that is faced by Quebec's municipalities: In 2006, only 8% of organic waste in the province was collected and composted (Ménard 2009).

To counter these poor results, the government of Quebec reconfirmed its commitment to organic waste recycling in its latest iteration of climate change legislation. In 2011, the *Politique québécoise de gestion des matières résiduelles* (Gouvernement du Québec 2011) was adopted with the goals of diverting waste

[2]*Éco-centers* are recycling centers that dealt generally with waste which cannot be disposed in regular recycling bins (such as batteries, paint, cars, etc.). There are several of them across Montreal in order to serve the population in different sectors of the city.

from landfills, contributing to the provincial climate change mitigation strategy, and empowering actors in the waste management sector. More specifically, the legislation sets out the following goals to be met by the end of 2015:

- Decrease to 700 kg/person the amount of waste send to landfills;
- Recycle 70% of paper, cardboard; plastic; glass; and metals;
- Recycle 60% of organic waste;
- Recycle 80% of concrete residue, bricks, and asphalt;
- Segregate at the source or send to a sorting center 70% of the residue coming from construction, renovation, and demolition coming from the building segment.

This time, the provincial government adopted financial incentives to encourage municipalities to increase their recycling rates and has committed to cover a large portion of the costs for the construction of organic waste recycling facilities. Under the compensation plan sets up in the 2011 provincial legislation, 70% of the costs of selective municipal waste collections on the CMM were covered by the province (CMM 2016).

Municipalities have done well in traditional recycling, where they are generally close to meeting the provincial targets. Progress has been slower for organic waste, and as of early 2017, several municipalities in Quebec still do not have such a program. Organic waste recycling remains unattractive to many who are put off by the potential odor. Table 5.2 provides an overview of recycling rates in the province and in the Montreal metropolitan area, demonstrating that Montreal is lagging behind the rest of the province (CMM 2012a, b). Figures 5.2 and 5.3 show the waste and organic waste recovery rates for the territories in the AM. The average rate of general recycling in the AM is 58%, while the organic waste recycling rate is 12%; both are lower than the provincial targets.

This discrepancy among Montreal's boroughs and related municipalities in the AM is the result of decentralized practices in the waste collection until 2015. Before 201, the boroughs and related municipalities provided curbside pickup programs while the AM oversaw waste disposal and coordinated recycling infrastructure. As a result, the services and programs offered by the boroughs and related municipalities differed greatly, especially for the organic waste collection. For instance, by the end of 2014, only eight of the 19 boroughs within the City of Montreal were offering a partial curbside pickup program for organic waste, and only one borough (Rosemont—La-Petite-Patrie) was able to implement a full municipal organic waste collection by the end of 2015 (Trottier 2015).

Table 5.2 Recovery Rates in the CMM Area (2012)

	Organic Recycling Rate (%)	Recycling Rate (%)
Provincial targets for 2015	60	70
Province actual rates	21	59
CMM actual rates	12	59
AM actual rates	12	58

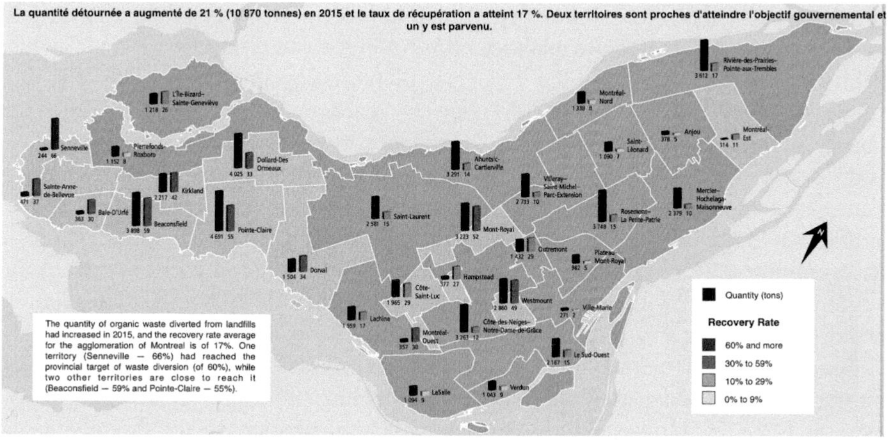

Fig. 5.2 Organic Recycling Rates on the Agglomeration of Montreal (Ville de Montréal 2015a). (Figures 5.2 and 5.3 come from the report *Bilan 2015 des matières résiduelles de l'agglomération de Montréal: Réduire pour grandir* done by the City of Montreal. The figures were modified in order to make their reading easier in a black and white print. Additionally, the legend was translated from French; and the text in the boxes have been translated and modified to put the emphasis on the information most relevant to our purpose.)

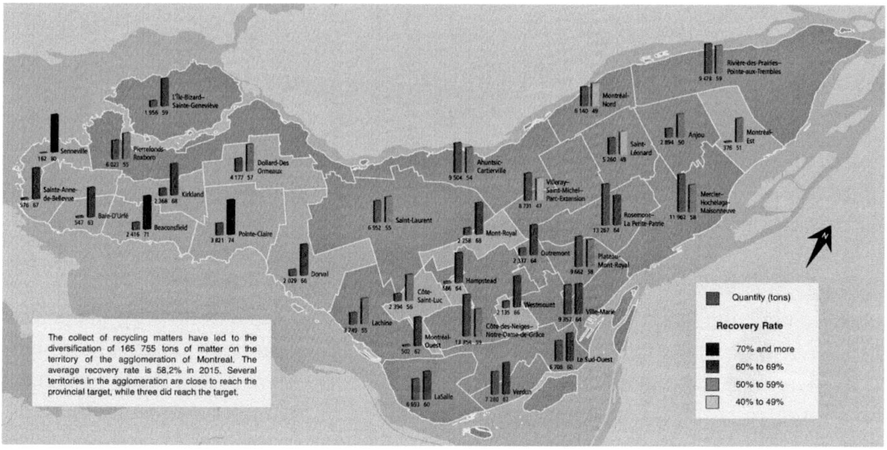

Fig. 5.3 Recycling Rates on the Agglomeration of Montreal (Ville de Montréal 2015a)

The municipalities in the region that are not part of the City of Montreal have made the most progress on organic waste recycling. Figures 5.2 and 5.3 show maps of the AM, including Montreal City (and its 19 boroughs), and the 15 related municipalities on the Agglomeration's territory. The boroughs are identified in dark gray, while the independent cities are identified in light gray. In Fig. 5.2, on each territory, the left column represents the quantity of organic waste produced (in tons), while the right column represents the organic recovery rate. In Fig. 5.3, the left column represents the quantity of recycling waste produced (in tons), while the right column represents the recovery rate of recycling matter. As Fig. 5.2 shows, the only territory in the AM that has reached the provincial target is the related municipality of Senneville, with an organic recovery rate of 66%. Additionally, the second and third best performing cities are two related municipalities that have almost reached the provincial 2015 targets: Beaconsfield (59%) and Pointe-Claire (55%). Even the borough of Rosemont—La-Petite-Patrie shows a disappointing 15% recovery rate. Thus, it seems that the good performance of the Agglomeration of Montreal is the result of the work done by the cities on its territory more than it was the work done by the City of Montreal.

Figure 5.3 demonstrates that municipalities implemented general recycling programs much sooner than organic waste recycling programs. Once again, however, the three territories showing the best recycling rates are the three same related municipalities that had the best organic waste recovery rates (although they do not hold the same positions within the top three): Senneville (80%), Pointe-Claire (74%), and Beaconsfield (71%). The general difference between the organic recovery rates and the recycling rates may be explained by the fact that organic waste curb pick programs are still at an early stage in Quebec while recycling pickup programs have been in place for years. As a result, we might expect that organic waste collection will improve over time, as did general recycling programs.

5.4 Centralization as an Urban Governance Response

Decision makers have tended to see decentralization in the metropolitan area as a roadblock to more efficient and equal service delivery. In the case of the City of Montreal, centralization of services allowed the city center to take progressive actions in the waste management sector and worked even further toward its goal of reducing greenhouse gas emissions. The reason centralization may work in this context may have less to do with economies of scale and more to do with the greater political opportunity to override household reluctance to participate in organic waste recycling schemes. The City of Montreal has the resources and legislative capabilities to require and incentivize participation in organic waste recycling, and, most importantly, had the political will to impose these measures despite residents' complaints about odors or inconvenience. Greater political distance from households may insulate decision makers from pushback and resistance to the program.

In January 2015, the City of Montreal was given several new responsibilities, including a waste collection that had previously been under the jurisdiction of the city's boroughs. This is a two-year experiment, after which a two-thirds vote by the Montreal city council is required to maintain the new arrangement (Cameron 2014, 2016). Following this reform, the Montreal city council has quickly adopted a regulation requiring an organic waste collection program for every Montreal household by 2019. Implementation of this program started in the fall 2016—offering for the first time a municipal organic waste collection to many residents living in Montreal (Cameron 2015). This new program has been accompanied by financial penalties to ensure citizen participation in the program. Indeed, the city can give fines up to $1000 to households who do not use the new green bin. However, it is not yet clear how the implementation of such a fine would work since it is not always apparent which bin belongs to which household.

In the years to come, the City of Montreal will take over waste collection contracts as the current contracts signed by the boroughs come to an end, putting the city in a better position to negotiate lower prices (Ville de Montréal 2015b). According to the city, centralization of the waste collection will engender economies of scale since Montreal will be in a better position to negotiate contracts and ensure equal services across its territory (Orfali 2014).

It still remains to be seen whether this centralization will provide the benefits the city claims. For instance, unified services coordinated at the city center might result in poor delivery at a more local scale. Each borough presents different characteristics, such as population density or household characteristics, which may influence the efficacy of service delivery in a sector such as waste management. Even within a single borough, those characteristics might be different from one street to another. The work of Chaput shows several difficulties that must be considered when dealing with waste management in the high-density population such as Montreal in order to increase efficiency (Chaput 2015). Such characteristics might appear quite trivial, but they are highly influential in the success of this type of policy. Criteria such as whether residents have the space to store three bins, for example, must be considered by municipalities when designing organic waste management programs (Chaput 2015). If the new organic program implemented by the city does not take into account those differences, there is a risk that residents will ignore the new policy put in place preferring using the system they are used to. To see the effect of the climate change innovative policy, the municipal government has to tailor service delivery, to some extent, to the need of each neighborhood as every citizen has a role to play in climate change mitigation.

Additionally, in terms of institutional structure, as Razin and Smith argue, "promises asserted in order to promote reform are nearly always exaggerated, unrealistic, or sometimes plainly wrong. Expected savings never fully materialize, if at all; unexpected extra costs tend to emerge; administrative chaos could accompany the transition period, and major territorial reforms frequently lead to substantial unintended and unanticipated consequences" (2006). For instance, the experience of the amalgamation process in Toronto did not provide the cost savings that were expected. On the contrary, Schwartz finds that the overall costs of the new city were higher following the amalgamation and that the cost of waste management services stayed largely the same (Schwartz 2009). In terms of the equalization of services, more time is needed before we can assess whether the implementation of the universal organic waste collection will be successful. So far, several boroughs have expressed concerns about the uniform waste collection that does take into account the particularity of each borough such as population density, type of households, and space for another collecting bin (Cameron 2014, 2015). The reform has also been critiqued for reducing the boroughs' budgets, many of whom claim they will not have enough money to provide the services that were left under their responsibility (Orfali 2014). For example, the borough of Outremont expects to close two ice rings due to lack of funding (Cameron 2014).

From a sustainability and climate change perspective, the centralization of waste management in the City of Montreal has so far been positive: it is likely that the new municipal organic waste legislation and services will increase waste diversion rates and decrease methane emissions in landfills. Indeed, the city plans to introduce an organic waste collection to 95,000 new households in 2017 and reach every household by 2019 (Ville de Montréal 2017). Moreover, given that the best organic waste recycling rates were achieved by related municipalities, and not by Montreal's boroughs, the centralization of powers will not prevent learning from best practices. In fact, the latest metropolitan waste management plan takes the related municipality of Beaconsfield as a success story (CMM 2016). In two years, Beaconsfield went from producing the second-highest quantity of waste destined to landfill to having the second best recycling rates. Indeed, through a new pricing scheme for its waste collection program—according to which households pay depending on the amount of waste they throw out—Beaconsfield has reduced is waste generation by 48% between 2014 and 2015 (Normandin 2016). The City of Montreal is currently considering this option, and learning from related municipalities may continue in the years to come.

As stated previously, it is too soon to provide a full assessment of the outcomes of the new waste management program in the City of Montreal as it is currently being implemented. If the organic waste recovery rates do increase, there may be interest in shifting other powers back to the City of Montreal. Before the implementation of the latest reforms in 2015, Montreal's boroughs enjoyed a level of autonomy that was unique in the province. As a result of this latitude, boroughs were able to develop innovative policies and programs, such as the participatory budgeting process adopted by Plateau—Mont-Royal's borough (Hamel 2009). Further centralization might take away the autonomy boroughs need in order to develop such innovative policies. Moreover, more centralization could shift the balance of powers between the city center and its boroughs and encourage boroughs to seek to leave the City of Montreal. The 2015 municipal reform was not well received by the boroughs, and the City of Montreal wants to go even further. The city has asked the provincial government to amend the city charter in order to make centralization even easier for the Montreal city council. As a result of this initiative, some borough mayors have suggested a second referendum round to leave the City of Montreal (Cameron 2016).

5.5 Going Forward: Grappling with Trade-offs

Debates about the benefits and challenges of decentralization in urban governance have a long and unresolved history, and the pursuit of climate change mitigation through service delivery reform adds another layer to the debate. As more and more cities around the world are pursuing climate change mitigation, it becomes more urgent to better understand how governance structures might enhance or hinder climate change action at the municipal level and the trade-offs they engender. In this chapter, we draw on and forward some of these debates in the context of waste management, and particularly efforts to introduce organic waste recycling programs to reduce urban GHG emissions.

The case of the City of Montreal, and its metropolitan area provides some evidence that centralization of services can facilitate progress on environmental aims in the waste management sector. Metropolitan restructuring is a common occurrence in Montreal as a response to policy and service challenges. In an effort to counter sluggish progress in the boroughs, the City of Montreal took on waste management and introduced measures to require and encourage organic waste recycling in the region. While the programs are still new, there are signs that they will improve organic waste recycling rates despite protests from the boroughs that centralized programs will be unresponsive to local needs. The City of Montreal has been committed to reducing GHG emissions since 2005 and has additionally pledged to reduce its GHG emissions by 80% by 2050 during the 2015 United Nations Climate Change Conference (COP 21) held in Paris (Ville de Montréal 2016).[3] To reach this target, the city has no choice but to take innovative actions, as the global experience shows us that "business-as-usual" is incompatible with meaningful climate change mitigation.

Whether or not centralization leads to policy innovation in the field of climate change mitigation, the case of Montreal reminds us of the broader factors and trade-offs at play when it comes to the centralization of metropolitan governance. While the centralization of powers in the city center in Montreal has resulted in policy innovation with promising outcomes from an environmental perspective, there are additional criteria against which urban governance should be—and is—evaluated. Indeed, efforts to move more authority into the hands of the City of Montreal are prompting some borough mayors to publicly criticize the city and even propose leaving the city structure altogether. Ongoing reforms can make it difficult for citizens and interest groups to know who to talk to, as responsibilities are continually shifting from one level to another. This may ultimately make programs like organic waste recycling more difficult because they require a level of buy-in and engagement that other reforms do not.

Indeed, the case of Montreal highlights the trade-offs that are inherent in designing urban governance arrangements able to facilitate innovations that reduce urban GHG emissions (see also Chap. 4). The more centralized authority may, in some cases, lead to more decisive action, equalization of services, and the political and financial resources needed to go forward. However, this may come at the cost of local control and input for decision-making, both of which are also valued features of urban climate change governance. Such trade-offs may have more than

---

[3]In 2005, the City of Montreal adopted its first GHG emissions reductions, aiming to reduce 30% of GHG emissions from 1990 levels by 2020 (Ville de Montréal 2016).

normative consequences. For instance, policies that are not tailored to different needs across the city may negatively affect citizen participation in climate change mitigation programs and thus hinder the city's ability to meet its GHG reduction goals. There are also signs in Montreal of decreasing stability in the more centralized system. While broader scholarship on metropolitan governance is aware of such trade-offs, it provides little guidance for how they might be effectively negotiated.

Based on these findings, we forward two recommendations that may alleviate some of the negative impacts of centralization for the City of Montreal and elsewhere. First is to increase participation spaces for citizens and smaller units of governments, such as the boroughs of Montreal. This would address some of the concerns raised by the boroughs about the extent to which local needs are incorporated in municipal or regional scale decision-making processes. Institutionalizing such spaces would allow citizens and small units to be more engaged and ensure their specific needs are met. This would also give boroughs the opportunity to adopt policies specific to their needs without sacrificing the benefits of centralization. Our second recommendation is to increase transparency in the decision-making process among and between the four levels of government in the Montreal region. In order for engagement processes to be effective, the public has to understand the role of each level of their regional government and to know where to address their concerns. In addition, transparency in the decision-making process is likely to reduce citizens' concerns over new policies, as they would be openly debated. These recommendations might not be enough to solve the current dissatisfaction with the governance structure in Montreal, but they are a first step to increase cohesion among the different levels of governments and between the elected officials and the general public.

Growing ambitions for climate change mitigation among urban policymakers may lead other metropolitan areas to reconsider their governance structures or, at least, confront the trade-offs they are facing in their current context. Scholars and practitioners of urban climate change governance should be mindful of the myriad ways multilevel governance arrangements shape local policy and political outcomes beyond policy innovations and reductions in GHG emissions. Urban climate change governance may benefit from the centralization of decision-making, but this may come at the expense of political and institutional stability and local engagement. Unwelcome reforms may hinder climate change mitigation innovations in the long term. Even if we might applaud the innovative policy adopted by the City of Montreal, the process underpinning it might ultimately undermine both its aims and the governance structure.

Urban climate change governance poses important challenges for municipal actors and comes with political trade-offs. Fostering innovation through institutional design must be balanced with local engagement and opportunities for input (See Chaps 6–9). To be effective in the long term, both policy innovations and new institutional designs must be sure to involve citizens and to incorporate their concerns. Despite having the best policy in place, if citizens are not participating in and benefitting from climate change programs, cities may be less likely to see major gains in the long-run. Urban climate change governance needs the involvement and engagement of all citizens for long-term results. Metropolitan centralization may be an effective way to compel municipalities to respond to climate change, but effective climate change policies and institutions must also have the capacity to respond to residents' concerns. An absence of tailored programs and local democracy might hinder policy innovation if not taken into account during institutional design. This is especially true at the local level where municipalities are asking citizens to change their daily habits—such as encouraging residents to take more often public transportation or to change their disposal habits by adopting often the third bin in their household—in order for the city to reach their climate change mitigation targets. Developing better metrics against which we may evaluate climate change mitigation innovations like organic waste recycling programs is an important area for future research.

References

Belley, S., Bherer, L., Chiasson, G., Collin, J.-P., Hamel, P., & Rivard, M. (2009). Quebec. In P. Dutil (Ed.), *Foundations of Governance: Municipal Government in Canada's Provinces* (pp. 70–137). Toronto: University of Toronto Press.

Boudreau, J-A. (2013). Penser la ville-région. Acteurs économiques et stratégies de développement à Montréal et à Toronto. In C. Lefèvre, N. Roseau, T. Vitale (Ed.), *De la ville à la métropole: les défis de la gouvernance* (pp. 77–97). Paris: Éditions L'OEil d'OR.

Brenner, Neil. (2002). Decoding the Newest 'metropolitan Regionalism' in the USA: A Critical Overview. *Cities, 19*(1), 3–21.

Cameron, D. (2014). Montréal: la réforme des arrondissements est adoptée. *La Presse*, November 25.

Cameron, D. (2015). Le compostage pour tous d'ici 2019 à Montréal. *La Presse*, April 28.

Cameron, D. (2016). Projet de loi 120: vers de nouveaux référendums de défusion à Montréal. *La Presse*, 22 November.

Chaput, N. (2015). *La gestion des matières résiduelles dans les milieux densément peuplés*. Sherbrooke: Université de Sherbrooke.

CMM. (2012a). *Taux de récupération des matières putrescibles, 2012*. Enquête sur la gestion des matières résiduelles effectuée en 2012 par la CMM auprès des 82 municipalités locales du Grand Montréal, Communauté métropolitaine de Montrléal (CMM).

CMM. (2012b). *Taux de récupération des matières recyclables, 2012*. Enquête sur la gestion des matières résiduelles effectuée en 2012 par la CMM auprès des 82 municipalités locales du Grand Montréal., métropolitaine de Montrléal (CMM).

CMM. (2015). *Territoire de la Communauté métropolitaine de Montréal*. Communauté métropolitaine de Montréal (CMM).

CMM. (2016). *Plan métropolitain de gestion des matières résiduelles (PMGMR) 2015–2020*. Communauté métropolitaine de Montrléal (CMM).

Collin J-P. (2002). La réforme de l'organisation du secteur municipal au Québec: la fin ou le début d'un cycle? *Organisations et territoires*: 5–13.

Collin, J.-P., & Robertson, M. (2005). The Borough System of Consolidated Montréal: Revising Urban Governance in a Composite Metropolis. *Journal of Urban Affairs, 27*, 307–330.

Gouvernement du Québec. (2000). *Politique québécoise de gestion des matières résiduelles 1998–2008*.

Gouvernement du Québec. (2011). *Politique québécoise de gestion des matières résiduelles*.

Hamel, P. (2006). Institutional Changes and Metropolitan Governance: Can De-amalgamation be Amalgamation? The Case of Montréal. In E. Razin & P. J. Smith (Eds.), *Metropolitan Governing: Canadian Cases, Comparative Lessons* (pp. 95–120). Jerusalem: The Hebrew University Magnes Press.
Hamel, P. (2009). Services centraux, services en trop: la dérive centrifuge des arrondissements de Montréal. In G. Sénécal, L. Bherer (Ed.), *La métropolisation et ses territoires*. Québec: Presses de l'Université du Québec.
Horak, Martin. (2013). State Rescaling in Practice: Urban Governance Reform in Toronto. *Urban Research & Practice, 6*(3), 311–328.
Institut de la statistique du Québec. (2016). *Le Québec chiffres en main*. Gouvernement du Québec.
Judd, D. R., & Swanstrom, T. (2012). *City Politics* (8th ed.). Glenview, IL: Pearson Education.
Keil, R., Boudreau, J-A. (2005). Arrested Metropolitanism: Limits and Contradictions of Municipal Governance Reform in Los Angeles, Montreal and Toronto. In H. Heinelt, D. Kübler (Ed.), *Metropolitan Governance: Capacity, Democracy and the Dynamics of Place* (pp. 100–116). London and New York: Routledge, Taylors & Francis Groupé.
Lafortune, M.-È., & Collin, J.-P. (2011). Building metropolitan governance capacity: The case of the Communauté métropolitaine de Montréal. *Canadian Public Administration, 54,* 42–399.
MAMOT. (2010). *Historique*. Ministère des Affaires municipales et de l'Occupation du territoire (MAMOT).
MAMOT. (2016). *L'organisation municipale au Québec en 2016*. Ministère des Affaires municipales et de l'Occupation du territoire (MAMOT).
Marsolais, M. (2014). La révision du financement des arrondissements à Montréal fait des mécontents. *Radio-Canada.*
Ménard, K. (2009). Un regard sur la Politique québécoise de gestion des matières résiduelles 1998-2008. *In L'État du Québec, 2009,* 238–244.
Normandin, P-A. (2016). Payer pour jeter: forte baisse des déchets à Beaconsfield. *La Presse*, 5 February.
Orfali, P. (2014). Le maire Coderre lance une importante réforme du financement des arrondissements. *Le Devoir*, 8 July.
Ostrom, Vincent, Tiebuot, Charles M., & Warren, Robert. (1961). The Organization of Government in Metropolitan Areas: A Theoretical Inquiry. *The American Political Science Review, 55*(4), 831–842.
Razin, E., & Smith, P. J. (2006). Metropolitan Governing: Been There, Done That, Where Are We Now? In E. Razin & P. J. Smith (Eds.), *Metropolitan Governing: Canadian Cases, Comparative Lessons* (pp. 3–25). Jerusalem: The Hebrew University Magnes Press.
Savitch, H., & Vogel, R. K. (2009). Regionalism and Urban Politics. In Jonathan S. Davies & David L. Imbroscio (Eds.), *Theories of urban politics* (pp. 106–124). London: SAGE Publications Ltd.
Schwartz, H. (2009). Toronto ten years after amalgamation. *Canadian Journal of Regional Science/Revue canadienne des sciences régionales, 32,* 483–494.
Swanstrom, T. (2001). What we argue about when we argue about regionalism. *Journal of Urban Affairs, 23*(5), 479–496.
Trottier M-C. (2015). Du compost presque partout à Rosemont. *TVA Nouvelles.*
Ville de Montréal. (2015a). *Bilan 2015 des matières résiduelles de l'agglomération de Montréal: Réduire pour grandir.*
Ville de Montréal. (2015b). *Nouveaux contrats pour l'élimination, la collecte et le transport des matières résiduelles—des économies substantielles pour la Ville de Montréal.*
Ville de Montréal. (2016). *Réduction des émissions de GES.*
Ville de Montréal. (2017). *Résidus alimentaires.*
Young, R. (2009). Canada. In N. Steytler, J. Kincaid (Ed.), *Local Government and Metropolitan Regions in Federal Systems* (pp. 107–135). Montreal Kingston: McGill-Queen's University Press.

Author Biographies

Eve Bourgeois is a doctoral student in the Department of Political Science at the University of Toronto in collaboration with the School of the Environment. Her research mainly focuses on the role of public-private partnerships in Canadian municipalities in the waste management sector. More particularly, she is interested in how coalitions within municipal governments influence their capacity to adopt and implement climate change policies in Canada. She is also working on cross-provincial initiatives regarding the carbon tax, carbon-trading system, and other actions taken by Canadian provinces to reduce their greenhouse gas emissions and mitigate climate change. Prior to her doctoral studies, Eve studied at the Université de Montréal, where she completed a B.Sc. and an M.Sc. in political science with a special interest in Canadian politics.

Sara Hughes is an Assistant Professor of Political Science at the University of Toronto and the University of Toronto-Mississauga. She is also a Senior Fellow at the Global Cities Institute and Bissell-Heyd Fellow at the Center for the Study of the United States at the University of Toronto. Her research focuses on the drivers of urban environmental politics and policy-making surrounding water resources, climate change mitigation and adaptation, waste management, and environmental justice.

Part II
Citizen Engagement Innovations

Chapter 6
Multi-level Governance, Civic Capacity, and Overcoming the Climate Change "Adaptation Deficit" in Baltimore, Maryland

Andrea Sarzynski

Abstract This chapter explores the policy and planning efforts of the city of Baltimore, Maryland, with respect to climate change adaptation using the institutional analysis and development framework. The city's innovative combined disaster preparedness and climate change adaptation plan was adopted in 2013 and situated within a complex, multi-level climate governance regime established in 2007. It's planning efforts have been recognized for their high quality from the federal government and nonprofit organizations. Additionally, city staff chose to build civic capacity on climate change resilience early in its implementation efforts, reaching more than one thousand residents to date. Yet civic dialogue around climate adaptation or private adaptive action has not emerged. Instead, adaptation efforts appear squarely rooted within the governmental realm and subject to resource constraints of its primary institutions, leaders, and staff. Thus, the Baltimore case reveals both the resilience of staff when conducting climate adaptation planning in an atmosphere of fiscal constraint, and the difficulties in fostering a community-wide sense of responsibility for climate adaptation action.

Keywords Climate change · Adaptation · Civic capacity · Baltimore · Governance

6.1 Introduction

The City of Baltimore, similar to other cities in the Mid-Atlantic region, is vulnerable to climate change hazards including sea level rise, chronic flooding, extreme heat, and worsening air quality. These challenges led the city to take stock of their situation and look for new ways to adapt and built-in mechanisms to create resilience.

A. Sarzynski (✉)
University of Delaware, Newark, DE, USA
e-mail: apsarzyn@udel.edu

S. Hughes et al. (eds.), *Climate Change in Cities*, The Urban Book Series,
https://doi.org/10.1007/978-3-319-65003-6_6

Mancur Olson's "logic of collective action" (1971) predicts no action on shared challenges faced by diverse groups with many participants, due to high transaction costs for coordination and a lack of tangible, short-term benefits. Urban climate change adaptation represents one policy area that may be more likely to fall prey to this logic, as the shared challenge is complex, the resolution costs are high and immediate while the benefits are less tangible and long-term, and coordination among the many diverse stakeholders is difficult to structure and maintain. Yet some cities have managed to make progress in this realm, motivating the question of how cities have successfully overcome the logic of collective action in adopting and implementing climate change adaptation policy (see Chap. 2: Homsy).

This chapter explores the decision process surrounding climate change adaptation policy in a large, diverse, and vulnerable legacy city—Baltimore, Maryland—using the institutional analysis and development framework (Ostrom et al. 2014). Baltimore adopted an innovative climate change adaptation and disaster risk reduction plan in 2013, following earlier efforts on sustainability and climate change mitigation. Baltimore's adaptation plan is nested within a dynamic, multi-level climate governance regime and coordinated by a new multi-stakeholder institution. As such, Baltimore's experience offers the opportunity to learn about the multi-level institutional arrangements and strategies that successfully enabled collective action in Baltimore and that could provide insight into climate adaptation strategies for other similarly situated cities with less-developed climate policies, such as Philadelphia. The case also provides insights into the study of civic capacity as a motivating factor for collective action on climate change adaptation, which so far has been limited in its application to a few policy areas including urban education and economic development (Briggs 2008; Stone et al. 2001).

6.2 Literature

Adaptive capacity is the concept used to denote "the ability of a system [such as a city] to adjust to climate change; it is thought to be determined by a range of factors including technological options, economic resources, human and social capital, and governance" (McEvoy et al. 2006). Scholars examine adaptive capacity with the often unstated presumption that limited capacity (especially of resources, both human and economic) is the primary barrier to action. Resources constitute a key set of barriers to climate adaptation at all levels of collective action (Ekstrom and Moser 2014; Moser 2009). Yet, not all communities with the resources to act do take action on climate adaptation, suggesting resources may be a necessary but insufficient precursor to action (Mackie 1980). Some scholars distinguish capacity from "willingness" to act (Heinrichs et al. 2013) while other scholars distinguish action from awareness of the problem or analysis of local impacts (Baker et al. 2012). Ekstrom and Moser (2014) describe the difference between potential capacity and adaptation action as "the adaptation deficit."

How do communities overcome this "adaptation deficit"? One possible explanation is the development of *civic capacity*, defined broadly as the institutional resources and shared understanding to act within and outside of government to solve collective problems (Stone et al. 2001). Such capacity develops over decades and is built on perceptions of benefits and costs of civic mobilization (as in Olson's classic "logic of collective action"), and the actual outcomes from prior collective problem solving (see Chap. 12: Brown).

Scholars have explored the importance of "social capital" in supporting collective problem solving (e.g., Maloney et al. 2000), although Briggs (2008) argues that civic capacity is a more expansive concept than social capital. Briggs distinguishes two dimensions of civic capacity: "the extent to which the sectors that make up a community are (1) *capable* of collective action on public problems (the resource dimension), given the norms and institutional arenas for local action; and (2) *choose* to apply such capability (the dimension of effort, will and choice, or "agency")" (2008, p. 13; emphasis in original).

Similarly, Mettler and Sorelle (2014) argue that policy outputs such as the provision of public goods and services, or payments to individuals, generates "resource effects" that then generate civic capacity and the civic predisposition (attitude or will) to work collectively to solve problems. Additionally, rules and procedures for civic engagement such as opportunities for public participation in policy-making generate "interpretive effects" of policy outputs, which also work to build civic predisposition to work collectively. In this way, prior public policy decisions and their outputs feed back into the political system by influencing the propensity and ability of individuals to participate further (Mettler and Sorelle 2014).

Briggs' second dimension calls to mind the classic sociological tension between structure and agency (i.e., Martin and Dennis 2010). Is it sufficient to provide the opportunity for collective action through structuring institutions and opportunities for interaction? Or is there an additional element necessary beyond structure when moving persons to act collectively? Shinn argues, "[o]bligation or responsibility [for collective action] comes from characteristics of society outside the formal political structure" (1999, p. 109). Similarly, an intermediate stage in the *climate adaptation pathway model* is perceiving a "sense of responsibility for developing a solution," which emerges after building "a clear understanding of climate change" generally and of your "own climate change vulnerability" specifically (Gardner et al. 2009, p. 19).

With this literature in mind, the following propositions are derived from the general elements necessary to build capability for collective action on climate change adaptation:

- P1: *A conductive institutional environment for deliberation* through public participation and repeated interaction;
- P2: *Various supports for action*, including financial and managerial resources, leadership, knowledge, legal authority, and public support (Jenkins-Smith et al. 2014); and
- P3: *A commitment to act collectively* in the best interests of the community, whether or not the participating individual will obtain tangible near-term benefits.

6.3 Methodology

6.3.1 Case Design and Selection

This analysis employs a single case, mixed method design (Yin 2009) to investigate the above-stated propositions in the context of Baltimore's climate adaptation decision-making. According to Yin, single case designs are appropriate where the motivation is to better understand a recent phenomenon of interest (an exploratory element) or to better understand the operations and causes behind a phenomenon of interest (an explanatory element). Single case designs can be used to better specify theory and, as such, generalize to theories rather than other cases (Yin 2009). The technique used here shares much similarity with historical analysis, as many of the relevant sources of information are archival in nature.

Baltimore was chosen for the case study for empirical reasons. It represents an instance of a still-relatively rare outcome of interest in a context of scholarly interest to urban and environmental researchers. The Center for Climate and Energy Solutions has so far documented 48 localities with some action on climate adaptation in the United States, including 20 localities with planning recommended in their general plans; 13 localities with planning in progress; 10 localities with completed adaptation plans; and 5 localities with adaptation actions but no formal plans (Center for Climate and Energy Solutions n.d.). A US Conference of Mayors survey from 2014 found 40% of their respondents (i.e., 81 of 203 cities) were engaged in climate adaptation planning and 40% were also taking adaptive actions (The United States Conference of Mayors 2014). These numbers are quite small given the approximately 36,000 municipal or town governments within the United States (as of the 2012 Census of Governments).

Other large, diverse, and physically vulnerable communities with adopted climate adaptation policies that could be studied include New York, Philadelphia, and Seattle (Center for Climate and Energy Solutions n.d.). New York City's climate policies have been the subject of numerous investigations (Rosenzweig et al. 2007; Rosenzweig and Solecki 2010); the other areas including Baltimore are ripe for study.

6.3.2 Analytic Approach

This paper uses the institutional analysis and development (IAD) framework to analyze the decision process involved in adapting to climate change in the Baltimore, Maryland metropolitan region. The IAD framework details the primary elements that are expected to be involved in a community's decision process. The framework itself does not imply the rationale motivating the process and its interactions, only the macro-structure of its elements (Ostrom et al. 2014).

The primary elements of the IAD framework are oriented around the "action arena" as the unit of analysis, which is "a social space where individuals interact, exchange goods and services, solve problems, dominate one another, or fight" (Ostrom et al. 2014, p. 271). The action arena is comprised of "action situations" where interactions occur and the actors participating in those situations. Results of interactions among actors can be examined for patterns and outcomes, both of which can be assessed by evaluative criteria such as efficiency or equity. The action arena itself is influenced critically by the local context, which includes attributes of the community, the resources available to the actors, and the institutional rules governing interactions. Ultimately, the outcomes of interactions within the action arena feed back into the local context, motivating further action (Mettler and Sorelle 2014).

In the Baltimore case, the unit of analysis is broadly conceived as the space, where collective decision-making happens concerning the community of Baltimore. The space is more encompassing than the formal governmental decision-making arenas, such as the City Council or Mayor's Office. Within each action arena, multiple action situations are identified corresponding to the instances when policy action was proposed, debated, or decided. Actors move in and out of action arenas depending on their interests and roles (e.g., the Mayor makes executive decisions relating to the management of the city bureaucracy but also approves legislative actions of the City Council). The key outcomes in the case include the adoption of public policies and implementation of actions regarding climate change adaptation within the region.

In the IAD framework, civic capacity can be considered prima-facie as an "attribute" of the community and part of the local context that feeds into the action arena. Civic capacity can be built up (or reduced) over time as a community-scale resource resulting from prior decisions and outcomes (Briggs 2008; Mettler and Sorelle 2014), and can be activated as situations warrant. In this analysis, civic capacity is presumed to influence the pattern of interactions among community members and the resulting policy choices that are made, although the mechanism of influence and its operation is underdeveloped theoretically.

6.3.3 Data Sources

The majority of information for the IAD analysis comes from archival research, including policy statements such as state and city plans, commissioned or staff-generated reports, public meeting notices and minutes, and testimony or written public comments on draft policies and plans. News articles and interest group or private citizen communications were also reviewed for relevant information. The archival research extended to the mid-2000s when the first calls for climate action in Baltimore were heard. The archival research generated both qualitative and quantitative information necessary for the IAD analysis, including a list of action situations (key events), actors (organizations and individuals), and sequences of interactions within each action arena. Additional information about the case context was obtained from secondary sources, such as from the U.S. Census Bureau, Vital Stats, and the State of Maryland.

6.4 Analysis

6.4.1 Case Context

Baltimore city is the largest principal city within the Baltimore-Columbia-Towson (Central Maryland) metropolitan statistical area and is the primate city in the state of Maryland. The metro area includes Baltimore city plus six surrounding Maryland counties: Anne Arundel; Baltimore; Carroll; Harford; Howard; and Queen Anne's. According to the U.S. Census Bureau, the metro area spans 6755 km^2 (2608 mi^2) and had a combined population of approximately 2.77 million in 2015. Baltimore city was home to approximately one fifth of the region's population in 2015. More than half of the metro area's population resided within 10 mi of downtown Baltimore in 2010, although the metropolitan area extends more than 40 mi from downtown at its furthest point in Queen Anne's county, across the Chesapeake Bay (Wilson et al. 2012).

The Port of Baltimore and the Inner Harbor are key features of Baltimore's geography and intimately connected to the regional economy. As a "legacy city" experiencing population decline since 1950 (Sampson 2014), Baltimore city has invested substantial time, energy, and public and private resources in revitalizing its downtown as a regional employment, educational, and entertainment center. Baltimore city now boasts nearly three times as many jobs as residents in its downtown (Berube and McDearman 2015), and many of those jobs are high paying. Nevertheless, mean per capita personal income of residents in the city of Baltimore was only 80% of the metropolitan area's average in 2014 according to the Bureau of Economic Analysis, and significant pockets of extreme poverty and social vulnerability persist to the west and east of downtown Baltimore, typically in majority black neighborhoods such as Cherry Hill (Berube and McDearman 2015; Bridges and Kaminowitz 2015).

The city's location along the Chesapeake Bay—with 98 km (61 mi) of coastline (Redding 2013)—introduces important climate hazards for the built environment, especially from sea level rise, storm surge, and higher precipitation. The Mid-Atlantic region experienced sea level rise of approximately 2.4–4.4 mm per year (1 ft in total) during the 20th century, due to a combination of global sea level rise and land subsidence (Titus et al. 2009). In Baltimore, tide gauges registered average sea level rise nearly twice as fast as the global average during the 20th century. The pace of sea level rise appears to be quickening, such that the region was projected to face sea level rise of 2–7 mm per year in the twenty-first century (Titus et al. 2009). More recent estimates indicate that Maryland will experience sea level rise of nearly 610 mm (2 ft) by 2050 (Boesch et al. 2013).

Baltimore is now one of ten U.S. cities most affected by "nuisance flooding," which occurs on a chronic basis and not as a result of a major weather event (Sweet et al. 2014). Baltimore experienced increases in its average nuisance flood days from 1.3 days in 1957–1963 to 13.1 days in 2007–2013 (occurring at 0.41 m or 1.35 ft above mean high water). Scholars estimate that Baltimore city will experience 63 nuisance flood days by 2030 and a remarkable 227 days by 2050 (Sweet et al. 2014). An estimated 2078 hectares (5136 acres) and 2800 buildings in Baltimore are presently located within the regulated floodplain (Baltimore Office of Sustainability 2014). Other high-risk areas in the region include Annapolis, Maryland, also located within the Baltimore metropolitan area as well as Atlantic City and Sandy Hook, New Jersey; Philadelphia, Pennsylvania; Washington, DC; and Norfolk, Virginia (Sweet et al. 2014).

The city is also vulnerable to extreme heat and worsening air quality in part from its urban heat island effect. In August 2001, researchers observed a decline of more than 10 °C in temperature from Baltimore's central city to its outlying suburbs (Earth Observatory 2008). Such an influential heat island will combine over time with any increase in average daily temperatures and/or humidity to increase the vulnerability of city residents and workers to heat-related injury or death, such as from asthma (Kenney 2016). The city of Baltimore committed to increasing tree canopy cover to 40% by 2037 (from 27% in 2007) to mitigate heat hazards and absorb hazardous air pollutants, among other reasons (O'Neil-Dunne 2009). The city also began analyzing the "heat burden in underserved neighborhoods" in 2014 to inform climate adaptation, as part of its *Growing Green Initiative*, with participation from Johns Hopkins University, Maryland Institute College of Art, and other community members (Baltimore Office of Sustainability n.d.-a). In this way, a preexisting policy on extreme heat mitigation was incorporated into the city's climate adaptation strategy.

6.4.2 Policy History

Climate adaptation policies for Baltimore city are nested within a multi-level governance regime, with relevant action dating back to 2007 from the U.S. federal government, the state of Maryland, and the Mid-Atlantic/New England region (see Table 6.1).

6.4.2.1 Relevant Federal Action

In 2007, the U.S. Congress commissioned the National Academy of Sciences (NAS) to examine the issues and opportunities for climate change response in the United States (Consolidated Appropriations Act 2008), following the publication of the fourth assessment report of the Intergovernmental Panel on Climate Change (Parry et al. 2007). The resulting NAS report recommended climate adaptation

Table 6.1 Major actions on climate adaptation relevant to Baltimore city

Date	Level	Actor	Action	Innovations
3/2007–4/2007	Nongovernmental	Intergovernmental panel on climate change	Issued fourth assessment report	The working group II highlighted local hazards such as flooding, extreme heat, and heavy precipitation
4/2007	State	Governor O'Malley	Signed executive order 01.01.2007.07	Established Maryland commission on climate change (new institution)
4/2007	State+Regional	Governor O'Malley	Signed memorandum of understanding with the regional greenhouse gas initiative	Established state climate mitigation goals
6/2007	City	Baltimore city Council/Mayor	Passed legislation	Established office of sustainability and commission on sustainability (new institutions)
12/2007	Federal	Congress/President	Passed legislation	Commissioned NAS panel on climate response
3/2009	City	Baltimore city council	Passed legislation	Adopted Baltimore sustainability plan as city ordinance
5/2009	State	General Assembly/Governor	Passed legislation	Established greenhouse gas reduction targets for state; direct agencies to develop plans to meet targets
10/2009	Federal	President Obama	Signed executive order 13,514	Directed federal agencies to lead on greenhouse gas reduction and prepare adaptation plans
10/2010	State	Department of natural resources	Issued policy statement	Guides decisions on vulnerable land purchases/management within state
12/2012	State	Governor O'Malley	Signed executive order	Directed all infrastructure investments to consider future climate impacts; initiates 'coast-smart' program
1/2013	City	Baltimore office of sustainability	Published plan	Adopted climate action plan

(continued)

Table 6.1 (continued)

Date	Level	Actor	Action	Innovations
11/2013	Federal	President Obama	Signed executive order 13,653	Established interagency task force to support community-scale climate preparedness
12/2013	City	Baltimore office of sustainability	Published plan	Adopted disaster preparedness and planning project plan
1/2014	State	Governor O'Malley	Signed executive order	Rescinds 2007 executive order; restructures and reorients MCCC
4/2014	City	Baltimore city council	Adopted ordinances	Updated building and floodplain regulations to obtain FEMA community rating system designation
4/2015	Nongovernmental	STAR communities	Makes award decision	Awards 5-STAR community certification to Baltimore city

planning by state, local, tribal, and nongovernmental entities irrespective of federal government action (Jacobs et al. 2010).

While the NAS panel was deliberating on their task, President Obama signed Executive Order 13,514 (2009), which had a two-pronged focus. Federal agencies were directed to lead-by-example on climate mitigation by reviewing and reducing their energy use and greenhouse gas emissions to the extent practicable. Agencies were also directed to lead on climate change adaptation by evaluating their assets and operations and engaging in cross-sector and multi-stakeholder planning to reduce vulnerabilities and improve resilience at the community scale (Executive Office of the President n.d.). The first agency plans were released in February 2013 (U.S. Global Change Research Program n.d.).

Federal law requires that public agencies seeking disaster assistance from the Federal Emergency Management Agency (FEMA) have an approved All Hazards Mitigation Plan (AHMP) and update it every five years (Disaster Mitigation Act of 2000).[1] In response to E.O. 13,514, FEMA encouraged its Disaster Mitigation Grant applicants to "incorporate climate change considerations in their project scoping and development… to support climate change preparedness and resilience" (Federal Insurance and Mitigation Administration n.d., p. 2). Corresponding changes to FEMA's state mitigation plan guidance were finalized in March 2015

[1]All coastal states are also required to assess coastal hazards every 5 years under Section 309 of the federal Coastal Zone Management Act of 1972 (16 U.S.C. §1456b).

and became effective in March 2016 (Federal Emergency Management Administration 2015).

In June 2013, the *President's Climate Action Plan* (PCAP) was released with Section 2 devoted to climate adaptation. The PCAP notes that "The Obama Administration will help state and local governments strengthen our roads, bridges, and shorelines so we can better protect people's homes, businesses and way of life from severe weather" (Executive Office of the President 2013, p. 5). This wording is important; the responsibility for action was placed on government entities to take adaptive actions that protect citizens, businesses, and private property.

Based on a recommendation in the PCAP, President Obama (2013) signed another executive order in November 2013 establishing an interagency task force specifically to support state, local, and tribal adaptation efforts. Working quickly, the task force issued recommendations in late 2014. Then Maryland Governor Martin O'Malley (former mayor and councilman of Baltimore) participated in the federal task force, providing representation for both his state and, informally, for its largest city.

6.4.2.2 Relevant State and Regional Action

In 2007, the state of Maryland also began to establish the institutional infrastructure for state-level climate adaptation policy beginning with Executive Order 01.01.2007.07 (O'Malley 2007), establishing the *Maryland Commission on Climate Change* (MCCC). The MCCC initially included four "Adaptation and Response Working Groups" tasked with preparing the state's climate adaptation policy, which was ultimately published as Chap. 8 in the state's *Greenhouse Gas Reduction Plan* (GGRP) (State of Maryland 2013). The GGRP plan was mandated as part of the state's Greenhouse Gas Emissions Reduction Act of 2009 and was coordinated with Maryland's participation in the Regional Greenhouse Gas Initiative (RGGI). The GGRP committed to a 25% reduction in greenhouse gas emissions below 2006 levels by 2020 and 40% reduction by 2030. Analysis of the GGRP implementation in 2015 found that the state was likely to exceed its 2020 reduction goal. The Greenhouse Gas Emissions Reduction Act was reauthorized in April 2016, retaining the commitments outlined in the GGRP (Maryland Climate Coalition 2016).

Additional state policies on climate adaptation came as a policy statement from the Maryland Department of Natural Resources in 2010, another Executive Order in 2012, and passage of House Bill 615 in 2014, variously directing public investments to minimize or adapt to future climate risk (Chapter 415: Coast Smart Council 2014; Maryland Department of Natural Resources 2010; O'Malley 2012).

Governor O'Malley (2014) followed up his participation in the federal interagency task force on community-scale adaptation with another executive order entitled "Strengthening Climate Action in Maryland." This action rescinded his 2007 executive order and restructured and reoriented the MCCC's mission, including reinstating one Adaptation and Response Working Group (ARWG) within MCCC to lead the state's adaptation policy efforts. MCCC's most recent report indicates "The ARWG has made it a priority to strengthen efforts in 2017 to place greater emphasis on supporting adaptation at the local level" (Maryland Commission on Climate Change 2016, p. A-1).

6.4.2.3 The City of Baltimore

The origins of Baltimore's climate policy also date from 2007, when the City Council and the Mayor's Office established two new institutional entities: (1) a *Commission on Sustainability* (COS), comprised of government and community leaders providing oversight of the city's policy development process, and (2) *Baltimore's Office of Sustainability* (BOS), initially comprised of former staff members from other city departments and providing support to the COS (Ewing and Knapp 2009). In 2007, Baltimore's mayor Sheila Dixon also signed the U.S. Conference of Mayor's *Climate Protection Agreement* and joined ICLEI's local government network, connecting Baltimore with a growing network of cities active in climate mitigation policy and planning.

In 2009, the city released its *Sustainability Plan* (hereafter the BSP), covering all environmental resources within the city (Baltimore Office of Sustainability 2009). The BSP was codified by the City Council as a city ordinance in March 2009. The BSP recommended the creation of a citywide *Climate Action Plan* (hereafter the CAP), which was subsequently released in 2012 (Baltimore Office of Sustainability 2013a). The CAP included some preliminary adaptation recommendations, although the report was oriented towards climate mitigation and greenhouse gas reduction, in line with the focus of the state's MCCC and GGRP implementation efforts.

Baltimore city's latest climate adaptation policy dates to 2013. At the time, the city was engaged in updating its All Hazards Mitigation Plan (AHMP). Recognizing substantive overlap in effort, the city of Baltimore used "data and community input from the AHMP [update process] as the foundation for creating a Baltimore Climate Adaptation Strategy" (Baltimore Office of Sustainability 2013a, Chap. 4). Baltimore's most recent AHMP—published in December 2013—is an integrated climate adaptation and hazard mitigation plan, termed the *Disaster Preparedness and Planning Project* (hereafter the DP3) (Baltimore Office of Sustainability n.d.-b). Baltimore is one of only a few public entities that have adopted such integrated and innovative hazard mitigation and climate change adaptation plans (Higbee 2014), although more are likely to emerge as localities update their AHMPs in line with FEMA's 2015 mitigation plan guidance.

Implementation of much of the DP3 is still in the early stages (Baltimore Office of Sustainability 2015), although the city successfully completed amendments to the city's building and floodplain regulations in April 2014. The amendments were subsequently recognized by FEMA for advancing beyond minimal standards of the National Flood Insurance Program, reaching "Class Five" designation in the Community Ratings System and reducing policyholders' insurance premiums

(Baltimore Office of Sustainability 2016b). Related, the city also received a 5-star designation from the nonprofit organization STAR Communities in April 2015, which covers sustainability generally. In particular, Baltimore city obtained high marks for its climate adaptation (15/15) and mitigation (20/20) efforts.

In late 2015, the city joined the U.S. Conference of Mayors, the National League of Cities, and several other cities in the "Local Government Coalition" in petitioning to participate as *amicus curiae* before the U.S. Court of Appeals in support of the federal Clean Power Plan and its climate policy more generally.[2]

The city recently endeavored to update its BSP with an explicit focus on social equity. The update process included hosting a Sustainability Town Hall in April 2016 to solicit community priorities, engaging neighborhood "ambassadors" to conduct community outreach and gather information, conducting an online survey of residents, and holding additional community meetings on the draft plan (Baltimore Office of Sustainability 2016a). An update to the DP3 is not expected until 2018.

6.4.3 Action Arenas

The preceding policy history reveals several dynamic action arenas relevant to Baltimore city's climate adaptation strategy, including the following:

- The city of Baltimore's formal action arena is comprised of its legislature and chief executive (i.e., city council and mayor), with multiple action situations leading up to adoption of the BSP, the CAP, and the DP3 and their implementation and updates.
- Within the state of Maryland's formal action arena, multiple action situations led to the development of the state's greenhouse gas reduction policy (which included a plan for adaptation) and the state's "coast smart" investment priorities.
- At the regional level, Maryland's participation in the 9-state Regional Greenhouse Gas Initiative (RGGI) is another relevant action arena. Despite RGGI's focus on climate mitigation and reduction of GHGs from the power industry, the state's involvement in RGGI improved access to knowledge, resources, and leadership about climate policy generally, including adaptation.
- At the federal level, the release of the PCAP and signing of executive orders 13,514 and 13,653 are relevant action situations within the chief executive action arena, with the agency and task force responses in the implementation of the orders as additional outputs.

---

[2]Their petition argued "The acute relevance of climate change to local governments' responsibilities and activities has led members of the Local Government Coalition to grasp both the need to adapt to climate change and the costs of failing to act to mitigate it. Prompted by lived experience and by the prospect of future impacts, they have made efforts both to adapt to their changing climatic circumstances and to slow or eliminate their greenhouse gas emissions" (*State of West Virginia* et al. *v. United States Environmental Protection Agency*, CA Case #15-1363, Document #1,589,943, p. 6).

The adoption of policy at each level reflects the operations of their unique action arenas, involving many differing action situations, participants, patterns of interactions, and contexts. The simultaneous operation of these action arenas exerted top-down pressure on Baltimore's climate adaptation policy but also established a political and institutional environment that was conducive to municipal policy development and change, in accordance with proposition 1.

6.4.4 *Participants*

Many actors and institutions were involved throughout the course of Baltimore city's climate policy history, from all levels of government and beyond.

6.4.4.1 Executive Leadership

Baltimore city's mayor and Maryland's governor initially provided high-level leadership on climate policy and sustainability more generally. The city of Baltimore has a strong-mayor/council form of municipal governance, where the mayor serves in the chief executive role much like the Maryland governor does for the state. On climate policy, both the Baltimore mayor and Maryland governor proposed policies to their legislative bodies and authorized executive actions when within their legal authority to do so. In 2007, Mayor Dixon signed the U.S. Conference of Mayors Climate Protection Agreement—committing the city to climate mitigation and targeting greenhouse gas reductions to 7% below 1990 levels by 2012—but left implementation to the BOS. Likewise, in 2007, Governor O'Malley committed Maryland to participate in the Regional Greenhouse Gas Initiative. But the details of the state's implementation in RGGI were only later promulgated through passage and implementation of the state's Greenhouse Gas Reduction Act of 2009.

Executive support for climate action in Baltimore city continued when Mayor Rawlings-Blake took office in 2010, providing policy consistency and fostering an environment conducive to ongoing conversations about sustainability. Additionally, Mayor Rawlings-Blake participated in the international Conference of Parties meeting in Paris in late 2015 on behalf of the U.S. Conference of Mayors and the city of Baltimore. Since then, however, executive leadership on climate policy has been elusive.[3]

---

[3]The city's new mayor, Catherine Pugh, began her term in late 2016 and has yet to establish a public position on climate change adaptation. Nevertheless, Pugh's background in the Baltimore City Council and state General Assembly implies familiarity with the climate change priorities of the state and city.

6.4.4.2 New Institutions

In both Maryland and the city of Baltimore, new institutional offices were established to develop climate adaptation policy. Governor O'Malley established the *Maryland Commission on Climate Change* in 2007 to coordinate development of the state's climate policies. Likewise, in 2007, the Baltimore City Council and Mayor established the multi-sector *Commission on Sustainability* and the *Office of Sustainability*. In these ways, interested parties at both the state and the city levels appeared to recognize the "need on the part of certain activities in the territorial system for a top leadership under whose auspices they can function" (Long 1958, p. 258). Having a single institutional home provides one venue in which to accrue resources (financial, human) and provide leadership and coordination among existing efforts, such as the tree canopy effort mentioned earlier. Their designation by the highest political authorities at each level helped to establish legitimacy for the new institutions and their efforts, building overall institutional capacity to address climate change within the region. The creation of these new institutional structures could also be seen as providing a window of opportunity for cross-sector dialogue on climate policy that had not existed previously (Kingdon 1995; Meyer and Minkoff 2004).

6.4.4.3 Staff Involvement

Government employees reside at the center of adaptation policy development and implementation in Baltimore, drawing upon their professional training, experience, and network connections. Aside from representatives of the new BOS and COS institutions, participants were drawn from across the city bureaucracy, including the Departments of Public Works, Transportation, General Services, Police and Fire Departments, as well as from the Mayor's Office of Emergency Management, and the City Council. Indeed, governmental employees or their contractors comprised 70% of the individually named participants from the climate policy and plan documents reviewed for this analysis.

Several champions of Baltimore's climate policy emerged from within the city's Department of Planning. For instance, Department Director Thomas Stosur served as member of the COS during the BSP process in 2009, Advisory Committee member during the CAP process in 2012, and co-chaired the Advisory Committee for the DP3 in 2013. Kristin Baja is perhaps the most well-recognized champion for climate policy in Baltimore today. She presently oversees the DP3 development and implementation, having begun as project manager and non-voting Advisory Committee member for DP3 in 2013, leading both the infrastructure and public services working groups. Baja supervises the city's floodplain regulation and all climate change-related communication and outreach.

Six additional persons were centrally involved in climate policy development both at the city and state levels, including four from state government agencies, one from the municipal electricity utility, and one from a private consulting business.

These persons share information and facilitate policy learning (Jenkins-Smith et al. 2014) within the region's multi-level climate governance regime.

Notably, the metropolitan planning organization for the Central Maryland region —the Baltimore Metropolitan Council (BMC)—played only a minor role in Baltimore's climate adaptation policy development. One employee of BMC served on the COS Advisory Committee that developed the city's DP3. In 2014, BMC produced a series of maps on sea level rise projections for 2050 and 2100 along the Chesapeake Bay, "developed to support the planners and leadership of the Maryland State Highway Administration (MDSHA) as they endeavor to mitigate or prevent the infrastructure impacts of sea level change resulting from land surface subsidence and rising sea levels" (Baltimore Metropolitan Council 2015b). BMC also produced various demographic and socioeconomic assessments and projections for the region to fulfill its technical assistance mission, including mapping social vulnerability that could be spatially associated with sea-level rise projections. To date, climate vulnerability assessments have not been published by BMC for the region or Baltimore city proper. The BMC has supported the development of a coordinated housing, transportation, and economic development plan within the region, although the plan does not overtly reference climate hazards or vulnerabilities (Baltimore Metropolitan Council 2015a).

6.4.4.4 Public Participation

Climate adaptation efforts worldwide exhibit a wide range and style of public participation, extending from traditional, government-led climate planning through to intensive public–private partnerships or entirely private actions (Sarzynski 2015). As with participation in environmental policy-making more broadly, each style of participation has a different breadth of participation and intensity of involvement (Dietz and Stern 2008).

A signature characteristic of Baltimore's climate policy strategy, as compared with 16 other climate active cities, has been its extensive and deliberate public consultation and outreach process (St. Juliana et al. 2016). While its policy development has been led and initiated by government staff, the process has many avenues for citizen input and consultation, such as during its annual town hall and other public meetings, online surveys, and the formal public comment process. These options provided ample opportunity for public input from persons motivated to participate, expanding the breadth of participation but with relatively low intensity of involvement (i.e., a couple hours of time while attending a meeting or preparing a written comment). More than 300 public comments were submitted on the original BSP and the city held 35 public meetings. The DP3 implementation process has so far included one large town hall meeting in April 2014 with 300 attendees and more than 40 public meetings "reaching more than 1450 community members" (St. Juliana et al. 2016, p. 69). BOS staff members deliberately sought to hold meetings in locations and at times convenient to community members, including at volunteers' homes or at regularly scheduled neighborhood association events.

Aside from serving on the Commission, a more intensive but less broad-reaching option for citizen involvement derives from "neighborhood ambassadors," who are volunteers trained to conduct public outreach and information gathering that feeds back into the larger policy development process. Over the years, Baltimore has engaged dozens of ambassadors who were individually recognized for their effort in the city's annual sustainability reports.

Related, the city's Office of Emergency Management oversees community emergency response teams (CERTs), including member recruitment and training. The CERTs provide citizen engagement and community outreach on all hazards including climate preparedness and resilience. Some CERT members participated in the DP3 development and implementation process, further cementing the connections between the city's disaster preparedness and climate change adaptation efforts.

Finally, representatives of private businesses or business organizations were less formally engaged in the DP3 policy development process than government staff. This situation contrasts somewhat with adaptation efforts in New York, where insurance companies, financiers, and property developers participated centrally in development of that region's adaptation plan (Rosenzweig and Solecki 2010). One of Baltimore's goals for DP3 implementation is to further engage with private businesses such as in the waterfront and vulnerable industrial areas (St. Juliana et al. 2016).

6.4.5 Resources

The city's extensive efforts to involve the public in development and implementation of the DP3 reflect various resource constraints facing the institutional actors, including financial, and a previously underdeveloped civic capacity for pursuing climate adaptation.

6.4.5.1 Financial

Although Baltimore city is home to many high-paying jobs within its metropolitan region, the city itself is cash-strapped and much policy attention is focused on the provision of public services such as safety and education. Plagued by limited financial resources with which to develop and implement climate policies, city staff looked elsewhere for financial support of its climate policy agenda. For instance, the city's CAP was first recommended in 2009 but was not completed until 2012, after receiving supplemental federal funding from the American Recovery and Reinvestment Act of 2009 (Baltimore Office of Sustainability n.d.-c). Likewise, the BOS originally sought funds from the Maryland Emergency Management Agency to prepare its DP3, but ended up also seeking funds from FEMA, the National Atmospheric and Oceanic Administration, and the Maryland Department of Natural

Resources (MDNR) before the plan could be completed (Baltimore Office of Sustainability 2013b, Chapter Appendix C).

In November 2014, the city received a grant from the MDNR to develop a data collection and tracking system of climate adaptation metrics that would allow the city to gauge its performance during DP3 implementation. The MDNR has also been funding floodplain management efforts in nearby jurisdictions within the Baltimore metropolitan area, including in Baltimore County, the city of Annapolis, Calvert County, and Talbot County (Skaggs 2015). While the city has been successful winning grants to support its climate work, future support is uncertain (St. Juliana et al. 2016).

6.4.5.2 Civic

In part due to financial uncertainty, city staff has sought to build civic capacity on climate adaptation. Indeed, a major prong of the city's DP3 implementation so far has been in raising awareness and providing tools that citizens could use to improve their own disaster preparedness and climate change resilience, including through the city's "Make a Plan, Build a Kit, Help Each Other" and "Resiliency Hubs" programs (St. Juliana et al. 2016). Notably, DP3 implementation frequently uses language around "resilience" or "preparedness" rather than "climate adaptation" per se, despite referencing actions that are also climate adaptive. These phrases directly relate to the language used by FEMA for its All Hazards Response Plan.

A search for news articles, letters to the editor, civic association messages, or related communications since 2005 revealed remarkably few making reference to climate change adaptation, and most all of those found were from the last year. Only three articles were found making any reference to adaptation and the DP3 (Cassie 2015; Lazarick 2013; Reutter 2016). Indeed, it seems that climate change was not a common topic for civic discourse within the city prior to governmental action in 2007. Even since then, city actions were not well covered by the print media and discussions that did occur were most likely held within government offices or stakeholder meetings. Sea level rise and the vulnerability of coastal areas did receive more news coverage, but without much of any discussion of what Baltimore residents or businesses should do, or how government employees are responding through policy or planning.

General civic engagement in the city of Baltimore is moderate. Voter turnout for the 2016 general election was 45%, up from 38% in 2014 but down from 51.5% in 2012. Approximately 75% of the eligible resident population is registered to vote. Prior assessments of social capital from 1997–2009 (Rupasingha et al. 2006 and online updates) found a relatively large number of non-profit organizations and membership associations within Baltimore city as compared to other counties throughout the nation. But when including its voter turnout rate, Census participation rate, and two organizational variables, the overall assessment of Baltimore city's social capital was below average compared to other counties nationwide. Its social capital increased some during the study period, although remained below average (Northeast Regional Center for Rural Development n.d.). The prevalence of all types of organizations (including civic, political, business, labor, and professional) declined in Baltimore city during the study period, excepting religious organizations.

While we cannot infer civic activity from the presence of organizations alone, the decline especially in civic and political organizations combined with the lack of print media coverage suggests that the civic capacity to address collective problems like climate change adaptation may be underdeveloped. Related, Baltimore city's 5-STAR community designation in April 2015 noted modest achievement on the civic engagement indicator, with no credit on outcomes regarding voting, volunteerism, and sense of empowerment, and half credit for its actions regarding diversity in governance, guidelines on engagement, partnerships, and training for government employees and citizens, including youth.[4]

Returning to the earlier-stated propositions, we see evidence of P1: *a conductive institutional environment for deliberation* established with the new institutional entities of the COS and BOS, which set the high-level policies (COS) and provide staff support and knowledge for implementation (BOS). We see evidence of P2: *various supports for action*, including early executive leadership, legal authority, and some financial resources, plus extensive public consultation that may equate to public support. We also see evidence of P3: *a commitment to act collectively* on the part of the participants involved most in policy development—the COS and BOS. Less clear evidence exists of a widespread, private commitment to act on climate adaptation, either from residents or businesses. In this way, grassroots mobilization appears lacking, with no clear civic organization activity emerging on climate adaptation.

If we recall the language from the policy history above, the responsibility for action was left to government, with the pressure moving from the federal government to state and local governments and tribes, and the pressure from the state government moving down to municipalities. In this way, Maryland and Baltimore city climate policy and plans have so far kept the onus for action on government actors with no stated responsibility for nongovernmental actors to build adaptive capacity on their own. BOS staff has been using some of the early DP3 implementation effort to help build citizen capacity to adapt but a widespread civic dialogue on roles and responsibilities and necessary actions in line with the full conception of civic capacity (Stone 2001) has yet to emerge in Baltimore city. Instead, the motivation for action taken so far in Baltimore appeared to lie with the executive leadership and senior agency staff, who used their existing authority to act when and how they could.

By combining effort across the disaster preparedness and climate change adaptation planning processes, and by seeking funding from higher governmental levels, the staff was able to move the policy process forward with fewer resources than otherwise. And by focusing some of their early implementation efforts on

[4]Details available at https://reporting.starcommunities.org/communities/68-maryland-baltimore.

building citizen capacity for adaptation, the staff are spreading awareness of the problem that may later feedback through Mettler's "interpretive effects" loop (2014) into higher civic predisposition to act on climate change adaptation. Only time will tell if this approach can yield the adaptive efforts needed to reduce vulnerability and exposure to the myriad climate hazards facing the city, while boosting resilient behavior.

6.5 Summary

This analysis examined the participatory process that Baltimore used on disaster preparedness, resilience, and climate change adaptation. The complex, multi-level governance regime in place on climate policy in Baltimore served to direction attention and limited resources down to municipal scale action, producing a conducive environment for policy change in the city. The creation of a new institutional home in which to concentrate resources, knowledge, and leadership has also facilitated further action on climate policy: first by creating the Sustainability Plan, then the mitigation-focused Climate Action Plan, and finally the Disaster Preparedness and Planning Project, as a combined all hazards mitigation plan and climate change adaptation plan. Implementation of the DP3 has sought to obtain further resources, both financial and civic, which will help to build additional public support and improve the effectiveness of the overall effort. To date, the city's adaptation planning efforts have been externally recognized for their quality by the federal government (FEMA) and the nonprofit STAR Communities.

Further attention appears warranted at this time to move beyond planning to action such that Baltimore's most vulnerable communities and properties are protected from harm. What will it take to fully implement the DP3 in Baltimore city? Is there sufficient civic capacity to take needed actions, in and out of government, to minimize harm from climate hazards? Or is there a "social limit" to what can be done (Adger et al. 2009), perhaps due to the community's values, norms, and expectations that climate action is the exclusive responsibility of government? If so, how can the government overcome resource constraints to implement its ambitious plans?

The present analysis was limited by the use of existing documents to trace the policy history, institutions, and actors, which were able to shed some light on the elements necessary to build effective collective action on climate change. The third proposition—*the will to act collectively*—appears within government staff but grassroots interest appears elusive. Further analysis should seek to tease out why adaptation policy action was taken in Baltimore and opportunities for further action through interviews with key participants from all sectors including private citizens (see Chap. 9: Iftikhar, Ali and Sarzynski).

Further analysis should also investigate the differences between Baltimore and a city like Philadelphia, which has a less well-developed climate adaptation strategy and which is located in a state with a notably different state environmental governance regime than Maryland, but which is located within the same region and faces many of the same climate hazards and socio-demographic conditions as in Baltimore. Such a comparison would help to improve our understanding of the structure and agency elements necessary for successfully overcoming the climate change adaptation deficit in diverse and vulnerable coastal cities.

References

Adger, W. N., Dessai, S., Goulden, M., Hulme, M., Lorenzoni, I., Nelson, D. R., et al. (2009). Are there social limits to adaptation to climate change? *Climatic Change, 93*(3–4), 335–354. doi:10.1007/s10584-008-9520-z.

Baker, I., Peterson, A., Brown, G., & McAlpine, C. (2012). Local government response to the impacts of climate change: An evaluation of local climate adaptation plans. *Landscape and Urban Planning, 107,* 127–136.

Baltimore Metropolitan Council. (2015a). *Baltimore regional plan for sustainable development.* Baltimore, MD: Baltimore Metropolitan Council. http://www.opportunitycollaborative.org/assets/RPSD_Final_June_2015.pdf. Accessed 1 April 2016.

Baltimore Metropolitan Council. (2015b, January 22). Environmental Coordination Mapping. http://baltometro.org/our-work/environmental-planning/environmental-coordination-mapping. Accessed 1 April 2016.

Baltimore Office of Sustainability. (2009). *The Baltimore Sustainability Plan.* Baltimore, MD: Baltimore Office of Sustainability. http://www.baltimoresustainability.org/wp-content/uploads/2015/12/Baltimore-Sustainability-Plan.pdf. Accessed 5 March 2016.

Baltimore Office of Sustainability. (2013a). *Baltimore Climate Action Plan.* Baltimore, MD: Baltimore Office of Sustainability. http://www.baltimoresustainability.org/wp-content/uploads/2015/12/BaltimoreClimateActionPlan.pdf. Accessed 16 February 2016.

Baltimore Office of Sustainability. (2013b). *Disaster preparedness and planning project: A combined all hazards mitigation and climate adaptation plan.* Baltimore, MD: Baltimore Office of Sustainability. http://www.baltimoresustainability.org/wp-content/uploads/2015/12/Executivesummary.pdf. Accessed 16 February 2016.

Baltimore Office of Sustainability. (2014). *2014 Annual Sustainability Report.* Baltimore, MD. http://www.baltimoresustainability.org/wp-content/uploads/2015/12/2014.pdf. Accessed 4 January 2017.

Baltimore Office of Sustainability. (2015). *2015 Annual Sustainability Report.* City of Baltimore. http://www.baltimoresustainability.org/wp-content/uploads/2015/05/UPDATED_CS_2015-Final-_June2016-min.pdf. Accessed 4 January 2017.

Baltimore Office of Sustainability. (2016a). Sustainability Plan—UPDATE. http://www.baltimoresustainability.org/plans/sustainability-plan/update/. Accessed 4 January 2017.

Baltimore Office of Sustainability. (2016b, June 17). Baltimore To Receive FEMA Floodplain Award, Residents To Receive Hundreds Of Thousands In Savings. http://www.baltimoresustainability.org/baltimore-to-receive-fema-floodplain-award-residents-to-receive-hundreds-of-thousands-in-savings/. Accessed 4 January 2017.

Baltimore Office of Sustainability. (n.d.-a). Urban Heat Island Sensors. http://www.baltimoresustainability.org/urban-heat-island-sensors/. Accessed 9 April 2016.

Baltimore Office of Sustainability. (n.d.-b). Disaster Preparedness Plan. http://www.baltimoresustainability.org/plans/disaster-preparedness-plan/. Accessed 16 February 2016.

Baltimore Office of Sustainability. (n.d.-c). Climate Action Plan. http://www.baltimoresustainability.org/plans/climate-action-plan/. Accessed 16 February 2016.

Berube, A., & McDearman, B. (2015, May 11). Good fortune, dire poverty, and inequality in Baltimore: An American story. *The Avenue| Rethinking Metropolitan America.* http://www.

brookings.edu/blogs/the-avenue/posts/2015/05/11-poverty-inequality-baltimore-berube-mcdearman. Accessed 12 March 2016.

Boesch, D. F., Atkinson, L. P., Boicourt, W. C., Boon, J. D., Cahoon, D. R., Dalrymple, R. A., et al. (2013). *Updating Maryland's Sea-level Rise Projections. Special Report of the Scientific and Technical Working Group to the Maryland Climate Change Commission*. Cambridge, MD: University of Maryland Center for Environmental Science. http://ian.umces.edu/pdfs/ian_report_413.pdf. Accessed 9 April 2016.

Bridges, J. M., & Kaminowitz, M. (2015). *Vulnerable Population Index* (White Paper). Baltimore, MD: Baltimore Metropolitan Council. http://baltometro.org/phocadownload/Publications/VPIWhitePaperFINAL.pdf. Accessed 1 April 2016.

Cassie, R. (2015, January 5). The Sea Also Rises. *Baltimore*. http://www.baltimoremagazine.net/2015/1/5/the-sea-also-rises. Accessed 24 February 2016.

Center for Climate and Energy Solutions. (n.d.). Climate Action| State and Local Climate Adaptation. http://www.c2es.org/us-states-regions/policy-maps/adaptation. Accessed 17 February 2016.

Chapter 415: Coast Smart Council., Pub. L. No. House Bill 615 (2014). http://mgaleg.maryland.gov/2014RS/Chapters_noln/CH_415_hb0615t.pdf. Accessed 5 March 2016.

Consolidated Appropriations Act. 2008. 121 STAT. 1844 (2007). https://www.gpo.gov/fdsys/pkg/PLAW-110publ161/pdf/PLAW-110publ161.pdf. Accessed 5 March 2016.

de Souza Briggs, X. (2008). *Democracy as problem solving: Civic capacity in communities across the globe*. Cambridge, MA: The MIT Press.

Dietz, T., & Stern, P. C. (Eds.). (2008). *Public participation in environmental assessment and decision making*. Washington, DC: National Academies Press. http://www.nap.edu/catalog/12434.html.

Disaster Mitigation Act of 2000., Pub. L. No. 106–390 (2000). http://www.fema.gov/media-library-data/20130726-1524-20490-1790/dma2000.pdf. Accessed 16 February 2016.

Earth Observatory. (2008, December 20). Urban Heat Island: Baltimore, MD. http://earthobservatory.nasa.gov/IOTD/view.php?id=36227. Accessed 10 March 2016.

Ekstrom, J. A., & Moser, S. C. (2014). Identifying and overcoming barriers in urban adaptation efforts to climate change: Case findings from the San Francisco Bay Area, California. *USA. Urban Climate, 9*(September), 54–74. doi:10.1016/j.uclim.2014.06.002.

Ewing, J., & Knapp, D. (2009). *Sustainability planning toolkit: A comprehensive guide to help cities and counties develop a sustainability plan*. Oakland, CA: ICLEI-Local Governments for Sustainability USA. http://portal.hud.gov/hudportal/documents/huddoc?id=20399_iclei_sustainabil.pdf. Accessed 5 March 2016.

Executive Office of the President. (2013). *The president's climate action plan*. Washington, DC: Executive Office of the President. https://www.whitehouse.gov/sites/default/files/image/president27sclimateactionplan.pdf. Accessed 9 April 2016.

Executive Office of the President. (n.d.). Climate Change Resilience. *The white house*. https://www.whitehouse.gov/node/7256. Accessed 16 February 2016.

Federal Emergency Management Administration. (2015). *State mitigation plan review guide* (No. FP 302-094-2). Washington, DC: Federal Emergency Management Administration. http://www.fema.gov/media-library-data/1425915308555-aba3a873bc5f1140f7320d1ebebd18c6/State_Mitigation_Plan_Review_Guide_2015.pdf. Accessed 9 April 2016.

Federal Insurance and Mitigation Administration. (n.d.). Resilience and Climate Change Adaptation. http://www.fema.gov/media-library-data/1424368115734-86cfbaeb456f7c1d57a05d3e8e08a4bd/FINAL_ResilienceClimateChange_JobAid_19FEB15_508_Complete_.pdf. Accessed 5 March 2016.

Gardner, J., Dowd, A.-M., Mason, C., & Ashworth, P. (2009). *A framework for stakeholder engagement on climate adaptation* (No. CSIRO Climate Adaptation Flagship Working Paper #3). South Victoria, Australia: Commonwealth Scientific and Industrial Research Organisation. http://www.csiro.au/resources/CAF-working-papers.html.

Heinrichs, D., Krellenberg, K., & Fragkias, M. (2013). Urban responses to climate change: Theories and governance practice in cities of the Global South. *International Journal of Urban and Regional Research, 37*(6), 1865–1878. doi:10.1111/1468-2427.12031.

Higbee, M. (2014). *Integrating hazard mitigation and climate adaptation planning: Case studies and lessons learned.* Oakland, CA: ICLEI- Local Governments for Sustainability USA. http://icleiusa.org/wp-content/uploads/2015/08/Integrating-Hazard-Mitigation-and-Climate-Adaptation-Planning.pdf. Accessed 16 February 2016.

Jacobs, K. L., Wilbanks, T. J., Baughman, B. P., Beachy, R. N., Benjamin, G. C., Buizer, J. L., et al. (2010). *Adapting to the impacts of climate change.* Washington, DC: National Academies Press. http://www.nap.edu/catalog/12783. Accessed 5 March 2016.

Jenkins-Smith, H., Nohrstedt, D., Weible, C. M., & Sabatier, P. A. (2014). The advocacy coalition framework: Foundations, evolution, and ongoing research. In P. A. Sabatier & C. M. Weible (Eds.), *Theories of the policy process* (3rd ed., pp. 183–223). Boulder, CO: Westview Press.

Kenney, J. (2016, February 3). Meet A CCANer: Baltimore Community Leader Keisha Allen. *Chesapeake Climate Action Network.* http://chesapeakeclimate.org/blog/meet-a-ccaner-baltimore-community-leader-keisha-allen/. Accessed 9 April 2016.

Kingdon, J. W. (1995). *Agendas, alternatives, and public policies* (2nd ed.). New York: Harper Collins College Publishers.

Lazarick, L. (2013, July 30). Rising seas 3: Weird weather and sea level inching to prompt Baltimore to confront future climate. *MarylandReporter.com.* http://marylandreporter.com/2013/07/30/rising-seas-3-weird-weather-and-sea-level-inching-up-prompt-baltimore-to-confront-future-climate/. Accessed 6 January 2017.

Long, N. E. (1958). The local community as an ecology of games. *American Journal of Sociology, 64*(3), 251–261. http://www.jstor.org/stable/2773192.

Mackie, J. L. (1980). *The cement of the universe: A study of causation.* Oxford: Oxford University Press.

Maloney, W., Smith, G., & Stoker, G. (2000). Social capital and urban governance: adding a more contextualized "top-down" perspective. *Political Studies, 48*(4), 802–820.

Martin, P. J., & Dennis, A. (Eds.). (2010). *Human agents and social structures.* Manchester, UK: Manchester University Press.

Maryland Climate Coalition. (2016). Greenhouse Gas Reduction Act. *Maryland Climate Coalition.* http://marylandclimatecoalition.org/factsheets/greenhouse-gas-reduction-act/. Accessed 4 January 2017.

Maryland Commission on Climate Change. (2016). *2016 Annual Report.* Annapolis, MD: State of Maryland. http://www.mde.state.md.us/programs/Marylander/Documents/MCCC/Publications/2016Report/MCCC_2016_final.pdf. Accessed 4 January 2017.

Maryland Department of Natural Resources. (2010, October 15). Building Resilience to Climate Change. http://www.dnr.maryland.gov/dnrnews/pdfs/climate_change.pdf. Accessed 5 March 2016.

McEvoy, D., Lindley, S., & Handley, J. (2006). Adaptation and mitigation in urban areas: Synergies and conflicts. *Proceedings of the ICE—Municipal Engineer, 159*(4), 185–191. doi:10.1680/muen.2006.159.4.185.

Mettler, S., & Sorelle, M. (2014). Policy feedback theory. In P. A. Sabatier & C. M. Weible (Eds.), *Theories of the Policy Process* (3rd ed., pp. 151–182). Boulder, CO: Westview Press.

Meyer, D. S., & Minkoff, D. C. (2004). Conceptualizing political opportunity. *Social Forces, 82*(4), 1457–1492. doi:10.1353/sof.2004.0082.

Moser, S. C. (2009). *Good morning, America! The explosive U.S. awakening to the need for adaptation.* Santa Cruz, CA: Susanne Moser Research & Consulting. http://www.preventionweb.net/files/11374_MoserGoodMorningAmericaAdaptationin.pdf. Accessed 22 July 2015.

Northeast Regional Center for Rural Development. (n.d.). Social Capital Variables for 1997, 2005, and 2009. http://aese.psu.edu/nercrd/community/social-capital-resources/social-capital-variables-for-1997-2005-and-2009. Accessed 6 January 2017.

O'Malley, M. (2007). Executive Order 01.01.2007.07: Commission on climate change. The State of Maryland. http://www.mde.state.md.us/assets/document/Air/ClimateChange/AppendixA_Executive_Order.pdf. Accessed 5 March 2016.

O'Malley, M. (2012). Executive order 01.01.2012.29: Climate change and "Coast Smart" construction. The State of Maryland. http://wetlandswatch.org/Portals/3/WW%20documents/sea-level-rise/exec_order.pdf. Accessed 5 March 2016.

O'Malley, M. (2014). Executive order 01.01.2014.14: Strengthening climate action in Maryland. The State of Maryland. http://climatechange.maryland.gov/wp-content/uploads/sites/16/2014/12/climate_change_commission_final_eo_01_01_2014_141.pdf. Accessed 5 March 2016.

O'Neil-Dunne, J. (2009). *A report on the City of Baltimore's existing and possible urban tree canopy*. Burlington, VT: University of Vermont. http://www.fs.fed.us/nrs/utc/reports/UTC_Report_BACI_2007.pdf. Accessed 9 April 2016.

Obama, B. (2009, October 8). Executive Order 13514—Federal Leadership in Environmental, Energy, and Economic Performance. U.S. Government Publishing Office. https://www.gpo.gov/fdsys/pkg/FR-2009-10-08/pdf/E9-24518.pdf. Accessed 5 March 2016.

Obama, B. (2013, November 1). Executive order—preparing the United States for the impacts of climate change. The White House Office of the Press Secretary. https://www.whitehouse.gov/the-press-office/2013/11/01/executive-order-preparing-united-states-impacts-climate-change. Accessed 5 March 2016.

Olson, M. (1971). *The logic of collective action: Public goods and the theory of groups*. Cambridge, MA: Harvard University Press.

Ostrom, E., Cox, M., & Schlager, E. (2014). An assessment of the institutional analysis and development framework and introduction of the social-ecological systems framework. In P. A. Sabatier & C. M. Weible (Eds.), *Theories of the Policy Process* (3rd ed., pp. 267–306). Boulder, CO: Westview Press.

Parry, M., Canziani, O., Palutikof, J., van der Linden, P., & Hanson, C. (Eds.). (2007). *Climate change 2007: Impacts, adaptation and vulnerability; contribution of working Group II to the fourth assessment report of the IPCC*. Cambridge, UK: Cambridge University Press. https://www.gpo.gov/fdsys/pkg/PLAW-110publ161/pdf/PLAW-110publ161.pdf. Accessed 5 March 2016.

Redding, L. (2013). To protect a city from rising seas: Build barriers or move. *Capital News Service|Sea Level Rise in Maryland*. http://cnsmaryland.org/sealevelrise/?p=66. Accessed 9 April 2016.

Reutter, M. (2016, November 4). The bureaucrats strike back. *Baltimore Brew*. https://www.baltimorebrew.com/2016/11/04/the-bureaucrats-strike-back/. Accessed 6 January 2017.

Rosenzweig, C., & Solecki, W. (2010). Introduction to climate change adaptation in New York City: Building a risk management response. *Annals of the New York Academy of Sciences, 1196*(1), 13–18. doi:10.1111/j.1749-6632.2009.05306.x.

Rosenzweig, C., Major, D. C., Demong, K., Stanton, C., Horton, R., & Stults, M. (2007). Managing climate change risks in New York City's water system: Assessment and adaptation planning. *Mitigation and Adaptation Strategies for Global Change, 12*(8), 1391–1409. doi:10.1007/s11027-006-9070-5.

Rupasingha, A., Goetz, S. J., & Freshwater, D. (2006). The production of social capital in US counties. *Journal of Socio-Economics, 35*, 83–101. doi:10.1016/j.socec.2005.11.001.

Sampson, N. H. K. (2014). Planning for climate change in legacy cities: The case of detroit, Michigan. *Michigan Journal of Sustainability, 2*. doi:10.3998/mjs.12333712.0002.004.

Sarzynski, A. (2015). Public participation, civic capacity, and climate change adaptation in cities. *Urban Climate, 14*(1), 52–67. doi:10.1016/j.uclim.2015.08.002.

Shinn, C. W. (1999). Civic Capacity: Theory, Research and Practice. *Administrative Theory & Praxis, 21*(1), 103–119. doi:10.1080/10841806.1999.11643352.

Skaggs, K. (2015, October 15). *CoastSmart communities program*. Presented at the Maryland Association of Floodplain and Stormwater Managers, Linthicum Heights, MD. http://mafsm.org/pdf/2015Conf/MAFSM_2015_Skaggs.pdf. Accessed 6 January 2017.

St. Juliana, A., Carney, K., & Vogel, J. (2016). Integrating climate change adaptation into an all-hazards mitigation plan. In *Climate adaptation: The state of practice in U.S. communities*. Kresge Foundation. http://kresge.org/sites/default/files/library/climate-adaptation-the-state-of-practice-in-us-communities-case-studies.pdf. Accessed 4 January 2017.

State of Maryland. (2013). *Greenhouse gas reduction plan; Chapter 8: adaptation*. State of Maryland. http://climatechange.maryland.gov/wp-content/uploads/sites/16/2014/12/chap8_adaptation_final_lowres1.pdf. Accessed 16 February 2016.

Stone, C. N. (2001). Civic capacity and urban education. *Urban Affairs Review, 36*(5), 595–619. doi:10.1177/10780870122185019.

Stone, C. N., Henig, J. R., Jones, B. D., & Pierannunzi, C. (2001). *Building civic capacity: The politics of reforming urban schools*. Lawrence, KS: University Press of Kansas.

Sweet, W., Park, J., Marra, J., Zervas, C., & Gill, S. (2014). *Sea level rise and nuisance flood frequency changes around the United States* (No. NOAA Technical Report NOS CO-OPS 073). Silver Spring, MD: National Oceanic and Atmospheric Administration. http://tidesandcurrents.noaa.gov/publications/NOAA_Technical_Report_NOS_COOPS_073.pdf. Accessed 10 March 2016.

The United States Conference of Mayors. (2014). *Climate mitigation and adaptation actions in America's cities: A 282-city survey*. Washington, DC. http://usmayors.org/pressreleases/uploads/2014/0422-report-climatesurvey.pdf. Accessed 17 July 2015.

Titus, J. G., Anderson, K. E., Cahoon, D. R., Gesch, D. B., Gill, S. K., Gutierrez, B. T., et al. (2009). *Coastal sensitivity to sea-level rise: A focus on the Mid-Atlantic region* (No. Synthesis and Assessment Product 4.1). Washington, DC: U.S. Climate Change Science Program, Environmental Protection Agency. http://downloads.globalchange.gov/sap/sap4-1/sap4-1-final-report-all.pdf. Accessed 20 July 2015.

U.S. Global Change Research Program. (n.d.). Federal adaptation resources. *GlobalChange.gov*. http://www.globalchange.gov/browse/federal-adaptation-resources. Accessed 4 January 2017.

Wilson, S. G., Plane, D. A., Mackun, P. J., Fischetti, T. R., & Goworowska, J. (2012). *Patterns of metropolitan and micropolitan population change: 2000–2010* (Special Report No. C2010SR–01). Washington, DC: U.S. Census Bureau. http://www.census.gov/prod/cen2010/reports/c2010sr-01.pdf. Accessed 20 February 2016.

Yin, R. K. (2009). *Case study research: Design and methods* (4th ed., Vol. 5). Thousand Oaks, CA: SAGE Publications.

Author Biography

Andrea Sarzynski is Associate Professor at the University of Delaware's School of Public Policy and Administration and the Department of Geography. Her research interests include urban governance, climate change policy, urban land use, shrinking cities/suburbs, and renewable energy policy. Prior to joining the faculty at Delaware, she worked at the George Washington Institute of Public Policy, the Metropolitan Policy Program at the Brookings Institution, the Rochester Institute of Technology, the White House Council on Environmental Quality, and the environmental law practice at Sidley Austin LLP. Andrea has a Ph.D. in public policy and public administration and a B.S. in natural resources from Cornell University.

Chapter 7
Enhancing Citizen Engagement in the Face of Climate Change Risks: A Case Study of the Flood Early Warning System and Health Information System in Semarang City, Indonesia

Aniessa Delima Sari and Nyoman Prayoga

Abstract This case study describes how two climate resilience action projects in Semarang City, Indonesia, were able to provide new mechanisms allowing better engagement between the Semarang city government and its citizens. With the introduction of the Flood Early Warning System (FEWS), flood-prone communities in the Beringin River Basin are now able to evacuate to safe shelters before flood incidents occur. Through the Health Information and Early Warning System (HIEWS), citizens can access real-time information related to dengue fever cases in the city. Although the focus areas are different, both projects aim to help communities become more resilient to the impacts of climate change, specifically floods and vector-borne disease. We find similar patterns in the two cases, in which efforts to enhance community participation are essential to guarantee the success of the projects. Enhanced community engagement is achieved through the thoughtful consideration of local knowledge and social networks, intensive assistance to increase awareness and motivation of the community, and understanding governance structures to ensure that funds are allocated through formal handover processes to continue and expand the results of the interventions. These findings are useful and important to guide any climate change adaptation projects toward better sustainability and ownership, especially in the application of an early warning system or information system that requires technology, sustainable budget allocation from the local government to operate and maintain the system, and buy-in from local communities.

A.D. Sari (✉) · N. Prayoga
Mercy Corps Indonesia, AD Premiere Building, 3rd Floor, Jl. Taman Margasatwa no 5, Ragunan, Jakarta 12550, Indonesia
e-mail: asari@id.mercycorps.org; aniessads@gmail.com

N. Prayoga
e-mail: iprayoga@id.mercycorps.org; prayoga.nyoman@gmail.com

S. Hughes et al. (eds.), *Climate Change in Cities*, The Urban Book Series,
https://doi.org/10.1007/978-3-319-65003-6_7

Keywords Flood early warning system · Disaster preparedness · Health information system · Climate change resilience · Climate change adaptation · Climate change governance

7.1 Introduction

The participation of citizens is required to make communities resilient and more inclusive (Chu et al. 2016). In Indonesia, national, regional, and local governments have an annual mechanism to ensure public participation to shape development pathways called *Musrenbang*, which stands for *Musyawarah Perencanaan Pembangunan* or Development Planning Deliberation. *Musrenbang* is an annual process during which residents meet to discuss the issues facing their communities and decide upon priorities (both infrastructure and nonstructural needs) for short-term improvements. At the city level, once a list of priorities or actions is agreed upon by communities, it is submitted to the local government planning department, Bappeda (Regional Development Planning Board), which will then assign resources to each neighborhood depending on the available funds and based on need. Actions identified and prioritized will be proposed to be funded by local government budget. The main objective of the *Musrenbang* is to help key stakeholders understand different challenges and development issues and to reach solutions through a consensus-based process. It aims to synchronize and reconcile "top-down" versus "bottom-up" approaches by having a community needs assessment, combined with technical assessment, to resolve conflicts between local government interests and nongovernment stakeholders in the local development agenda (such as between needs and fiscal capacity), and to synergize budget allocations (Ministry of Home Affairs 2007).

Although the mechanism is available, in many cases, the *Musrenbang* process is hardly effective. Most development proposals from the community tend to be limited to infrastructure needs, which are repeated every time a *Musrenbang* is held. Not everyone is able to voice their ideas or requests as the elected representatives who attend *Musrenbang* rarely propose activities or development projects that genuinely reflect the needs of their constituents. The activities resulting from *Musrenbang* are usually implemented by local governments and contractors in ways that are inflexible or strictly regulated. For example, the procurement system that is in place to select a particular implementer or vendor is often determined based on the lowest price and not on highest quality. Therefore, the city government is often unable to adapt if the vendor is underperforming or if there is new information from either the communities as beneficiaries or from direct observations to improve the quality of the work. Furthermore, *Musrenbang* currently lacks an information platform that can provide information to citizens to help track their proposed actions (i.e., whether it is approved, already implemented, still planned, or rejected). As a result, any proposed ideas run the risk of being duplicated in the following year, which leads to a general inability to meet the long-term needs of the community. Finally, even when communities are dissatisfied with the results of projects, citizens often do not know where and how to report complaints, as well as whether their complaints will be handled properly by the relevant authorities.

This chapter describes how two climate resilience action projects in Semarang City, Indonesia—namely the Flood Early Warning System (FEWS) and the Health Information and Early Warning System (HIEWS)—were able to provide new mechanisms, beyond the *Musrenbang*, to facilitate better engagement between the city government and its citizens. Although the concept of deliberative planning is not new in the Indonesian context, this chapter notes that an ad hoc multi-stakeholder team overseeing the entire process of planning, budgeting, implementing, and coordinating climate change actions—especially a process that complements the *Musrenbang*—is a governance innovation. More specifically, in the case of the Flood Early Warning System (FEWS), seven Disaster Community Groups in Beringin River Basin are now able to safely evacuate themselves to safe shelter before the arrival of a flood incident, and they are able to communicate cases of flooding and needs of resources or logistics directly to the Disaster Management Board. The FEWS project used known and trusted local technologies, such as bamboo slit drums (*kentongan*) and mosque loudspeakers, to alert the community when the early warning system was activated. In the case of Health Information System, a participatory online system allows community cadres, student doctors,[1] and teachers to report-free larvae index numbers, which can then indicate the risk of Dengue Hemorrhagic Fever (DHF) incidences in the area. This information is processed through the Health Early Warning and Information System (HIEWS) along with climate data and DHF cases reported by local hospitals and local clinics. All the data is then synthesized and disseminated as early warning signals to different sub-district chiefs so that they can quickly respond and work with communities to prevent DHF cases and to remove mosquito larvae from the local environment.

The analysis of Semarang's experience highlights how digital technology, used in a clearly designated and coordinated manner, can improve multilevel communication pathways within cities, especially when dealing with climate impacts on local public health and environmental quality needs. The case study also highlights the importance of coupling the use of technology with a community engagement and deliberative governance approach, which includes utilizing local knowledge as reliable (as well as a "backup") mechanisms for communities dealing with recurring climate change and disaster incidents.

---

[1]Student doctors—or locally known as *Dokter Kecil*—are groups of elementary school students appointed based on their leadership qualities and proactive attitude to promote healthy lifestyles.

7.2 Case Study of Climate Resilience Action Projects in Semarang City

7.2.1 Case Study: Flood Early Warning System in Beringin River Basin

Due to the city's natural conditions, coastal flooding is a major problem in Semarang and is posing increasing challenges for the city government (Marfai and King 2008a, b). The topographic characteristic of Semarang is divided into a northern region, which is a flat lowland area and a southern highland area. This topography results in particular challenges for managing floods because of the possibility of flash flood events when upstream areas receive heavy rain. Flash flooding has become recurrent on an annual cycle, so the city government has put particular focus on this issue. The high level of loss and damage resulting from flood is one of the consequences of not having a warning system, as well as a lack of preparedness within local communities. In addition, the Semarang Disaster Management Board was only established in 2011. Prior to that, there were no preexisting community disaster groups to address flooding, no flood shelters and evacuation routes in areas prone to flooding, and there were weak coordination mechanisms between stakeholders involved in disaster management. This included a lack of coordination mechanisms available to connect communities in the upstream and downstream areas, as well as between communities and different government agencies related to flood management.

Semarang was one of the pilot cities in the Rockefeller Foundation's Asian Cities Climate Change Resilience Network (ACCCRN) program (Taylor and Lassa 2015). Semarang's Vulnerability Assessment (2010) and Semarang City Resilience Strategy (2010) showed that the city is prone to flooding, so the ACCCRN program encouraged the government of Semarang to implement a Flood Early Warning System (FEWS) project. The project was implemented by the BINTARI Foundation in partnership with the Semarang City Government and the University of Diponegoro, with close coordination by Mercy Corps Indonesia. The project was conducted between 2012 and 2014, and was located in Beringin watershed (see Fig. 7.1). The project aimed to reduce vulnerability to flood risk and impacts of flood disasters by building preparedness capacity of both the most vulnerable communities and the local government through the development of an early warning system and evacuation strategies. The resulting project, known as the Flood Early Warning System (FEWS), worked to reduce vulnerability, loss, and casualties attributed to floods by strengthening community and government response. This project also strengthened communities and government preparedness through the development of a flood information system, an evacuation strategy, as well as the identification of evacuation shelters for those who are most affected.

The pilot projects focused on the Beringin watershed because it is prone to floods from the overflowing of the Beringin River as well as tidal inundation. Under these conditions, the projects aimed to improve preparedness to flood risks as a means of disaster risk management in communities to reduce overall climate

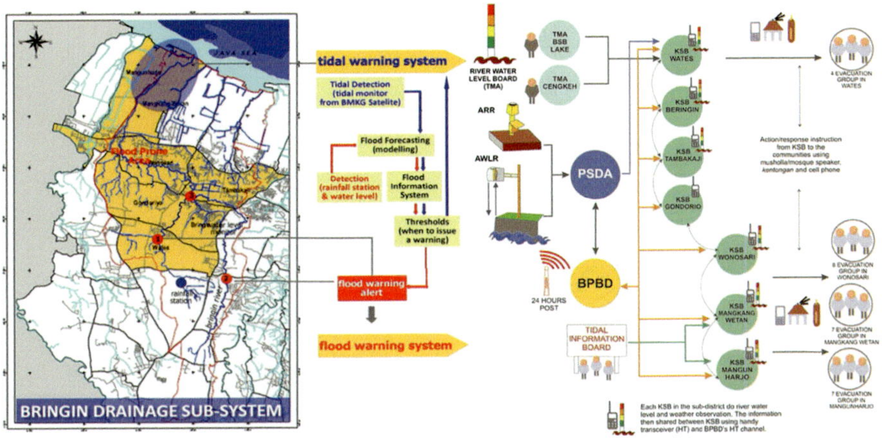

Fig. 7.1 FEWS information pathway within the system

Fig. 7.2 Scenes of the community actively participating within the KSB

vulnerability in Semarang City. The interventions were implemented across seven sub-districts in the Beringin watershed, and helped to develop a flood forecasting and early warning system that considers climate scenarios for future land use changes. As part of the overall management of the early warning system, the city's water resource management agency—the PSDA (*Pengelolaan Sumber Daya Air, Energi, dan Sumber Daya Mineral*)—is responsible for maintaining the drainage system and monitoring rainfall. In the case of flooding, the PSDA provides flood warning information to disaster preparedness groups at the community level,

Fig. 7.3 Scenes of the community actively participating within the KSB

which are called *Kelompok Siaga Bencana* (KSB) or sometimes *Kelompok Masyarakat Peduli Bencana* (KMPB). These community groups respond quickly and disseminate information to the public (see Figs. 7.2 and 7.3). The information is also communicated to the local disaster management agency, called the *Badan Penanggulangan Bencana Daerah* (BPBD), so that they can also prepare if there is a possibility of flooding.

Flood Early Warning System (FEWS) activities are carried out through three components, which consist of acknowledging flood pattern, providing flood evacuation shelter, and the involvement of stakeholders such as the municipal government and nongovernmental organizations. Local communities—as represented by different disaster preparedness groups (*Kelompok Siaga Bencana*) spread across seven sub-districts—play an important role as local agents together with the Semarang City government's BPBD and PSDA, who also play vital roles in disaster management. The PSDA has responsibility for managing the early warning system tools, which include an Automatic Water Level Recorder and Automatic Rainfall Recorder. The BPBD has the responsibility of establishing and maintaining the different disaster preparedness groups in communities by periodically increasing the capacity and awareness through socialization and training simulations.

To reflect on these developments, in January and February 2014, a severe flooding event occurred in and around the Beringin watershed. The average height of the floodwaters was around 40 cm with a receding rate of 30 min to an hour. Local residents claimed that there were some changes in how they coped with floods in 2010 compared to 2014, which can be attributed to the FEWS intervention project. In 2010, floods in the northern area of Semarang City caused seven deaths and hundreds of injuries; but in 2014 there were no casualties, though some injuries were inevitable.

7.3 Case Study: Health Information and Early Warning System in Semarang City

According to various studies, changes in climate cycles are proven to increase the likelihood and risks of vector-borne diseases, especially Dengue Hemorrhagic Fever (DHF) (Case et al. 2007, Anyamba et al. 2006, and Sofyan, 2010). The dengue virus is sensitive to climate change because of changes in average temperature and changing rainfall. This trend tends to be exacerbated in breeding areas of the *Aedes Aegypti* mosquito, especially in dense neighborhoods and places with many water containers.

Based on records from 2008 to 2012, Semarang ranked as having the highest number of dengue cases in Central Java. However, this data is based on the number of dengue fever cases reported by hospitals that treat patients, and is often not on time nor accurate as it was reported over an average of 20 days instead of within 24 h. Information coming from the Health Agency is considered more reliable. One reason that data reports from clinics and hospitals tends to be late is because the communication mechanisms are primarily manual, operated by government officials who wait for telephone calls or visits from health workers. Data reports are then received late, which result in weak interactions between local governments and the hospitals and clinics. In addition, there is no warning system for dengue fever provided by the local government that can be accessed by communities to help them prepare. Additionally, there is no rigorous documentation about the impacts of climate change toward incidences of dengue infection in Semarang City. As a result, the city government is often not aware that climate change exerts additional risks for dengue prevention efforts across Semarang. The level of awareness across communities was also a problem, as the design and the tools for dengue prevention were considered boring and not attractive. Local community leaders were therefore not motivated to distribute dengue prevention awareness to the wider public.

Starting in 2013, Mercy Corps Indonesia—through the ACCCRN program and in collaboration with the Semarang City Government—sought to tackle the increasing rates of Dengue Hemorrhagic Fever (DHF) attributed to climate change. The ACTIVE (Actions Changing Incidence of Vector-Borne Disease Endemic) program attempts to build collaboration between stakeholders across citizen, school, and government levels to jointly increase the city's resilience and to reduce the risk of Dengue Hemorrhagic Fever. The ACTIVE program was held in six sub-districts, six public health clinics, and 19 primary schools across five districts in Semarang (see Fig. 7.4 for an illustration of the program design). Dengue risks must be addressed before it emerges as a disease outbreak; so in this context, preventive efforts took a significant role with the development of the Health Early Warning System (HEWS) and Health Information System (HIS).

Fig. 7.4 Activity design of the dengue prevention program in Semarang

As part of the dengue prevention program, a well-designed Health Information and Early Warning System (HIEWS) was developed.[2] The program intended to increase the Semarang City Government's capacity to generate reliable public information and a warning and alert system to dengue fever risk to the citizens, thereby increasing the possibility of actively preventing dengue fever transmission early on. All warning systems, however, simple they may be, are based on an analysis of scientific data and trends. The adequate prediction of dengue fever incidences becomes one key factor, which has proved to be very useful in preventing dengue transmission. Since the transmission of dengue fever is strongly influenced by climate change, predictions are conducted based on climate data that is supported by robust statistical models, forecasts of dengue-climate change interactions, as well as vulnerability cluster mapping across Semarang City. The ACTIVE program tested and documented the application of different information and communication technologies, which helped to integrate climate information and to allow the system to provide early warning and public information about dengue fever incidents across Semarang City. The early warning system provides public health officials and the general public with as much advance notice as possible about the likelihood of a disease outbreak in a particular location, thus widening the range of feasible response options.

The Health Information and Early Warning System (HIEWS) in Semarang City was developed through intensive collaboration with the Semarang City Health Agency. The climate-related dengue fever prediction system uses rainfall, humidity, and air temperature data supplied by the National Meteorology, Climatology, and Geophysics Agency (BMKG—*Badan Meteorologi, Klimatologi, dan Geofisika*). These climate parameters are used as inputs to dengue forecasting as formulated by the Climate Change Center of the Bandung Institute of Technology (Nuraini, et al. 2014). The HIEWS helps the City Health Agency in documenting dengue cases and reports findings from the community, schools, public health centers, and hospitals. The ACTIVE Project Team trained school teachers, health workers, and community cadres to understand the impacts of climate change on dengue incidences, to be able

[2]Accessible at http://www.dinkes-kotasemarang.go.id/simdbd/.

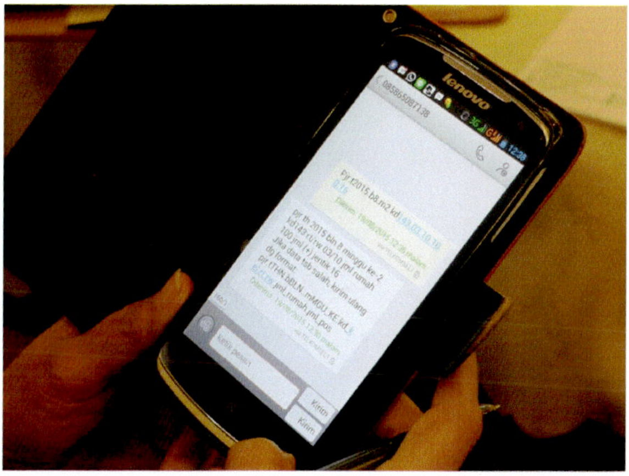

Fig. 7.5 Community and student doctors perform mosquito larvae monitoring and send the report to the City Health Agency through SMS as part of the HIEWS mechanism

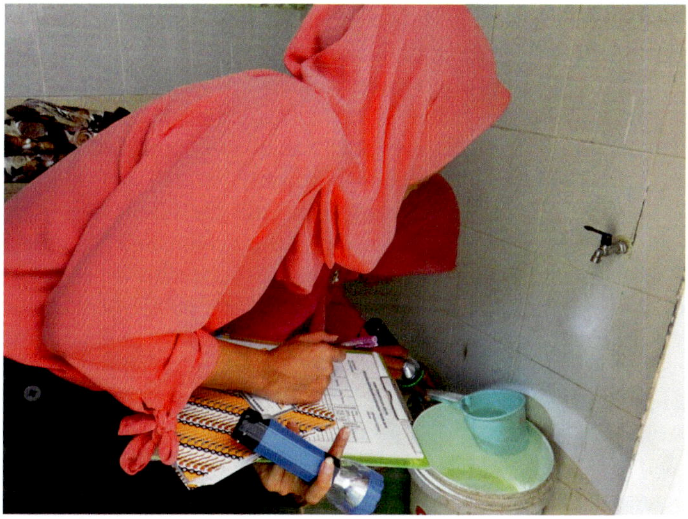

Fig. 7.6 Community and student doctors perform mosquito larvae monitoring and send the report to the City Health Agency through SMS as part of the HIEWS mechanism

to report to the HIEWS system, and to deliver training to the wider community. Thus, hospitals are now able to report incidences of dengue online, which enables timely sharing of dengue-related information (see Figs. 7.5, 7.6, and 7.7 for illustrations). The communities and student doctors are also able to report Free Larvae Index collected from the neighborhood using SMS text. The SMS system is linked to key stakeholders, such as the head of the district or sub-district and neighborhood leaders, who have the authority to act if there are dengue cases or outbreaks

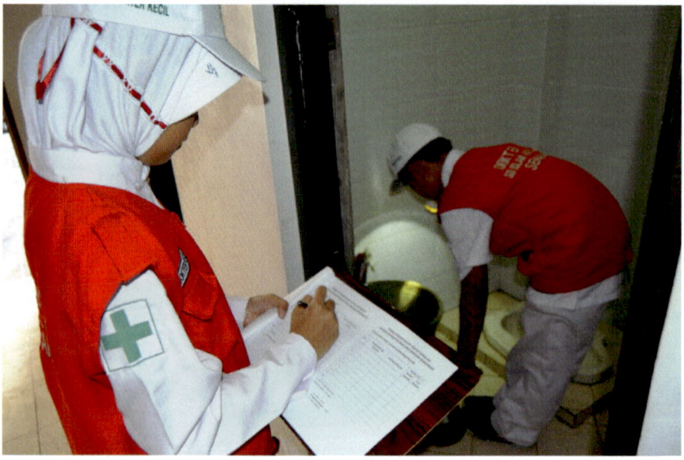

Fig. 7.7 Community and student doctors perform mosquito larvae monitoring and send the report to the City Health Agency through SMS as part of the HIEWS mechanism

identified. At the same time, the system provides a digital account for public health centers and hospitals to complete online reports.

The ACTIVE program also increased partnerships among scientific researchers of dengue fever and climate change vulnerability and adaptation. In particular, the program benefitted from the rigorous documentation of current and future trends of dengue fever that consider climate change impacts, as well as from the improved design of dengue awareness tools, such as the Behavior Change and Communication Campaign. The campaign was conducted by Mercy Corps Indonesia in partnership with the Semarang City Health Agency to increase knowledge, skills, and the motivation of households, public health workers, and elementary schools to identify, manage, and to regularly and accurately report dengue fever data. Research from the Semarang City Health Agency in 2015 showed that 29% of dengue fever patients were children of 5 to 9 years old, making children the most vulnerable group to the dengue virus. In addition, the campaign also increased access to information about dengue fever by using key messages and action cues among households and elementary schools. With increased knowledge and access to information, the ACTIVE program expects to improve adaptive practices and behavior toward dengue fever in Semarang City.

7.4 Community Participation as a Key Element Within Early Warning System Implementation

From both case studies, we found that the community plays a significant role in the process of building local resilience to climate change and disaster risks. For example, the Disaster Preparedness Group (*KSB—Kelompok Siaga Bencana*) plays a core coordinating function within the Flood Early Warning System (FEWS) and community champions and student doctors provide dengue-related information to the Health Information and Early Warning System (HIEWS). To promote resilience to climate change and to other community stressors, a phased course of action is required to facilitate community-based adaptation strategies that engage stakeholders in a proactive problem-solving process, thereby enhancing social capital across local and national levels (see Chap. 9: Iftikhar, Ali and Sarzynski).

Community-based adaptation to climate change is a community-led process based on the community's priorities, needs, knowledge, and capacities, and should empower people to plan for and cope with the impacts of climate change (Dodman and Mitlin 2013; Ebi and Semenza 2008; Magee 2013). Stakeholder engagement is important to ensure that messages designed to reach vulnerable groups provide the information and motivation necessary for individuals to make appropriate choices (Anguelovski et al. 2014; Moser and Pike 2015; Nay et al. 2014). In addition, working with stakeholders can help ensure individual and community acceptance of the intervention, along with reducing constraints to implementation (Lassa and Nugraha 2014). However, in Semarang, those who are affected by impacts of climate change—such as flood and dengue outbreaks—are also left with an increased need to build their capacity in disaster management, especially related to prevention and preparation. Community members, particularly those who are most vulnerable, should be actively involved in activities or initiatives for building resilience, especially with issues that directly relate to them, and with the expectation that what they are doing differently can contribute to a better situation in the future.

7.4.1 Benefit to the Community by Taking Part in Urban Climate Change Resilience

Demonstrating economic or livelihoods benefits is an effective way to invite people to build resilience (Forsyth 2013). In a disaster context, having an early warning system helps the community prevent higher losses in the event of a disaster. It is important to ensure that communities understand the different outcomes that result from having an early warning system versus not having one. Based on the experiences of ACCCRN interventions in Semarang, both case studies explained in this chapter show the need to reduce vulnerability and the impacts of flood disasters and dengue outbreaks by building the preparedness capacity of vulnerable communities. For example, in the flooding context, with better preparation prior to a disaster event, communities can reduce potential material losses by receiving early warning information about disaster risks, which gives them time to protect their important property. In the health context, people understand that the cost of dengue treatment is expensive and vulnerable community members tend to not allocate budget for health needs. As a result, as seen through the Health Information and Early Warning

System (HIEWS) case, by having more real-time data that is reported using a bottom-up approach, the response to dengue cases as well as future preventative actions will be better managed. Further, the no or low-cost direct actions that community members can engage into protect their health—such as increasing abilities for dengue fever prevention and mosquito monitoring and an increased focus on clean and healthy living—have resulted in a dramatic decrease in the number of DHF cases, the number of lives saved, as well as lowering the costs to the city for medical services.

7.4.2 Building Community Ownership

When community members are given clear roles and responsibilities, they feel that they are acknowledged and have an important contribution to the project (see Chap. 4: Boswell and Mason). Each community member must understand each specific role and should discuss together what they can each contribute. For example, in the case of the Disaster Preparedness Group, the different responsibilities can be shared amongst group members in order to make them more manageable. There are people who are in charge of monitoring the water levels of the river, disseminating information through their communication systems, giving warning to the neighborhood through loudspeakers of local mosques, and also those who are responsible during an evacuation. Another example comes from the implementation of the HIEWS, where the student doctors from elementary schools feel that they contribute to the protection of the school and neighborhood by regularly monitoring mosquito larvae. This is also applied to community champions, who send monitoring reports to the HIEWS because they feel appreciated when receiving confirmation messages from the City Health Agency when their report has been received.

7.4.3 Impacts of Community Projects on Urban Climate Change Governance

The case studies of the FEWS and HIEWS have similarities in terms of the project design, in which both put careful considerations on the role of governments, communities, and how governments and communities interact in the context of climate change resilience efforts. Development projects at the community level often only focus on the community, thereby neglecting to consider the role and involvement of the government as part of the process.

In the case of the Flood Early Warning System, the establishment of the disaster preparedness groups becomes a platform where government and communities interact. The Flood Early Warning System sent information to the PSDA and BPBD

—the local disaster management agencies—which was then forwarded to the head of the local disaster preparedness group. However, all of the parties realized that the early warning system is not the only way to evaluate the risks of flooding. In particular, the local governments rely on disaster preparedness groups to ensure that the system is working well and is valid. For example, group members in the upstream area report rainfall data to downstream group members who manually monitor the water level of the river via hand-held communication devices (colloquially known as *handy talkies*). This is a backup in case the equipment runs out of battery power, breaks or malfunctions, or experiences interferences that cause delays or false recording of the data. For instance, large piles of trash can cause the Automatic Water Level Recorder (AWLR) to mistake it for high water levels. In addition, the use of *handy talkies* that were distributed to all disaster preparedness groups (KSB) and connected to the local disaster management agency (BPBD) allowed communities to engage more regularly and report their needs to the local governments directly. This type of engagement model is now promoted by other local municipal officials in their engagements with community residents and programs for flood management in other areas of the city.

In the case of the ACTIVE program, the Health Information and Early Warning System (HIEWS) is the platform that connects local government (such as the City Health Agency) to citizens, public and private hospitals, and clinics. The City Health Agency, which was previously only responsible for storing data, has now expanded its role to include storing, managing, and sharing data. Communities are using their own funds to send SMS texts to inform the Health Information and Early Warning System's real-time free larvae index. Citizens are now also able to access this real-time information, either through the City's Health Agency's website or SMS text, which are now sent to the various chief of districts and sub-districts, who will then issue a dengue incidence warning to local communities through community-based Women's Welfare Groups, local leaders, as well as through traditional media outlets. The impacts have been beneficial for both the community and government. Communities can now report cases and receive warnings from the Health Information and Early Warning System about potential and existing dengue cases in a timely fashion, which means that the government can also quickly respond and take necessary action.

7.5 Challenges and Opportunities in the Application of Warning Systems and Stakeholder Engagement

The link between the application of early warning systems and the process of engaging stakeholders is strong. Projects in the community should not only involve the community, but also many relevant actors including the government. It is also expected that once programs or projects are finished, the government can continue to support the initiative in the community as necessary. In reflecting on the implementation of the Flood Early Warning System (FEWS) and the Health Information Early Warning System (HIEWS) projects in Semarang, there are a number of lessons on the challenges and opportunities that can be learned.

The first challenge is that to make the system work, the program needs strong participation from both community and government (as well as other relevant stakeholder groups). It is important to find key individual actors within the stakeholder groups to help the implementers accelerate and strengthen efforts by including them as a notable agents or project champions. As a result, such community leaders can help to convince and bring the community together to actively engage with the process, or to bridge the communication divide and bureaucratic processes within the local governance systems (Carmin et al. 2012, 2013). Finding key actors may be tricky in the beginning of the process, but once they are on board, they can be a great value for the process.

Second, the development of the early warning system is closely linked to the availability and utilization of technology. However, this can be difficult for communities that lack familiarity with information and communications technology (ICT). One option may be to persistently teach and assist community members in using the technology or to better acknowledge their preferences in alternative communication methods. For example, since many community beneficiaries are actually elderly people who are often not familiar with digital operating system-based applications, an alternative system that is more convenient may be one that uses SMS technology instead. However, the main objective should be to find the most effective method—which may include existing or traditional community communication systems—for disseminating knowledge to allow community members to participate in the early warning system.

Third, a significant challenge for the early warning system project is the maintenance of the system infrastructure, especially once external funding sources conclude. This could be anticipated by ensuring proper ownership of the project deliverables. For example, there should be a clear catalogue of action items delineating the actors responsible for maintaining the infrastructure and for obtaining funding to upgrade the system in the future. This could also be strengthened by utilizing support from higher level government officials and community leaders. Support from the leadership is important to ensure the prioritization and continuity of climate-related pilot projects. This is also related to the challenge of understanding and influencing how the city government allocates resources and institutionalizes new approaches within existing institutional arrangements.

In terms of opportunities, the proper implementation of different early warning system projects can lead to better coordination between the government and community, as well as between governments and between communities. Establishing an effective communication system is essential to help with opening and bridging the access of involved stakeholders, and to be better able to dissemination information and knowledge amongst community members (see Chap. 8: Enberg). An interesting example can be learned from the Flood Early Warning System (FEWS) project. In this case, disaster preparedness groups at higher elevations are able to communicate to disaster preparedness groups at lower elevations about rainfall patterns, thereby enabling them to alert others of the possibility of flash flooding events.

A second opportunity is that this participatory approach may become a trend in many community projects within and beyond Semarang. The case studies of the Flood Early Warning System (FEWS) and the ACTIVE programs highlight the importance of engagement processes to gain support from local leaders as well as the need for intensive assistance to motivate stakeholders. For example, such processes of intensive assistance included introducing interactive methods, fun learning, and creative facilitation workshops for dengue prevention and surveillance. In addition, the community empowerment officers also motivated the community cadres by instilling an understanding of why the cadres' roles are important and how they are part of a positive and collaborative movement, which is then further acknowledged by the municipal health agency through a variety of appreciative gestures. However, it must be acknowledged that behavior change is neither an easy nor short process, but the efforts to encourage better participation by everyone must be persistent. In both cases, community volunteers took up practical action with direct and meaningful benefits. Behavioral change can be a gradual shift as long as community members have a good sense of ownership, built throughout the process by intensively involving them in many activities and also by designating specific roles and responsibility.

Finally, the early warning projects provide an opportunity to raise awareness of the public and to put more consideration into overall climate change protection mechanisms. People are increasingly realizing that disaster is not a problem that can be solved in a short term, as no one can guarantee when or where natural disaster and disease outbreaks will take place. However, we can help mitigate the risk or even reduce the impact of disaster through a well-coordinated, participatory, and inclusive early warning system. In this context, early warning systems will help to anticipate and cope with risks in pre-disaster, during disaster, and post-disaster phases, as well as to prevent losses when a disaster event occurs. This also involves interaction with different stakeholders, which facilitates knowledge exchange and experiential learning. It is expected that this can influence government actors—both within and beyond Semarang—to better prioritize and put more consideration toward climate and disaster risk prevention activities instead of focusing their budget on post-disaster recovery.

References

ACCCRN Program (Asian Cities Climate Change Resilience Network). (2010). Vulnerability and Adaptation Assessment to Climate Change in Semarang City. Semarang, Indonesia: Mercy Corps Indonesia.

ACCCRN Program (Asian Cities Climate Change Resilience Network). (2010). City Resilience Strategy: Semarang's Adaptation Plans in Responding to Climate Change. Semarang, Indonesia: Government of Semarang City.

Anguelovski, I., Chu, E., & Carmin, J. (2014). Variations in approaches to urban climate adaptation: Experiences and experimentation from the global South. *Global Environmental Change, 27,* 156–167. doi:10.1016/j.gloenvcha.2014.05.010.

Anyamba, A., Chretien, J.-P., Small, J., Tucker, C. J., & Linthicum, K. J. (2006). Developing Global Climate Anomalies Suggest Potential Disease Risks for 2006-2007. *International Journal of Health Geographics, 5,* 60.

Carmin, J., Anguelovski, I., & Roberts, D. (2012). Urban Climate Adaptation in the Global South: Planning in an Emerging Policy Domain. *Journal of Planning Education and Research, 32*(1), 18–32. doi:10.1177/0739456X11430951.

Carmin, J., Dodman, D., & Chu, E. (2013). *Urban Climate Adaptation and Leadership: From Conceptual Understanding to Practical Action* (No. 2013/26). Paris, France: Organisation for Economic Co-operation and Development (OECD). doi:10.1787/5k3ttg88w8hh-en.

Case, M., Ardiansyah, F., & Spector, E. (2007). *Climate Change in Indonesia: Implications for Humans and Nature*. Jakarta, Indonesia: WWF-Indonesia.

Chu, E., Anguelovski, I., & Carmin, J. (2016). Inclusive approaches to urban climate adaptation planning and implementation in the Global South. *Climate Policy, 16*(3), 372–392. doi:10.1080/14693062.2015.1019822.

Dodman, D., & Mitlin, D. (2013). Challenges for Community-Based Adaptation: Discovering the Potential for Transformation. *Journal of International Development, 25*(5), 640–659. doi:10.1002/jid.1772.

Ebi, K. L., & Semenza, J. C. (2008). Community-based adaptation to the health impacts of climate change. *American Journal of Preventive Medicine, 35*(5), 501–507. doi:10.1016/j.amepre.2008.08.018.

Forsyth, T. (2013). Community-based adaptation: A review of past and future challenges. *Wiley Interdisciplinary Reviews: Climate Change, 4,* 439–446. doi:10.1002/wcc.231.

Lassa, J. A., & Nugraha, E. (2014). From shared learning to shared action in building resilience in the city of Bandar Lampung, Indonesia. *Environment and Urbanization*. doi:10.1177/0956247814552233.

Magee, T. (2013). *A Field Guide to Community Based Adaptation*. London and New York: Routledge.

Marfai, M. A., & King, L. (2008a). Coastal flood management in Semarang. *Indonesia. Environmental Geology, 55*(7), 1507–1518. doi:10.1007/s00254-007-1101-3.

Marfai, M. A., & King, L. (2008b). Potential vulnerability implications of coastal inundation due to sea level rise for the coastal zone of Semarang city. *Indonesia. Environmental Geology, 54* (6), 1235–1245. doi:10.1007/s00254-007-0906-4.

Ministry of Home Affairs. (2007). Decree No.050-187/Kep/Bangda/2007 on Musrenbang. Jakarta, Indonesia: Government of Indonesia.

Moser, S. C., & Pike, C. (2015). Community engagement on adaptation: Meeting a growing capacity need. *Urban Climate, 14,* 111–115. doi:10.1016/j.uclim.2015.06.006.

Nay, J. J., Abkowitz, M., Chu, E., Gallagher, D., & Wright, H. (2014). A review of decision-support models for adaptation to climate change in the context of development. *Climate and Development, 6*(4), 357–367. doi:10.1080/17565529.2014.912196.

Nuraini, et al. (2014). *Climate Models on Vector Borne Diseases in Semarang City in conjunction with Health and Climate Vulnerability Assessment*. Semarang: Mercy Corps Indonesia and Bandung Institute of Technology (ITB)-Climate Change Center.

Sofyan, A., Sukowati, S., & Slamet, J. S. (2010). *Indonesian Climate Change Sectoral Roadmap (ICCSR) - Health Sector*. Jakarta: Bappenas (National Development Planning Agency).

Taylor, J., & Lassa, J. A. (2015). *How can Climate Change Vulnerability Assessments best impact policy and planning?: Lessons from Indonesia*. London: International Institute for Environment and Development (IIED).

Author Biographies

Aniessa Delima Sari has over 7 years of experience in leading and implementing a wide range of urban resilience work in Indonesia. For the past 3 years, she has worked as a Program Manager in Mercy Corps Indonesia's Resilience Portfolio, which include the Asian Cities Climate Change Resilience Network (ACCCRN) Indonesia, the Zurich Flood Resilience Program, 100 Resilient Cities support to Semarang City, and the recently launched Global Resilience Partnership TRANSFORM Flood Resilience Program. She works with a wide variety of stakeholders from government, NGO, community, business, and academic institutions to increase the capacity of Indonesian cities to effectively implement climate change resilience strategies.

Nyoman Prayoga is a Member Relations Manager for the Asian Cities Climate Change Resilience Network (ACCCRN) at Mercy Corps Indonesia. He holds a Master of Engineering in Urban and Regional Development from Diponegoro University. From 2014 to 2016, he helped manage resilience projects in ACCCRN cities across Indonesia, where he was responsible for communication, generating knowledge, and building awareness on urban climate change resilience issues. At Mercy Corps Indonesia, he is responsible for ensuring active participation from members in the ACCCRN network, attracting new membership, maintaining member engagement, generating knowledge and resources for the network, and encouraging and influencing the ACCCRN community to connect and collaborate on building urban climate change resilience.

Chapter 8
Climate Adaptation and Citizens' Participation in Denmark: Experiences from Copenhagen

Lars A. Engberg

Abstract In a Danish perspective, climate adaptation and investments in new infrastructures for storm water management open up new possibilities for citizens' engagement in urban spatial improvements. Climate adaptation projects are developed in a multilevel governance framework that balances cross-sectoral approaches with participation mechanisms. The chapter accounts for the Copenhagen climate adaptation case, and it explores how the city strategically meta-govern the boundaries between the expert governed large-scale water management scheme against small-scale place-based bottom-up projects in collaboration with citizens and other place-based stakeholders.

Keywords Climate adaptation · Participation · Urban governance · Consensus-steering · Urban renewal · Place-making

8.1 Introduction

When the City of Copenhagen experienced cloudbursts with flooding and damage to property in 2011, and torrential downpours in 2014 and 2015, climate adaptation became a political priority in the city. Copenhagen is, alongside most other Danish cities, concerned about the sustainable transition to the zero carbon society. The city promotes energy efficiency in buildings and numerous other mitigation initiatives, but the cloudburst event was the first reminder that the day-to-day life of Copenhageners is directly and fundamentally challenged by extreme weather. Instead of initiating a costly overhaul of the sewer system, the City Council decided to go for a surface solution, and combine it with the construction of four large underground pipes. The surface solution implies that some streets are modified to channel water in extreme weather situations, in a specific city-wide hydraulic pattern. The Copenhagen climate adaptation strategy therefore relies on a process of

---

L.A. Engberg (✉)
Danish Building Research Institute, Aalborg University Copenhagen, Copenhagen, Denmark
e-mail: lae@sbi.aau.dk

S. Hughes et al. (eds.), *Climate Change in Cities*, The Urban Book Series,
https://doi.org/10.1007/978-3-319-65003-6_8

public–private 'co-creation'; the participation of citizens, small businesses and other nonpublic actors in designated areas to remodel the city scape, adapting squares, streets, lakes, and parks to better cope with future cloudbursts.

Urban governance refers to the strategic ordering of public action that influences or regulates the actions of nonpublic actors in networks between public, private and civil actors (Jacquier 2005). The fragmented and multilevel characteristics of the urban governance system reflect the silo-based structures of the public sector, and the tendency to include nonpublic actors in "co-production" of welfare services (Meijer 2016). For three decades, practitioners in the Danish urban regeneration policy field have introduced mechanisms of public–private co-creation. The term co-creation refers to the notion that a plurality of public and private ideas, resources and knowledge is merged in a joint creation of public solutions (Torfing et al. 2016, p. 3). In the Danish tradition, City Councils initiate and steer urban regeneration initiatives and programs where planners and local stakeholders co-create integrated urban space projects. As public governance actors, planners and civil servants use a toolbox of meta-governance mechanisms to enable and encourage this co-creation, a process by which nonpublic actors influence, define, concretize and prioritize projects that improve their neighborhoods in collaboration with city officials (Fig. 8.1).

The chapter provides an account of how the City of Copenhagen changed an Integrated Urban Renewal project (IUR) in the Skt. Kjelds District into 'The Copenhagen Climate District' after the 2011-flooding event. The district became the urban lab for innovation of concrete solutions and inspired the development of a new city-wide water handling structure. Today, the city administration works together with local committees, IURs, housing associations, owners associations, citizens' groups, businesses and other local stakeholders to develop the new surface-based water handling structure. The city surface cannot be transformed without the consent and participation of residents in their roles as owners, users, and local decision-makers (Fig. 8.2).

Because climate adaptation measures cut across policy boundaries and different bureaucratic domains, they need to be developed in a complex multilevel steering-

Fig. 8.1 Citizens and planners transformed a school yard at a public school into both school yard and public park. Integrated Urban Renewal (IUR) project 2012–2016. Courtesy of Amager Fælled Skole

Fig. 8.2 Flooding in Copenhagen, June 2011. Courtesy of the Copenhagen City Council

and decision-making context. When Copenhagen currently develops its response to extreme weather events, the city is facing a type 3 wicked problem (Head and Alford 2013, p. 7) where both problem definition and solutions are initially unclear and require extensive discussion and learning for all concerned. The response to the climate challenge is innovated in a fuzzy domain of competing modes of technical, social, and embedded or practical knowledge and operational experience. Some technical solutions are givens that define the scope for citizens' projects, and some solutions are only possible when bottom-up inputs are developed and incorporated into city-wide systems. The notion of co-creation captures this open-ended innovation process, where a priori distinctions between specialized knowledge and context-relevant inputs and solutions are hard to draw.

8.2 Co-creation and Meta-Governance

A key question in the volume is how to foster and encourage innovation in a multilevel climate governance context. In this chapter, climate governance is conceptualized as meta-governance of top-down steering of bottom-up participation in co-creation processes at both neighborhood and municipal levels. To clarify this analytical perspective, the key arguments are introduced in the following.

First, the participatory arenas in the urban regeneration policy field integrate dialectics of top-down structural policies and bottom-up approaches rooted in local uniqueness and diversity (Albrechts 2005). In this form, the classical organizational hierarchy and a less coordinated self-organizing "heterarchy" are at play within the same program (Jessop 1998). In a Danish context, the multilevel governance dimension primarily refers to intra-municipal coordination between hierarchically organized city departments and professional teams that operate in collaborative horizontal settings with a plurality of local stakeholders.

Second, public–private co-creation in a public governance context can be defined as a process that enhances the production of public value through improvements in outputs and outcomes, or through innovative[1] changes that transform problem understandings and devise new ways to solve problems (Torfing et al. 2016, p. 8). To stimulate co-creation, public administrations should engage in "constructive collaboration with relevant and affected actors who can help to define and solve the shared problems and common tasks" (p. 6) and create synergy across the individual, institutional, and epistemic boundaries that divide public agencies, users, citizens, firms, civil society organizations, etc. (ibid). Following Torfing and colleagues, there is a close relationship between co-creation and innovation in a public policy context and the main criteria are that the parties somehow contribute to the development of public value. Inspired by Arnstein's ladder (Arnstein 1969) the authors discuss what motivates citizens to take part in public value creation, and they formulate a 'co-creation ladder' that differentiates between five dimensions of citizens' influence that spell out different dimensions of co-creation (pp. 10–11): Citizens can co-create (1) the services they are affected by; (2) value for others; (3) inputs to new tasks and solutions; (4) the design of new tasks and solutions; and (5) collaborative innovations at the institutional level.[2]

Third, co-creation as creative exchange and innovation of ideas and preferences occurs in a steering context of conflicting interests and turf fights[3] in the urban regeneration field. When nonpublic stakeholders participate as co-producers of public policies and programs, they also become co-producers of new organizational forms of government (Hastings and McArthur 1995, in Jacquier 2005). It is a classical challenge to create solutions that to some extent accommodate the preferences of all parties in a multilevel governance context with parties that represent different constituencies with partly overlapping partly conflicting interests (Baldersheim and Ståhlberg 2002, p. 78). People bring ideas and interests to the table, but being there, new things emerge. In the urban regeneration field they do so in open network relations based on voluntary participation where roles and positions of authority are subject to negotiations and conflict. Public steering of participation processes (top-down bottom-up organizing) is ripe with conflicts and boundary negotiations at all levels. Some power games can be related to observable conflicts, while others are manipulated so that conflicts do not arise in the first place through processes of definition, selection, and exclusion of issues.

---

[1]Innovation defined as the "development and realization of new disruptive ideas" (Torfing et al., p. 8).

[2]Actors' motivations for taking part in co-creation can be theoretized, e.g., (Alford 2013) the point here is to use the ladder as a heuristic that identifies empirical dimensions in specific examples.

[3]In network governance "*patterns of inclusion and exclusion and the constitutional rules that regulate networks are settled not in advance but as an integrated part of the policy process*" (Sørensen and Torfing 2007, p. 244).

Fourth, mechanisms of meta-governance (governance of governance) (Jessop 1998; Sørensen 2005; Christopoulos et al. 2012) directly and indirectly steer the framework conditions under which governance takes place in public–private co-creation processes. Examples of meta-governance mechanisms are discursive framing; strategic guidance; direct involvement; enabling self-governance; defining the rules of the game; and using fear (Sørensen 2005; Sehested 2009; Nederhand et al. 2015). Meta-governance mechanisms provide a sense of direction and policy integration in a steering context characterized by actors who have partly overlapping partly conflicting interests, who are endowed with asymmetrical distributions of power, knowledge and resources and who, considering the likely costs and lack of clear incentives, hesitate to take part in public value creation. Typically, it is necessary to work with the framework conditions for co-creation to govern the ambivalent control of the implementation process as both ante innovation efforts and ex post error correction.

In the Copenhagen case below, the analytical focus highlights how the polycentric multilevel governance context contextualizes climate adaptation at the level of the city district. The approach is 'meta-steering of the learning by doing process in ambivalent networks'. The case analysis focuses on the specific organizing mechanisms of the participation process, and how these relate to the substantial co-creation dimensions as suggested by Torfing et al. in the five dimensions of the co-creation ladder. Two research questions structure the analysis: (1) What are the public governance strategies to enable/persuade residents to contribute resources (ideas, time, money) to climate adaptation solutions? And (2) Does the governed participation process reflect the five dimensions of influence in the co-creation ladder?

The co-creation paradigm is a heuristic only; different administrative paradigms (NPM, WM, NPG) co-exist as layered realities for politicians and managers in contingent governance contexts (Hartley 2005, p. 29). Empirically, the case is based on desktop research combined with 10 qualitative interviews conducted in 2012 (5) and 2016 (5) with planners from the Skt. Kjelds IUR, other IURs, and the Copenhagen Climate Adaptation Office. Interviews with 9 active residents (Kvorning Boysen 2016) also inform the analysis.

8.3 Public Participation in the Nordic Countries

According to Sellers and Lidström (2007), the Nordic welfare states combine the most decentralized governance systems in the advanced industrialized world with the most universalistic and egalitarian welfare systems. The Nordic model is characterized by localism, central-local integration and "guided democracy" (Baldersheim and Ståhlberg 2002, pp. 76–77) where central-local relations are coordinated in a system of state-supervision, earmarked grants, and legal review of local decisions in a context of relative trust and shared norms. Governance relations are conditioned by a pragmatic consensus culture where informal coordination is stimulated by shared normative and cultural orientations. The culture stimulates experimental methods of consensus-building and joint search for solutions through experiments (soft governance). In all of the Nordic countries, two tendencies stand out (Löfgren and Ringholm 2009, p. 513): Inclusion of nonpublic actors, and focus on democratic values in policy-making and service delivery. The current neoliberal political trend and the economic crisis with growing fiscal pressures and expectation overload create tensions in both consensus culture and central-local relations, and local governments are under pressure to rationalize their welfare systems. Still, in international comparisons, Nordic countries continue to rank at the top in terms of low levels of corruption, high levels of trust, and government effectiveness (Haveri 2015, p. 136).

The Danish welfare state is essentially local.[4] City councils handle environmental control, adult education, public schools, an employment system, and specialized social services. A basic principle in the Danish model is that money follows tasks. The primary revenue comes from income tax and property tax. Local tax levels vary, and city councils receive an annual block grant from the state. The block grant system transfers money from the rich to the less affluent local governments. The consensus culture is related to a high degree of integration of societal interests in policy-making processes. The degree of political participation in Denmark is high, especially at the local level (Andersen et al. 2000, Chap. 11). Almost 50% of all citizens have been involved in some kind of political activity within a year, and about half of all parents have been actively involved in activities in day care institutions, school boards, etc. Two out of three is active in one or more voluntary associations. At the same time, voter behavior is socially biased. Analyses of voter turnout in municipal elections document two groups in terms of voter behavior: A democratic A team with a high turnout, and a B team that stays at home. The A team consists of the middle-aged, women, the highly educated, higher income groups, and ethnic Danish. The B team consists of the young (19–29 years) males, low educated, the lowest income group, the socially marginalized, and immigrants and their descendants (Bhatti and Møller Hansen 2010).

8.4 Citizens' Participation and Co-creation in Danish Urban Regeneration

The Danish urban regeneration tradition is tied into the development of the social democratic welfare state in the post-war period. The governance context has evolved in three historical phases. In the first top-down phase (50s to mid-80s) the state initiated large-scale, top-down organized slum clearance and inner city sanitation programs creating "better and brighter" inner cities. The Urban Renewal Act of 1983 decentralized urban renewal to local authorities, strengthened procedures for public involvement, and introduced specified technical standards. In the mid-80s

[4]In 2007, the number of Danish local governments was reduced from 275 to 98.

the top-down approach was challenged, especially in Copenhagen, when residents fought the bulldozer tactics of the City Council. The legislative body regulating urban regeneration has been continuously reformed (Andersen et al. 2001; Ærø et al. 2008) reducing public expenditures while mobilizing private resources in a co-creation framework (Norvig Larsen and Engberg 2011).

In the second collaborative phase (90s to 00s) law-makers introduced collaborative mechanisms to better plan and negotiate large-scale renewals with residents and local stakeholders. State and local governments co-fund regeneration activities comprising housing retrofitting, IURs, and green refurbishment of inner city courtyards. The area-based approach is initiated and organized by city planners, who set up a consensus-based steering framework in a designated neighborhood selected according to measurable criteria of social disadvantage (Engberg et al. 2000). The IUR model is the key vehicle for co-creation. It combines public meetings, working groups, and a steering committee consisting of local stakeholders (individual citizens, shop-keepers, neighborhood associations, NGOs, social housing representatives, others) and city professionals. Planners make detailed place analyses and suggest ideas for regeneration projects, but citizens define and prioritize the content of regeneration activities in collaboration with planners. The collaborative fora produce a District Plan, which is consequently approved by the City Council. IURs typically channel citizens' wishes for urban space improvements like parks, new community buildings, playgrounds, traffic-reduction projects, climate and energy efficiency initiatives, etc. Urban space improvements are linked to individual building renewal projects under a building renewal scheme, where building owners receive a public subsidy for building renovation. The IURs are one-third funded by the state and two-third by the municipality. The 90s saw marked reductions in previously plentiful public budgets, and focus narrowed to city centers and socially distressed neighborhoods. The collaborative turn was not without frictions and coordination conflicts, and local governments approached these in terms of local conflicts at the level of the neighborhood rather than systemic issues to be explicitly dealt with at city hall (Engberg et al. 2000).

In the third strategic phase (2000s–present), city administrations increasingly approach place-based coordination in urban regeneration as a new public governance challenge also to be dealt with at the level of the city administration (Engberg and Norvig Larsen 2010). As a consequence of budget cuts in the 2000s and 2010s, a growing number of local governments rethink regeneration as "smart money" to catalyze integrated sustainable growth agendas involving citizens in collaborative efforts to create better urban spaces. Urban renewal programs continue to target the bottom of the housing market in disadvantaged neighborhoods, integrating top-down steering of bottom-up processes. But planners and decision-makers also seek to integrate the urban regeneration domain with other policy areas as a general strategy to promote integrated and sustainable urban development.

8.4.1 Consensus-Steering in Urban Regeneration

From empirical research, we know that public agents who organize participation in Danish urban regeneration contexts combine two basic mechanisms: (1) They build up local alliances understood as groups that work together in relation to one or several projects, gradually developing mutual trust and a capacity for collaborative behavior; (2) They formalize these alliances as informal partnerships representing the neighborhood using consensus-steering as a key meta-governance mechanism (Engberg et al. 2000, pp. 15–22). The rule 'mandatory local consensus for the process to go on' surfaces as steering mechanism to manage both local conflicts between stakeholders, and intra-municipal boundary conflicts in the informal decentralization process. In all instances, urban regeneration stirs some level of disagreement and conflict horizontally and vertically between local stakeholders and between city officials. Alliances typically emerge at the intra-municipal and local level for projects to succeed. Whenever local parties conflict, city councils threaten to (or do in fact) abolish steering committee arrangements, pulling back the process to the level of the city administration. Planners and civil servants see themselves as 'neighborhood advocates', but they also emphasize an 'emergency brake function' when they feel that local stakeholders go beyond politically and administratively sanctioned boundaries. The consensus-steering mechanism is an authoritatively enforced necessary consensus at the local level for the process to move on. The mechanism operates as a friendly threat to reinsert government in governance relations if network actors do not reach agreement within the relative autonomy of the network in the context of institutional and political limitations as defined by public stakeholders in the specific case.

The consensus culture in urban regeneration entails an institutional tendency to suppress disagreement and local turf fights. It is clear that the apparent non-coercive deliberative dialogue is embedded in processes of exclusion of difference and antagonism (Hillier 2002; Mouffe 1999, 2000, 2005). As meta-governance mechanism, the tendency is that consensus-steering legitimizes a top-down performance regime by depoliticizing struggles that are confined to the particularist concerns of neighborhoods (Davies 2011, in Blanco et al. 2014, p. 3135). Steering committee members have some degree of organizational skills (Agger 2005) and they are typically from the A team. At the same time, IURs organize many projects that target groups in the locality that do not typically take part in regeneration activities, and IUR activists support a rhetoric of citizens' empowerment and bottom-up logics to include the plurality of needs and preferences of the citizenry.

8.5 Case: Climate Adaptation in Copenhagen

The City of Copenhagen aims to become the first carbon neutral capital in the world by 2025.[5] Ten percent of the projected total CO_2 reduction is realized through construction and retrofitting projects, equivalent to 50,000 tons of CO_2. The city enforces energy efficiency requirements (energy savings, use of renewable energy) in construction and renovation of buildings which the municipality owns, rents, or supports financially, like urban renewal projects. Using state subsidies to urban renewal, Copenhagen has refurbished the main bulk of its building stock in the post-war period, resulting in relatively high housing standards. Today, about 11% or 33,000 homes of the city's 295,000 housing units are considered to be of inadequate standard, lacking basic amenities as district heating, toilet, and/or a bath.

In 2012, Danish local governments agreed to formulate climate adaptation plans on a voluntary basis (Ministry of Finance 2012). The plans describe risk analyses and suggest adaptation measures to counteract anticipated extreme weather risks (flooding). Climate Adaptation Plans are part of Municipal Plans.[6] The state supports local governments with technical expertise (risk assessments, GIS information, flooding maps, data), a task force, and the necessary legal changes to pave the way for local adaption measures. By 2015, 95 of 98 local governments had adopted CA plans (http://www.klimatilpasning.dk). With the 2011 flooding incident in Copenhagen, climate adaptation became an important policy problem and a political priority. The same year the City Council adopted a Climate Adaptation Plan[7] that sets the framework for how the city can protect itself against climate change. Adaptation is implemented through the Municipal Plan and Municipal Strategy, enforced by the Financial Department and organized by the Technical and Environmental Department (Zandersen et al. 2014). Following the projections of the UN Intergovernmental Panel on Climate Change (IPCC), the city estimates damage costs in Copenhagen over the next 100 years to reach DKK 16 billion.[8] The plan proposes a new structure that combines cloudburst management solutions on the surface with four large cloudburst pipes below ground that discharge water to Copenhagen Harbor and the city lakes acting as retention basins.

---

[5]Copenhagen City Council: Copenhagen Climate Plan, https://www.energycommunity.org/documents/copenhagen.pdf, downloaded September 14, 2016.

[6]The Planning Act regulates national, regional and local spatial plans. Local governments are responsible for a comprehensive land use regulation, with legally binding guidelines for property owners. In the first part of the 4-year election period, the 98 City Councils adopt a Municipal Plan. The Planning Act stipulates minimum rules for public participation in the planning process, with a mandatory 8-week consultation period.

[7]The City Council estimates that the plan requires municipal investments of EUR 360 million and public and private "green" investments totaling EUR 3,6 billion in the period up to 2025, resulting in 35,000 new jobs.

[8]Copenhagen City Council (2015) Climate change and adaptation and investment statement, part 1, p. 6.

The 2012 Cloudburst Management Plan divides the city into seven catchment areas, and details a new infrastructure for storm water management. In the worst-case scenario; a 100-year rainfall event, the maximum accepted road surface water level is 10 cm (3.94 inches). The key principles in the plan are (1) as much rain water as possible is managed at the surface (retention and surface channeling) and (2) solutions on the surface are treated as new recreational urban spaces in the city, developed together with citizens. The City Council has formulated 300 cloudburst management projects to be developed in cooperation with citizens. Each year, the city council selects a number of projects. The projects are funded by the city and the water utility company[9] (water charges). The plan is to be implemented over the next 20–30 years. Because pipe construction takes 7–9 years, some surface projects are treated as stand-alone projects[10,11].

8.5.1 Context: The Multilevel Governance System in Copenhagen

The governance model in Copenhagen is based on a modified Committee rule with shared responsibility for the city administration. The City Council with 55 members elects members to seven Standing Committees according to the principle of proportional representation. Each committee chairman is appointed mayor, and the chairman of the most powerful Finance Committee is the Lord Mayor. The Finance Committee comprises the mayors and six other city council members. The principle of proportional representation implies that a mayor from a small party does not have the ruling majority per se in the committee that she or he chairs. The Lord Mayor cannot instruct the six mayors, and the seven mayors are on equal footing. Mayors negotiate all issues and political disagreements on a case by case basis. Further, the model is based on joint administrative governance: Each administration is subject to the authority of both mayor and Standing Committee, jointly responsible for running services.

As a rather complex multilevel governance model, the Copenhagen system is known to generate tensions and conflicts within and between administrations that hamper cross-departmental coordination (Engberg and Norvig Larsen 2010). Most importantly, minority mayors may be responsible for executing policies that he or she disagrees with. Also, when sectoral committees have a joint say in running departments, it blurs the division of power between execution and legislative

[9]The utility company (HOFOR) in charge of water handling, energy, and waste.

[10]Cloudburst management roads that transport the water in the event of torrential rain; retention roads, which retain the water; green roads, typically side streets, that retain and store water; retention spaces, which store the water in the event of torrential downpours.

[11]Copenhagen cooperates with New York City, transferring ideas and results from the Climate Resilient Neighbourhood in Østerbro to a district in New York, and with the city of Beijing on solutions for management of everyday rain and torrential downpours.

scrutiny (OECD 2009, p. 230). The structure creates bottlenecks that restrict the flow of information to vertical lines of control even though many areas of local government responsibility are cross-sectoral in nature. Civil servants negotiate complex ties of loyalty in a difficult hierarchical structure, the misreading of which harms career prospects resulting in a defensive and risk-averse culture (ibid.). From a citizens' perspective, the system allows for minority influence on the day-to-day decisions, but it also reduces transparency and blurs lines of accountability.

8.5.2 The Climate District Skt. Kjelds (2010–2015)

The City of Copenhagen has initiated IUR projects every year in relatively disadvantaged neighborhoods for about two decades. Each project pursues an area-based approach, combing a steering committee structure with public meetings and citizens' working groups. Citizens and planners formulate a district plan in a bottom-up process, consequently approved by the City Council. The plan is a framework for a plethora of co-created activities to improve the neighborhood. In the Skt. Kjelds project, the initial focus was broad[12]: Collaboration, urban space improvements, better conditions for pedestrians, a cultural project at the local school, etc. The 2011-flooding event happened in the vicinity of the neighborhood, and together with the City Council the IUR partnership decided to focus on climate adaptation as the main topic. The City Council, the Climate Adaptation Unit, the city utility company, and the IUR formed a broader partnership focusing on climate adaptation. This partnership set forth to analyze and implement climate adaption solutions at a city scale while developing these in the collaborative planning context of the Skt. Kjelds IUR. Also, in 2012 the Technical and Environmental Administration (organizing IURs) pursued an NPG-innovation strategy, using 'urban design labs' working with co-creation and user-generated innovation to promote 'livability' and sustainable urban development in the city (Fig. 3).

> We described our IUR-approach with a cupcake metaphor: The utility company was the cake (funding of new infrastructure); the city administration the icing (public resources to invest in green solutions) and Skt. Kjelds IUR the cherry, developing resources for urban space projects while strategically combining the three funding streams. As an IUR we had the means to facilitate the innovation process while developing the recipe so to speak (city planner).

As it happened, the Skt. Kjelds IUR emerged as the Copenhagen laboratory for developing the specifics of the supposedly climate-resilient neighborhood. The main public discourse was 'urban nature' as solution strategy, targeting problems of pollution, overheating and cloudburst management while offering citizens new recreational facilities. The discourse was supported by the fact that a hydraulic

---

[12]Skt. Kjelds Distric Plan (2011) http://kk.sites.itera.dk/apps/kk_pub2/pdf/809_xwILhbTCFi.pdf.

Fig. 8.3 Taasinge Square in the Skt. Kjelds District, before project start, 2010. Courtesy of the Copenhagen City Council

surface solution was considerably cheaper than modernizing the city's sewer system.

The Skt. Kjeld's process was a key to develop the city's approach to climate adaptation; new solutions would integrate residents' ideas and wishes while the greening strategy was made concrete at the city level. Also, citizens would benefit from new social ties in the neighborhood developing green squares, green inner courtyards, green buildings, and green streets together with planners, architects, and contractors.

An initial discussion was about 'cloud burst roads'; how should existing roads in the district be modified[13] to channel rain burst water on the surface, and how much water is realistic to manage in extreme weather situations? City experts decided not to alter roads or elevate curbs. Instead they opted for cross-cutting sidewalks at intersections to keep water on the designated road. This technical choice was not subject to citizens' influence, but it meant that water levels could not rise more than 10 cm on average in this scenario. The norm 'an acceptable threshold of 10 cm water in a 10-year occurrence' was accepted in debates and integrated in district plan and city guidelines. Another topic was a 'first flush' technology letting the first part of (polluted) rainwater into the sewers and the rest onto the streets, a technology still to be developed. Rainwater detention is a complex health issue, and residents discussed issues of use and maintenance of new green areas considering how pollution accumulates in the soil over time. The process made it clear that the notion of a Sustainable Drainage System is still in the making. Solutions for issues like salt on the roads in the winter time, pollution, water separation, water quality and health issues still need to be adequately dealt with. These technical issues were part of a broader place-making activity in the district. Planners and contracting

[13]A recess in the middle of the road? Elevated curbs? Elevated sidewalks at intersections?

architects visualized the green narrative (while arguing that realistic outcomes would hardly be identical to the projected images) and a series of workshops and public meetings were organized to concretize the imagined future (Fig. 8.4).

The IUR hired an urban consulting firm to build a temporary urban space in one of the squares in the district, creating a tangible visualization of the climate adaptation potentials together with the local community (Fig. 8.5). The idea was that citizens through engagement in specific initiatives get to see the potential benefits of the greening approach and become encouraged to also engage in the long-term effort. The mock up was a success in terms of participation, and after the event the temporary installation quickly deteriorated. Also, citizens could apply for 50% funding of projects in their private courtyards or in public spaces in the neighborhood. Projects ranged from raised garden beds and rainwater barrels to small green roofs and rainwater activities for children (Kvorning Boysen 2016, p. 14).

Fig. 8.4 Taasinge Square after climate adaptation as visualized by contracting architects. Courtesy of the architectural firm Tredje Natur [Third Nature]

Fig. 8.5 Taasinge Square as temporary urban space in 2013. Courtesy of the Copenhagen City Council

The greening strategy highlighted a basic disagreement between professionals in the city administration. Planting trees in Copenhagen is costly in most places because of soil conditions and underground infrastructure, the cost of planting one tree can add up to 30,000 euros. A planner from the Skt. Kjelds project describes his position on the issue like this:

> If we spend 1 million on a square that attracts 100 people on a daily basis, or 2 million on a square that attracts a thousand, we should opt for the latter. But some colleagues disagree and argue for the low-cost approach. At the end of the day it is a subjective assessment: Can we legitimize added costs on the basis of how people experience an urban space? What *is* an attractive urban space? (planner).

In the Skt. Kjelds neighborhood, most roads are private common roads where homeowners have an obligation to cover minor road maintenance. According to a recent law, private landowners can apply municipal utility companies for funding of climate adaptation projects (covered by water charges) when organized in land lot associations. One-third of all climate projects are planned on private roads where house owners have a joint maintenance responsibility. The city and the utility company negotiate in partnership cloudburst management initiatives with owners. The IUR managed to persuade homeowners around Taasing Square to set up a neighborhood association to maintain the new square (Fig. 8.6). All owners' associations irrespective of size agreed to pay an annual lump sum of EUR 800 toward general maintenance of the new square:

> It was a sustained effort to build trust amongst local residents towards the City Council and our intentions with the climate project at Taasinge Square. We labored quite intensively to persuade the different owners' associations to support the project financially, talking to both elected representatives and individual residents at the estate level. In this regard, we work with heterogeneous groups to build a bottom-up demand for climate adaptation solutions. In fact, we have to stimulate and in some sense create people's love for green urban spaces (planner).

Owners unanimously voiced their interest in keeping the same number of parking lots after the renovation of the square, and the IUR accepted this demand. To do the puzzle of relocating parking lots necessary to design the square, the IUR organized public meetings, workshops, etc. People active at these events influenced the pattern of relocation. When more people became aware of the new traffic plan, some of those benefitting the least protested. The IUR responded to the criticism by refuting it, arguing that the outcome was legitimized by the consensus process leading up to the plan. Some residents voiced frustration that their individual climate projects were too costly, and inadequately supported by the IUR and the city (Kvorning Boysen 2016, pp. 71–78). Also, residents criticized the IUR for being less supportive of their own bottom-up initiatives, and more responsive to horizontal coordination and innovation in the context of the city's own plans and agendas (ibid).

Fig. 8.6 Realized sustainable urban drainage system and park project, Taasinge Square, 2016. Courtesy of the Copenhagen City Council

8.5.3 The Cloudburst Management Plan (2012–Present)

The Skt. Keld process showed the city that it could create an integrated cloudburst system on the surface, using the micro-context of the district to innovate and standardize solutions in the larger planning context of the entire city. The IUR articulated the greening strategy as a new city discourse, it clarified a number of technical issues, and it was a citizens' participation arena that inspired intra-departmental development of a (more) integrated approach to climate adaptation. In 2012, the City Council adopted the Cloudburst Management Plan, and some years later it set up a green think tank to inspire green solutions for cloudburst management projects to boost urban nature in Copenhagen. The plan identifies 300 hydraulic projects in the seven water catchment areas. The geographical location of projects sets the frame for the type of participation process relevant to the implementation of the project. Some climate projects are situated in areas with IURs, and all Copenhagen IURs now integrate climate adaptation as a key topic.

Each year, the Technical and Environmental Committee (TEC) in the City Council decides on new projects in the water catchment areas. Each project is analyzed, and in some instances TEC funds additional urban space improvements. Climate projects are integrated with other relevant projects and services wherever possible, e.g., roads and maintenance, constructions, park management, and urban space improvements, but also with municipal projects and services in the social and cultural policy domains. Before the Climate Adaptation Office as gatekeeper submits the annual project package to the TEC, the office consults with other city departments about their policy agendas, and it consults with local committees[14] in the seven catchment areas, IURs, and other local stakeholders. City departments pursue a strict policy of cost reduction wherever possible, and the climate office negotiates each project. If residents are aware of such ongoing budget negotiations, they can (and sometimes do) put a direct pressure on councilors to override the internal budget process. Some municipal professionals argue against citizens' participation in the implementation of the climate plan. The Climate Adaptation Office in charge of implementation maintains that participation is necessary for two reasons in particular: 1) The city needs to inform about the consequences of the cloud burst plan to increase knowledge and accept of the new city wide infrastructure; and 2) It needs to better understand and analyze local problems and synergies together with local stakeholders to improve solutions. The plan specifies projects in the hydraulic structure to make sure that hydraulic flows are adequate, but it does not specify the design and recreational functions of individual projects.

> It is not up for discussion whether a specific road is turned into a cloud burst road, you do not get to decide were the water will flow. But you influence what it looks like, how it works when it is dry, and when the water is there. Implementing CMP we have relatively well-defined participation arenas: Local committees in the seven catchment areas, IURs if present in the area, owners associations, social housing associations, and insurance companies (the latter group conducts campaigns; owners are responsible for water in their basements). But also, it's an open agenda: Whom might be relevant to include? (planner).

According to the Climate Adaptation Office, Copenhageners generally support climate initiatives they see as urban space improvements: "At this point I have yet to experience direct refusal of our projects. In one case we met local disagreement about the specifics of the plan, something which we then negotiated" (planner). In this example, local resistance resulted in a modification of the municipal plan which was consequently adopted. The main idea of the project was to turn the local lake into a 'rain water park' as a more effective retention basin. After a public consultation phase and public meetings, co-chaired with four local committees representing wards adjacent to the lake, one local committee in particular criticized the plan. The city planners hired anthropologists to investigate the local opposition. On the basis of a survey and interviews with stakeholders and residents, they found that local people generally agreed with the critique.

> From the councils perspective, we did not see the local opposition as NIMBYism, people supported an adaptation intervention, but they disagreed with our project. People simply did not want to alter the lake features. The TEC did not want the public conflict and we accepted a compromise, a patch-work solution, even though the initial project was economically and technically better (planner).

---

[14]In 2005, the city introduced a system of 12 local committees with 23 elected members, of which 15 are active in associations in the local district, and eight are members of political parties. The committees have no prerogatives in relation to the municipal budget, but they influence budget negotiations and constitute the "neighbourhood voice" in relation to the different policy areas. The committees support civil society activities, and citizens can apply for funding of community projects.

8.6 Discussion

From a primary focus on mitigation, the 2011-flooding introduced climate adaptation as a relevant and workable policy problem in Copenhagen to be dealt with through government action. The municipal R&D process identified a cost-effective solution with four pipes and surface channeling of cloudburst water as preferable to a renovation and enlargement of the sewer system. The approach depended upon citizens' consent and participation, especially on privately owned public roads. The basic catalogue of solutions in the new urban drainage system was developed and tested in the participatory framework of the Skt. Kjelds IUR, and exported in a parallel process to the rest of the city. Currently, the Climate Adaptation Office operates in primarily five participation domains: (1) Dialogue and consultation with local committees; (2) Campaigns directed at house owners encouraged to organize and engage in climate projects under a funding scheme; (3) Engagement in larger area-based partnership processes (IUR or the like) that focus on squares, roads, and buildings; (4) Collaboration with social housing associations that administer estates in catchment areas; and (5) Collaboration with insurance companies to make house owners aware of liabilities. The intra-municipal process continues to be linked to co-creation at the neighborhood level where some local solutions and design concepts are developed and tested to be used in other parts of the city.

Because the City Council's adaptation plan depends upon the consent of house owners and other local stakeholders, the city invests in co-creation and the forming of collaborative relations based on reciprocity and trust between citizens and city officials. The Skt. Kjelds IUR was used as governance vehicle for building necessary trust and engagement amongst residents to test the realism of the surface solution in the context of the larger scheme:

> To sum it up: Water runs down. We had to see the larger picture in our IUR approach. We learned that it is easier to carry out climate projects on municipal land and roads. But we also proved that the city can co-create solutions on private roads, and that extra time and resources invested in collaboration with owners is necessary. But climate adaptation continues to be an open-ended agenda: How do we create most value in terms of recreational urban space? And how much money can and should we spend on this? (planner).

The positive framing of the climate agenda was the key meta-governance mechanism in the IUR context. Architect visualizations of the green and blue climate adapted cityscape were combined with a positive rhetoric of the 'green city' and the 'blue city', and temporary climate adaptation installations created a tangible and positive narrative in relation to the climate adaptation agenda. To increase credibility planners critically distanced themselves from the idealized narratives hereby further strengthening the meta-mechanism when residents defined the scope for co-creation as narrowing down the distance between perceived and actualized reality. The storytelling also created an atmosphere of positive expectations in a situation where public works would cause a nuisance in everyday life in the district for longer periods of time. As meta-governance mechanism, the discursive framing mobilized engagement and relational coordination of loosely coupled networks of experts and lay people that worked toward the same ends across public–private boundaries in the district.

Multiple barriers exist to intra-municipal and cross-sectoral policy integration. Despite frictions and shortcomings in terms of resources and decision-making power vis-à-vis sector administrations, the IUR framework has a track record of relative success as a hybrid organizational form that meta-govern across different boundaries: It is a framework that can be used to mediate and negotiate functional organizational logics with place and context orientation (tailoring of solutions); to handle the area lobbyism that generates tensions in the city administration; to build operational consensus across multiple administrative levels and domains in dialogue with nonpublic actors; and to merge rational–technical logics with the more complex dialogues between proponents of competing value frameworks. Considering it is important to identify actors in charge of linking between government sectors and citizens in adaptation projects (Cloutier et al. 2015, p. 47), the Copenhagen case highlights a hybrid organizational form that enables this process.

Suggesting a 'co-creation ladder', Torfing et al. (2016, pp. 10–11) identify the rungs of the ladder as citizens' influence on (1) the services they are affected by; (2) value created for others; (3) inputs to new tasks and solutions; (4) design of new tasks and solutions; and (5) collaborative innovations at the institutional or system level. To what extent does the participation process as described above reflect these different dimensions of influence? A first observation is that the climate district rhetorically merges all five dimensions in the framing of the participation narrative. Residents have to co-create the global solution (and the tailored specifics) for it to work (see Chap. 7: Sari and Prayoga). Further, the process of concretizing this participation narrative is meta-governed: Public officials possess the power to discursively frame the innovation process in the deliberative and semi-transparent collaborative context. Consensus-steering as governance strategy puts pressure on citizens to 'climb the rungs' in terms of developing their private needs and preferences into collective solutions and engage in public value creation. For instance, the IUR incentivizes stakeholders to convert private domains into publicly accessible domains to receive support and funding (Fig. 8.7). Also, a dominant public perspective on climate governance as co-creation is to persuade residents to cofinance place-making projects, for instance, setting up associations that cover running costs of new green squares. To sum up, the influence dimensions of the co-creation arena resembles less a structured ladder and more a staged play of symbolic, discursive, and material negotiations of the key participation narrative and the cost-sharing strategies necessary for its implementation. If anything, the governance strategy conflates the five influence dimensions rhetorically (if not practically) to persuade citizens to engage. The strong participation narrative in the IUR framework effectively mediates a number of key binaries (complex to simple, intangible to tangible, at risk to resilient, atomized private to collective public, us vs. them, from costs to values) in a situation with limited public resources for expensive climate adaptation measures.

Fig. 8.7 Before and after: New private front gardens are open to people in the community, *Source* http://klimakvarter.dk/projekt/aabne-haver/, downloaded January 11, 2017

There is a growing need for building increased capacity for public engagement in climate change understood as the deliberative process of involving the public in matters of public concern and decision-making (Moser and Pike 2015 pp. 112–113). Meaningful citizens' participation in urban climate governance may be difficult to obtain "given the technical knowledge needed to understand the complex problems associated with climate change, the uncertainty and risks involved in global environmental systems, and how global changes translate into local hazards" (Sarzynski 2015, p. 63). The Copenhagen case shows that collaborative climate governance in this instance is about creating a participation context that exemplifies and concretizes the public value perspective, linking individual initiatives with integrated solutions that are part of a larger structure (see Chap. 12: Brown). The IUR framework (with its pros and cons) offered decision rules that structured the co-creation processes, building bridges between technical issues, individual projects, communal initiatives, and the larger design choices at the district and city levels.

The climate district process was relatively devoid of (discernable) antagonistic conflict, partly due to the historically institutionalized consensus steering mechanism, partly due to the fact that adaptation measures represented a plus-sum game where residents would benefit from partly subsidized urban space improvements. In relation to the city-wide process currently underway, planners confirm the picture:

> We meet two 'religions'; those who believe in pipes, and those who believe in trees. Most citizens believe in trees, and often we have to emphasize that we plan for extreme cloudburst situations that require more than trees…In one instance residents fought to prevent some old trees to be cut down, and they succeeded. Apart from that we have seen very few conflicts (planner).

In terms of intra-municipal coordination, citizens' participation is supported by but also supports the development of a framework for practical policy integration and governance innovation in the case context (see Chap. 16: Ninomiya and Burch). As such, co-creation in this context does bridge a perceived performance gap (Voorberg et al. 2015, p. 1349). From a public governance perspective, the co-creation with citizens functions as an indirect means to align public agencies in a more integrated approach to the wicked problem issue. The co-creation process in the IUR district links the impact of public policies to the urban environment and integrates different policy priorities embedded in practical designs (accessibility, relocation of parking lots, balancing greening costs and opportunities, aesthetic design principles, acceptable maintenance costs, etc.) with water handling solutions. The process is not a panacea, but the tendency to integrate solutions to the wicked problem issue is discernible.

In a survey of Danish local governments (Jensen et al. 2016) there is an absolute consensus amongst practitioners and decision-makers that climate adaptation is an interdisciplinary and interdepartmental task. But respondents do not agree which part of the city government should take the lead, or how policy innovation should be organized. The same realism underlies the Copenhagen context, where many boundary spanning issues relating to water management still need to be innovated:

> We invite citizens into the machinery of administrative and political decision making in a process that we have not yet fully defined. Obviously this creates a sense of insecurity. There is a lack of knowledge of what works, and of the overall direction of the professional decision making process, which is not fully clear to people. Citizens do not demand storm water management solutions. And on top of that, we ask people to provide inputs to a content that the larger system is only beginning to define in more specific terms (planner).

If more rainwater is to stay longer on the surface for recreational purposes, health and safety issues need to be dealt with. It is expensive to plant trees in Copenhagen, and economic limits to the greening strategy will become more manifest in other city districts. New hydraulic functions on the surface fixate the urban space in relation to future uses, and they are easily blocked; they are rarely used, and people tend to forget about them (Lund 2016, p. 30).

Using the urban regeneration policy field and the Copenhagen case on climate adaption in the Skt. Kjelds district as markers, there is a clear tendency that the hybridization of the multilevel governance context for climate adaptation policy actually delivers in the Danish context. As a public–private hybrid and informal collaborative framework, the IUR model exemplifies a more layered institutional landscape acting as an intermediary or brokerage mechanism forging collaborative ties between City Council and the local community. It is a two-way road: Citizens influence the concretization of policy outcomes and improve their neighborhood, and public policy agents access nonpublic resources and integrate and legitimize policy delivery (see Chap. 9: Iftikhar, Ali, and Sarzynski). Yet, considering current budget pressures in Copenhagen and other Danish cities, the economically justified NPM efficiency agenda and the politically justified NPG integration agenda (Valkama 2016) are co-existing realities in Danish climate adaptation. So, when the weather sets in, planners and citizens alike continue to be charged with the fundamental challenge of co-creating better and cheaper solutions to a very wicked problem.

References

Agger, A. (2005). Demokrati og deltagelse—et borgerperspektiv på kvarterløft. Statens Byggeforskningsinstitut.

Albrechts, L. (2005). Creativity as a drive for change. *Planning Theory, 4*(3), 247–269.

Alford, J. (2013). The multiple facets of co-production: Building on the work of Elinor Ostrom. *Public Management Review, 16*(3), 299–316.

Andersen, J. G., Torpe, L., & Andersen, J. (2000). Hvad folket magter. Demokrati, magt og afmagt [The powers of the people. Democracy, power and public alienation] Demokratiprojektet, Jurist- og Økonomforbundets Forlag.

Andersen, H. S., Vestergaard, H., & Gottschalk, G. (2001). Det danske boligmarked-udvikling i boligforsyning og boligønsker [The Danish Housing Market: Trends in Housing Supply and Housing Preferences] Copenhagen, AKF, 2001.

Arnstein, S. R. (1969). A ladder of citizen participation. *Journal of the American Institute of Planners, 35*(1969), 216–224.

Baldersheim, H., & Ståhlberg, K. (2002). From guided democracy to multi-level governance: Trends in central-local relations in the Nordic Countries. *Local Government Studies, 28*(3), 74–90. doi:10.1080/714004149.

Bhatti Y., & Møller Hansen K. (2010). Valgdeltagelsen ved kommunalvalget 17, November 2009. Beskrivende analyser af valgdeltagelsen baseret på registerdata, Københavns Universitet, Institut for Statskundskab Arbejdspapir 2010/03.

Blanco, I., Griggs, S., & Sullivan, H. (2014). Situating the local in the neoliberalisation and transformation of urban governance. *Urban Studies, 51*(15), 3129–3146.

Christopoulos, S., Horvath, B., & Kull, M. (2012). Advancing the governance of cross-sectoral policies for sustainable development: A metagovernance perspective. *Public Administration and Development, 32*, 305–323.

City of Copenhagen. (2015, October). *Climate Change. Adaptation and Investment Statement*. Copenhagen: City of Copenhagen.

Cloutier, G., Joerin, F., Dubois, C., Labarthe, M., Legay, C., & Viens, D. (2015). Planning adaptation based on local actors' knowledge and participation: A climate governance experiment. *Climate Policy, 15*(4), 458–474. doi:10.1080/14693062.2014.937388.

Davies, J. S. (2011). *Challenging governance theory: From networks to hegemony*. Bristol: Policy Press.

Engberg, L. A., Bayer, S., & Tarnø, C. S. (2000). Konsensus-Styring i Kvarterløft: Kommunernes Erfaringer med Organisering af Kvarterløft [Consensus-Steering in "Neighbourhood Uplift". Organisation of Area-Based Interventions in Local Government] (By og Byg Resultater, 002). Hoersholm: Statens Byggeforskningsinstitut.

Engberg, L. A., & Larsen, J. N. (2010). Context-oriented meta-governance in Danish urban regeneration. *Planning Theory & Practice, 11*(4), 549–571.

Ærø, T., Andersen, H. S., Gottschalk, G., Jensen, J. O., Engberg, L. A., Norvig Larsen, J., et al. (2008). Evaluering af lov om byfornyelse ogudvikling af byer. Sammenfatning [Evaluation of law on urban renewal. Summary], SBI 2008:02.

Hajer, M., Nilsson, M., Raworth, K., Bakker, P., Berkhout, F., de Boer, Y., et al. (2015). Beyond cockpit-ism: Four insights to enhance the transformative potential of the sustainable development goals. *Sustainability, 7*, 1651–1660.

Hartley, J. (2005). Innovation in governance and public services: Past and present. *Public Money & Management, 25*(1), 27–34. doi:10.1111/j.1467-9302.2005.00447.x.

Haveri, A. (2015). Nordic local government: A success story, but will it last? *International Journal of Public Sector Management, 28*(2), 136–149.

Head, B. W., & Alford, J. (2013). Wicked problems: Implications for public policy and management. *Administration & Society*, published online 28 March 2013.

Hillier, J. (2002). *Shadows of power: An allegory of prudence in land-use planning*. London: Routledge.

Jacquier, C. (2005). On relationships between integrated policies for sustainable urban development and urban governance. In *Tijdschrift voor economische en sociale geografie, 96*(4), 363–376.

Jessop, B. (1998). The rise of governance and the risk of failure: The case of economic development. *International Social Science Journal, 50*(1), 29–45.

Jensen, A., Nielsen, H. Ø., & Nielsen, M. L. (2016). *Climate adaption in local governance: Institutional barriers in Danish municipalities.* Aarhus University, DCE—Danish Centre for Environment and Energy, 102 pp. Scientific Report from DCE—Danish Centre for Environment and Energy No. 104. http://dce2.au.dk/pub/SR104.pdf.

Kvorning Boysen, P. (2016). *Empowerment and co-creation: Integrated urban renewal in Skt. Kjelds Kvarter.* Master Thesis, Sustainable Cities, Aalborg University Copenhagen, Copenhagen.

Lund, D. H. (2016). Implementering af 10 klimatilpasningsplaner—aktiviteter, udfordringer og gode oplevelser. IGN Rapport Marts 2016, Institut for Geovidenskab og Naturforvaltning, Frederiksberg. 56 s. ill.

Löfgren, K., & Ringholm, T. (2009). Introduction: New network modes of Nordic local governance. *Local Government Studies, 35*(5), 505–514. doi:10.1080/03003930903227337.

Meijer, A. (2016). Coproduction as a structural transformation of the public sector. *International Journal of Public Sector Management, 29*(6), 596–611.

Ministry of Finance. (2012). National agreement on municipal finances. Sustainable transition and development of welfare. Ministry of Finance https://www.fm.dk/nyheder/pressemeddelelser/2012/06/aftale-om-kommunernes-oekonomi-for-2013_groen-omstilling-og-udvikling-af-velfaerd. Downloaded October 25, 2016.

Moser, S. C., & Pike, C. (2015). Community engagement on adaptation: Meeting a growing capacity need. *Urban Climate, 14*(2015), 111–115.

Mouffe, C. (1999). Deliberative democracy or agonistic pluralism? *Social Research, 66*(3), 745–757.

Mouffe, C. (2000). *The democratic paradox.* London: Verso.

Mouffe, C. (2005). *On the political.* London: Routledge.

Nederhand, J., Bekkers, V., & Voorberg, W. (2015). Self-organization and the role of government: How and why does self-organization evolve in the shadow of hierarchy? *Public Management Review, 18*(7), 1063–1084. doi:10.1080/14719037.2015.1066417.

Norvig Larsen, J., & Engberg, L. A. (2011). Organisational change and knowledge management in urban regeneration planning. *Landinspektoeren. Tidsskrift for kortlaegning og Arealforvaltning, 46*(1), 114–126.

OECD. (2009). OECD Territorial Reviews: Copenhagen, Denmark. doi:10.1787/19900759.

Pløger, J. (2004). Strife: Urban planning and agonism. *Planning Theory, 3*(1), 71–92.

Sarzynski, A. (2015). Public participation, civic capacity, and climate change adaptation in cities. *Urban Climate, 14*(2015), 52–67.

Sehested, K. (2009). Urban planners as network managers and metagovernors. *Planning Theory & Practice, 10*(2), 245–263.

Sellers, J. M., & Lidström, A. (2007). Decentralization, local government, and the welfare state. *Governance: An International Journal of Policy, Administration, and Institutions, 20*(4), 609–632.

Sørensen, E. (2005). Metagovernance: The changing role of politicians in processes of democratic governance. *The American Review of Public Administration, 36*(1), 98–114.

Sørensen, E., & Torfing, J. (Eds.). (2007). *Theories of democratic network governance.* Basingstoke: Palgrave Macmillan.

Torfing J., Sørensen E., & Røiseland A. (2016). Transforming the public sector into an arena for co-creation: Barriers, drivers, benefits, and ways forward. *Administration and Society*, 1–31. doi:10.1177/0095399716680057.

Valkama, A.-V. A. P. (2016). Post-NPM-style service integration: Partnership-based brokerage in elderly care. *International Journal of Public Sector Management, 29*(7), 675–689.

Voorberg, W. H., Bekkers, V. J. J. M., & Tummers, L. G. (2015). A systematic review of co-creation and co-production: Embarking on the social innovation journey. *Public Management Review, 17*(9), 1333–1357. doi:10.1080/14719037.2014.930505.

Zandersen, M., Jensen, A., Termansen, M., Buchholtz, G., Munter, B., Bruun, H. G., & Andersen, A. H. (2014). *Ecosystem based approaches to Climate Adaptation—Urban Prospects and Barriers*. Aarhus University, DCE—Danish Centre for Environment and Energy, 94 pp. Scientific Report from DCE—Danish Centre for Environment and Energy No. 83. www.dce2.au.dk/pub/SR83.pdf.

Author Biography

Lars A. Engberg is Senior Researcher at the Danish Building Research Institute, Aalborg University Copenhagen. In his research, Engberg looks at collaborative governance mechanisms in complex and hard to govern sustainable urban transitions. Working from a New Public Governance perspective, Engberg focuses on social housing issues, climate mitigation and adaptation in the built environment, urban renewal, smart cities, spatial planning, and the relationships between area-based policies to combat social segregation and current public service innovations in local government. Engberg is a member of CONCITO, a Danish green think tank.

Chapter 9
Community–Government Partnership for Metered Clean Drinking Water: A Case Study of Bhalwal, Pakistan

Muhammad Naveed Iftikhar, Shujat Ali and Andrea Sarzynski

Abstract The case study explores the design and operation of a community–government partnership initiative for clean drinking water called *Changa Pani* (Clean Water) Program in Bhalwal, a small city of Pakistan. It is a joint initiative by the city government of Bhalwal and residents, facilitated by a civil society intermediary organization that has succeeded in developing and operating a viable water distribution system to supply metered clean water to residents after purely government-organized efforts did not deliver. The *Changa Pani* Program (CPP) appears as a manifestation of community-based adaptation to climate change, given rising temperatures, depletion/contamination of groundwater, increased spread of waterborne disease, and childhood stunting. Considering the formidable challenges of access to clean drinking water and sanitation for Pakistan and other countries working to achieve the Sustainable Development Goals, lessons from successful grass root level projects like CPP can be broadly valuable. The case study analyzes determinants of the initiative's success and the opportunities and challenges for replicating the CPP for other types of infrastructure and/or service in other locations, in Pakistan, and in other rapidly urbanizing nations.

Keywords Drinking-water · Sustainability · Community

M.N. Iftikhar (✉) · A. Sarzynski
University of Delaware, Newark, DE, USA
e-mail: cheema@udel.edu

A. Sarzynski
e-mail: apsarzyn@udel.edu

S. Ali
Ministry of Finance, Islamabad, Pakistan
e-mail: ch.shujat.ali@gmail.com

S. Hughes et al. (eds.), *Climate Change in Cities*, The Urban Book Series,
https://doi.org/10.1007/978-3-319-65003-6_9

9.1 Background

Pakistan is becoming one of the most rapidly urbanized countries in South Asia. The urban population was 17% in 1951, 32% in 1998, and now it is close to 40%. More than half of the population in Pakistan will urbanize in the next 10 years (Planning Commission 2015; Hussain 2014). The official urbanization statistics in Pakistan may be under-reported in order to present the country and its economy as rural (Haque 2015).

Urbanization provides a range of opportunities for lifting people out of poverty by pulling them from low productivity agriculture to high-value urban sectors. Nevertheless, urbanization poses challenges ranging from law and order to provision of basic services to the citizens. Some of these services include education, water, infrastructure, and health. Water and sanitation are the most vulnerable services in this regard, due to their cross-cutting impact on human health and productivity (Glaeser 2012).

Only half of the current population of 180 million in Pakistan has access to safe drinking water. Water availability in Pakistan has declined from approximately 5000 cubic meters per capita in 1950 to around 1300 cubic meters today. Scholars estimate this "water-stressed" situation may fall further to "water scarcity" (1000 m^3 per capita) in a decade due to several factors including population growth, climate changes, and inefficient water usage (Kamal 2009).

If we take into account issues related to water quality and reliability, the situation is even worse (Farooq et al. 2008). An intermittent supply of water in the pipes often leads to cross contamination from sanitation pipes, which severely damages water quality. A considerable number of deaths of children and adults are associated with waterborne disease in Pakistan. Moreover, health problems strain meager resources of marginalized segments of society. This situation points to the need for urgent action by the government and communities.

Other developing countries are also facing similar challenges in the drinking water sector. Over 100,000 people die in India annually due to waterborne diseases, and 70% water is polluted with sewage effluents. The groundwater in one-third of India's 600 hundred districts is not suitable for drinking due to hazardous concentration that is higher than prescribed tolerance standards (Chabba 2013). A similar situation could be found in developed countries in previous centuries. It is reported that 11,000 died in London's cholera epidemic of 1854, and the same number of people died in New York City in 1866 (Condran 1995, pp. 29–35). However, these countries made a significant investment in the water sector and developed better governance systems to address these issues (Glaeser 2011). Even so, the recent water crisis in Flint, Michigan revealed the vulnerability of the United States' aging water infrastructure and raises concerns about equity and affordability (Kane and Puentes 2016). Despite high payoff of clean water and reliable sanitation services, governments usually do not give high priority to this sector (Glaeser and Sims 2015). Other mega-scale projects such as roads and power generation often get budgetary priority over drinking water.

The purely government-run service delivery organizations in Pakistan have been facing tremendous challenges in terms of their ability to ensure professional governance, maintain public assets, and meet growing service delivery needs of people at large. In an environment of clientelism, nonprofessional governance structures, and lack of capital investment it becomes difficult for these government organizations to keep a balance between commercial sustainability and vested interests by different powerful stakeholders (Iftikhar 2014, 2015). A recent article highlighted the plight of citizens of Karachi, one of the world's largest cities and adding over half million people annually: "options for access to water include erratic supply (where pipelines exist), collusion with area gangs who puncture water mains for commercial sales, boring and the use of semi-brackish water and purchasing tanker loads" (Ahmad 2016, para 5). Felbab-Brown (2017, para. 5) explains further:

> In developing countries, such water loss amounts to some 30 to 50 percent of all treated water. In Karachi and Delhi, such water leakages from old water pipes and outright water theft amounts to water loss of some 30 to 35 percent in each city. Water mafia networks divert water through illegal pipelines by drilling holes into official pipelines or setting up illegal hookups to water canals. In Karachi, water mafias also illegally tap into water hydrants and set up illegal wells and pipelines for industries. They fill up trucks and sell the water to industries and large businesses—some 70 percent of the stolen water!—as well as to the poor, at inflated prices. Illegal traders who operate 30 to 40 tankers earn as much as $16,000 a day."

Despite an apparent threat of a water crisis, Pakistan continues to have inefficient use of water in agriculture, industry, and households. Residential water charges are minimal and embody flat rates based on house size instead of quantity used, signifying a poorly designed pricing scheme to accomplish conservation goals. One government official explained that "many folks [in Karachi] considered urban piped water a gift of nature that does not need any monetary compensation!" (Ahmad 2016, para 7). Evidence suggests that people are willing to pay higher water prices, contrary to general expectations, as they realize the importance of clean drinking water. For instance, a survey of residents of Lahore city in Pakistan found that people were willing to pay $7.50–$9 per month for clean piped drinking water, which is comparable to the monthly expenditures on in-house water treatment and is about three to four times the average monthly water bill being paid (Akram and Olmstead 2011).

One important consideration for clean water provision is not to consider water as a traditional market good but as a collective resource available to the community. The market failure in the use of a natural resource arises when everyone tries to maximize its use for a short-term gain without realizing negative consequences in the long run for the community and society as a whole (Hardin 1968). Traditional market interventions of incentives, taxes, and fines do not work properly in many developing countries where formal institutions are not well developed and enforcement is weak (Ashraf et al. 2016).

The self-governance and polycentric framework for management of common pool resources explained by Ostrom and Gardner (1993) is relevant in this context of inefficient pricing and weak institutions. The authors found that irrigation infrastructure in Nepal was developed by public investment but the government could not create or enforce rules for efficient and level-playing use of irrigation water. The farmers closed the gap through the formation of collaborative arrangements and continuous evolution in rules in response to changing needs. Thus, a partnership evolved between government and self-organized farmers. Ostrom and Gardner emphasized the importance of self-organization and building consensus to develop institutions and rules that helped in better management of collective water resources. The complexity of service delivery in the water sector requires polycentric governance—interdependence, contractual and cooperative partnerships, and conflict resolution by independent center of decision-making (Ostrom 2010).

The design and operations of *Changa Pani* Program (CPP) in Bhalwal city of Pakistan are a similar response to the lack of access to clean drinking water. Climate change is driving up temperatures and humidity in the central Punjab Province, and the associated lack of access to clean drinking water has largely contributed to childhood stunting and waterborne diseases. Through a cross-country analysis, Fay et al. (2005) have found that provision of basic infrastructure such as piped water, sanitation, and electricity reduce incidence of childhood stunting and child mortality rates. In these circumstances, the response of the residents and government of Bhalwal city in the form of CPP is an innovative form of community-based adaptation (CBA) to climate change and service delivery constraints. The apparent success of the CPP model has triggered a debate about its core features, scaling, and replication to other situations.

9.2 The Research

The case study here reports on the innovative community–government partnership implemented for clean, metered drinking water in Bhalwal, Pakistan called the *Changa Pani* Program (clean water), whose lessons may also be relevant for other settings and contexts.

9.2.1 Research Questions and Design

This analysis employs a single case study, aimed at exploring the design and implementation of community–government partnerships for the provision of clean drinking water and climate change adaptation. The case specifically reviews (1) the design and implementation of CPP in Bhalwal; (2) the key factors contributing to success of CPP; and (3) the challenges in expansion of CPP. A case study design is most appropriate in this context given that the research is exploratory and the researcher has no control of behavioral aspects of the phenomenon being studied

(Yin 2013). Additionally, the phenomenon is contemporary and key persons are available for interviewing.

Single case studies face validity threats especially externally, where the results may be difficult to generalize beyond the studied location (for a good discussion of the threats, see Flyvberg 2006). Nevertheless, this study can assist in our understanding of an innovative program design and implementation plus provide insight into community engagement, the power of self-governance, and partnership dynamics.

9.2.2 Information Sources

Qualitative information for this case study was compiled from a focus group, consumer interviews, and document analysis. An hour-long focus group was conducted in January 2017 by the lead author in Bhalwal city. Participants included officials from the local government (*Tehsil Municipal Administration*, TMA) and community organization managing this initiative. Focus groups help researchers understand the collective perspective of participants through discussions of motivations, priorities, experiences, and reflection (Berg and Lune 2012; Morgan 1996). Focus groups may also help to gain an understanding of perspectives through interaction amongst participants, which may not be obtained through individual interviews.

Five semi-structured interviews were also conducted in January 2017 with consumers—identified by the intermediary organization *Anjuman Samaji Behbood* (ASB). The focus of these interviews was to understand the consumer perspective about CPP, the importance of clean water, the impact of metering on water consumption, and water pricing. It was useful to have an exchange and interaction with consumers, although the number was limited due to the scarcity of time and resources.

Last, documents were requested from ASB including the partnership agreements between the community and government, other reports and independent evaluations of the project, water bills, and a video documentary by Sustainable Development TV Channel (Satellite).

9.2.3 Case Selection

Bhalwal was selected for study due to the relative rarity of the phenomenon of interest—the CPP—and the potential benefits of this partnership model for clean water provision, participatory governance, and climate change adaptation.

Bhalwal is an agro-based *Mandi* (Market) Town located close to the city of Sargodha, within the Sargodha district. It has a population of over 100,000 with a rural, village-based hinterland peopled by settler families who migrated to this area from neighboring older districts of the Punjab. The settlers came here to benefit from the various land colonization and development schemes that accompanied the advent of the new irrigation canal and railway infrastructure in the early years of the twentieth century. Bhalwal is now the administrative center of the Bhalwal Sub Division, the lowest regional tier of the provincial administrative hierarchy. The town is also a municipality responsible for delivery of basic public services to the town's population.

According to the Multiple Indicator Cluster Survey (MICS 2014) by the Government of the Punjab Province (with the support from UNICEF Punjab), only 2.8% of the Sargodha's district household population has piped access to water, compared to 11.6% in Punjab Province. By contrast, 73.4% of the district's water comes from handpump, compared to 30.6% in Punjab Province. Infant and under-5 mortality rates in Sargodha District are 84 and 107 as compared to the Punjab Province's average of 75 and 93. And the district's number of stunted children are 34.1 and 11.7 for "moderate and severe" and "severe", respectively, as compared to Punjab's average of 33.5 and 13.3%. The survey also reported that the percentage of children who in the last two weeks had an episode of diarrhea is 20.5% and an episode of fever is 21.6% as compared to Punjab's average of 17.4 and 20.8%, respectively. Literacy and school attendance rates in Sargodha are relatively better than Punjab Province's average. The indicators for Bhalwal may be worse than the district average as Sargodha is relatively a large city and has better socioeconomic development than its sub-divisions.

Groundwater of Bhalwal city is brackish, which is not suitable for drinking or irrigation except in a few locations. Water contamination due to the sugar industry and heavy use of pesticides and fertilizers in citrus orchards has also contributed to various waterborne diseases in Bhalwal. The city is situated in central Punjab, which has a hot and humid environment. This leads to dehydration and other health complications. The temperature touched 52 °C (125 F) in the summer of 2016. The decrease in rain levels and rising temperatures have further enhanced the importance of the provision of clean drinking water to the public in this region. These factors point to the significance of the case of Bhalwal as an innovative governance model for climate change adaptation. Moreover, a traditional approach could not succeed to complete the water supply scheme which was later revived and completed through a community–government partnership initiative—facilitated by the intermediary civil society organization (Sahi 2014).

Pakistan has traditionally followed a hierarchical and top-down development approach with fewer episodes of local governments coupled with nonrepresentative governments at the center (Cheema et al. 2006). This context highlights centralization that may not allow citizen participation and decision-making at a local level. This factor implies a need to carry out devolution of resources and decision-making to the local level. It also signifies the importance of studying the success of partnership-based CPP in such a centralized governance structure.

9.3 Findings

9.3.1 The CPP Model

CPP is a component sharing model of clean drinking water provision, developed by Faisalabad (one of the largest metropolitan cities in Punjab Province) based on ASB. It was implemented in Faisalabad in the 1990s and later in Lahore (provincial capital of Punjab Province) in early 2000s. The model embodies participatory planning and an integrated approach for 24/7 water and sanitation system governance.

The process starts with the realization by the community and then intermediary organization ASB mobilizes the community to negotiate a partnership with the government. ASB provides guidance and training to local people to carry out such projects starting from a basic understanding of engineering design of the project to operations and maintenance of the water distribution system. Under the partnership, government finances infrastructure for tube wells, overhead tank, disposal systems, treatment systems, and main pipes (6 in.), which make up almost 80% of the total cost of the project. The community participates in the procurement process and contributes financially to water supply pipes (3 in.), sewerage pipes (9 in.), and water meters. The project's management is ultimately handed over to a local community organization that is composed of community members and local government officials.

9.3.2 Bringing CPP to Bhalwal

Before CPP, groundwater of the locality was brackish, so people were relying on water being transported on donkey carts and bicycles from the nearby canal. Water quality was poor and it was also expensive due to high charges by the water transporters.

The government initiated a piped drinking water scheme in 2006 in one of Bhalwal city's three union councils (UCs). But after laying substandard pipes in wider streets and getting payments against those, the contractor ran away, leaving the schemes half way completed.

Some of the members of the community realized the necessity of additional efforts and started looking for solutions. They came to know about ASB and CPP in 2011 and requested ASB to work with them to revive the failed drinking water project. After a number of discussions with the community, ASB started working to convince people about the CPP approach. It was not a smooth process, as people were used to looking toward governments for solutions instead of making collective efforts and negotiating active partnerships with the government, as the CPP model requires.

The founder of ASB and CPP, Malik Nazir Ahmad Wattoo, noted that some of the community members originally opposed the effort. Mr. Wattoo said "*It is good to have opponents as it is usually a test that you are doing a good work. Moreover, if an initiative is opposed, it spreads more.*" He had long discussions with the community about the importance of clean water, even at a price. The local people were resisting it due to the potential financial burden on them and the new water metering system. With the new system, residents would pay for their actual water consumption instead of property size. Without water meters, residents throughout the country are charged solely based on lot size.

With his personal commitment and understanding about community mobilization, Mr. Wattoo continued meetings with the government officials at local, district, and provincial levels. Government offices were reluctant to transform their earlier failed project into the CPP model, as they were afraid of exposure of deficiencies in their earlier work. Mr. Wattoo said that he had the contacts in the bureaucracy at different levels due to his long experience of this type of work in other locations. He used all those contacts to arrange meetings between local community members and government officials. The purpose of these meetings was to agree on a common agenda for starting work on the CPP model and handing over its control to the local community.

This effort succeeded in 2012 when there was a considerable consensus of the community to adopt the CPP model. A Memorandum of Understanding (MoU) was signed between ASB as an intermediary organization and TMA Bhalwal. Since the MoU was signed, 1720 connections have been provided out of 2600 potential connections. A progress report dated on January 9, 2017 mentioned complaints of water quality and leakages, which were being addressed. After the successful launch of the drinking water initiative, the community began to negotiate with the government to start a sanitation scheme as well.

9.3.3 Benefits of CPP in Bhalwal

It is important to understand the success of community–government partnership-based initiative in Bhalwal, Pakistan called the *Changa Pani* Program (clean water) in the situation where the performance of the traditional government-managed scheme was not up to par. Debates on water availability are often focused either on pure market solutions or government run schemes. *Changa Pani* Model provides a quasi-market, partnership-based solution for cities that enable both government and community members to collaborate and take ownership of this important aspect of human life. The model also emphasizes the importance of local self-governance for climate change adaptation by residents of urban areas in developing countries.

9.3.3.1 Procurement Improvements

In traditional local water schemes, the provincial Department of Public Health Engineering issues procurement tenders for contractors and consultants. One of the city government officials reported that no classification of contractors exists at the local level in case of different types of development work like sewerage, water supply, buildings, roads, and firefighting, among others. Therefore, a fundamental problem in the procurement process is that often contractors who do not have specialization are hired to develop water supply schemes. One of the officials explained that "*the consultant's job was to oversee the contractor and to carry out quality control tests, but there was weak accountability of consultant in the failed project.*" The focus group discussion also revealed that pipe manufacturing companies are registered in provincial headquarters and Chief Engineer of Public Health Engineering Department is responsible for material testing. In this case, local government officials have little direct responsibility for the quality of work being carried out by the contractors and pipe manufacturers. According to the focus group participants, community engagement and partnership in the procurement process from the beginning of the project could help to avoid many of the problems that occurred during projects in a traditional government-centric model.

It was determined that the pipe manufacturing company, in connivance with the contractors and consultants, provided substandard pipes for the project before it was handed over to the ASB and local people. Moreover, the contractor did not have a map or layout of pipes in the streets. So the first challenge for the community was to figure out the layout of pipes, and then carry out testing of pipes. Laboratory tests revealed that the substandard material used in the pipes could not carry water at a certain pressure. Accordingly, people of the locality went to the pipe manufacturing company and negotiated with them to replace the pipes. The government officials explained that they were not allowed to negotiate with the pipe manufacturers under the government's procurement rules, but the community could do it once the project had been handed over to them under a partnership framework. Fearing a bad reputation for the company, the pipe manufacturer agreed to replace 975 m (3200 ft) of pipes with the better standard material. Moreover, the community applied stringent tests to the water quality as they knew that they would be drinking it in the future. The cost of the project was rationalized considering that the community would pay for all expenses not covered by the government.

9.3.3.2 Better Service Delivery

Consumer interviews provided insights about service delivery. They were happy to have the piped water supply for the first time in that locality's history. Moreover, there was recognition that ASB initiated this project from very low performance levels. The focus groups participants explained that they were making efforts to provide 24/7 water in order to eliminate chances of contamination in empty water pipes. However, their efforts are often undermined due to electricity load shedding.

Load shedding hurts extraction of clean water from the tube wells, which are 6 km away from the locality. Second, some of the pipes are still the ones laid out by the contractor, which cannot bear the water pressure, creating frequent leakages. The consumers confirmed that they were not getting a full-time supply of water. However, they were satisfied with the quality of water and convenience of piped availability instead of other expensive modes. It was realized after discussion with consumers that the organizing committee and management would have to make further efforts to improve service delivery and undertake communication with consumers to convey the challenges that they were facing.

9.3.3.3 Water Conservation and Pricing

As discussed earlier, water bills in Pakistan are charged based on lot sizes and are minimal, hindering efficient use of water. Even in this situation, traditional government-run schemes are unable to recover the bills. Recovery rates are usually below 50% as compared to over 90% in CPP schemes. In this locality of Bhalwal, water charges were introduced based on consumption starting from Pakistan Rupees (PKR) 50 per 1000 gallons to PKR 70 for an additional 1000 gallons and PKR 100 for each additional 1000 gallons. The water bill also includes fixed charges of PKR 50 per month. After introducing the new water pricing in Bhalwal, they observed that per person per day water consumption declined by more than 50%.

However, it was informed that water pricing was still a tricky issue due to the applicable legal framework. The administrative authority of the province has to approve prices after a number of layers of approvals by the different governmental offices and committees. Additionally, one consumer revealed that he was getting a very high bill for water and it was beyond his capacity to pay. The management explained that there were about 30 members of this household living in a small house, constituting three families. The consumers had a wrong impression of water billing based on house size instead of actual consumption. This issue highlights the importance of communication with consumers for promoting the efficient use of water.

9.3.4 Challenges for Expanding CPP

CPP was first implemented in Faisalabad, later in Lahore, and now in one of the Union Councils (UCs) in Bhalwal. After the success of CPP in one UC in Bhalwal, ASB has been requested by the city government of Bhalwal to take this model to the remaining two UCs of the city. Similarly, ASB has received requests from other nearby cities, which require new conversations on the appropriate adaptation of this approach for those contexts. An international organization has also recently approached ASB to collaborate with them to revive redundant water supply schemes in one of the cities in South Punjab.

Thus, CPP appears to be following the process of diffusion of innovation (Rogers 2003), in which ideas travel from one place or situation to another. Such diffusion of innovation is challenging. Every city and community must consider the local needs and capabilities for designing such partnerships. The spectrum of engagement, investment, and control by both public sector and community may vary under different circumstances and contexts. Any replication of CPP to another situation or place must take into account the capacity for engaging in such partnerships especially in societies where contract enforcement and other formal institutions are not strong and credible.

Here, we identify some of the challenges in replication and expansion based on our analysis of the model and personal insights into the political economy and urban governance in Pakistan.

9.3.4.1 Need for a Skilled Policy Entrepreneur

One important challenge is to mobilize the community to develop a critical mass and contribute a financial investment to negotiate a partnership framework with the government. The intermediary organization has a central role in this model as they have to work with both sides to help them to reach an agreeable framework. Both sides pose unique challenges. The government is often not willing to devolve governance and share powers with communities, and the local communities may not be willing to contribute financially to such projects.

Kingdon (1984) explained that policy entrepreneurs played a crucial role in connecting problems to the potential solutions. By exploiting the window of opportunity, the policy entrepreneurs present new ideas and pursue policy goals with innovative approaches. A similar phenomenon has been observed in the case of CPP. The charismatic personality of Mr. Wattoo, the 75-year-old founder of ASB and CPP, had a critical role in the success of the project. We asked him "*is it possible to do it all without you?*" He replied that he has trained local people to manage and govern this initiative while pointing to Mr. Imtiaz Ahmad, the head of the organizing committee. However, we understand that it would not be an easy task to undertake such a complex partnership framework without the support and guidance of an experienced and dedicated social mobilizer.

9.3.4.2 Urban Political Economy

Rapid urbanization (around 3% at the moment) has enhanced ethnic and cultural diversity in Pakistan in many ways. Even so, cities can best be characterized as contested spaces (Harvey 1997). Local government officials, political activists, influential groups, and mafias want to have their own say into service delivery of water sector and other public goods as well. Perhaps, this is one of the reasons that

CPP has yet not been adopted throughout any large city. It has been carried out in semi-urban and quasi-planned urban settlements only.

Last year, local government elections took place after an interval of almost a decade. Newly formed local governments have limited mandate and resources to deliver services, as most of those are vested with provincial governments. The CPP team in Bhalwal explained that after the local government was installed the newly elected officials pushed to provide free water to please their constituents and build political support. The officials did not foresee their inability to realize the cost of water extraction, transmission, distribution, and maintenance. Such decisions reinforce the prevailing view that the provision of clean drinking water is the responsibility and burden of government, and undercut the foundation of shared governance established by CPP.

9.3.4.3 Nascent Civil Society

Additionally, community organizations are best placed to carry out advocacy and education of the stakeholders involved to ensure achievement of mutually beneficial goals and avoid retrenchment. Yet, grassroots community organizations, having strong local ties and funding mechanisms, are relatively scarce in Pakistan. Most of the local community organizations rely on internationally funded projects with outside interests. Without the engagement of a local intermediary organization like ASB, a CPP-style project may achieve suboptimal results.

9.4 Discussion

9.4.1 CPP as a Participatory Governance Innovation

The CPP is a multilevel and multi-sector governance arrangement to address drinking water infrastructure development, operations, maintenance, and service delivery. The approach can be characterized as a form of New Public Governance, which entails participatory governance, citizen engagement, and networks in the public sector (Osborne 2010). Notably, "governance entails a move away from traditional hierarchical forms of organization and the adoption of network forms. It also entails a revision of the relationship between the state and civil society in a more participatory direction. Governance is said to be responsible for shifting the emphasis away from statute law to more flexible forms of regulation and implementation" (Pollitt and Bouckaert 2011, p. 21, citing Bellamy and Paulmbo 2010). Participatory modes of governance arguably provide more flexibility and customization for the achievement of service delivery goals.

CPP is based on a partnership between local government and community residents, who jointly finance, own, and manage the drinking water and sanitation infrastructure. As a form of "Arm's Length Governance", partnerships are sometimes criticized for taking control away from the democratically elected leaders to new players in public governance. Some scholars argue that involvement of non-governmental players makes it difficult for the governments to assess consumer preferences and deploy accountability framework. Sometimes, such models are also construed as a deviation of the government from its core responsibility. Notably, local government officials in Faisalabad had not given positive feedback about CPP despite the failure of their own schemes and poor quality of service delivery (Farooqi 2013).

CPP is different from traditional public–private partnership models such as outsourcing, privatization, and governmental grants. It is an inclusive partnership. The government has to make a significant upfront public investment, determines water pricing, and has representation on the management committee. The community develops rules for itself and then negotiates with the government to create a unique governance structure for the project. The community can better understand its needs, can cooperate with each other to develop mutually beneficial rules, and can revisit the arrangement as need arises.

Thus, community engagement in public service delivery is a democratic process. Through community engagement, local people are allowed and encouraged to participate in managing and controlling public assets instead of attending mere public participation forums (Durose et al. 2015). Most of the time, if citizens are involved in such forums, participation remains at the level of information sharing, which raises questions regarding the effectiveness of citizen participation (Arnstein 1969). Participatory initiatives like CPP ultimately contribute to democratic accountability, legitimacy, and policy effectiveness (Fung 2006).

Additionally, effectiveness here was aided by a local social mobilizer (Mr. Wattoo) who understood the context and tailored the CPP model to local needs and behaviors. This model contrasts with prior models developed by academicians, scholars, and international development experts that rely on top-down governance despite weak institutional capacity. Easterly (2013) eloquently advocated for this type of spontaneous development and evolutionary learning for problem-solving in developing countries. Moreover, there is some discussion in the literature that decentralization reforms succeed if there is a sufficient capacity of collective action in the communities. Gonzelez-Rivas (2014) found that decentralization of water service delivery in Mexico only succeeded in communities that had a history of collective action. Stone (2001) has also emphasized the importance of the civic capacity to develop an agenda for a collective action and then to implement it successfully.

9.4.2 CPP as a Climate Adaptation Strategy

Despite facing severe consequences of climate change in Pakistan especially from rising temperatures and humidity, there is little recognition of the need for adaptation at the community level. Most discussions on climate change are taking place at federal and provincial levels. In such state of affairs, it is challenging to initiate adaptation initiatives at the city level.

Nevertheless, the CPP can be seen as a climate adaptation effort along the lines of other community-based adaptation (CBA) projects aimed at simultaneously improving livelihoods and climate adaptive capacity from the ground up (Archer et al. 2014) (see Chap. 7: Sari and Prayoga). CBA is "a community-led process, based on communities, needs, knowledge and capacities, which should empower people to plan for and cope with the impacts of climate change" (Reid et al. 2009, p. 13). Under CBA, the community is the central agent in the climate governance process. Experiences with CBA in developing nations have been examined for insights and lessons (e.g., Archer et al. 2014; Soltesova et al. 2014).

The CPP's community–government partnership marks it as a form of co-production of climate adaptation, where the community is empowered to make the decisions and control the public resources (see Chap. 8: Enberg). This type of adaptation strategy contrasts with many other examples of local climate adaptation, where citizen participation is limited and government typically leads the planning and manages any implementation (Sarzynski 2015).

Despite its successes in several contexts, CBA creates complications in upscaling of the approach (Ayers and Forsyth 2009). Moreover, such models are not easy and smooth. They face a range of issues of integrity, credibility, and professionalism. Reid (2016) highlights the challenges faced by CBA approaches regarding integration with the political structures and mainstream governance processes. Mansuri and Rao (2012) highlight that government and market failure does get attention of scholars and policy–makers; however, civil society failure is relatively less researched. These considerations point to the challenges faced by community-led initiatives. It is important to appreciate that CPP embodies a multilevel and multi-sector governance approach that helps to overcome the constraints common in CBA initiatives by formally integrating the effort through the partnership agreement.

9.5 Next Steps

Further research is needed to understand the relevance of the CPP model for addressing governance challenges in other settings and other parts of the world. The irrigation sector in Pakistan faces similar challenges as drinking water. Considering the rapid climate changes in Pakistan and nonresponsiveness of traditional governance structures, it appears beneficial to examine the possibilities of CPP for irrigation.

On drinking water, a database of various provision models being pursued by the public and community organizations is needed with an aim to formulate key lessons from local experiences. Similarly, it is imperative to compare the outcomes and sustainability of CPP with other governance models to properly guide future efforts.

Notably, none of the stakeholders involved in this research identified CPP as a climate change adaptation strategy, marking the need for further discussion and communication of the ways in which various efforts to provide clean drinking water can also serve multiple sustainable development goals (SDGs) including improving health and well-being, building climate-resilient infrastructure, promoting economic growth, and promoting democratic governance and inclusive institutions. Indeed, the provision of clean drinking water has a cross-cutting impact on the SDGs; without clean drinking water, many other goals cannot be achieved.

References

Ahmad, N. (2016, December 30). The politics of urban water. *Dawn*. https://www.dawn.com/news/1305255.

Akram, A. A., & Olmstead, S. M. (2011). The value of household water service quality in Lahore, Pakistan. *Environmental & Resource Economics, 49*(2), 173–198.

Archer, D., Almansi, F., DiGregorio, M., Roberts, D., Sharma, D., & Syam, D. (2014). Moving towards inclusive urban adaptation: Approaches to integrating community-based adaptation to climate change at city and national scale. *Climate and Development, 6*(4), 345–356. doi:10.1080/17565529.2014.918868.

Arnstein, S. R. (1969). A ladder of citizen participation. *Journal of the American Institute of Planners, 35*(4), 216–224.

Ashraf, N., Glaeser, E. L., & Ponzetto, G. A. (2016). *Infrastructure, incentives, and institutions*. Working Paper 21910. National Bureau of Economic Research, Cambridge, MA.

Ayers, J., & Forsyth, T. (2009). Community-based adaptation to climate change. *Environment: Science and Policy for Sustainable Development, 51*(4), 22–31.

Bellamy, R. P., & Palumbo, A. (Eds.). (2010). *From government to governance*. Farnham: Ashgate.

Berg, B., & Lune, H. (2012). *Qualitative research methods for the social sciences*. London: Pearson Education Inc.

Bureau of Statistics Punjab and UNICEF Punjab. (2014). *Multiple indicator cluster survey*. Planning and Development Department, Government of Punjab.

Chabba, A. P. S. (2013). Water-born diseases in India. RESET, Smart Approaches to Sustainability. https://en.reset.org/blog/water-borne-diseases-india. Accessed May 12, 2016.

Cheema, A., Khwaja, A. I., & Qadir, A. (2006). Local government reforms in Pakistan: Context, content and causes. In P. Bardhan & D. Mookherjee (Eds.), *Decentralization and local governance in developing countries: A comparative perspective* (pp. 257–284). Cambridge: MIT Press.

Condran, G. A. (1995). Changing patterns of epidemic disease in New York City. *Hives of sickness: Public health and epidemics in New York City*, 27–41.

Durose, C., Justice, J., & Skelcher, C. (2015). Governing at arm's length: Eroding or enhancing democracy? *Policy & Politics, 43*(1), 137–153.

Easterly, W. (2013). *The tyranny of experts: Economists, dictators, and the forgotten rights of the poor*. New York: Basic Books.

Farooq, S., Hashmi, I., Qazi, I. A., Qaiser, S., & Rasheed, S. (2008). Monitoring of coliforms and chlorine residual in water distribution network of Rawalpindi, Pakistan. *Environmental Monitoring and Assessment, 140*(1–3), 339–347.

Farooqi, S. A. (2013). *Assessment of decentralized service delivery arrangements and institutional performance: The case of Pakistan Local Government Reforms* (Doctoral dissertation, University of Stirling, UK).

Fay, Marianne, Leipziger, Danny, Wodon, Quentin, & Yepes, Tito. (2005). Achieving the child-health-related millennium development goals: The role of infrastructure. *World Development, 33*(8), 1267–1284.

Felbab-Brown, V. (2017). *Order from chaos: Stealing water*. Brookings Institution, DC. https://www.brookings.edu/blog/order-from-chaos/2017/03/23/stealing-water/. Accessed April 1, 2017.

Flyvbjerg, B. (2006). Five misunderstandings about case-study research. *Qualitative Inquiry, 12* (2), 219–245.

Fung, A. (2006). Varieties of participation in complex governance. *Public Administration Review, 66*(s1), 66–75.

González Rivas, M. (2014). Decentralization, community participation, and improvement of water access in Mexico. *Community Development, 45*(1), 2–16.

Glaeser, E. (2011). *Triumph of the city*. New York: Penguin Random House.

Glaeser, E. L. (2012). The challenge of urban policy. *Journal of Policy Analysis and Management, 31*(1), 111–122.

Glaeser, E., & Sims, H. (2015). *Contagion, crime, and congestion: Overcoming the downsides of density*. London: IGC growth brief series.

Haque, N. U. (2015). *Flawed Urban development policies in Pakistan* (No 2015:119). Islamabad: Pakistan Institute of Development Economics.

Hardin, G. (1968). The tragedy of the commons. *Science, 162*(3859), 1243–1248.

Harvey, D. (1997). Contested cities: Social process and spatial form. *Transforming Cities*, 19–27.

Hussain, I. (2014). *Urbanization in Pakistan*. Karachi: Institute of Business Administration.

Iftikhar, M. N. (2014). Corporate governance of state-owned enterprises (SOEs): The case of Pakistan. *Board Leadership* 2014 (134).

Iftikhar, M. N. (2015). *State-owned enterprises in Pakistan: The need for corporate governance and private investment*. Islamabad: Policy Research Institute of Market Economy.

Kamal, S. (2009). *Pakistan's water challenges: Entitlement, access, efficiency, and equity. Running on Empty*. Washington, DC: Wilson Center.

Kane, J., & Puentes, R. (2016). *Flint's water crisis highlights need for infrastructure investment and innovation*. Washington, DC: Brookings Institution.

Kingdon, J. W. (1984). *Agendas, alternatives, and public policies*. Boston: Little, Brown & Company.

Mansuri, G., & Rao, V. (2012). *Localizing development: Does participation work?* Washington: World Bank Publications.

Morgan, D. L. (1996). *Focus groups as qualitative research*. Thousand Oaks: Sage publications.

Osborne, S. P. (Ed.). (2010). *The new public governance: Emerging perspectives on the theory and practice of public governance*. UK: Routledge.

Ostrom, E. (2010). Beyond markets and states: Polycentric governance of complex economic systems. *American Economic Review 100* (June 2010), 641–672.

Ostrom, E., & Gardner, R. (1993). Coping with asymmetries in the commons: Self-governing irrigation systems can work. *The Journal of Economic Perspectives, 7*(4), 93–112.

Planning Commission. (2015). *Pakistan Vision 2025*. Islamabad: Government of Pakistan.

Pollitt, C., & Bouckaert, G. (2011). *Public Management Reform: A comparative analysis-new public management, governance, and the Neo-Weberian state*. Oxford: Oxford University Press.

Reid, H. (2016). Ecosystem-and community-based adaptation: Learning from community-based natural resource management. *Climate and Development, 8*(1), 4–9.

Reid, H., Alam, M., Berger, R., Cannon, T., Huq, S., & Milligan, A. (2009). Community-based adaptation to climate change: An overview. *Participatory Learning and Action* 60, 11–33. London: IIED.

Rogers, E. M. (2003). *Diffusion of innovations*. New York: The Free Press.

Sahi, A. (2014, March 2). Water in the pipeline. *The News*. http://tns.thenews.com.pk/water-pipeline-bhalwal/#.WSHthWjyvIU.

Sarzynski, A. (2015). Public participation, civic capacity, and climate change adaptation in cities. *Urban Climate, 14,* 52–67.
Soltesova, K., Brown, A., Dayal, A., & Dodman, D. (2014). Community participation in urban adaptation to climate change: Potential and limits for community-based adaptation approaches. In E. L. F. Schipper, J. Ayers, H. Reid, S. Huq, & A. Rahman (Eds.), *Community based adaptation to climate change: Scaling it up* (pp. 214–225). New York: Routledge.
Stone, C. N. (2001). Civic capacity and urban education. *Urban Affairs Review, 36*(5), 595–619.
Yin, R. K. (2013). *Case study research: Design and methods.* Thousand Oaks: Sage Publications.

Author Biographies

Muhammad Naveed Iftikhar has worked for the public sector, academia, research centers, and international development organizations in the fields of public policy, governance, sustainable development, and economic reforms. Naveed's current research focuses on entrepreneurship, public sector governance, cities, and sustainable economic development. As a Governance Specialist in the Economic Reforms Unit of Pakistan's Ministry of Finance, he contributed to the design and implementation of policy reform initiatives in key sectors of the economy. Naveed has consulted for international development organizations including The World Bank, Asian Development Bank, United Nations Development Program, and Japan International Cooperation Agency among others. Naveed holds M.Phil. degree in Economics and is currently a Ph.D. Candidate in Urban Affairs and Public Policy at the University of Delaware.

Shujat Ali is a former civil servant from the Pakistan Administrative Service. He has served in various capacities in the districts and provincial and federal governments. He has vast experience of stints in the provincial finance and planning departments and in the Ministry of Finance in the federal government. His last assignment in government was as Secretary of the Statistics Division in the Ministry of Finance. He holds an undergraduate degree from the London School of Economics and a doctorate in economics from Simon Fraser University Canada. He has patronized and supported a number of community participation initiatives in public service delivery in various parts of Pakistan over the years. His research interests lie primarily in the areas of agriculture productivity, climate change, and models for service delivery.

Andrea Sarzynski is an associate professor at the University of Delaware's School of Public Policy and Administration and the Department of Geography. Her research interests include urban governance, climate change policy, urban land use, shrinking cities/suburbs, and renewable energy policy. Prior to joining the faculty at Delaware, she worked at the George Washington Institute of Public Policy, the Metropolitan Policy Program at the Brookings Institution, the Rochester Institute of Technology, the White House Council on Environmental Quality, and the environmental law practice at Sidley Austin LLP. Andrea has a Ph.D. in public policy and public administration and a B.S. in natural resources from Cornell University.

Part III
City Networks Innovations

Chapter 10
Connecting the Dots: The Politics of Governing Urban Climate Adaptation Innovations Through Transnational Municipal Networks

Ryan G. Bellinson

Abstract Since Agenda 21 was ratified at the Rio Earth Summit in 1992, a number of transnational municipal networks (TMNs) have emerged to promote a variety of sustainability and climate change actions. Despite research on the role of TMNs in facilitating urban climate action, there have been few reflexive inquiries regarding how governance actors, institutions, and processes have transformed resulting from local governments' participation in TMNs. In this chapter, I investigate how TMNs have promoted urban adaptation actions as well as their effects on local government climate policy processes. I first draw on the theories of multi-level governance, urban planning, and social innovation in the context of climate change to develop a framework for governing adaptation innovations through a model of contentious urban politics. I then apply this framework to explore climate adaptation processes in Rotterdam (The Netherlands) and Berkeley (California, United States), and examine the actors, resources, and networks involved in motivating, sustaining, and prioritizing adaptation across competing sectoral agendas. The results note that the successful and eventual uptake of adaptive innovations requires communicating competing interests, confronting conflicts, and building cross-sectoral coalitions. I conclude by arguing that a model of contentious politics, especially as advocated by TMNs, is more appropriate to understand how governance innovations can be harnessed to promote more climate adaptive cities.

Keywords Transnational municipal networks · Mutli-level governance · Climate adaptation · Urban innovation

R.G. Bellinson (✉)
Department of Urban Studies and Planning, University of Sheffield, Sheffield, United Kingdom
e-mail: rbellinson@gmail.com

S. Hughes et al. (eds.), *Climate Change in Cities*, The Urban Book Series,
https://doi.org/10.1007/978-3-319-65003-6_10

10.1 Introduction

Though cities around the globe confront place-based climate change risks and vulnerabilities such as sea level rise, flooding, droughts, landslides, and heatwaves, they are nevertheless organizing within transnational municipal networks (TMNs) to exchange knowledge, build capacity, and collectively plan for their climate challenges (Bulkeley 2013). TMNs have become important political actors in the arena of climate change policy and have successfully advanced innovative adaptive planning agendas such as harnessing new program funding mechanisms, translating the most current science into policy, advancing cross-sectoral collaborative decision-making, and promoting policy co-benefits (Boyd and Osbahr 2010; Dzebo and Stripple 2015).

Current research on TMNs mainly investigates their origins, potential benefits, and role within global environmental governance, most extensively within the context of mitigation (Fünfgeld 2015; Andonova et al. 2013; Lee 2013). In this chapter, I apply these insights to recent adaptation actions and utilize these theories to examine the attitudes and outcomes of municipalities participating in 100 Resilient Cities (100RC), a TMN funded by the Rockefeller Foundation. As innovation is communicated and subsequently interpreted differently across cities, this chapter asks: how are cities building new multi-level governance arrangements to internalize emerging discourses of *innovation* in adaptation planning as advocated by TMNs? And relatedly, how are cities communicating the need for *innovative* adaptive approaches across governance scales?

To answer these questions, I draw on theories of urban innovation and multi-level governance in the context of climate change. I then apply this framework to two case studies to illustrate the governance pathways of adaptation innovations. To conclude, I affirm that TMNs are indeed aiding cities to some degree by developing procedural innovations to address climate change, but contemporary theoretical discourses of policy innovation must additionally consider the inherent *political* nature of urban governance, which has significant implications on how adaptation innovations are negotiated, designed, and implemented.

10.2 Theories of Climate Adaptation, Multi-level Governance, and Innovation

In this section, I draw upon the theories of multi-level governance, urban planning, and social innovation to construct a conceptual framework for governing climate adaptation innovations (see Fig. 10.1). The framework indicates that the potential of cities to expand their adaptive capacity rests on their ability to integrate urban political and economic interests, climate risk awareness, and external capacity and resource support. From this starting point, cities enter into what Sørensen and Torfing (2011) describe as an “innovation cycle”. The results of this innovation

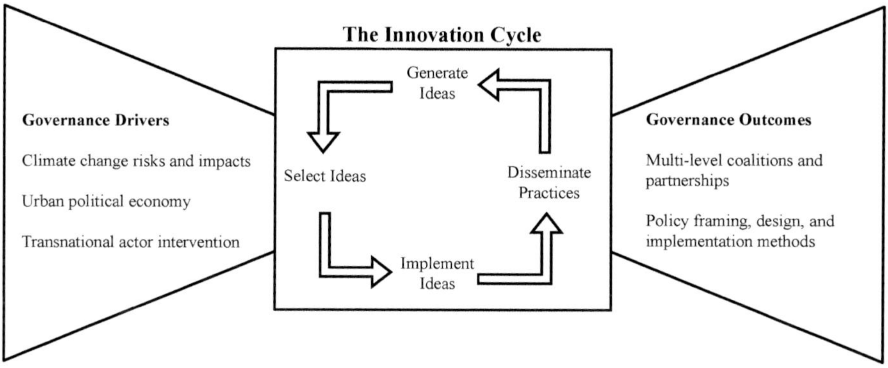

Fig. 10.1 Conceptual framework of the innovation cycle in the context of urban climate change governance

cycle—particularly within the context of adapting to unpredictable climate risks—are a set of procedural innovations that attempt to more effectively govern urban political and economic interests.

10.2.1 Urban Climate Adaptation and TMNs

Cities often struggle with political fragmentation between local government, non-governmental actors in the public and private sectors, the general public, and higher levels of government that can lead to political stagnation, particularly in multi-sectoral policy arenas such as climate change (Chu and Schenk 2017; Blanco 2013; Innes 1998). TMNs can address this fragmentation by aligning institutions and actors across differing political scales and by fostering new collaborative planning and decision-making pathways (Coen and Thatcher 2008; Bogason and Musso 2006). Traditionally, TMNs are nonhierarchical (Blanco 2013; Andonova and Mitchell 2010; Provan and Kenis 2008), and facilitate enhanced learning, improve competitiveness, more efficient uses of resources, and improve capacity to plan for and confront complex problems (Bäckstrand 2008; Provan and Kenis 2008; Bogason and Musso 2006; Giest and Howlett 2013).

As emblematic forms of multi-level governance, TMNs facilitate both vertical and horizontal collaboration between all levels of government, private actors, and civil society organizations (Gordon 2013; Measham et al. 2011; Nyseth 2008; Healey 1998). Additionally, TMNs help ground scientific projections and models, enable sharing of practices, facilitate inclusive and equitable adaptation processes, and stimulate collaboration across their local government members (Betsill and Bulkeley 2004; Leck and Simon 2012). Present scholarship on urban regimes posits that robust local governance demands dynamic collaboration between private and public actors operating at all political levels to mobilize resources and capacity

(Blanco 2013; Leck and Simon 2012), but this frequently proves cumbersome for climate adaptation efforts due to competing sectoral interests, unsupportive political ideologies, or unclear mandates (Carmin et al. 2013). For instance, clashing planning agendas—such as between economic development and environmental conservation—can cause misuse of already-finite resources (Leck and Simon 2012; Betsill and Bulkeley 2004; Chu et al. 2016).

Despite their acknowledged advancements, TMNs have also been criticized for not assisting cities develop corresponding multi-level governance architectures to help limit political conflicts arising from institutional confusion and lack of communication (Jordan and Huitema 2014; Bulkeley 2013; Anguelovski and Carmin 2011). TMNs have also been critiqued for being homogenous networks of exclusive "elite", larger cities at the exclusion of "second tier" cities. (Andonova and Mitchell 2010; Kern and Alber 2009; Bulkeley and Betsill 2013). TMNs display this membership bias because the networks themselves have an incentive to overly highlight their positive outcomes to improve access to resources and favorable recognition from the international community (Kern and Alber 2009; Bäckstrand 2008; Betsill and Bulkeley 2004).

10.2.2 Governing Innovations in Cities

Innovation is an "intentional and proactive process that involves the generation and practical adoption and spread of new and creative ideas which aim to produce a qualitative change in a specific context" (Sørensen and Torfing 2011: 849). Governance innovation often demands collaborative interactions between private and public actors as local government resources—such as political capacity, knowledge supply, and finance—are limited and frequently insufficient for meeting the cross-sectoral pressures of adaptation (Chu et al. 2016; Anguelovski and Carmin 2011; Sørensen and Torfing 2011; Bulkeley and Betsill 2013).

Innovation takes place through a cycle with four distinct stages (see Fig. 10.1), of which TMNs engage in the last three. During the initial stage, policy ideas are generated to address problems illuminated during the agenda setting process (Sørensen and Torfing 2011). During the second stage, TMNs bring together policy-makers with diverse backgrounds and expertise to collaboratively assess risks and benefits of prospective policies that were conceived during the policy generation process (Anguelovski et al. 2014; Sørensen and Torfing 2011). Afterwards, TMNs help to diminish implementation barriers by organizing adaptive efforts between member cities, supporting the orchestration of efficient, new internal multi-level governance architectures, and mobilizing financial and human resources (Sørensen and Torfing 2011; Andonova 2013). Lastly, TMNs establish member-to-member connections that help disseminate innovative policies through peer exchange events (Anguelovski and Carmin 2011; Sørensen and Torfing 2011; Bulkeley 2013).

Cities often face difficulties in sparking and sustaining innovation due to a lack human capacity, financial resources, and political support (Gordon 2013;

Anguelovski and Carmin 2011; Innes 1998). To respond, TMNs can focus on addressing the structural governmental constraints on cities by mediating knowledge and communication dissemination pathways that span the siloed character of urban planning (Jordan and Huitema 2014; Giest and Howlett 2013; Measham et al. 2011; Hartley et al. 2013). TMNs can act as instigators of innovation by decoupling city governments from their traditional planning tracks, and moving towards alternative decision-making pathways through which bureaucratic constraints can be negotiated (Gordon 2013; Sørensen and Torfing 2011; Okereke et al. 2009).

TMNs build upon the innovation cycle within cities by enhancing multi-level governance relationships between actors, fostering new cross-sectoral policy decision pathways, and facilitating increased inter-municipal collaboration (Massey et al. 2014; Bulkeley and Betsill 2013). Although the role of TMNs as political agents involved in governing climate adaptation is clear, there remains uncertainty regarding potential rescaling of municipal governance structures in reaction to TMN engagement as well as the repercussions for how innovation manifests within cities. Particularly, there have yet to be empirical studies illustrating the nature by which cities capture innovative planning techniques advocated by TMNs (Measham et al. 2011). Moreover, existing strategies of innovation and experimentation are frequently restricted when addressing trans-boundary multi-scalar challenges like climate change (Pattberg and Stripple 2008; Bulkeley and Broto 2015; Bulkeley and Tut 2013). In response, I seek to contribute to this knowledge gap by arguing that local governments must transform institutionally to fully internalize discourses of innovation within the contentious, political domain of planning to inclusively and equitably implement emerging adaptation innovations.

10.3 Methods

Two case studies were analyzed to illustrate how cities have internalized different concepts and methods of adaptive innovation. The cities of Berkeley (California, United States) and Rotterdam (The Netherlands) were selected because both became members of 100RC during the primary round of member selection in December 2013. Second, both cities are governed by mayoral city council executive branches charged with setting local policy agendas, representing the public's interest, approving budgets, and maintaining legislative oversight. Third, Berkeley and Rotterdam face shared climate risks including storm surges, sea level rise, and saltwater intrusion. Lastly, Rotterdam and Berkeley are economically significant nodes placed within robust regional metropolitan areas—the Randstad and the Bay Area—both with relatively substantial regional GDPs.

The methodology includes policy document analysis, semi-structured interviews, and analysis of public meeting records. Recent climate change policies from each city were thematically coded to identify climate adaptation procedural and policy outcomes (see Table 10.1). In identifying and evaluating various process innovations, I investigated each policy's motivations, funding sources, significant

Table 10.1 Key adaptation policy and planning actors and outcomes in Berkeley and Rotterdam

City	Policy/Plan	Involved scoping actors	Main planning outcomes
Berkeley	*Climate action plan* (2009)	City office of energy and sustainable development; City manager's office; Berkeley city council; HUD; ICLEI civil society	• Create regional and state partnerships for infrastructure adaptation projects • Create neighborhood level climate vulnerability impact assessment
	Climate action plan 2015 update (2015)	City office of energy and sustainable development; City manager's office; The ecology center; City of Berkeley sustainability working group; Civil society	• Increase horizontal integration within city government • Increase connectivity with regional government agencies
	Berkeley resilience strategy (2016)	100RC; City office of energy and sustainable development; City of Oakland; City of San Francisco; Association of bay area governments; Berkeley city council; Civil society	• Require land use planning decisions to be based off of the most current science available • Work towards building regional resilience
Rotterdam	*Rotterdam climate proof* (2009)	Regional water boards; Rotterdam climate initiative; Port of Rotterdam authority; Erasmus University; TU Delft; TNO	• Build knowledge and innovation on water-related planning sectors • Market innovation to other cities
	Rotterdam climate city mitigation action program (2010)	City of Rotterdam; Rotterdam climate initiative; Port of Rotterdam authority; Ministry of economic affairs; Ministry of the environment; DCMR; Erasmus University; Civil society	• Develop and implement sustainable building codes for residential and business purposed • Optimize energy and infrastructure • Promote sustainable business management
	Rotterdam adaptation strategy (2013)	City of Rotterdam; Port of Rotterdam authority; Climate-proof Rotterdam; Rotterdam city council; TNO; DCMR; Civil society	• Fortify outer and inner dike flood barriers • Strive towards developing regional adaptiveness
	Rotterdam resilience strategy (2016)	City of Rotterdam; Port of Rotterdam authority; Utility companies; Ministry of the environment; TNO; DCMR; 100RC; Rotterdam CRO strategy team	• 68 action items under an umbrella of seven resilience themes

actors, and key recommendations. The semi-structured interviews were conducted with leading members of each cities' environmental, planning, and climate change departments, as well as officials from pertinent regional agencies (see Appendix A). These interviews concentrated on the themes of procedural change, long-term agenda setting, the adaptation policy cycle, and TMN governance outcomes. Ten total interviews were held with policy-makers, five for Rotterdam and five for Berkeley. Furthermore, video recordings of public meetings were analyzed to identify discourses and outcomes during the scoping process of climate change policies (see Appendix B). Public meeting records were only analyzed in Berkeley because Rotterdam did not involve the public during the scoping process of its climate policies.

10.3.1 100RC

In December of 2013, 100RC—a TMN funded by the Rockefeller Foundation—named its first group of 32 member cities. Subsequently, 100RC has had two rounds of membership expansion and, as of early 2017, has a network of 100 member cities. Upon admittance into 100RC, members are given four main resource pathways.

The first and most prominent resource provided by 100RC is a 2-year grant to fully fund a Chief Resilience Officer (CRO), who is a senior local government policy entrepreneur and decision-maker typically closely aligned with the Mayor's or City Manager's Office. The primary role of the CRO is to coordinate within local government and to oversee the city's resilience efforts. An essential responsibility of the CRO is to work across governmental departments in order to improve internal government collaboration and dissolve sectoral planning silos to more effectively implement resilience policies. The CRO's close ties to the city's executive branch allow him/her to harness political leadership in order to cultivate interdepartmental and multi-level cooperation, dialogue, and coordination. The second resource 100RC provides is their City Resilience Framework—a methodology through which members develop their resilience strategies. Third, 100RC members have access to "Platform Partners", which helps them connect to different NGO and private sectors organizations that dispense tools and services for developing and implementing resilience strategies. Lastly, member cities have access to 100RC's other members through a digital database, which can facilitate peer learning opportunities and knowledge exchange.

10.4 Comparative Approaches to Governing Innovation for Urban Climate Adaptation

Rotterdam and Berkeley have distinct experiences with climate change adaptation. Berkeley has a background of developing a climate adaption agenda through community-driven directives while Rotterdam has taken a more top-down approach. In this section, I detail each city's recent history of climate policy and highlight processes of governance innovation.

10.4.1 Berkeley

Historically, Berkeley has enjoyed a politically active citizenry and so the city has made it a priority to be publicly inclusive (Lipset and Albach 1966). In 2006, Berkeley's voters overwhelming passed Ballot Measure G, mandating the city to reduce its greenhouse gas emission by 80% below 2000 levels by 2050, which led Mayor Tom Bates to begin developing a plan to reach the clear target along with other climate priorities (City of Berkeley 2009). Measure G was codified through the resolution of Berkeley's Climate Action Plan (CAP) in 2009. Although the CAP was mostly conceptualized as a climate mitigation policy, there were components that addressed adaption efforts, such as one broadly defined goal of community resilience and preparedness (City of Berkeley 2009). The development of the CAP was facilitated by the Office of Energy and Sustainable Development (OESD), in close coordination with other municipal departments, regional agencies, and TMNs (City of Berkeley 2009).

Although the CAP was developed through cross-departmental coalitions and with multi-level support, it never lost its community-inspired ethos since it continuously focused on bottom-up initiatives such as public meetings and citizen participatory committees (City of Berkeley 2009). To this effect, an Association of Bay Area Governments (ABAG) official who was involved with the CAP noted,

> The level of community engagement Berkeley has been able to harness in the face of the potentially intersectional, competing conflicts that this can ignited, is remarkable (Interview 2016).

The CAP exemplified a form of a multi-level governance as seen by the variety of involved government actors, broad range of local stakeholders, and active community members that were represented throughout the process. The subsequent range of perspectives led to contentious political negotiations that were integral to developing the ultimately ratified CAP while simultaneously meeting the original emissions reduction target called for by Measure G.

Once Berkeley's CAP had been in place for 4 years, the City Manager's Office and the City Council commissioned the CAP 2015 Update, which was to be coordinated by the OESD. The principle decision-making pathways of the CAP

2015 Update were reached through the cross-departmental City of Berkeley Sustainability Working Group (BSWG), a volunteer citizen committee, and public meetings (City of Berkeley 2015). Public commentary was a mechanism that the BSWG utilized to foster political contention and new ideas (City of Berkeley 2015). At a City Council meeting in which the BSWG and OESD presented a full draft of the CAP 2015 Update, one citizen proposed that,

> We should align city budget with the goals of the CAP…The budget needs to reflect our emissions reductions priorities (Berkeley City Council Meeting 2015).

After the public commentary period, the mayor closed the City Council meeting by echoing the constituent's sentiment, and requested the OESD to work to incorporate this viewpoint into the finalized policy. As instructed, the OESD work closely with the public and ultimately codified stricter emissions standards for buildings constructed with public funds, as well as more stringent emissions standards for vehicles added to the public fleet which was realized in large part through state funding (City of Berkeley 2015).

The main finding of the CAP 2015 Update was an acknowledgement that greater funding was required to fulfill the CAP's adaptation-oriented programs. Additionally, collaboration between the City Council, the OESD, involved municipal departments, and higher levels of government was necessary to increase implementation capacity for programs (City of Berkeley 2015). From these findings, Berkeley made a strategic policy process accommodation for climate adaptation and institutionalized greater coordination between the City Council and BSWG (City of Berkeley 2015). By strengthening the working ties between these cross-institutional actors, the city believed it could more efficiently implement its adaptation priorities and improve overall governance effectiveness (City of Berkeley 2015).

Berkeley published its most recent major climate policy, the *Resilience Strategy*, in 2016, which further advanced municipal climate mitigation and adaptation priorities. The *Resilience Strategy* was motivated by Berkeley's membership into 100RC (City of Berkeley 2016). The OESD was tasked with overseeing the scoping, design, and implementation of the *Resilience Strategy* (City of Berkeley 2016). Based upon their findings from the CAP 2015 Update, the OESD focused on developing the *Resilience Strategy* through cross-departmental coalitions with additional support from regional, state, and federal government agencies, as well as public stakeholders (City of Berkeley 2016).

Over one thousand local residents provided input during the Resilience Strategy's scoping process through events hosted by community organizations such as Livable Berkeley, League of Women Voters, Bicycle Friendly Berkeley Coalitions and the Sierra Club, as well as events hosted by the city including City Council meetings, and community workshops (City of Berkeley 2016). Furthermore, the private sector, civil society organization, and a citizen volunteer committee were also invited to provide input to be included in the Resilience Strategy's scoping process. Robust ties between various city departments, the constellation of community-wide actors, and regional collaboration coordination through ABAG created a process of political negotiation that informed specific actions and broad goals articulated in the Resilience Strategy (City of Berkeley 2016). This hallmark of community-based policy-making was again impactful, but the levels of involved stakeholders, the extent to which public input was sought, and the number of consulted regional agencies and organizations institutions were all new procedural advancements.

After the OESD had spent over a year developing a draft of the Resilience Strategy, Berkeley's CRO formally presented a draft at a June 2015 City Council meeting before receiving the public's response. The public raised several points that would ultimately be incorporated into the ratified Resilience Strategy, including providing disaster preparedness education opportunities to the public, addressing cascading disaster events, and revising city ordinances and building codes to reflect newly adopted resilience frameworks (Berkeley City Council Meeting 2015). One citizen who provided comments during this period raised several concerns within the Resilience Strategy draft but also articulated encouraging remarks regarding Berkeley's resiliency efforts in collaboration with 100RC,

> I think all the issues we (the public) raised are important and need to be addressed and I look forward to seeing the progress we're able to make with help from 100RC and Rockefeller (Berkeley City Council Meeting 2015).

Berkeley's Resilience Strategy, as well as the adaption programs within it, were conceptualized through a collaborative multi-level process utilizing cross-departmental coalitions like the BSWG and partnerships with community organizations like ABAG and the Ecology Center. This form of enduring multi-level decision-making was a key contributor to Berkeley's Resilience Strategy.

10.4.2 Rotterdam

As a delta city, water is viewed as both an asset and climate risk in Rotterdam (City of Rotterdam 2009). Rotterdam has consistently remodeled their city governance in response to flooding and other water related risks by developing new municipal departments, working groups, and offices. Over the last two decades, decision-makers have sharply transitioned from conceptualizing water issues as being a distinctive policy arena to now viewing it as a multi-sectoral, holistic, decentralized policy sphere. An important policy and structural development that exemplifies this transition was the creation of Rotterdam Climate Proof (RCP) in 2009, which was designed to build upon the progress of Rotterdam Climate Initiative (RCI), the most significant climate mitigation policy in Rotterdam to date (City of Rotterdam 2009).

RCP was established from collaborations between local public authorities such as the City of Rotterdam and the Port of Rotterdam Authority, regional government agencies and institutions, and Delta Cities (an international network of coastal cities). The main objective of RCP was to help construct a climate-proof Rotterdam by 2025, and also sought to further the mitigation efforts from RCI, strengthen Rotterdam's climate adaptation actions, and develop clear institutional structures for climate policy between local actors, stakeholders, and the local government (City of Rotterdam 2009). In addition to the overarching goal of establishing a climate-proof Rotterdam, RCP outlined three ancillary goals: to develop knowledge and create innovation around the intersection of water and spatial development, create innovative forms of adaptive development and build green infrastructure projects, and promote RCP's innovations to other cities (City of Rotterdam 2009).

In 2013, Rotterdam ratified the Rotterdam Climate Change Adaptation Strategy, which was a policy designed to centralize municipal authority and increase the city's adaptive capacity (City of Rotterdam 2013). The Adaptation Strategy also had the overarching goal of advancing the role of regional authorities from "guardians of public affairs" to "facilitators and initiators" (City of Rotterdam 2013: 26). To implement this goal, the City of Rotterdam worked closely with several regional authorities including the DCMR, which is the environmental agency for the Rijnmond region (City of Rotterdam 2013). For example, one of the focuses of the Adaptation Strategy was the further development of regional dike infrastructure that protects Rotterdam from rising sea levels and storm surges (City of Rotterdam 2013). This was achieved through working closely with regional government agencies to increase political and funding capacities.

The Adaptation Strategy was designed to improve upon the progress, both in terms of environmental policy and structural advancements, from RCI. An important procedural development that stemmed from the Adaptation Strategy was the blending of policy arenas and the integration of previously separated actors, institutions, and stakeholders. As the DCMR Deputy Director of Environment stated,

> Rotterdam is the largest, most influential municipality within our regional jurisdiction. When they (Rotterdam) asked us to work closely with them on the Adaptation Strategy, we were excited because not only would we be able to help Rotterdam but it presented an opportunity to help our smaller municipals advance their adaptation programs (Interview 2016).

The RCP coordinated the scoping process of the Adaption Strategy but, as originally intended, included a broad swath of multi-level partners such as the Port of Rotterdam Authority, DCMR, TNO (the Netherlands Organization for Applied Scientific Research), local government cross-departmental working groups, and the public in developing the finalized strategy (City of Rotterdam 2013).

The Rotterdam Resilience Strategy, published in 2016, was the next major climate policy with a broad adaptation component. As with Berkeley, Rotterdam's Resilience Strategy was largely prompted by the city's recent membership in 100RC (City of Rotterdam 2016). The Resilience Strategy was an innovative policy for a few reasons. First, the Resilience Strategy was intended to connect previously detached actors and further integrate levels of government in an effort to holistically plan for and develop policy mechanisms and processes to most efficiently prepare for shocks and stresses. Rotterdam's CRO elaborated on this objective stating that,

> If you want to become a resilient city, you have to go outside of the way you developed policy in the past, and develop meaningful relationships with the widest variety possible of new stakeholders and institutions (Interview 2016).

From its earliest stages, the Resilience Strategy was conceptualized differently compared to previous mitigation and adaptation policies in Rotterdam due to its wide range of community partners and representation of regional governments and private actors in the steering committee (City of Rotterdam 2016). Therefore, one can see that resilience was not conceptualized a discrete policy arena concerning a singular city government department, or City of Rotterdam, but was actually an emerging policy sphere deeply entangled throughout in the city's broad political and economic interests.

One explanation for why Rotterdam approached resilience differently than previous climate policies may have been their position within 100RC network. Unlike their role in other TMNs, Rotterdam was able to utilize their 100RC membership to extract knowledge and best practices from other cities to develop their own innovative policy advancements. As Rotterdam's CRO explained,

> Because of 100RC, I have new counterparts to collaborate with because it is difficult to invent the wheel all by yourself. It's so much easier, so much more productive to share ideas and create on topics together (Interview 2016).

Another procedural innovation resulting from the development and implementation of the Resilience Strategy was the variety and scope of incorporated cross-department municipal government actors and multi-level institutions. Almost every city government department had some involvement in the scoping or implementing Resilience Strategy, numerous regional government agencies and offices were consulted in scoping the strategy, the private sector was closely involved through the designing process, and the general public was also included. A policy advisor described this multi-level collaborative attitude and noted,

> I have no doubt that through developing our Resilience Strategy we've strengthened our regional governance commitments with the help of 100RC. Their methodology has made a big difference in strengthening our relationships with regional government agencies in every aspect (Interview 2016).

From Rotterdam's experience, one can see that the city has long relied on durable coalitions between local government and broad stakeholders—such as the local private sector, the Port of Rotterdam Authority, and different regional government agencies like DCMR and TNO—to cooperatively work towards determined priorities by developing processes to harness local innovation. These unique multi-level ties have contributed to Rotterdam's decision-making and governance restructuring within the context emerging climate vulnerabilities and risks.

10.5 Patterns of Governance Innovation in Urban Climate Adaptation

In the cases of Rotterdam and Berkeley, adaptation innovations were supported by increased multi-level structures, which in these instances signified the inclusion of new horizontal and vertical political actors, collaborative and cross-sectoral networks, harnessing new forms of knowledge, and the use of new participatory policy pathways. For Rotterdam, the CRO team located within the department of city development developed greater cross-departmental communication mechanisms while fostering increased involvement from local stakeholders, vertical government agencies, and TMNs in the decision-making process. The CRO-led OESD in Berkeley catalyzed increased cross-department collaboration, robust public engagement, and coordination with regional and state government to broaden the political discourse. In Rotterdam and Berkeley, the urban political economic context, climate change risks and impacts, and TMN resource and capacity support affected the local governance context within which the innovation cycle and policy process takes place. The multi-level intersection of differing priorities, interests, actors, and mandates contributed to several procedural innovations and challenges. In this section, I expand on Fig. 10.1 and analyze the different factors that have led to governance innovations in Rotterdam and Berkeley.

10.5.1 Drivers of Innovation

Local governance contexts often dictate or constrain approaches to policy innovation (Gordon 2013). For urban climate adaptation policy, the practical point of departure is profoundly affected by the local political economy, location-dependent climate change impacts and risks, and transnational actor interventions (Kern and Bulkeley 2009; Healey 2004; Gordon 2013). Rotterdam and Berkeley both confront well-understood climate change impacts and risks that are paired with robust political discourses and high public awareness of the documented climate change scenarios. Both Rotterdam and Berkeley are located within large economic regions and are closely tied to local knowledge institutions, including technology firms, research organizations, and universities. Both have utilized this background of innovation from the local industrial, economic, educational, and service sectors to provide a framework for translating innovation into the governance of infrastructure, sustainability, and climate change.

10.5.2 Innovation Outcomes

From analyzing the climate adaptation experiences of Rotterdam and Berkeley, two main observations of innovative outcomes surfaced (see Table 10.2). The first

Table 10.2 Summary of innovation drivers and outcomes in Berkeley and Rotterdam

City	Innovation drivers	Innovation outcomes
Berkeley	• Strong public perception of climate risks • Deep economic ties to other local governments in the region • Political desire to work on policy at a regional level • History in TMNs as active participant	• Cross-governmental coalitions (City of Berkeley Sustainability Working Group) • Citizen stakeholder committee • New policy pathways that disintegrate planning silos and deepen regional government agencies involvement
Rotterdam	• High perception of climate risks by local government • Strong economic ties to other local governments in the region • Leading city in the region, smaller cities in the region following actions • Long history within TMNs as knowledge distributor	• Citywide, cross-actor Resilience strategy team • 100RC help mold definition of "resilience" on a policy level • New policy pathways that disintegrate planning silos and deepen regional government agencies involvement

observation is the development of new multi-level governance coalitions and partnerships, which describes informal relationships that TMNs help foster between constellations of local institutions and wider stakeholders and government agencies that share comparable priorities. There are several notable examples of TMNs spurring internal coalitions in both Rotterdam and Berkeley, including the citywide Resilience Strategy Team in Rotterdam and Berkeley's cross-departmental BSWG. Both examples help build their city's capacity to innovate by facilitating coordination between disparate local government factions, as well as in positively utilizing "groupthink" to tackle adaptation obstacles collaboratively. By creating a more diverse political coalition to govern adaptation through the involvement of local and regional actors, TMNs can help cities broaden the scope of the political discourse, cultivate greater public inclusivity at all stages of the policy process, and negotiate potentially conflicting interests (Provan and Kenis 2008; Anguelovski and Carmin 2011).

The other major procedural innovation is new methods for policy framing, design, and implementation. In particular, tools and resources that TMNs provide to their members to support innovation include opportunities for harnessing policy entrepreneurs and multi-level coalitions, tools for assessing climate adaptation requirements, methodologies for prioritizing adaptation options, and identifying synergistic sectoral policy entry points. For instance, in Rotterdam and Berkeley, decision-makers described resources dispensed by 100RC—such as Platform Partners, 100RC's curated database of resources offered to cities—that have contributed to the disintegration of traditional planning silos that inhibit adaptation efforts (Interview 2016). The new institutional structure of the CROs and their team also contributes to this procedural innovation. In other words, this new construction reinforces innovative methods of policy framing, design, and implementation by effectively internalizing and reproducing drivers of innovation.

However, while there are clear advantages to the CRO position, there are potential uncertainties that merit further exploration. 100RC member cities are provided with 2 years of funding through the Rockefeller Foundation to finance the CRO (Berkowitz 2015). As funding support is time-bound, there is an underlying question of the sustainability and longevity of the position's long-term impact. Furthermore, the hiring of a CRO can raise legitimacy concerns should this high-ranking official be hired externally and being new to a local government because the position is authoritative, requires strong relationships, leadership skills, and the ability to bridge departmental divides.

10.5.3 Implications for the Innovation Cycle

The literature notes that urban climate adaptation innovations can develop when cities are able to internalize and translate TMN resources using a multi-level context within which implementation takes place (Giest and Howlett 2013; Søresen and Torfing 2011; Okereke et al. 2009). TMNs have structurally intertwined themselves within the field of contentious urban adaption politics, as demonstrated with the case of 100RC providing funding to create an entirely new policy entrepreneur position in local government. Building upon Sørensen and Torfing's (2011) cycle of innovation, I further note that processes of innovation are additionally reliant upon intrinsically political variables. For instance, the ability of cities to increase adaptive capacity rests on their ability to channel urban political and economic interests, climate risk awareness, and external resource and capacity support. Therefore, the literature must additionally unpack issues of ideology, conflicting interests, and uneven political power. In this chapter, I consequently argue that the result from the innovation cycle—particularly in the framework of adapting to uncertain climate vulnerabilities and risks—is a succession of procedural innovations that attempt to contain often-conflictual urban economic and political interests.

One inherently political dimension of innovation is the topic of leadership, which in the context of 100RC manifests as the CRO. As the CRO's main objective is to assist local governments in creating more effective collaboration across departmental divides by aligning divergent sectorial interests, confronting competing departmental agendas, and by working across departments to facilitate co-beneficial policies, the CRO's main duty is contentious by nature. However, there are uncertainties regarding the long-term stability of the position once the two years of 100RC funding concludes. An additional concern of the CRO position is rooted in the top-down nature of 100RC. New members hiring a CRO adhere to 100RC's strict member-to-network organization that places rigid guidelines from which the position is created. The lack of flexibility to adapt the CRO's installation to fit the local context additionally contributes to potentially problematic outcomes.

A second implication is that innovative policies, particularly when scaled to the urban level, often face place-based bureaucratic and jurisdictional limitations. Designing and implementing adaptation policy can be difficult as decision-making is constrained by spatial and capacity mismatches, higher levels of government, and the complexity of urban systems (Carmin et al. 2013; Bai et al. 2010). Governance innovations advanced by TMNs materialize at differing scales, at different junctures, and within dynamic political contexts (see Chap. 3: Kemmerzell). In certain situations, TMNs can help facilitate innovation through supporting community involvement and cross-departmental coalitions, as seen in the case of Berkeley. In different circumstances, TMNs can help form coalitions and foster multi-sectoral collaboration to drive new policy processes, such as in Rotterdam. The top-down leadership style of 100RC, however, may constrain potential innovations by limiting organizational flexibility, instrumental for climate adaptation planning across many contexts.

10.6 Conclusion

Cities are continually innovating in an effort to address their emerging climate adaptation challenges (Andonova et al. 2013). Through TMNs such as 100RC, cities are constructing new multi-level governance arrangements in which climate adaptation innovations are pursued (Chu 2016; Bulkeley 2013). As shown in the cases of Rotterdam and Berkeley, multi-level governance approaches are a crucial component of advancement towards desired adaptation policy outcomes, particularly as local governments are required to wade through complex political contexts made up of diverse stakeholders with different interests and resources, while constantly attempting to traverse spheres of authority (Bulkeley and Betsill 2013).

Cities can leverage their TMN relationships to help advance towards more collaborative institutional forms that are more conducive to adaptation innovations compared to traditional siloed forms of governing. Meanwhile, TMNs also need to reflexively reorganize their own governing methods to provide sufficient autonomy and flexibility to suit the individualized needs of their members. TMNs could additionally focus their efforts on reinforcing a multi-level political apparatus that comprehensively guides the work of the CRO. By doing so, the inherently political factors of urban governance can be better planned for and arbitrated at all stages of the innovation cycle and across different political levels. TMNs are predominately focused on developing strategies to achieve outcomes in specified planning areas; but by doing so, they provide cities with tools that help integrate and unify their overall governance structures and individual actors.

Appendix A: List of Interviews

Assistant Director of Planning and Research. Association of Bay Area Governments. 15 December 2016.

Chief Resilience Officer. Department of City Development. City of Rotterdam. 24 August 2016.

Climate Action Program Manager. Energy and Sustainable Development Department. City of Berkeley. 18 August 2016.

Deputy Director of the Environment. DCMR. 9 December 2016.

Environmental Policy Advisor. Department of City Development. City of Rotterdam. 26 August 2016.

Innovation and Change Program Advisor. Department of Engineering. City of Rotterdam.17 October 2016.

Planner, Energy and Sustainable Development Department. City of Berkeley. 24 August 2016.

Resilience Planner. Association of Bay Area Governments. 6 December 2016.

Soil and Ground Water Division Manager. DCMR. 6 December 2016.

Sustainability Program Manager, Energy and Sustainable Development Department. City of Berkeley. 22 August 2016.

Appendix B: List of Public Meetings

Climate Action Plan Drafting Special Meeting. City Council Meeting. City of Berkeley. 9 January 2008.

Climate Action Plan Update & Resilience Strategy Planning Special Meeting. City Council Meeting. City of Berkeley. 9 June 2015.

Resilience Strategy Update Special Meeting. City Council Meeting. City of Berkeley. 3 November 2015.

References

Andonova, L. B. (2013). Boomerangs to partnerships? Explaining state participation in transnational partnerships for sustainability. *Journal of Chemical Information and Modeling, 53*(9), 1689–1699.

Andonova, L. B., Betsill, M. M., & Bulkesley, H. (2013). Transnational climate governance. *Journal of Chemical Information and Modeling, 53*(9), 1689–1699.

Andonova, L. B., & Mitchell, R. B. (2010). The rescaling of global environmental politics. *Annual Review of Environmental Resources, 35*, 255–282.

Anguelovski, I., & Carmin, J. (2011). Something borrowed, everything new: Innovation and institutionalization in urban climate governance. *Current Opinion in Environmental Sustainability, 3*(3), 169–175.

Anguelovski, I., Chu, E., & Carmin, J. (2014). Variations in approaches to urban climate adaptation: Experiences and experimentation from the global south. *Global Environmental Change, 27,* 156–167.

Bäckstrand, K. (2008). Accountability of networked climate governance: The rise of transnational climate partnerships. *Global Environmental Politics, 8*(3), 74–102.

Bai, X., McAllister, R. R. J., Beaty, R. M., & Taylor, B. (2010). Urban policy and governance in a global environment: Complex systems, scale mismatches and public participation. *Current Opinion in Environmental Sustainability, 2*(3), 129–135.

Berkowitz, M. (2015). What a chief resilience officer does. http://www.100resilientcities.org/blog/entry/what-is-a-chief-resilience-officer1#/-_/.

Betsill, M., & Bulkeley, H. (2004). Transnational networks and global environmental governance: The cities for climate protection program. *International Studies Quarterly*, 471–493.

Blanco, I. (2013). Analysing urban governance networks: Bringing regime theory back in. *Environment and Planning C: Government and Policy, 31*(2), 276–291.

Bogason, P., & Musso, J. A. (2006). The democratic prospects of network governance. *American Review of Public Administration, 36*(1), 3–18.

Boyd, E., & Osbahr, H. (2010). Responses to climate change: Exploring organizational learning across internationally networked organizations for development. *Environmental Education Research, 16*(December), 629–643.

Bulkeley, H. (2013). Reconfiguring environmental governance: Towards a politics of scales and networks. *Journal of Chemical Information and Modeling, 53*(9), 1689–1699.

Bulkeley, H., & Betsill, M. M. (2013). Revisiting the urban politics of climate change. *Environmental Politics, 22*(1), 136–154.

Bulkeley, H., & Broto, C. V. (2015). Government by experiment? Global cities and the governing climate change. *Transactions of the Institute of British Geographers, 38*(3), 361–375.

Bulkeley, H., & Tuts, R. (2013). Understanding urban vulnerability, adaptation and resilience in the context of climate change. *Journal of Chemical Information and Modeling, 53*(9), 1689–1699.

Carmin, J., Dodman, D., & Chu, E. (2013). *Urban climate adaptation and leadership: From conceptual understanding to practical action.* OECD Publishing, (26).

Chu, E. (2016). The governance of climate change adaptation through urban policy experiments. *Environmental Policy and Governance, 26*(6), 439–451.

Chu, E., Anguelovski, I., & Carmin, J. (2016). Inclusive approaches to urban climate adaptation planning and implementation in the global south. *Climate Policy, 16*(3), 372–392.

Chu, E., & Schenk, T. (2017). Communicating about climate change with urban populations and decision-makers. In M. C. Nisbet (Ed.), *Oxford research encyclopedia of climate science,* Oxford: Oxford University Press.

City of Berkeley. (2009). Department of planning and development. Climate action plan.

City of Berkeley. (2015). Office of energy and sustainable development. Climate action plan progress update.

City of Berkeley. (2016). Office of energy and sustainable development. Berkeley resilience strategy.

City of Rotterdam. (2009). Rotterdam climate proof. Rotterdam climate proof: The Rotterdam challenge on water and climate adaptation.

City of Rotterdam. (2013). Rotterdam climate proof. Rotterdam climate change adaptation strategy.

City of Rotterdam. (2016). Rotterdam climate initiative. Rotterdam resilience strategy: Ready for the 21st century.

Coen, D., & Thatcher, M. (2008). Network governance and multi-level delegation: European networks of regulatory agencies. *Journal of Public Policy, 28*(1), 49–71.

Dzebo, A., & Stripple, J. (2015). Transnational adaptation governance: An emerging fourth era of adaptation. *Global Environmental Change, 35,* 423–435.

Fünfgeld, H. (2015). Facilitating local climate change adaptation through transnational municipal networks. *Current Opinion in Environmental Sustainability, 12*(February), 67–73.

Giest, S., & Howlett, M. (2013). Comparative climate change governance: Lessons from european transnational municipal network management efforts. *Environmental Policy and Governance, 23*(6), 341–353.
Gordon, D. J. (2013). Between local innovation and global impact: Cities, networks, and the governance of climate change. *Canadian Foreign Policy Journal, 19*(3), 288–307.
Hartley, J., Sørensen, E., & Torfing, J. (2013). Collaborative innovation: A viable alternative to market competition and organizational entrepreneurship. *Public Administration Review, 73*(6), 821–830.
Healey, P. (1998). Building institutional capacity through collaborative approaches to urban planning. *Environment and Planning A, 30*(9), 1531–1546.
Healey, P. (2004). Creativity and urban governance. *Policy Studies, 25*(2), 87–102.
Innes, J. E. (1998). Information in communicative planning. *Journal of the American Planning Association, 64*(1), 52–63.
Jordan, A., & Huitema, D. (2014). Innovations in climate policy: The politics of invention, diffusion, and evaluation. *Environmental Politics,* 4016 (November 2014), 1–20.
Kern, K., & Alber, G. (2009). Governing climate change in cities: Modes of urban climate governance in multi-level systems. *Competitive Cities and Climate Change,* 171–196.
Kern, K., & Bulkeley, H. (2009). Cities, Europeanization and multi-level governance: Governing climate change through transnational municipal networks. *JCMS. Journal of Common Market Studies, 47*(2), 309–332.
Leck, H., & Simon, D. (2012). Fostering multiscalar collaboration and co-operation for effective governance of climate change adaptation. *Urban Studies, 50*(May), 1221–1238.
Lee, Taedong. (2013). Global cities and transnational climate change networks. *Global Environmental Politics, 13*(1), 108–127.
Lipset, S. M., & Altbach, P. G. (1966). Student politics and higher education in the United States. *Comparative Education Review, 10*(2), 320–349.
Massey, E., Biesbroek, R., Huitema, D., & Jordan, A. (2014). Climate policy innovation: The adoption and diffusion of adaptation policies across Europe. *Global Environmental Change, 29,* 434–443.
Measham, T. G., Preston, B. L., Smith, T. F., Brooke, C., Gorddard, R., Withycombe, G., et al. (2011). Adapting to climate change through local municipal planning: Barriers and challenges. *Mitigation and Adaptation Strategies for Global Change, 16*(8), 889–909.
Nyseth, T. (2008). Network governance in contested urban landscapes. *Planning Theory & Practice, 9*(4), 497–514.
Okereke, C., Bulkeley, H., & Schroeder, H. (2009). Conceptualizing climate governance beyond the international regime. *Global Environmental Politics, 9*(1), 58–78.
Pattberg, P., & Stripple, J. (2008). Beyond the public and private divide: Remapping transnational climate governance in the 21st century. *International Environmental Agreements: Politics, Law and Economics, 8*(4), 367–388.
Provan, K. G., & Kenis, P. (2008). Modes of network governance: Structure, management, and effectiveness. *Journal of Public Administration Research and Theory, 18*(2), 229–252.
Sørensen, E., & Torfing, J. (2011). Enhancing collaborative innovation in the public sector. *Administration & Society, 43*(8), 842–868.

Author Biography

Ryan Bellinson is currently an urban studies and planning Ph.D. candidate at the University of Sheffield. He holds a Research Masters degree in Urban Studies from the University of Amsterdam in the Netherlands. His research primarily focuses on the procedural impacts of transnational municipal networks on their member local governments. Prior to his graduate studies, Ryan worked as a land use policy fellow in the Natural Resource Office of former Governor of Oregon, John Kitzhaber, and as a field coordinator in the field office of United States Senator Jeff Merkley (D-OR). He holds a B.S. in environmental studies with minors in urban planning and political science from the University of Oregon.

Chapter 11
Climate Resilient Smart Cities: Opportunities for Innovative Solutions in India

Umamaheshwaran Rajasekar, Soumita Chakraborty and Gopalkrishna Bhat

Abstract India's overall exposure to climate-induced risks—such as cyclones, extreme rainfall, riverine floods, heat waves, and disease outbreaks—is causing large-scale damage to many states as well as unprecedented economic loss in cities. Since the late 2000s, the Government of India has been actively trying to incorporate climate change within the country's development goals. Inclusion of the Climate Change Department within the Ministry of Environment and Forestry is one such initiative. This initiative has had some success in preserving biodiversity, but has contrastingly led to limited actions in cities, which function under the Ministry of Urban Development. In cities, most urban administrators have a short tenure ranging from 1 to 3 years, while political representative such as the Mayor has a maximum term of 4 years, thus making it difficult for them to deal with challenges that are probable 10–20 years in the future. Further, the magnitude of developmental challenges given rapid urban migration has led urban managers to address the immediate basic needs of citizens rather than plan for future eventualities, both in terms of climate change adaptation/mitigation and future population growth. The lack of inclusion within government programs, lack of incentives, increasing exposure, and spending for coping has created opportunities for individuals, communities, and institutions to design and implement resilience measures to address the problems of shocks and stresses that arise from climate extremes. This chapter highlights some of the wicked challenges faced by smart city initiatives in the context of climate change and offers innovative solutions to address them across scales and in time.

Keywords Climate change · Resilience · Smart cities · Urban development · India

---

U. Rajasekar (✉) · S. Chakraborty · G. Bhat
TARU Leading Edge, M6, Aurobindo Marg, Hauz Kaus, New Delhi 110016, India
e-mail: mrajasekar@taru.org

S. Chakraborty
e-mail: schakraborty@taru.co.in

G. Bhat
e-mail: gkbhat@taru.org

S. Hughes et al. (eds.), *Climate Change in Cities*, The Urban Book Series,
https://doi.org/10.1007/978-3-319-65003-6_11

11.1 Introduction

The Smart Cities Mission (MOUD 2015) is the first step towards building infrastructure and services within Indian cities to attract people and investments for urban growth and development. The initiative focuses on area-based development in selected cities, whose good practices and successes can be replicated in other parts of the same city for an overall improvement of the city's "smartness" (MOUD 2015). It involves three types of interventions:

- City improvement (retrofitting of more than 500 acres)
- City renewal (redevelopment of more than 50 acres)
- City extension (green field development in previously vacant or unused land of more than 250 acres).

Pan-city solutions are additional interventions undertaken by the Smart Cities Mission to ensure the inclusiveness of all residents of the city. The Ministry of Housing and Urban Poverty Alleviation referred to the "4S and 4P formula" to make Indian cities smart, which include smart leadership, smart governance, smart technologies, smart people, and public–private people partnerships. Some of the major areas of concern that have been proposed for development and implementation under this program are water supply and sanitation, solid waste management, energy supply, transit system, affordable housing for the poor, robust IT connectivity and digitization, governance, sustainable environment, citizen's safety, health, and education. This chapter analyzes and highlights initiatives undertaken in the water supply, sewerage, waste management, energy efficiency, urban mobility, and land use sectors across the 20 Indian cities selected in first round of the Smart Cities program.

11.2 Smart Initiatives: Correspondence or Discrepancy?

The Smart Cities Mission is funded partly by the central government of India and partly by different state and urban local bodies (ULB). The remaining amount of the proposed cost is acquired from other funding sources, such as through government projects and schemes. All cities have stated their individual proposal costs based on different levels of city improvement, renewable, and extension initiatives. Figure 11.1 illustrates the total cost of projects proposed for the Smart Cities Mission. For example, the per person proposal cost in metropolitan cities is Rs. 1719 in the New Delhi Municipal Corporation (NDMC), Rs. 2940 in Chennai, and Rs. 4105 in Ahmedabad, which are lower than that of smaller cities such as Belagavi (Rs. 72,404), Kakinada (Rs. 61,138), Bhubaneswar (Rs. 53,794), Ludhiana (Rs. 41,818), and Jabalpur (Rs. 37,881). This gap can be attributed to the population difference between cities.

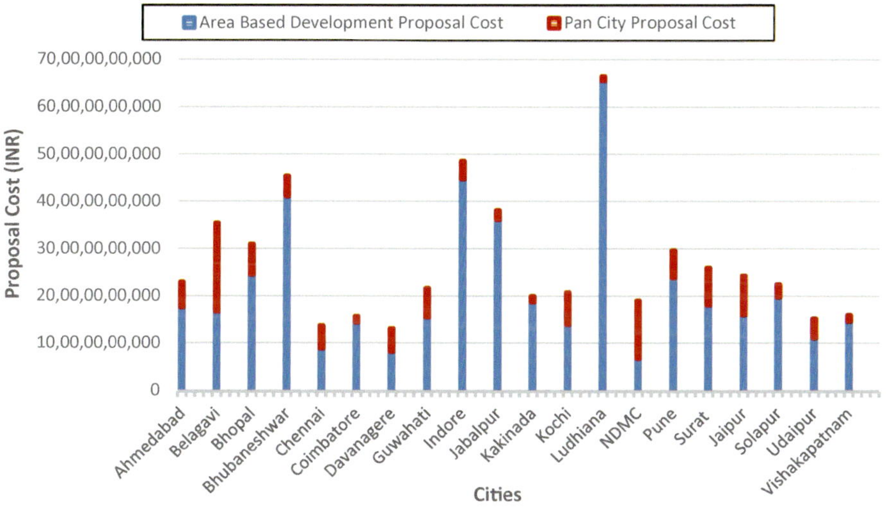

Fig. 11.1 Proposed project costs under Smart Cities Mission (2015). *Source* Authors' own analysis

A comparative study of the proposal costs of the selected cities indicate that a majority of the implementation and investments are focused on area-based development, with a negligible amount of the total being spent on strategic pan-city interventions, such as traffic management, transit-oriented development to increase efficiency, energy efficient street lighting, smart parking, smart water solutions, common card payment system, integrated solid waste management, surveillance camera network, smart unified governance, command and control center establishment, central monitoring, and e-governance. However, Belagavi and NDMC are two exceptions where the proposed initiatives have equal costs for both area-based and pan-city developments. For example, the NDMC proposed Rs. 1228.17 crore[1] worth of initiatives for e-governance, waste management, energy efficiency, smart water solutions, education and health improvements.

11.3 Current Indian Scenario

More than half of the world's population now lives in urban areas. Cities and towns have become the engines for economic growth. By 2030, urban areas are expected to house around 40% of India's population and contribute to 75% of the nation's GDP (MOUD 2015). These ever-increasing numbers exert a maximum amount of pressure on water supply systems, solid waste management systems, energy networks, transportation, and land availability.

[1]A crore is equal to 10,000,000 Indian Rupees (Rs.).

11.3.1 Climate Change

In Indian cities, urbanization, overcrowding, lack of inclusive growth and planning, service failure, difficulty in maintenance of aging infrastructure, the adverse effects of infrastructural expansion, and a growing gap between higher and lower income groups are aggravated by the additional stresses of climate change. Direct impacts of climate change are seen in the form of slow onset, unidirectional phenomenon (e.g., sea level rise), slow onset periodic phenomenon (e.g., drought and heat wave), and sudden, high intensity phenomenon having large-scale effects (e.g., flood and cyclone). All of these are responsible for the disruption of city life and economic growth. Indirect impact includes health consequences and additional stress on the energy systems due to temperature fluctuations (Bhat et al. 2013).

Simulations to understand future monsoon trends in India indicate an increase in rainfall on a seasonal mean and area-average basis.[2] This is due to increasing land–sea thermal contrast and increased warming over the Indian Ocean, which allows more moisture to be carried by southwest monsoons. In spite of different patterns emerging from different climate simulations, an average change of 5–10% in the monsoon remains constant. Indian monsoons have been remarkably stable, so even small variations can have tremendous effects on agricultural production, stocks and commodity markets (Turner n.d.). For example, heavy intermediate showers followed by a week or more break in the monsoon has resulted in flooding, water-logging, and inundation in parts of the country while leading to water shortage and agricultural drought in others. This interannual variability is increasing and will continue to do so in the future.

Trend of climate variability over the last two and half decades is presented in Table 11.1. An analysis of the selected smart cities for the period of 1990–2013 indicates that the number of days with extreme rainfall (>100 mm) increased over time in Belagavi, Bhubaneswar, Kochi, Pune, and Surat. It remained more or less constant in Chennai, Jabalpur, Udaipur, and Visakhapatnam. However, temperature trend portrays a different picture, where majority of the cities are showing a substantial increase in the number of days with temperature soaring above 35 °C. However, exceptional cities like Belagavi, Coimbatore, Davanagere, Guwahati, and Kochi are experiencing either a reduction or consistency in number of extreme temperature days. Temperature trends in these cities also clearly indicate a rise in the average and maximum temperatures over time. This situation worsens in coastal cities like Bhubaneswar, Chennai, Kochi, Surat, and Vishakhapatnam, where high temperature coupled with humid weather increases the level of discomfort for citizens in summer. These cities are also severely affected during monsoons, when an increase in precipitation disrupts normal functioning of the city due to riverine floods, water logging or tidal floods caused by cyclonic depressions. High or rising temperature is an impending threat for landlocked cities like New Delhi, Bhopal,

[2]*Source* Gridded rainfall data as procured from Indian Meteorological Department for the period of 1950–2013.

Table 11.1 Smart cities extreme event scenario (Post 2000 Climate Analysis)

S. No.	Name of state/territory	Smart cities	Extreme rainfall (Days >100 mm rainfall)	Rainfall trend	Extreme temperature (No. of Days >35 °C)	Temperature trend
1	Gujarat	Ahmedabad	Decrease	Increase	Increase	Increase
2	Karnataka	Belagavi	Increase	Increase	Constant	Increase
3	Madhya Pradesh	Bhopal	Decrease	Decrease	Increase	Increase
4	Odisha	Bhubaneswar	Increase	Decrease	Increase	Increase
5	Tamil Nadu	Chennai	Constant	Increase	Increase	Constant
6	Tamil Nadu	Coimbatore	Decrease	Increase	Decrease	Constant
7	Karnataka	Davanagere	Decrease	Increase	Constant	Constant
8	Assam	Guwahati	Decrease	Decrease	Decrease	Increase
9	Madhya Pradesh	Indore	Decrease	Constant	Increase	Increase
10	Madhya Pradesh	Jabalpur	Constant	Constant	Increase	Increase
11	Rajasthan	Jaipur	Decrease	Decrease	Increase	Increase
12	Andhra Pradesh	Kakinada	Decrease	Constant	Increase	Increase
13	Kerala	Kochi	Increase	Constant	Decrease	Constant
14	Punjab	Ludhiana	Decrease	Increase	Increase	Increase
15	New Delhi	New Delhi	Decrease	Decrease	Increase	Constant
16	Maharashtra	Pune	Increase	Constant	Increase	Constant
17	Maharashtra	Solapur	Decrease	Decrease	Increase	Increase
18	Gujarat	Surat	Increase	Increase	Increase	Increase
19	Rajasthan	Udaipur	Constant	Increase	Increase	Increase
20	Andhra Pradesh	Visakhapatnam	Constant	Constant	Increase	Increase

Source Authors' own analysis

Ahmedabad, Indore, Jabalpur, Jaipur, Kakinada, Ludhiana, Solapur, and Udaipur, who experience a continental climate with already extreme climatic conditions in both summers and winters.

India—being a highly vulnerable country to different types of hazards—requires shifts grounded in policy, regulations, fiscal/financial, and institutional instruments in order to adapt to climate change. However, we see a disconnect across Indian metropolises between policies with an objective vision and projects, which are the means of realizing these policies (Alankar 2015). Thus, the ever-increasing resource needs along with a growing population highlight the requirement for preparing Indian cities for unpredictable and uncertain climate impact and to lower anthropogenic contribution of greenhouse gases (Satterthwaite 2011, 2008; Bicknell et al. 2009).

11.4 Water Supply

Increased water supply and productivity require modifications to water economics, formulation of fact-based water policies, development of efficient allocation and investment patterns, inter-stakeholder coherence, prediction of future demand trend, future supply possibility and identification of technical options. Growing competition for water resources poses considerable challenges for related resources (e.g., food security and energy sustainability) and has increased business risk (Addamset et al. 2009).

11.4.1 Current Scenario

The 2011 census data indicates that only 49% of the Indian population has access to piped water supply within their premises. In addition to the quantity of water supplied, intermittent supply and questionable quality are two major areas of concern for this sector. Most cities have the norm of 4–5 hours of water supply in a day, with the average being 2–3 hours (MOUD n.d.). It is worse for the poor who need to spend more time and money to obtain water due to lack of in-house connections (Bhat et al. 2013). Cities like Bhubaneswar and Kochi have continuous water supply for more than 18 h, whereas Davanagere, Indore, Jabalpur, Kakinada, Surat, and Udaipur have intermittent supply of less than 5 h per day. Table 11.2 highlights the status of water supply across Indian cities.

Moreover, contamination in the water supply network due to pipe leakage and mixing of sewer systems are also critical issues demanding attention. While there are several challenges in the service standards, coverage expansion and distribution loss due to decrepit pipes, lack of maintenance and incomplete metering and billing continue to pose problems. Although Ahmedabad, Davanagere, and Indore do not have metered water connection, cities like Udaipur and Kochi have metered more than 70% of the water supply connections. Belagavi, Ahmedabad, Bhopal, Bhubaneswar, Indore, Jabalpur, Kakinada, Kochi, Pune, Solapur, Surat, and

Table 11.2 Service level benchmarking indicators for water supply

Indicator	Unit	Benchmark	Median	Average
Coverage of connections	%	100	53	50.2
Per capita supply	Lpcd	135	69	69.2
Metering of connections	%	100	0	13.3
Non-revenue water	%	20	29	32.9
Continuity of supply	Hours	24	2	3.1
Quality and treatment	%	100	94	81.7

Source Ministry of Urban Development (2010)

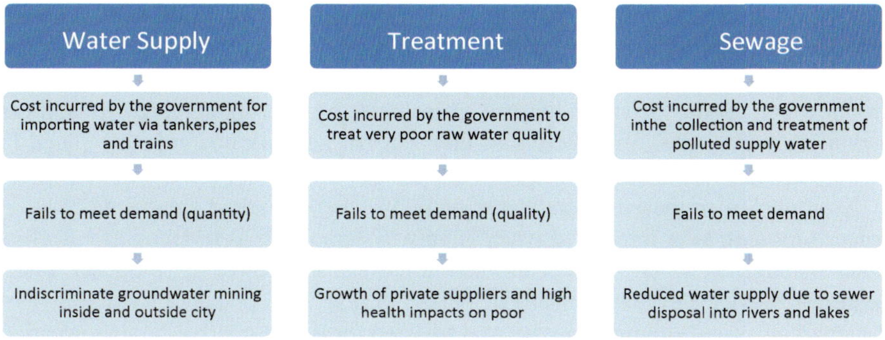

Fig. 11.2 Current urban water supply paradigm

Udaipur have high (more than 90%) quality and treatment of supply water. Cost recovery is more than 80% for cities like Ahmedabad, Kakinada, Solapur, and Surat. The cause and effect of urban water supply and sewage are presented in the Fig. 11.2.

At the city level, the biggest concerns are high distribution loss and non-revenue water. Jawaharlal Nehru National Urban Renewal Mission (JNNURM), a flagship program launched by the Government of India, spent one-half to two-thirds of the funding on expansion and construction of water supply systems and a quarter was used for the augmentation of supply (Wankhade et al. 2014).

Since storm water drainage and sewage systems are two essential components of effective and clean water supply systems, the focus has been to enhance and increase efficiency of both. Household-level coverage of storm water drains is more than 50% in Belagavi, Bhopal, Davanagere, Jabalpur, Pune, Surat and Udaipur, whereas connection efficiency is more than 80% in Bhubaneswar, Kochi, and Surat. More than 70% of Ahmedabad, Indore, Pune, Surat, and Udaipur is covered by a sewage network system, while Bhopal, Davanagere, Indore, Pune, and Surat have adequate sewage treatment capacities in place. The quality of sewerage treatment is 100% in Ahmedabad, Indore, and Udaipur.

11.4.2 Developmental Plans

To ensure a smooth and effective water supply system, most of the cities focused on retrofitting the existing network, whereas pan-city proposals for the same were limited to only a selected few including Ahmedabad, Chennai, Jabalpur, Kochi, NDMC, Pune, Solapur, and Belagavi. The augmentation or rehabilitation of sewage network and existing pumping stations have been proposed and initiated by Chennai, Coimbatore, Davanagere, Guwahati, Indore, NDMC, Pune, and Solapur. Cities like Chennai, Coimbatore, and Vishakhapatnam have addressed this issue while keeping in mind their area specific needs and limitations. While Chennai focused on disposal arrangements, flood management, primary drain encroachment and sensor-based water-level monitoring, Coimbatore proposed an integrated development of the urban lake network to augment water source and recharge groundwater. Some of the initiatives prioritized by the smart cities to address prevailing water crisis, irrespective of their current network coverage are stated in Table 11.3.

Examples of proposed initiatives at pan city level include a smart unified governance looking after integrated water supply system (such as in Bhopal), effective water management, and intelligent water solutions. To realize the proposed interventions, Service Level Improvement Plans (SLIPs) were prepared by majority of the smart cities and submitted to the Atal Mission for Rural to Urban Transformation (AMRUT) Scheme. The Scheme primarily focuses on areas not covered by city sewerage network. Along with AMRUT, many of the smart city water solutions are integrated with national-level schemes and programs such as JNNURM, the Swachh Bharat Mission, and the Storm Water Drainage Scheme (Tables 11.4 and 11.5).

Table 11.3 Water supply interventions proposed by percentage of cities

80%	Retrofitting water supply network for adequate supply, including dual piping system
80%	Provision of water 24×7 with smart meters and e-billing by Municipal Corporations to reduce non-revenue water at community or household level
70%	Integrated water management through rainwater harvesting, productive landscaping and green infrastructure. Development of proper drain channels to help collect and store rainwater to be reused for landscaping irrigation and other purposes
40%	Lake rejuvenation/neutralization along with primary and secondary drain layout in surrounding areas for water cycle restoration to ensure groundwater recharge
35%	Smart water solutions including governance and service provision on demand along with better access through common service delivery outlets to ensure efficiency, transparency and reliability at affordable costs
30%	Inclusion of pressure and flow monitoring system along with special emphasis on customer interactions and cost recovery

Table 11.4 Storm water management interventions proposed by percentage of cities

95%	Increase storm water networks
25%	Integrated storm water management: Sensor-based flow measurements helping to identify flood prone areas, surveillance and alert system to monitor the water level in reservoirs

Table 11.5 Sewerage treatment interventions proposed by percentage of cities

90%	Construction of sewage treatment plants and rehabilitation of old sewer lines. Construction of new ones under Comprehensive Sewerage Scheme wherever available
50%	Waste water treatment at modern plants and their reutilization by industries

11.4.3 Impact of Climate Change

The most threatening and prominent impact of climate change on cities will be water stress caused by extreme events. Rising temperatures can lead to water scarcity, particularly in cities like Jaipur, Kakinada, Belagavi, and Davanagere. Soaring temperature will reduce the level of both groundwater as well as surface water bodies. It will also increase the demand for water due to its usage for a wide range of activities, both domestic as well as commercial. Extreme rainfall, on the other hand, resulting in water logging or urban floods, riverine floods and tidal upsurge in coastal cities, can lead to overflowing of municipal sewer drains, thus contaminating the drinking water supply network with untreated sewage and storm water. Waterlogging in specific pockets coupled with high humidity—particularly in riverine and coastal cities like Visakhapatnam, Chennai, Kochi, and Surat—can spread water and vector-borne diseases. It will increase water pollution and contaminate both ground as well as surface water. Coastal cities using freshwater aquifers to supply drinking water will need to increase investment for desalination in case of storm surges. This, in turn, can increase the need for water purification and enhancement of drinking and sewerage treatment capacity. Such additional infrastructural requirement will inevitably raise the cost of supplying water. Moreover, extreme precipitation can damage or render water supply, sewerage, as well as storm water drainage network useless.

Recommendations:

- Improve water use efficiency through increase in water storage capacity, recycling, and employing rainwater harvesting techniques.
- Undertake infrastructural modification to include revamping of underground pipes and networks for separation of rainwater drains from sewerage and water supply network.
- Develop watershed approach for micro planning wherever possible.
- Desalinate seawater in coastal cities.
- Expand use of economic incentives such as water metering and pricing to encourage water conservation.

11.5 Waste Management

Responsibility of collecting, processing, transporting and disposing the waste generated by cities in India lies with the Urban Local Bodies (ULBs). A majority of cities are struggling to provide efficient waste management services, with most facing low waste collection coverage and efficiency, widespread collection in unsafe and unsanitary manner (such as via open trucks), limited waste recovery, processing and open dump disposal of the wastes for leachate treatment (HPEC 2011). Such inefficient disposal of waste has long-term effects on the health of the citizens, environment, and quality of ground and surface water, which also facilitates disease vectors like rats, flies, and mosquitoes. Moreover, open burning of waste and landfill fires contribute to about 20% of the pollution due to particulate matter, carbon monoxide, and hydrocarbons.

11.5.1 Current Scenario

Most Municipal Corporations face community protests when they intend to setup waste management facilities within or near the city. Municipal officials repeatedly stress the issue of scarcity of land for waste disposal, which leads to overflowing dump sites and waste treatment facilities receiving more waste than what they were designed for (Clean India Journal 2013).

Municipal Solid Waste (Management and Handling) Rule 2000 mandates the collection, segregation, storage, transportation, processing, and disposal of municipal waste. However, most Indian ULBs find it difficult to comply with and implement these rules at the city level, due to lack of required scientific, technical, managerial and financial structures in place. Others lack the willingness or awareness to engage with the private sector and concerned officials to address solid waste management challenges. To facilitate this process, between 2006 and 2009, JNNURM funded 49 solid waste management (SWM) projects across various cities (MOUD n.d.). Ahmedabad, Belagavi, Bhubaneswar, Surat, and Davanagere have 70% household level coverage for solid waste collection, with their efficiency exceeding 80%. Moreover, Ahmedabad, Belagavi, and Pune have 100% scientific disposal of the collected solid waste. Solid waste management is also a key initiative under the National Mission on Sustainable Habitat (Government of India 2010) and Swachh Bharat Mission (2014) (NIUA 2015).

11.5.2 Developmental Plans

In line with the Municipal Solid Waste Rule 2000, the Smart Cities Mission proposed interventions under the Swachh Bharat Mission to enhance solid waste collection coverage, transportation, and processing. Some cities like Ahmedabad, Chennai, Coimbatore, Kochi, NDMC, Belagavi, Ludhiana, Pune, Surat, Udaipur, and Vishakhapatnam already have a high collection coverage ranging from 90 to 100%. However, the segregation and treatment of collected waste is currently being carried out only by Belagavi, Bhopal, Chennai, Kochi, and Solapur. Majority of cities—with the exception of Bhopal, Bhubaneswar, Coimbatore, and Guwahati—have proposed integrated solid waste management initiatives through retrofitting existing structures that are already in place. Surat, who is undertaking redevelopment of 1406 acres of land to ensure high quality of life, proposed scientific treatment through the construction of a waste to energy (RDF) and bio-methanation

Table 11.6 Waste management interventions proposed by percentage of cities

50%	Undertaking information, education, and communication (IEC) campaigns to create public awareness regarding solid waste disposal
45%	Providing infrastructure for 100% door to door collection
40%	Area-Based Development (ABD) to integrate public conveniences like toilets with retail facilities and public spaces to ensure public areas are free of open defecation. This maintenance is to be entrusted with private retailers. The Central Control Room would release Collection Vans at optimal intervals accordingly
35%	Waste segregation at source, use of weight sensors, and innovative waste processing techniques
20%	Geographical Positioning System (GPS) installation in vehicles to monitor collection
20%	Following the 4R (reduce, reuse, recycle, respond/refuse) principle at building level
5%	A pneumatic waste collection system (including sensor-based waste bins) which would negate the need for door to door collection
5%	Improve service through a location based grievance system

plant. Cities like Chennai, on the other hand, focused on the usage of compost for green rooftop and radio frequency identification system (RFID) to improve efficiency, economy, and traceability of waste collection and transportation. Some of the initiatives prioritized by the cities to monitor and improve solid waste network and infrastructure are listed in Table 11.6.

11.5.3 Impacts of Climate Change

Although solid waste treatment is one of the primary contributing factors to climate change, the rate of decomposition and treatment, in turn, is impacted by changing temperature and rainfall, induced by the former. However, adaptation strategies to reduce potential negative impacts of the latter on treatment facilities and landfill sites have received much less attention. Climate change impacts urban waste management systems primarily through the disruption of supporting infrastructure such as roads and railways used for waste collection and transportation. Also, an increase in the number of days with extreme temperature can alter biological composting (anaerobic digestion) at the landfill site, site hydrology, and leachate production. Increasing temperature also increases vermin and pests in surrounding areas thus making solid waste management sites the producer and transmitter of vector-borne diseases. Extreme precipitation, on the other hand, will inundate waste management facilities and contaminate groundwater and surrounding surface water bodies. A sudden increase in rainfall can also alter decomposition rates and site hydrology. Storm surge in coastal cities like Vishakhapatnam, Bhubaneswar, and Kochi can cause potential damage to buildings.

Recommendations:

- Assess and examine waste management facilities and processes that are more vulnerable to climate change and extreme events, such as low-lying coastal sites.
- Awareness generation among waste collection and disposal authorities on issues arising during and in the aftermath of extreme events in the region, which can influence service delivery responsibilities and contractual commitment.

11.6 Energy Efficiency

As a prerequisite for economic development, the energy sector has witnessed rapid growth in India. However, to keep pace with this growing population, there is a need to augment energy generation and supply in the country (TERI n.d.).

11.6.1 Current Scenario

Although around 92% of the population have access to electricity (2011 Census), energy provision in Indian cities is characterized by frequent power cuts. According to studies, by 2030, energy usage in urban India itself will increase four to six times due to space cooling requirements. Thus, critical issues plaguing these cities are frequent power cuts and low quality electricity supply (Bhat et al. 2013).

The massive renewable energy potential of the country can form the backbone of energy efficient India. However, of all the untapped renewable energy sources in the country, installed capacity has only reached 12.5%. Thus, in spite of being one of the largest energy consumers, national energy shortage and inadequate energy infrastructure have pushed India to perpetuate energy poverty. Summer heat, frequent labor strikes, and natural disasters have made it harder to meet the market demand of coal, which remains the country's staple energy source. This can also lead to increases in coal imports. Falling production of natural gas within the country, excess reliance on import from other nations, and their failure due to geopolitical issues are critical areas of concern in India (TERI n.d.). To meet India's growing demand, increase production, profit and overall economic growth of the industrial sector, improvement of infrastructure, and other initiatives have been undertaken to address energy poverty, logistical problems, and effective implementation (Durns 2014).

11.6.2 Developmental Plans

Being endowed with only 1% of the oil and natural gas reserves of the world, India is faced with the need for energy security. Hence, energy efficiency and sustainable resources are two main focus areas of smart initiatives for Indian cities. Ahmedabad, Belagavi, Bhubaneswar, Chennai, Coimbatore, Davanagere, Guwahati, Indore, NDMC, Pune, Surat, and Vishakhapatnam have identified solar, wind, hydro, and biomass (based on their respective geographical location) as renewable energies which can be harnessed in the near future. Some of the measures proposed by smart cities for energy efficiency have been listed in Table 11.7.

The need for such measures arises due to frequent outages, including in Ahmedabad, Belagavi, Bhubaneswar, and Jaipur. However, unscheduled shut downs have been reduced in Indore, Coimbatore, Jabalpur, Ludhiana, Udaipur, and Vishakhapatnam. Chennai, NDMC, and Pune, on the other hand, no longer experience scheduled outages. Retrofitting and redevelopment interventions identified by cities essentially aim to strengthen the power supply network, smart metering, better access to Pipelined Natural Gas (PNG) and Compressed Natural Gas (CNG) through infrastructural development, and increase in energy generation through renewable sources such as roof top solar panels. Initiatives for augmenting and improving the energy supply network throughout has been proposed by only NDMC and Belagavi. NDMC focused on power management, demand forecasting (including peak load management), outage management system, substation automation, asset monitoring, short-circuit analysis, and relay coordination across the metropolitan area.

To realize the mentioned interventions, most smart cities proposed convergence with national and state-level schemes such as the Integrated Power Development Scheme (IPDS) fund for strengthening electricity transmission and distribution networks, the National Solar Mission and Energy Efficiency Services Ltd. (EFSL) for conversion of street and domestic lighting into LED, MNRE Solar City Mission, and the Jawaharlal Nehru National Solar Mission (JNNSM).

Table 11.7 Interventions to enhance energy efficiency proposed by percentage of cities

100%	Assured electric supply with a sizeable portion generated by renewable sources
85%	Installation of solar panels on all government and public buildings
70%	Supervisory control and data acquisition (SCADA) for maintenance and management activities
60%	Rehabilitation and strengthening of transmission and distribution network in urban areas through underground cabling, elimination of overhead network, and use of smart meters and sensors
50%	100% energy efficiency using green building concept and through implementation of Net Metering Policy
30%	Retrofitting and redevelopment of Compressed Natural Gas (CNG) and Pipelined Natural Gas (PNG) distribution lines

11.6.3 Impact of Climate Change

Changes in temperature, precipitation, sea/river/lake water level, and frequency or severity of extreme events impact the generation, delivery, and consumption of energy. Increases in urban temperatures will increase energy demand for cooling, with the peak demand in summer months of May and June requiring additional investment for new energy generation and distribution infrastructure. Power plants will need more water for cooling, but will face shortages due to rising temperature. Moreover, energy intensive methods such as pumping, irrigation or water distribution over longer distances—particularly across drier parts of the country such as Belagavi, Jaipur, Davanagere, and Solapur—will increase both demand and costs. Hydroelectric power plants that are sensitive to volume and timing of stream flows will face problems during heavy rainfall. Extreme precipitation will also lead to increased risks of energy supply disruption particularly in coastal cities where energy supply facilities can be uprooted by storm surge. Urban, riverine, and coastal flooding, in turn, can damage electricity distribution lines and equipment. It can also delay the repair and maintenance procedure of the same for reestablishing normal energy supply system.

Recommendations:

- Reinforce walls and roofs, construct dikes, and undertake measures for structural improvement of transmission assets.
- Deploy energy efficient technologies in end-use facilities, power generation, transmission, and transmission. It will help cope with increasing demand with rise in temperature.
- Undertake demand response programs aimed at peak loads in summer season to counteract the increase in peak demand and avoid additional power plant installations.
- Increase the use of cool roof technology to reduce ambient temperature.
- Undertake water efficiency programs to address climate impacts on water resources and reduce energy use for pumping and treating water.

11.7 Urban Mobility

India has witnessed a rapid increase in vehicle ownership over the years, with it being higher for four-wheeler ownership (Schafer and Victor 2000). An increase in motorization resulted in rising greenhouse gas emissions, local air pollution, congestion, and noise. The success of a city depends on the ease of business, comfort of life, and reduced cost in connecting people through agglomeration, productivity, efficiency, and basic amenities, all of which are directly or indirectly affected by transportation system and technology (Mittal and Singh 2015).

11.7.1 Current Scenario

Policies such as the National Highway Development Program, National Urban Renewal Mission, and National Urban Transport Policy have led to construction of flyovers and widening of roads. Though such efforts initially improved traffic, in the long run, it increased latent demand for vehicles (Mittal and Singh 2015). Inadequate and inappropriate public transport systems, unscientific route rationalization, and slow progress of construction of metro rails have increased congestion in many large Indian cities. All these problems can be attributed to a disproportionate allocation of funds for road infrastructure, which could otherwise be met by a cost-effective public transit system (Padam and Singh n.d.). City planners have allotted funds for road development but left urban mobility to be determined by market forces. This increased the use of private vehicles, which are convenient, high status, and time saving (Kahn 2014). More private vehicles, mainly cars, result in more space consumed inefficiently, which in turn increases road congestion (Glaeser 2011).

11.7.2 Developmental Plans

Public transportation systems have a mere 50% coverage in cities like Coimbatore, Guwahati, Indore, and Jaipur. Currently, a Bus Rapid Transit System (BRTS) is operational only in Ahmedabad, Bhopal, Bhubaneswar, Indore, Ludhiana, Surat, Jaipur, and Vishakhapatnam. Smart city proposals highlight the need to ensure decongestion, manage lanes, increase convenience for private vehicle commuters, identify smart parking solutions, focus on road safety, reinforce traffic regulations, and develop a fleet management system, passenger information system, and common electronic payment mechanism. Coimbatore proposed the development of a 30 km eco-mobility corridor linking vantage points with dedicated walking or biking tracks, with the possibility of conversion of select streets into non-motorized transit streets. Bhubaneswar, on the other hand, having less than 75% of area covered by public transport and poor supporting infrastructure, proposed first and last mile connectivity through the implementation of a pilot public bus system scheme, e-rickshaws, city-wide pink auto service for children's safety, and the redevelopment of 12 acres at the railway terminal. Similarly, to reduce pollution and congestion and improve health and safety, Ludhiana proposed the operationalization of GPS enabled smart e-rickshaws over the next 5 years.

At the city level, Ahmedabad identified initiatives for an Intermodal Transportation System-based development, along with installation of Intelligent Red Light Violation System, Area Traffic Control System and Variable Message Signs at key junctions. Some of the key interventions prioritized by the Smart Cities initiative are to ensure eco-friendly mobility, reduce road congestion, and better access, connectivity, and reduced travel time (see Table 11.8).

Table 11.8 Interventions to enhance urban mobility proposed by percentage of cities

75%	Integrated road network redesigned for seamless mobility. Improved walkability and mobility of the area through pedestrian walkways, cycle tracks, central and city bus terminals, flyovers, underpass, and improved road junctions, enhanced bus connectivity, and efficiency through provision of new bus routes, and additional bus stops and redevelopment of existing old bus terminal into a state-of-the-art central transit hub
75%	Decongestion of roads in planned manner and ensuring compliance. Integrated City Operations and Management Center for area-based traffic control, traffic flow monitoring, video surveillance, incident management, video analytics supported traffic violation detection, and e-*challan*
65%	Smart parking, nonvehicle streets, and zones
45%	Increased use of public transport through provision of crucial last mile connectivity, thereby expanding the catchment area of the region's transit system

However, Ahmedabad, Belagavi, Chennai, Davanagere, Indore, Jabalpur, Kakinada, Kochi, Ludhiana, NDMC, Pune, Jaipur, Solapur, Udaipur, and Vishakhapatnam are yet to realize the need for smart parking (on and off street). Ahmedabad, Belagavi, Bhubaneswar, Coimbatore, Guwahati, and Udaipur also need to check pollution along with improving the transit system. In addition to area-based development, city solutions for smart traffic and integrated transit management system have been proposed by Ahmedabad, Belagavi, Bhubaneswar, Chennai, Coimbatore, Davanagere, Guwahati, Indore, Ludhiana, Kakinada, Pune, Surat, Jaipur, and Udaipur. Intelligent Transit System including non-motorized transport (cycle tracks, safe route to schools, pedestrian facilities, smart parking, and bike sharing facilities) are planned under convergence with AMRUT and other state government funds.

11.7.3 Impact of Climate Change

Urban mobility and climate change are closely connected. Increases in number of days with extreme temperature in cities like Ahmedabad, Jaipur, and Chennai will increase heat exposure and hence damage roads, bridges, and diesel engines. Soaring temperature will also slow down and hinder transport service provision as well as urban mobility. Extreme rainfall, on the other hand, often results in urban flooding, particularly in the riverine and coastal cities such as Chennai, Kochi, Pune, Surat, and Visakhapatnam. Waterlogging not only threatens railways by washing out railway beds, but it can also inundate urban road networks. Structural damages will be incurred in the form of potholes, which will subsequently disrupt transport in the event of extreme rainfall. It can also undermine support structures such as bridge footings. Storm surges in coastal cities can cause physical damages to road and rail networks.

Recommendations:

- Use more heat resistant materials such as asphalt for roads and heat tolerant metals for railway tracks.
- Coastal, lake, and riverine cities like Ahmedabad, Bhopal, Bhubaneswar, Chennai, Guwahati, Indore, Jabalpur, Surat, and Visakhapatnam need to use remote sensing and GIS to detect damaging water levels and trends which can help develop better management of drainage systems.
- Undertake smart growth approaches that stress high urban density with mixed land use close to public transport nodes and walkable communities. These strategies will help obtain less car-use for short trips, more transit efficiency, and reduction in greenhouse gas emissions.

11.8 Land Use

In India, the rate of land acquisition has been increasing in order to ensure development and cater to the demands of the increasing population. Shortage of housing in urban areas reached 26.53 million dwelling units during the period of 2007–2012 (MOUD n.d.).

11.8.1 Current Scenario

The land use scenario in India is characterized by inadequate housing, overcrowding, urban sprawl, and squatter settlements. These are the output of unmet demand for housing, attributable to inadequate or improper land use and lack of inclusive urban planning. Whereas there are numerous middle and high-income residential units lying vacant or unsold, it is the poor who are often pushed towards informal solutions due to unaffordable housing costs (Bhat et al. 2013). Therefore, uncontrolled land prices, inadequate infrastructure, and high cost of extending basic services are some of the pressing issues faced by Indian cities. They have direct impact on the environment and health of the population.

11.8.2 Developmental Plans

According to many critics, land is the primary face of smart urbanism in India. The proposals recommend extraction of surplus value from unproductive land and transform it into real estate to deal with the land crisis in urban India. For example,

Ahmedabad proposed redevelopment of the existing Wadaj slum, resulting in efficient infrastructure and housing provision free of cost. This includes rehabilitation of slum dwellers to multi-storied buildings developed at the same location as that of the existing slums (under *Mukhya Mantri Gruh Yojana* program) and improvements in housing conditions of these residents. This proposal includes the provision for establishing well-defined and compact urban form with smart features across the city. Bhopal intends to provide 20% affordable houses with a mix of rental public housing for low-income groups and inexpensive studios for sale. Bhubaneswar, on the other hand, proposed the redevelopment of 40 acres of underutilized government land into institutional space promoting commerce, education, and business, along with providing 6000 affordable houses. Ahmedabad, Bhubaneswar, Coimbatore, Indore, Kakinada, Kochi, Pune, Surat, Visakhapatnam, Davanagere, Jabalpur, Kakinada, and Jaipur have been retrofitting existing infrastructure to ensure affordable housing, development of intermodal hub, and densification of existing residential and commercial areas. Some of the key interventions identified by the smart Indian cities have been listed (Table 11.9).

The proposed interventions are likely to be implemented in convergence with *Pradhan Mantri Awas Yojana* (PMAY) under the "Housing-for-All" mission. This primarily focuses on construction of houses for a slum-free city. However, land crises will be addressed only in specific pockets of these cities, thus citywide interventions need to be undertaken in future. For example, in relation to the growing urban population and eviction of informal residents, there is a need to evaluate the public–private partnerships proposed for developing sound social housing policies. When interventions are driven by private actors instead of social welfare organizations, it is usually the urban poor who suffer due to relocation to remote areas isolated from livelihood, school, health care, and community networks. Moreover, the issue of land crisis in India can only be solved after addressing national-level agrarian reforms, land rights, and rural development. In the absence of the latter, unskilled migratory population will continue to increase and the newly developed affordable housings in selected areas of the smart cities will fail to accommodate them.

Table 11.9 Interventions for land use efficiency proposed by percentage of cities

50%	Affordable housing
50%	Mixed use and compactness of land. Plan encourages a robust and effective utilization of available land for environmental regeneration, development of intermodal hub, improved connectivity and residential development to attract new residents
30%	Revitalizing economic precincts in respective cities, transforming the area into a vibrant economic hub
30%	Institutional core promoting commerce, business, and education
15%	Delineate priority commercial streets in Local Development Plan
5%	Redevelopment of public land to create shared public open spaces (both neighborhood and sector level)

11.8.3 Impact of Climate Change

Recent climate change analyses show temperatures frequently rising above 40 °C in Ahmedabad, Bhopal, Chennai, Indore, Jaipur, Jabalpur, Kakinada, Ludhiana, NDMC, Solapur, Surat, Udaipur, and Vishakhapatnam. Such extreme temperature events will lead to outbreak of invasive species and diseases, which in turn can damage or degrade the micro ecosystems, particularly those having a coastal, riverine, or lake ecosystem in the vicinity. However, changes in land use and land cover pattern also alter the local climate. Urbanization is a major causal factor of climate change, with local environmental condition of cities being warmer due to increased heat release in a confined area. This over time has and will continue to increase the number of urban heat islands adversely affecting the lifeline services and facilities.

Recommendations:

- Increase green space to counteract local warming effects through increases in carbon sinks.
- Vegetation and soil information needs to be monitored and made accessible to city planners. Instead of type of soil, land use and land cover need to be assessed based on soil density, conductivity, and heat capacity.

11.9 Discussion

11.9.1 Key Urban Trends and Challenges

The largest challenge is the increasing role of real estate developers who are circumventing planning bylaws especially within new developmental areas. This has resulted in peri-urban areas undergoing ad hoc development. Increasing real estate prices have resulted in the transition from low-rise high-density settlements to high-rise high-density settlements. In cities, old low-rise buildings are bought and converted into multi-storey flats. These are points of high demand for water, energy, and road infrastructure, which often are not well integrated within city plans. As a result, water and energy demand increases beyond the planned demand. Water scarcities and power outages become common due to increasing demands, as well as increasing pressures on sewerage and waste disposal systems.

The rapid rate of urban growth often hides the financial poverty of municipalities. Local governments are unwilling to increase tariffs for urban services while users facing scarcities are willing to pay for better services. Because of financial constraints, municipalities are forced to rely on national flagship programs such as the Smart Cities Mission (MOUD 2015) or the Jawaharlal Nehru National Urban Renewal Mission (2005–2014). These programs come with attached generic policies and guidelines, which often do not address local contexts or are not informed by the lessons learned from previous experiences.

Before the 74th Amendment, municipalities were mandated to manage basic services. Capital expenses and maintenance of some functions were done by state departments, while most municipalities have shortage of staff—especially in the middle and upper managerial levels—and they rely on administrative leaders for decision-making. Modern tools such as GIS are necessary to manage large cities, but are mostly lacking. City administrations are mandated to act on social development—especially focused on poor and vulnerable sections—but very few cities have professionals with social development backgrounds. Instead, most municipalities are saddled with a huge burden of semi-skilled workers, with very few managers and planners.

Smart city management and implementation require interdisciplinary knowledge and skills. The current plans also lack an anticipatory culture such as forecasting future population growth and options for introducing disruptive planning innovations. The status of lifeline services within many cities are decrepit and will require a considerable amount of human resources to bring change. A history of capacity building within these organizations has shown that most of these efforts have yielded less than expected results. Since most of the senior staff are busy addressing day-to-day problems, they are often unable to engage in short to medium term capacity building programs.

As the population across India is increasing, natural resources (especially land and water) are facing major competition between the traditional users and the cities. Major interstate and rural–urban water conflicts are disrupting city water resources, but city planners continue to extract and plan augmentation projects based on these contested resources. Climate change related uncertainties in water resources can further amplify the scarcities. Unfortunately, with limited knowledge and a lack of anticipatory culture, Indian cities continue to plan and manage urban systems based on old planning and management paradigms. For example, urban development authorities in India still practice land use or land cover based approaches to planning. Such approaches work well in low-rise lowdensity scenarios but do not address the complexity of high-rise and high-density environments where network-based planning plays a key role in the design of horizontal and vertical systems (both below and above the ground).

Since cities have major knowledge gaps and face financial constraints, they are forced to resort to knee-jerk measures and only work from day to day. These constrain their capacity to plan, especially in the world of limited per capita resources, growing uncertainties, and the lack of innovations that can transcend current infrastructural deficiencies to cater to the need of citizens. To plan and manage a city of the future, anticipatory culture—informed by multidisciplinary inputs—is essential.

11.9.2 Solutions for Building Resilience

Indian cities have inherited many planning paradigms developed by and for European and American cities. A variety of land use-based planning approaches was experimented in India. In 1960s, low-rise high-density settlement planning and single-use zoning was common. In the context of mobility and availability of land, these frameworks have created more problems that the Smart Cities Mission (MOUD 2015) is expected to solve without altering the planning process. In particular, resilient solutions in Indian cities face a number of key disruptors.

First, cities must ensure network integrity across seasons and emergencies. Network informed master plans are needed to ensure the availability of different lifeline services, even with changing land uses. Transit oriented development is one of such efforts. Major disruptions are expected in the mobility sector with increasing penetration of electrical and hybrid cars, taxi aggregators, and self-driving cars. These can reduce the number of vehicles required to provide sufficient mobility across the city, and can potentially result in many flyovers and other road infrastructure redundant.

Second, water recycling can be a disruptor, which can reduce dependency on the external sources and provide a certain degree of autonomy to neighborhoods. Combined with rainwater harvesting and groundwater management, one can hope to increase cellular autonomy at neighborhood levels.

Third, solar energy and battery technology are two major disruptors, which can reduce the dependency on regional grid-based electricity supply. Over the next decade, it is expected that the local electricity generation from solar energy may be cheaper than the transmission and distribution costs of regional grid-based electricity. Mixed grids of electricity will be able to power the cities. Major advances in energy efficiency of devices and appliances can further reduce the dependency of regional grids, which can threaten the financial viability of energy utilities.

In response, innovations that target the aforementioned disruptions (i.e., disruptive innovations) in water, energy, food, and transportation systems can provide opportunities for increasing the autonomy of communities in meeting a significant part of their lifeline resources and service needs. Despite the high-density of urban settlements, local natural resources can meet a significant part of the demand by disruptive technologies such as water recycling and solar energy. As we are transforming into an intelligent and knowledge-based society, there are huge potentials for transforming from the daily commuting work styles to "work from home" lifestyles. Also, the forthcoming disruptions in the mobility sector can potentially reduce dependencies on a large number of vehicles, reduce the area under roads and transportation infrastructure, and create greener and more self-sufficient communities.

Further, cellular autonomy can create green neighborhoods and improve the quality of life. A judicious mix of non-motorized transport in local areas and faster transportation corridors can create "rurban" lifestyles for the citizens. A significant proportion of water, food, and energy can be produced locally, which can create larger number of green jobs. Urban local bodies in India must devolve some of the functions to these cells/communities, such as managing local water bodies, energy, green areas, and local non-motorized transport systems. The principle of subsidiarity can be effectively used to identify the devolution subjects and empowerment of the cells.

Indian cities must also integrate water, energy, and food (biomass) cycles based on principles of urban metabolism in order to limit carbon, energy and water footprints. Urban planners need to be taught to integrate multiple sectors and urban local bodies—rather than individual departments—should develop mechanisms to manage urban metabolism. As the availability of per capita resources may reduce in the future and climate change is likely to increase their uncertainty, we must learn to achieve synergy across sectors to increase the efficiency of the urban systems. We need an anticipatory culture to understand these complex issues unfolding into the future and learn to live in the age of limited resources and uncertainties.

The disruptive innovations described here have the potential to convert current consumers into "prosumers" (consumer who also is a producer) at household and neighborhood levels. When these disruptive innovations are grounded, we need to build synergy across different scales ranging from household to cities and regions. This will ensure that actions at one scale will not create dysfunctionality at lower or higher scales. For example, neighborhood-level water recycling can build up nutrients in groundwater, which needs to be flushed out during monsoons through drainage systems at city and regional levels. Similarly, excessive use of solar and the biomass-based energy can result in solar energy sprawls, which can reduce food-growing areas. Therefore, it is necessary to connect with the lower and higher scales so that there is coherence and synergy of actions across scales (see also Chaps. 2 and 3).

Urban systems become more complex as they grow and draw natural and human resources, while also creating large amounts of products and wastes. So far, Indian cities have been externalizing the responsibilities of managing land, water, energy, and transportation to municipalities and utilities. As a result, cities have been seeking bureaucratic solutions to complex urban challenges. Unless there is synergy across stakeholders, the best intentions of one set of stakeholders cannot make changes at the city scale. Further, we are witnessing the shift of control on urban development from municipal bodies to real estate and other private sector players. Currently the total investments on urban development by the private sector is much higher than the municipal and national government program investments. Since smart city interventions address predominantly area-based measures, such innovations can act as demonstrators for medium to large level projects coming up in peri-urban areas.

11.10 Conclusion

Climate change poses serious threat to infrastructure in cities, quality of life, and the entire urban system. Extreme events can also potentially lead to social unrest and stresses due to shortage and price hikes, mass migration, unemployment, disrupted food distribution, energy provision, water supply, and waste removal. Thus, smart city initiatives need to build on current experiences, with municipal capacity and management services undertaking measures to ensure effective support systems that can aid in recovering from extreme events and slow onset challenges such as high temperature, increased precipitation, changing local weather patterns, and lack of water.

However, in the context of the Smart Cities Mission (2015), there are a number of critical governance challenges facing cities in India. In particular, urban local bodies manage services through multiple departments, with limited coherence between them. For example, water supply, sewerage, and drainage are managed by separate departments and/or organizations. Parastatals such as water supply and sewerage boards manage the part of the water cycle while drainage is managed by municipalities in many cities. As a result of the fragmentation and overlapping roles between different departments and organizations, cities are unable to adequately manage natural resources such as water, and end up importing water from distance sources, even though wasting rainwater and sewage creates more problems for downstream communities.

With the various governance challenges and opportunities (i.e., in the form of disruptive innovations) in mind, this chapter analyzed the interventions proposed by the 20 cities selected as candidate cities in the first round of the Smart Cities Mission (2015). While some targeted slum redevelopment and development of underutilized government land, others stressed decentralizing their inner-city areas to create commercial corridors and new growth centers with mixed land use, or strategized to ensure sustainable economic development and high quality of life using smart interventions. Cities dependent on tourism, on the other hand, proposed enhanced tourism potential and local ecosystem, among other interventions.

Some of the identified shortcomings of the Smart Cities Mission are listed below.

- Limited clarity on private–public partnership and lack of interdepartmental coordination risks the entire implementation mechanism;
- Prioritization of initiatives from different departments that are currently planned under different plans can jeopardize the proposals;
- Local political opposition and difficulty in removing encroachments, hawkers, and vendors for infrastructure and service development or expansion;
- No indicators for assessment of success or failure of the smart interventions at the city level;
- Lack of a well-structured and detailed implementation plan. In absence of it, the selected smart cities may become centers where corporations sell their products, or even worse, acquire land with the help of government;
- In spite of the current global climate change scenarios and increasing climate change awareness, other than a few cities proposing mitigation measures to a selection of climate-induced disasters, broad climate change action has not been addressed by any of the future smart cities.

For proper implementation of the smart city interventions, there needs to be a holistic approach for timely and orderly collaboration of Special Purpose Vehicle (SPV) with the municipal corporations of the respective cities. Inter-departmental cooperation, coordination, and compliance with SPV is critical for implementation of the projects through infrastructure sharing, knowledge sharing, training, and capacity building. In cases of noncompliance, there is a need for the SPV to improve their negotiation skills through the provision of better incentives.

In sum, there is tremendous potential for creating constructive urban development pathways by engaging with multiple stakeholders. We need to create platforms for engagement and identify the rules for collective action through building synergy across the stakeholders. We must bridge interests between stakeholders, develop common visions, and develop individual or collective roadmaps for realizing smart, sustainable, and climate resilient cities.

References

Addams, L., Boccaletti, G., Kerlin, M., & Stuchtey. M. (2009). Charting our water future: Economic frameworks to inform decision-making. New York: McKinsey & Company.

Alankar. (2015). India's Megacities and climate change: Explorations from Delhi and Mumbai, STEPS Working Paper 79, Brighton: STEPS Centre.

Bhat, G. K., Raghupathi, Usha, Rajasekar, Umamaheshwaran, & Karanth, Anup. (2013). *Urbanization-poverty-climate change (Synthesis Report)*. Gurgaon, India: TARU and NIUA.

Bicknell, J., Dodman, D., & Satterthwaite, D. (2009). *Adapting cities to climate change: Understanding and addressing the development challenges'*. London: Earthscan.

Clean India Journal. (2013, January 18). Indian Crisis Scenario—A Report. Retrieved on 1st August, 2016 from http://www.cleanindiajournal.com/indian_crisis_scenario_a_report/.

Durns, S. (2014, March 20). 5 energy problems confronting India. Retrieved on 1st August, 2016 from http://globalriskinsights.com/2014/03/5-energy-problems-confronting-india/.

Glaeser, E. (2011). The death and life of cities. Harvard University and NBER report.

Government of India. (2010). *National Mission on Sustainable Habitat*. New Delhi: Ministry of Urban Development.

HPEC. (2011). *Report on Indian Urban Infrastructure and Services*. Ministry of Urban Development: High Powered Executive Committee. New Delhi.

Kahn, M. E. (2014), Sustainable and smart cities. UCLA and National Bureau of Economic Research.

Mittal, T., & Singh, C. (2015). *Smart Urban mobility: Road less travelled* (Working Paper No. 492). Bangalore: IIM. Retrieved on 1st August, 2016 from https://www.iimb.ernet.in/research/sites/default/files/WP%20No.%20492_0.pdf.

MOUD. (n.d.). *Annual Report 2014-2015*. New Delhi. Retrieved on 1st August, 2016 from http://moud.gov.in/sites/upload_files/moud/files/English%20Annual%20Report.pdf.

MOUD. (n.d.). *Annual Report 2015-2016*. New Delhi. Retrieved on 1st August, 2016 from http://moud.gov.in/sites/upload_files/moud/files/AR2015-16ENGLISH.pdf.

MOUD, Government of India. (2010). *Strategic Plan (2011-2016)*. New Delhi: Second Draft.

MOUD. (2015). *Smart Cities Mission*. New Delhi: Ministry of Urban Development.

NIUA. (2015). *Urban solid waste management in Indian cities*. UMC Pvt Ltd. Retrieved on 1st August, 2016 from http://pearl.niua.org/sites/default/files/books/GP-IN3_SWM.pdf.

Satterthwaite, D. (2011). How urban societies can adapt to resource shortage and climate change. *Philosophical Transactions of the Royal Society A, 369*, 1762–1783.

Satterthwaite, D. (2008). Cities' contribution to global warming: notes on the allocation of greenhouse gas emissions. *Environment and Urbanization, 20,* 539.

Schafer, A., & Victor, D. G. (2000). The future mobility of the world population. *Transportation Research Part A: Policy Practice, 34,* 171–205.

Padam, S., & Singh, S. K. (n.d.). *Urbanization and urban transport in India: The sketch for a policy*. Pune: Central Institute of Road Safety. Retrieved from http://www.seas.harvard.edu/TransportAsia/workshop_papers/Padam-Singh.pdf.

TERI. (n.d.). India Energy Portal. Retrieved on 1st August, 2016, from http://www.indiaenergyportal.org/about.php.

Turner, A. (n.d.). The Indian monsoon in a changing climate. Retrieved August 3, 2016, from https://www.rmets.org/weather-and-climate/climate/indian-monsoon-changing-climate.

Wankhade, K., Balakrishnan, K., & Vishnu M. J. (2014). *Sustaining policy momentum: Urban water supply & sanitation in India*. Bangalore: IIHS. Retrieved on 1st August, 2016 from http://iihs.co.in/knowledge-gateway/wp-content/uploads/2015/08/RF-WATSAN_reduced_sized.pdf.

Author Biographies

Umamaheshwaran Rajasekar is Director at TARU Leading Edge. He is a disaster management and climate change expert with over 15 years of experience in developing risk reduction strategies, climate change analysis, hazard modeling, institutional analysis, and adaptation and mitigation plans for governments, INGOs, bilateral, and multilateral agencies. His work has been published in peer reviewed journal articles, technical reports, and has been presented at various international and national forums. His recent work on the development and implementation of Urban Health Services Monitoring System has been recognized as an intervention to watch by the UN Secretary General's global pulse initiative.

Soumita Chakraborty is a project associate at TARU Leading Edge. She completed a postgraduate degree in disaster management from the Tata Institute of Social Sciences in Mumbai, India, and graduated with a geography degree from Calcutta University. She has also previously worked with the UNDP office in New Delhi as well as the National CSR Hub in Mumbai. Her recent projects include conducting social impact assessments, disaster management planning, and participatory vulnerability analyses across India.

Gopalkrishna Bhat is the Chairman of TARU Leading Edge Pvt. Ltd., India. He started his career as a geologist in the Ladakh Projects of the Geological Survey of India after obtaining a postgraduate degree from IIT Roorkee, India. He has over three decades of experience in addressing water issues in rural and urban environments. Over last 5 years, he has been working on climate change risk assessment and resilience building activities in urban areas. Recently, he published *Road to Resilience*, a book providing a menu of technologies for improving resilience at household and neighborhood levels.

Chapter 12
Visionaries, Translators, and Navigators: Facilitating Institutions as Critical Enablers of Urban Climate Change Resilience

Anna Brown

Abstract For cities endeavoring to shore up their resilience to a range of climate change impacts, the approach to do this can be overwhelming. The concept of resilience is complex and is not easily understood or absorbed without opportunities for iterative learning and review. Furthermore, the measures needed to build resilience often cut across multiple sectors and city departments, which can challenge norms and institutional incentives and practices. Experience from The Asian Cities Climate Change Resilience Network (ACCCRN)—a 9-year initiative supported by The Rockefeller Foundation—has highlighted the important role that facilitating institutions can play in accompanying cities on their resilience building journey. These intermediary organizations are able to engage credibly with a range of stakeholders in cities while also bringing in technical and process expertise that is sustained over time. Experience in six countries has revealed insights into some of the qualities that make facilitating organizations effective in helping cities make progress toward resilience building. This experience has also shed light on a set of principles that these important institutions have adhered to, which has resulted in robust and high-quality interactions with cities and, more importantly, a high level of local ownership and buy-into the urban climate change resilience agenda.

Keywords Resilience · Climate change · Institutions · Asia · Facilitation · Urban

12.1 Introduction

"Urban climate change resilience" is not a term that rolls off the tongue with many audiences, including among city government officials. It is jargon and complex. (see for example Bahadur and Thornton 2015; Friend and Moench 2013; Meerow et al. 2016; Ziervogel et al. 2017). Each of the words brings certain connotations,

A. Brown (✉)
Aequita Consulting, 29 Pine St., Freeport, ME 04032, USA
e-mail: annalibbybrown@gmail.com

S. Hughes et al. (eds.), *Climate Change in Cities*, The Urban Book Series,
https://doi.org/10.1007/978-3-319-65003-6_12

none of which tend to embody the fullness of the meaning. It does not feel immediate or pressing, in contrast to many other palpable demands such as access to work or affordable safe drinking water or health care. This obtuseness was even more the reality in 2007, when The Rockefeller Foundation launched a 9-year initiative—the Asian Cities Climate Change Resilience Network (ACCCRN)—which was an program that engaged more than 60 cities in six countries: Bangladesh, India, Indonesia, the Philippines, Thailand, and Vietnam (Brown et al. 2012; Carmin et al. 2013; Chu 2016; Karanth and Archer 2014; Kernaghan and da Silva 2014; Opitz-Stapleton et al. 2011; Tyler et al. 2010). More significantly, the notion of what building urban climate change resilience looks like in practice—and how to do it—was an unwritten book.

In spite of this, over the course of the initiative, a range of innovations have emerged through ACCCRN that have better enabled the people, institutions, and systems of cities to be more resilient to climate related shocks and stresses. It has been the hard work of a set of organizations that have, over time, partnered with cities and a range of other actors at different scales to enable this new body of grounded urban climate change resilience practice to emerge. Organizations such as the Institute for Social and Environmental Transition (ISET) in Vietnam—which initially worked through ACCCRN in Can Tho, Da Nang, and Quy Nhon—have provided technical assistance and know-how throughout a process that included everything from wrestling with definitions (where "resilience" rarely translates well into other languages), to understanding climate change impacts in the context of specific sectors, and to designing targeted interventions to strengthen the capacity of localities to weather climate change impacts (Moench et al. 2011). ISET Vietnam has played the role of a facilitating organization by guiding cities and building a process that enables joint learning among actors that are positioned to translate the work into changes in decisions and practices locally (Friend et al. 2016; Orleans Reed et al. 2013). They have also worked with provincial stakeholders and national ministries—such as the Urban Development Agency inside the Ministry of Construction—to identify opportunities to support the broader scale-up of resilience building nationally. The approaches used by facilitating institutions like ISET Vietnam have not only brought their own expertise to bear on the contexts in which they work, but, more importantly, they have helped build local ownership for the agenda of building climate resilience (Anguelovski et al. 2014; Chu et al. 2016). In turn, what was once seen as inaccessible has been translated into something discernable for city actors—and a basis for action.

12.2 Setting the Context: The Asian Cities Climate Change Resilience Network is Seeded

In 2008, The Rockefeller Foundation kicked off The Asian Cities Climate Change Resilience Network (ACCCRN) as part of a global US$70 million "Building Climate Change Resilience" initiative. After a Board of Trustees approved extension and expansion, the initiative totaled $59 million over 9 years. ACCCRN focused explicitly on secondary cities, which are rapidly growing in Asia both in terms of geographic size and population. Unlike megacities where there is a greater need for retrofitting and updating past decisions, smaller growing cities present the opportunity to inform forward-looking investments in infrastructure and land development (Shi et al. 2016). The overarching vision of the ACCCRN initiative was to catalyze attention, funding, and action to build the resilience of cities to climate change impacts. The initiative focused on strengthening the resilience of the city as a whole while also emphasizing that the needs of poor and marginalized communities require deliberate focus. The program sought to achieve three primary outcomes:

1. *Capacity*: The capacity to plan, finance, coordinate, and implement climate change resilience strategies within ACCCRN cities is strengthened;
2. *Knowledge and engagement*: Shared practical knowledge to build urban climate change resilience deepens the quality of awareness, engagement, demand, and application by ACCCRN cities and other stakeholders; and
3. *Money and leverage*: Urban climate change resilience (UCCR) is expanded with ACCCRN and new cities taking action through existing and additional support (finance, policy, and technical) generated by a range of actors.

While the program expanded programmatic work to six countries and about 60 cities, the initial phases of work focused deeply on ten cities in India, Indonesia, Thailand, and Vietnam, each with a lead facilitating partner (see Table 12.1) and in some instances additional technical assistance providers.

The core methodology of ACCCRN followed a common process though each facilitating organization adapted the approach according to the context of activity. The Institute for Social and Environmental Transition (ISET) International played a formative role in the early stages of ACCCRN to help build a methodology and provide support to the lead facilitating partners in each county (Moench et al. 2011). Figure 12.1 provides a visual representation of the overarching steps taken in each city beginning with stakeholder engagement (A), moving through a process of urban and climate assessments (B), synthesizing the range of analyses and perspectives into a City Resilience Strategy (C), and using a process of multi-stakeholder engagement and shared learning dialogues (F). As part of the strategy developed for each city, a set of specific interventions were developed for support from The Rockefeller Foundation (D). Learning, synthesis, and documentation (E) take place throughout this iterative process. A more expanded set of steps is captured in Box 12.1.

Table 12.1 ACCCRN lead facilitating institutions by country and city

Country	City	Lead facilitating organization
Initial cities involved in ACCCRN that have developed city resilience strategies and have undertaken resilience building projects		
India	Indore, Surat	TARU Leading Edge (TARU)
India	Gorakhpur	Gorakhpur Environmental Action Group (GEAG)
Indonesia	Bandar Lampung, Semarang	Mercy Corps Indonesia
Thailand	Chiang Rai, Hat Yai	Thailand Environment Institute (TEI)
Vietnam	Can Tho, Da Nang, Quy Nhon	Institute for Social and Environmental Transition (ISET) Vietnam
Replication and expansion cities that have been involved in ACCCRN[1]		
Bangladesh	Singra, Barisal, Rajshahi, Mongla, Kushtia, Sirajganj, Dinajpur	ICLEI South Asia
India	Shillong, Leh, Panaji, Dharamshala, Mandi, Dehradun, Kurseong, Gangtok, Nasik, Nainital, Keylong, Kochi, Siliguri, Bhubaneswar, Mysuru, Shimla	ICLEI South Asia
India	Guwahati	The Energy and Resources Institute (TERI)
India	Basirhat, Saharsa, Jorhat	Gorakhpur Environmental Action Group (GEAG)
Indonesia	Sukabumi, Cimahi, Bandung, Bogor, Jambi, Balikpapan, Tanjung Pinang, Tangerang Selatan, Jakarta, Bekasi,	ICLEI Oceania
Indonesia	Blitar, Probolinggo, Cirebon, Tarakan, Pekalongan and Palembang	Mercy Corps Indonesia
Philippines	Quezon City, Makati City, Marikina City, Tuguegarao, San Fernando, Bacnotan, Baguio, Tublay, La Trinidad, Catbalogan, Borongan, Naga City, Batangas, Bohol Province, Santa Rosa	ICLEI Southeast Asia
Thailand	Phuket, Udon Thani	Thailand Environment Institute (TEI)
Vietnam	Hue, Lao Cai	Institute for Social and Environmental Transition (ISET) Vietnam

[1]This list is not comprehensive of all of the cities that have engaged with ACCCRN, but rather focuses on those where involvement has been more multi-pronged and has involved multiple points of support

Box 12.1 Typical steps used by facilitating organizations in ACCCRN

- Partners initially scoped the city to determine the demand from government stakeholders and to gauge their interest in a longer process of developing climate change resilience. [A in Fig. 12.1]
- A multi-stakeholder Shared Learning Dialogue (SLD) was generated within each city. This mechanism cultivated trust among individuals and groups who may not have engaged with each other much, or who may have opposed each other on specific issues in the past. [F in Fig. 12.1]
- Within and outside the SLD, there was extensive discussion on the term 'resilience', which led to a more aligned understanding of what it means and how it might mitigate the impacts of future events. [F in Fig. 12.1]
- Much time was expended to collectively analyze and understand urbanization and climate science, with a particular focus on what future threats might be and the levels of uncertainty surrounding those potential threats. [B in Fig. 12.1]
- Each group of city stakeholders developed a vulnerability assessment to better understand who might be exposed to various types of future threats and in what ways. Sector studies were conducted to deepen the understanding of what priority sectors would experience in terms of climate change and urbanization. [B in Fig. 12.1]
- Each city then developed a City Resilience Strategy (CRS), based upon the collective analysis and holistic thinking regarding how a city might approach future threats. Within the strategy, each city prioritized short- and longer term activities to build resilience. This prioritizing ultimately led to a short list of projects that the Rockefeller Foundation then funded. [C in Fig. 12.1]
- Those projects were then conceptualized in more detail, including a plan for implementing them. [D in Fig. 12.1]
- Each project's activities are monitored and their progress documented, including reflections by city stakeholders that point to the links between concrete action and the broader objective of building resilience. The project documentation should promote more sophisticated dialogue within each city in the future. [E in Fig. 12.1]

Source Adapted from "10 Cities, 4 Countries, 5 Years: Lessons on the Process of Building Urban Climate Change Resilience." 2015. https://assets.rockefellerfoundation.org/app/uploads/20150102000721/Ten-Cities-Four-Countries-Five-Years-Lessons-on-the-Process-of-Building-Urban-Climate-Change-Resilience.pdf

The purpose of the "Shared Learning Dialogue" approach, which sat at the heart of the ACCCRN process, was originally envisioned by ISET International to: (1) Engage key actors, identify critical climate risks and evolving potential

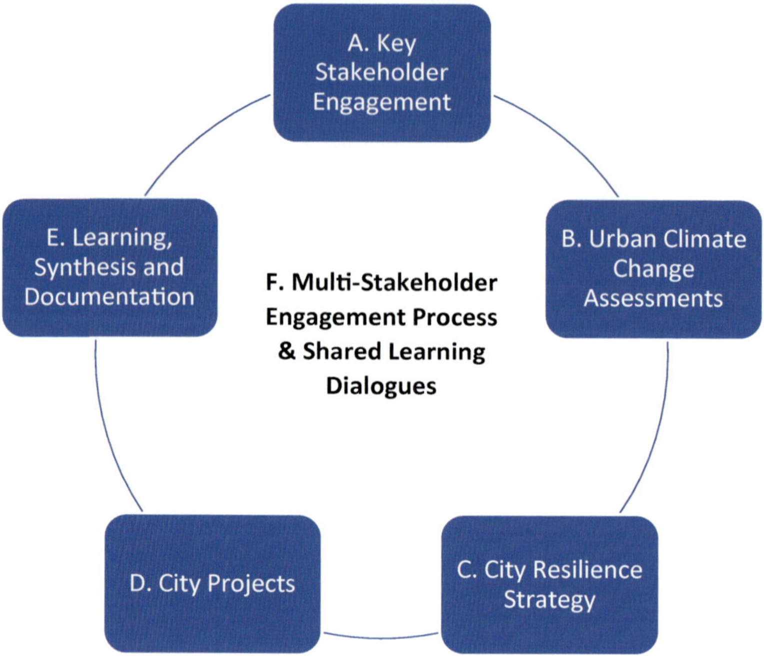

Fig. 12.1 Urban climate change resilience building process used in ACCCRN

responses that build resilience; (2) Build ownership among stakeholders necessary for successful implementation; (3) Overcome knowledge systems divides and coordinate across scales and sectors; (4) Engage vulnerable groups; (5) Build understanding of divergent interests of stakeholders; and (6) Compile and make accessible relevant local information sources (Moench et al. 2011).

In addition to the city-level work, the ACCCRN program also focused on different opportunities to scale-up urban climate change resilience efforts. This included work led by the same facilitating organizations as well as a collection of other actors at state, provincial, and national levels through training efforts, research, and policy levers. Likewise, The Rockefeller Foundation helped design and launch a $160 million multi-donor partnership, the Urban Climate Change Resilience Trust Fund, to support urban climate change resilience across Asia.

This chapter will provide an overview of the role that facilitating organizations have played in introducing urban climate change resilience to cities and will include some reflections on the qualities displayed by the facilitating organizations (see Table 12.1 for a list). The approaches taken in different countries and cities varied in part because of the distinctions in each context in terms of governance, political economy, and what resonated with local actors. Facilitating organizations helped navigate these contextual realities to lead a process that worked in each place.

Finally, the chapter discusses some of the principles that facilitating partners utilized in the implementation of ACCCRN work.

12.3 Insights from Indore, India

Beginning in 2008, The Rockefeller Foundation partnered with TARU Leading Edge Private Limited, a transdisciplinary institution formed in the 1990s to address India's development challenges. TARU initially conducted an analysis for The Rockefeller Foundation to identify a range of possible focal cities in India for deeper engagement. After the Rockefeller Foundation had selected the initial cities of Indore, Surat, and Gorakhpur and TARU was identified as the lead institution to facilitate the work in Indore and Surat, the engagement with city stakeholders got underway in 2009.

Indore is the largest city in Madhya Pradesh and as of 2011 had a population of nearly 2 million. An educational, trade, and medical hub in the western part of the state, Indore is growing rapidly. The population in 2001 was 1.64 million and the city is on a trajectory for 4 million people by 2030. Like many second-tier cities in India, Indore faces numerous pressing challenges, including access to reliable, safe drinking water, health problems, and traffic congestion. And also like other secondary cities, Indore is faced with insufficient resources to address the spectrum of development needs and aspirations (Chu 2017; Indore City Resilience Strategy 2012). Much of the industrial base is located outside the city limits, for example, meaning that the Indore Municipal Corporation (IMC) has struggled to generate a strong enough tax and resource base to handle operations and maintenance expenses, much less new investments. This reality means that any issue that gets raised with city leaders is in competition with a long list of other pressing challenges. From a governance perspective, Indore has developed a patronage system of administration dating back to before and during British colonial rule. This has continued to influence the political system to some degree such that the systems of redress and grievances in the city often contribute to lower expectations in terms of leadership accountability.

12.3.1 The Resilience Building Process in Indore

Beginning in early 2009, TARU took steps to build a process of engagement around urban climate change resilience (UCCR) that generally aligned to those outlined in Box 12.1. The City Advisory Committee (CAC) formed by mid-2009, and included representatives from the Indore Municipal Corporation, City Development Agency, nongovernmental organizations (NGOs), academic institutions, private sector, and ISET and TARU (Indore City Resilience Strategy 2012).TARU launched a process that drew on the perspectives of different stakeholders in the city, initially to build a grounded understanding of how the city is urbanizing, who and what in the city is vulnerable to, and how climate change affects this whole picture. With guidance from the CAC that TARU helped to convene, Indore undertook a set of studies looking at vital sectors of the city. The teams that conducted the studies were identified by the CAC and covered water security, energy security, urban environment and health, green buildings, and urban transport.

In spite of some early champions in Indore, including the City Planner of the Indore Municipal Corporation, it took time for TARU to build enough interest in the city beyond the small City Advisory Committee (CAC). This was in part due to time constraints and the lower perceived importance of the resilience agenda on the part of municipal leaders. However, through steady engagement and adaptive approaches—in part aided by the presence of a TARU staff member based in Indore—gradually UCCR gained greater purchase in the city, including among constituents that typically have not shown active participation in governance, such as community members and fishermen (Chu 2017).

One of the adaptive approaches that TARU introduced to help make the issues of urban climate resilience more palpable for Indore stakeholders was the introduction of climate science and scenario exercises in April and May 2010. TARU conducted a set of consultations—titled "Risk to Resilience" workshops—for CAC members and other city stakeholders. In the initial meeting, participants developed future urban development scenarios. In a follow-up workshop, participants explored future growth trajectories for the city, considering issues like resource shortages/scarcity, governance, infrastructure services, connectivity, and migration. TARU introduced climate information to anticipate the implications of changed rainfall patterns, which would likely mean more intensive periods of rainfall for the city. TARU also helped bring information about changing temperature patterns into the consultations with different stakeholders. In Indore, climate models point to temperature increases of two degrees Celsius by 2030—and three to four degrees increase by 2080. The urban heat island effect, which makes cities hotter than surrounding less developed areas, would add a further two to four degrees on top of that. This could mean maximum temperatures of close to 50° Celsius, rendering outside work highly dangerous and adding impetus for greater energy inputs for space cooling (especially for those that could afford such an expenditure). By framing these projections around the context of *existing challenges* of waterlogging, poor drainage, and health (including water-borne and water-related disease as well as incidents of malaria, dengue, and other vector-borne diseases), TARU was able to more effectively reach different city constituents.

TARU worked with city leaders and other stakeholders in the business and civil society sectors to digest the findings and generate a synthesis of how urbanization, climate change, and poverty and vulnerability would affect the city's future. Through this, TARU led a process to prioritize actions that could be taken on a short- and medium-term time horizon to strengthen the resilience of the city. Compiled, this comprised Indore's City Resilience Strategy (CRS) (see Table 12.2 for details). The Rockefeller Foundation funded a subset of these actions.

Table 12.2 Examples of short- and medium–long-term measures for Indore to incorporate climate change risks

Sector	Needs	Short term	Medium–long term
Water	• Ensure sufficient water availability • Prevent leakage and revenue loss • Decrease dependency on long distant sources • Protect local groundwater resources	• Assess local water resources and upgrade existing infrastructure • Comprehensive water management planning, such as leak detection, preventing illegal water withdrawal, and repairing local pipelines • Institute community-level conjunctive water management practices • Support rainwater harvesting and wastewater recycling/reuse technologies	• Strengthening existing distribution pipelines to cover the entire city • Develop a citywide water metering and audit system • Conduct groundwater surveys to locate aquifers and their recharge areas • Dedicate open space for rainwater harvesting facilities • Restrict overexploitation of groundwater
Energy	• Cope with increasing urbanization and population pressures • Ensure energy reliability • Build redundancies to meet energy demands and price requirements	• Implement energy demand and supply management system • Promote household electricity metering • Upgrade urban electricity transmission network • Promote energy efficiency products and passive cooling for buildings • Develop guidelines for environmentally sustainable building design and construction	• Map the urban electricity distribution network using GIS • Protect infrastructure during climatic events • Promote solar energy for household • Ensure that all new buildings adhere to energy efficiency building codes
Disaster management	• Reduce risk exposure of poor communities • Improve disaster warning and forecast • Strengthen city disaster management plan	• Develop a storm water drainage master plan • Improve disaster response plans, including evacuation of citizens from waterlogging zones • Undertake floodplains risk zoning with advance warning system	• Map floodplains for identification of waterlogging prone area of the city
Health	• Decrease cases of vector-borne disease	• Undertake disease monitoring system and Health GIS with epidemiological research	• Develop scenarios for temperature and rainfall changes that can be integrated with a

(continued)

Table 12.2 (continued)

Sector	Needs	Short term	Medium–long term
		• Consider a SMS-based health monitoring system	GIS-based health monitoring system
Sewage/waste management	• Address inadequate waste storage • Poor management of waste management system	• Improve public awareness over the segregation of waste • Modernize waste management system with route planning, waste transfer systems, and integrated processing facilities • Strengthen household garbage collection • Use biodegradable, recyclable, and inert waste to reduce dependence on landfill	• Initiate drainage cleaning programs and evaluate resettlement options for communities along the natural drainage systems • Assess climate change impact on the decomposition of disposed waste • Design a municipal waste processing facility with a goal of converting 90% of solid waste into reusable products

Source Indore City Resilience Strategy (2010)

12.3.2 Water Resilience and the Case of Lake Restoration

One of the salient issues that surfaced through the multi-stakeholder process that TARU led was the challenges of water extremes in Indore. The studies undertaken revealed poor drainage in much of the city, due in part to limited investments to expand coverage and to poor maintenance of existing systems. Waterlogging was flagged as a regular problem, especially during and after the rainy season, in light of low soil permeability. Urban development and the increase in paved surfaces as well as improper solid waste management has exacerbated the problem for this growing city. Although flooding and waterlogging affect Indore, the city also struggles with water scarcity (Chu 2017). In 2010, only 54% of the population had access to piped water supply (on alternate days and only for one hour) and the rest of the population relied on ground water or water tankers. Illegal connections meant that approximately 30–50% of the water supply was unaccounted for or lost. Findings captured in an internal ACCCRN status brief in 2010 reported that residents tended to feel more comfortable with informal methods of acquiring water rather than relying on the municipal system. These methods include digging pits to collect water seeping from pipes, the construction of sump pumps, and installing pumps to draw from municipal lines. Without water metering, water pricing was based on a flat tariff, curtailing incentives for conservation. Water scarcity concerns have also given way to protests and road blockages in Indore on occasions. With this context, in the early stages of ACCCRN involvement, many stakeholders were hopeful that the new planned water supply from the Narmada River would decrease scarcity concerns. However, this perspective was controversial in particular because the system is inherently energy intensive and financially costly as it requires water to be transported 70 km after being pumped up a 500 m vertical.

As such, it is not surprising that water emerged as a priority resilience action area for Indore (Chu 2016). The Rockefeller Foundation funded a total of four city projects in Indore (see Table 12.3 for an overview of the selected projects). One of the resilience building options that emerged in the City Resilience Strategy was to increase the redundancy of the water available by enabling access to an emergency supply (Indore City Resilience Strategy 2012)—something that could be vital given water demand projections in the face of climate change and growth. Results from one of the sector studies estimated that local reservoirs had lost about 25% of their capacity due to silting and eutrophication and improper waste management. Indore is comprised of a large number of urban lakes, which range in size from 2.5 to 428 acres. TARU's analysis revealed 26 lakes within the Municipal Corporation's jurisdiction. Many of these have become overgrown with vegetation and are subject to wastewater disposal, rendering the water quality very poor.

With funding from The Rockefeller Foundation, TARU designed a project targeting peri-urban lake restoration for emergency water supplies (Chu 2016). TARU conducted a stakeholder analysis to understand the user base as well as the stakes that people have in these urban waters. When the project started, there were no concrete efforts by communities to conserve urban lakes, in spite of regular use by fishermen, local households, and religious institutions. In TARU's early analysis and discussion with representatives from the Municipal Corporation, there was a sense that the city could provide capital investment for restoration efforts, but that the required maintenance and operation demands would not realistically be met by the city. The project that TARU designed focused on two lakes—Lasudia Mori and Khajrana Lake—to demonstrate both a range of technical options for restoration and wastewater treatment, as well as approaches to governance that would allow for long-term management of these resources.

Through this lake restoration project, several multi-stakeholder actions have been taken with participation by the city government as well as civil society and community members. Restoration approaches have included:

- Solid Waste Management: The Water Conservation and Management Committee (WCMC) has led an intensive effort of solid waste management at Lasudia Mori to both change practices related to dumping waste and to clean up the lake. The WCMC worked with local youth leaders and the Indore Health Department on a lake clean-up effort. Local fishermen supported this clean-up effort as well. The WCMC and other community leaders have adopted regular monitoring practices to ensure that the lake remains clean.
- Eco Idol Immersion campaigns: TARU worked with the Municipal Corporation to organize a campaign to prevent lake pollution from the customary practice of immersing religious idols in the lake water at the time of festivities. A 2015 'Eco Immersion Campaign' inspired the Municipal Corporation to expand the activity to all other lakes in the city. As a result, more than 67,000 idols were collected.

Table 12.3 Urban climate change resilience projects funded by the Rockefeller Foundation under the ACCCRN Initiative

ACCCRN City projects in Indore	Land use & urban planning	Drainage, flood & solid waste management	Water demand & conservation systems	Emergency management & early warning systems	Responsive health systems	Resilient housing and transport systems	Ecosystem service strengthening	Diversification & protection of affected livelihoods	Education & capacity building of citizens	Institutional coordination & capacity support
Testing and promoting decentralized systems for differential water sources and uses			√			√				
Strengthening vector-borne disease surveillance and response systems				√	√					
Cool roof and passive ventilation promotion for low-income housing (also implemented in Surat)						√				
Indore urban lake restoration for emergency water provision	√		√				√			

(Adapted from Brown, Dayal and Rumbaitis del Rio, 2012)

As an alternative, this provided an artificial tank for idol immersion. This effort drew on the combined efforts of many actors including lake committees NGOs, city councilors, and social workers.
- Artificial Floating Islands: As a technique to improve the water quality of the lake, provide bird habitat, and to improve the aesthetic quality of the lake, TARU worked with city stakeholders to design and construct an artificial floating island in Annpurna Lake. The Municipal Corporation provided regular monitoring and maintenance of this construction.

The projects, which have relied on multiple partnerships between the Municipal Corporation and a range of community groups, have generated results in terms of improved water quality as well as shifts in practice and expectations among community members (Chu 2017). Water Conservation and Management Committees (WCMCs) have formed around the lakes and these entities take on a regular role of monitoring waste disposal and discouraging dumping into the water bodies. The WCMC of Khajrana Lake, for example, conducts weekly monitoring walks on Sundays, using these meetings to monitor as well as reach out to community members to raise awareness about the value of keeping the lakes clean. They also use these walks to plan upcoming activities. In a final grant project report in November 2016, TARU reported that the areas adjacent to Lasudia Mori lake were waste-free. On the governance side, two new community associations, Rangvasa and Talawali Chanda, were formed and registered under the Society Registration Act in 2016. These groups have taken on ownership of the lake restoration agenda, marking a shift in norms and expectations within the community around lake protection.

Awareness of the urban lakes and recognition of the value that they provide to Indore communities has increased, as has a joint commitment from the Municipal Corporation and community members to maintain their health. In response to the ACCCRN lake conservation activities, the Nai Dunia Newspaper Group initiated a lake conservation awareness campaign using the two Indore lakes, Khajrana and Lasudia Mori. The campaign has the goal of reviving 1100 traditional water bodies across all of Madhya Pradesh. This campaign has been joined not only by members of the Khajrana WCMCs, but also by other local NGOs, city Councilors, and Indore Municipal Corporation officials and administrators.

Indore's journey to arrive at the City Resilience Strategy and proposed projects did not happen quickly. Initial engagement did not galvanize traction beyond the small group of committed constituents involved with the City Advisory Committee. Over time, however, TARU was able to cultivate interest across both technocrats and political leaders in the city and Indore has taken several measures to strengthen their resilience to climate change impacts. The next section probes further the qualities that TARU—and indeed other facilitating institutions in ACCCRN—demonstrated to enable this traction for urban climate change resilience measures to take hold, even among historically less active members of society.

12.4 Qualities and Competencies of Organizations Facilitating Urban Resilience

The fact that Indore was able to mobilize around the seemingly nonurgent agenda of climate change resilience begs a question about how this arose. TARU played an important role in building awareness and understanding over time and providing sustained support. As mentioned earlier in this chapter, the facilitating organizations that have partnered with ACCCRN cities have followed the same general process to generate a city resilience strategy and a set of possible actions (see Fig. 12.1 and Box 12.1). Even with this common approach, each facilitating organization has brought a particular orientation, which reflects comparative strengths, skills, and networks, and a theory of change. For example, the Gorakhpur Environmental Action Group (GEAG), which played a facilitating role in Gorakhpur, in the Indian state of Uttar Pradesh, drew on their strengths in working directly with neighborhood groups. GEAG led robust processes that deeply engaged some of the poorest and most marginalized residents of that city. Likewise, the head of the organization has a strong background in systems thinking, which was reflected clearly in the risk framework that GEAG developed (Gorakhpur Environmental Action Group n.d.).

GEAG has also worked closely with stakeholders within the municipal government, building awareness and interest, but they have not waited on government-led action in particular because decentralization has not borne out in Gorakhpur or for other local bodies in Uttar Pradesh state. The approach taken by TARU, on the other hand, has been informed by their robust technical capacities, including innovative analytical methods, deep knowledge of urban development and related sectors (e.g., water supply and sanitation), strong command of technology, and positive reputation among government actors. This—as well as the diversity in contexts—has resulted in variation in terms of how ACCCRN has been rolled out in each city. Despite variations in core strengths, a set of some common qualities and competencies, captured below, cut across the different facilitating organizations that have guided the city-level work in ACCCRN.

12.4.1 Credibility Within Local and National Context

The Rockefeller Foundation learned about TARU's work through their strong reputation among international NGOs, multi-laterals, and bi-laterals and from the research community. However, in terms of credibility among actors relevant to decisions affecting Indore, TARU needed to demonstrate knowledge of local context—including physical, institutional, political, and cultural dimensions—as well as an ability to work in relevant local languages. TARU had also earned a strong reputation within India among different levels of government—including the National and State Disaster Management Authorities, the National Institute of Urban Affairs (NIUA), and with research and technical institutions such as the

Indian Council for Research on International Economic Relations (ICRIER)—for their grounded technical expertise and ability to provide innovative and creative solutions. TARU has been one of the institutions backing a Sanitation Innovation Accelerator effort in India, for example, which has aimed to generate affordable and sustainable sanitation solutions for rural communities. This grounded expertise not only helped open doors for initial meetings with city stakeholders, but through it they continued to demonstrate value that sustained and expanded engagement.

The same has been true for other facilitating organizations. The Institute for Social and Environmental Transition (ISET) Vietnam is another good example. While ISET is a known—and credible—international research institution, and one that played a pivotal role in developing the process for building shared learning among city stakeholders involved in ACCCRN, ISET Vietnam formed only in 2010 to lead the ACCCRN work in Vietnam. ISET Vietnam grew out of engagements and partnerships that ISET International had forged in Vietnam, drawing on a mix of expertise from inside and outside of the country. The very first staff member of ISET Vietnam proved immensely strategic given her strong operational knowledge of how to get things done, project management expertise, and local, provincial and national networks. Very swiftly, this first staff person was able to build a team of strong subject matter experts, which has continued to deliver on the complex and multi-pronged programming needs across the country in a range of contexts. The team has built up positive partnerships with staff at key government agencies, like the Urban Development Agency, which sits under the Ministry of Construction.

At the city level, ISET Vietnam has also earned the trust of local players, including the Da Nang Women's Union with whom they partnered on a credit and housing upgrading scheme under ACCCRN. This trust has enabled them to lead processes that sustain interest and activity over time. ISET Vietnam has also been able to work with cities in a manner that has built local ownership and engagement over the UCCR agenda. Like other ACCCRN institutions that played the role of facilitating organizations, the analyses conducted were in part generated by local and national actors, and not only by international institutions. Furthermore, the products generated, including the city resilience strategies, were developed and legitimized by city actors. Part of this meant that the reports generated were prepared—and shared—in local languages. It may seem a modest step, but it is not a given that such analyses make their way to local constituents.

12.4.2 Adaptive, Learning, and Flexible

The organizations that have provided support and guidance to cities on urban climate change resilience have had to demonstrate flexibility, adaptiveness, and an ability to apply learning quickly given the unchartered nature of the work—all characteristics of resilience. In different ways, each of the facilitating organizations had to learn how to introduce the topic in a manner that would generate enough interest to participate in a process. Several facilitating organizations quickly learned—for example in TARU's experience in India—that they could generate more interest and engagement from city stakeholders by framing conversations not around climate change, but around current urban challenges (Anguelovski et al. 2014; Chu 2016). While some Indore city stakeholders understood that climate change did represent a long-term risk to the city, TARU was able to capture more interest in conversations framed around water scarcity given the more urgent nature of this concern.

In the initial interactions with cities, partners tended to line up meetings and conversations using climate change as the entry point. However, by listening to the concerns and challenges articulated by different representatives from the city—and experiencing the challenge of getting access to meetings—partners like TARU learned that they needed to instead start with questions related to an urban systems analysis used in the ACCCRN conceptual framework (da Silva et al. 2012). That is, how well are the city's infrastructure (e.g., roads, energy systems, water, and sanitation systems) and institutions (e.g., service providers, government bodies, knowledge and innovation centers) functioning to support the well-being of its inhabitants? Are the ecosystems and hinterlands upon which city dwellers depend for basic needs like food and water adequately protected? How is the city changing in terms of population and land expansion? What will this mean for infrastructure and service provision needs? ACCCRN facilitators have learned to raise questions about climate change only after understanding the current and projected future urban realities. Then questions related to how direct and indirect climate impacts may play out tend to garner more concrete ideas about impacts.

The scenario exercise that TARU designed with stakeholders in India represented an adaptation in approach to stimulate traction in the city and build joint vision for the future of the city. The output and the process used to generate it provided a concrete basis for building engagement and sharing information around potential future climate resilience pathways for the city. In Indore, this exercise created an opportunity to cultivate common interests around water conservation and the city's lakes. Importantly, it also helped generate more robust interest in the city from influential actors in the city. In Surat, that included members of the business community and chamber of commerce (Chu 2016). In Indore, that included prominent citizens and university professors. In Gorakhpur, the city government is relatively weak, in part because decentralization has not fully been implemented in this state. In spite of positive engagements with the Municipal Corporation, the Gorakhpur Environmental Action Group (GEAG) adapted the approach of engagement to focus on resilience building measures that could be taken up by civil society and non-state actors. These included a community-based resilience building process, which mobilized a neighborhood level solid waste management plan that resulted in massive reductions in waterlogging in this part of the city. While GEAG has been able to generate support from government, the ability to tap resources from the local government emerged after having provided some tangible proof points in terms of actions and impacts.

12.4.3 Open Communication

A listening and learning mode, which is valuable for the resilience building journey, requires humility to be open to new information, perspectives, and experiences from different sources. In different ways, the facilitating partners involved in ACCCRN demonstrated this quality by engaging in a process of cogeneration with city stakeholders (Chu and Schenk 2017). The process led by ISET Vietnam in the three Vietnam cities, for example, engaged a range of local government department in the process of information sharing, building collaboration and generating consensus. This cross-sectoral process is highly unusual and seen as innovative the Vietnam context (Moench et al. 2011). Rather than assume that they held all of the answers, ISET Vietnam and other facilitating organizations recognized the need to work with local constituents to develop a grounded understanding of both the potentials risks of climate change and urban development as well as what measures could help strengthen resilience—given the particularities of each given city context.

12.4.4 Connecting Resources Locally, Nationally, and Globally

One of valuable functions that facilitating actors have played has been to build connections—be they knowledge, tools, training or financial resources—among different scales of actors. In Vietnam, for example, ISET brought in international expertise in climate science, including a recognition of the fundamental uncertainties associated with climate change and the resulting challenges for decision makers. While ISET did play a role of providing some climate information to some cities (in Vietnam and elsewhere), they also made a point of tapping key national, provincial, and local actors. For example, in the case of climate projections, ISET interacted closely with Vietnam's Institute for Meteorology, Hydrology and Environment (IMHEN) to vet the climate information findings, particularly given the role of National government data authorization in Vietnam. In other instances, ISET partnered with local institutions and experts like the DRAGON Institute of Can Tho University to conduct research and analysis related to water management and saline intrusion.

This role of connecting across scales and domains of knowledge and resources was also demonstrated by each of the other facilitating organizations in ACCCRN. In Indonesia, Mercy Corps was able to not only stitch together expertise that came from their international networks, including from their resilience advisors, but they could also tap into opportunities within the Indonesia context. Mercy Corps leveraged funds from Zurich Insurance, for example, to scale up a project initially supported by The Rockefeller Foundation, which developed a flood early warning system for the city of Semarang, in Central Java (see Chap. 7: Sari and Prayoga).

With the additional resources, the project was able to expand the reach to the entire river basin. In another case, the Thailand Environment Institute (TEI) was able to draw on their strong private sector networks nationally to generate interest in the urban climate change resilience agenda. This included helping link private sector leaders from the city of Hat Yai to national dialogues focused on generating more interest and commitment from the private sector to climate change resilience. TEI also drew on a strong network of expertise nationally, including in the domains of climate science and urban planning.

12.5 Core Principles of Facilitating Organizations

In spite of slight variations in the approaches that different facilitating organizations have used to guide cities along their climate change resilience journey, each of these actors has followed a set of common principles. Since the methodologies and practices deployed in ACCCRN evolved over the course of the program, one can retroactively distill these common principles to shed light on which variables contribute to effective implementation.

12.5.1 Centrality of a Multi-stakeholder Process

Without a doubt, the combined impacts of climate change and urbanization on a city are mediated by a diverse range of factors, including physical realities (e.g., the geography of the city, including low-lying and hilly areas, coastlines and waterways, and soil composition), capacities of individuals and institutions, the existence of inclusive policies and practices, levels of governance autonomy, and the fiscal and budgetary priorities and processes. Climate change will affect different individuals and interests in the city unevenly, so it would be insufficient, incomplete, and inequitable to generate a vision and plan for how the city would go about strengthening resilience without a multi-stakeholder process (Chu et al. 2016). Urban resilience planning processes are designed to enable diverse actors to come together over a sustained period of time, providing a vehicle to counter top-down approaches to urban management where government leaders make decisions without drawing on input or perspectives from those whom the decisions will affect. A well-designed urban resilience planning process can also help surface opportunities and risks for businesses and industries and, in so doing, can engage them by creating incentives to make internal resilience investments that benefit them directly (or may have other positive benefits). The approach also helps to develop a joint understanding of potential climate impacts and how these relate to existing and anticipated developmental challenges of the city in ways that are relevant to key municipal departments as well as to the business, academic, and civil society sectors.

Each of the ACCCRN facilitating partners put multi-stakeholder engagement in the heart of processes of working with cities through using guidance by ISET, who had developed a methodology of "shared learning dialogues" (SLDs). These SLDs, though adapted by facilitators, became vehicles for joint learning and bringing new information into a common space for consideration and reflection (Orleans Reed et al. 2013). The multi-stakeholder approach used for urban resilience planning also has the benefit of helping to identify discreet solutions that can achieve multiple outcomes, thereby improving resource efficiency and avoiding investments that work at cross purposes (Chu et al. 2017). For example, in Quy Nhon, Vietnam, the city prioritized a project to restore mangroves in an urban lagoon. In addition to the benefits of providing a physical buffer to prevent erosion and impact from storms and high water levels, the project helped prompt a wider debate on the need to cement a moratorium on development in the floodplains of the city, which is adjacent to the lagoon. This pause on development arose in response to severe flooding in 2009 when Typhoon Mirinae caused unprecedented flooding. Eventually, in 2013, the Prime Minister of Vietnam called for the revision of the Quy Nhon master plan to protect the flood plain—a highly unusual move that signaled an interest in revising land use planning and urban development practices more widely in the country.

12.5.2 Working with Local Experts in a Locally Grounded Process

In order to build resilience and not generate solutions that are off-base, ungrounded, or maladaptive, each of the ACCCRN facilitating partners came to appreciate the importance of working with local experts. In each city, early in the process of context analysis—which was foundational to developing the City Resilience Strategy—local experts were engaged often through the working group established or as experts leading on a sector study or another component of analysis. In Indore and Surat, for example, the City Advisory Committee helped identify the actors who would undertake the sector studies. In contrast to some other development practices, the role of local experts was to weigh in and, in some cases, generate research products that would be consumed locally, as well as by actors at other scales. It is not uncommon for international development actors to commission or solicit research that draws on local expertise through contracting and sub-contracting. What emerged as distinctive in ACCCRN—and what proved important to being able to work toward the desired outcomes of a more resilient city—was the role of these local experts in the broader process of generating a vision and strategy. That is, often bids are issued for a scope of research that local institutions undertake, but the extracted products may not necessarily feed into an ongoing and iterative learning journey that involves other local actors, be they elected or appointed officials, technocrats, or leaders in civil society and research.

In some cases, the reports may not even get translated to local languages. When many local actors, including knowledge experts, are involved it creates potential to knit a fabric of engagement and shared vision and purpose, especially when brought into a process of shared learning and reflection (Bartlett and Satterthwaite 2016).

12.5.3 Combination of Hard and Soft Measures

The resilience of a city requires functioning physical infrastructure—even in the face of stresses and shocks. However, the resilience of these "hard" systems in the city is a necessary but not sufficient ingredient. The facilitating partners involved in ACCCRN understood the importance of social capital: including knowledge of who lives in different communities, what specific needs may be, and the strength of connections within and across groups. They understood that awareness and knowledge are important requirements to strengthen capacity, which is needed both at the levels of individuals and households, but also of the institutions who set rules, provide goods, support and services, and who in turn build the capacity of others. These facilitators recognized that shifts in policy and practice could significantly affect the development trajectory of a city, as well as the degree to which it moves on a pathway toward resilience or builds in risks and vulnerabilities.

Out of nearly 40 projects (for a total of approximately US$15.5 million) that The Rockefeller Foundation funded in the initial ten ACCCRN cities, only a few examples focused on physical infrastructure. Of these, most honed in on the "soft" side of these efforts, like research and development and the creation of finance instruments. For example, TARU led a project for Indore and Surat to identify and promote a range of low-cost housing solutions for thermal comfort, including technologies for passive ventilation and cool roofs. ISET Vietnam worked with the Da Nang Women's Union to develop a credit scheme for low-income families to upgrade their homes to be resilient to storms. Several examples of "green infrastructure" also emerged, like the ecological restoration of a portion of an urban river in Chiang Rai, Thailand, which was facilitated by the Thailand Environment Institute (TEI). Other "soft" measures included mechanisms to improve coordination—like the End-to-End Early Warning System in Surat—or the establishment of cross-departmental coordination offices in Can Tho, Da Nang, and Quy Nhon, Vietnam, to navigate the typical silos of planning, decision-making, and implementation.

The heavy emphasis on soft measures was in part a result of the scale of funds that were available from The Rockefeller Foundation for resilience building projects. It also arose, however, from the experience and knowledge of the facilitating organizations, who understood the potential of non-infrastructure agendas to contribute to urban climate resilience. Better land use planning and coordination of information and investments can, for example, help align agendas and resources across different sectors to achieve multiple outcomes (see Chap. 13: Cook and Chu).

12.5.4 Linking Current Problems to Longer Term Visions

Facilitating organizations working with cities to advance urban climate change resilience quickly learned that to gain an entry point with cities amidst a crowded set of pressures, there was a need to connect the longer term vision of a resilient city to the current issues cities wrestle with daily (Carmin et al. 2012). It is not surprising that water featured strongly in Indore's City Resilience Strategy and in the prioritized actions laid out in the document given the pressures on the water systems (Chu 2017; Indore City Resilience Strategy 2012). In addition to the project focused on restoring peri-urban lakes for an emergency water supply, TARU worked with the city to develop a separate project on decentralized water solutions for a diverse set of neighborhoods (Chu 2016).

By listening to what is salient to the different voices in the city today and using that as a jumping off point for a longer term resilience agenda, it has made that task of aligning with the energy, focus, and interests of the city less Sisyphean. Surat, the most flood-prone city in Gujarat and also a thriving industrial hub, had undergone a major flood event in 2006. At that time, an emergency release of the upstream Ukai dam during a period of heavy rainfall resulted in 75% of the city being flooded—lasting for a week in some areas. The damages and losses from business grinding to a halt are estimated at about $4.5 billion. When TARU started to engage with different actors in the city in 2009, flooding was understandably a dominant agenda. By listening and helping to build an understanding of what such a flood might look like in the future with added pressures from urbanization, like the loss of permeable surfaces, and with the shifts in precipitation that are projected with climate change, it helped build momentum for engagement and action toward a more resilient city (Bhat et al. 2013; Karanth and Archer 2014). Indeed, TARU worked with the Municipal Corporation as well as a set of other actors at different scales, including the Narmada Water Resources, the Gujarat State Disaster Management Authority, the Surat Urban Development Authority, to design an end-to-end early warning system that relied on cross-scalar information and coordination around rainfall and water levels in the Ukai reservoir—and informing procedures for release. This project has proven important to improving water management. In 2013, it played a pivotal role in averting another flood event that could have hit the city even harder than the 2006 floods. By meeting the city where they were on flood-related concerns, TARU was able to help Surat develop a cross-jurisdictional approach that has improved water management, as well as the ability to handle water-related stresses and shocks (Bhat et al. 2013).

In the cases of Surat and Indore, water concerns provided an entry point for a deeper analysis and exploration of potential ways to strengthen resilience. But in both instances—as was the case with other cities involved in the initial phase of ACCCRN—the conversation and domains for action broadened. Both cities, for example, came to grasp the risks of heat stress and other health-related concerns connected to a changing climate and introduced measures to improve the resilience of the health system. These domain of impact, were not immediately obvious or salient to city stakeholders, but this perception and understanding shifted over time as knowledge and engagement with the resilience agenda deepened.

12.5.5 A Focus on Vulnerability and Poor People

Climate change impacts—and even measures to build resilience—will play out differently on different geographies and populations of the city. In some cases, there will be winners and losers—with the most marginalized and poor segments of society bearing the brunt of climate impacts (Ayers and Dodman 2010; Shi et al. 2016). Given this, ACCCRN partners have included a strong focus on understanding the vulnerabilities and capacities of poor and vulnerable populations. TARU developed a methodology that involved surveying households to create a set of geo-referenced indices on education, income, and social capacity. The Gorakhpur Environmental Action Group (GEAG) in India, which already had close ties with many poor and marginalized communities, was able to undertake a deeper analysis than in many other cities using community surveys and focus groups. The process used by GEAG also helped provide community members with information that they used in reaching out to elected officials. It is not sufficient to consider what hazards and risks must be anticipated, but ACCCRN facilitators have also asked whose resilience will be built through different proposed measures.

12.6 Conclusion

The field of urban climate change resilience has advanced significantly since 2007, with considerably more conceptual knowledge and practical know-how in existence today. Cities interested in shoring up their resilience to climate change impacts now are able to draw on many more resources to inform their journey and the materials available continue to expand (see Chap. 10: Bellinson). Central to the resilience building process of ACCCRN cities has been the steady, consistent, and grounded presence of facilitating organizations who have served as guides, technical assistance providers, connectors, and translators. Organizations like Mercy Corps Indonesia, Thailand Environment Institute, ISET Vietnam, Gorakhpur Environmental Action Group, and TARU Leading Edge have each developed deep subject matter expertise on urban resilience, but equally they have become process experts in understanding the importance of building buy-in and ownership from city stakeholders if this work is to endure and deepen. As there is never a "resilience" finish line to be crossed at which point all desired outcomes are achieved, building local knowledge, capacity, and investment in the agenda is critical.

While the resilience building approaches have varied some across the ACCCRN cities, the facilitating partners have relied on similar qualities. These include having credibility in both local and national contexts, approaching the work in an adaptive, learning, and flexible manner, using open communication and being connected to resources at different scales. The body of ACCCRN work has revealed that certain core principles are also important to the process of building urban resilience. These principles, which each ACCCRN facilitating partner adopted in different ways, were visible when looking comparatively across the different examples of implementation. These include using a multi-stakeholder processes, a strong focus on poor and vulnerable people, relying (at least in part) on local experts to conduct the research and analysis that fed into the resilience strategy building process (see Chap. 8: Engberg), and focusing on both hard and soft measures for building climate change resilience.

The impacts of climate change are likely to quicken and grow more severely in the future, affecting more and more urban communities. The urgency for climate change resilience building will increase and so too will experienced actors who can support cities and communities in their efforts to improve their ability to handle shocks and stresses. Out of ACCCRN emerged an experienced set of facilitators that have navigated the messy and contentious politics of cities to help urban communities become more resilient to climate change. Organizations like ICLEI have also been important partners to ACCCRN, using their approaches and networks to reach larger numbers of cities. Still more actors are needed. Looking ahead, innovations are needed to enable a much larger number of cities to benefit from resilience building guidance and accompaniment. The Rockefeller Foundation is supporting other efforts to generate greater resilience building capacity as a whole, including the piloting of a Global Resilience Academy. The challenge will be to develop ways for larger numbers of cities to benefit while also retaining the important lessons around sustained guidance from facilitators that help cultivate local expertise, leadership, and ownership over the climate change resilience agenda.

References

Anguelovski, I., Chu, E., & Carmin, J. (2014). Variations in approaches to urban climate adaptation: Experiences and experimentation from the global South. *Global Environmental Change, 27,* 156–167. doi:10.1016/j.gloenvcha.2014.05.010.

Ayers, J., & Dodman, D. (2010). Climate change adaptation and development I: The state of the debate. *Progress in Development Studies, 10*(2), 161–168. doi:10.1177/146499340901000205.

Bahadur, A. V., & Thornton, H. (2015). Analysing urban resilience: A reality check for a fledgling canon. *International Journal of Urban Sustainable Development, 7*(2), 196–212. doi:10.1080/19463138.2015.1060595.

Bartlett, S., & Satterthwaite, D. (Eds.). (2016). *Cities on a finite planet: Towards transformative responses to climate change*. Routledge.

Bhat, G. K., Karanth, A., Dashora, L., & Rajasekar, U. (2013). Addressing flooding in the city of Surat beyond its boundaries. *Environment and Urbanization, 25*(2), 429–441. doi:10.1177/0956247813495002.

Brown, A., Dayal, A., & Rumbaitis Del Rio, C. (2012). From practice to theory: Emerging lessons from Asia for building urban climate change resilience. *Environment and Urbanization, 24*(2), 531–556. doi:10.1177/0956247812456490.

Carmin, J., Anguelovski, I., & Roberts, D. (2012). Urban climate adaptation in the Global South: Planning in an emerging policy domain. *Journal of Planning Education and Research, 32*(1), 18–32. doi:10.1177/0739456X11430951.

Carmin, J., Dodman, D., & Chu, E. (2013). *Urban climate adaptation and leadership: From conceptual understanding to practical action* (No. 2013/26). Paris, France: Organisation for Economic Co-operation and Development (OECD). doi:10.1787/5k3ttg88w8hh-en.

Chu, E. (2016a). The political economy of urban climate adaptation and development planning in Surat, India. *Environment and Planning C: Government and Policy, 34*(2), 281–298. doi:10.1177/0263774X15614174.

Chu, E., Anguelovski, I., & Carmin, J. (2016). Inclusive approaches to urban climate adaptation planning and implementation in the Global South. *Climate Policy, 16*(3), 372–392. doi:10.1080/14693062.2015.1019822.

Chu, E., Anguelovski, I., & Roberts, D. (2017). Climate adaptation as strategic urbanism: assessing opportunities and uncertainties for equity and inclusive development in cities. *Cities, 60*, 378–387. doi:10.1016/j.cities.2016.10.016.

Chu, E. K. (2016b). The governance of climate change adaptation through urban policy experiments. *Environmental Policy and Governance, 26*(6), 439–451. doi:10.1002/eet.1727.

Chu, E. K. (2017). Urban climate adaptation and the reshaping of state–society relations: The politics of community knowledge and mobilisation in Indore, India. *Urban Studies*, 4209801668650. doi:10.1177/0042098016686509.

Chu, E., & Schenk, T. (2017). *Communicating about climate change with urban populations and decision-makers*. In: M. C. Nisbet (Ed.), *Oxford Research Encyclopedia of Climate Science* (vol. 1). Oxford and New York: Oxford University Press. doi:10.1093/acrefore/9780190228620.013.413.

da Silva, J., Kernaghan, S., & Luque, A. (2012). A systems approach to meeting the challenges of urban climate change. *International Journal of Urban Sustainable Development, 4*(2), 125–145. doi:10.1080/19463138.2012.718279.

Friend, R. M., Anwar, N. H., Dixit, A., Hutanuwatr, K., Jayaraman, T., McGregor, J. A., et al. (2016). Re-imagining inclusive urban futures for transformation. *Current Opinion in Environmental Sustainability, 20*, 67–72. doi:10.1016/j.cosust.2016.06.001.

Friend, R., & Moench, M. (2013). What is the purpose of urban climate resilience? Implications for addressing poverty and vulnerability. *Urban Climate, 6*, 98–113. doi:10.1016/j.uclim.2013.09.002.

Gorakhpur Environmental Action Group. (n.d.). *Rooting resilience in ODDS: Asian cities climate change resilience network (ACCCRN)—Gorakhpur*. Gorakhpur, India: Gorakhpur Environmental Action Group. http://geagindia.org/pdf/Rooting-Resilience-in-ODDS.pdf.

Indore City Resilience Strategy. (2012). *Indore city resilience strategy for changing climate scenarios*. In: G. K. Bhat, V. P. Kulshreshtha, U. A. Bhonde, U. Rajasekar, A. K. Karanth, & M. K. Burvey (Eds.), Indore, India: TARU Leading Edge.

Karanth, A., & Archer, D. (2014). Institutionalising mechanisms for building urban climate resilience: Experiences from India. *Development in Practice, 24*(4), 514–526. doi:10.1080/09614524.2014.911246.

Kernaghan, S., & da Silva, J. (2014). Initiating and sustaining action: Experiences building resilience to climate change in Asian cities. *Urban Climate, 7*, 47–63. doi:10.1016/j.uclim.2013.10.008.

Meerow, S., Newell, J. P., & Stults, M. (2016). Defining urban resilience: A review. *Landscape and Urban Planning, 147*, 38–49. doi:10.1016/j.landurbplan.2015.11.011.

Moench, M., Tyler, S., & Lage, J. (Eds.). (2011). *Catalyzing urban climate resilience: Applying resilience concepts to planning practice in the ACCCRN Program (2009-2011)*. Boulder, CO: Institute for Social and Environmental Transition (ISET).

Opitz-Stapleton, S., Seraydarian, L., Macclune, K., Guibert, G., Reed, S., Uennatornwaranggoon, F., et al. (2011). Building resilience to climate change in Asian cities. In K. Otto-Zimmermann (Ed.), *ReSILIENT CITIes: Cities and Adaptation to Climate Change Proceedings of the Global Forum 2010* (pp. 401–409). Dordrecht, Netherlands: Springer. doi:10.1007/978-94-007-0785-6.

Orleans Reed, S., Friend, R., Toan, V. C., Thinphanga, P., Sutarto, R., & Singh, D. (2013). "Shared learning" for building urban climate resilience—experiences from Asian Cities. *Environment and Urbanization, 25*(2), 393–412. doi:10.1177/0956247813501136.

Shi, L., Chu, E., Anguelovski, I., Aylett, A., Debats, J., Goh, K., et al. (2016a). Roadmap towards justice in urban climate adaptation research. *Nature Climate Change, 6*(2), 131–137. doi:10.1038/nclimate2841.

Shi, L., Chu, E., & Carmin, J. (2016). Global Patterns of Adaptation Planning: Results from a Global Survey. In K. C. Seto, W. D. Solecki, & C. A. Griffith (Eds.), *The Routledge Handbook of Urbanization and Global Environmental Change* (pp. 336–349). Routledge.

Tyler, S., Reed, S., MacClune, K., & Chopde, S. (2010). *Planning for Urban climate resilience: Framework and examples from the asian cities climate change resilience network* (No. ISET Working Paper 3). Boulder, CO.

Ziervogel, G., Pelling, M., Cartwright, A., Chu, E., Deshpande, T., Harris, L., et al. (2017). Inserting rights and justice into urban resilience: a focus on everyday risk. *Environment and Urbanization, 29*(1), 123–138. doi:10.1177/0956247816686905.

Author Biography

Anna Brown is an interdisciplinary practitioner with more than 15 years of experience in international development, philanthropy, and sustainable development. This includes 10 years working on urban climate change resilience. Until December 2016, Anna served as a Senior Associate Director with The Rockefeller Foundation (RF). Over her 10 years with RF, including nearly 7 based out of Bangkok, Anna played a role in several major efforts, including management of the Asian Cities Climate Change Resilience Network (ACCCRN), one of the Foundation's flagship initiatives. In 2017, Anna formed Aequita Consulting and focuses on equity, resilience, and environmental stewardship. Anna's past work includes the MIT-USGS Science Impact Collaborative, the Quaker United Nations Office, and UNESCO. Anna earned a bachelor's degree in Environmental Studies from Brown University and a master's in City Planning from the Massachusetts Institute of Technology.

Chapter 13
Between Policies, Programs, and Projects: How Local Actors Steer Domestic Urban Climate Adaptation Finance in India

Mitchell J. Cook and Eric K. Chu

Abstract In light of the slow progress in mobilizing international finance for climate adaptation in developing countries, a growing body of research promotes the idea of *pooling* blended forms of climate finance to leverage limited financial flows and enhance domestic control over allocation and accountability. Yet the constraints imposed by weaknesses in existing urban public finance institutions complicates perspectives on climate finance that envision the smooth pooling of blended finance from multiple sources across multiple scales. This chapter presents the case of Surat in India to illustrate how city governments can sustain an innovative approach to local climate adaptation while switching between various sources of funding. In this chapter, we ask two research questions: How do local governments that wish to pursue climate adaptation, often outside a comprehensive planning framework, steer these efforts around the numerous institutional, operational, and political constraints at the local level? Second, when the pursuit of climate adaptation is supported by time-bound external funding sources, how do local governments sustain action while switching between different types of financial flows? In the case of Surat, we find that the city is beginning to intentionally draw on intergovernmental fiscal transfers and, increasingly, their own local revenue resources. Such an implementation pathway emerges from the local government's ability to innovatively identify specific adaptation and development co-benefits and to exploit this with projects funded by existing and forthcoming streams of public revenue. Surat's experience suggests that, in contrast to *pooling*, cities will have to strategically *steer* climate adaptation action around local fiscal constraints created by the different governance logics associated with policies, programs, and projects.

M.J. Cook (✉)
Department of Urban Studies and Planning, Massachusetts Institute of Technology, Cambridge, MA, USA
e-mail: mjcook@mit.edu

E.K. Chu
School of Geography, Earth and Environmental Sciences, University of Birmingham, Edgbaston, Birmingham, UK
e-mail: e.chu@bham.ac.uk

S. Hughes et al. (eds.), *Climate Change in Cities*, The Urban Book Series,
https://doi.org/10.1007/978-3-319-65003-6_13

Keywords Climate change adaptation · Municipal public finance · Multilevel governance · Innovation · Urban politics · Fiscal planning · India

13.1 Introduction

Currently, there is no global consensus on the annual costs for cities in developing countries to adapt to climate change, nor an agreement on what material or institutional attributes constitute adaptation finance. The United Nations Framework Convention on Climate Change (UNFCCC) estimates that annual global adaptation costs could range from US$27 to $66 billion (UNFCCC 2007), while the World Bank puts costs slightly higher between US$70 and $100 billion (World Bank 2010). More recently, the United Nations Environment Programme (UNEP) projected the costs could rise to between US$280 and $500 billion annually by 2050 under the worst-case scenarios for mean global temperature increase (UNEP 2016). While there may be no consensus on adaptation costs or the classification of adaptation finance, there is growing agreement that international climate finance flows will not be enough to meet the adaptation needs of vulnerable cities in developing countries (UNFCCC 2016).

In light of slow progress in mobilizing resources and concerns over accountability for the use of global climate funds, various development organizations as well as a growing body of research promote the idea of *pooling* different sources of finance to leverage limited international public climate finance flows (Flynn 2011; Brugmann 2012). These sources range from official development aid (ODA) and other forms of development assistance in the broader global climate finance regime, private finance such as bank lending, bonds, and commercial equity, and domestic tax revenues and proceeds from natural resource extraction. Pooling or blending climate finance enhances domestic ownership and control over the design of adaptation interventions, while potentially unlocking private finance for adaptation by simultaneously supporting improvements in project execution and specific policy reforms (Irawan et al. 2012; Smallridge et al. 2013; Flynn 2011). Some developing countries have responded by forming new institutions and policies to structure and manage climate finance at the national level, such as in the case of Bangladesh's Climate Change Trust Fund and Climate Change Resilience Fund (Ayers et al. 2014). The formation of national climate funds and experimentation with national climate fiscal frameworks suggests that domestic public finance institutions are increasingly concerned with *pooling* climate finance flows that operate along diverse, and sometimes conflicting, governance logics. For instance, while allocation mechanisms in different global climate finance funds generally adopt a needs-based framework, they disburse according to different logics such as existing technical capacity or "readiness" to deliver benefits at scale (e.g., GEF), opportunities to leverage private investment (e.g., GCF and CIF), or even willingness to accommodate return seeking investors (e.g., G-20) (UNFCCC 2016).

Few studies have examined how—and to what effect—cities in developing countries combine or shift dependence on different sources of domestic and external finance to support climate adaptation (see also Chap 14). In contrast to ideas of *pooling* or *blending* climate finance by national governments, this chapter builds on the idea of *steering* climate action through different forms of domestic and external assistance to pursue adaptation and development co-benefits. The notion of steering is based on a bottom-up perspective on urban public finance in developing countries. A bottom-up perspective foregrounds the challenges of moving between time-bound external funds for mainstreaming adaptation and enhancing local capacity, which often tie revenue to the production of strategic plans or planning capacity, and forthcoming funding streams associated with domestic intergovernmental grants for urban infrastructure and services, which often tie funding for capital investments to specific local governance reforms.

In the long run, mobilizing finance to sustain support for climate adaptation planning and consolidate the benefits of material investments will inevitably require predictable growth in the revenue base of municipalities. In an encouraging move, early forms of multilateral and bilateral funding that focus on mainstreaming climate adaptation into local development planning are increasingly targeting reforms to fiscal policy frameworks at the national level. Yet, even as slow progress is made in mainstreaming climate adaptation into national budgets, the resource base over which municipal governments exercise control remains hampered by a range of factors associated with the uneven and partial implementation of fiscal decentralization reforms and limited capacity at the local level (Smoke 2015). Because of these organizational and institutional weaknesses, the capacity of local governments in developing countries to *pool* resources from different domestic and external (public and private) sources is limited. Thus, adopting a bottom-up perspective on urban public finance raises questions about how local governments pursuing adaptation options can sustain action even as reliance on different sources of funding changes, for instance, from external organizations to the local tax base.

This chapter presents a case study of the city of Surat in India to illustrate the motivations behind as well as the policy approaches to mobilizing public revenue to support and implement urban climate adaptation. We ask the following questions: How did the local government plan across various forms of public revenues to mobilize domestic public finance for urban climate adaptation? What were the implications of these fiscal strategies for planning and implementing climate adaptation interventions? This chapter contributes to the urban adaptation literature and practice in two ways. First, we detail the relationship between local fiscal dynamics and existing programmatic and institutional constraints around how climate adaptation can be mainstreamed into urban planning and development. Second, we relate these fiscal dynamics to the particular pathways available to cities to sustain climate adaptation actions following sponsorship by external organizations—in this case the Rockefeller Foundation's Asian Cities Climate Change Resilience Network (ACCCRN). Rather than focusing on the political-economic dynamics surrounding the allocation of adaptation finance at the international level, we examine bottom-up public sector approaches to financing local adaptation action. In the case of Surat, we find that the city is beginning to intentionally draw on intergovernmental fiscal transfers and, increasingly, their own local revenue resources to support adaptation. Such an innovative implementation pathway, switching between external philanthropic funding and domestic public finance, emerges from the local government's ability to incrementally identify specific adaptation and development co-benefits and to exploit this nexus with projects funded by existing and forthcoming streams of public revenue.

13.2 Urban Adaptation in Developing Countries: Understanding the Municipal Finance Dimension

The literature reviewed in the following sections relates to two primary knowledge gaps. First, even as many developing countries have begun *pooling* financial flows at the national scale to enhance the level and effectiveness of existing adaptation finance, the question of how local governments *steer* climate adaptation actions under conditions of institutional ambiguity in the municipal finance system remains open. Second, in contrast to the idea of *pooling*, the nature of municipal finance institutions in developing countries suggests the challenges of financing adaptation are more related to the complexities of *steering* plans and executing climate adaptation actions while *switching* between time-bound international, national, and local sources of funding. The following sections relate local climate adaptation actions to the dynamics of urban development and, more specifically, to local taxes and other sources of public finance available to local governments.

13.2.1 Urban Development and Climate Adaptation

The global discourse on climate adaptation has focused on recognizing the distributional inequities associated with climate impacts and seeking to remedy them by transferring resources, capacities, and technologies between industrialized and developing countries (Ciplet et al. 2013; Roberts 2009). Many scholars have identified barriers to effective adaptation in the urban context, which include the absence of science on climate hazards and their impacts, lack of understanding on how broader socioeconomic processes influence urban vulnerabilities, difficulties in integrating risk projections into local planning agendas, and lack of suitable governance and fiscal frameworks for risk management (Bulkeley 2010; Corfee-Morlot et al. 2010; Hunt and Watkiss 2011; Huq et al. 2007; Romero Lankao and Qin 2011). Despite the creation of dedicated climate funds under the UNFCCC—such as the Global Environment Facility Trust Fund's Strategic Priority on Adaptation, the Least Developed Countries Fund, the Special Climate Change Fund, and the Adaptation Fund—much of this money is channeled through national institutions and is not directly accessible to cities (Ciplet et al. 2013; Harris and Symons 2010; Schlosberg

2012; Shepard and Corbin-Mark 2009). Although there is likely some overlap in the priorities of national governments and the diverse needs of municipal governments, financial interventions solely at the international level likely will not suffice to address the imminent climate concerns of local governments (Kern and Bulkeley 2009).

The concentration of people and infrastructure at the urban scale provides distinct opportunities for reducing vulnerability and promoting adaptation. Adaptation is critical for communities in developing countries because of their disproportionate exposure to impacts and lower capacities to respond to extreme events (IPCC 2014). Local governments, in particular, will be at the forefront of responding to climate impacts due to the locally and contextually specific nature of risks and vulnerabilities (Carmin et al. 2012; Hallegatte et al. 2010; Hunt and Watkiss 2011). For most cities, key adaptation goals are reducing climate risks as well as securing people's well-being, enhancing adaptive capacity of the poor, and reforming land use policies that produce more vulnerability (Dodman and Satterthwaite 2009).

Cities pursue various adaptation planning approaches that fit with their existing political-institutional contexts and vulnerabilities (Anguelovski et al. 2014; Anguelovski and Carmin 2011). Cities tend to formalize adaptation planning in order to strengthen the legitimacy and facilitate implementation and coordination across sectors and departments (Carmin et al. 2012). Mainstreaming proposes the integration of adaptation priorities into citywide development planning (Bowen et al. 2012; Huq and Reid 2004; Ayers and Dodman 2010; Pasquini et al. 2015). Important reasons for integrating adaptation with development are to streamline decision-making (Smit and Wandel 2006), to accommodate the newly recognized climate agenda without "reinventing the wheel", to articulate co-benefits with other development priorities, and to reduce future remedial costs from climate variability (Agrawala and van Aalst 2008; Doll et al. 2013; Mercer 2010; Puppim de Oliveira 2013). Recently, climate action has begun to incorporate actors beyond the state, including civil society and private businesses (Bulkeley and Betsill 2013; Anguelovski et al. 2014; Chu 2017).

13.2.2 Urban Public Finance for Adaptation in the Global Climate Finance Landscape

Challenges associated with the long-term planning and implementation of adaptation for most cities in developing countries are compounded by more than just the scarcity of financial resources at the national level (Carmin et al. 2013). Rather, policy and administrative ambiguity in urban public finance institutions exert severe constraints over operationalizing innovative urban climate adaptation investments and services (Chu 2016). Efforts over the past three decades to strengthen the fiscal position of municipalities in developing countries by decentralizing tax authority have stalled, leaving city governments with a mismatch between available resources and spending obligations due to rapid urbanization. Potentially productive revenue streams to support implementation of local adaptation measures, such as the local property tax, continue to be inhibited by antiquated tax structures, poor administration, and weak enforcement (Bahl et al. 2013; Bahl and Wallace 2010; Mathur et al. 2006).

These features of municipal finance in developing countries complicate perspectives on climate finance that envision the smooth pooling of blended finance from multiple sources at the national and subnational scales (Brugmann 2012). For instance, research often points to a decisive role for municipal credit markets, particularly bond markets, in attracting private finance for adaptation. Yet in countries like India, critical aspects of bank lending and the deepening of municipal bond markets—such as accurate debt instrument pricing and repayment risk management—are hindered by the structural mismatch between local taxing authority and spending obligations (Pethe and Lalvani 2006), volatility in annual local tax yields (Mohanty et al. 2007), skills deficiency among staff overseeing local finances, and the extensive reliance on obsolete information technologies (Sheikh and Asher 2012). Moreover, some financial instruments—such as revenue or land-secured municipal bonds—are biased towards investment projects that can quickly generate positive cash flows in the form of user charges. Consequently, specific urban adaptation investments and services for which it is difficult to assign user charges for cost recovery, such as storm water drains or emergency shelters for the urban poor, are inevitably excluded.

Thus, given both the slow progress in meeting insufficient climate finance pledges and problems in the allocation mechanisms for distributing existing international climate finance for local adaptation action, domestic public finance institutions (i.e., both local and intergovernmental) should receive more consideration as potential resource bases for innovative strategies.[1] Table 13.1 lists four different categories of funding for adaptation from the perspective of municipal governments and their potential linkage to urban adaptation expenditures. This section restricts discussion to the local tax base and intergovernmental fiscal transfers, as these were crucial to how Surat sustained adaptation action at the conclusion of the external sponsorship by the Rockefeller Foundation. Spending on urban adaptation can come from different local revenue instruments, such as property taxes, user charges, impact and betterment fees, or earmarked and conditional fiscal transfers. Though often challenging to administer, property taxes can be buoyant sources of general revenue income. Buoyant property tax revenue affords local governments autonomy and flexibility to finance climate-related research, operations, infrastructure maintenance, small capital investments, and to service debt obligations (Bahl and Bird 2008; Oates 2001).

[1]Because of space constraints, we limit our discussion to various sources of tax, fee, and intergovernmental transfer revenue for direct public spending by general purpose municipal governments. The discussion is intended to be indicative and not comprehensive. This means we do not cover how local tax instruments, such as the property tax, could be designed to pursue climate adaptation objectives. Nor do we go into extensive detail about sector-specific intergovernmental transfers, such as for education or health.

Table 13.1 Potential sources of revenue and their uses for urban climate adaptation

Level of government	Revenue sources	Potential linkage to urban adaptation expenditures
Municipal	Property taxes User charges Impact fees Betterment levies Land value capture Vehicle taxes Licenses/registrations	• Maintaining infrastructure assets and retrofitting small capital investments • Setting rates to cover differential costs of adaptation expenditures • Covering the cost of incremental extension or upgrading of infrastructure and services • Cover personnel costs for research and advisory services
Intergovernmental	Earmarked grants Conditional grants Shared taxes	• Tying grants to programs and giving cities flexibility to decide how they wish to spend funds • Making grant disbursement conditional on reforms to local public administration and adaptation policy, programs, and expenditures
International public finance	International donor funds Official development assistance Technical and capacity support	• Providing concessional loans through national and/or regional governments for local infrastructure and service delivery improvements • Providing direct technical assistance to city governments to mainstream climate adaptation into urban planning
Private	Loans Municipal bonds Private investment Philanthropy Insurance/re-insurance Microfinance	• Supporting private investment or public-private partnerships in local adaptation services • Long-term finance for green infrastructure • Soliciting donations for targeted urban adaptation capacity programs for local officials and communities • Covering the costs of future welfare losses from weather or other climate-related events

User charges can be applied to public services for which local governments can impose direct consumption charges. If prices are set equal to the marginal cost of supplying the service and are perceived as fair by consumers, user charges can lead to efficient provision of services, improved service quality, and full cost recovery (Bahl and Linn 1992; Bird and Tsiopoulous 1997). Impact fees can generate upfront revenues to cover specialized adaptation expenditures, particularly in urban areas that might otherwise not be covered given existing revenue constraints. Local governments can also use betterment levies and other land-based taxes to capture a portion of the "unearned" increment in land values stemming from specific neighborhood investments, especially when adaptation investments enhance land values by reducing vulnerabilities to location-specific climate stressors (Fensham and Gleeson 2003; Smith and Gihring 2006). Lastly, earmarked or conditional transfers from higher levels of government could be used to encourage local governments to make investments that support objectives outlined in national climate action plans.

Though Table 13.1 lists many potential revenue instruments to finance adaptation, in reality, most cities often have limited autonomy over modify existing tax structures or diversify the local tax base. For instance, higher levels of government often restrict local tax rates to a predetermined range, require approval for increases in user charges, or outright prevent local governments from levying certain taxes. Regional governments often selectively intervene in local land management functions—ranging from property tax assessment policy to control over specific land use decisions—thus limiting the scope of local land value capture initiatives. Additionally, the formation of interim bureaucracies to implement programmatic elements of national urban policy in cities, such as project approval and oversight committees, generates strong centralizing pressures within the intergovernmental fiscal system (Baindur and Kamath 2009).

It is important to acknowledge that the technical, operational, and political complexities associated with the various local revenue instruments listed in Table 13.1 create opportunities and constraints for adaptation at the local level. For instance, while there is a strong economic case to support the position that the costs of regular infrastructure maintenance and upgrading should be covered by beneficiaries through local property taxes, many governments in developing countries resist levying property taxes where vulnerability is most concentrated—such as in informal settlements—in order to avoid giving the poor a formal claim on services (Smolka and Cesare 2012). Therefore, the conditions that establish the need to *steer* various forms of climate finance arise not only because of time-bound engagements funded through external climate finance, but also from the particular mix of technical, operational, and political factors associated with different revenue streams.

13.2.3 Adaptation and Domestic Public Finance in Indian Cities

India's *National Action Plan on Climate Change* (2008) outlined goals for addressing climate change at the national level, while also noting various responsibilities for state and local governments (Government of India 2008; Thaker and Leiserowitz 2014). Key climate impacts in India include rising surface temperatures, increasingly erratic rainfall leading to changes in monsoon, changing water availabilities, rising sea levels, and increasing cyclone intensities (Brenkert and Malone 2005; Dhiman et al. 2010; Prabhakar and Shaw 2007; Ranger et al. 2010; Revi 2008). Climate risks in Indian cities are typically associated more with exacerbated vulnerability than hazard exposure (Malone and Brenkert 2008; Revi 2008; Sharma and Tomar 2010). These vulnerabilities accentuate and further degrade the resilience of poor communities through loss of livelihoods, loss of community social safety nets, and reduced coping capacity to impacts (Kruks-Wisner 2011).

Several philanthropic foundations, NGOs, and development agencies have stepped in to support adaptation in cities, but these interventions are often limited to technical guidance or capacity development (Anguelovski and Carmin 2011; Bulkeley 2005). Though some cities are making use of these emerging opportunities, many are also discovering that they lack staffing and technical capacities to fulfill complex monitoring, reporting, and evaluation requirements set forth by funders and grant administrators (Carmin et al. 2013). As a result, many local governments are actively identifying alternative climate adaptation options that can be financed with their own local revenue income or with domestic intergovernmental revenue transfers.

In India, local governments are commonly assigned jurisdiction over roads, water, waste treatment, and public transportation (Bagchi and Chattopadhyay 2004). Despite these spending responsibilities, the annual revenue available to most Indian cities severely constrains the scale and scope of their activities. For example, in the 2007–2008 fiscal year, the share of local revenue in total public sector revenue was only 1.7% (ADB 2013). Most municipalities in India can levy some combination of property taxes, vehicle taxes, professional taxes, impact fees and betterment charges on construction, and user charges for solid waste, water, and electricity services. Consequently, cities continue to be highly dependent on intergovernmental transfers to fulfill their expenditure obligations and make long-term capital investments (HPEC 2011). Intergovernmental transfers typically include conditionalities related not only to the product of government expenditure (i.e., what the funds can be used on) but also to local governance *processes* (i.e., targets for tax collection, use of certain tax bases, the production of plans, etc.). As a result, relative changes in local dependence on different revenue instruments (i.e., taxes, intergovernmental transfers, or borrowing) are likely to generate considerable spatial (regional and intra-urban) and temporal variations in functional strategies to adapt to climate change, particularly if there are requirements for certain groups to benefit or pay for climate adaptation actions.

Municipalities receive fiscal transfers directly from state and national governments through the Central Finance Commission and Planning Commission for objectives outlined in each *Five-Year Plan*, as well as through participating in Centrally Sponsored Schemes and Central Sector Schemes with conditions determined by the various national ministries (Planning Commission 2011). In 2006, the Government of India introduced the Jawaharlal Nehru National Urban Renewal Mission (JNNURM), a conditional intergovernmental financing scheme that made available matching grants for urban infrastructure and low-income housing to 71 participating cities across the country. From 2006 to 2013, the national government sought to mobilize US$20 billion in public and private finance for urban infrastructure and low-income housing (Sivaramakrishnan 2011). JNNURM was designed to facilitate local institutional reforms in planning and fiscal management to enhance the capacity for maintaining public infrastructure assets (Kundu 2014).

Under the scheme, participating cities were required to match 50% of project costs covered by the national government with 25% from their own revenue (the other 25% was matched by state government revenue). Consequently, some of the most important reforms under JNNURM were in the areas of property tax administration and user charges for urban services.

In summary, this review has pointed to two primary knowledge gaps. The first gap relates to various institutional dynamics at the urban scale that render the strategy of resource *pooling* less feasible at the local level compared to other scales. How do local governments that wish to pursue urban adaptation actions, often outside a comprehensive planning framework, steer these efforts around the numerous local institutional, operational, and political constraints? Second, when the pursuit of climate adaptation actions is supported by time-bound external funding sources, how do local governments sustain action while switching between different types of financial flows? This chapter, therefore, seeks to trace how Surat has navigated these complexities to tap into different sources of public finance to achieve climate adaptation and development co-benefits. These lessons can then potentially be instructive to other cities in developing countries that are trying to pursue adaptation in the face of ongoing institutional, political, and fiscal constraints.

13.3 Methodology

This chapter presents a case study of climate adaptation planning in the city of Surat, with a particular focus on how local fiscal dynamics intersect with project framing and implementation. Surat was selected because the city has a long history of engagement with and has received support from key international programs, including the Rockefeller Foundation's Asian Cities Climate Change Resilience Network (ACCCRN) (see Table 13.2 for details). Surat was also selected based on an expectation of high "information content" (cf. Flyvbjerg 2006) related to the process of *transitioning* to a reliance on domestic public finance to sustain urban climate adaptation actions. Data for this chapter was collected in New Delhi (India's capital city) and Surat between January 2012 and May 2014. The methodology included semi-structured interviews with national policymakers and key actors of the city's adaptation process, observations of planning meetings, and an analysis of municipal budgets and plans.

Table 13.2 Key Climate and Development Indicators for Surat

State	Gujarat
Population (2011)	4,500,000
Climate impacts	Changing precipitation, increasing storm surges, and changing disease patterns
Development pressures	Rapid urbanization, migration, urban infrastructure deficits
Key adaptation needs	Improving disaster warning, vector-borne and gastrointestinal disease control, protecting vulnerable populations against climate stressors, reducing economic losses, improving housing quality, increasing public awareness of hazard impacts
Select adaptation interventions	Forming climate watch group, installing early warning system, developing disaster management plans, improving disease surveillance, monitoring public health, building skills in disaster response, training citizen groups, and building community awareness over climate impacts
Key city institutions	Surat Municipal Corporation, Southern Gujarat Chamber of Commerce and Industry, Surat Climate Change Trust
External actors	Asian Cities Climate Change Resilience Network (ACCCRN), Rockefeller Foundation, 100 Resilient Cities

13.4 Financing Urban Adaptation in Surat

Surat, in the western state of Gujarat, is projected to experience increasing climate impacts such as river flooding and vector-borne diseases (ACCCRN 2011). Since the 1960s, Surat has experienced nearly 60% decadal population growth mainly due to an influx of migrants (Kantor et al. 2006; Shah 1994). In 2006, unusually high rainfall around the city produced high discharges from the Ukai Dam, which is situated upstream from Surat on the Tapi River. During this particular episode, 75% of the city's built-up area was flooded, leading to high incidences of gastrointestinal and vector-borne diseases (Bhat et al. 2013; Karanth and Archer 2014). Coming off the heels of this disaster, Surat was selected to be a part of Rockefeller Foundation's Asian Cities Climate Change Resilience Network (ACCCRN) in 2008, which directed the city in producing vulnerability and impact assessments, designing pilot projects, and drafting a city resilience strategy (Brown et al. 2012; ACCCRN 2011; Kernaghan and da Silva 2014; Chu 2016). The resulting *Surat City Resilience Strategy*, published in June 2011, identified adaptation strategies across all municipal departments and sectors.

A particularly unique characteristic in Surat is the systematic effort to involve representatives of the private sector in the governance of economic and industrial policy (Bhat et al. 2013; Dholakia 2000; Yagnik and Sheth 2005). The state-level Gujarat Industrial Development Corporation provides comprehensive infrastructural facilities to industrial enterprises (Dholakia 2000) while representatives from the Southern Gujarat Chamber of Commerce and Industry also play a crucial role in facilitating trade, attracting investment, and supporting action around climate adaptation (Interview 2013). Because of Surat's status as an industrial hub, these private actors have historically exerted strong influence on policymaking in the city, hence the participation of the Chamber of Commerce in adaptation processes (Chu et al. 2016).

Financial support for Surat from the ACCCRN program concluded at the end of 2015. However, the city believed that climate adaptation was critical to its continued development, is an important mechanism for preparing the growing urban population against projected climate impacts, and would serve to raise the profile of Surat in the international arena. As a result, the Surat Climate Change Trust was established in June 2012. Instead of being an informal gathering of experts, the Surat Climate Change Trust would now be able to pursue structured and sustained adaptation projects. The institutionalization of the adaptation agenda through a public–private body outside the city government facilitated a dedicated focus on building the adaptive capacity of particular vulnerable sectors to articulate concrete channels of funding support and to ensure the continuity of the adaptation agenda in urban planning (Anguelovski et al. 2014).

The Surat Climate Change Trust generates income through project grants from the Surat Municipal Corporation and through soliciting funds from international organizations, including through the Rockefeller Foundation's new 100 Resilient Cities program. A number of managerial and institutional benefits underpinned the formation of the Surat Climate Change Trust. First, as it is outside the direct control of the Surat Municipal Corporation, the Surat Climate Change Trust would be able to bypass funding restrictions to solicit support from external sources as a private entity. One particular municipal officer stressed this issue,

> Cities cannot directly take money from external agents and funding agencies. It has to come to the central government, then to the state government, then to the cities… It is difficult for [cities] to then make use of this money for implementable and fully financed projects (Interview 2013).

Second, since the Surat Climate Change Trust exists separately from the bureaucratic and political organs of city administration, the Surat Climate Change Trust will be relatively immune to any changes in political or administrative direction. Lastly, as a formal legal entity devoted to charitable causes, the original objectives of the trust can never be changed or redirected.

Since its inception, the Surat Climate Change Trust has primarily worked on two adaptation projects: the End-to-End Early Warning System and the Urban Health and Climate Resilience Center, which cost US$509,000 and US$521,800 respectively, and were supported by seed funds from the Rockefeller Foundation (ACCCRN 2013). The Early Warning System provides a mechanism for integrating existing hydrological, climate, and urban development models into one comprehensive database (Interview 2013). The Urban Health and Climate Resilience Center, launched in June 2013, is aimed at building on the knowledge and operating procedures of the city's existing public health facilities.

On top of the pilot projects overseen by the Surat Climate Change Trust, in 2013, the Surat Municipal Corporation formally introduced a new line item in the municipal budget for funding climate-related studies and capital investments. The Surat Municipal Corporation budgeted a total of 20 million rupees (approximately US$300,000) to cover both recurrent and capital expenditures (Interview 2013). This new budget line item is not just a symbolic recognition of the importance of climate issues to the overall development trajectory of the city, it is also intended to complement and build upon existing urban infrastructure upgrading and service enhancement efforts, such as programs for slum relocation and rehabilitation, road infrastructure improvement, flood control, water distribution, and wastewater management (Interview 2013). Introducing a climate adaptation line item potentially provides long-term security, financial stability, and enables longer term planning without having to rely on often fickle and unpredictable external support.

Finally, while noting the projected economic costs of climate impacts, the Surat Municipal Corporation partnered with the Surat Climate Change Trust to implement specific upgrades and improvement projects to its existing wastewater and sewage treatment facilities using funds from JNNURM. As one municipal officer remarked,

> [Cities] have the National Urban Renewal Mission for improving infrastructure and services… Cities have been attaching a lot of adaptation projects to this program because there is no other channel to get resources to implement adaptation projects. The program only supports certain urban sectors or areas, for example water supply, drainage, or sewage. So, apart from this piecemeal approach, and if there are any cross-sectoral needs, cities are at a loss in terms of how to actually look for resources to implement projects (Interview, 2013).

Since adaptation priorities in Surat are often framed around disaster risk reduction and infrastructure protection, combining adaptation needs with current infrastructure upgrading needs made financial sense. For example, the Anjana Sewage Treatment Plant in the east of the city, which was heavily damaged in the 2006 floods, was selected for reconstruction and renovation under JNNURM. Direct capital investments in reconstruction and improvements of the Anjana Plant entailed many different types of expenditures, some of which were classified as adaptation. An example of such improvements included raising critical instrumentation above a flood line marked in the midst of the previous flooding. Municipal officials noted that there was both an explicit and implicit recognition of projected climate impacts even though much of the spending entailed standard upgrades given the availability of new technology (Interview 2013).

Much like the various Surat Climate Change Trust pilot projects that made use of dedicated adaptation funds, the Anjana Plant example highlights similar municipal approaches to innovatively navigating available urban financial resources to produce both urban adaptation and developmental co-benefits. In Anjana, though costly in relation to the Municipal Corporation's own revenues, total project financing was covered through combining existing local revenue with funds from JNNURM that were originally earmarked for general urban development, infrastructure upgrade, and public service provisions in the sewage sector.

13.5 Assessing the Nexus of Climate Adaptation and Urban Public Finance

The examples from Surat highlight the various pathways through which the city government has achieved adaptation benefits while shifting dependence from external philanthropic to domestic public funds (such as local revenues and intergovernmental transfers). The following section draws on these experiences to distill two implications for climate adaptation in the context of municipal finance. First, the institutional and policy changes that took place in Surat over the past decade show that more attention needs to be paid to how adaptation actions are sustained within the institutional, operational, and political dynamics of urban fiscal adjustment, particularly in resource-constrained contexts. Second, even under such challenging conditions, adaptation projects in Surat were realized through reframing and complementing existing programs for urban development, which have led to incremental institutional reform and innovation in light of the municipal budget cycle and existing multilevel resource dynamics.

13.5.1 From New Opportunities to Old Constraints in Financing Adaptation

The reality in Surat is that external resource and capacity interventions provided by actors such as ACCCRN are time-bound. When time-bound external interventions run up against intergovernmental transfer schemes with different governance logics, the environment in which local climate action takes place can quickly shift from one of new opportunities to old constraints. The objective of the external funding to Surat was to help the city government integrate adaptation into their day-to-day operations. Even though Surat has experienced some success in implementing climate adaptation measures, the local government is still positioned in an evolving multilevel fiscal system that constrains the long-term implementation and institutionalization of such measures. In the context of chronic underfunding for infrastructure and public services, Surat and other cities across India face a mismatch between growing expenditure responsibilities and limited control over revenue resources stemming from uneven implementation of the constitutional, legal, and administrative dimensions of fiscal decentralization.

In Surat, the local government has strived to adjust to fluctuations in the availability of different revenues sources by intensifying the use of existing tax sources and channeling intergovernmental transfers to investments with adaptation benefits. These maneuvers allow the city to tap into new fiscal opportunities to pursue adaptation objectives. As one municipal officer noted,

> [Our] major sources of income [for climate adaptation] are our internal sources: our tax revenues and non-tax revenues. The tax revenue consists mainly of property taxes, tax on vehicles, and provisional taxes. Non-tax revenue consists of rent from other municipal

assets - these are some of the internal sources. External sources are granted from state government and other grants we receive from central ministry (Interview 2013).

In addition to changes in this fiscal landscape, the overall structure of own-source revenues in the city has changed significantly since 2006. In particular, as a condition of participating in JNNURM, the city was required to abolish the *Octroi* tax on inter-jurisdictional commerce levied at entry points into the municipality. The abolishment of *Octroi*, in effect, reduced municipal revenue autonomy and negatively impacted subsequent capacities to maintain or upgrade urban infrastructure and basic services (Interview 2013). The Gujarat state government introduced a new grant transfer to compensate for the loss of *Octroi* revenue, but the transfer is ad hoc and did not match the size of the foregone revenue.

As a condition of receiving grants sanctioned by JNNURM, Surat had to offset the loss of *Octroi* revenue by reforming property taxes and user charges—with the goal of increasing property tax coverage to 90% and user charge collection rates to 85%—to increase local revenues. Between 2006 and 2013, the city increased revenues from property taxes, user charges, betterment charges, and impact fees on new construction and received additional intergovernmental grant transfers (see Table 13.3). Other direct taxes increased from 684,000 rupees in 2005–06 to over 22 million rupees in 2012–13. Even though the combination of these reforms supplied much-needed revenue to the local government, total revenues have never recovered to the levels they would have had Surat still had access to *Octroi* revenue.

In Surat, climate adaptation and disaster resilience only became priority issues after experiencing the plague in 1994 and severe floods in the summer of 2006. Financial and technical supports from ACCCRN starting in 2008 catalyzed further adaptation action in the city, which coincided with municipal revenue shortfalls after abolishing the *Octroi* tax in 2007. The convergence of all these events led the city to reframe climate adaptation as a component of infrastructure upgrading and service improvement. This reframing, in effect, institutionalized adaptation into a particular development strategy under the umbrella of a much larger, longer term project of governance reform in Surat. The transformation of adaptation into a

Table 13.3 Surat Municipal Revenues 2006–13 (in 100,000 Indian Rupees)

Revenue Income	2005–06	2006–07	2007–08	2008–09	2009–10	2010–11	2011–12	2012–13
Property tax	10,793	12,417	14,765	13,066	14,715	14,715	18,531	21,938
User charges	3,555	3,446	4,459	5,011	5,615	5,615	6,755	7,293
Other direct taxes	684	1150	596	9208	10,640	11,639	20,533	22,489
Octroi	40,520	48,559	34,734	0	0	0	0	0
Grant transfers	5509	8826	25,985	60,652	62,316	60,365	65,232	61,095
Total income	67,347	80,244	89,657	95,602	100,032	101,812	123,721	151,928

Source Surat Municipal Corporation

public good, as a result, allowed these projects to utilize public funds that were otherwise not specifically earmarked for adaptation. The Surat Climate Change Trust, the climate protection line item in the municipal budget, and the Anjana Sewage Treatment Plant upgrading project are all examples of adaptation projects that were facilitated through innovatively utilizing such public funds.

13.5.2 Between Policies, Programs, and Projects: Implementing Adaptation Within Local Fiscal Constraints

As we noted earlier, the approach to climate adaptation in Surat depended on the ability to find complementarities between available and forthcoming revenue streams and different infrastructure development needs. However, as one municipal corporation officer remarked:

> In case of development priorities, most cities have to draft city development plans that are required under the National Urban Renewal Mission. Under those terms, [cities] have to identify their development priorities. So initially, people asked how climate change actually aligns with their development priorities. This sort of disjoint in their understanding was there (Interview 2013).

In this context, Surat has pursued mainstreaming through slowly framing adaptation and development co-benefits, identifying complementary projects, creatively making use of available public finance, and relating adaptation objectives to a variety of funding sources. The Surat Climate Change Trust is one such platform upon which adaptation projects are framed and where climate adaptation planning authorities are devolved to a nonprofit institution with public and private representation (Chu et al. 2016). This is an innovative strategy that coincides with the general culture of governance reform promoted by all levels of governments in India. The creative navigation of existing and pipeline funds, primarily in the form of intergovernmental transfers, resulted in the inclusion of adaptation as one of the key components of the overall urban development vision.

Though Surat's experience with innovatively steering multiple public finance streams into adaptation actions is somewhat unique, the basis of this incremental approach to adaptation is not surprising in contemporary India. The national trend under JNNURM (and beyond) has shown that cities are indeed embarking on major public capital investments. While the political and economic contexts in which India's national urbanization transition is occurring has raised the profile of urban investment projects, the design of these major urban investment schemes has failed to pressure cities to integrate these projects within an urban governance framework that encourages participation and comprehensive planning. With limited local revenues and operating under conflicting objectives, project design and execution are often not fully linked to larger infrastructure networks or policy frameworks (Mahadevia 2011).

The Surat case also emphasizes how the mainstreaming of adaptation into overall urban development is determined by a project-based approach increasingly promoted by all levels of government in India (Chu et al. 2017). To some extent, reinforcing a project-based approach to local reform has not been the intention of policymakers responsible for operationalizing national urban policy in the design of fiscal transfers. As described by an official in the Planning Commission that oversaw the design of JNNURM, the aim of the scheme was misunderstood at the local level:

> While state and local governments felt that the money was coming for projects, the Government of India felt that the money was being given for putting in place a system (Interview 2012).

Nevertheless, the design of national urban policy with regards to intergovernmental transfers forces the linking up of revenue sources with specific projects and reform measures. Since the existing local bureaucratic and administrative culture is project-oriented, adaptation planning in Surat, unsurprisingly, has also been pursued on a project-by-project basis.

Lastly, the squeezing of local revenue autonomy in Surat reveals the city's significant reliance on intergovernmental fiscal transfers to support day-to-day urban services and infrastructure maintenance and upgrading functions. This reliance has both institutional and structural implications. The first points to the slow pace of improving local tax and user charge administration, such as making regular revisions to tax assessment roles and systematically enforcing penalties for non-compliance, which would set local revenue systems on a more sustainable path. The second relates to the sheer size of the challenge of building, maintaining, and expanding public capital investments in the context of rapid urbanization, population growth, and environmental change. Even with the emergence of many public–private partnerships targeting infrastructure and real estate development, the income raised from local tax bases is insufficient to meet the growing investment demand in Surat and other cities across India.

13.6 Conclusion

Surat's experience exemplifies the fiscal realities in many developing country cities. As urban governments continue to operate with low levels of fiscal autonomy, finance for implementing climate adaptation will necessarily take on a polycentric character because of the need to combine resources from multiple levels of government. This was observed clearly through examples of the various ACCCRN-supported pilot projects and other capital investments, which were subsequently expanded and implemented using matching intergovernmental grants. Under JNNURM, Surat was required to match 25% of the project costs out of its own revenue. Even though awareness of climate impacts is still quite low among municipal officers, the expectation is that future projects on adaptation and risk management will increasingly rely on public finance sources.

Consequently, even if international and domestic public finance flows at the national level can be strategically channeled into local climate adaptation investments, the expectation that cities match portions of national government funding therefore requires continued improvements in tax administration. To the extent that cities rely on intergovernmental transfers for investments in infrastructure and basic services, this structural reform agenda forms the basis of a largely hidden association between local fiscal reform and urban climate adaptation. The challenge then is to frame adaptation needs around existing development objectives and financial streams, and to facilitate adaptation actions around these constraints and opportunities. Through conceiving projects that further both adaptation and urban development goals, Surat has reframed adaptation into a public good and, thus, has established a budgetary basis that can make use of various funding streams that tie into existing funds for service provision, infrastructure upgrading, and institutional reform.

Cities around the world are beginning to realize that existing sources of international finance will not be enough to support the wide range of local adaptation efforts needed. As a result, some cities are actively identifying public sources of revenues—such as taxes, user charges, and intergovernmental transfers—to address climate risks and to improve the resilience of vulnerable populations. Nevertheless, public finance is not distributed evenly throughout a country's multilevel fiscal system. Cities have control over a narrow resource base while national and state governments retain control over the most productive taxes. Furthermore, some cities like Surat have the capacity to mobilize local tax revenues while others remain highly dependent on intergovernmental transfers.

These realities point to the complexities of *steering* climate action when revenue resources are distributed between policies, programs, and projects with distinct multilevel governance logics. While blending and pooling financial flows can enhance resource mobilization and improve utilization of funds, an urban public finance perspective on climate action reveals the process is more about *steering* activities into funding streams with different time horizons. Research on the nexus of climate adaptation and public finance is in its infancy, and the findings of this chapter should be interpreted in that context. Even though external assistance from philanthropic organizations and international networks has improved local capacity to plan for climate change and adapt to a broader range of impacts in the near future, the perspective of municipal finance in developing countries should receive more attention if adaptation is to be implemented, sustained, and institutionalized in the long-term.

References

ACCCRN. (2011). *Surat City resilience strategy*. Surat, India: Asian Cities Climate Change Resilience Network (ACCCRN) and TARU-Leading Edge.

ACCCRN. (2013). *ACCCRN City Projects: Asian Cities Climate Change Resilience Network. May 2013*. Bangkok, Thailand: Asian Cities Climate Change Resilience Network (ACCCRN), The Rockefeller Foundation, Asia Office.

ADB. (2013). *Municipal Finance Matters: India Municipal Finance Report (IMFR)*. New Delhi, India: Asian Development Bank (ADB) and National Institute of Public Finance and Policy (NIPFP).

Agrawala, S., & van Aalst, M. K. (2008). Adapting development cooperation to adapt to climate change. *Climate Policy, 8*(2), 183–193. doi:10.3763/cpol.2007.0435.

Anguelovski, I., & Carmin, J. (2011). Something borrowed, everything new: Innovation and institutionalization in urban climate governance. *Current Opinion in Environmental Sustainability, 3*(3), 169–175. doi:10.1016/j.cosust.2010.12.017.

Anguelovski, I., Chu, E., & Carmin, J. (2014). Variations in approaches to urban climate adaptation: Experiences and experimentation from the global South. *Global Environmental Change, 27*, 156–167. doi:10.1016/j.gloenvcha.2014.05.010.

Ayers, J., & Dodman, D. (2010). Climate change adaptation and development I: The state of the debate. *Progress in Development Studies, 10*(2), 161–168. doi:10.1177/146499340901000205.

Ayers, J. M., Huq, S., Faisal, A. F., & Hussain, S. T. (2014). Mainstreaming climate change adaptation into development: A case study of Bangladesh. *Wiley Interdisciplinary Reviews: Climate Change, 5*(1), 37–51. doi:10.1002/wcc.226.

Bagchi, S., & Chattopadhyay, S. (2004). Decentralised urban Governance in India: Implications for financing urban infrastructure. *Economic and Political Weekly, 39*(49), 5253–5260.

Bahl, R. W., & Bird, R. M. (2008). Subnational taxes in developing countries: The way forward. *Public Budgeting & Finance, 28*(4), 1–25. doi:10.1111/j.1540-5850.2008.00914.x.

Bahl, R. W., & Linn, J. F. (1992). *Urban public finance in developing countries*. Oxford and New York: Oxford University Press.

Bahl, R., & Wallace, S. (2010). A new paradigm for property taxation in developing countries. In R. Bahl, J. Martinez-Vazquez, & J. Youngman (Eds.), *Challenging the conventional wisdom on the property tax* (pp. 165–201). Cambridge, MA: Lincoln Institute of Land Policy.

Bahl, R. W., Linn, J. F., & Wetzel, D. L. (Eds.). (2013). *Financing Metropolitan Governments in Developing Countries*. Cambridge, MA: Lincoln Institute of Land Policy.

Baindur, V., & Kamath, L. (2009). *Reengineering Urban infrastructure: How the World Bank and Asian Development Bank Shape Urban Infrastructure Finance and Governance in India*. New Delhi, India: Bank Information Centre, South Asia.

Bhat, G. K., Karanth, A., Dashora, L., & Rajasekar, U. (2013). Addressing flooding in the city of Surat beyond its boundaries. *Environment and Urbanization, 25*(2), 429–441. doi:10.1177/0956247813495002.

Bird, R. M., & Tsiopoulous, T. (1997). User charges for public services: Potentials and problems. *Canadian Tax Journal, 45*(1), 25–86.

Bowen, A., Cochrane, S., & Fankhauser, S. (2012). Climate change, adaptation and economic growth. *Climatic Change, 113*(2), 95–106. doi:10.1007/s10584-011-0346-8.

Brenkert, A. L., & Malone, E. L. (2005). Modeling vulnerability and resilience to climate change: A case study of India and Indian States. *Climatic Change, 72*(1–2), 57–102. doi:10.1007/s10584-005-5930-3.

Brown, A., Dayal, A., & Rumbaitis Del Rio, C. (2012). From practice to theory: Emerging lessons from Asia for building urban climate change resilience. *Environment and Urbanization, 24*(2), 531–556. doi:10.1177/0956247812456490.

Brugmann, J. (2012). Financing the resilient city. *Environment and Urbanization, 24*(1), 215–232. doi:10.1177/0956247812437130.

Bulkeley, H. (2005). Reconfiguring environmental governance: Towards a politics of scales and networks. *Political Geography, 24*(8), 875–902. doi:10.1016/j.polgeo.2005.07.002.

Bulkeley, H. (2010). Cities and the governing of climate change. *Annual Review of Environment and Resources, 35*(1), 229–253. doi:10.1146/annurev-environ-072809-101747.

Bulkeley, H., & Betsill, M. M. (2013). Revisiting the urban politics of climate change. *Environmental Politics, 22*(1), 136–154. doi:10.1080/09644016.2013.755797.

Carmin, J., Anguelovski, I., & Roberts, D. (2012). Urban climate adaptation in the Global South: Planning in an emerging policy domain. *Journal of Planning Education and Research, 32*(1), 18–32. doi:10.1177/0739456X11430951.

Carmin, J., Dodman, D., & Chu, E. (2013). *Urban Climate Adaptation and Leadership: From Conceptual Understanding to Practical Action* (No. 2013/26). Paris, France: Organisation for Economic Co-operation and Development (OECD). doi:10.1787/5k3ttg88w8hh-en.

Chu, E. (2016). The political economy of urban climate adaptation and development planning in Surat, India. *Environment and Planning C: Government and Policy, 34*(2), 281–298. doi:10.1177/0263774X15614174.

Chu, E. K. (2017). Urban climate adaptation and the reshaping of state–society relations: The politics of community knowledge and mobilisation in Indore, India. *Urban Studies*. doi:10.1177/0042098016686509.

Chu, E., Anguelovski, I., & Carmin, J. (2016). Inclusive approaches to urban climate adaptation planning and implementation in the Global South. *Climate Policy, 16*(3), 372–392. doi:10.1080/14693062.2015.1019822.

Chu, E., Anguelovski, I., & Roberts, D. (2017). Climate adaptation as strategic urbanism: Assessing opportunities and uncertainties for equity and inclusive development in cities. *Cities, 60,* 378–387. doi:10.1016/j.cities.2016.10.016.

Ciplet, D., Roberts, J. T., & Khan, M. (2013). The politics of international climate adaptation funding: Justice and divisions in the greenhouse. *Global Environmental Politics, 13*(1), 49–68. doi:10.1162/GLEP_a_00153.

Corfee-Morlot, J., Cochran, I., Hallegatte, S., & Teasdale, P.-J. (2010). Multilevel risk governance and urban adaptation policy. *Climatic Change, 104*(1), 169–197. doi:10.1007/s10584-010-9980-9.

Dhiman, R. C., Pahwa, S., Dhillon, G. P. S., & Dash, A. P. (2010). Climate change and threat of vector-borne diseases in India: Are we prepared? *Parasitology Research, 106*(4), 763–773. doi:10.1007/s00436-010-1767-4.

Dholakia, R. H. (2000). Liberalisation in Gujarat: Review of recent experience. *Economic and Political Weekly, 35*(35/36), 3121–3124.

Dodman, D., & Satterthwaite, D. (2009). Institutional capacity, climate change adaptation and the urban poor. *IDS Bulletin, 39*(4), 67–74. doi:10.1111/j.1759-5436.2008.tb00478.x .

Doll, C. N. H., Dreyfus, M., Ahmad, S., & Balaban, O. (2013). Institutional framework for urban development with co-benefits: the Indian experience. *Journal of Cleaner Production, 58,* 121–129. doi:10.1016/j.jclepro.2013.07.029.

Fensham, P., & Gleeson, B. (2003). Capturing value for urban management: A new agenda for betterment. *Urban Policy and Research, 21*(1), 93–112. doi:10.1080/0811114032000062164.

Flynn, C. (2011). *Blending climate finance through national climate funds: A guidebook for the design and establishment of national funds to achieve climate change priorities*. New York: United Nations Development Programme.

Flyvbjerg, B. (2016). Five misunderstandings about case-study research. *Qualitative Inquiry, 12* (2), 219–245. doi:10.1177/1077800405284363.

Government of India. (2008). *National Action Plan on Climate Change*. New Delhi, India: Prime Minister's Council on Climate Change, Government of India.

Hallegatte, S., Henriet, F., & Corfee-Morlot, J. (2010). The economics of climate change impacts and policy benefits at city scale: A conceptual framework. *Climatic Change, 104*(1), 1–12. doi:10.1007/s10584-010-9981-8.

Harris, P. G., & Symons, J. (2010). Justice in adaptation to climate change: Cosmopolitan implications for international institutions. *Environmental Politics, 19*(4), 617–636. doi:10.1080/09644016.2010.489716.

HPEC. (2011). *Report on Indian Urban Infrastructure and Services*. New Delhi, India: High Powered Expert Committee (HPEC) for Estimating the Investment Requirement for Urban Infrastructure Services, Ministry of Urban Development, Government of India.

Hunt, A., & Watkiss, P. (2011). Climate change impacts and adaptation in cities: A review of the literature. *Climatic Change, 104*(1), 13–49. doi:10.1007/s10584-010-9975-6.

Huq, S., Kovats, S., Reid, H., & Satterthwaite, D. (2007). Reducing risks to cities from disasters and climate change. *Environment and Urbanization, 19*(1), 3–15. doi:10.1177/0956247807078058.

Huq, S., & Reid, H. (2004). Mainstreaming adaptation in development. *IDS Bulletin, 35*(3), 15–21. doi:10.1111/j.1759-5436.2004.tb00129.x.

IPCC. (2014). Climate change 2014: Impacts, adaptation, and vulnerability. In C. B. Field, V. R. Barros, D. J. Dokken, K. J. Mach, M. D. Mastrandrea, T. E. Bilir, et al. (Eds.), *Contribution of Working Group II to the Fifth Assessment Report of the Intergovernmental Panel on Climate Change*. Cambridge and New York: Cambridge University Press.

Irawan, S., Heikens, A., & Petrini, K. (2012). *National climate funds: Learning from the experience of Asia-Pacific countries. Working paper*. New York: United Nations Development Programme.

Kantor, P., Rani, U., Unni, J. (2006). Decent work deficits in informal economy: Case of Surat. *Economic and Political Weekly, 41*(21), 2089–2097.

Karanth, A., & Archer, D. (2014). Institutionalising mechanisms for building urban climate resilience: Experiences from India. *Development in Practice, 24*(4), 514–526. doi:10.1080/09614524.2014.911246.

Kern, K., & Bulkeley, H. (2009). Cities, Europeanization and multi-level governance: Governing climate change through transnational municipal networks. *JCMS: Journal of Common Market Studies, 47*(2), 309–332. doi:10.1111/j.1468-5965.2009.00806.x.

Kernaghan, S., & da Silva, J. (2014). Initiating and sustaining action: Experiences building resilience to climate change in Asian cities. *Urban Climate, 7,* 47–63. doi:10.1016/j.uclim.2013.10.008.

Kruks-Wisner, G. (2011). Seeking the local state: Gender, caste, and the pursuit of public services in Post-Tsunami India. *World Development, 39*(7), 1143–1154. doi:10.1016/j.worlddev.2010.11.001.

Kundu, D. (2014). Urban development programmes in India: A critique of JnNURM. *Social Change, 44*(4), 615–632. doi:10.1177/0049085714548546.

Mahadevia, D. (2011). Branded and renewed? Policies, politics and processes of urban development in the reform era. *Economic and Political Weekly, 46*(31), 56–64.

Malone, E. L., & Brenkert, A. L. (2008). Uncertainty in resilience to climate change in India and Indian states. *Climatic Change, 91*(3–4), 451–476. doi:10.1007/s10584-008-9472-3.

Mathur, O. P., Thakur, D., & Rajadhyaksha, N. (2006). *Urban property tax potential in India*. New Delhi, India: National Institute of Public Finance and Policy.

Mercer, J. (2010). Disaster risk reduction or climate change adaptation: Are we reinventing the wheel? *Journal of International Development, 22*(2), 247–264. doi:10.1002/jid.1677.

Mohanty, P. K., Misra, B. M., Goyal, R., & Jeromi, P. D. (2007). *Municipal finance in India. Study 26*. Mumbai, India: Reserve Bank of India.

Oates, W. E. (Ed.). (2001). *Property taxation and local government finance*. Cambridge, MA: Lincoln Institute of Land Policy.

Pasquini, L., Ziervogel, G., Cowling, R. M., & Shearing, C. (2015). What enables local governments to mainstream climate change adaptation? Lessons learned from two municipal case studies in the Western Cape, South Africa. *Climate and Development, 7*(1), 60–70. doi:10.1080/17565529.2014.886994.

Pethe, A. & Lalvani, M. (2005). *A comparative study of municipal finances in maharashtra: Patterns, problems and prospects. Working paper no. 16 in urban economics and regional development*. Mumbai, India: University of Mumbai.

Planning Commission. (2011). *Report of the Committee on Restructuring of Centrally Sponsored Schemes (CSS)*. New Delhi, India: Planning Commission, Government of India.

Prabhakar, S. V. R. K., & Shaw, R. (2007). Climate change adaptation implications for drought risk mitigation: A perspective for India. *Climatic Change, 88*(2), 113–130. doi:10.1007/s10584-007-9330-8.

Puppim de Oliveira, J. A. (2013). Learning how to align climate, environmental and development objectives in cities: Lessons from the implementation of climate co-benefits initiatives in Urban Asia. *Journal of Cleaner Production, 58,* 7–14. doi:10.1016/j.jclepro.2013.08.009.

Ranger, N., Hallegatte, S., Bhattacharya, S., Bachu, M., Priya, S., Dhore, K., et al. (2010). An assessment of the potential impact of climate change on flood risk in Mumbai. *Climatic Change, 104*(1), 139–167. doi:10.1007/s10584-010-9979-2.

Revi, A. (2008). Climate change risk: An adaptation and mitigation agenda for Indian cities. *Environment and Urbanization, 20*(1), 207–229. doi:10.1177/0956247808089157.

Roberts, J. T. (2009). The International dimension of climate justice and the need for international adaptation funding. *Environmental Justice, 2*(4), 185–190. doi:10.1089/env.2009.0029.

Romero Lankao, P., & Qin, H. (2011). Conceptualizing urban vulnerability to global climate and environmental change. *Current Opinion in Environmental Sustainability, 3*(3), 142–149. doi:10.1016/j.cosust.2010.12.016.

Schlosberg, D. (2012). Climate justice and capabilities: A framework for adaptation policy. *Ethics & International Affairs, 26*(4), 445–461. doi:10.1017/S0892679412000615.

Shah, G. (1994). Economy and civic authority in Surat. *Economic and Political Weekly, 29*(41), 2671–2676.

Sharma, D., & Tomar, S. (2010). Mainstreaming climate change adaptation in Indian cities. *Environment and Urbanization, 22*(2), 451–465. doi:10.1177/0956247810377390.

Sheikh, S., & Asher, M. (2012). A case for developing the municipal bond market in India. *ASCI Journal of Management, 42*(1), 1–19.

Shepard, P. M., & Corbin-Mark, C. (2009). Climate justice. *Environmental Justice, 2*(4), 163–166. doi:10.1089/env.2009.2402.

Sivaramakrishnan, K. C. (2011). *Re-visioning Indian cities: The urban renewal mission.* New Delhi, India: Sage.

Smallridge, D., Buchner, B., Trabacchi, C., Netto, M., Lorenzo, J., & Serra, L. (2013). *The role of national development banks in catalyzing international climate finance.* Washington, D.C.: Inter-American Development Bank.

Smit, B., & Wandel, J. (2006). Adaptation, adaptive capacity and vulnerability. *Global Environmental Change, 16*(3), 282–292. doi:10.1016/j.gloenvcha.2006.03.008.

Smith, J. J., & Gihring, T. A. (2006). Financing transit systems through value capture. *American Journal of Economics and Sociology, 65*(3), 751–786. doi:10.1111/j.1536-7150.2006.00474.x.

Smoke, P. (2015). Managing public sector decentralization in developing countries: Moving beyond conventional recipes. *Public Administration and Development, 35*(4), 250–262. doi:10.1002/pad.1736.

Smolka, M., & Cesare, C. M. D. (2012). Property tax and informal property. In W. J. McCluskey, G. C. Cornia, & L. C. Walters (Eds.), *A primer on property tax* (pp. 265–286). West Sussex: Blackwell Publishing Ltd.

Thaker, J., & Leiserowitz, A. (2014). Shifting discourses of climate change in India. *Climatic Change, 123*(2), 107–119. doi:10.1007/s10584-014-1059-6.

UNEP. (2016). *The Adaptation Finance Gap Report 2016.* Nairobi, Kenya: United Nations Environment Programme (UNEP).

UNFCCC. (2007). *Climate change: Impacts, vulnerabilities and adaptation in developing countries.* Bonn: United Nations Framework Convention on Climate Change.

UNFCCC. (2016). *2016 Biennial Assessment and Overview of Climate Finance Flows Report.* Bonn: United Nations Framework Convention on Climate Change.

World Bank. (2010). *Development and climate change: Stepping up support to developing countries.* Washington, DC: The World Bank.

Yagnik, A., & Sheth, S. (2005). *The shaping of modern Gujarat: Plurality, Hindutva and beyond.* New Delhi, India: Penguin Books India.

Author Biographies

Mitchell Cook is a development consultant and Ph.D. candidate in the Department of Urban Studies and Planning at the Massachusetts Institute of Technology. His research examines the politics and practice of municipal finance reform, particularly in the areas of urban climate adaptation finance, digital information systems for government fiscal operations, and local tax policy and administration. As a recipient of Fulbright-Nehru Fellowship in 2014–2015, he studied the development and implementation of innovative public finance technology and urban governance in Bengaluru (Bangalore), India. Mitchell has held various research consultancy positions in the private sector and with the Asian Development Bank, the World Bank, and UNICEF.

Eric K. Chu is a Lecturer in Planning and Human Geography in the School of Geography, Earth and Environmental Sciences at the University of Birmingham in the UK. His research is on the politics of climate change governance in cities, with particular emphasis on the globally comparative perspectives of socio-spatial change, development planning, policy reform, and local environmental justice.

Part IV
Drivers and Obstacles of Multi-Level Innovations

Chapter 14
Multilevel Governance and Innovations in the Financing of Urban Climate Change Strategies

Jacqueline Peterson

Abstract Cities have become key policy actors in the global fight against climate change. Local climate change initiatives, however, generally come with a cost and for policies to be successful, appropriate revenue sources and financing mechanisms are necessary. Decentralization theories and municipal finance literature provide guidance on how most municipal services should be financed, yet encounter challenges when it comes to the policy area of climate change. Models based on aligning the benefits of a given municipal service with those who pay for it are of limited use, as the beneficiaries of urban climate change efforts are simultaneously local and global. Cities have been able to tailor traditional financing mechanisms (such as property taxes and user fees) to incentivize citizens to reduce their carbon footprints while generating the resources necessary to fund local climate change projects. However, given the unconventional and global nature of climate change as an urban policy area, cities will likely benefit by looking beyond the formal intergovernmental institutions that structure cities' financing abilities. This chapter advocates applying a multilevel governance lens that recognizes the importance of new, outside actors that operate at both the local and global scales to climate change financing solutions. By acknowledging the important relationship between municipal finance and the broader multilevel frameworks that govern climate change policy, cities have the opportunity to pursue innovations in municipal finance to support their climate change objectives.

Keywords Municipal revenue · Climate finance · Financial tools · Urban finance · Federalism

J. Peterson (✉)
University of Toronto, Toronto, Canada
e-mail: jacqueline.brown@mail.utoronto.ca

S. Hughes et al. (eds.), *Climate Change in Cities*, The Urban Book Series,
https://doi.org/10.1007/978-3-319-65003-6_14

14.1 Introduction

Cities across the world often face challenges in securing the funding needed to adequately deliver the most standard of city services. The revenue generating tools available to cities are limited not only by local political and economic realities but —importantly—are inevitably affected by policies set by other levels of government (DuPuis and McFarland 2016). All cities must navigate within multilevel governance frameworks that include not just other levels of government, but non-governmental actors as well (Bulkeley and Betsill 2003) such as private interests, international organizations and transnational networks. These factors can affect both cities' policy obligations *and* fiscal health. As cities become key players in the fight against climate change (Gore and Robinson 2009), the question must be not only which climate change strategies to adopt, but how to pay for them. The latter question is often ignored in the climate change policy literature, although it is widely acknowledged that fiscal capacity is a critical determinant of how successful local climate change policy can be (Lubell et al. 2009; Bulkeley 2010; Homsy and Warner 2015). The traditional revenue sources available to local governments—such as property taxes, user fees and intergovernmental transfers—can certainly be used to fund projects directed at climate change mitigation and adaptation. However, given the unconventional and global nature of climate change as an urban policy area, many cities are sharpening their tool kits and recognizing the need to innovate and adapt their financing mechanisms to encourage behavioural change and achieve their policy objectives. Because climate change is multi-jurisdictional and outside actors have such a large impact on municipalities' resources, many of the innovations in municipal finance are reflective of changes in multilevel governance and the institutional arrangements governing urban climate change policy.

Municipal finance is not generally considered a policy area with high levels of innovation. Given the complexity of the policy area and the importance of reliable revenue sources to the provision of municipal goods and services, municipal finance mechanisms tend to be predictable, uncontroversial and slow to change. Moreover, much of the municipal finance literature highlight the formal, hierarchical government institutions and the impact that higher levels of government and intergovernmental relations can have on urban policy outcomes, rather than the informal horizontal networks that can also impact local capacity. Indeed, cities *are* able to make small changes to existing finance mechanisms within these institutional arrangements to advance their environmental aims. Taxes, grants and fees can be tailored to incentivize citizens to act in a more environmentally conscientious manner and can also be used to raise the revenue necessary for climate change projects. However, climate change presents a unique policy challenge for cities: while solutions (and expenditures) are local, the impacts are global. As a result, cities will likely benefit by drawing on new, innovative financing strategies involving outside actors that operate at both the local and global scales. Policymakers will benefit by applying a multilevel *governance* lens that looks beyond formal intergovernmental structures and recognizes the importance of new, outside actors in climate change governance and the finance opportunities they represent.

In the sections that follow, I first highlight the theoretical dilemmas posed by situating urban climate change policy within the study of municipal finance. Theory suggests which level of government ought to offer a given service and which tax and revenue instruments should be used to finance the service. However, the financial tools available to cities are often inadequate to address the problem of climate change, requiring us to revisit the multilevel arrangements through which urban climate change policy is financed and governed. To understand how innovations in multilevel finance can better equip cities with the mechanisms needed to fight climate change, I provide an overview of cities' main revenue sources: the property tax, user fees, sales taxes, income taxes, intergovernmental transfers and municipal borrowing. I assess the potential for each instrument to finance climate change programmes and review the ways these revenue and financing tools have been adapted to incentivize behavioural change and advance urban climate change objectives. In the fourth section, I identify new tools and innovations that become available when cities recognize changing multilevel governance dynamics and involve outside actors whose roles and resources are better equipped to meet the unique demands of climate change. The chapter concludes by suggesting further avenues of inquiry into the role of municipal finance innovations on urban sustainability. While I focus primarily on the financing mechanisms available to cities in North America, these mechanisms are commonly used in cities across the world. As such, this chapter has relevance beyond North America and I frequently draw on global cases to highlight alternative practices, perspectives and opportunities for innovation.

14.2 Theory and the Dilemma of Urban Climate Change Finance

The question of how cities should finance climate change strategies presents a challenge for municipal finance scholars. Climate change is global problem, and climate protection is a global public good (Krause 2012). Greenhouse gas (GHG) emissions cross-borders and the effects of environmental policy—both good and bad—in one jurisdiction inevitably 'spill over' into other jurisdictions (Kunce and Shogren 2005). Climate change mitigation, however, requires localized action and cities are key policy actors (Bulkeley and Betsill 2003). It has been estimated that up to 75% of GHG emissions come from urban areas (Bulkeley 2010). Cities are usually responsible for land use and development patterns, which can have a significant environmental impact, and a number of municipal goods and services—such as transit, utilities and waste and recycling collection—have direct environmental consequences (Robinson and Gore 2005). Thus, while the problem of climate change is global, policy solutions are often local, highlighting the need for multilevel policy analysis. Like climate change, municipal finance must also be analyzed through a multilevel governance lens that recognizes the many actors—both governmental and non-governmental—that can have a direct influence on cities' finances.

Fiscal decentralization theories inform our understanding of how government services should be distributed and financed among different levels of government, and specifically that whenever possible, services should be governed and financed by local government when the benefits and geographic scope of a given policy area are contained. Two principles underpin decentralization theories of local government responsibilities and revenue sources. First, Oates' 'Subsidiarity Principle' (1972) theorizes 'that government services should be provided by the lowest level of government that can do so efficiently' (Alm 2015, 233). Second, Olson's 'equivalence principle' (1969) argues that the level of government responsible for a given programme or policy area should align with the geographic area and electorate impacted by the policy (Ahmad and Brosio 2015). This principle complements the 'benefits model' of local finance: wherever possible, those that benefit from a given service should be those that pay for it (Kitchen 2006). According to fiscal federalism perspectives, local governments might be better positioned to administer programmes that require local knowledge and evaluation, such as local or land-based programmes (Shah and Shah 2006). However, to promote economic efficiency and national equity, higher levels of government should manage redistributive taxes and taxes on mobile goods and services (Shah and Shah 2006).

Economists argue that revenue source and expenditures should be linked; different tax and financing tools are appropriate for different types of expenditures (Kitchen and Slack 2016). To increase accountability, governments should access own-source revenue as much as possible to fund their expenditure needs (Shah and Shah 2006). Also, as per the benefits model, whenever it is possible for the users of a given service or piece of infrastructure to be identified and charged for it, 'user fees' should be levied (Bazel and Mintz 2014). Many expenditures, however, are public goods that are not exclusive and may be better financed through land-based taxes such as property taxes (Kitchen and Slack 2016). Transfers from higher levels of government may be appropriate when benefits from a given policy 'spill-over' beyond a given jurisdiction (Slack 2011). Furthermore, borrowing can be an appropriate tool to finance large capital projects that will benefit future generations (Shah and Shah 2006; Alm 2015).

These theoretical models, however, only go so far in explaining how urban policies are *actually* financed. First, the division of fiscal capacities advocated by fiscal federalism is an extremely political process, as such divisions help to shape the power distribution between different levels of government. The use of different municipal finance tools—such as taxes and fees—to finance different services can be similarly political and not always reflective of economic principles (Pagano and Perry 2008). Regardless of the economic efficiencies associated with a given financing mechanism or distribution of responsibilities, governments may leverage the tools and powers available to them to advance their policy preferences within—and across—jurisdictions. As a result, decentralization approaches such as fiscal federalism are often charged with being overly normative (Ahmad and Brosio 2015); they explain how services *ought* to be governed and financed between different levels of government but are less adept at explaining *actual* policy arrangements.

Second, these approaches focus largely on formal government institutions and less on the informal institutions and private actors involved in urban governance. Over the past 20 years, there has been a conceptual shift among urban scholars away from government towards *governance* (Rhodes 1997; Stoker 1999). Governance involves the 'rise of networks' and the 'blurring of public–private boundaries' (Kjaer 2009, 138). Urban policy isn't simply defined within the confines of government institutions; rather, urban governance theory highlights the ways that norms, economic conditions and different types of actors and institutions interact to create policy objectives (Pierre 2011). A multilevel governance approach looks beyond the formal hierarchical intergovernmental relations and recognizes the many vertical and horizontal relationships and opportunities for collaboration that shape cities' policy preferences and capacities (Peters and Pierre 2012). Multilevel governance approaches recognize that certain public goods and services may not be efficiently provided by formal 'general purpose' governments, and that more flexible 'task specific' governance arrangements may be necessary to address problems that are not confined within a given jurisdiction (Hooghe and Marks 2003).

The disconnect between municipal finance theory and actual urban governance dynamics is particularly acute in the context of climate change policy. The equivalence principle suggests that climate change, as a policy issue, ought to be the responsibility of the highest level of government. The spillover effects associated with greenhouse gas emissions assume the need for national, if not international, policy solutions and financing mechanisms. Intergovernmental grants are typically considered appropriate financing instruments for services that are non-exclusionary and inter-jurisdictional, which define many urban climate change programmes. However, action at the international and state level has been slow and financial support for climate change initiatives have been inconsistent, inadequate and highly political (Gore and Robinson 2009). Additionally, many cities have experienced the downloading of responsibilities from higher levels of government without the resources to finance them (Pagano and Perry 2008; Abels 2014). As such, cities are increasingly tackling climate change themselves (Bulkeley 2010) with the local financing mechanisms at their disposal. While the financing tools traditionally available to cities can be adapted and used to incentivize behaviour change and reduced emissions at the local level, the disconnect between the *global* nature of climate change and the need for *local* policy solutions challenges existing models of municipal finance and highlights the need for innovation in urban climate change finance and creative multilevel solutions.

Climate change governance is complex; theory offers no parsimonious explanation of how urban climate change strategies are (or should be) financed. Theory does, however, provide conceptual markers that help us evaluate municipal financing mechanisms in terms of their ability to advance sustainability aims. It allows us to assess which policy areas (e.g. transit, building regulations and water infrastructure) are best served by different financing mechanisms, and whether the appropriate actors and levels of government are involved. Recognizing the complex and multilevel nature of climate change and municipal finance allows us to identify the ways cities can work with different actors to improve these tools and develop new ones.

14.3 Taxes, Fees, Transfers and Borrowing: Can Cities Adapt Existing Finance Mechanisms to Advance Climate Change Policy?

Cities generally draw from a limited number of revenue sources to fulfil their everyday service responsibilities. While these revenue sources can vary, they typically consist of a mixture of property, sales and income taxes, user fees and intergovernmental transfers (OECD 2010). Additionally, cities can finance capital investments by borrowing funds through practices such as issuing municipal bonds (Alm 2015). These tax and revenue tools can be used both to fund cities' climate change initiatives as well as incentivize more sustainable behaviour. Municipal finance mechanisms are not 'climate neutral'; how and whether certain instruments are used can affect cities' ability to achieve their sustainability goals (Kamal-Chaoui and Robert 2009). In this section, I consider each of these mechanisms individually, looking at the ability of each instrument to advance cities' climate change objectives and the impact of other actors on the efficacy of these tools. I include examples of innovation and highlight ways that each of these common financing mechanisms has been adapted to better meet the needs associated with urban climate change objectives.

14.3.1 Property Tax

In many countries, the property tax remains the most important revenue source available to cities (Chernick et al. 2011; OECD 2010). Because the tax is levied on a fixed property, the tax base is 'immovable' and an excellent tool for cities to finance local public goods (Slack 2011). Moreover, although both the tax rate and assessed value of properties can fluctuate with economic and political changes, property taxes provide cities with an overall reliable revenue stream with which to finance most municipal services (Kitchen and Slack 2016).

Cities can—and do—use revenue generated from property taxes to help realize their sustainability objectives. The property tax is an appropriate funding source for local environmental initiatives whose benefits are public and non-exclusive, which may range from bike lanes to rain gardens to environmental building standards. The stability of the property tax as a revenue source provides cities with secure and predictable funding for these and other 'green' infrastructure and services.

While property tax revenues may help fund a number of climate change initiatives, as it is commonly implemented it is not always useful for incentivizing sustainable behaviour by residents. Property assessments do not always reflect a property's 'ecological footprint' and the amount of infrastructure required to service a certain property. Properties in automobile-dependant suburban areas often have effective tax rates lower than central, multi-family mixed use properties in denser areas (Kamal-Chaoui and Robert 2009). To address these biases and incentivize more sustainable development patterns, cities may choose to reform their property taxes and explore options such as split-rate property taxes (e.g. Sydney, Hong Kong, Pittsburgh and Denmark), differential taxation (Austin) and preferred rates for multi-family units (Denmark) (Kamal-Chaoui and Robert 2009, 120–121). Cities can also provide green property incentives and offer rebates for environmental retrofits, such as California's residential turf rebate, which encourages natural landscaping (California Department of Water Resources).

Although the property tax is usually levied at the local level, cities do not always have full control of the tax and the revenue generated from it. Higher levels of government usually set the terms and rates under which property taxes can be collected and changed; and reforming property tax models would require changes in national or subnational legislation. Also, property tax increases are often highly political; local policymakers may try to limit increases to property tax rates by turning to other revenue sources or limiting expenditures that are not considered urgent, such as climate change mitigation projects.

14.3.2 Fees

Cities are increasingly turning to user fees as a way to limit property tax increases and better align ratepayers and the beneficiaries of a given municipal service (Pagano and Perry 2008). User fees are appropriate when services are excludable and can be tied to a specific user. This is usually the case with utilities, transit and use of recreation facilities, for example. Like property taxes, user fees can generate a reliable revenue stream for cities. Moreover, they are often considered fair and efficient, as users are charged directly for the services provided to them (Slack 2011). At the same time, user fees can also be politically unpopular, especially when some services have traditionally been 'free' (Slack 2014). Also, cities often have limited flexibility as to how revenue generated from user fees can be spent; these fees are typically earmarked and tied to specific expenditures. As with property taxes, higher levels of government often impact the ability for municipalities to charge user fees and the conditions under which they are able to do so.

User fees can be effective tools for cities to finance sustainability targets. Fees can incentivize behavioural change by charging individuals for practices that increase GHG consumption (Kamal-Chaoui and Robert 2009). Metering water and charging users for the amount of water used has been proven to encourage water conservation; similarly, charging a waste collection fee per bag has shown to reduce waste to landfills (Kitchen 2006, 16). Innovative utility pricing mechanisms, such as increasing water block rates ('progressive rate structures') or 'bi-directional' power can affect how utilities are consumed, and, in some cases, produced (Mullin 2008; Braun and Hazelroth 2015).

Many user fees with environmental impacts are those imposed on transportation use. Toll roads can be used to put a price on the automobile use and help finance associated infrastructure costs (Bazel and Mintz 2014); they can also reduce GHG emissions by pricing negative externalities and encouraging more 'efficient' transportation choices (Slack 2014). Other 'paid-use' models include the new Vehicle Miles Traveled (VMT) fee piloted by Oregon and California (DuPuis and McFarland 2016, 13), Singapore's Electronic Road Pricing system (Kitchen 2006) and congestion charges used in London, Stockholm and Milan (Kamal-Chaoui and Robert 2009). Improved technology has facilitated this, allowing cities to match infrastructure and service use to specific individuals in ways that would not have been feasible earlier (Pagano and Perry 2008; Bazel and Mintz 2014).

Impact fees (also known as 'development charges' or 'developer levies') have also become popular for cities trying to mitigate the economic and environmental costs of sprawl. While not considered a user fee in the typical sense, impact fees may be placed on new developments to help cover the extra costs associated with growth and incentivize more efficient growth patterns (Kitchen 2006; Lee et al. 2014).

In addition to incentivizing behavioural change, user fees can support cities' climate change initiatives if cities choose to channel expenditures generated from user fees to encouraging sustainable practices and subsidizing the development of new technologies in a given service area. For example, solid waste collection fees in Minnesota are collected by the state and then redistributed by counties to municipalities to support their recycling and organics recycling initiatives (Peterson and Hughes, in press).

14.3.3 Sales Tax

Many cities also have the option of gathering revenue through a sales tax, which can be applied throughout a municipality on all—or select—goods and services. Like both property taxes and user fees, sales taxes produce a reliable revenue stream for cities. Moreover, cities often have discretion as to how this revenue is spent; although it can be tied to specific expenditures, this need not always be the case. Sales tax revenue often feeds into cities' general funds, from which a variety of projects (including climate change initiatives) can occur.

Cities that have the authority to impose a tax on the sale of specific goods or services can use the tax to incentivize climate-conscious behaviour when the tax is applied to products associated with greenhouse gas emissions, for example. Moreover, as is the case with user fees, revenue generated from some sales taxes can be earmarked for specific climate change initiatives. For example, Missouri diverts one-eighth of every cent of sales tax to fish, wildlife and forest conservation efforts (Ziehmer 2010). Fuel taxes may be considered a form of sales tax, and are often exacted by higher levels of government and then shared with local governments to fund transportation projects (Kitchen 2006). Fuel taxes are a relatively popular form of sales tax to finance urban climate change initiatives, and it is common for revenue generated from fuel taxes to be earmarked for transit, as is the case in Victoria and Vancouver, Canada (TD Economics 2004).

However, local sales taxes can be politically volatile, and may cause consumers to cross-jurisdictions for products if the tax is not consistent throughout a metro area. Moreover, the ability of cities to set and collect sales taxes is almost always determined by higher levels of government. Many municipalities do not have the legal authority to collect sales taxes. For other cities, higher levels of government will either set the tax rate or set a limit for it. For example, the state of Texas allows cities to impose a sales tax of up to 2% (although it is rare for cities to actually set their sales tax this high) (Texas Comptroller of Public Accounts).

14.3.4 Income Tax

While extremely rare in North America (New York City is a notable exception), many European cities—such as Copenhagen, Helsinki and Stockholm—are financed primarily through income taxes (OECD 2010). Income taxes are useful mechanisms for redistributing resources, but may not be a useful revenue mechanism for localized, land-based expenditures. The ability for cities to exact income taxes is dependent on legislation made by higher levels of government, which generally restricts local bodies from charging income taxes.

For cities that do have access to income tax revenue, resources can be directed towards climate change programmes. There is little incentive for citizens, however, to reduce their environmental impact through this mechanism, as tax owed is a reflection of income, not behaviour. That being said, some cities have sought to use income taxes to address service distortions caused in large metro areas where the place of employment differs from place of residence and suburban communities benefit from investments made by the center city (Slack 2011). A 'commuter tax' has been employed in both Ohio and Kentucky, where cities can impose a 'uniform income tax rate' on residents and nonresidents alike; however, these commuter taxes can often lead to 'protracted political battles', as has been the experience in New York City (Pagano and Perry 2008, 27). It is not clear what the climate change benefits of a 'commuter tax' may be, but access to these funds may discourage driving, incentivize suburban municipalities to provide more employment opportunities and reduce the amount of commuters into central areas.

14.3.5 Intergovernmental Transfers and Grants

Cities across the world differ in terms of their reliance on transfers from higher levels of government, but nearly all do to some extent. Intergovernmental transfers are an appropriate way for state and subnational governments to redistribute funding to capture inefficiencies associated with 'spillovers' from cross-jurisdictional policy areas, such as transportation and environmental protection. A large proportion of intergovernmental transfers to cities come in the form of grants, which can be either conditional or unconditional in nature (Tindal and Tindal 2009). Grants are a useful way for cities to finance capital projects with significant upfront costs, including large investments in 'green' infrastructure (ex. transit lines, flood mitigation infrastructure, etc.). The US federal government, for example, has encouraged local energy efficiency projects through the Energy Efficiency and Conservation Block Grants (EECBGs) (Krause 2012).

However, grants are not always a reliable revenue source for cities. They often reflect the political priorities of other levels of government, and in most cases are earmarked for specific projects or purposes (Tindal and Tindal 2009). Moreover, changes in administration at other levels of government can lead to changes in the availability of grants and distribution of funds. This has implications for cities that depend on intergovernmental grants to reach their sustainability objectives. Cities have less autonomy with respect to fulfilling their climate change objectives, and goal alignment between different levels of government may be necessary for significant investments in 'green' infrastructure to occur.

At the same time, however, grants may be a way for higher levels of government to incentivize action if local leadership is unwilling to invest in climate change strategies. Conditional funding may help national governments steer urban policy and incentivize climate change initiatives at the local level (Tindal and Tindal 2009). Germany, Portugal and Brazil all consider environmental indicators in their grants to local governments (OECD 2010). The increased use of sustainability criteria represents a shift from previous intergovernmental transfer practices and demonstrates the impact that financing mechanisms can have on urban climate change strategies. However, cities that receive transfers or grants from higher levels of government often experience accountability issues, and may even be incentivized to 'overbuild' certain infrastructure (Kitchen 2006). Also, intergovernmental transfers can impact cities' spending preferences and lead to inefficient and inequitable expenditures (Bazel and Mintz 2014).

14.3.6 Borrowing and Municipal Bonds

Cities also borrow money to finance large capital projects. Borrowing is well suited to finance projects with large upfront costs and whose benefits are distributed over the long term (Alm 2015). As such, borrowing may be an appropriate way for cities to support climate change infrastructure projects that help encourage 'inter-generational equity' (Slack 2014).

One common borrowing mechanism is the issuance of municipal bonds on private capital markets (Alm 2015). Issuing municipal bonds is especially popular in the United States (Pagano and Perry 2008). Bonds can take the form of either 'revenue bonds', which are backed by a dedicated revenue source, such as tolls, utilities, etc., or 'general obligation bonds', for which state governments often require public approval of municipal bond packages. Public approval requirements may cause policymakers to favour uncontroversial projects with immediate impact over climate change projects with future benefits (Peterson 2017). Moreover, while borrowing does allow cities to acquire capital to finance the large infrastructure projects necessary for a sustainable future, municipal credit ratings are generally detached from the environmental performance. As such, borrowing and borrowing opportunities rarely directly incentivize cities to undertake climate change projects; however, a few exceptions exist. In recent years, some cities have looked towards 'green bonds' as a way to finance their sustainability projects. Increased demand from investors has caused this part of the municipal bond market to grow, with bonds being primarily directed towards transportation and water infrastructure projects (Saha 2016).

In some countries outside the United States, cities have the option to borrow through specialized institutions [i.e. Dexia in Belgium and France (OECD 2010)] or through subnational or national governments in order to capitalize on lower interest rates (Hanniman 2015). Some cities have access to infrastructure banks or 'revolving funds' through which they can borrow funds from higher levels of government to finance environmental projects. California's Infrastructure and Economic Development Bank allows borrowing for 'environmental mitigation measures', and the Oregon Transportation Infrastructure Bank (OTIB) considers 'innovation and environmental sustainability' in their selection process (DuPuis and McFarland 2016, 15).

As will be discussed in the following section, while cities' borrowing regulations are almost always determined by higher levels of government, increased investment opportunities from private and international actors may provide cities with innovative financing mechanisms for their climate change programmes.

14.4 Multilevel Governance and Innovations in Urban Climate Change Finance

As demonstrated in the previous section, cities have indeed found modest ways to advance their urban climate change initiatives by adapting the financing mechanisms traditionally available to them. These changes have generally occurred within existing intergovernmental arrangements, however, and involve little innovation with respect to the multilevel arrangements that govern climate change. As a result, the financing of urban climate change initiatives remains an area largely defined by traditional government institutions and hierarchies, which may be contributing to the slow pace of implementation (Bulkeley and Betsill 2013). While taxes and fees have been modified to incentivize behavioural change and finance certain 'green' services and infrastructure at the local level, they have limited ability to address the problem of 'spillovers' and the global nature of climate change in any broad and concerted fashion.

Traditional municipal finance mechanisms are designed to address local problems by funding local services and infrastructure that benefit residents. Climate change presents a new policy challenge for cities. The problem is global in nature, and while local residents will generally benefit from climate change strategies, they are not the sole beneficiaries of these programmes. Still, cities are responsible for a large percentage of GHG emissions, and it is incumbent upon cities to develop local solutions and take local action. There are also significant local co-benefits to action on climate change, such as improved air quality and energy savings. To finance their urban climate change strategies, cities may benefit by better aligning their financing mechanisms with multilevel governance principles.

As previously noted, urban policy scholars have advocated a conceptual shift away from formal government institutions towards *governance*. Governance is a dynamic process where power and policy are shaped by both public and private interests within formal and informal institutions alike. Multilevel governance, in particular, involves recognizing different configurations of authority and the multiple actors and relationships which govern a given policy area (Peters and Pierre 2012). This is an especially relevant framework for understanding urban climate change policy, as

> Urban climate governance is a complex process driven by the intersection of the specific challenges of the issue itself and the reconfiguration of political authority across multiple levels and between public and private actors. (Bulkeley 2010, 231)

Multilevel governance approaches may be necessary to account for potential externalities and spillovers (such as GHG emissions or water run-off) between municipalities and overcome the coordination challenges and capacity constraints associated with urban climate change action (Homsy and Warner 2015).

Despite the increasing popularity of multilevel governance frameworks in urban policy research, they are rarely applied in the subfield of municipal finance. Municipal finance literature stresses the impact of higher levels of government on local resources and the regulatory and fiscal constraints facing cities. Indeed, this is generally a suitable conceptual framework, as most local services and infrastructure can be adequately financed with local revenue sources available to cities through legislation set by higher levels of government. However, given the global nature of climate change and unique policy challenges it presents, cities must recognize the relevance of outside actors (including non-governmental actors) to their financial opportunities and constraints, and the applicability of multilevel governance principles when considering how to finance their climate change strategies. Collaborating with international organizations, transnational advocacy networks and private actors may provide cities with more effective and efficient financing mechanisms for urban climate change strategies. Public–private partnerships, climate funds and climate and carbon financing mechanisms are examples of how cities may collaborate with new sets of actors to access new revenue sources. Although the impact and effectiveness of these tools remain under evaluation, it is evident that there is potential for further innovation in financing mechanisms to occur when municipal finance arrangements reflect the multilevel governance dynamics of a given policy area.

14.4.1 Public–Private Partnerships

Over the past few decades, cash-strapped cities have turned to public–private partnerships (PPPs or P3s) as a means of building and managing local infrastructure. Private actors may find PPPs to be attractive investment opportunities, especially when the project has a dedicated revenue stream upon completion (i.e. toll roads). PPPs can be a way for cities to finance projects that have long lifecycles and require large upfront capital investments, and may be drawn on to advance urban climate change projects. In theory, the competitive process through which PPPs are awarded will incentivize cost savings and innovation (Kitchen 2006; Slack 2014), including potential innovations in green technology and practices (Koppenjan 2015). Many private actors have global experience, technical expertise and finance opportunities that may benefit local projects.

However, Koppenjan (2015) demonstrates that using PPPs for green infrastructure results in numerous 'tensions and challenges'. PPPs cannot necessarily be considered an effective means of advancing climate change initiatives, especially if cost and environmental incentives do not align. There are cases of private firms imposing internal environmental standards, with a few even going as far as putting an internal price on carbon (IISD/SDG Knowledge Hub). Other private firms, such as HSBC, have recognized the impact cities can have on climate change and have targeted their climate efforts accordingly (Bulkeley 2010). However, proactive actions and investments by private firms are by far the exception than the norm. For PPPs to be a viable financing option for urban climate objectives, appropriate government incentives and standards will be necessary.

14.4.2 Climate Funds and International Grant Programs

International and national networks have the potential to encourage local action and be important actors in cities' climate change initiatives (Gore and Robinson 2009). Given the global nature of climate change and the need for local participation in climate change initiatives, it is perhaps unsurprising that international organizations, non-governmental actors and advocacy networks are increasingly supporting climate change programmes at the local level. Often, this assistance has taken the form of international organizations working in coordination with national governments to support local initiatives. However, mayors and local policymakers are calling on international organizations to facilitate direct access for cities to international climate funds such as the Green Climate Fund, the Global Environment Facility and the Adaptation Fund and create ‘city-specific mitigation and adaptation funds’ geared to cities’ needs and capacities (Eduardo 2016).

Transnational municipal networks provide cities with a venue to collaborate directly with other cities at both the national and international levels. C40, 100 Resilient Cities and ICLEI-Local Governments for Sustainability are examples of prominent transnational municipal networks. These international networks often partner with national municipal networks such as the US Council of Mayors, the Federation of Canadian Municipalities (FCM) and the European Green Cities Network to support local climate change initiatives (Gore and Robinson 2009). These collaborations at both the national and international scales help cities design global solutions through local action. They support capacity building in cities through technical support, knowledge sharing, and, in some cases, grants and resources. ICLEI’s *Cities for Climate Protection* (CCP) campaign has assisted the sustainability programmes of over 1000 municipalities (ICLEI), and its *Transformative Actions Program* (TAP) seeks to ‘catalyze and improve capital flows to cities’ for sustainability projects (TAP). In Canada, FCM manages the federal governments’ ‘Green Municipal Fund’, which provides support for local green infrastructure projects (Peck 2000; Robinson and Gore 2005; FCM 2017). Allowing a non-governmental organization to administer funds may reduce politics and provide cities with additional capacity-building supports for their climate change programmes.

While climate funds and international grants present new opportunities to cities to finance their climate change strategies, there remain challenges with respect to how reliable these funding sources are, especially over the long term. International funds are typically administered by national governments, which adds additional uncertainty for cities. Serious shortfalls and gaps in these international funds have been noted (Ayers 2009). Additionally, accountability problems may arise when cities draw on outside funds.

14.4.3 Climate Finance: Carbon Finance and Access to Capital Markets

Some cities are turning to global financial markets and institutions to support their climate change agendas (OECD 2010). ‘Climate finance’ allows private investors to support climate change programmes and promote socially responsible investing through market mechanisms (Richardson 2014). Private climate finance can take various forms, including investments, carbon market payments and voluntary funding (Stadelmann et al. 2013). Many cities have sought direct access to these markets in recent years, but often encounter challenges given their relatively low expertise and administrative and technical capacities vis-à-vis higher levels of government (Eduardo 2016). In addition to capital markets, some cities have entered carbon offset and carbon emissions markets (OECD 2010). Chicago, Los Angeles, Santiago and Tokyo have become involved in emissions trading through cap-and-trade programmes. There is also the potential for cities to generate revenues from emission reduction projects through the two carbon offset mechanisms created through the Kyoto protocol: the Clean Development Mechanism (CDM) and Joint Implementation (JI) (OECD 2010). So far, however, very few cities have done so, with Bogota, Delhi and select metro areas in France and Germany being notable exceptions (OECD 2010).

Cities will likely encounter difficulties when pursuing climate and carbon finance. So far, there has been limited buy-in from cities, who face large upfront costs and high administrative and organizational requirements in order to enter global markets and attract private investment. Moreover, these may not be a reliable means of financing urban climate change policy, as investments may privilege one-off projects over comprehensive climate action plans (Kamal-Chaoui and Robert 2009; OECD 2010). The issuance of green bonds, as summarized in the previous section, may allow cities to overcome these hurdles and access private capital in a more consistent manner and at a more modest scale. Also, cities will still have to work together with their counterparts at higher levels of government, as state regulations impact the effectiveness of the climate finance 'marketscape' (Richardson 2014). To date, there is little data on how much investment has been generated through climate finance (Stadelmann et al. 2013) as well as how effective such investments have been at advancing environmental outcomes. Moreover, as is the case with PPPs, it remains to be seen how well urban environmental objectives can ultimately align with the priorities of private investors.

14.5 Conclusion

As cities seek to finance their climate change strategies, they will likely encounter frustrations and inefficiencies if they rely solely on the traditional financing mechanisms available to them through hierarchical intergovernmental institutions. Instead, cities will benefit from adopting a multilevel governance approach that promotes collaboration with both governmental and non-governmental actors alike, at both the local and global levels. By recognizing the important relationship between municipal finance and the broader multilevel frameworks that govern climate change policy, cities have the opportunity to pursue innovations in municipal finance to support their climate change objectives (see Chap. 15: Bausch, Eakin & Lerner). Higher levels of government can encourage these innovations by fostering a supportive regulatory environment and helping municipalities develop the networks and capacity necessary to benefit from outside revenue sources (Williams and McNutt 2013; Eduardo 2016).

Many of these financing tools are still in their infancy and have yet to be widely used or evaluated; nevertheless, they demonstrate the potential for cities to work with new actors to find financing solutions that may be better aligned with the global policy imperatives of climate change.

This chapter presents early attempts to bridge municipal finance and urban climate change literature, and understand how different municipal finance tools affect cities' climate change strategies. The impact of municipal finance is often overlooked in climate change policy literature. Scholars agree that fiscal capacity and local revenues are important indicators of local climate change initiatives (Krause 2012), but there is little attention on how revenue is generated and the impact of different tools on policy success. Additional quantitative and qualitative research highlighting these causal connections would be valuable for cities seeking to design—or improve—their climate change action plans. Further research is also necessary to the social impacts and equity concerns associated with different financing models of climate change. Moreover, the financing innovations introduced here raise broader questions about accountability and policy responsiveness, as the involvement of global actors in financing urban climate change strategies may impact (or bypass) conventional political accountability mechanisms at the local level. Ultimately, researchers must move beyond descriptive reports that detail the financing options available to cities. These tend to be highly normative, outlining which revenue mechanisms cities ought to employ for different expenditures. Instead, more research is needed on the political dynamics that define when and how different financing mechanisms are actually employed and their consequences for global climate change response.

References

Abels, M. (2014). Strategic alignment for the new normal: Collaboration, sustainability, and deliberation in local government across boundaries. *State and Local Government Review, 46* (3), 211–218. doi:10.1177/0160323X14551179.

Ahmad, E., & Brosio, G. (2015). *Handbook of multilevel finance.*

Alm, J. (2015). Financing urban infrastructure: Knowns, unknowns, and a way forward. *Journal of Economic Surveys, 29*(2), 230–262. doi:10.1111/joes.12045.

Ayers, Jessica. (2009). International funding to support urban adaptation to climate change. *Environment & Urbanization., 21*(1), 225–240. doi:10.1177/0956247809103021.

Bazel, P., & Mintz, J. (2014). The free ride is over: Why cities, and citizens, must start paying for much-needed infrastructure. *University of Calgary, School of Public Policy SPP Research Papers,* 7(14).

Braun, G., & Hazelroth, S. (2015). Energy infrastructure finance: Local dollars for local energy. *The Electricity Journal, 28*(5), 6–21. doi:10.1016/j.tej.2015.05.008.

Bulkeley, H., & Betsill, M. M. (2003). *Cities and climate change: Urban sustainability and global environmental governance*. London: Routledge.

Bulkeley, H. A., & Betsill, M. M. (2013). Revisiting the urban politics of climate change. *Environmental politics., 22*(1), 136–154.

Bulkeley, H. (2010). Cities and the governing of climate change. *Annual Review of Environment and Resources, 35,* 229–253.

California Department of Water Resources. (2017). California turf replacement initiative. http://www.water.ca.gov/turf/. Accessed 23 Feb 2017.
Chernick, H., Langley, A., & Reschovsky, A. (2011). Revenue diversification in large U.S. cities. *University of Toronto, IMFG Papers on Municipal Finance and Governance, 5*.
DuPuis, N., & McFarland, C. K. (2016). *Paying for local infrastructure in a new era of federalism: A state-by-state analysis*. Washington: National League of Cities.
Eduardo, P. (2016). Finance Systems for Sustainable Cities. *Climate and Clean Air Coalition*. http://www.ccacoalition.org/en/blog/finance-systems-sustainable-cities.
FCM. (2017). Green municipal fund. http://www.fcm.ca/home/programs/green-municipal-fund.htm. Accessed 23 Feb 2017.
Gore, C. D., & Robinson, P. (2009). Local government response to climate change: Our last, best hope? In H. Selin & S. D. VanDeveer (Eds.), *Changing climates in North American politics: Institutions, policymaking, and multilevel governance* (pp. 137–158). Cambridge, MA: MIT Press.
Hanniman, K. (2015). A good crisis: Canadian municipal credit conditions after the Lehman brothers Bankruptcy. *University of Toronto, IMFG Papers on Municipal Finance and Governance, 22*.
Homsy, G., & Warner, M. (2015). Cities and sustainability: Polycentric action and multilevel governance. *Urban Affairs Review, 51*(1), 46–73. doi:10.1177/1078087414530545.
Hooghe, L., & Marks, G. (2003). Unraveling the central state, but how? Types of multi-level governance. *American Political Science Review, 97*(2), 233–243. doi:10.1017/S0003055403000649.
IISD/SDG Knowledge Hub. (2016, July 5). June 2016 climate finance update: Funds meet, non-state actor engagement accelerates. http://sdg.iisd.org/news/june-2016-climate-finance-update-funds-meet-non-state-actor-engagement-accelerates/. Accessed 23 Feb 2017.
Kamal-Chaoui, L., & Robert A. (eds). (2009). Competitive Cities and Climate Change, Regional Development Working Papers No 2, OECD.
Kitchen, H. M. (2006). *A state of disrepair how to fix the financing of municipal infrastructure in Canada*. Toronto: C.D. Howe Institute.
Kitchen, H., & Slack, E. (2016). New tax sources for Canada's largest cities: What are the options? *University of Toronto, IMFG perspectives,* 15
Kjaer, A. M. (2009). Governance and the urban bureaucracy. In J. S. Davies & D. L. Imbroscio (Eds.), *Theories of urban politics* (pp. 137–152). Los Angeles: Sage.
Koppenjan, J. (2015). Public-Private Partnerships for green infrastructures. Tensions and challenges. *Current Opinion in Environmental Sustainability 12*, 30–34. http://dx.doi.org.myaccess.library.utoronto.ca/10.1016/j.cosust.2014.08.010.
Krause, R. M. (2012). Political decision-making and the local provision of public goods: The case of municipal climate protection in the US. *Urban Studies, 49*(11), 2399–2417. doi:10.1177/0042098011427183.
Kunce, M., & Shogren, J. F. (2005). On interjurisdictional competition and environmental federalism. *Journal of Environmental Economics and Management, 50*(1), 212–224. doi:10.1016/j.jeem.2004.11.004.
Lee, Y., Feiock, R., & Lee, I. (2014). A multilevel governance and growth management: The use of impact fees by Florida cities. *Economic Development Quarterly, 28*(4), 328–338. doi:10.1177/0891242414548510.
Lubell, M., Feiock, R., & Handy, S. (2009). City adoption of environmentally sustainable policies in California's Central Valley. *Journal of The American Planning Association, 75*(3), 293–308. doi:10.1080/01944360902952295.
Mullin, M. (2008). The conditional effect of specialized governance on public policy. *American Journal of Political Science, 52*(1), 125–141. doi:10.1111/j.1540-5907.2007.00303.x.
Oates, W. (1972). *Fiscal federalism*. New York: Harcourt Brace Jovanovich.
OECD. (2010). *Financial instruments and funding new expenditure needs, in cities and climate change*. Paris: OECD Publishing.
Olson, M. (1969). The principal of "fiscal equivalence": The division of responsibilities among different levels of government. *American Economic Review, 59,* 479–487.

Pagano, M., & Perry, D. (2008). Financing infrastructure in the 21st century city. *Public Works Management & Policy, 13*(1), 22–38. doi:10.1177/1087724X08321015.
Peck, S. (2000). Green infrastructure funds. *Alternatives Journal, 263,* 39.
Peters, B. G., & Pierre, J. (2012). Urban governance. *Oxford Handbooks Online.* Retrieved 24 Feb 2017, from http://www.oxfordhandbooks.com.myaccess.library.utoronto.ca/view/10.1093/oxfordhb/9780195367867.001.0001/oxfordhb-9780195367867-e-5.
Peterson, J. (2017). Do municipal bond elections privilege certain infrastructure investments over others? The case of San Antonio, Texas. In *Paper Presented at the 47th Annual Urban Affairs Association Conference*, Minneapolis, April 19–22.
Peterson, J., & Hughes, S. (in press). *Governing garbage: Advancing urban sustainability in the context of private service delivery.* Cities.
Pierre, J. (2011). *The politics of urban governance.* Houndmills: Palgrave Macmillan.
Richardson, B. J. (2014). The evolving marketscape of climate finance. *Climate Law, 4,* 94–106.
Rhodes, R. A. W. (1997). *Understanding governance.* Buckingham: Open University Press.
Robinson, P. J., & Gore, C. D. (2005). Barriers to Canadian municipal response to climate change. *Canadian Journal of Urban Research, 14*(1), 102–120.
Saha, D. (2016, Oct 25) Green bonds take root in the U.S. municipal bond market. *The Brookings Institute.* https://www.brookings.edu/blog/the-avenue/2016/10/25/green-bonds-take-root-in-the-u-s-municipal-bond-market/.
Shah, A., & Shah, S. (2006). The new vision of local governance and the evolving roles of local governments. In A. Shah (Ed.), *Local governance in developing countries* (pp. 1–46). Washington: The World Bank.
Slack, E. (2011). Financing large cities and metropolitan areas. *University of Toronto, IMFG Papers on Municipal Finance and Governance,* 3.
Slack, E. (2014). Own-source revenues for metropolitan cities. In *Presentation to Seminario Internacional sobre alternativas de financiamiento de ciudades*, Bogota, Columbia, 18 November 2014.
Stadelmann, Martin, et al. (2013). Difficulties in accounting for private finance in international climate policy. *Climate Policy, 13*(6), 718–737.
Stoker, G. (1999). *The new management of British local governance.* Basingstoke: Macmillan.
TAP. (2017). Transformative Actions Program. http://tap-potential.org/about-tap/. Accessed 23 Feb 2017.
TD Economics. (2004, May 20). *Mind the gap: Finding the way to upgrade Canada s aging public infrastructure.*
Texas Comptroller of Public Accounts. (2017). Local Sales and Use Tax. http://www.trcc.state.tx.us/taxinfo/local/city.html Accessed 23 Feb 2017.
Tindal, C. R., & Tindal, S. N. (2009). *Local government in Canada.* Toronto: Nelson Education.
Williams, R. A., & McNutt, K. (2013) Climate change adaption and policy capacity in the Canadian finance sector: A meso analysis. *Review of Policy Research,* 30(1).
Ziehmer, R. (2010). Missourians are Committed to Conservation. *Missouri Department of Conservation* https://mdc.mo.gov/conmag/2010/06/note-our-readers. Accessed 23 Feb 2017.

Author Biography

Jacqueline Peterson is a Ph.D. candidate at the University of Toronto in the Department of Political Science. Her research interests include municipal finance, urban infrastructure policy, and waste management. She has a M.I.A. from Columbia University and a B.A. from McGill University.

Chapter 15
Adaptation for Whom to What? Challenges and Opportunities in Agriculture-Urban Collaboration for Climate Change Adaptation

Julia C. Bausch, Hallie C. Eakin and Amy M. Lerner

Abstract As leaders in climate change governance, urban governments have the opportunity to interact, coordinate, and collaborate with different sectors and actors to set and pursue both private and public adaptation goals. Urban and peri-urban agriculture (growing and raising food and non-food products within or at the periphery of an urban area) is recognized as both vulnerable to climate change and as a potential strategy for urban climate adaptation and mitigation. However, few cities have formally incorporated it into their climate change policies. Mexico City is one exception. It's 2014–2020 Climate Action Plan—*El Programa de Acción Climática: Ciudad de México* (PACCM)—outlines actions and programs to benefit the city's peri-urban agrarian communities and farmers. This chapter examines the PACCM to explore the drivers, obstacles, and opportunities of agriculture-urban collaboration for climate change adaptation. We examine: (1) how and why agriculture became part of the PACCM; (2) the stressors and vulnerabilities that the PACCM's agrarian actions and programs seek to mitigate, for private and/or public benefit; and, (3) the barriers to and opportunities for this collaboration. We analyzed the PACCM programs that target agrarian actors, activities, and lands, and interviewed government officials, PACCM coordinators and authors, agrarian community leaders, and farmers about the Plan's development and implementation. We found that the PACCM implicitly considers peri-urban agrarian actors as private providers of public adaptation benefits for the city, through measures intended to also benefit agrarian actors. However, the Plan does not articulate how agrarian actors and lands fit into the city's larger vision for adaptation, nor does it adequately address the specific vulnerabilities and socioeconomic dynamics shaping agrarian

J.C. Bausch · H.C. Eakin
Arizona State University, Tempe, AZ, USA
e-mail: jcbausch@asu.edu

H.C. Eakin
e-mail: hallie.eakin@asu.edu

A.M. Lerner (✉)
Universidad Nacional Autónoma de México, Mexico City, Mexico
e-mail: amy.lerner@iecologia.unam.mx

S. Hughes et al. (eds.), *Climate Change in Cities*, The Urban Book Series,
https://doi.org/10.1007/978-3-319-65003-6_15

actors' decisions, which may undermine the Plan's success. The results suggest several guidelines to promote the private provision of public adaptation in the context of social-ecological change. First, governments and private providers must explicitly communicate their needs and expectations for the collaboration so that the needs of both parties can be addressed during policy development. This also requires consideration of the effects of social, economic, and environmental changes on the private providers and beneficiaries. Secondly, to encourage private actors to provide specific public adaptation benefits, governments must develop policy mechanisms that explicitly and directly promote the desired benefits, ideally in collaboration with private providers. Finally, policy processes and outcomes that promote private provisioning of public adaptation benefits warrant close attention to how winners and losers, and synergies and trade-offs are mediated.

Keywords Adaptation · Climate change · Public goods · Private provision · Agriculture · Cities · Peri-urban · Mexico

15.1 Introduction

Cities play an important role in climate change adaptation,[1] both as material entities that contribute to climate change and as political entities leading efforts to reduce carbon emissions and adapt to the effects of climate change. On both fronts, cities are making new governance arrangements and bringing new actors and sectors into the sphere of urban politics. These changes may provide new opportunities for adaptation to and mitigation of climate change and other risks.

One new arrangement is a collaboration with the agricultural sector (de Zeeuw et al. 2011; Lwasa et al. 2015; Simon 2012; United Nations Human Settlements Programme 2009). Because agriculture, forestry, and other land use activities contribute 24% of global net greenhouse gas (GHG) emissions (Intergovernmental Panel on Climate Change (IPCC) 2014), it is widely acknowledged that farming practices must be changed as part of mitigation efforts (Food and Agriculture Organization (FAO) 2008). However, agriculture can also help urban areas adapt to climate change. Urban and peri-urban agriculture[2] has been found to provide climate adaptation through urban greening, microclimate mediation, nutrient cycling, water infiltration, flood mitigation, and enhanced biodiversity and ecosystem

---

[1]Climate change adaptation can be defined as "the processes and actions that enable people to cope better with increasingly challenging weather and climatic conditions" (Tompkins and Eakin 2012 p. 3).

[2]Urban and peri-urban agriculture includes the diverse set of activities for growing and raising food and non-food products within or at the periphery of an urban area. It is practiced by diverse actors, often on small parcels or in confined spaces, and can include many different crops and products. It is closely interconnected with the urban system, and complements, rather than replaces, rural production and imported foods (McIntyre et al. 2009; Mougeot 2000).

services. Agricultural activities near cities can also provide GHG mitigation by reducing greenhouse emissions from food production and transportation and methane emissions from landfills (Lwasa et al. 2015).

However, peri-urban agriculture is also *vulnerable* to climate change and other stressors (Eakin and Appendini 2008; Eakin et al. 2010). Climate change increases climate variability and changes crop suitability, production potential, and the characteristics and availability of agricultural land (Fischer et al. 2002; Misselhorn et al. 2010). Farming households face crop failure, market shocks, and declining livelihoods (Morton 2007). Their vulnerability may translate into reduced food security at all scales (FAO 2008). Climate change also threatens agriculture's capacity to provide ecosystem services (Misselhorn et al. 2010). Farmers and agrarian communities face many stressors and vulnerabilities beyond climate change, including economic, social, and environmental stressors (Eakin 2005; Eriksen and Silva 2009; O'Brien and Leichenko 2000; Taylor 2013). Urbanization threatens agricultural land use in peri-urban areas, but it also creates opportunities for farmers to move into non-farm economic activities (Simon 2008; Eakin et al. 2010).

In Mexico City, agriculture has long had an uneasy relationship with urban development and planning. Over the last century, farmers have lost land and voice to urbanization; the future viability of farming hinges less on adaptation to climatic threats than it does on the process of urbanization itself. However, Mexico City has included peri-urban agricultural activities in its Climate Action Plan (PACCM, *Programa de Acción Climática Ciudad de México* 2014–2020). In fact, Mexico City's is one of the few urban climate plans that include programs explicitly for agrarian actors, activities, and lands; the inclusion demonstrates both the potential connections and tensions between urban climate policy and peri-urban agriculture (*Gobierno del Distrito Federal* (GDF) 2014).[3]

This chapter explores the drivers, obstacles, and opportunities of agriculture-urban collaborations for climate change adaptation using the case of Mexico City's plan, offering an empirical case of public funding for the private provision of public adaptation benefits, of which there are few (Kotchen and Moore 2007). It identifies the motivations for and assumptions behind the inclusion of agrarian actors and rural land uses in the city's climate plan, and describes the mechanisms the city government uses to promote adaptive actions by agrarian actors. It analyzes the obstacles to peri-urban agrarian actors serving as private providers of public adaptation benefits for the city and explores opportunities for improving climate adaptation through collaboration with the agricultural sector.

[3]The formal name of the administrative unit of Mexico City was changed on January 29, 2016 from the *Distrito Federal*, or Federal District, to the *Ciudad de México*, or Mexico City. Thus, in the text we refer to the city government as the Government of Mexico City (GMC); however, policy documents dating from before this change still refer to the *Gobierno del Distrito Federal* (GDF).

15.2 Private Provision of Public Adaptation

Both governments and private actors adapt to stressors, vulnerabilities, and disaster events, but often for different reasons. Governments are mandated to provide and protect the public good, which includes mitigating risks and responding to disasters associated with climate change. Private actors, in contrast, are assumed to adapt if they perceive a private benefit, following the logic of economic rationality. In other words, private actors will adapt if they believe the benefits exceed the costs, and if they know they will experience the direct benefits of their actions. Regardless of who adapts and why, other actors will be affected, for good, bad, or both: there is interplay between private and public actions and responsibilities (Mees et al. 2012; Milman and Warner 2016; Tompkins and Eakin 2012). Public adaptation can benefit some private actors more than others; for example, the public sector can subsidize or provide disaster insurance for specific populations (Erdlenbruch et al. 2009). Likewise, there are cases of private adaptation for public benefit; for example, farmers who contribute to urban flood risk management by retaining or slowing runoff on their lands through farming practices (e.g., reduced grazing), green infrastructure (e.g., hedgerows, ponds), or hard infrastructure (e.g., retention walls) (Posthumus et al. 2008; Milman and Warner 2016).

Private adaptation for public benefit can be supplied accidentally or deliberately. Accidental provision occurs as a positive externality of actions carried out by private actors for other purposes or motivations (e.g., increased profitability of an agroforestry product stimulates reforestation, and incidentally results in greater carbon sequestration and lower probability of landslides). There is little empirical evidence on deliberate private provisioning of public adaptation. Deliberate providers are theorized to have three motives for supplying public adaptation goods: (1) they value the good more than the cost and are therefore willing to supply it to others (known as Olsonian privilege groups); (2) they have altruistic motivations; or, (3) they are motivated by profit or maximizing their welfare in other ways (Tompkins and Eakin 2012).

Obtaining public adaptation benefits from private providers is challenging because the provision of a benefit can occur at a different spatial and/or temporal scale from where/when the benefit manifests. In addition, a minimum number of private actors may need to participate for the benefit to be felt. In addition, a provider's interests may not align with public interests, especially where the provider incurs new risks, costs, or perceives few direct benefits. Thus, appropriate institutional mechanisms (e.g., incentives, rules) may be necessary to encourage private provision of public adaptation. Mechanisms used to motivate providers who seek increased profit or welfare include compensatory payments (e.g., grants for infrastructure installation and maintenance, land purchases), public markets for public goods (e.g., payments for ecosystem services), regulations (e.g., a logging ban), and reassigning property rights (Mees et al. 2012; Tompkins and Eakin 2012). Over time, an accidental private provider might evolve into a deliberate private provider, such as when a landholder who has accidentally provided ecosystem services by using her land for grazing animals is confronted with new incentives to urbanize her land. In these cases, where the provider's role in adaptation may not have been articulated and did not previously require institutional mechanisms, as a deliberate provider, his or her role needs to be made explicit and supported through institutional mechanisms.

Like all policy, climate adaptation policies and plans are inherently political: they mediate winners and losers among actors and synergies and tradeoffs among different goals. They emerge in a specific context that shapes climate adaptation decision processes and outcomes (Næss et al. 2015), including which collaborations emerge, what policy goals and mechanisms are selected, who is considered vulnerable, and who benefits. These elements warrant special attention where a public authority is incentivizing private investment in adaptation for private and/or public benefit. The collaboration between the Government of Mexico City (GMC) and peri-urban farmers, formalized in the PACCM, is one such example.

15.3 Agriculture, Conservation, and Climate Change in Mexico City

Mexico City urbanized rapidly over the twentieth century to become the megacity it is today (Aguilar 2008; see Table 15.1). Concerns about aquifer recharge and the degrading effects of urbanization on the environment led the urban government to establish a contiguous conservation area in 1992, known as the *Suelo de Conservación*, or Conservation Zone (SC) (Sheinbaum Pardo 2008). The SC covers 59% of the formal territory of Mexico City, essentially the southern half. It contains forests (53%), grasslands (10%), wetlands and water bodies (1.5%), agricultural lands (20%), and human settlements (12%) (GDF 2012). The Ecological Ordinance of the Federal District (*Ordenamiento Ecológico del Distrito Federal*) (GDF 2000) zoned agricultural, forest, and urban land uses in the SC and established regulations for those land uses. Seventy-one percent of the SC is in the hands of 44 agrarian communities (*ejidos*[4] and indigenous communities), who share management responsibilities with the city's Secretariat for the Environment (SEDEMA) (GDF 2012), and participate in activities to maintain environmental quality, prevent and respond to forest fires, and manage natural resources. The Ordinance severely restricts and regulates land-use activities in the SC.

The SC provides ecosystem services such as carbon sequestration, biodiversity conservation, climate regulation, and agricultural production (GDF 2013). Of particular importance is aquifer recharge: 60–70% of the city's water comes from

[4]An *ejido* is an area of communally held land, established following the 1910 Mexican Revolution via land redistribution to smallholders and indigenous farmers. *Ejido* members have rights to use the land for individual cultivation, residential settlement, and communal use. In 1992, *ejidos* were given the right to privatize their lands; however, many *ejidos* have persisted.

Table 15.1 Characteristics of Mexico City and the Agrarian Sector

Description	Data	Reference
Area of Metropolitan area (2010)	7866.1 km^2 (3037.12 mi^2)	Félix Guerra et al. (2012)
Metropolitan population (2010)	20,116,842	Félix Guerra et al. (2012)
Area of Mexico City proper	1485.49 km^2 (573.55 mi^2)	SIAP (2014a)
Population of Mexico City proper	8,874,724 people	SIAP (2014a)
GDP	2,204,492 million Mexican pesos (MXN)*; 17.1% of national GDP**	SIAP (2014a)*; GDF (2013)**
Agriculture sector as % of city's GDP	0.1%	SIAP (2014a)
Area of the SC	85,554 ha (211,408.54 acres); 59% of city area	GDF (2012)
SC population	53,248 people, 0.6% of city's population	SIAP (2014a)
# Agrarian communities	44 *ejidos* & indigenous communities	GDF (2015)
# Agrarian subjects	22,500 people (*ejidatarios* & *comuneros)*	GDF (2015)
Participants in primary sector	Estimates range from 16,000* to 35,000**	*Quintanar (2014); **SEDEREC (2014)
Active productive units in SC (2007)	11,881	INEGI (2009)
Area planted (2014)	17,607.73 ha (43,509.65 acres) 11.9% of land in city proper	SIAP (2014b)
% Rain-fed production	89.6% of production in SC	SIAP (2014a)

groundwater (GDF 2012). To protect ecosystem services, the government regulates land use, builds and funds infrastructure for soil and water retention, and distributes payments for ecosystem services to agrarian communities (Schteingart and Salazar 2010). Despite these measures, urban expansion into the SC continues, although it has slowed since the 1970s (Aguilar 2008). Illegal urbanization in the SC is due, in large part, to the lack of affordable housing. By 2010, there were 859 irregular settlements occupying 2800 ha (6919 acres) of the SC (GDF 2013).

The economic viability of agriculture in Mexico City began to decline in the 1960s, which led farmers to abandon agriculture for urban employment (Pensado Leglise 2001). According to official census data, the number of farming units in Mexico City fell 37% from 1960 to 2007 (Secretaría de Industria y Comercio 1965; INEGI 2009). Area planted fell nearly 90% over nearly the same period, from its peak of 165,657 ha in 1960 (Secretaría de Industria y Comercio 1965) to 17,608 ha in 2014 (SIAP 2014a).[5] Between 1995 and 2014 alone, the area planted dropped 33.9% (*Servicio de Información Agroalimentaria y Pesquera* (SIAP) 2014b). Today the SC is a peri-urban space with both rural and urban characteristics and activities. Much of its population is poor, although poverty rates have gone down in recent decades. A recent study speculated that the reduction in poverty rates in the SC between 1990 and 2000—from 77.3% to 51.6%—likely reflects the shift of the economically active population in the SC from agricultural activities to better-paying urban employment (Aguilar 2008). Agricultural abandonment is associated with urban expansion on *ejido* land (Cruz Rodríguez 1995), as well as the decline of ecosystem services (Aguilar 2008).

Currently, only 0.6% of the population within the city proper lives in the SC and roughly 0.3% of the city's population participates in primary sector activities (i.e., agriculture and forestry) (SIAP 2014a; see Table 15.1). Many farmers are affiliated with agrarian or indigenous communities (Dieleman 2016; Losada Custardoy et al. 1998). Like the rest of Mexico, the farmer population is aging: nationally, 59% of farmers are over the age of 50[6] (Secretaría de Agricultura Ganadería Desarrollo Rural Pesca y Alimentación (SAGARPA) 2014). Many of those who practice agriculture in Mexico City do so as a secondary economic activity, for subsistence, or as a hobby (Cruz Rodríguez 2001; Losada Custardoy et al. 1998).

15.3.1 Climate Change Policy in Mexico City

Climate change has been on the policy agenda in Mexico City since 2000 (Hughes and Romero-Lankao 2014; GMC 2016b). The Secretariat for the Environment (SEDEMA, *Secretaría del Medio Ambiente*) published the city's first climate change strategy in 2004, as part of the city's Environmental Protection Plan, which focused on air quality, ecological restoration, and energy-saving measures (GDF 2004). With funding from the World Bank and strong support from Mayor Marcelo Ebrard Casaubón (2007–2012), the 2004 strategy was revised and expanded in 2008 to address cultural changes, financing, and technology adoption to reduce emissions and mitigate risks associated with climate change (GDF 2008). It included adaptation actions directly related to agriculture, such as protecting native maize varieties, managing soil and water resources, and promoting organic farming practices.

---

[5]Agricultural land use in the territory of Mexico City at any point in the twentieth century is hard to determine because of the rapid rate of urbanization over that period (see Ezcurra et al. 1999; Ward 1990). We provide the official 1960 census data as a point of reference, with the caveat that it may not accurately reflect the reality of land use at that time.

[6]Mexico City was not included in the cited study; no specific data on the age of farmers in Mexico City could be found. Thus, we extrapolate from the national and regional data that the majority of farmers in Mexico City are also over the age of 50.

In 2011, the GMC institutionalized climate change as a policy area by passing the Law for Mitigation and Adaptation to Climate Change and Sustainable Development of the Federal District (*Ley de Mitigación y Adaptación al Cambio Climático y Desarrollo Sustentable del Distrito Federal*). The legislation authorized the government to develop political and funding instruments for climate change adaptation and mitigation, and established a legal mandate for SEDEMA to implement climate actions regardless of the political party in office (Hughes and Romero-Lankao 2014).

In 2014, again with strong support from the mayor (Miguel Ángel Mancera, 2013–2018), SEDEMA published a new Climate Action Plan for Mexico City (the PACCM) that expanded the previous plan. The new plan encourages, coordinates, and integrates other city agencies into climate change adaptation and mitigation. SEDEMA involved government functionaries, researchers, experts, and NGOs in the plan's development, and did an online consultation with the public. The resulting plan increased the number of climate actions to 69: 38 new, 15 continued, and 12 revised from the 2008 plan. Each action is the responsibility of a city agency; two of these agencies work closely with the agrarian population—the Secretariat for Rural Development and Equity for Communities (SEDEREC, *Secretaría de Desarrollo Rural y Equidad para las Comunidades*) and the Natural Resources Commission of the Secretary for the Environment (CORENA, *Comisión de Recursos Naturales de la Secretaría del Medio Ambiente*).

The main objective of the 2014 plan is to "increase the quality of life and sustainable development with low carbon intensity in Mexico City" (GDF 2014, p.73). Mexico City emitted 31 tons of CO_2 equivalents in 2012, 80% of which came from energy consumption for transport, industry, commerce, services, and electricity. Agriculture is estimated to contribute less than 1% of the city's total emissions. Collectively, agriculture, forestry, and other land uses constitute an estimated 2% of the city's emissions (GDF 2014). The government does not provide an estimate of how much SC lands and activities, agricultural or otherwise, contribute to the city's climate *mitigation* efforts. The city's 2016 report on climate adaptation identifies Mexico City's principal climate risks as floods and mudslides, heat waves and high temperatures, vector-transmitted diseases, and drought in the SC (GMC 2016b).

Although the 2004 plan discussed agriculture primarily as a *source* of GHG emissions and environmental degradation, the 2014 plan explicitly characterizes agriculture as *vulnerable* to climate change. It identifies rural lands, forests, and agricultural activities as especially vulnerable to climatic change and extreme climate events, including hydrological stress on crops, increased pest incidence, reduced yields from rain-fed production, and altered phenology of some species. The PACCM also recognizes the economic stressors of low profits and low employment facing the agrarian population (GDF 2014).

15.4 Methods

We analyzed the 2014 PACCM document, focusing on the 15 actions (programs; see Appendix 1) directly relevant to agrarian actors (farmers, landholders, and representatives and members of agrarian communities) and land uses (i.e., agriculture, natural resource management), conducted semi-structured interviews with PACCM coordinators, government officials, and agrarian actors, and used text analysis on both sets of data (Bernard 2011). Notes from the interviews were coded inductively, with codes emerging from the sources, with the support of MAXQDA 12 (VERBI GmbH 2016). Though our sample is not representative, for reference we provide the number of participants who spoke about a given issue (see Appendix 2). All the interviews were audio-recorded. Direct quotes were transcribed from the audio recording of the interview in question.

We conducted interviews with 33 agrarian actors in Mexico City from July 2014 to June 2015. Participants were farmers, representatives of farmer organizations, and agrarian community leaders. They were identified at rural development events and meetings, and by recommendation (snowball sample). Most participants were from the heavily forested borough of La Magdalena Contreras; the other six were from other boroughs. The majority (25 of 33) had off-farm employment or was retired from off-farm employment. Interviews lasted between 1 and 4 hours. The participants discussed their objectives for and challenges in agriculture and/or conservation activities, their concerns for the sector, their perception of and interactions with existing government agencies and programs for rural development and conservation, and their expectations for agriculture in the SC in the future.

Seven urban government officials and coordinators involved in the plan's development and implementation were also interviewed. Participants were selected based on information from government reports, websites, and recommendations from other interviewees. These interviews were conducted in person, via phone, or via Skype, from June 2015 to January 2017, and lasted 45 min to an hour. Participants responded to questions about: (1) how the themes and mechanisms related to agrarian actors and lands emerged in the PACCM, (2) what role agrarian actors played in the plan's development, (3) the objectives and intended beneficiaries of the actions relevant to agrarian actors and lands, and (4) the greatest successes of and barriers to realizing the plan to date.

To identify the motivations behind, objectives of, and barriers to the agrarian programs, we inductively coded the PACCM (GDF 2014) and notes from interviews with government officials and contractors. To identify the GMC's understanding of the public and private benefits provided, and the beneficiaries of those actions, we relied on the PACCM document and the interviews with government officials as data sources (Appendix 1). We identified the benefits and beneficiaries and assessed their relative importance through deductive coding, by identifying how many of the PACCM actions explicitly identified them at least once. Finally, we contextualized our results with secondary data about government programs.

15.5 Results

15.5.1 Drivers

According to interviews with people involved in the development of 2014 PACCM, one of the motivations for involving agrarian actors in the Plan was recognition of the vulnerability of the agricultural sector to climate change. In the participatory stage of the plan's development, plan coordinators, together with environmental NGOs, researchers, and government functionaries who represented agrarian interests in the plan's development, recognized agriculture's vulnerability to climate change. Farmers and agrarian communities did not participate in the plan's development (see Sect. 15.5.5).

Other motivations for including programs for agrarian actors and lands were utilitarian and pragmatic. One interviewee explained that government agencies were motivated to participate in the PACCM because it was a way to improve the chances that the Legislative Assembly of Mexico City (ALDF) would approve their budget requests; in other words, it helped secure their annual operating budget for programs marked as climate change adaptation actions. Including agrarian programs in the PACCM was due in part to the fact that city agencies had existing programs for natural resource management, conservation, and agriculture. These programs recognized agrarian actors as land and resource managers, and could easily be reframed as climate change adaptation programs. Therefore, in the development stages of the PACCM, SEDEREC and CORENA proposed a number of actions for the plan based on their existing programs. In sum, the inclusion of agrarian actors and lands in the PAACCM in part reflects SEDEREC's and CORENA's efforts to solidify and justify their programs and budgets by framing agriculture and forests as vulnerable to climate change, as well as providing adaptation benefits to the city at large.

15.5.2 Objectives

Despite the fact that most of the PACCM's agrarian programs originated in existing policies and programs, it was clear from the PACCM document and interviews with PACCM coordinators that the intention was to include actions that would produce mutual benefits for urban and rural sectors. However, the explicit goal of most of these actions is to mitigate climate-related stressors for the public in general rather than for farmers in particular (Appendix 1). These include reducing GHG emissions (five actions), diminishing the threat of extreme precipitation and consequent erosion, landslides, and floods (three actions), and managing the groundwater supply. The main benefits mentioned in the PACCM agrarian programs are ecosystem services (13 actions, including soil retention, water retention, aquifer recharge, green space, carbon capture, air quality, biodiversity, recreation, climate regulation, and nutrient cycling), and quality of life and well-being of urban residents (4 actions). These benefits are largely derived from agricultural lands, rather than agricultural activities per se; in other words, their provision depends on farmers maintaining their farmland and forests in non-urban uses. The benefits to farmers are identified as local economic development and livelihood stability (5 actions), and maize genetic diversity (2 actions). Other benefits mentioned (in only one action) included local organic food production, crop pest management, and public health.

One SEDEMA official elaborated on the link between economic benefits for farmers in the SC and ecosystem services for the city:

> There are two levels: one is the city level, and the importance in terms of environmental services that maintain a quality of life for those of us who live here. However, it is also at a more local level, with these communities: they are the owners of this territory [the SC], and it is a material element from which they subsist. Therefore, we have two interests to confront: the need to conserve this space for environmental services for Mexico City, and on the other hand the need for development of these *pueblos* and communities, which have been limited by the conservation of this space. (Authors' translation)

15.5.3 Mechanisms

The mechanisms used by the GMC to encourage adaptive actions among agrarian actors include expanding channels for local commercialization of farm products (a financial incentive); payments for ecosystem services (a market for a public good); financial supports for water and soil-retention infrastructure (compensation); land use regulation and monitoring; and training in natural resource management, agroecological practices, and meeting commercial quality standards (Appendix 1). The GMC's program offering payments for ecosystem services (action ENVS5) is currently only available to agrarian communities, not individual landholders. The payment program provides a minimal incentive of 400 Mexican pesos per hectare per year, equivalent to one person's labor for one week at minimum wage, and requires significant transaction costs (i.e., compiling application materials, attending mandatory training sessions, and accounting) of the communities' leaders (GDF 2015). Land use regulation and monitoring limit agrarian actors' land use choices without offering any incentives or compensation. Training programs provide new information to encourage agrarian actors to change their practices, on the assumption that they will be motivated by potential yield improvements and increased product value (e.g., organic produce fetches a higher market price than conventional produce; soil conservation techniques might result in higher yields).

15.5.4 Adaptation Providers and Beneficiaries

In most of the PACCM actions we analyzed, it is difficult to distinguish between who is the intended provider and who is the intended beneficiary of adaptation: the government, the urban public or agrarian actors. Two PACCM coordinators saw the actions relevant to agrarian actors as being designed to help agrarian actors adapt to climate change rather than to contribute to the city's adaptation. For example, in actions SC5 and ENV4 (as referenced in Appendix 1), the urban government provides private adaptation benefits for farmers and food consumers by monitoring heirloom maize for contamination from transgenic maize, and by expanding the city's seed bank to conserve genetic diversity of heirloom maize.

However, in most of the Plan's actions, agrarian actors are implicitly presented as private providers of public adaptation benefits, with the government providing some support for adaptive actions. These public adaptation benefits can be summarized as environmental services (especially soil and water retention to reduce the risk of floods and landslides, aquifer recharge for the city's water supply, and biological and genetic diversity conservation to adapt to an uncertain future), and reduced GHG emissions from the purchase of local, sustainable products. The PACCM identifies three ways that farmers would privately benefit from their provision of these public benefits: yield gains from improved soil fertility and moisture; economic gains related to increased demand and market opportunities for their products; and the maintenance of cultural traditions strongly linked with heirloom maize varieties. It implies that these benefits would enhance farmers' and the city's resilience to climatic and environmental change. For example, in action SC4 (microbasin management for rural development and soil and water conservation on agricultural lands), the government provides supports for soil- and water-retention infrastructure to improve landholders' resource management. The desired outcomes are improved agricultural production at the farm level, and the public benefits of soil and water retention, aquifer recharge, and reduced risk of landslides. In this program, farmers would be providing public *and* private adaptation benefits with government support.

Agrarian actors recognized that their activities and lands provide private and public benefits. For example, one interviewee who farms and works in government at the borough level explained:

> [Farming] is an opportunity for self-employment, to be able to develop more, and to even have the opportunity to create jobs… I'm referring to two points, and everything follows from this, which is to conserve what we have in the city in terms of natural resources… agricultural activities are part of the periphery of the city, where fortunately we still have natural resources that are fundamental for quality of life which is at minimum necessary for mental health, air, water… it's an area for a big percentage of aquifer recharge… I identify a lot with nature, and it seems to me that we have to continue taking care of this for everything it means: carbon absorption, water capture, [and] maintaining important biodiversity. (Authors' translation)

Though many interviewees identified their motivations for farming as economic (e.g., income, self-employment), participants also mentioned tradition, culture and enjoyment (e.g., love of the countryside—"*amor al campo*"), household food security, self-reliance, and health (Appendix 2). None of the participants were unwilling to provide public benefits; in fact, the majority identified environmental stewardship and/or ecosystem services as a motivation for their agricultural activities.

15.5.5 Obstacles

Interviews with agrarian and government actors revealed four key obstacles to realizing the objectives of the PACCM for agrarian lands, actors, and activities (Appendix 2). First is the economic stress of low profits from agricultural activities. Most of the farmer interviewees have continued to practice agriculture and/or participate in conservation activities as a secondary economic activity. Though they enjoy this work, many described it as an unreliable or insufficient source of income. Some said they thought it might be possible to earn sufficient income from farming if they pursued it full-time, switched crops (e.g., to vegetables or livestock), had the infrastructure to help increase their production (e.g., greenhouses), or combined it with other economic activities such as tourism. Still, many felt that accessing high-end markets, and transporting their produce to markets and/or clients were key challenges. Farmers cited other limits to and stressors on their activities, including urban expansion, water access, infrastructure needs (e.g., irrigation systems, greenhouse repair), limited space, and government regulations. Yet the PACCM expects farmers to develop their farms into profitable commercial enterprises (e.g., actions SC1, SC4). While the PACCM acknowledges both the economic stressors facing agrarian actors and the pressures of urbanization, it ignores the fact that most agrarian actors have more gainful employment opportunities in the city.

A second obstacle is that the incentives and compensation the GMC offers agrarian actors are small but carry big transaction costs. Agrarian actors explained that the government only awards part of the amount requested or needed, and many farmers do not have enough financial capital to complete their projects, especially when the activity is only marginally profitable, if profitable at all. These actors described strict requirements, complex bureaucracy, and major time investments, just to *apply* for government supports. Several complained that many government employees are incompetent, and/or rarely follow up on farmers' requests for help or information. The majority of agrarian actors in the study perceived the distribution of government supports for environmental and agricultural activities to be motivated by politics and nepotism, tainted by corruption and mediated by local political leaders (*caciques*). For example, a farmer from the borough of Tláhuac explained:

> Unfortunately, SEDEREC [the rural development agency] programs are centralized, they are labeled. They are for a certain person who will receive the economic resources… they arrive to *campesinos* [peasant farmers] but only to the friend, or mate, or relative [of powerful/influential people] so they have more resources… As a *campesino*, I can't easily access [program benefits]. (Authors' translation)

A third obstacle is the lack of political participation of agricultural actors and representation of their concerns in urban decision-making processes. Interviews with PACCM coordinators revealed that PACCM decision makers chose not to involve farmers or agrarian communities directly in the plan's development. Instead, their interests were represented by government agencies and environmental NGOs. From the decision makers' perspective, the complexity of the SC context, and the diversity of problems and needs among agrarian actors (and an apparent lack of a unified voice or agenda from agrarian actors) were barriers to even inviting agrarian actors to contribute. One interviewee involved in the plan's development explained:

> Unfortunately, our participation with the people associated with the agrarian communities or also with the forest, with conservation… was representative. It was never direct with *ejidatarios* or *comuneros*, because, first, it was already hard to find the entry point for us to understand what was happening in the area [the SC]… During our work meetings, we [the PACCM coordinators, SEDEREC agents, researchers from the University of Chapingo, environmental NGOs] arrived at the conclusion that it was going to be very difficult to integrate [agrarian actors and communities] as a direct voice because there are a lot of them, and among them there are many social problems, with land tenure, economic problems, et cetera… We decided together that we were going to work in a representative manner, and look for support directly from SEDEREC with the people who already work in the countryside and with the NGOs that were already working there. (Authors' translation)

As a result, farmers and agrarian communities did not have the opportunity to present their needs and interests relevant to climate change or other issues, or to discuss whether and how they would be willing to participate in climate change adaptation. Nor were they able to propose actions for or changes to the Plan. Many farmers stated they had little political representation in urban politics, and few opportunities to voice their opinions and needs, and many cited examples of being ignored when they did so. For example, in July 2014, the ALDF organized the first *Campesino* Parliament of the Federal District (*Primer Parlamento Campesino del Distrito Federal*), in which Mexico City's farmers were invited to the legislature building to discuss the challenges they face. Two farmers whom we interviewed told us that by the end of the event; only one high-level government official had stayed to listen to the farmers, which made the farmers feel like the event had been a waste of their time.

A fourth, longer-term obstacle that both farmers and government actors acknowledged is cultural and generational: the aging farmer population, loss of rural tradition and knowledge, and lack of interest among and/or economic opportunity for youth in farming and natural resource management. One PACCM coordinator said:

> We are talking about these generations… who were born, learned [farming], and it is what they know. Their kids and grandkids are of the generation that studied, or already have a taxi, or already have a more urban perspective. They aspire to having a car like the guy who

> lives in Condesa [the city center]. When these generations start to become the landholders in this territory [the SC], primary activities are going to decline. For 1500 pesos, it's better to drive a taxi and sell the land. (Authors' translation)

Some farmers lamented that other farmers do not want to work, are unwilling to dedicate enough time to make farming more profitable, or will eventually lose interest in farming altogether. Describing her concerns about development in the area, one farmer from the borough La Magdalena Contreras was worried:

> …People are going to lose interest in the countryside. Because, if there are no people who like it, or who see it as difficult, or who don't put in much effort, well, [they think] I would be better off going to work in the United States, right? Alternatively, go work in an office, and leave the countryside because it doesn't provide enough to live from. So, well, we are going to die of hunger, because who is going to plant? Because these people see [agriculture] as ugly, or dirty. Therefore, it is difficult that people who are interested in [farming] have their wings clipped, it would be chaos. There is no other way besides planting, and the number one most basic thing is food, so it worries me. (Authors' translation).

While these concerns cause many participants to feel pessimistic about the future of agriculture in the city, others were optimistic. These optimistic participants reasoned that agriculture would continue because there will always be food demand, and that niche markets (i.e., organic, local) will provide economic opportunities.

15.5.6 Opportunities

Government and agrarian actors acknowledged the need for policy changes to improve both economic development in the SC and agrarian actors' capacity to provide public benefits, to achieve mutual benefits, or win-wins. PACCM coordinators acknowledged that the economic needs of the SC population must be met and that doing so will likely involve economic diversification. However, as one participant put it, the city's environmental objectives present opportunity costs for economic development, and finding a solution that meets both environmental and economic objectives would merit the "Nobel Prize for ecology." Many agrarian actors felt that government agents should spend more time in the field to familiarize themselves with the SC context and the challenges farmers face there, provide more technical assistance, and develop programs more appropriate for smallholders in the SC. While in the field, government agents would also see who is actively farming, and could distribute supports accordingly. Others recommended giving bigger financial awards to farmers to help them develop their farming projects. A few participants recommended improvements in planning, to integrate urban and rural land use plans and policies, and coordinate across objectives for conservation, ecosystem services, and economic development. One interviewee suggested an alternative management scheme for the SC modeled after the United States National Park Service, in which the government could fund conservation, agricultural, and tourism activities, and pay a professional staff to maintain environmental quality, ideally comprised of agrarian community members.

15.6 Discussion

Private provisioning of public adaptation is complex because disparate actors have different expectations and understandings of what their roles are, how they should fulfill those roles over time, what the outcomes should be, and what mechanisms should be put in place to ensure that certain benefits are provided. Adaptation has costs and risks, so the role of private provider of public adaptation benefits may not be one that everyone is willing or able to fill. For the agricultural sector, climate change adaptation is also complicated by other stressors, including socioeconomic change in the short and long term (Eakin 2005; O'Brien and Leichenko 2000). Facing economic stress and urbanization pressure, many peri-urban farmers have diversified their livelihoods with urban employment, reflecting a new rurality in which farmers' lifestyles and livelihoods are a hybrid of urban, rural, and international influences (Lerner and Eakin 2011).

Our results from Mexico City show that the GMC has little choice but to work with agrarian actors to ensure the provision of desired public adaptation benefits that result from agrarian lands and their management. The GMC has limited funds to invest in public benefits and is constrained by corruption within the government and within agrarian communities. These constraints might mean that mechanisms requiring less government involvement are more attractive to decision makers (Tompkins and Eakin 2012). However, to achieve its adaptation goals, the GMC may find more success by developing a genuine partnership with agrarian actors and being explicit about its expectations of those actors. It needs to take into account the specific vulnerabilities facing the peri-urban agrarian sector and the socio-economic changes that sector is undergoing. It needs to explore the circumstances under which agrarian actors are willing and able to provide public benefits. Finally, it should treat agrarian actors as *deliberate* adaptation providers—perhaps even more so than as providers of food and fiber—by offering them appropriate, direct incentives and compensation for the public benefits they provide.

The GMC has implicitly expected peri-urban agrarian actors to be private providers of adaptation benefits because it assumes that agrarian actors are willing to commit to and rely on agricultural production for their livelihoods, and thus will be motivated to take adaptive actions to increase production. Yet, to sustainably provide adaptation benefits, agrarian actors would either have to earn an adequate living from their agricultural activities such that they would be individually vested in the adaptive benefits promised through the PACCM (representing Olson's notion of a privileged group), or they would have to have a secure enough off-farm income that they could afford to be altruistic (Tompkins and Eakin 2012). However, agrarian actors in the SC are highly heterogeneous: some provide public benefits accidentally, others seek the private adaptation benefits associated with participating in PACCM programs (e.g., improved soil quality), and still others seek full compensation for any public benefits they are expected to provide (e.g., full compensation for installing and maintaining water-retention infrastructure on their lands). For many, urbanization has provided new economic opportunities through urban employment and (illegal) land sales (Aguilar 2008), both of which are associated with the decline of agriculture.

A second assumption is that relatively little public investment is necessary to make agriculture viable in the SC. The GMC's efforts to support development and adaptation in the SC are commendable, as are its efforts to create adaptation strategies that are mutually beneficial to city dwellers and agrarian actors. However, the PACCM mechanisms that support agrarian actors are insufficient, and the long-term result will likely be the under-provisioning of public adaptation benefits. This dilemma resembles a poverty trap (Banerjee and Duflo 2011): neither agrarian actors nor government is willing or able to invest enough to make agricultural activities profitable or to realize the desired public and private benefits. If the government provided agrarian actors no additional support for adaptation, many adaptation services would still be provided accidentally, though perhaps not as efficiently as they would through coordinated efforts with agrarian actors and other landholders, with economic compensation for the public benefits they provide. Over time, urbanization will continue to threaten these lands and undermine adaptation services.

Direct compensation for the provision of public benefits would be fair, considering that the GMC's land use regulations and conservation measures have created considerable opportunity costs for agrarian actors by constraining their economic development options for the benefit of the city's water supply and risk management, and also considering the high transaction costs of participating in government programs for private provisioning of public adaptation. Agrarian actors should be compensated for the time they spend providing public adaptation at a level at least comparable to urban employment (e.g., as a taxi driver, construction worker, etc.), and should also be compensated for the rent of their lands. Adaptation providers would not need to depend *solely* on this compensation for their livelihoods but would be better able to provide public adaptation services in ways compatible with off-farm employment. This approach might help those interested in farming to pursue or maintain those activities and make illegal land use changes less appealing for landholders. Other measures would also be needed to address the shortage of affordable housing that drives the demand side of the urbanization process.

The case of Mexico City suggests three guidelines for government efforts to promote private provisioning of public adaptation benefits. First, to achieve adaptation goals in collaboration with private providers, governments should provide mechanisms that explicitly and directly promote the desired outcomes, rather than assuming that the desired outcomes can be achieved indirectly through particular livelihood activities and economic development initiatives, such as agriculture. Direct compensation mechanisms (e.g., grants for infrastructure installation and maintenance, land purchases), and/or markets for public adaptation benefits (e.g., payments for adaptation services, payments to compensate resource managers for opportunity and transaction costs incurred from providing public benefits) may be effective mechanisms for achieving the desired adaptation outcomes (Tompkins and Eakin 2012), together with appropriate technical training. In Mexico City, for example, the government could extend conservation and natural resource management programs (i.e., supports for soil-retention infrastructure, infiltration ponds; payments for ecosystem services)—currently available only to agrarian communities with large landholdings—to individual smallholders. The government could also amend its land use plans to allow alternative uses for lands currently zoned for agriculture that benefit the landholder while also providing the adaptive benefits the government seeks. The GMC has the opportunity to do so now, as it revises its land use plan (see GMC 2016a).

Secondly, whether it is accidental or deliberate, private provisioning of public adaptation benefits requires engagement with the providers themselves, and explicit communication of needs and expectations on both sides. Developing mechanisms for private provisioning in consultation, if not collaboration, with private providers is likely to yield more sustainable and mutually beneficial outcomes (see Chap. 14: Peterson). In Mexico City, the government's recognition of the potential adaptation gains from agriculture-urban collaboration is progressive and commendable. However, not including agrarian actors in the PACCM's development, and relying on existing but insufficient programs, were missed opportunities to co-develop appropriate policy mechanisms, increase agrarian actors' participation in urban political processes, and improve trust between government and agrarian actors.

Thirdly, cases of private provisioning require closer attention to the politics of adaptation. Demands for private provisioning could reinforce historical power imbalances between the government and potential adaptation providers. Working toward a genuine partnership could provide opportunities to correct historical power imbalances, and result in truly mutual benefits for adaptation.

In cases where agrarian actors serve as private providers of public adaptation benefits, compensating them directly for desired public benefits rather than for agricultural production might appear to ignore or undermine food system adaptation and sustainability goals. In Mexico City, these goals include supporting Mexican smallholders, traditional farming practices, and conserving maize genetic diversity. However, farming households, the city, and the food system adapt and develop differently. To pursue synergies for adaptation and development across these systems, the objectives for each, and the resources and actions needed to achieve those objectives should first be evaluated separately. In some cases, the objectives for these systems should be addressed separately, and actions to achieve these objectives compensated for separately. Doing so would make policy objectives clear, and facilitate the creation of policy mechanisms that adequately and appropriately respond to the specific vulnerabilities and stressors affecting both adaptation providers and beneficiaries. It would allow more space for adaptation and transformation while striving for development and adaptation objectives.

15.7 Conclusion

Peri-urban agriculture is recognized as vulnerable to climate change and as a potentially strategic sector for urban adaptation to climate change. Yet little research has been done on how urban governments have collaborated with agrarian actors or how adaptation has been conceptualized in these arrangements. Likewise, few have asked whom the adaptations are intended to benefit. Our analysis of Mexico City's Climate Action Plan reveals the urban government's perception that agrarian actors are largely accidental providers of public benefits: that is, the public benefits are positive externalities of adaptive actions that agrarian actors would take anyway to improve their farms. However, short-term economic pressures and urbanization are driving many agrarian actors into urban employment. Many of these actors maintain agricultural and conservation activities as a secondary source of income or hobby. If the government wants these actors to provide public adaptation benefits, they will have to approach them as *deliberate*, profit-maximizing adaptation providers, and provide them with direct compensation and/or incentives for the desired benefits, independent of any particular livelihood activity.

To collaboratively advance climate change adaptation, urban decision makers will need to establish a mutual understanding with their collaborators of who is adapting for whom, why, and how. Both adaptation providers and beneficiaries will have to explicitly communicate what they need and expect from the collaboration so that specific vulnerabilities and socioeconomic changes may be taken into account. The government should develop mechanisms that directly promote the desired adaptation benefit, ideally in collaboration with private providers. To promote more productive collaborations, research could explore the politics of adaptation inherent in such collaborations; for example, how histories of social, political, economic, and environmental interactions affect institutional structures, trust, and conflict today, and how policy processes and outcomes mediate winners and losers, and synergies and tradeoffs between development and adaptation goals. This case study of Mexico City illustrates both the potential for adaptation and the pitfalls of such private–public partnerships. Genuine collaboration over time, with clearly defined roles, benefits, and costs may increase the capacity of actors in agriculture-urban collaborations to advance adaptation to climate change.

Appendix 1

Actions of 2014 PACCM pertinent to peri-urban agrarian lands and actors

Action/program	Objective (actor)	Benefit created (target beneficiary)	Policy mechanism	Cost to farmers and landholders
IPU1	Integrate environmental and urban planning policies (government)	Urban quality of life (urban residents)	Land use planning, regulation	Limits land use and development options
EVI1	Increase green space per capita (landholders)	Urban quality of life (urban residents)	Reforestation, environmental management	Transaction costs; time investment in project implementation and maintenance; potential yield and income losses; opportunity costs in other employment
SC1	Increase organic farming practices (farmers)	Environmental quality, local food (urban residents); economic development (farmers)	Training in farming practices; incentive of possible government purchasing	Transaction costs; time investment in training, transitioning farm to organic, establishing market relationships; opportunity cost of other employment
SC2	Evaluate the logging ban and forest quality (government)	Carbon sequestration and other ecosystem services (urban residents)	Regulation, evaluation	Limits land use and development options
SC3	Integrate local farmers as providers of the "green purchases" program (government)	Reduced GHG emissions (urban residents); sustainable economic development (farmers)	Agreements to realize guidelines for government purchases of local produce	Transaction costs; time to transition to meet commercial standards, establish market relationships

(continued)

(continued)

Action/program	Objective (actor)	Benefit created (target beneficiary)	Policy mechanism	Cost to farmers and landholders
SC4	Conserve soil and water on agricultural land (farmers)	Ecosystem services; reduced risk of landslides and floods (urban residents)	Economic supports for infrastructure, management plan	Transaction costs; time to implement and maintain conservation projects; potential financial investment; opportunity cost of other employment
SC5	Conserve and protect native maize varieties (government)	Maize genetic diversity (farmers)	Monitoring transgenes, regulation, training, and evaluation	Limits crop choice and development options
SC6	Conserve soil, water, and ecosystems in the SC (agrarian communities)	Ecosystem services; reduced risk of landslides and floods (urban residents)	Financial supports for infrastructure; monitoring	Transaction costs; time to implement and maintain conservation projects; potential financial investment
SC7	Conserve biodiversity through milpa agriculture (government)	Food diversity, ecosystem services, local food (urban residents); livelihoods (farmers)	Demonstration plots, technical assistance, maize samples collected for seed bank	Time in training; opportunity costs for economic development
SC8	Increase carbon capture in SC (government, landholders, women's groups)	Ecosystem services, quality of life (urban residents); employment (agrarian actors)	Land use plans, project implementation, and maintenance on abandoned agricultural land	Transaction costs; limits land use and development options
SC9	Rainwater capture for irrigation (farmers)	Increased water supply (urban residents); improved water management (farmers)	Design and implement plan for eco-technologies for water capture	Transaction costs; time in training with new technology
ENV3	Restore environmental quality in natural	Ecosystem services (urban residents)	Management plans	Transaction costs; time to implement and maintain conservation projects

(continued)

(continued)

Action/program	Objective (actor)	Benefit created (target beneficiary)	Policy mechanism	Cost to farmers and landholders
	protected areas (landholders)			
ENV4	Conserve maize diversity and biodiversity (government)	Conserve genetic diversity (farmers) and biodiversity (urban residents)	Create lab, reactivate seed bank	None
ENV5	Create new ecological reserves in agrarian communities (agrarian communities)	Ecosystem services (urban residents)	Market for public good (ecosystem services)	Transaction costs; time in labor, training, implementing and maintaining projects
F15	Contain urban expansion (government)	Urban quality of life (urban residents)	Land use planning	Limits land use and development options

Appendix 2

Analysis of Interviews with Agrarian Actors (N = 33)

Motivations to farm

- Tradition, culture, enjoyment: N = 20
- Economic: N = 18
- Environmental stewardship and ecosystem services: N = 17
- Household food security and self-reliance: N = 11
- Health: N = 7

Obstacles to success of agrarian climate actions

- Economic barriers to farming: N = 20
  - Insufficient or unreliable income: N = 16
  - Limited market access: N = 10
  - Profits are limited but could improve: N = 9
- Other stressors and limits on farming activities: N = 25
  - Urbanization: N = 17
  - Infrastructural needs: N = 9

  - Water supply/access: N = 7
  - Space, capacity limits: N = 7
  - Regulations: N = 6
- Government supports: N = 24
  - Insufficient: N = 14
  - Transaction costs: N = 13
  - Strict requirements: N = 15
  - Access biased/corruption, nepotism: N = 18
- Limited political participation and representation: N = 15
- Lack of interest in farming: N = 14

Opportunities for agricultural development in the SC and private provision of public benefits

- Field presence: N = 12
- Better Funding: N = 12
- Tech support: N = 6
- Appropriate programs: N = 5
- Integrated urban–rural planning: N = 4

References

Aguilar, A. G. (2008). Peri-urbanization, illegal settlements and environmental impact in Mexico City. *Cities, 25*(3), 133–145.

Banerjee, A. V., & Duflo, E. (2011). *Poor economics: A radical rethinking of the way to fight global poverty*. New York, NY: PublicAffairs.

Bernard, H. R. (2011). *Research methods in anthropology: Qualitative and quantitative approaches* (5th ed.). New York: Altamira Press.

Cruz Rodríguez, M. S. (1995). Políticas agrarias en la periferia ejidal de la ciudad de México. *Sociológica, 10*(29), 16.

Cruz Rodríguez, M. S. (2001). *Propiedad, poblamiento y periferia rural en la Zona Metropolitana de la Ciudad de México*. México: Universidad Autónoma Metropolitana.

de Zeeuw, H., Van Veenhuizen, R., & Dubbeling, M. (2011). The role of urban agriculture in building resilient cities in developing countries. *Journal of Agricultural Science, 149*(S1), 153–163.

Decreto de Programa General de Ordenamiento Ecológico del Distrito Federal, 90 pp. (August 1, 2000).

Dieleman, H. (2016). Urban agriculture in Mexico City: Balancing between ecological, economic, social and symbolic value. *Journal of Cleaner Production*. doi:10.1016/j.jclepro.2016.01.082.

Eakin, H. (2005). Institutional change, climate risk, and rural vulnerability: Cases from Central Mexico. *World Development, 33*(11), 1923–1938.

Eakin, H., & Appendini, K. (2008). Livelihood change, farming, and managing flood risk in the Lerma Valley, Mexico. *Agriculture and Human Values, 25*(4), 555–566.

Eakin, H., Lerner, A. M., & Murtinho, F. (2010). Adaptive capacity in evolving peri-urban spaces: Responses to flood risk in the Upper Lerma River Valley, Mexico. *Global Environmental Change, 20*(2010), 14–22.

Erdlenbruch, K., Thoyer, S., Grelot, F., Kast, R., & Enjolras, G. (2009). Risk-sharing policies in the context of the French Flood Prevention Action Programs. *Journal of Environmental Management, 91*(2), 363–369.

Eriksen, S., & Silva, J. A. (2009). The vulnerability context of a savanna area in Mozambique: Household drought coping strategies and responses to economic change. *Environmental Science & Policy, 12*(1), 33–52.

Ezcurra, E., Mazari-Hiriart, M., Pisanty, I., & Aguilar, A. G. (1999). *The Basin of Mexico: Critical environmental issues and sustainability*. Tokyo: United Nations University Press.

Félix Guerra, H., Topelson Fridman, S. H., de la Garza Estrada, G., Poiré Romero, A., Mohar Betancourt, G., Fernández Varela, F. V., et al. (2012). Delimitacion de las zonas metropolitanas de Mexico 2010. México: Secretaría de Desarrollo Social (SEDESO), Consejo Nacional de Población (CONAPO), Instituto Nacional de Estadística y Geografía (INEGI).

Fischer, G., Shah, M., & Velthuizen, H. v. (2002). Climate Change and Agricultural Vulnerability (pp. 152). Johannesburg: International Institution for Applied systems Analysis.

Food and Agriculture Organization (FAO). (2008). Climate Change and Food Security: A framework Document (pp. 93). Rome.

Gobierno de la Ciudad de México (GMC). (2016a). Plan CDMX: Proyecto de PGDU para Consulta Pública (pp. 251). Ciudad de México: Gobierno de la Ciudad de México, Secretaría de Desarrollo Urbano y Vivienda.

Gobierno de la Ciudad de México (GMC). (2016b). Programa de Acción Climática de la Ciudad de México, Informe de avances al 2016 (pp. 116). México, D.F.: Secretaría del Medio Ambiente, C40.

Gobierno del Distrito Federal (GDF). (2004). Estrategia Local de Acción Climática del Distrito Federal (pp. 123).

Gobierno del Distrito Federal (GDF). (2008). Programa de Acción Climática de la Ciudad de México 2008-2012. México, D.F.: Secretaría Del Medio Ambiente del Distrito Federal, World Bank.

Gobierno del Distrito Federal (GDF). (2012). Atlas Geográfico del Suelo de Conservación del Distrito Federal (pp. 100). México, D.F.: Secretaría del Medio Ambiente, Procuraduría Ambiental del Ordenamiento Territorial del Distrito Federal, México, D.F.

Gobierno del Distrito Federal (GDF). (2013). Programa General de Desarrollo del Distrito Federal 2013-18 (pp. 160). Ciudad de Mexico: Jefatura de Gobierno, Administración Pública del Gobierno del Distrito Federal.

Gobierno del Distrito Federal (GDF). (2014). Programa de Acción Climática: Ciudad de México 2014-2020 (pp. 383). México, DF: SEDEMA, Centro Mario Molina para Estudios Estratégicos sobre Energía y Medio Ambiente, A.C.

Gobierno del Distrito Federal (GDF). (2015). *Tercer Informe de Gobierno del Distrito Federal* (p. 621). Mexico City: Secretaría Particular del Jefe de Gobierno.

Hughes, S., & Romero-Lankao, P. (2014). Science and institution building in urban climate-change policymaking. *Environmental Politics, 23*(6), 1023–1042.

Instituto Nacional de Estadística y Geografía (INEGI). (2009). Censo Agropecuario 2007, VIII Censo Agrícola, Ganadero y Forestal: Distrito Federal. Aguascalientes.

Intergovernmental Panel on Climate Change (IPCC). (2014). Climate Change 2014: Synthesis Report. Contribution of Working Groups I, II, and III to the Fifth Assessment Report of the Intergovernmental Panel on Climate Change. In R. K. Pachauri & L. A. Meyer (Eds.), (pp. 151). Geneva, Switzerland: IPCC.

Kotchen, M. J., & Moore, M. R. (2007). Private provision of environmental public goods: Household participation in green-electricity programs. *Journal of Environmental Economics and Management, 53*(1), 1–16.

Lerner, A. M., & Eakin, H. (2011). An obsolete dichotomy? Rethinking the rural-urban interface in terms of food security and production in the global south. *The Geographical Journal, 177* (4), 311–320.

Ley de Mitigación y Adaptación al Cambio Climático y Desarrollo Sustentable para el Distrito Federal 18 (June 16, 2011).

Losada Custardoy, H., Martínez, H., Vieyra, J., Pealing, R., Zavala, R., & Cortés, J. (1998). Urban agriculture in the metropolitan zone of Mexico City: Changes over time in urban, suburban and peri-urban areas. *Environment and Urbanization, 10*(2), 37–54.

Lwasa, S., Mugagga, F., Wahab, B., Simon, D., Connors, J. P., & Griffith, C. (2015). A meta-analysis of urban and peri-urban agriculture and forestry in mediating climate change. *Current Opinion in Environmental Sustainability, 13,* 68–73.

McIntyre, B. D., Herren, H. R., Wakhungu, J., & Watson, R. T. (2009). *Agriculture at a Crossroads: Global Report* (pp. 590). Washington DC: International Assessment of Agricultural Knowledge Science and Technology for Development (IAASTD).

Mees, H. L. P., Driessen, P. P. J., & Runhaar, H. A. C. (2012). Exploring the scope of public and private responsiblities for climate adaptation. *Journal of Environmental Policy & Planning, 14* (3), 305–330.

Milman, A., & Warner, B. P. (2016). The interfaces of public and private adaptation: Lessons from flooding in the Deerfield River Watershed. *Global Environmental Change, 36,* 46–55.

Misselhorn, A., Eakin, H., Devereux, S., Drimie, S., Msangi, S., Simelton, E., et al. (2010). Vulnerability to what? In J. Ingram, P. Ericksen, & D. Liverman (Eds.), *Food security and global environmental change* (pp. 87–114). Washington, DC: Earthscan.

Morton, J. F. (2007). The impact of climate change on smallholder and subsistence agriculture. *PNAS, 104*(50), 19680–19685.

Mougeot, L. J. A. (2000). Urban agriculture: definition, presence, potentials and risks. In N. Bakker, M. Dubbeling, S. Gündel, U. Sabel-Koschella & H. d. Zeeuw (Eds.), *Growing cities, growing food: Urban agriculture on the policy agenda. A reader on urban agriculture* (pp. 99–117). Feldafing, Germany: DSE/ETC.

Næss, L. O., Newell, P., Newsham, A., Phillips, J., Quan, J., & Tanner, T. (2015). Climate policy meets national development contexts: Insights from Kenya and Mozambique. *Global Environmental Change, 35*(2015), 534–544.

O'Brien, K. L., & Leichenko, R. M. (2000). Double exposure: Assessing the impacts of climate change within the context of economic globalization. *Global Environmental Change, 10*(3), 221–232.

Pensado Leglise, M. d. R. (2001). El desafío rural del siglo XXI para el Distrito Federal. *Revista Estudios Agrarios, 16*, 39–66.

Posthumus, H., Hewett, C. J. M., Morris, J., & Quinn, P. F. (2008). Agricultural land use and flood risk management: Engaging with stakeholders in North Yorkshire. *Agricultural Water Management, 98*(7), 787–798.

Quintanar, E. (2014). *Diagnostico: La Agricultura Urbana y Peri-Urbana de la Ciudad de México* (p. 76). México, D.F.: UNEP/UNDAP.

Schteingart, M., & Salazar, C. E. (2010). *Expansión urbana, sociedad y ambiente*. México, DF: El Colegio de México.

Secretaría de Agricultura Ganadería Desarrollo Rural Pesca y Alimentación (SAGARPA). (2014). Estudio sobre el envejecimiento de la población rural en México. Ciudad de México.

Secretaría de Desarrollo Rural y Equidad para las Comunidades (SEDEREC). (2014). Segundo Informe de Gobierno 2013-2014 (pp. 97). Mexico City.

Secretaría de Industria y Comercio, D. G. d. E. (1965). IV Censos Agricola, Ganadero, y Ejidal. 1960. México, D.F.

Servicio de Información Agroalimentaria y Pesquera (SIAP). (2014a). Infografía Agroalimentaria del Distrito Federal 2014 (pp. 64). México, D.F.

Servicio de Información Agroalimentaria y Pesquera (SIAP). (2014b). SIACON. http://www.gob.mx/siap/acciones-y-programas/produccion-agricola-33119?idiom=es. Accessed October 14, 2015.

Sheinbaum Pardo, C. (2008). *Problemática ambiental de la Ciudad de México*. México, D.F.: Editorial LIMUSA, S.A. DE C.V.

Simon, D. (2008). Urban environments: Issues on the peri-urban fringe. *Annual Review of Environment and Resources, 33,* 167–185.

Simon, D. (2012). Climate and environmental change and the potential for greening African cities. *Local Economy, 28*(2), 203–217.
Taylor, M. (2013). Climate change, relational vulnerability and human security: Rethinking sustainable adaptation in agrarian environments. *Climate and Development, 5*(4), 318–327.
Tompkins, E. L., & Eakin, H. (2012). Managing private and public adaptation to climate change. *Global Environmental Change, 22*(1), 3–11.
United Nations Human Settlements Programme. (2009). Planning Sustainable Cities: Global Report on Human Settlements 2009 (pp. 306). Sterling, VA.
VERBI GmbH. (2016). *MAXQDA, software for qualitative data analysis*. Berlin, Germany: Sozialforschung GmbH.
Ward, P. M. (1990). *Mexico City: The production and reproduction of an urban environment*. Boston: G.K. Hall & Co.

Author Biographies

Julia C. Bausch recently earned her Ph.D. in Sustainability from Arizona State University. Her research focuses on how governance and policy for agriculture are evolving to address sustainability and adaptation concerns for both agriculture and the social-ecological systems in which it is embedded. She explores how priorities, tradeoffs and synergies across different functions of agriculture are mediated for the benefit of different actor groups and systems.

Hallie C. Eakin is an Associate Professor in Sustainability Science with the School of Sustainability, Arizona State University where she teaches food system sustainability. Her research focuses on the adaptability, resilience and sustainability of smallholder farmers and food systems in the face of global environmental and socioeconomic change.

Amy M. Lerner is an assistant research professor in the National Laboratory of Sustainability Science (LANCIS), within the Institute of Ecology at the National Autonomous University of Mexico (UNAM). She has a Ph.D. in Geography from the University of California, Santa Barbara, a Master's in Environmental Studies from Brown University, and an undergraduate degree in Human Biology from Stanford University. Her research focuses on the human dimensions of global environmental change, particularly in the context of rural to urban transitions, sustainable agricultural production, and urban resilience in Latin America, and specifically how residents and land stewards contribute and react to global change.

Chapter 16
Beyond "The Business Case": The Emerging Role of Entrepreneurs in the Multilevel Governance of Urban Decarbonization in Canada

Scott Morton Ninomiya and Sarah Burch

Abstract In 2015, at the 21st Conference of the Parties to the United Nations Framework Convention on Climate Change (UNFCCC), global leaders agreed that urgent, collaborative action on climate change is not only an environmental and social imperative but also represents a significant economic opportunity. The transition to a decarbonized global economy presents a challenge of unprecedented scale, yet a growing body of research identifies local level energy transition processes as an effective locus for action and change. There is an identified lack of systematic learning about transitions at the municipal level and this presents an ideal opportunity to address it through collaborative cross-sectoral research. This paper analyzes the case of Waterloo, Canada, through the lens of socio-technical transitions to assess potentially innovative pathways to a more sustainable (low carbon) future. In particular, we examine a new participatory process called Decarbonize Waterloo Region, a forum within which local stakeholders, scholars, and practitioners can envision low carbon futures and negotiate paths toward them. Participation in the forum as embedded scholars and the administration of post-forum surveys delivered insights into the effectiveness of this process, the partnerships, and policies at various levels of government that might be required to accelerate this sustainability transition, and the roles that entrepreneurs can play in driving innovation. Waterloo Region presents an interesting case as a municipality with a history as a hub of technological innovation, in which local entrepreneurs have played a significant part. The analysis of the Decarbonize Waterloo Region process will investigate how entrepreneurs can now be engaged to articulate and implement a vision to create a local hub of transition innovation. New actors and participatory processes are of particular interest. Our investigation probes the role of entrepreneurs in experimentation, a crucial part of sustainability transition processes. Innovative low-carbon energy solutions require testing before they can be

S.M. Ninomiya (✉) · S. Burch
University of Waterloo, Ontario, Canada
e-mail: sdmorton@uwaterloo.ca

S. Burch
e-mail: sarah.burch@uwaterloo.ca

S. Hughes et al. (eds.), *Climate Change in Cities*, The Urban Book Series,
https://doi.org/10.1007/978-3-319-65003-6_16

scaled up, and local entrepreneurs have the potential to play a key role in the development, roll out and evaluation of those early experiments. Ultimately, we find that a clear picture of local energy systems and distinct set of challenges (as daunting and complex as they are) provided local stakeholders with places to start and arenas for ongoing innovation. The forum process also strengthened bonds among an expanding network of stakeholders who can now move forward with a collective desire to take on those challenges and transform their local energy picture.

16.1 Introduction

Global emissions of greenhouse gases (GHG) have pushed atmospheric concentrations of carbon dioxide consistently above 400 parts per million (ppm)—a level unprecedented over the course of human history, and clearly linked to the escalating consumption of fossil fuels. Even so, in 2014 and 2015, global greenhouse gas emissions leveled off, indicating the possible decoupling of economic growth from emissions (IEA 2016). Level (i.e., not increasing) flows of greenhouse gases, however, will still lead to an increasing stock of carbon in the atmosphere (especially if the capacity of oceans and terrestrial ecosystems to absorb carbon reaches its threshold), and ultimately produce dangerous shifts in the global climate. Thus, two conclusions are becoming apparent: without dramatic and accelerated reductions in greenhouse gas emissions, communities around the globe (especially those in high latitudes and impoverished regions) will experience significant and detrimental climate change impacts; and dogged, ambitious efforts from the global to the local scale to reduce greenhouse gas emissions over the last two decades may, in fact, be bearing fruit.

Climate change has traditionally been the domain of international policy-making, with state-to-state negotiations yielding the Kyoto Protocol (1997) and more recent Paris Agreement (2015). These agreements may spur a global conversation about the necessity of climate change mitigation and adaptation, but any treaty must ultimately be legislated and implemented domestically. Canada unveiled a Pan-Canadian Framework on Clean Growth and Climate Change in 2016. While the total contribution of Canada to total global emissions is rather small (less than 2%), Canadians emit over 20 tons per capita annually—one of the highest rates in the world (Environment and Climate Change Canada 2016). In response to this challenge, the federal government has pledged to reduce its emissions by 30% from 2005 levels by 2030 and is developing several pan-Canadian climate policy tools aimed at decarbonization (Government of Canada 2016), which have yet to bear fruit.

Given the ways in which jurisdiction over sources of greenhouse gas emission is shared among municipalities, provinces, and the federal government in Canada, a coordinated multilevel governance approach is required to create substantial reductions in emissions. In multilevel governance, various levels of authority are implicated in a policy domain, in this case local climate action. Accordingly, a number of actors, organizations, and agendas must be coordinated to achieve coherent policies. Individual provinces and municipalities had been implementing a vast array of climate policies for at least two decades (Burch 2010; Castan Broto and Bulkeley 2013; Robinson 2005). This distributed decision-making on climate change is, in fact, a prerequisite for success: the federal government has jurisdiction over only a portion of the greenhouse gas emissions produced, while provinces (or territories) and municipalities hold sway over the rest.

Furthermore, individual citizens, private sector actors, and civil society groups also make crucial decisions and possess particular capacities to address climate change. Biermann et al. (2009, p. 4), for instance, defined earth system governance as "the interrelated and increasingly integrated system of formal and informal rules, rule-making systems, and actor-networks at all levels of human society (from local to global) that are set up to steer societies towards preventing, mitigating and adapting to global and local environmental change." Emerging governance arrangements among state and non-state actors may "support an appropriate balance between permanence and change" (Pahl-Wostl 2009, p. 358) in markets and bureaucracies—an important attribute in the face of the uncertainty and complexity associated with climate change, for instance (Burch et al. 2014). So, the decarbonization challenge is one that ultimately takes place on the ground in communities but nested within intersecting policies and priorities, changing the way that individuals live, play, and work (Scannell and Gifford 2013). Consequently, the local scale is emerging as a key locus for action on climate change (Broto and Bulkeley 2013) and local government leadership and collaboration on climate action has matured significantly in the past decade (Kinley 2016).

The desire for tangible, local action has also fuelled growing scholarly and policy interest as well as financial investment in clean technology solutions and the entrepreneurs who create them (McCrone and Molseneer 2016; Loorbach and Wijsman 2013; Klewitz and Hansen 2014). Bulkeley (2016) conceives of this shift toward a local locus for climate action as inverting a pyramid to form a funnel. Communities are increasingly taking local action instead of waiting for the impact of international climate treaties to trickle down to them. As millions of citizens and entrepreneurs are mobilized in thousands of cities, the cumulative impact is felt significantly at national and global levels in the form of social, technical, and policy innovations. The dynamics of multilevel climate governance are being flipped on their head: actors at the local level are claiming their climate agency and feeding the funnel with innovative ideas and actions (Bulkeley 2016).

Despite this proliferation of local innovation, greenhouse gas emissions grow out of deep roots: the design of communities (the proximity of residential areas to workplaces, for instance, and the availability, diversity, and abundance of mass or active transportation options) leads to an over-abundance of personal vehicles and single-family detached homes (Burch 2011; Burch et al. 2014). A transformative shift, or a radical, nonlinear divergence (Pelling et al. 2015)[1], in the nature of this underlying development path may be required to reach the ambitious climate change targets laid out both in the Paris Agreement and in the thousands of local climate change action plans around the world.

If transformative change is required, then decarbonization is not simply the task of government. Regulations and incentives play a crucial role, but many of the technical and social innovations that hold the most promise will likely come from the private sector. The economy and private sector have been recognized as important arenas for sustainability research, yet little empirical work has been conducted (see Driessen et al. 2015; Pereira et al. 2015) that sufficiently characterizes the varied motivations, capacities, and transformative potential of entrepreneurs and the businesses of which they are a part. Furthermore, little work has been done to interrogate the nature of relationships between government and entrepreneurs that might trigger or enable sustainability-oriented innovation. Local authorities, where many decarbonization policies will be implemented, often rely on external organizations to provide advice and assistance to businesses (Bradford and Fraser 2008; Klewitz et al. 2012). External organizations appear to fall into two groups: (1) intermediaries, organizations such as NGOs which engage with businesses with the express purpose of improving some aspect of their operations; and (2) networks, collaborative spaces wherein businesses engage with other organizations and ideas freely evolve. The intention of both is the same, however: to provide businesses with the knowledge resources and financial capacity to overcome the characteristics (such as a lack of technical capacity, poor return on investment, or low likelihood of market penetration) that discourage environmental investment (Klewitz et al. 2012; Parker and Rowlands 2007).

Employing a multilevel governance lens, this chapter explores initial steps towards decarbonization in practice. The Decarbonize Waterloo forum was a participatory visioning process held in 2016 that aimed to gather representatives from government, industry, and academia to explore pathways to full decarbonization by 2050. Observational and survey results underscore the importance of forum design, with a particular focus on the capacities and barriers faced by entrepreneurs.

16.2 Local Participatory Transition Processes

Decarbonization is a multifaceted, iterative, and deeply value-laden process that is only just beginning in many nations around the world. Not simply a challenge of technological innovation, it is also a social and political project. This suggests the need for innovative participatory processes that allow multiple stakeholders to

[1]Following on Pelling et al. (2015), transformation can be thought of as a fundamental shift in the structure or function of a social-ecological system, including a rethinking of core values and relationships. Whether this is the result of accumulated incremental action or a more radical, disruptive shock, remains an issue of discussion and debate.

envision desirable futures (Shaw et al. 2009), voice concerns about the direction of change, and engage actively in the implementation of any plans. Engaging with multiple stakeholders offers opportunities to learn more about the interwoven components of a complex system, but also to elicit the values that are implicit in a transition to a decarbonized community. Inevitably, some groups will benefit from the decarbonization pathway that is chosen, while others may suffer; participatory processes allow for these contentious implications to be raised and addressed.

Much of the early participatory processes literature utilizes Habermasian ideas surrounding competent discourse to analyze methods of citizen participation. The central tenets of this literature hold that cooperative decision-making processes allow for the incorporation of different ways of knowing, but challenges arise in ensuring that all participants are equally skilled at the methods of communication needed, and thus equally represented (Renn et al. 1995). Although it has traditionally been thought that the most authentic deliberation occurs within the formal institutions of the government, alternative forms of deliberation, from story-telling and greeting (Dryzek 2000) to citizen juries and study groups, might be better placed to produce relevant, competent outcomes (Renn et al. 1995) (see Chap. 8: Engberg). In recent years, these methods have evolved dramatically to include participatory visioning (Sova et al. 2012) and geo-visualization (Shaw et al. 2009) and a host of other tools. Even with more inclusive methods, however, it is possible that participants will be self-selected, with vested interests (Pidgeon et al. 2005)

A suite of international research projects is currently taking place that aims to identify the various actors that play a role in urban sustainability transitions, focusing on different geographical areas and actor dynamics. Governing Urban Sustainability Transitions (GUST), for instance, is developing the concept of Urban Living Labs as experimental sites that design, test, and share lessons about both social and technical innovation (Bulkeley et al. 2015). The Governing and Accelerating Transformative Entrepreneurship (GATE) project focuses explicitly on the transformative potential of entrepreneurs, and the governance challenges inherent in engaging small businesses to innovate on sustainability (Burch et al., Forthcoming). Taken together, these and other projects illustrate that decarbonization at the local scale is not simply a technical task involving the pursuit of renewable energy production and reduced energy consumption, but a deeply value-laden process that will shape the look, feel, and function of future communities. Furthermore, it is a moving target: as new technologies, preferences, and socioeconomic realities emerge, the policies considered most desirable and feasible by communities will change. As such, there is a need for iterative and inclusive participatory processes that enable creative visioning, the sharing of expertise, and the deepening of partnerships to enable implementation. The process described in the sections that follow represents a nascent attempt to do this.

16.3 Decarbonize Waterloo Region: A Participatory, Multilevel Process

The Decarbonize Waterloo Region process takes place in an evolving regulatory and political context. As mentioned, there is increasing interest and investment in climate action at the federal level in Canada, represented by the introduction of carbon pricing and the extensive nation-wide consultations that informed the recent Pan-Canadian Framework on Clean Growth and Climate Change (Government of Canada 2016). The Province of Ontario also released its own five-year Climate Action Plan in June 2016, including a goal to reduce GHG emissions to 80% below 1990 levels by 2050. Early analysis of the Ontario Climate Action Plan has noted that many of the actions and goals in the plan are tied directly to steps that can only be taken by municipalities—particularly land use planning (Burch 2016). In Waterloo Region, the Waterloo Region Climate Action Plan was developed in 2013 by local civil sector actors in collaboration with local governments but offered only a modest near-term goal of 6% emissions reductions by 2020.

The Decarbonize Waterloo Region forum was initiated in this context as a participatory process, with the goal of bringing together the pluralistic knowledge and capacities of many different local actors. The process was initiated by researchers at the University of Waterloo, who sought out a broad range of experts from the local civil sector, academia, all levels of government, energy sector leaders, as well as local entrepreneurs. These actors were convened for a two-day invitation-only forum, hosted at the Balsillie School of International Affairs in Waterloo. The local government and civil sector leaders of the Waterloo Region Climate Action Plan assisted in the development of the list of participants, based on the significant groundwork they had already done to engage local expertise and build partnerships. That ground work was led by local civil sector organizations like Sustainable Waterloo Region and REEP Green Solutions when they called together many actors to create the Climate Action Waterloo Region plan that was launched in 2013. This plan is unique in Canada in that it was initiated and is still led by civil sector actors. The plan is implemented by a joint management team and three committees that include diverse stakeholders from all sectors. The goals of the plan are relatively modest, compared to the broader goal of decarbonization, but the Plan's unique governance structure and the connections that it has created set the table for deeper decarbonization (Waterloo Region, 2013).

On November 17 and 18, 2016, over 50 participants gathered for the two-day workshop. The forum organizers articulated ambitious goals: namely, the development of scenarios for fully decarbonized local energy systems and pathways to achieve those scenarios. The scenarios were focused on the three energy use sectors with the highest GHG emissions, as identified in the Waterloo Region Climate Action Plan—transportation, residential, and industrial/commercial. Scenarios were also to be developed for the transformation of three broad energy source categories, including electricity, fuels, and thermal energy. The forum intentionally excluded agriculture and waste because in the local context these represented relatively small proportions of total emissions. In addition to the development of scenarios and pathways, the Decarbonize Waterloo Region forum also included the development of policy recommendations for federal, provincial, and municipal governments.

Participants at the forum sought to quantify the complexity of the carbon challenge in their own Canadian municipality and devise specific strategies that might trigger and accelerate the transition to a zero-carbon local economy. The forum brought a diversity of voices to the local transition planning table, and the list of the unusual suspects included a small but vocal handful of local entrepreneurs including engineering, construction, and clean tech firms.

16.4 Methods: An Embedded Scholarship Approach

The Decarbonize Waterloo Region forum was designed to be a reflexive process, and as such an embedded scholarship approach was integrated from the earliest planning stages. The authors of this chapter took part in the forum as both participants and as researchers analyzing the process to determine what lessons can be learned. These lessons will be applied to ongoing decarbonization efforts in Waterloo Region and disseminated widely to inform future decarbonization initiatives in communities across Canada. This embedded scholarship approach is emerging as an important aspect of transition initiatives. Wittmayer and Schapke, for instance, summarize some of the new roles that researchers are exploring in transition processes: "they guide collective learning processes, mediate between different frames, commit themselves to transforming reality, and put sustainability into action" (Wittmayer and Schapke 2014, p. 483).

The role of the embedded scholars was disclosed and explained to participants by means of a letter emailed to them in advance of the forum. This letter also outlined the voluntary nature of the study and made a commitment to refraining from attributing remarks by the participants in any published material. At the forum, the embedded scholars also gave brief introductory remarks reiterating their role. Throughout the two-day forum, the scholars took part in discussions and took notes about how the process was unfolding.

Immediately after the forum, a 12-question survey was administered electronically to all forum participants. It asked questions about their experience in the forum process: what worked well, what did not work well, stakeholder interactions and contributions to the process, and what lessons can be learned as the decarbonization process moves forward. Survey participants were recruited by means of a letter that was sent to all forum participants. Participation in the survey was optional. Thirty-five out of fifty participants responded to the survey, yielding a 70% response rate. The results of the survey have been used to inform this chapter, and will also be used to write an analysis of the key lessons learned in the forum process.

Decarbonize Waterloo Region is an ongoing process. The next step will involve convening a series of small focus groups consisting of forum participants in winter of 2017. The purpose of these focus groups is to determine the next steps for innovation and action to achieve the goal of decarbonizing local energy systems. These groups will be convened in the first half to 2017. Due to time restrictions, results from these focus groups cannot be included in this chapter.

16.5 Observations from a Participatory Process: Entrepreneurs and Innovation at the Decarbonize Waterloo Region Forum

Observations from discussion at the forum and the follow-up survey supplied rich insights into barriers and opportunities for decarbonization innovation in Waterloo Region, as well as shedding light on the emerging potential roles for entrepreneurs in that innovation.

16.5.1 Innovation Stifled: Barriers to Decarbonization

Many barriers to innovation were illuminated in both the forum discussion and in the follow-up survey. The sheer magnitude of the task was noted by many participants. Statistics on local energy use and sources compiled for the forum made it clear that Waterloo Region's energy systems are extremely carbon intensive. Over 77% of energy consumed in the residential, transportation, and commercial/industrial sectors (combined) come from fossil fuels. After 2 days in a room full of local experts, elegant scenarios and straightforward pathways to decarbonization did not materialize. Many participants noted that it was difficult to get their head around the problem—one survey respondent said: "I knew it would be complex—but not this complex."

Participants also noted that several key stakeholders were missing from the discussion, including energy end users with low incomes and rural stakeholders. Participants also noted disappointment the paucity of representation from key stakeholders like electrical and gas utilities, energy intensive industries, and Metrolink (the commuter train and bus system). Some participants noted a lack of expertise on specific energy technologies and social scientists with expertise on behavior change. The fact that no elected leaders attended was noted with dismay by some. Without these key actors on board, it will be difficult to move innovative solutions forward, so engagement of the missing actors is an area for ongoing work.

The realization that many of the 2016 forum's municipal policy recommendations were already listed in the Region's 2013 Climate Action Plan—but not yet accomplished—was named as a low point by several participants. There was disappointment at how slowly the local policy environment is evolving to support climate innovation, despite consensus by all local governments on the Climate Action Plan goals. This indicated that perhaps more concrete mechanisms must be put in place to monitor specific progress on identified local policy goals.

Negative local attitudes were also named as a barrier to the innovation that is needed to move the decarbonization process forward. At one point in the forum, a local entrepreneur interjected in a group discussion: "I'm sick and tired of being told that decarbonization can't be done here. I see so many things happening in other places." This powerful statement is double edged. It indicates an awareness of, and deep frustration with, local attitudes that stifle local innovation that could contribute to decarbonization. However, it also points to a strong desire to move beyond those attitudes and an awareness of viable alternatives to the Region's current energy systems which rely heavily on fossil fuels. Further investigation and understanding of the stifling attitudes that motivated this double-edged statement is an important area of future local research.

16.5.2 Innovation Fostered: Forum Outputs Present Opportunities

Though many barriers to innovation were identified, and no clear scenarios for decarbonization were produced, several significant outputs of the forum present opportunities for innovation going forward.

In preparation for the forum, statistics on present and future (2050 projections) energy use and sources for the Region were compiled into a series of diagrams and displayed in terms of petajoules. These diagrams provoked a great deal of discussion at the forum (as well as surprise and dismay). Survey responses revealed that many participants found this visualization very helpful for conceptualizing the challenge of decarbonization. In addition to the diagrams, the discussion that they generated helped to make the Region's energy systems—which are normally in the background of people's consciousness—more visible. It also provides a rough projection of how much energy will be needed going forward and notions about where reductions are required. Though the Region's energy future is complex and concomitantly uncertain, these diagrams provide a concrete starting place for ongoing and evolving action. As one participant remarked in the survey, "I am able now to see the beginnings of what the path might look like."

As part of making the beginnings of the path to decarbonization clearer, the first day of the forum included several presentations by local experts on emerging tools that organizers evaluated as promising innovations for the decarbonization of the Region. These included social innovations like coordinated neighborhood retrofits for district energy, financial innovations such as impact investment, and technical innovations such as geothermal energy systems. Surveys revealed that participants found this part of the forum enlightening, useful, and inspiring(Fig. 16.1).

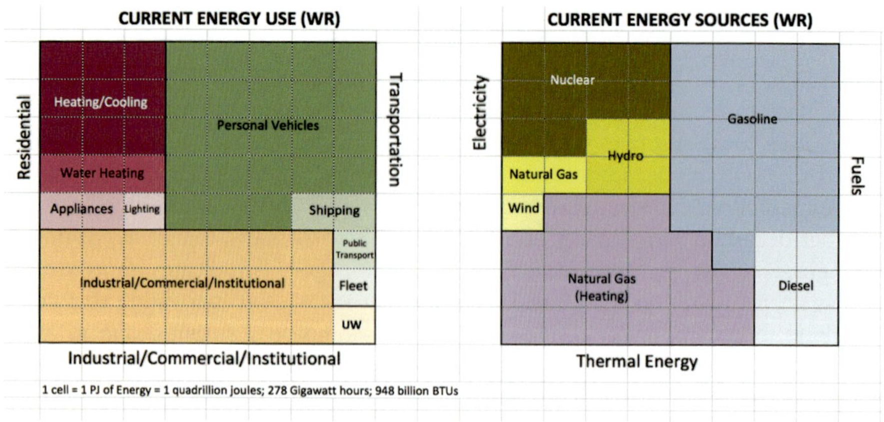

Fig. 16.1 Current energy use and sources in Waterloo region (in petajoules)

The forum process also helped to identify four major challenges which must be addressed by 2050 if full decarbonization is to be achieved in the Region:

Challenge 1 Dramatically reducing energy needs of the built environment
Challenge 2 Maximizing local renewable power generation
Challenge 3 Eliminating fossil fuels-based transportation
Challenge 4 Replacing natural gas as a source of energy for heating

While these challenges are undoubtedly daunting, moving forward, these challenges can be shaped into arenas for research and experimentation by scholars, entrepreneurs, and other local stakeholders.

> Creating a favorable policy environment was also identified as a prerequisite for climate innovation in the Region. A list of policy recommendations for federal, provincial, and local governments was generated by forum participants in a session designed specifically for that purpose on day two. This discussion included early ideas on how to coordinate policies across the jurisdictions to promote a rapid local transition to decarbonized energy systems, including building codes, infrastructure renewal, and many others.

Finally, participants also recognized that innovation for decarbonization will only be fostered in Waterloo Region when there is public support for it. A survey respondent made this clear: "The only way we're going to get community buy in is if we approach decarbonizing as a co-benefit." A session on day two of the forum was dedicated to the articulation of co-benefits that will accrue from efforts to decarbonize the Region's energy systems (in addition to the direct impact of reducing local GHG emissions). The long list included social, economic, and environmental co-benefits, including growth of the local clean tech sector and attraction and retention of young talent.

16.5.3 An Emerging Network: Entrepreneurs as Key Players

In an informal plenary session during the forum, one participant congratulated the organizers for starting "an adult conversation" about energy realities and the potential for innovation. The participant remarked that often discussions about energy get bogged down in political or ideological debates and consequently fail to articulate the concrete challenges and opportunities that lie ahead. Survey responses reflected that many participants shared the sentiment that this forum set the stage for ongoing constructive dialogue that will lead to action. One survey respondent enthused: "There is such a knowledgeable network emerging with this focus."

The post-forum survey also asked forum participants to identify MVPs—a participant who made an extraordinarily positive contribution to the forum. Interestingly, half of the respondents who identified a specific individual as an MVP named entrepreneurs, though they only made up a handful of the group. The input of these entrepreneurs was characterized by fellow forum participants as practical, eloquent, and visionary. Several of the entrepreneurs present at the forum expressed a strong desire to move decarbonization forward at various points in the collective discussion. This suggests an untapped potential for local entrepreneurs as passionate and innovative leaders.

Among the most important intangible outcomes of the forum was the fact that in addition to bringing together a large group of local experts, it also fostered a collective desire to move forward on the complex task of decarbonizing the region. Surveys showed clearly that participants were willing to stay involved in a variety of ways. In addition, 100% of survey respondents expressed willingness to take part in follow-up focus groups to explore next steps in the decarbonization process.

16.6 Beyond the Business Case: Emerging Roles for Entrepreneurs

A broadening dialogue has begun to take root in Waterloo Region through the Decarbonize forum, raising the following questions that pertain to where innovation is needed, where it is stifled, and how it can be fostered.

Against the backdrop of shifting multilevel governance dynamics, the local participatory process unfolding in Waterloo Region provides a rich context to explore where transformative innovation is needed, stifled, and fostered, as well as the role that entrepreneurs can play in that innovation.

Waterloo Region has a strong local narrative of innovation, built in part around the Region's entrepreneurs who imagined technology (like the smart phone) that has had global transformative impact. As mentioned, the Decarbonize Waterloo Region forum identified many current opportunities for technological, financial, and social innovations that can contribute to the transformation of local energy systems. In light of the transformative potential demonstrated historically by local entrepreneurs, it is interesting to consider the role of current entrepreneurs to catalyze the technical, social, and financial innovations required to transform local energy systems.

A starting place for this engagement was introduced at the forum when the potential economic co-benefits of decarbonization were outlined. Statistics presented at the forum showed that of the $2.1 billion spent locally on energy per annum, $1.8 billion leaves the Region (85%). It, therefore, follows that efforts to decarbonize the region through the development of local, renewable energy systems will help to stem the outflow of money and benefit the region's economy. There is also a growing amount of literature that corroborates the effectiveness of engaging the public on the economic aspects of decarbonization. Westley et al. (2011) point out that, "recent studies of developing and developed economies, and oil-based and non-oil-based economies, give evidence that citizen support for renewable energy can be garnered through linking it to jobs" (Westley et al. 2011, p. 769).

16.6.1 Entrepreneurs Help to Reimagine the Good Life

Emphasis on the economic co-benefits of decarbonization is only a starting place. As Westley et al. (2011) further argue, "invention and innovation [by entrepreneurs] have served as a means of economic value creation, rather than a means to reduce our impact on the biosphere—our life supporting environment." (Westley et al. 2011, p. 763). Westley and her coauthors warn that emphasis solely on economic value creation can lead to unintended consequences, path dependency, and lock-in—and the creation of energy systems that are no more sustainable than the current carbon-based systems (Davies and Oreszczyn 2012). This points to a role for entrepreneurs that goes beyond simply making the business case for the economic benefits of decarbonization.

Furthermore, decarbonization is not simply a technical matter of fuel switching but rather a "social, economic, and deeply political" process (Burch 2010), where "transformative shifts thus require communities to be imaginative, radical and ambitious, pursuing sustainability as a complex set of value propositions about what defines a good life" (Burch 2016). It became evident to participants in the Decarbonize forum that it is difficult to even imagine a "good life" that is not deeply rooted in the present highly carbonized, invisible energy infrastructure.

The task of reimagining the "good life" may present another ideal entry point for entrepreneurs to engage the broader public in decarbonization dialogue. To date, the "deeply political" discourse on decarbonization is framed in such a way that it is disempowering to the general public. A growing body of academic literature emphasizes the importance of moving away from jargon-laden or alarmist appeals for action on climate change, to an approach that is more place-based and rooted in free-choice learning (Schweizer et al. 2013). As noted above, participants in the forum characterized the discussion at the forum as a refreshing "adult conversation" that successfully transcended the disempowering aspects of broader public debate that often lead to unproductive ends.

The Decarbonize Waterloo Region forum provided a platform for the voices of some local entrepreneurs, who appear ready and willing to join the public discourse on decarbonization in the Region. As mentioned earlier, one of those entrepreneurs said that he is "sick and tired of being told [decarbonization] can't be done." This comment suggests that innovation which might contribute to the decarbonization of local energy systems is being stifled in the Region. Survey respondents demonstrated a keen awareness that decarbonization efforts will face public resistance. For example, in Ontario's current political climate, rising energy costs have been closely linked to the provincial government's efforts to expand renewable energy. This is a formidable barrier to political and public support for any initiatives promoting energy alternative. One survey respondent articulated the anticipated resistance starkly, warning that, "this will be a very strong uphill battle."

Experience at the forum points to the potential of local entrepreneurs as helpful navigators of that resistance. As author Bell Hooks said, "if I do not speak a language that can be understood, there is little chance for dialogue" (Hooks 1989). The "adult conversation" that started at the forum could be a starting point in the effort to reframe public discourse into a more constructive dialogue that can be understood by the broader public as a call to creative action. As mentioned, survey respondents identified local entrepreneurs at the forum as eloquent and influential MVPs of the process.

The potential roles of entrepreneurs are not limited to reimagining the good life and include articulating the new visions that are generated, as well as experimentation, implementation, and integration of innovations that can transform energy systems.

16.6.2 Entrepreneurs Foster Social, Technical and Financial Innovation—Through Experiments

A key step in transforming energy systems is the development of alternatives. The diverse set of participants at the Decarbonize Waterloo Region forum explored the growing panoply of technical, financial, and social innovation in the field of local energy solutions. New innovations are making it increasingly feasible for communities to take steps toward the *adjacent possible*, which then opens a host of new possibilities for innovations that can drive system change. Panelists and other participants at the forum brought forward innovations in technical, social, and financial realms that have emerged strongly in the past decade, including electric vehicle infrastructure (technical), neighborhood energy strategies (social), and impact investment (financial), among many others. Entrepreneurs can play an important role in the social, financial, and technical innovations that were brought forward in the Decarbonize Waterloo Region forum.

The development of transition experiments is a crucial first step for moving these innovations ahead and here entrepreneurs can play a key role as well. Westley et al. (2011) assert that "experiments can help prepare for a transformation by beta testing alternative policy options." Waterloo Region's history of entrepreneurial innovation provides the narrative that can help to build support for experimentation among policy makers and the public.

Partnerships between entrepreneurs and other stakeholders are crucial to success and experiments present a natural milieu for the development of mutually beneficial collaboration. Partnerships with academics in the development and evaluation of experiments are one example. Local governments can also collaborate with entrepreneurs by helping to fund experiments using new financial tools such as green bonds. Civil sector organizations can partner with entrepreneurs who are pursuing social innovations such as neighborhood energy plans.

Respondents to the post-forum survey all expressed interest in being part of follow-up focus groups after the forum, which will provide an opportunity to discuss concrete ideas for potential transition experiments and the specific partnerships that could power them. Fostering and coordinating productive partnerships is a key aspect for moving urban decarbonization forward, which leads to the question of governance.

16.6.3 Governance that Supports Entrepreneurial Leadership

Early in the Decarbonize forum, a civil society leader from Waterloo Region asserted that "people go where systems lead them." Creating new energy systems means creating new governance systems that engage and empower entrepreneurs as key actors in urban decarbonization. The Decarbonize Waterloo Region forum provided an early glimpse of what innovative and effective governance for urban decarbonization might look like going forward. The event was organized by social scientists who propounded the concepts of uncertainty and complexity, but some participants were vocal about their struggle to think in those realms. One forum participant put forward the thesis that we require both linear and divergent thinking to drive the local transition to decarbonized energy systems. Going forward, governance structures must synergize different kinds of thinking to develop technically, socially, and politically feasible means to make the shift to decarbonized local energy systems (see Chap. 8: Engberg). As Burch et al. state: "Such shifts also rest on the model of governance that is participatory, and effectively integrates the often divergent and contested knowledge and capacities of civil society, technical experts, indigenous communities, the private sector, and decision makers." (Burch 2016, p. 4).

Decarbonization is also facilitated by the creation of a culture and governance structures that permit and encourage experimentation as a tool of learning and reflexivity. Bulkeley et al. (2013) encourage "governance by experiment" in the realm of local level sustainability transitions. Entrepreneurs can play a crucial role in fostering a culture of governance by experimentation. As Kerr et al. (2014) assert, "entrepreneurship is fundamentally about experimentation because the knowledge required to be successful cannot be known in advance or deduced from some set of first principles."

If experimentation is central to transformative innovation for decarbonization—and entrepreneurship is "fundamentally about experimentation" and "a voyage into the unknown"—then it makes sense to create more opportunities for entrepreneurs to help communities to navigate the complex uncertainties of decarbonization processes. The Decarbonize Waterloo Region forum points to the potential for entrepreneurs as drivers and MVPs with a largely untapped capacity to inspire, innovate, experiment, and collaborate with a broad range of other local level actors.

16.7 Conclusions and Directions for Future Research

Shifting multilevel governance dynamics of climate action have caused the locus of control to move from the international level to the local level, and beyond government to include a host of other actors, including entrepreneurs. Participatory, multi-actor processes have sprung up around the world, and the Decarbonize Waterloo Region process presents an interesting opportunity for learning and innovation. The ambitious goal of full decarbonization by 2050 will require hard work and creativity from all the members of the emerging local network. The failure to produce clear or straightforward scenarios for decarbonization in the November 2016 forum was disheartening for the participants, but many positive outcomes emerged nonetheless. A clear picture of local energy systems and distinct set of challenges (as daunting and complex as they are) provided local stakeholders with places to start and arenas for ongoing innovation. The forum process also strengthened bonds among an expanding network of stakeholders who can now move forward with a collective desire to take on those challenges and transform their local energy picture.

The emergence of local entrepreneurs as vocal, visionary, and practical MVPs bodes well for their potential as leaders of ongoing decarbonization efforts. This development in Waterloo Region suggests that further action research into the emerging role of local entrepreneurs in urban energy transitions is both an imperative and an interesting prospect.

In the Decarbonize Waterloo Region forum, a community got a glimpse of the leadership potential of local entrepreneurs. There are many latent possibilities for entrepreneurs to be much more than the inventors of the next great eco-widget. This glimpse into the adjacent possible of entrepreneurial potential gives rise to some exciting questions:

- What would happen if communities explored tools beyond the carrot of incentives and the stick of regulations in their efforts to engage entrepreneurs in the enterprise of urban decarbonization?
- How could entrepreneurs move from forum invitees to become more fully engaged as visionary leaders of urban decarbonization processes? In other words, how could they take the wheel as opposed to being brought along for the ride?
- How can participatory processes be designed to better engage entrepreneurs who run small businesses that demand a great deal of their time and focus?
- If "people go where systems lead them", then how can the emerging lessons from the Decarbonize Waterloo Region inform the creation of local innovation systems that engage entrepreneurs alongside other key stakeholders to foster truly transformative action for decarbonization?

As communities across Canada move forward with their own decarbonization initiatives, researchers need to explore how entrepreneurs can help to reimagine "the good life" without fossil fuels, and foster social, technical, and financial innovation through experimentation. Research is also needed to develop innovative governance approaches that support entrepreneurial leadership in the complex and uncertain endeavor of urban decarbonization.

References

Biermann, F., Betsill, M., Gupta, J., Kanie, N., Lebel, L., Liverman, D., et al. (2009). Earth system governance: people, places, and the planet. *Science and Implementation Plan of the Earth System Governance Project, Earth System Governance Project Report, 1.*

Bradford, J., & Fraser, E. D. (2008). Local authorities, climate change and small and medium enterprises: Identifying effective policy instruments to reduce energy use and carbon emissions. *Corporate Social Responsibility and Environmental Management, 15,* 156–172.

Bulkeley, H. (2016). STEPS 2016 annual lecture at the University of Sussex. Accessed at: https://www.youtube.com/watch?v=KATUvvZ8fiw.

Bulkeley, H., & Castán Broto, V. (2013). Government by experiment? Global cities and the governing of climate change. *Transactions of the Institute of British Geographers, 38*(3), 361–375.

Bulkeley, H., Breitfuss, M., Coenen, L., Frantzeskaki, N., Fuenfschilling, L., Grillitsch, M., et al. (2015). Theoretical framework working paper on Urban living labs and urban sustainability transitions. Lund University, Lund.

Burch, S. (2010). In pursuit of resilient, low carbon communities: An examination of barriers to action in three Canadian cities. *Energy Policy, 38,* 7575–7585.

Burch, S. (2011). Development paths: Investigating the roots of responses to climate change. *Sustainable Development, 19,* 176–188.

Burch, S. (2016). Will Ontario's climate change action plan transform communities? Policy Brief No. 90, Centre for International Governance Innovation. Available at: https://www.cigionline.org/sites/default/files/documents/PB%20No.90web_0.pdf.

Burch, S., Andrachuk, M., Carey, D., Frantzeskaki, N., Schroeder, H., Mischkowski, N., et al. (Forthcoming). Governing and accelerating transformative entrepreneurship: Exploring the potential for small business innovation on urban sustainability transitions. *Current Opinion in Environmental Sustainability*.

Burch, S., Shaw, A., Dale, A., & Robinson, J. (2014). Triggering transformative change: A development path approach to climate change response in communities. Climate Policy 14, 467–487. business in sustainability transitions. *Journal of Cleaner Production, 45,* 20–28.

Castan Broto, V., & Bulkeley, H. (2013). A survey of urban climate change experiments in 100 cities. *Global Environmental Change, 23,* 92–102.

Davies, M., & Oreszczyn, T. (2012). The unintended consequences of decarbonising the built environment: A UK case study. *Energy and Buildings, 46,* 80–85.

Driessen, P. P. J. (2015). Societal transformations in the face of climate change: Research priorities for the next decade. JPI Climate [online].

Dryzek, J. S. (2000). *Deliberative democracy and beyond: Liberals, critics, contestations*. Oxford: Oxford University Press.

Environment and Climate Change Canada. (2016). Canadian environmental sustainability indicators: Greenhouse gas emissions. Consulted on Feb 6, 2017. Available at: www.ec.gc.ca/indicateurs-indicators/default.asp?lang=en&n=FBF8455E-1.

Government of Canada. (2016). *Pan-Canadian framework on clean growth and climate change: Canada's plan to address climate change and grow the economy*. Ottawa: Government of Canada.

Hooks, B. (1989). *Talking back: Thinking feminist, thinking black*. New York: South End Press.

IEA. (2016). Decoupling of global emissions and economic growth confirmed. International Energy Agency.

Kerr, W. R., Nanda, R., & Rhodes-Kropf, M. (2014). Entrepreneurship as experimentation. *Journal of Economic Perspectives, 28*(3), 25–48.

Kinley, R. (2016). Climate change after Paris: From turning point to transformation. *Climate Policy*, 1–7.

Klewitz, J., & Hansen, E. G. (2014). Sustainability-oriented innovation of SMEs: A systematic review. *Journal of Cleaner Production, 65,* 57–75.

Klewitz, J., Zeyen, A., & Hansen, E. G. (2012). Intermediaries driving eco-innovation in SMEs: A qualitative investigation. *European Journal of Innovation Management, 15,* 442–467.

Loorbach, D., & Wijsman, K. (2013). Business transition management: Exploring a new role for business in sustainability transitions. *Journal of Cleaner Production, 45,* 20–28.

McCrone, A., & Moslener, U. (2016). Global trends in clean energy investment. *United Nations Environment Program and Bloomberg New Energy Finance*.

Pahl-Wostl, C. (2009). A conceptual framework for analyzing adaptive capacity and multi-level learning processes in resources governance regimes. *Global Environment Change, 19,* 354–365.

Parker, P., & Rowlands, I. H. (2007). City partners maintain climate change action despite national cuts: Residential energy efficiency programme valued at local level. *Local Environment, 12,* 505–517.

Pelling, M., O'Brien, K., & Matyas, D. (2015). Adaptation and transformation. *Climatic Change, 133*(1), 113–127. doi:10.1007/s10584-014-1303-0.

Pereira, L., Karpouzoglou, T., Doshi, S., & Frantzeskaki, N. (2015). Organising a safe space for navigating social-ecological transformations to sustainability. *International Journal of Environmental Research and Public Health, 12,* 6027–6044.

Pidgeon, N. F., Poortinga, W., Rowe, G., Horlick-Jones, T., Walls, J., & O'Riordan, T. (2005). Using surveys in public participation processes for risk decision making: The case of the 2003 British GM nation? Public debate. *Risk Analysis, 25*(2), 467–479.

Renn, O., Webler, T., & Wiedemann, P. (Eds.). (1995). *Fairness and Competence in Citizen Participation*. Dordrecht: Springer Netherlands. doi:10.1007/978-94-011-0131-8.

Robinson, P. (2005). Barriers to Canadian municipal response to climate change. *Canadian Journal of Urban Research, 14,* 102–120.

Scannell, L., & Gifford, R. (2013). Personally relevant climate change the role of place attachment and local versus global message framing in engagement. *Environment and Behavior, 45*(1), 60–85.

Schweizer, S., Davis, S., & Thompson, J. L. (2013). Changing the conversation about climate change: A theoretical framework for place-based climate change engagement. *Environmental Communication: A Journal of Nature and Culture, 7*(1), 42–62.

Shaw, A., Sheppard, S., Burch, S., Flanders, D., Weik, A., Carmichael, J., et al. (2009). Making local futures tangible—Synthesizing, downscaling, and visualizing climate change scenarios for participatory capacity building. *Global Environmental Change, 19,* 447–463.

Sova, C.A., Chaudhury, A., Helfgott, A., & Corner-Dolloff, C. (2012). Community-based adaptation costing: An integrated framework for the participatory costing of community-based adaptations to climate change in agriculture, CCAFS working paper 16. CCAFS, Denmark, Copenhagen.

Westley, F., Olsson, P., Folke, C., Homer-Dixon, T., Vredenburg, H., Loorbach, D., et al. (2011). Tipping toward sustainability: Emerging pathways of transformation. *Ambio, 40*(7), 762–780.

Wittmayer, Julia M., & Schäpke, Niko. (2014). Action, research and participation: Roles of researchers in sustainability transitions. *Sustainability Science, 9*(4), 483–496.

Author Biographies

Scott Morton Ninomiya is a sustainability scholar and a proud native of Waterloo Region, Canada. His current research and practice are focused on engaging entrepreneurs, governments, civil sector, and scholars in the transformative work of creating low carbon communities in Canada. At the University of Waterloo, he led a participatory action research project investigating a local decarbonization project in Waterloo Region, funded by Energy Council of Canada. He is also a research team member of the *GATE project*, (*Governing and Accelerating Transformative Entrepreneurship*), a collaboration of leading international scholars investigating the transformative potential of entrepreneurs in sustainability governance. In September 2017, he will begin pursuing a Ph.D. of Global Governance in Environment at Balsillie School of International of International Affairs in Waterloo.

Sarah Burch is an Assistant Professor and Canada Research Chair in Sustainability University of Waterloo. She is also a Senior Fellow, Centre for International Governance Innovation and a Fellow at the Balsillie School of International Affairs. Her research focuses on governing response to climate change (both adaptation and mitigation) in urban spaces. She explores triggers of transitions toward more sustainable development pathways, and strategies for engaging a variety of actions in conversations about desirable futures.

Chapter 17
Multilevel Governance of Sustainability Transitions in Canada: Policy Alignment, Innovation, and Evaluation

Ann Dale, Sarah Burch, John Robinson and Chris Strashok

Abstract Local communities are on the front line of climate action, mitigation, and adaptation implementation. This chapter explores the research outcomes of a tri-university five-year research collaboration studying local climate innovators in the province of British Columbia. At the time the research began, there was a unique opportunity to study multilevel governance between the province and local governments albeit in a national vacuum. Lessons learned from the first phase and preliminary analysis from the second phase are then applied to the province of Ontario poised to take province-wide action. Ontario's case is different in that there is now alignment between the federal and provincial levels, but less engagement to date with local governments. Our research shows that the active engagement of local communities is essential for accelerating climate innovation and multilevel governance.

Keywords Climate change adaptation and mitigation · Multilevel governance · Climate innovation · Policy strategies

---

A. Dale (✉) · C. Strashok
School of Environment & Sustainability, Royal Roads University, Victoria, Canada
e-mail: ann.dale@royalroads.ca

C. Strashok
e-mail: chris.strashok@techsightconsulting.com

S. Burch
Department of Geography and Environmental Management, University of Waterloo, Waterloo, Canada
e-mail: sburch02@gmail.com

J. Robinson
Munk School of Global Affairs and School of the Environment, University of Toronto, Toronto, Canada
e-mail: johnb.robinson@utoronto.ca

S. Hughes et al. (eds.), *Climate Change in Cities*, The Urban Book Series,
https://doi.org/10.1007/978-3-319-65003-6_17

17.1 Introduction

In light of particularly wicked socio-ecological problems such as climate change, calls have been made for new forms of governance that allow for a range of actors, flexible partnerships, and creative co-production of knowledge to enable transitions to more sustainable development pathways. Twenty years ago at the 1992 United Nations Conference on Environment and Development, Agenda 21 recommended new governance arrangements between governments at different scales, with non-governmental organizations (NGO) and local authorities. Since then, the need for systemic change has become more urgent with climate change adaptation and mitigation imperatives and the requisite transition to a low-carbon economy. Many scholars and organizations are calling for a transformative change in current development paths (Burch et al. 2014; Westley et al. 2011). The degree, timing, and nature of this change are socially contested and evoke vested interests, making consensus at this time elusive. To further complicate matters, the low-carbon economy transition is not simply the task of formal government but rather a shifting constellation of private and public actors, through formal and informal mechanisms, investments and the acceleration of innovations by local governments across the country.

While the accumulation of human-induced greenhouse gas emissions is a phenomenon occurring at the global scale, emissions stem from local contexts (Guston 2008; Ibrahim et al. 2012). In this way, climate change requires integrated governance that bridges social-ecological, temporal, and jurisdictional scales (Adger et al. 2005; Bulkeley and Betsill 2005; Bulkeley and Castan Broto 2013). Furthermore, governance is not simply the domain of formal government; it encompasses all of the processes and interactions aimed at solving a collective problem (Bevir and Rhodes 2006).

Greenhouse gas emissions trajectories are clearly shaped by fossil fuel-based technologies, but perhaps even more important are the social, political, and economic contexts underpinning the use of these technologies (Burch 2010; Shaw et al. 2014). In other words, in order to build communities that are sustainable, resilient, and low carbon, a deeper shift in the logic of economies, and the values that reinforce them, must inevitably occur. These transformative shifts thus require communities to be imaginative, radical, and ambitious, pursuing sustainability as a complex set of value propositions about what defines a "good life" (Burch 2016). Such shifts also rest on a model of governance that is participatory, and effectively integrates the often divergent and contested knowledge and capacities of civil society, technical experts, Indigenous communities, the private sector, and decision-makers (while of course recognizing that these groups are not mutually exclusive). Jurisdiction over greenhouse gases overlaps, so it is crucial that municipal, provincial, and federal policies are congruent rather than contradictory (Dale 2008; Shaw et al. 2014). These overlapping responsibilities draw our attention to the governance dynamics that are at play in the design and implementation of climate change policy.

Such considerations suggest the importance of a multilevel governance approach. While the urban or community scale is an important context within which to explore sustainability transitions, such an approach highlights the dynamic interactions amongst scales (Bulkeley and Betsill 2005), mirroring the systems-based approach of the sustainability transitions literature. Furthermore, it highlights the potential influence of "fluid, issue-oriented alliances" among levels of government and various actors (a polycentric model) in contrast to a more hierarchical model in which competencies are distributed rather than overlapping (Bulkeley and Betsill 2003; Hooghe and Marks 2003).

Understanding the relationships among actors, the distribution of power [viewed as "the capacity of actors to mobilize resources to realize a certain goal" (Avelino and Rotmans 2011)] is central to an exploration of governance that has sustainability as its explicit goal (Bulkeley et al. 2015). This enlarged and expanded sphere of stakeholders (civil society, researchers, practitioners, and private sector leaders) demands an unprecedented level of collaboration between governments and civil society. Some are calling for a collective intelligence model taking advantage of new digital technologies that convene large groups—a community, region, city or nation—to think and act intelligently in a way that amounts to more than the sum of their parts (Saunders and Muligan 2007).

In the next sections, we present the results of a five-year research project in British Columbia, intended to explore the dynamics of innovative local responses to climate change in BC and their application locally, regionally and nationally to changing development paths. As climate leadership in British Columbia wanes and other provinces take center stage, we speculate about the implications of these research outcomes in the context of Ontario.

From the point of view of the governance of climate policy, British Columbia and Ontario represent two variants of emerging multilevel governance. In the case of British Columbia during the first phase of our research, there was strong provincial and local government alignment, within the context of a federal leadership vacuum. Presently, Ontario is benefitting from launching its extensive climate action plan in the context of complementary federal leadership and its recent announcement of a national carbon tax. It is now moving forward to incent local municipalities to accelerate their take-up of climate actions.

17.2 British Columbia: Provincial Leadership Spurring Municipal Innovation

17.2.1 Case Context

British Columbia (BC) presented a unique opportunity to explore the implementation of climate innovations as beginning in 2008, there was unprecedented provincial leadership and local government cooperation happening with respect to climate change adaptation and mitigation, but in a national vacuum. The BC Climate Action Secretariat (CAS) had begun a series of orchestrated and coordinated steps to accelerate the take-up of climate action across the province (Dale 2015). These included a strong legislative framework to stimulate climate change adaptation and mitigation innovation designed to create a level playing field for local governments. The Carbon Tax Act, introduced on July 1, 2008, started to phase in an escalating revenue-neutral carbon tax, where one hundred percent of the revenue from the tax was returned to taxpayers through reductions in other provincial taxes, with built-in protection for lower income British Columbians.

A key policy instrument, the BC Climate Action Charter, complemented this legislative innovation, and as of today, 180 of BC's 188 local governments have signed the charter. The province also mandated carbon neutrality and reporting across all public sector organizations including government offices, schools, post-secondary institutions, Crown corporations, and hospitals, to measure operational GHG emissions, reducing those where possible, offsetting the remainder, and demonstrating leadership through public reporting. Since 2010, British Columbia has achieved carbon neutrality each year across its entire provincial public sector (BC Government website, accessed March 5, 2017).

A suite of reporting and accountability measures, on the premise that "what is measured is managed," underpinned these legislative and policy innovations. Financial incentives were also put in place as well as tools to accelerate policy implementation. Across the province, and in the face of acute economic constraints, local governments have reduced GHG emissions, developed local projects to balance emissions, purchased offsets to compensate for emissions, and, in many cases, developed financing innovations ranging from carbon funds to regional offset strategies (Burch et al. 2014; Shaw et al. 2014).

The evolution and drivers of climate change responses in local governments in BC from 2010 to the present have been extensively studied in Meeting the Climate Change Challenge (MC^3), a tri-university research project involving over fifteen research partners from civil society, public sector, and quasi-institutional organizations in the province of British Columbia. The first phase of the project, 2012–2014 explored eleven local governments[1] in the province that were identified as climate innovators. The second phase, begun in 2015, builds on this previous research looking at current development paths in the original case study communities, what has changed, if anything, and what could be described as transformative change.

The second phase of our MC^3 research is gathering new evidence about whether or not mitigation and adaptation innovations in the 11 local governments from the first phase resulted in transformative changes toward more sustainable paths at the local level. Part of our second research phase involves developing a coherent theory

[1]Used interchangeably with communities, although interviews were confined to local government officials and not the wider community.

of development path change, the conditions under which development paths[2] can be transformed, including the drivers and barriers to action, as well as key indicators associated with such a shift. We have recently concluded a series of interviews with a sub sample from the first set of interviews and our preliminary analysis is discussed below.

17.2.2 Key Governance Factors

Research from the first MC^3 phase gave rise to several major findings. *First* among these was the importance of leadership at multiple levels. Provincial leadership and the Charter, in particular, were crucial in moving local decision-makers toward accelerated investments in on-the-ground climate action and innovation. One outcome from the concluding peer-to-peer learning exchange (which brought together all the case study interviewees with the research team) was that a provincial Charter 3.0 should be put in place. Participants argued it should have even stronger targets and timelines to accelerate more local innovation and to create another higher level playing field (Dale 2015). On the other hand, a group of mayors and ex-mayors convened toward the end of Phase 1 indicated that they thought that provincial leadership had been essential but that many communities would now act on their own initiative. Therefore, while local government staff from diverse departments were strongly of the view that the Charter had been critical in convincing political decision-makers to support and move on climate change, the need for such leadership in the future was less clear. This became a central question in the Phase 2 analysis.

Second, not surprisingly, the most innovative local governments were those with political and staff alignment. In other words, when municipal staff and municipal politicians had the same agenda, much was possible. Conversely, if there was a conflict between the two levels, little was accomplished. *Third*, the capacity for cross-departmental (horizontal) planning processes was essential to climate action and sustainability. In particular, systematic frameworks for policy-making, such as a consistent reporting requirement, fostered inter-departmental collaboration and inter-sectoral cooperation. *Fourth*, provincial leadership resulted in a majority of the case study local governments integrating climate change into broader sustainability planning. The embedding of climate innovations into existing policies and programs within a larger sustainability agenda, for example, Official Community Plans and Integrated Community Sustainability Plans was essential to sustaining momentum between electoral swings and transforming current development paths (Dale et al., forthcoming). Related to this embedding process is the

[2]For the purposes of this research, a development path consists of social systems (formal and informal rules, habits, and norms), networks amongst actors, diverse technologies and ecological systems (Burch et al. 2014), which governs and shapes how individuals and organizations act in a given jurisdiction.

institutionalization of sustainability in the organizational structure and function of municipal government. *Finally*, all of the case study communities took advantage of new partnerships and strategic alliances because of having greater access to networks stimulated by the leadership of the Climate Action Secretariat. This involved exploiting windows of opportunity for collaboration and framing climate change in a way that is synergistic with other pressing priorities.

Preliminary findings from Phase 2 of this research confirm the importance of the findings introduced above, but further (ongoing) research will elucidate the ways in which these dimensions are evolving as climate change responses mature in British Columbia. The following sections describe our main findings in Phase 2 expanding on the findings from the first phase.

17.2.3 Emerging Lessons from Phase 2

17.2.3.1 Leadership

Preliminary analysis from the second phase indicates that the majority of the case study communities are still engaged in climate action using a systems-oriented sustainability mandate but tailored to their specific context. All communities still credit the provincial government's 2008 Climate Action Charter (CAC) for either legitimizing or incentivizing climate mitigation efforts occurring within their communities and draw on the funding they receive by fulfilling their CARIP reporting requirements. This reinforces the Phase 1 finding that leadership at multiple levels is important. However, the provincial regime has changed considerably since our initial interviews and British Columbia is no longer leading in the same ways that it was during the first phase of our interviews, particularly the leadership from the Climate Action Secretariat. The province has not increased the carbon tax as was originally planned, and its next iteration of the Charter has been criticized as conservative and not accelerating the necessary conditions to build upon earlier innovations (Campbell 2016). As a result, local governments are no longer anticipating or waiting for renewed provincial government leadership in order to act and some see the province as now actually in their way. For example, while many are calling for strong provincial building codes to incentivize municipal reductions in building energy use, others are striving for standards that are more progressive than what the province has proposed and some see the province as a hindrance to achieving their objectives around sustainability and climate change adaptation. This suggests that provincial leadership—essential in the early stages of climate response at the municipal level in BC—is now still desirable but perhaps not required for communities to continue and further develop their climate policies.

17.2.3.2 Political and Staff Alignment

Phase 2 data reinforced the finding that alignment between political and staff agendas is crucial for continuity and momentum between changes in administration for sustaining and building further local action, even in the most innovative communities. Communities without this alignment have stalled and in some cases lost, their initial momentum observed in the first phase of our research. This is a further rationale for embedding and institutionalizing policies and programs to "ride out" large swings in political mandates and avoid losing momentum in innovative practices, which may become even more important as many EU countries move to more stringent carbon neutral targets than Canada. Equally key is policy alignment within local government departments and policy congruence between levels of governments.

17.2.3.3 Embedding and Institutionalizing Sustainability

The majority of local governments are now strengthening policy alignment between departments, moving from stand-alone sustainability units toward institutionalization of sustainability goals in existing departmental mandates. The latter extends beyond one or two departments, to cross-departmental implementation as part of every department's mandate, where their executives are accountable and responsible for its achievement, and it is an operational line item in departmental budgets. The potential benefits of this "institutionalization" are increased access to more diverse resources, augmented collaboration on sustainability/climate projects, and broader, integrated horizontal planning.

It is also possible, however, that such processes will result in a loss of identity and momentum for sustainability initiatives that now have lost institutionally distinct champions and visibility, now buried in departments with quite different priorities. As a result, this integration may be most fruitful once a community has made significant progress towards sustainability, such that it has become an uncontroversial part of the identity of the municipality; in the early stages of this process, when this is not the case, an identifiable champion and concentration of sustainability expertise may be more important.

Whatever the outcomes of such institutional changes, the existence of dedicated human resources, including new building energy management positions for both the public and corporate sectors, is contributing to continued climate innovation, and the continual build out of resources.

17.2.3.4 Partnerships and Strategic Alliances

In addition, central in all the local government cases was continuing and enlarging upon strategic partnerships and alliances in the broader community, particularly with the business community; which can be key to increasing access to diverse resources. All local governments had evidence of accessing additional resources outside their community, many of which provided investments in intellectual capital, for example, the BC Hydro energy managers program. This was a key leverage point for smaller communities as it provided a staff person who was responsible for creating new programs, but more importantly implementing measuring and monitoring systems. Equally, partnerships with quasi-institutional organizations such as the Fraser Basin Council and the Columbia Basin Trust also served to accelerate innovations, in addition to nationally, the Federation of Canadian Municipalities (FCM). This level of collaboration, for example, supported the adoption and public support for Vancouver's Renewable City Strategy aimed at getting to 100% renewable energy usage by 2050.

Data from Phase 2 suggest that strategic partnerships are enhanced by the recognition and exploitation of windows of opportunity for collaboration. When strategic priorities align, traditionally oppositional frameworks can be bridged to keep accelerating momentum and enhancing ambition. Those local governments further along the innovation curve still emphasize the importance of how issues are framed, and taking advantage of extreme events in local contexts to capitalize on linking leading-edge science and research outcomes to the climate action imperative. Initially, many framed the issue more narrowly as energy efficiency and then after realizing the benefits of acting in this narrower domain, built on their successes to adopt a wider sustainability frame.

17.2.3.5 Financing

In Phase 2, the importance of new financing mechanisms emerged as a central theme. In addition to embedding policies and programs into existing mandates, it is important to establish innovative financing solutions to continue funding larger and more difficult innovations. In British Columbia, over 40 local governments now have green revolving funds and even very small communities have established energy revolving funds.

17.2.3.6 Knowledge Transfer

Our research in both phases demonstrates that best practices sharing among community innovators and with climate scholars is a major driver accelerating climate action. Peer-to-peer learning exchanges, face-to-face between local government staff and researchers have been important in building new networks of collaboration that accelerate the take-up of climate innovations. We have only anecdotal evidence; however, that the virtual meetings we held with elected officials also resulted in one or two local governments beginning to adopt climate change through access to the knowledge sharing with peers and the research team.

A summary chart of the differences between the two research phases for the three largest urban centers (Surrey, Vancouver, Victoria) can be found in Appendix A.

17.3 Emerging Leadership in Ontario: Lessons from the MC^3 Project

Since the early leadership on climate policy has shown in British Columbia, the climate governance context has shifted substantially. There is now a distinct national presence to climate change implementation, starting with the federal government signing the 2016 COP 21 agreement committing the world's nations to limiting temperature increases to 1.5°, and the more recent announcement of a national climate action plan and a national carbon tax. Against this backdrop, Ontario (the most populous province in Canada) has announced a carbon cap-and-trade system, and one of the more ambitious climate action plans in the country. This cap and trade is intended to finance the vast majority of plan implementation and incentivize local innovation. If successful, it may prove to be an effective strategy for continuous iterative investments in local government innovations independent of which administration is in power, unlike the situation in British Columbia. The province is now starting to work with local governments to accelerate climate action and innovation. So, what are the key lessons learned from our research in British Columbia that can be applied to governance in other jurisdictions?

Released in June of 2016, Ontario's Five Year Climate Change Action Plan represents a controversial and ambitious effort to de-link economic growth from fossil fuel consumption, stimulate the uptake of renewable energy technologies, and apply a price to carbon that begins to capture the true costs of carbon-intensive communities and lifestyles. This requires spending between $5.9 to $8.3 billion over the next 5 years, which would come from the revenues generated by auctioning off carbon emissions credits as part of the cap-and-trade market that Ontario will join (along with Quebec and California) (Province of Ontario 2016).

The Ontario Five Year Climate Change Action Plan (2016–2020) is comprised of eight action areas: transportation, buildings and homes, land-use planning, industry and business, collaboration with indigenous communities, research and development, government, and agriculture, forests, and lands. Each action area consists of a number of proposed actions, specific targets, and estimated costs. In this it is not dissimilar from provincial and municipal climate change action plans developed across Canada and elsewhere, but a number of dimensions of this plan distinguish it: the central position of a *cap-and-trade system* in order to put a price on carbon, the extremely *short time frame* of the action plan, and the level of *ambition* of both the targets and the proposed actions.

The province has set greenhouse gas reduction targets of 15% below 1990 levels by 2020, 37% by 2030, and 80% by 2050. This action plan takes the province to the first of its goals, and should set the stage for the increasingly transformative medium- and long-term targets (for which specific actions have yet to be assigned). As such it is important to iteratively take stock of the progress that specific actions and policies will make while keeping in mind the potential for these (and additional) actions to ultimately yield exponentially increasingly greenhouse gas reductions.

The main sources of emissions and the stated reduction targets suggest that efforts to densify communities, improve public transit, shift homes away from a reliance on natural gas, and accelerate a transition toward electric cars (since the vast majority of electricity in Ontario is produced by hydropower) will yield significant results for Ontario.

Many of the action areas and goals, especially those related to land-use planning in communities, however, are tied directly to steps that can only be taken by municipalities. While the Province can require municipalities to embed climate change considerations in their official plans, and send a clear signal that climate change is a priority at the provincial level, municipalities have control over how communities are designed (such as the proximity of work to home and play, which affects commuting distances and viability of active/mass transportation), water and waste management, parks, and economic development (Province of Ontario 2001). All of these domains have direct implications for reaching provincial greenhouse gas reduction targets, and so provincial policies must reinforce (rather than contradict) municipal climate change actions.

We started our research in 2012 at the local level because local governments are on the front line of delivering climate action on the ground. They have direct control of critical sources of emissions (Bulkeley and Betsill 2005) and are the scale at which the potentially catastrophic impacts of climate change will play out (Wilbanks and Sathaye 2007). While BC concentrated its efforts on provincial/municipal coordination, policy and program congruence, and the creation of incentives, the current plan in Ontario seems to focus mainly on provincial action, with little explicit attention paid to engaging actively with municipalities, a traditional hierarchical top down approach. Our research in BC suggests that Ontario's plan is unlikely to be successful unless municipalities are deeply engaged in local actions through innovative policies and incentive programs designed to accelerate the take-up and knowledge transfer of successful innovations between local governments.

Our research demonstrates what can happen with a multilevel governance approach between two levels of government, in partnership with numerous civil society groups and researchers. Ontario now has a golden opportunity to capitalize on a complementary federal/provincial landscape to adopt this approach to build congruence between all three jurisdictions.

There are many lessons that are transferable to both urban and smaller governments across the country that also highlight the efficacy of moving to a multi-level governance system for addressing implementation gaps and fast-tracking climate action locally. The greatest potential for accelerating shifts in current development paths occur when the three levels of government are working congruently with one another, coupled with internal *alignment* within and across each level of government. Perhaps even more important is alignment between political and official staff as demonstrated by the leading-edge climate innovators in British Columbia.

Another central lesson for the Ontario government, if it wishes to become a climate innovator, is the importance of provincial leadership, but with policies and programs designed to accelerate local climate action. Furthermore, *embedding* and *institutionalizing* policies and programs into existing departments is an important driver for continuing the momentum between administrations, especially if another less favorable government replaces it.

A complementary suite of policy instruments and incentives should accompany the legislative framework. British Columbia limited its Climate Action Charter to the public sector. Based on its effectiveness[3] in accelerating local government action in British Columbia, we recommend that Ontario implement a Climate Action Charter that commits all public sector organizations, including crown corporations, to carbon neutrality with mandatory targets and timelines that also includes the *industrial sector*. These policy instruments should be accompanied by monitoring and measuring incentives that build on the BC experience, such as the CARIP program.

Identifying and costing wherever possible the co-benefits of climate change adaptation and mitigation, sustainable development, the green economy and green jobs—including health outcomes, infrastructure, operational savings, and household energy savings will also lead to greater acceleration and take-up. This links to the key finding from MC^3, which suggested that a broader *framing* of the climate issue to connect with related sustainability priorities might foster accelerated action.

Finally, the Ontario case highlights a linkage between two key findings from MC^3 research in British Columbia: the province can play a key *leadership* role in brokering *strategic alliances and partnerships* for local governments, particularly smaller to mid-sized communities, to move ahead in climate actions, ensuring no community is left behind. As demonstrated in British Columbia, successful models of innovative partnerships and community engagement exist that illustrate the importance of sharing responsibility for climate action across different levels of governance (Dale et al. 2013, 2015). Such partnerships are necessary because effective climate take-up rests on the integration of the divergent and contested knowledge and capacities of civil society, technical experts, Indigenous communities, the private sector and decision-makers. Addressing asymmetries of scale and resources, including intellectual capital, through innovative incentives such as BC's energy manager program, has proven to be a key factor in climate action locally. Government leadership is also essential for the transition to more sustainable development paths that simultaneously restrains energy demand (despite population growth), drives the production of low-carbon energy sources and designs complete and compact neighborhoods and communities that create alternative forms of transport and encourage multi-use development.

[3]Our historical dataset of local government GHG emissions for the years 2010 and 2015 reveals that corporate emissions have decreased in all but three of the case study communities, and some significantly.

17.4 Conclusion

Given the distribution of authority throughout Canada's federal system and shifting political priorities at all levels of government, progress on climate change mitigation and adaptation has come in fits and starts. The recent surge in federal support for climate change, however, represents an important opportunity for provinces and municipalities to align goals, learn from past mistakes, and accelerate greenhouse gas reductions. Our findings suggest that leadership at multiple levels, policy alignment, embedding and institutionalizing sustainability and creating strategic partnerships and alliances based on a broader sustainability framing are crucial. In particular, municipal experiences in British Columbia indicate that an explicit multilevel governance approach, which actively engages and provides an incentive for, strong local government action is a desirable and perhaps necessary component of a successful provincial climate policy.

Ultimately, it is early days for Ontario's climate change planning—cap and trade has only recently come into effect, and none of the vehicle electrification or residential renewable incentives has had time to bear fruit. If policy learning is to occur, there must be a conscious effort by policy-makers to draw on the experience of different jurisdictions to ascertain the most effective policies for achieving a particular objective (Peter 1992). In particular, our research shows that innovative financing mechanisms, capitalizing on provincial leadership to create policy alignment, deep partnerships with civil society organizations, and ultimately embedding and institutionalizing climate change concerns throughout the day-to-day operations of municipalities hold significant potential to trigger and sustain significant greenhouse gas reductions and more inclusive and effective multilevel governance. Delivering co-benefits for other community priorities, such as social justice, environmental integrity, and community cohesion may serve to deepen the resilience of climate change policies, ensuring their survival even as the political winds shift.

Appendix A

Community	Approach	Initial driver of climate response	First MC^3 Phase (2011–2012)	Second MC^3 Phase (2015–2017)	Governance/Organizational changes
Surrey	Sustainability focus	District energy aligned priorities of Surrey (increased tax base via density) and B.C. Hydro (reduce energy consumption) aligned, Community Energy Manager role (partially funded by B.C. Hydro) was crucial Climate Action Charter motivator to develop more integrated, forward-looking approach to sustainability. Competition between municipalities to be innovative	Climate change framed as co-benefit in order to create mandate acceptable to business community, residents, and political leadership; mitigation and adaptation were originally integrated, however, mitigation has been discarded and the focus is on an ICLEI adaptation pilot	Finalized climate action strategy and updated the sustainability charter, once again combining mitigation and adaptation into planning. Working to adapt to sea level rise through coastal flood protection strategy. Expanding district energy policy with the construction of a biofuel facility. Completing biodiversity and conservation strategy	Stable governance environment and support since the first phase. Incorporating sustainability across the organization by embedding the sustainability department into other functional departments, starting with the planning and development group, and then moving to a new department every 2 years (with the flexibility of staying longer if needed)
Vancouver	Sustainability focus	CO_2 emissions reductions from policy perspective began with 1990 Clouds of Change report. Creation of Sustainability Support Group in 2002	Extensive mitigation underway: e.g., new buildings to be LEED Gold standard. Vancouver first city to implement adaptation strategy focused on storm water management, urban forest planning, and projections of sea level rise. Dedicated mitigation and adaptation staff in Sustainability Office. Long-term investment in climate change since 1990	Through the city's work over the years, staff have gained a better understanding of what is working and what is not as part of the Greener City Action Plan	Stable governance environment and support since the first phase
					Sustainability has infiltrated all of the city's departments and is now part of now a new department called Planning, Urban Design and Sustainability
				Started Renewable City Strategy, which has put even more focus on energy and greenhouse gas emissions	The city has collaborated with businesses and community organizations on dozens of initiatives generating a lot of confidence throughout the organization and community—success has bred success
				Looking at not just climate change adaptation and mitigation but moving toward exploring what is resilience	
				Focusing on current and future building stocks and creating advanced energy efficiency targets for building codes	

(continued)

(continued)

Community	Approach	Initial driver of climate response	First MC^3 Phase (2011–2012)	Second MC^3 Phase (2015–2017)	Governance/Organizational changes
Victoria	Sustainability focus	An overall 2008 sustainability framework and the initiation of the 2009 Sustainability Department. Came to climate change late but put resources to it and took it beyond the expectations of the Climate Action Charter (CAC) (including community emissions and adaptation planning)	Integrated corporate (Carbon Neutral Plan) and community emissions planning (Climate, Energy, and Resilience). Adaptation planning at the community scale. Climate change response integrated throughout Official Community Plan in land-use, transportation, and infrastructure and food security. Storm water utility builds business case for permeable surfaces, minimizing future infrastructure expenditures and emissions	Reworking Climate Action Plan to better address energy, waste management, transportation, infrastructure and building energy requirements	Stable governance environment and support since the first phase
				Performing heat	Exploring the idea of hiring more support in the sustainable community and planning departments so they really are the keepers of the development policy process
				Island sea level rise mapping to understand where to taking action to mitigate and adapt to climate change.Exploring the elimination or reduction of waste streams as well as supporting the construction of sewage waste and waste water treatment facility within the district	

References

Adger, W. N., Arnell, N., & Tompkins, E. (2005). Successful adaptation to climate change across scales. *Global Environment Change Part A, 15,* 77–86.

Avelino, F., & Rotmans, J. (2011). A dynamic conceptualization of power for sustainability research. *Journal of Cleaner Production, 19,* 796–804.

Bevir, M., & Rhodes, R. (2006). *Governance Stories*. London: Routledge.

British Columbia Government Website. (2017). Accessed March 5.

Bulkeley, H., & Betsill, M. M. (2003). *Cities and climate change: Urban sustainability and global environmental governance*. London: Routledge.

Bulkeley, H., & Betsill, M. (2005). Rethinking sustainable cities: Multi-level governance and the 'urban' politics of climate change. *Environmental Politics, 14,* 42–63.

Bulkeley, H., & Castan Broto, V. (2013). Government by experiment? Global cities and the governing of climate change. *Transactions of the Institute of British Geographers, 38,* 361–375.

Bulkeley, H., Breitfuss, M., Coenen, L., Frantzeskaki, N., Fuenfschilling, L., Grillitsch, M., & Hartmann, C. (2015). *Theoretical framework working paper on urban living labs and urban sustainability transitions*. Lund: Lund University.

Burch, S. (2010). In pursuit of resilient, low-carbon communities: An examination of barriers to action in three Canadian cities. *Energy Policy, 38,* 7575–7585.

Burch, S. (2016). *Will Ontario's climate change action plan transform communities?*. Waterloo, ON: Centre for International Governance Innovation.

Burch, S., Shaw, A., Dale, A., & Robinson, J. (2014). Triggering transformative change: A development path approach to climate change response in communities. *Climate Policy, 14,* 467–487.

Campbell, I., Edwards, M., Pedersen, T., Horne, M., Smith, M., & Berman, T. (2016). Letter to premier christy Clark from climate leadership team members. Pembina Institute.

Dale, A. (2008). Governance for sustainable development: As if it mattered? In G. Toner & J. Meadowcroft (Eds.), Innovation, Science and Environment 2009–2010. Special edition—charting sustainable development in Canada 1987–2007 (pp. 54–71). Montreal: McGill-Queen's University Press.

Dale, A. (2015). Prioritizing policy: Protecting nature by ensuring that the law is for the land. *Alternatives Magazine, 41*(1), 77–79.

Dale, A., Shaw, A., & Newell, R. (2013). MC^3 final progress report: Submitted to the Pacific Institute for Climate Solutions.

Dale, A., Robinson, J., Herbert, Y., & Shaw, A. (2015). *Meeting the Climate Change Challenge (MC^3). Submission to the BC Government*. http://www.mc-3.ca.

Guston, D. H. (2008). Innovation policy: not just a jumbo shrimp. *Nature, 454*(7207), 940–941.

Hooghe, L., & Marks, G. (2003). Unraveling the Central State, but how? Types of multilevel governance. *The American Political Science Review, 97,* 233–243.

Ibrahim, N., Sugar, L., & Hoornweg, D. (2012). Greenhouse gas emissions from cities: Comparison of international inventory frameworks. *Local Environment, 17,* 223–241.

Peter, J. M. (1992). Policy Learning and Failure. *Journal of Public Policy, 12*(4), 331.

Province of Ontario. (2001). Municipal Act, 2001, S.O. 2001.

Province of Ontario. (2016). *Ontario's climate change action plan 2016–2020*. Toronto: Government of Ontario.

Saunders, T., & Mulgan, G. (2007). *Governing with Collective Intelligence*. UK: NESTA

Shaw, A., Burch, S., Kristensen, F., Robinson, J., & Dale, A. (2014). Accelerating the sustainability transition: Exploring synergies between adaptation and mitigation in British Columbian communities. *Global Environmental Change, 25,* 41–51.

Wilbanks, T. J., & Sathaye, J. (2007). Integrating mitigation and adaptation as responses to climate change. *Mitigation and Adaptation Strategies for Global Change, 12*, 957–962.
Westley, F., Olsson, P., Folke, C., Homer-Dixon, T., Vredenburg, H., Loorbach, D., et al. (2011). Tipping toward sustainability: Emerging pathways of transformation. *Ambio, 40,* 762–780.

Author Biographies

Ann Dale is a Professor in the School of Environment and Sustainability at Royal Roads University. Professor Dale held her university's first Canada Research Chair in sustainable community development (2004–2014), is a Trudeau Alumna (2004), and a Fellow of the World Academy of Art and Sciences. Her book, *At the Edge: Sustainable Development in the 21st Century*, received the 2001 Policy Research Initiative Award for Outstanding Research Contribution to Public Policy. She is the principal investigator for MC3: Meeting the Climate Change Imperative and her research focuses on social strategies for climate innovation, sustainable community development and governance. She is deeply committed to novel research dissemination tools and techniques using social media channels and is currently exploring the use of contemporary museum curatorial practices for research knowledge transfer.

Sarah Burch is an Assistant Professor and Canada Research Chair in Sustainability University of Waterloo. She is also a Senior Fellow, Centre for International Governance Innovation and a Fellow at the Balsillie School of International Affairs. Her research focuses on governing response to climate change (both adaptation and mitigation) in urban spaces. She explores triggers of transitions toward more sustainable development pathways, and strategies for engaging a variety of actions in conversations about desirable futures.

John Robinson is a Professor at the Munk School of Global Affairs, and the School of the Environment, at the University of Toronto; an Honorary Professor with the Institute for Resources, Environment & Sustainability at The University of British Columbia; and an Adjunct Professor with the Copenhagen Business School. At the University of Toronto, he is also Presidential Advisor on the Environment, Climate Change and Sustainability. Prof. Robinson's research focuses on the intersection of climate change mitigation, adaptation and sustainability; the use of visualization, modeling, and citizen engagement to explore sustainable futures; sustainable buildings and urban design; the role of the university in contributing to sustainability; creating partnerships for sustainability with non-academic partners; and, generally, the intersection of sustainability, social and technological change, behaviour change, and community engagement processes.

Chris Strashok is a researcher and instructor in the School for Environment and Sustainability at Royal Roads University with a background in chemical engineering and simulation modelling. Chris works on developing creative and interactive ways of modelling dynamic systems and has developed and implemented various simulation models in the area of community planning around energy and emissions as well as exploring how a community's health is influenced by their built environment. The goal of Chris' modelling work is to share feedback about the space we are living in and build a culture of knowledge, sharing and collaboration.

Part V
Conclusion

Chapter 18
Conclusion: Multilevel Governance and Climate Change Innovations in Cities

Eric K. Chu, Sara Hughes and Susan G. Mason

Abstract This book explored the extent to which cities across the global North and South are able to harness the innovative and creative potential of multilevel pathways of resource, capacity, and authority to support climate change action. While the potential of cities and city regions to offer innovative strategies for climate change mitigation and adaptation is well recognized, the work of realizing this potential is still in its infancy. To address this gap, scholars and practitioners must interrogate the political, economic, and social perquisites of urban innovation pathways; the multilevel opportunities and trade-offs associated with "urban" governments that are increasingly taking on nontraditional forms; and finally, recognize that different ideologies, interests, and authorities are mediated in the process of governing climate change innovations in cities. Without addressing these issues head-on, cities will be unable to realize the full potential of innovations toward more climate resilient, inclusive, and socially just urban futures.

Keywords Cities · Climate change · Multilevel governance · Innovation · Urban politics

E.K. Chu (✉)
University of Birmingham, Edgbaston, Birmingham, UK
e-mail: e.chu@bham.ac.uk

S. Hughes
University of Toronto, Toronto, Canada
e-mail: sara.hughes@utoronto.ca

S.G. Mason
Boise State University, Boise, ID, USA
e-mail: susanmason@boisestate.edu

S. Hughes et al. (eds.), *Climate Change in Cities*, The Urban Book Series,
https://doi.org/10.1007/978-3-319-65003-6_18

18.1 Revisiting the Premise of the Volume

In the introductory chapter, we posited that climate change solutions in cities have two central characteristics. First, we argued that inclusive and equitable mitigation and adaptation actions require and reflect political, institutional, economic, spatial, and social innovation. Addressing climate change in a meaningful way requires the development of new patterns and processes of political engagement, finance, and collaboration. Second, we posited that urban climate change actions are frequently embedded in and (re)produced by multilevel governance contexts, such as through intergovernmental arrangements or transnational municipal networks. In this sense, governing cities is a collective endeavor, and climate change solutions are particularly likely to straddle political, jurisdictional, and socio-ecological boundaries.

From this starting point, the chapters in this book explored the extent to which cities across the global North and South are able to harness the innovative and creative potential of multilevel pathways of resource, capacity, and authority to support climate change action. The different chapters drew upon theories in the fields of public policy, urban planning and administration, multilevel governance, socio-ecological systems, and environment and society to interrogate the complex governance structures and processes through which climate change innovations arise (or not) in cities across different contexts. In this concluding chapter, we will first revisit the primary research questions posed in the introduction. Then, we compare and synthesize key lessons from across the 16 empirical chapters. Finally, we reflect on our contributions to better understand multilevel governance and climate change innovations in cities and present several fruitful opportunities for future research.

This volume highlights that while the importance of cities to tackling climate change is well established, the work of realizing the potential of cities and city-regions to offer meaningful and sustained innovative strategies for climate change mitigation and adaptation is still in its infancy. Therefore, the chapters in this volume help us evaluate climate change responses in cities, what is missing, what it takes to realize the potential of cities, as well as the factors and processes at play that are shaping the outcomes we see. Our hope is that these theoretical and empirical insights will pave the way for more inclusive, equitable, and transformative mitigation and adaptation actions in urban areas around the world.

18.2 Revisiting the Motivating Questions

This volume began with a set of questions about the relationship between urban climate change innovations and multilevel governance. In particular, we first asked how multilevel governance arrangements relate to innovations for urban climate change governance across different contexts. Then, we asked where is the greatest need for innovation, where is innovation difficult or stifled, and how innovations can be fostered and encouraged. Each chapter contributed to different aspects of these central questions, and collectively provide guidance on where the opportunities and obstacles to greater innovation lie. In this section, we reflect on these insights and begin to draw out the larger patterns and conclusions that can be found from the collection.

18.2.1 How Do Multilevel Governance Arrangements Relate to Innovation for Urban Climate Change Governance?

The chapters reveal two distinct dimensions of the relationship between multilevel governance arrangements and innovation for urban climate change governance. First, multilevel governance arrangements can themselves serve as important innovations for facilitating urban climate change governance. New decision-making venues can provide an opportunity for new voices to be heard and for new ideas to come forward. For example, Ninomiya and Burch (Chap. 16) highlighted the role of a new participatory process in Waterloo, Canada, in engaging local entrepreneurs to develop decarbonization strategies for the region. Similarly, both Pakistan and Indonesia have been experimenting with new citizen engagement procedures and forums as a means of enhancing and promoting urban climate change adaptation (Ifitkhar et al.; Sari and Prayoga, Chaps. 7 and 9). In both examples, the engagement processes are facilitated by open channels of communication, committed public servants, the targeting of key stakeholders in the community, and—in some instances—new forms of digital technologies.

New roles and strategies for and within city governments can also be important innovations for urban climate change governance. Engberg (Chap. 8) examined the way the City of Copenhagen took on a new role of "meta-governance," effectively linking the expert-led large-scale water management scheme and the small-scale place-based projects in collaboration with local stakeholders to facilitate adaptive measures in water management. Bourgeois and Hughes (Chap. 5) follow the re-centralization process in Montreal, Canada, and highlight the trade-offs that arose between greater policy effectiveness and local accountability. In their review of best practices in Canadian cities, Dale et al. (Chap. 17) find that there are several institutional strategies that facilitate innovation and effectiveness, including institutionalizing climate change across municipal departments and developing systematic policy-making frameworks. Boswell and Mason (Chap. 4) see potential in California's strategy of incentivizing land use and transportation planning at the regional scale, especially when large and mid-size cities engage in activities that support California's Sustainable Communities Strategies.

Multilevel governance arrangements also structure innovation supply and demand for urban climate change governance. They create a need for more innovative approaches, present opportunities for new practices to emerge, help to dictate the terms and set the boundaries of local action, and in some cases, act as obstacles to greater innovation. One clear example of this is the intergovernmental relationships that structure local decision-making. Homsy (Chap. 2), Kemmerzell (Chap. 3), and Rajasekar et al. (Chap. 11) all find that vertical intergovernmental dynamics play a significant role in shaping local decision-making about climate change. For Kemmerzell, cities that participate in European Union projects have more comprehensive climate change programs. Homsy similarly finds that U.S. cities are more likely to have climate change policies of their own when higher levels of government have already acted. This is particularly true for smaller cities that may have fewer opportunities and resources with which to innovate on their own. Finally, Rajasekar et al. noted the importance of national schemes such as the Smart Cities Mission in structuring many governance opportunities and constraints to embedding smart urbanism efforts into current climate change resilience programs across Indian cities.

Multilevel governance arrangements extend beyond vertical intergovernmental dynamics and there are additional ways they structure innovations. For example, Cook and Chu (Chap. 13) find that in Surat, India, the shifting and complex climate change finance landscape has fostered innovative strategies at the city level for accessing funds and steering adaptation actions to reflect and adjust to local fiscal constraints. Peterson (Chap. 14) similarly highlights the need and opportunities for innovative financing strategies as cities confront changing fiscal landscapes. Cities are increasingly participating in and embedded within transnational municipal networks and must negotiate between emerging best practices and expectations from such networks with the contentious and local politics of urban climate change governance. Bellinson (Chap. 10) draws a clear picture of the actors, resources, and networks involved in navigating this boundary, and highlights the importance of confronting conflicts and building cross-sectoral coalitions in response to these tensions arising from multilevel governance arrangements. Such networked approaches to financing climate change action and sharing mitigation or adaptation knowledge illustrate the role of emerging actors—whether within the public sector or beyond—that are contributing to governance innovation in cities.

18.2.2 Where Is the Greatest Need for Innovation?

The chapters in this volume present some of the most up-to-date experiences of cities as they pursue climate change governance in multilevel governance contexts. In doing so, they highlight emerging innovations and exciting experiments in climate change governance. They also provide insight into where there is a need for further innovation going forward or places where we are only beginning to understand the role of innovative practices. We see three areas for greater innovation, drawing on the lessons from the different chapters: financing urban climate change governance, designing democratic and inclusive institutions, and building intergovernmental relationships that are able to balance efficiency, effectiveness, and equity in urban climate change governance outcomes.

Urban climate change finance stands out to us as an important arena for greater innovation going forward. The chapters by Bausch et al. (Chap. 12), Cook and Chu (Chap. 13), and Peterson (Chap. 14) all highlight the importance of urban climate change finance and the diverse but often inadequate funding options city governments have available to them. While these dynamics have in some cases led cities to be quite innovative in how they pursue climate change projects (e.g., Surat in the Cook and Chu case), many chapters have highlighted the constraining role of a lack of financial resources (e.g., Sarzynski, Homsy, and Dale et al.). We know there is the potential for innovation in urban climate change finance, but that not all cities are being innovative in this area. Greater attention to financial innovations going forward would be an important contribution to knowledge on urban climate change governance.

The second area for further innovation is in designing democratic and inclusive institutions. Several chapters in this volume focus on new and emerging institutional forums for stakeholder and citizen participation, and highlight the increased awareness of the importance of these forums and processes for urban climate change governance (e.g., in the cases of Iftikhar, Sari and Prayoga, Ninomiya and Burch, Engberg, and Sarzynski). In particular, many notes that democratic and inclusive innovations are required to recognize the context-dependence of many climate risks and vulnerabilities, the need to integrate climate change efforts into existing development or livelihood strategies, as well as the importance of accounting for equity, justice, and community rights in any potential climate change actions. As a result, as noted by Brown (Chap. 12), greater innovation is needed to scale these initiatives up and out in a way that can facilitate meaningful shifts and catalyze greater effectiveness in urban climate change governance.

Finally, the chapters in this volume demonstrate that vertical intergovernmental arrangements clearly matter for the success of urban climate change innovations (e.g., in the cases of Kemmerzell, Homsy, Bourgeois and Hughes, Bellinson, Boswell and Mason). However, they also highlight the need for arrangements that are better able to account for a range of conditions, local priorities, and political contexts. Innovations are needed in the way such vertical relationships are structured and in their ability to maintain consistency and provide direction while also incorporating flexibility and providing space for local priorities.

18.2.3 Where Is Innovation Difficult or Stifled?

Even though there are many opportunities for governance innovation, the chapters in this volume show that innovative efforts are in fact incredibly complex, politically contentious, and time-consuming. For some, the difficulty stems from an operational bias towards short-term strategic projects in cities. For example, as described by Brown (Chap. 12), cities often experience budget and capacity shortfalls, which inhibit their ability to make long-term commitments to climate change policies that require extensive financial, political, and institutional support. As a result, many cities opt for co-beneficial actions that yield immediate economic and strategic development benefits, such as in terms of real estate, financial investment, or property tax gains. Furthermore, Bellinson (Chap. 10) noted that the complexity of different urban interests set the stage for contentious policy discussions around which projects to prioritize when confronted with constrained bureaucratic and financial resources. Again, such politically contentious discussion often end up favoring short-term economic goals that yield immediate investment opportunities or electoral gains. Such forms of short-sightedness may stifle more comprehensive efforts to reform governance practices in cities, which are required to address the long-term implications of climate change.

Many chapters noted that multilevel governance contexts can stifle innovative behavior in cities when jurisdictional boundaries are uncertain. In this case, multilevel interactions can serve as a deterrent for cross-sectoral partnerships, inter-municipal coordination, or intergovernmental support because they lack clearly codified procedures (both legally and informally) for ensuring equitable distribution of capacities and resources between different actors. For example, Boswell and Mason (Chap. 4) noted that California's climate change strategy does not articulate clear, legally embedded guidelines for local-level actions, which leads to some cities committing to climate action while others lag behind.

A third difficulty encountered by cities is the interaction between resource intensity and institutional lethargy. Governance innovations require long-term political commitment, dedicated arenas for policy deliberation, and clear guidelines for assessing impacts, which imply that the amount of resources and dedication needed to successfully foster innovation can be significant. As a result, cities can be resistant to fostering climate innovations outside of the occasional demonstration/pilot project or when experiments are fully supported by external donors. For example, Sari and Prayoga (Chap. 7) noted that innovative disaster resilience strategies required investments from nongovernmental partners such as Mercy Corps to help organize community engagement activities, pilot studies, and project assessment exercises. Brown (Chap. 12) similarly noted that cities that participated in the Asian Cities Climate Change Resilience Network benefited from a structured "shared learning dialogue" engagement process, which facilitated close partnerships between the Rockefeller Foundation, intermediary organizations, and local community beneficiaries. In both cases, this facilitation—and, at times, hand-holding—offered by nongovernmental actors ensured the sustainability of resources, time, and effort required to foster innovative approaches to urban climate action. In cases where cities offer such resources internally, such as in the case of Montreal's waste management policy (see Bourgeois and Hughes, Chap. 5), some boroughs in the city actually saw a decline in resources available to provide other essential public services, and so have sought to cede from the city as a result of the centralization of services. Aside from the example of the Surat Climate Change Trust (see Cook and Chu, Chap. 13), few cities have built corresponding institutional structures to support the time, resources, and effort required to sustain innovative climate actions across time.

18.2.4 How Can Innovation Be Fostered and Encouraged in a Multilevel Governance Context?

The chapters in this volume offer numerous strategies to encourage innovation in different multilevel contexts, which are essential for ensuring the procedural inclusiveness and equity of potential climate change actions. However, this simultaneously poses serious strains on cities that are already financially stressed. To overcome this, some chapters have highlighted examples where institutional leaders, facilitators, or entrepreneurs have managed to navigate the complex governance arenas on behalf of particular climate change priorities. For example, Bellinson (Chap. 10) noted the important role played by City Resilience Officers (CROs) within cities participating in the 100 Resilient Cities program. These policy entrepreneurs help to identify, connect, and facilitate relationships between different municipal actors, as well as to channel necessary resources to implement certain pilot projects. Similarly, Brown (Chap. 12) illustrated incidences where local facilitating organizations in South and Southeast Asia have helped to build collaborative platforms for identifying and prioritizing climate resilience interventions. Finally, Dale et al. (Chap. 17) also argued for the strong role of facilitating organizations that are more amenable to collaboration and experimentation. Not only are policy facilitators important for bridging potentially disparate policy domains, it also helps to put a "human face" on certain agenda items and may help to build social and political capital in preparation for complex and contentious policy negotiations. Furthermore, facilitating actors tend to be more agile, flexible, and adaptive in the face of political (and environmental) uncertainty, which allows for the testing of prospective interventions before they are fully applied in public policy.

Some chapters have noted the opportunities presented by inclusive and deliberative arenas that allow for the presentation and assessment of disparate climate change needs. Since climate change will benefit and impact urban sectors, communities, and environments differently, inclusive decision-making arenas help to foster political debate and to ensure the recognition of minority needs. For example, in the case of climate adaptation in Mexico City, Bausch et al. (Chap. 15) illustrated the need to account for opportunities presented by agricultural communities. Similarly, Sari and Prayoga (Chap. 7) and Iftikhar et al. (Chap. 9) both show how community beneficiaries can be effective drivers of climate resilience programs if given the recognition and opportunity. Deliberation can encourage climate innovations not just because participatory arenas facilitate more diverse voices, it can also promote urban equity and inclusiveness as different interests are taken into account in the policy process.

Finally, many chapters point to the importance of building higher level structures to offer a broader policy direction. Such forms of meta-governance allow for more structured political leadership, communication pathways, and the distribution of tasks between actors participating in climate change governance. For example, Engberg's discussion of climate change adaptation in Copenhagen (Chap. 8) illustrated the importance of structuring a hybrid organizational form that enables intra-municipal and cross-sectoral policy integration. This way, the advances in decision-making produced gains for both the community and the city. In Boswell and Mason's discussion of the case of California (Chap. 4), the meta-governance structure for regulating greenhouse gas emissions was not strong enough to direct actions from cities across regions. Conversely, Rajasekar et al.'s discussion of the Smart Cities Mission in India (Chap. 11) did offer a strong policy structure in which resource pathways were clearly delineating, thus providing clarity for cities who wish to participate and creatively think through how smart urbanism priorities can be integrated into existing climate adaptation and disaster resilience efforts in cities.

18.3 Innovation in the Context of Multilevel Climate Governance: Towards New Insights

The previous section highlighted the procedural complexity, political "wickedness", and sheer scale of the climate change challenge facing cities across the global North and South. In this section, we turn our attention to distilling new insights as we relate these findings to current theories in multilevel climate change governance, innovation, and the role of urban actors. Specifically, the following sections illustrate three key theoretical contributions offered by the volume, where we seek to illuminate the political, social, and economic perquisites of urban innovation pathways; the opportunities and trade-offs associated with "urban" governments that are increasingly taking on nontraditional forms; and finally, we end with a call to resist the *post-political turn* of climate change governance innovations in cities.

We find that innovative strategies to either mitigate or adapt to climate change will only succeed in multilevel governance arenas if we take the *politics* of innovation seriously. In other words, we must confront head-on the questions of how are narratives and strategies of innovation being constructed by political actors; how do directives, mandates, and resources supplied by actors from both within and beyond cities interact with existing municipal regimes, special interests, and resources; who and according to what criteria are innovative strategies being implemented, prioritized, and evaluated in cities; and finally, how are different ideologies, interests, and authorities mediated in the process of governing climate change innovations in cities.

18.3.1 The Institutional Foundations for Urban Innovations

Many of the chapters in this volume note that progressive climate change action in cities depends on a baseline level of knowledge, capacity, and resources. As a result, understanding the foundations of and motivations behind innovative behavior is critical to shedding light on what types of mitigation and adaptation actions eventually transpire. The literature on climate change mitigation, for example, has noted that innovations must be situated in particular spatial, political, social, and economic contexts (Bulkeley et al. 2015; Castán Broto and Bulkeley 2013). Emerging climate change adaptation efforts also require a baseline understanding of projected climate risks—such as sea level rise, extreme precipitation, urban heat islands, etc.—as well as a conducive political environment that allows for cross-sectoral, comprehensive, and forward-thinking policies and plans (Carmin et al. 2013).

For both mitigation and adaptation, cities—meaning urban environments, urban governments, and urban communities in particular—are thought of as particularly effective spaces for experimentation and innovation (Bulkeley and Castán Broto 2013; Chu 2016; Evans 2011). As shown in this volume, many cities across the global North and South find experimentation attractive because it promotes innovation through improving overall decision-making efficiency, effectiveness, inclusiveness, and responsiveness. Procedurally, experiments are able to support evidence-based policy-making by giving robust and timely advice for evaluating and redesigning existing approaches. In this sense, experiments can be seen as "laboratories" of learning and best practices (see for example Engberg, Chap. 9), which allow diverse actors, networks, and policy pathways to embed emerging needs and priorities into urban policies and plans (Karvonen and van Heur 2014). When applied to climate change, for instance, experiments allow cities to flexibly frame objectives, implement trials and pilot projects, and monitor and evaluate project outcomes. They also allow local governments to test implementation pathways, help prioritize options, and evaluate overall project benefits in the face of uncertain climate futures and highly decentralized or fragmented governance arenas (Anguelovski et al. 2014). Although some have challenged the external validity and replicability of experiments, this methodology has been shown to be a good arena for intensive dialogues and small-scale innovations. The interactive quality of many experiments facilitates deliberative processes that may even improve the social inclusiveness of policy outcomes.

In this vein, the chapters in this volume have noted that innovations in climate governance require informed, effective, and accountable institutions in order for particular experimental strategies to take hold. For example, some chapters in this volume touched upon the importance of local civic capacity to facilitate climate awareness (see Sarzynski, Chap. 6) or to form the basis for sustained grassroots actions (see Sari and Prayoga, Chap. 7). Others noted the importance of community networks (see Iftikhar et al., Chap. 9) and bottom-up, deliberative, and co-creative processes (see Engberg, Chap. 8) for rendering climate change more relevant for urban development policies and spatial plans. Further, some authors have noted the importance of strong political leaders and entrepreneurs (see Chaps. 12 and 17) or effective policy facilitators and translators who can manage their substantive expertise against local bureaucratic constraints (see Cook and Chu, Chap. 13). These observations mirror those highlighted in the literature, where scholars note the importance of knowledge capacity, local social and political awareness, as well as popular ownership over programs and projects as key foundational requirements for effective and equitable climate change actions in cities. Our volume, therefore, offers numerous empirical examples of the different sorts of institutional pre-requisites to innovative and experimental behavior in urban climate governance across the global North and South.

18.3.2 Unbounding the Urban in Climate Governance

One central premise of the literature on the multilevel governance of climate change notes that cities are increasingly critical actors because they are often more attuned and responsive to local environmental conditions and socioeconomic needs (Betsill and Bulkeley 2006; Rosenzweig et al. 2010). Furthermore, many cities oversee primary responsibility for managing infrastructure and social services that are essential for promoting good living standards, social inclusiveness, and the reduction of vulnerability to many environmental hazards. However, several chapters in this volume have noted that the leadership role played by cities may be more nuanced—cities are often dependent on state-level policy direction (see Homsy, Chap. 2) or are led by regional or supra-national mandates (see Kemmerzell, Chap. 3). In this sense, urban authorities are typically jurisdictionally and legally confined, which constrains the autonomy of cities to pursue locally innovative mitigation and adaptation strategies. Yet, in other cases, cities are simply nudged to follow the policy recommendations of region-level actions (see Boswell and Mason, Chap. 4).

The chapters in the volume, therefore, all note that even though cities are critical sites for innovative climate change governance, many of the policy tools, planning strategies, and social and political networks employed to facilitate innovative behavior extend beyond the jurisdictional confines of many municipal governments. The "urban", in other words, is not necessarily solely a spatially and legally defined sub-national entity; rather, it is an assemblage of public, private, and civil society actors all exerting their interests via particular procedural channels while simultaneously interacting with complex political networks, economic flows, social identities, and environmental ideologies. For example, this volume noted that though the sites of innovation are often cities, the key facilitating actors vary from traditional municipal public sector institutions to supra-national governments (see for example Kemmerzell, Chap. 3), regional or metropolitan agencies (see for example Boswell and Mason, Chap. 4, and Bourgeois and Hughes, Chap. 5), or different transnational networks (see for example Bellinson, Chap. 10). Furthermore, these external actors are often complemented (or contested) by strong grassroots community coalitions, including those documented in Part II of this volume. As a result, in order to critically evaluate the true sources, pathways, and implications of innovative climate change governance in cities, one must *unbound the urban* and look beyond the traditional spatial and political confines of urban arenas.

This need to more critically evaluate the "urban" in climate governance mirrors recent developments in the literature (see Angelo and Wachsmuth 2015; Brenner and Schmid 2015; Roy 2016). Wachsmuth et al. (2016) have raised doubts about the true role of cities in solving sustainability and climate change challenges given inaccurate accounts and assessments of cities' environmental footprints. Furthermore, many cities are fragmented, with municipal boundaries dividing what are otherwise contiguous urban regions. For example, changing urban mobility behaviors by incentivizing public transportation usage or transit-oriented development is critical for reducing emissions, but such actions rely on coordinated policies across regions, as transportation networks and urban agglomerations transcend political boundaries (Chu and Schenk 2017). Furthermore, many of the chapters in the volume note the scalar dependency of urban climate actions, such as in the case of overcoming the "adaptation deficit" in Baltimore, USA (see Sarzynski, Chap. 6) or in the case of assessing whether city-regions or agglomerations of smaller cities in the United States actually respond to higher level climate change mandates or policy incentives (see Homsy, Chap. 2).

In addition to the need *to unbound* the scalar and spatial aspects of urban climate governance, many chapters in this volume speak to the need *to unbound the politics of the urban*. In fact, urban climate mitigation and adaptation actions are no longer under the sole jurisdiction of municipal public sector agencies as they increasingly rely on transnational, regional, nongovernmental, and private sector actors for resources, capacity, and knowledge. This can be seen in the case of community-based partnerships over metered drinking water in Bhalwal, Pakistan (see Iftikhar et al., Chap. 9), as well as in the example of engaging private sector entrepreneurs and leaders in Waterloo, Canada (see Ninomiya and Burch, Chap. 16). Furthermore, many chapters in this volume note that cities are increasingly nodes for transnational cooperation, where networks help to bridge fragmented authority through their ability to align actors and institutions, as well as by engaging and fostering new cooperative planning and decision-making pathways (Andonova et al. 2009; Fünfgeld 2015). This point is especially stressed by advocates who suggest that city-to-city networks have become a governance arena that is parallel—and gaining importance—to nation-state-based geopolitical spaces (Andonova and Mitchell 2010; Toly 2008). Networks such as C40, ICLEI, and the Rockefeller Foundation's 100 Resilient Cities program facilitate horizontal dialog and cooperative relationships between local government, civil society, and private actors (see for example Brown, Chap. 12). Finally, as illustrated in the examples from Berkeley and Rotterdam (see Bellinson, Chap. 10), networks also share best practices, distribute scientific projections and models, and foster collaboration across different cities.

In sum, the power of a multilevel conceptual approach—especially one that unbounds the politics of urban actions to include global, transnational, and local-level interactions—is to allow for deep, empirical, and multifaceted interrogations of the intersecting global-local, public-private, institutional-spatial, and hegemonic-grassroots interactions of political power in the context of climate change action within, between, and amongst cities.

18.3.3 Resisting the Post-Politics of Climate Innovations

The previous two sections illustrated the importance of engaging with the institutional foundations of urban innovation and the need to unbound the "urban" politics of climate change governance, which both point to the reality that innovation—i.e., the process of generating, adopting, and disseminating new and creative ideas which aim to produce a qualitative change in a specific context (Sørensen and Torfing 2011)—is in fact intensely political and contentious. Innovation, as the chapters in this volume all note, represents iterative, co-creative, and uncertain processes that are deeply mired in conflicts between urban regime interests, multilevel political and economic forces, as well as local environmental needs and development priorities. In other words, climate change innovations are not simply technological, infrastructural, or engineering fixes to urban systems that are either transitioning to low-carbon economies or are needing to adapt to extreme disaster risks; instead, governance innovation is a deeply political exercise that embodies the variegated aspirations for, ambitions of, and pathways toward climate resilient urban futures—ones that have real consequences for long-term urban equity, inclusion, and justice.

Though there is a long history of innovation scholarship in the domains of national economic and industrial policy (see for example Freeman 1991; MacCullum et al. 2009; Nelson 1993), there has been relatively little on innovation at the urban scale, especially beyond the field of economic geography or territorial/spatial development (see for example Moulaert and Sekia 2003; Walker and Storper 1989). The chapters in this volume, therefore, contribute to recent advancements in theorizing the genesis and implications of urban innovations in the context of climate change, particularly through the lens of experimentation or urban "living labs" (Bulkeley et al. 2015; Evans and Karvonen 2014) and through the participation of non-state actors (Chu et al. 2016). In terms of substance, this recent literature on urban climate governance has focused mainly on three categories of innovation—first, the opportunities for citizen-led democratic innovations (see for example Seyfang and Haxeltine 2012; Wamsler 2016); second, the procedural aspects of policy innovation (see for example Jordan and Huitema 2014; Krause 2011; Massey et al. 2014); and finally, the opportunities for innovative socio-technical arrangements that facilitate urban climate transitions (see for example Geels 2004; Rutherford and Coutard 2014). Although interrogations into these three distinct domains are important, the literature must also critically assess the conditions in which different categories and pathways of innovation emerge, interact, and take-hold in cities. This is especially important because cities are only one actor within multilevel governance systems, and thus do not hold sole authority over determining who participates in innovation, which innovations are promoted, and how the outcomes of innovations are evaluated. Furthermore, innovation requires collaborative interactions between different urban actors because municipal resources are finite and often insufficient to meet the cross-sectoral demands of mitigation and adaptation (Bulkeley and Betsill 2013; Chu et al. 2016).

One main contribution of this volume is, therefore, a call to reinsert the *lived experiences* of urban political actors in the process of articulating, designing, and implementing innovative climate change governance strategies. The politics of the *lived experience* is critical because climate change priorities are complex, interrelated, and embedded in urban norms, ideologies, and power differentials. For example, internal bureaucratic constraints in Indian cities regulate the way in which external strategies for climate change finance are applied to urban spatial plans (see Cook and Chu, Chap. 13). Similarly, the degree to which a "business case" for climate change adaptation can be crafted determines the extent to which private actors participate in urban decarbonization efforts (see Ninomiya and Burch, Chap. 16).

This focus on the lived experience, therefore, relates to more critical works by urban geographers and sociologists who note that the overwhelming focus of climate change governance on scientific expertise, bureaucratic rule-making, legal construction, and epistemic knowledge renders it "sanitized" and apolitical (see Braun 2014; Swyngedouw 2010). This post-political critique of climate governance asserts that the lack of negotiation and deliberation between different urban interests actually results in governance outcomes that codify or entrench hegemonic—often neoliberal—forms of knowledge (see Clarke 2012; Swyngedouw 2005), or worse, lead to the active disenfranchisement of minority voices and needs (Ziervogel et al. 2017). A post-political turn in the study of urban climate change governance, therefore, implies the lack of attention paid to the equity, inclusiveness, and justice dimensions of potential mitigation and adaptation strategies (Anguelovski et al. 2016; Hughes 2017; Shi et al. 2016). In response, the findings of this volume contribute to calls for resisting the post-politics of urban climate governance by empirically showing the inherently political—and *lived*—ramifications of innovation across the global North and South.

18.4 Ways Forward for Research and Practice

As cities increasingly realize the opportunities and costs associated with prospective mitigation and adaptation strategies, the rhetoric of governance innovation has emerged to help cities conceptualize the different sources of political support, finance, and institutional capacity that are available across local, metropolitan/regional, national, and transnational governance actors. On the one hand, governance innovations can help address the structural constraints in cities by facilitating communication, resource, and knowledge dissemination pathways that bridge the siloed nature of urban planning and policy-making. Innovative behavior can also uncouple cities from their traditional planning models—such as those based on the logic of real estate development or strategic mega-projects—to experiment with alternative learning and implementation pathways through which bureaucratic constraints can be negotiated. However, on the other hand, governance innovation is a much more politically complex, contentious, and context-dependent phenomenon, especially when considering that cities are only one actor in the multilevel political reality that we all live in today. Climate change vulnerabilities and impacts also affect communities and environments without regard to political or jurisdictional boundaries. Therefore, innovation often requires dedicated political attention, resource support, communication channels, and community coalitions to sustain processes across time and across space.

The chapters in this volume have all empirically illustrated the practical realities of facilitating and sustaining innovative urban climate change mitigation and adaptation strategies across different contexts. Our examples have ranged from evaluating climate policy uptake in German cities, including community-based early warning systems in Indonesia, opportunities for embedding local entrepreneurs in decarbonization programs in Canada, and many more. These examples offer in-depth, timely, and nuanced understandings of how climate governance innovations emerge and take root given different political opportunities, resource constraints, and participatory arrangements. In response to these lessons, we offer three directions for future research and policy development.

First, we must further interrogate the ideological basis for climate governance innovations in cities. Many of the chapters in this volume have noted that the lived experience of mitigation or adaptation innovations must balance the procedural or practical dimensions of the policy process with overall normative or ethical aspirations of the policy designs. The latter refers to issues such as the political rhetoric behind climate change action—i.e., where such discourses come from and how they are framed—or the equity and distributive implications of climate change projects —i.e., which actors are included and how underrepresented, minority, or vulnerable communities are recognized in the process. Although the literature has been quite thorough in assessing the procedural aspects of facilitating and implementing climate change actions in cities—which include testing, trials, and mainstreaming (see for example Carmin et al. 2013; Noble et al. 2014)—we do not have a good understanding of how these practical institutional opportunities or constraints interact with grounded ideologies, interests, ethics, and biases within particular urban contexts. This is a particularly relevant critique because urban climate change innovations are increasingly taking on neoliberal, speculative, and exclusionary characteristics. As a result, we must ask: how do governance innovations contribute to broader visions of climate resilient urban futures through navigating competing ideals and aspirations of "what ought to be"?

Second, and building off of the first question, many chapters in this volume (particularly Part II) advocate for more inclusive, participatory, and representative decision-making when designing and implementing climate governance innovations in cities. The argument is that more inclusive processes can take into account the differentiated needs and interests of citizens who may have been neglected or marginalized in the past. This is particularly important as climate change tends to impact those who are less able to cope (Shi et al. 2016). However, we also see that innovation requires dedicated political attention and resource support, which means that public sector and private actors will continue to be drivers of many projects. As such, given the differentiated and complex social, political, and economic interests that are at play, how do we ensure the equitable distribution of decision-making power and the benefits of innovation across different contexts? Furthermore, based on the lessons learned, are there opportunities for developing methods or metrics to assess the conditions within which innovations are successful, ineffective, or even counterproductive? Finally, how can these lessons be reflected upon comparatively and longitudinally—e.g., to what extent can experience from Copenhagen or Semarang be applied to Baltimore or Bhalwal?

Third, and in a more practical sense, many of the chapters in this volume highlighted the importance of institutional and financial resources for kick-starting and sustaining climate governance innovations in cities. However, many chapters also noted that the availability of resources can be a double-edged sword. For example, Climate Resilience Officers have a lot of power in leading climate policy in cities, but this position is time-constrained and may encounter legitimacy issues along the way (see Bellinson, Chap. 10). In other cases, donor funds earmarked for climate change projects may be lumped together with existing intergovernmental grants to fund infrastructure or development projects that were already in the books (see for example Cook and Chu, Chap. 13). The particular roles of resources—either as facilitating or steering agents—is therefore unclear. In response, we must ask: How do strong leaders or policy entrepreneurs navigate complex urban governance arrangements to sustain and disseminate climate innovations across time? Under what circumstances will financial resources redirect climate change innovations in favor of more speculative or exclusionary—i.e., neoliberal—strategies that do not offer benefits for all?

In summary, these three directions for future research all seek to bridge the empirical illustrations of governance innovations with more critical theories from the fields of political economy, environment and development, urban geography, city planning, and public policy. This is a much-needed advancement because we find that innovative strategies to either mitigate or adapt to climate change will only succeed in multilevel governance arenas if we take the *politics* of innovation seriously. Given the prevalence of promoting experimentation and innovation in climate change governance today (see Bulkeley et al. 2015; Jordan and Huitema 2014), the chapters in this volume collectively point to a need for future scholarship that further interrogates the political, economic, and social perquisites of urban innovation pathways; the multilevel opportunities and trade-offs associated with "urban" governments that are increasingly taking on nontraditional forms; and finally, recognizes that different ideologies, interests, and authorities are mediated in the process of governing climate change innovations in cities. Without addressing these issues head-on, cities across the global North and South will be unable to realize the full potential of innovations towards more climate resilient, inclusive, and socially just urban futures.

References

Andonova, L. B., Betsill, M. M., & Bulkeley, H. (2009). Transnational climate governance. *Global Environmental Politics, 9*(2), 52–73. doi:10.1162/glep.2009.9.2.52.

Andonova, L. B., & Mitchell, R. B. (2010). The rescaling of global environmental politics. *Annual Review of Environment and Resources, 35*(1), 255–282. doi:10.1146/annurev-environ-100809-125346.

Angelo, H., & Wachsmuth, D. (2015). Urbanizing urban political ecology: A critique of methodological cityism. *International Journal of Urban and Regional Research, 39*(1), 16–27. doi:10.1111/1468-2427.12105.

Anguelovski, I., Chu, E., & Carmin, J. (2014). Variations in approaches to urban climate adaptation: Experiences and experimentation from the global South. *Global Environmental Change, 27,* 156–167. doi:10.1016/j.gloenvcha.2014.05.010.

Anguelovski, I., Shi, L., Chu, E., Gallagher, D., Goh, K., Lamb, Z., et al. (2016). Equity impacts of urban land use planning for climate adaptation: Critical perspectives from the global north and south. *Journal of Planning Education and Research, 36*(3), 333–348. doi:10.1177/0739456X16645166.

Betsill, M. M., & Bulkeley, H. (2006). Cities and the multilevel governance of global climate change. *Global Governance: A Review of Multilateralism and International Organizations, 12*(2), 141–159.

Braun, B. P. (2014). A new urban dispositif? Governing life in an age of climate change. *Environment and Planning D: Society and Space, 32*(1), 49–64. doi:10.1068/d4313.

Brenner, N., & Schmid, C. (2015). Towards a new epistemology of the urban? *City, 19*(2–3), 151–182. doi:10.1080/13604813.2015.1014712.

Bulkeley, H., & Betsill, M. M. (2013). Revisiting the urban politics of climate change. *Environmental Politics, 22*(1), 136–154. doi:10.1080/09644016.2013.755797.

Bulkeley, H., & Castán Broto, V. (2013). Government by experiment? Global cities and the governing of climate change. *Transactions of the Institute of British Geographers, 38*(3), 361–375. doi:10.1111/j.1475-5661.2012.00535.x.

Bulkeley, H., Castán Broto, V., & Edwards, G. A. S. (2015). *An urban politics of climate change: Experimentation and the governing of socio-technical transitions.* New York and London: Routledge.

Carmin, J., Dodman, D., & Chu, E. (2013). *Urban Climate adaptation and leadership: From conceptual understanding to practical action* (No. 2013/26). Paris, France: Organisation for Economic Co-operation and Development (OECD). doi:10.1787/5k3ttg88w8hh-en.

Castán Broto, V., & Bulkeley, H. (2013). A survey of urban climate change experiments in 100 cities. *Global Environmental Change, 23*(1), 92–102. doi:10.1016/j.gloenvcha.2012.07.005.

Chu, E., Anguelovski, I., & Carmin, J. (2016). Inclusive approaches to urban climate adaptation planning and implementation in the Global South. *Climate Policy, 16*(3), 372–392. doi:10.1080/14693062.2015.1019822.

Chu, E. K. (2016). The governance of climate change adaptation through urban policy experiments. *Environmental Policy and Governance, 26*(6), 439–451. doi:10.1002/eet.1727.

Chu, E., & Schenk, T. (2017). Communicating about climate change with Urban populations and decision-makers. In M. C. Nisbet (Ed.), *Oxford Research Encyclopedia of Climate Science* (Vol. 1). Oxford and New York: Oxford University Press. doi:10.1093/acrefore/9780190228620.013.413.

Clarke, N. (2012). Urban policy mobility, anti-politics, and histories of the transnational municipal movement. *Progress in Human Geography, 36*(1), 25–43. doi:10.1177/0309132511407952.

Evans, J., & Karvonen, A. (2014). "Give Me a Laboratory and I Will Lower Your Carbon Footprint!"—Urban laboratories and the governance of low-carbon futures. *International Journal of Urban and Regional Research, 38*(2), 413–430. doi:10.1111/1468-2427.12077.

Evans, J. P. (2011). Resilience, ecology and adaptation in the experimental city. *Transactions of the Institute of British Geographers, 36*(2), 223–237. doi:10.1111/j.1475-5661.2010.00420.x.

Freeman, C. (1991). Networks of innovators: A synthesis of research issues. *Research Policy, 20* (5), 499–514. doi:10.1016/0048-7333(91)90072-X.

Fünfgeld, H. (2015). Facilitating local climate change adaptation through transnational municipal networks. *Current Opinion in Environmental Sustainability, 12,* 67–73. doi:10.1016/j.cosust. 2014.10.011.

Geels, F. W. (2004). From sectoral systems of innovation to socio-technical systems. *Research Policy, 33*(6–7), 897–920. doi:10.1016/j.respol.2004.01.015.

Hughes, S. (2017). The politics of urban climate change policy. *Urban Affairs Review, 53*(2), 362–380. doi:10.1177/1078087416649756.

Jordan, A., & Huitema, D. (2014). Innovations in climate policy: The politics of invention, diffusion, and evaluation. *Environmental Politics, 23*(5), 716-734. doi:10.1080/09644016. 2014.923614.

Karvonen, A., & van Heur, B. (2014). Urban laboratories: Experiments in reworking cities. *International Journal of Urban and Regional Research, 38*(2), 379–392. doi:10.1111/1468-2427.12075.

Krause, R. M. (2011). Policy innovation, intergovernmental relations, and the adoption of climate protection initiatives by U.S. Cities. *Journal of Urban Affairs, 33*(1), 45–60. doi:10.1111/j. 1467-9906.2010.00510.x.

MacCullum, D., Moulaert, F., Hillier, J., & Haddock, S. V. (Eds.). (2009). *Social Innovation and Territorial Development*. London: Routledge.

Massey, E., Biesbroek, R., Huitema, D., & Jordan, A. (2014). Climate policy innovation: The adoption and diffusion of adaptation policies across Europe. *Global Environmental Change, 29,* 434–443. doi:10.1016/j.gloenvcha.2014.09.002.

Moulaert, F., & Sekia, F. (2003). Territorial innovation models: A critical survey. *Regional Studies, 37*(3), 289–302. doi:10.1080/0034340032000065442.

Nelson, R. R. (Ed.). (1993). *National Innovation Systems: A Comparative Analysis*. Oxford and New York: Oxford University Press.

Noble, I. R., Huq, S., Anokhin, Y. A., Carmin, J., Goudou, D., Lansigan, F. P., et al. (2014). Adaptation needs and options. In C. B. Field, V. R. Barros, D. J. Dokken, K. J. Mach, M. D. Mastrandrea, T. E. Bilir, et al. (Eds.), *Climate Change 2014: Impacts, Adaptation, and Vulnerability. Part A: Global and Sectoral Aspects. Contribution of Working Group II to the Fifth Assessment Report of the Intergovernmental Panel on Climate Change* (pp. 833–868). Cambridge, UK and New York, NY: Cambridge University Press.

Rosenzweig, C., Solecki, W., Hammer, S. A., & Mehrotra, S. (2010). Cities lead the way in climate–change action. *Nature, 467*(7318), 909–911. doi:10.1038/467909a.

Roy, A. (2016). What is urban about critical urban theory? *Urban Geography, 37*(6), 810–823. doi:10.1080/02723638.2015.1105485.

Rutherford, J., & Coutard, O. (2014). Urban energy transitions: Places, processes and politics of socio-technical change. *Urban Studies, 51*(7), 1353–1377. doi:10.1177/0042098013500090.

Seyfang, G., & Haxeltine, A. (2012). Growing grassroots innovations: Exploring the role of community-based initiatives in governing sustainable energy transitions. *Environment and Planning C: Government and Policy, 30*(3), 381–400. doi:10.1068/c10222.

Shi, L., Chu, E., Anguelovski, I., Aylett, A., Debats, J., Goh, K., et al. (2016). Roadmap towards justice in urban climate adaptation research. *Nature Climate Change, 6*(2), 131–137. doi:10. 1038/nclimate2841.

Sørensen, E., & Torfing, J. (2011). Enhancing collaborative innovation in the public sector. *Administration & Society, 43*(8), 842–868. doi:10.1177/0095399711418768.

Swyngedouw, E. (2005). Governance innovation and the citizen: The janus face of governance-beyond-the-state. *Urban Studies, 42*(11), 1991–2006. doi:10.1080/00420980500279869.

Swyngedouw, E. (2010). Apocalypse forever?: Post-political populism and the spectre of climate change. *Theory, Culture & Society, 27*(2–3), 213–232. doi:10.1177/0263276409358728.

Toly, N. J. (2008). Transnational municipal networks in climate politics: From global governance to global politics. *Globalizations, 5*(3), 341–356. doi:10.1080/14747730802252479.

Wachsmuth, D., Cohen, D. A., & Angelo, H. (2016). Expand the frontiers of urban sustainability. *Nature, 356*, 391–393. http://www.nature.com/news/expand-the-frontiers-of-urban-sustainability-1.20459#/b3.

Walker, R., & Storper, M. (1989). *The Capitalist Imperative: Territory, Technology and Industrial Growth*. Wiley-Blackwell.

Wamsler, C. (2016). From risk governance to city-citizen collaboration: Capitalizing on individual adaptation to climate change. *Environmental Policy and Governance, 26*(3), 184–204. doi:10.1002/eet.1707.

Ziervogel, G., Pelling, M., Cartwright, A., Chu, E., Deshpande, T., Harris, L., et al. (2017). Inserting rights and justice into urban resilience: A focus on everyday risk. *Environment and Urbanization, 29*(1), 123–138. doi:10.1177/0956247816686905.

Author Biographies

Eric K. Chu is a Lecturer in Planning and Human Geography in the School of Geography, Earth and Environmental Sciences at the University of Birmingham in the UK. His research is on the politics of climate change governance in cities, with particular emphasis on the globally comparative perspectives of socio-spatial change, development planning, policy change, and local environmental justice.

Sara Hughes is an Assistant Professor of Political Science at the University of Toronto and the University of Toronto-Mississauga. She is also a Senior Fellow at the Global Cities Institute and Bissell-Heyd Fellow at the Center for the Study of the United States at the University of Toronto. Her research focuses on urban politics, institutions, and policymaking for water resources, climate change mitigation and adaptation, waste management, and environmental justice.

Susan G. Mason is Emeritus Faculty in the School of Public Service at Boise State University. She was previously Director of Strategic Initiatives for the School of Public Service and Professor and Founding Director of the Department and Master and Certificate programs in Community and Regional Planning at Boise State University. Her research focuses on regional governance, urban and community development, and sustainability.

GPSR Compliance

The European Union's (EU) General Product Safety Regulation (GPSR) is a set of rules that requires consumer products to be safe and our obligations to ensure this.

If you have any concerns about our products, you can contact us on ProductSafety@springernature.com

In case Publisher is established outside the EU, the EU authorized representative is:

Springer Nature Customer Service Center GmbH
Europaplatz 3
69115 Heidelberg, Germany

Batch number: 10756376

Printed by Printforce, the Netherlands

Lecture Notes in Mechanical Engineering

ABCM Series on Mechanical Sciences and Engineering

This series publishes selected papers as well as full proceedings of events organized and/or promoted by the Brazilian Society for Mechanical Sciences and Engineering (ABCM) on an international level. These include the International Congress of Mechanical Engineering (COBEM) and the International Symposium on Dynamic Problems of Mechanics (DINAME), among others.

Paulo Kurka · Milton Pereira
Editors

Advances in Structural Vibration

Selected works from COBEM 2023

ABCM
Associação Brasileira de
Engenharia e Ciências Mecânicas

Springer

Editors
Paulo Kurka
Secretary of the Committee of Dynamics of ABCM
Universidade Estadual de Campinas
Campinas, São Paulo, Brazil

Milton Pereira
Chair of the Program Committee of COBEM 2023
Federal University of Santa Catarina
Florianópolis - SC, Brazil

ISSN 2195-4356 ISSN 2195-4364 (electronic)
Lecture Notes in Mechanical Engineering
ISSN 2662-3021 ISSN 2662-303X (electronic)
ABCM Series on Mechanical Sciences and Engineering
ISBN 978-3-031-71539-6 ISBN 978-3-031-71540-2 (eBook)
https://doi.org/10.1007/978-3-031-71540-2

This Springer imprint is published by the registered company Springer Nature Switzerland AG
The registered company address is: Gewerbestrasse 11, 6330 Cham, Switzerland

ABCM—Brazilian Society of Mechanical Sciences and Engineering

Foreword

Mechanical engineering is at the core of industrial development as it provides the knowledge and tools to design, manufacture, analyze, operate and maintain industrial equipment and systems, domestic, commercial, healthcare and communication devices and appliances, energy systems, vehicles, motors, actuators and robots, sensors and instruments, appliances, implants and prostheses and many other applications that support and facilitate our lives. COBEM is the most important congress promoted by the Brazilian Society of Mechanical Sciences and Engineering-ABCM and the largest technical meeting for mechanical engineering in South America. COBEM 2023 took place in Florianópolis, Brazil, from December 4 to 8, 2023, and I had the honor to be its president. During the event, eighteen symposia were held simultaneously, with 180 oral and 45 poster sessions, in which 801 papers and 444 posters were presented. The event was attended by 1419 participants. The congress not only provided a forum for the latest research in the specialist areas but was also intended to provide fertile ground for discussion, planning and new collaborations between complementary fields. It was one of the most exciting service activities I have experienced in my life as a university professor. The opportunity to provide a high-level technical environment to colleagues and students in the field of mechanical engineering and then to meet all the organizers, supporters, sponsors and participants of COBEM 2023 was a rewarding and memorable experience.

The ABCM Technical Committee on Dynamics is one of the longest established ABCM Technical Committees, dating back to the founding of ABCM in 1975. The Symposium on Dynamics, Control, Vibration and Acoustics was one of the 4 symposia that received the most submissions at COBEM 2023. There were 132 papers submitted in final format, of which 89 were presented orally and 43 in poster format. This is an exciting field, both from a scientific and applied perspective, which is present in various aspects of our society and makes a major contribution to our lives. The papers selected for this Symposia Series represent the quality, impact and novelty of the state-of-the-art contributions in this field. I would like to thank the coordinators of the symposium and the editors of this volume for their excellent work.

On behalf of the organizers of COBEM 2023, I would like to thank the ABCM members, staff and board for giving us this opportunity and supporting the organization of the event. Special thanks also go to the UFSC volunteers who did considerable work as local coordinators of symposia, as well as the student volunteers. Our sincere thanks also goes to the secretaries of the technical committees and the chairs of the technical sessions, without whose hard work and diligent cooperation the congress could not have taken place. Last but not least, we would like to thank the authors of the contributed papers—we sincerely hope that they benefited from the event.

Florianópolis, SC, Brazil
June 2024

Amir A. M. Oliveira Jr
President of COBEM 2023

Preface

Congresso Brasileiro de Engenharia Mecânica **(COBEM)** is a biennial conference organized by the Brazilian Association of Engineering and Mechanical Sciences (ABCM). The conference covers a wide range of topics related to mechanical sciences and engineering, as addressed by all the technical committees within ABCM.

During the 2023 COBEM meeting, the Committee of Dynamics of ABCM took the initiative to compile and publish a selection of works in book format. These works specifically focus on dynamic problems in mechanics. The chosen works demonstrate updated techniques and practical applications related to dynamics and vibration.

The aim of the book is to provide ready-to-use knowledge directed at scientists, engineering professionals and research students in the field of dynamics, vibration, signal processing and nonlinear analysis.

We hope this work contributes to the development of new ideas and applications. We encourage direct consultation with the authors of the book's sections for the exchange of technical information and the acquisition of research experience.

Campinas, Brazil — Paulo Kurka
Florianópolis, Brazil — Milton Pereira

Contents

Contributors

Jose Manoel Balthazar Department of Mechanical Engineering, Engineering School, São Paulo State University (UNESP), Bauru, Brazil;
Federal University of Technology—Paraná (UTFPR), Ponta Grossa, Brazil

Cláudio Henrique Cerqueira Costa Basquerotto Instituto de Geociências e Engenharias, Faculdade de Engenharia Mecânica, Universidade Federal do Sul e Sudeste do Pará (Unifesspa), Marabá, Pará, Brazil

Fernando Luiz Martinechen Beghetto UTFPR, Federal University of Technology—Paraná, Curitiba, Brazil

João Bezerra School of Mechanical Engineering, University of Campinas (UNICAMP), Campinas, Brazil

Luís Armando Ferreira Bispo Department of Mechanical Engineering, School of Engineering, UNESP, Bauru, Brazil

Vinícius Gabriel Peixoto Borges Department of Mechanical Engineering, University of Brasília, Brasília, Brazil

Michael John Brennan Department of Mechanical Engineering, School of Engineering, UNESP, Bauru, Brazil

Douglas D. Bueno São Paulo State University, Ilha Solteira, São Paulo, Brazil

Solidônio R. Carvalho Heat and Mass Transfer Laboratory, School of Mechanical Engineering, Federal University of Uberlandia, Uberlandia, Minas Gerais, Brazil

Daniel A. Castello Department of Mechanical Engineering, Universidade Federal Do Rio de Janeiro, Cidade Universitária—Ilha Do Fundão, Rio de Janeiro (RJ), Brazil

Adhemar Castilho LEDAV/COPPE/UFRJ, Federal University of Rio de Janeiro, Rio de Janeiro, Brazil

Marcelo S. Castro School of Mechanical Engineering, University of Campinas, Campinas-SP, Brazil;
Center for Petroleum Studies, University of Campinas, Campinas-SP, Brazil

Juliana R. Cenzi School of Mechanical Engineering, University of Campinas, Campinas-SP, Brazil

Hernán Darío Cerón-Muñoz São Carlos School of Engineering, University of São Paulo, São Carlos, Brazil

Rafael A. Cordeiro Electrical Engineering Department, Federal University of Espirito Santo (UFES), Vitória, Brazil

Vinicius da Silva Almeida Academic Unit of Mechanic Engineering, Federal University of Campina Grande, Campina Grande, Brazil

André Victor da Silva Castilho CEFET/RJ, Centro Federal de Educação Tecnológica Celso Suckow da Fonseca, Rio de Janeiro, Brazil;
UERJ, Universidade do Estado do Rio de Janeiro, Rio de Janeiro, Brazil

Nícolas da Silva Dias Department of Mechanical Engineering, University of Brasília, Brasília, Brazil

Agnaldo Antônio Moreira Teodoro da Silva University of Brasília, Brasília, DF, Brazil

Daniely A. das Neves Department of Mechanical Engineering, University of Brasília, Brasília-DF, Brazil

Felipe Lima de Abreu Federal University of Grande Dourados, Dourados, Mato Grosso do Sul, Brazil

Fabrício César Lobato de Almeida Department of Mechanical Engineering, School of Engineering, UNESP, Bauru, Brazil

Fernanda Carolina de Almeida Department of Mechanical Engineering, Faculty of Engineering, UNESP, Ilha Solteira, Brazil

Henrique Edno Leoncini de Carvalho Department of Mechanical Engineering, Faculty of Engineering, UNESP, Ilha Solteira, Brazil

Alexandre de Macêdo Wahrhaftig Department of Construction and Structures, Polytechnic School, Federal University of Bahia (UFBA), Salvador, Brazil

Marcus Vinicius Girão de Morais University of Brasilia, Gama, Brazil

Sillas de Oliveira Cézar Department of Mechanical Engineering, School of Engineering, UNESP, Bauru, Brazil

Vinicius Pires de Oliveira Doca Universidade Federal Do Paraná, Curitiba, Brazil

Eduardo Márcio de Oliveira Lopes Federal University of Paraná (UFPR), Curitiba, Brazil

Ely de Paiva School of Mechanical Engineering, University of Campinas (UNICAMP), Campinas, Brazil

Thiago de Paula Sales Mechanical Engineering Division, ITA—Aeronautics Institute of Technology, São José Dos Campos, SP, Brazil

Aline Souza de Paula Department of Mechanical Engineering, Faculty of Technology, University of Brasília, Brasília, Brazil

Igor Jordan Guilherme de Sena Academic Unit of Mechanic Engineering, Federal University of Campina Grande, Campina Grande, Brazil

Daniel Henrique de Sousa Obata Department of Mechanical Engineering, Faculty of Engineering, UNESP, Ilha Solteira, Brazil

Gabriel Ruggiero do Amaral Federal University of Paraná (UFPR), Curitiba, Brazil;
Mechanical Testing Laboratory, Institute of Technology for Development—Lactec, Curitiba, Brazil

Paulo Victor Gomes dos Santos CEFET/RJ, Centro Federal de Educação Tecnológica Celso Suckow da Fonseca, Rio de Janeiro, Brazil

Pedro Henrique Gomes dos Santos Rheimmetall, Rome, Italy

Nathali Dreher School of Mechanical Engineering, University of Campinas (UNICAMP), Campinas, SP, Brazil

Adriano Todorovic Fabro Department of Mechanical Engineering, University of Brasília, Brasília-DF, Brazil

Raul Dario Durand Farfan University of Brasilia, Gama, Brazil

Jhonny Feldhaus Norte Energia S.A.—NESA, Vitória do Xingu, Pará, Brazil

Murilo Cesar Filipus Federal University of Grande Dourados, Dourados, Mato Grosso do Sul, Brazil

Jeferson Camargo Fukushima Department of Mechanical Engineering, Faculty of Engineering, UNESP, Ilha Solteira, Brazil

Andressa Caroline Rodrigues Garcia São Carlos School of Engineering, University of São Paulo, São Carlos, Brazil

Gabriel Guimarães de Souza Braga de Albuquerque Department of Mechanical Engineering, University of Brasília, Brasília, Brazil

Maurício Kiotsune Iwanaga Department of Mechanical Engineering, Faculty of Engineering, UNESP, Ilha Solteira, Brazil

Mauro F. Koyama School of Mechanical Engineering, University of Campinas (UNICAMP), Campinas, Brazil

Flávio Issao Kubota Universidade Federal Do Paraná, Curitiba, Brazil

Tiago Machado School of Mechanical Engineering, University of Campinas (UNICAMP), Campinas, SP, Brazil

Niederauer Mastelari School of Mechanical Engineering, University of Campinas (UNICAMP), Campinas, Brazil

Vinícius Mauro de Souza Santos Mechanical Engineering Division, ITA—Aeronautics Institute of Technology, São José Dos Campos, SP, Brazil; SUPMICROTECH, Institut FEMTO-ST, Université de Franche-Comté, CNRS, Besançon, France

Vítor S. Medeiros School of Mechanical Engineering, Federal University of Uberlandia, Uberlandia, Minas Gerais, Brazil

Juan Antonio Flores Mezarina São Carlos School of Engineering, University of São Paulo, São Carlos, Brazil

Ulisses A. V. Monteiro LEDAV/COPPE/UFRJ, Federal University of Rio de Janeiro, Rio de Janeiro, Brazil

Suzana Regina Moro Universidade Estadual de Campinas, São Paulo, Brazil

Morvan Ouisse SUPMICROTECH, Institut FEMTO-ST, Université de Franche-Comté, CNRS, Besançon, France

Amarildo Tabone Paschoalini Department of Mechanical Engineering, Faculty of Engineering, UNESP, Ilha Solteira, Brazil

Thomas Paulsen Department of Civil and Mechanical Engineering, Technical University of Denmark (DTU), Lyngby, Denmark

João Antonio Pereira Faculdade de Engenharia, Universidade Estadual Paulista (UNESP), Ilha Solteira, São Paulo, Brazil

Luiz A. Vaz Pinto LEDAV/COPPE/UFRJ, Federal University of Rio de Janeiro, Rio de Janeiro, Brazil

Matheus Silva Proença Department of Mechanical Engineering, Faculty of Engineering, UNESP, Ilha Solteira, Brazil

Kevin Mauricio Menon Ribeiro Federal University of Paraná (UFPR), Curitiba, Brazil

Thiago G. Ritto Department of Mechanical Engineering, Universidade Federal Do Rio de Janeiro, Cidade Universitária—Ilha Do Fundão, Rio de Janeiro (RJ), Brazil

Marcela Rodrigues Machado Department of Mechanical Engineering, University of Brasília, Brasília, Brazil

Ilmar Santos Department of Civil and Mechanical Engineering, Technical University of Denmark (DTU), Lyngby, Denmark

Marcelo B. Santos Mechanical System Laboratory, School of Mechanical Engineering, Federal University of Uberlandia, Uberlandia, Minas Gerais, Brazil

Rodrigo B. Santos Federal University of Grande Dourados, Dourados, Mato Grosso do Sul, Brazil

Oscar Scussel Department of Mechanical Engineering, Faculty of Engineering, UNESP, Ilha Solteira, Brazil

Richard Senko Academic Unit of Production Engineering, Federal University of Campina Grande, Campina Grande, Brazil

Antônio Almeida Silva Academic Unit of Mechanic Engineering, Federal University of Campina Grande, Campina Grande, Brazil

Antonio C. Troyman LEDAV/COPPE/UFRJ, Federal University of Rio de Janeiro, Rio de Janeiro, Brazil

Antônio Eduardo Turra Faculdade de Engenharia, Universidade Estadual Paulista (UNESP), Ilha Solteira, São Paulo, Brazil

Sandro C. Valente Department of Mechanical Engineering, Universidade Federal Do Rio de Janeiro, Cidade Universitária—Ilha Do Fundão, Rio de Janeiro (RJ), Brazil

Felipe W. Varga School of Mechanical Engineering, University of Campinas (UNICAMP), Campinas, Brazil

Saon C. Vieira PETROBRAS, Santos-SP, Brazil

Laís Bitencourt Visnadi Department of Mechanical Engineering, University of Brasília, Brasília, Brazil

Pedro Falcomer Pontes Viégas University of Brasilia, Gama, Brazil

Gabriel Weiss Mattioli Universidade Federal Do Paraná, Curitiba, Brazil

Chapter 1
Investigation of a Metastructure for Vibration Attenuation in a Crankshaft with Non-constant Inertia

Nícolas da Silva Dias and Aline Souza de Paula

Abstract Torsional vibrations are inherent in the operation of reciprocating machines, which is one of the leading causes of mechanical failures. This work analyzes the torsional vibration attenuation of a crankshaft of a single-cylinder reciprocating compressor when identical resonators are attached. The shaft is modeled by a continuous nonlinear system, considering the inertia variation of the crank-slider mechanism with rotation and under a concentrated harmonic excitation. Two main configurations are presented, without and with resonators. At first, the nonlinear and linearized systems are compared for two different crank length values. Results show that the systems present similar behaviors for a low value of the crank length. On the other hand, when the crank lengths are raised, the resonance frequencies are slightly shifted backward and the response exhibits nonlinear effects. The second case presents the system response when the resonators are attached. The performance of the resonators is evaluated for two crank length values as in the previous analysis. In both cases, the resonators can reduce the vibration amplitudes in the neighborhood of the target frequency, even when the nonlinearity is increased in the system response.

Keywords Torsional vibration · Vibration attenuation · Identical resonators · Non-constant inertia

N. da S. Dias (✉) · A. S. de Paula
University of Brasília, Faculty of Technology, Department of Mechanical Engineering, Brasília-DF, Brazil
e-mail: nicolas-silva.ns@aluno.unb.br

A. S. de Paula
e-mail: alinedepaula@unb.br

P. Kurka and M. Pereira (eds.), *Advances in Structural Vibration*, Lecture Notes in Mechanical Engineering, https://doi.org/10.1007/978-3-031-71540-2_1

1.1 Introduction

Vibrations are an intrinsic phenomenon in rotating machines (Chandrashekar et al. 2018). Typically, these vibrations manifest during startup, continuous operation, or shutdown (Wachel and Szenasi 1993), being the leading causes of coupling failures, broken shafts, and worn gears, among others (Corbo and Melanoski 1996). They are caused by the excessive forces and moments that vary cyclically with the periodicity given by the rotational speed of the system (Wachel and Tison 1994). In this context, vibration control is one important subject, and it is studied extensively by the scientific community to ensure a long service life of these mechanical equipment.

Over the years, several researchers have proposed dynamic models for a better understanding of the bending and torsional vibrations in rotating systems. In the case of machines with crankshafts, such as internal combustion engines and reciprocating compressors, the torsional vibration analysis is more relevant. This is because the rotation of the shaft initiates the compression inside the cylinder performed by the crank sliding mechanism, which undergoes a periodic reciprocating motion resulting in vibration and noise (Liu et al. 2019).

The first available studies in the literature consider discrete linear models with lumped parameters (Li et al. 2022). However, as they are simplified, they often cannot adequately describe the system dynamics. It is common to assume that the crank-slider mechanism has constant inertia, but this approximation can lead to significant errors (Pasricha and Carnegie 1979). The variable inertia of the mechanism enhances the realism in representing dynamic phenomena, as stated by the authors Pasricha and Carnegie (1976, 1979), Brusa et al. (1997), Metallidis and Natsiavas (2003), Huang et al. (2012), and results in nonlinear equations of motion. This nonlinearity gives rise to the emergence of a second resonance caused by the coupling of natural modes of the system (Draminsky 1965; Hesterman and Stone 1994, 1995).

In general, dynamical systems can be described by discrete or continuous models. The continuous modeling of the torsional vibration of crankshafts can be achieved by considering a uniform shaft, as defined by the classical problem (Inman 2007), or non-uniform, assuming the variation of torsional stiffness along its length (Li 2003; Wu 2007). The crank-slider mechanism can be modeled as concentrated inertia elements introduced into the problem by boundary conditions (Qiao et al. 2002) or directly in the governing equation (Gürgöze 1984; Wu and Lin 1990; Peng and Yan 2022). Nevertheless, there is a lack of research covering the nonlinear model with variable inertia. Typically, the system is linearized to simplify the dynamic analysis (Koser and Pasin 1995, 1997).

This work presents an analysis of vibration attenuation of the torsional vibration of a single-cylinder reciprocating compressor's crankshaft using metamaterials, showing the effects of the nonlinearity of inertia of the crank sliding mechanism. This paper is structured as follows. The next section describes the equations of motion of the crankshaft with the attached concentrated inertia, and its analytical solution uses the modal expansion theory. Two configurations are presented: without and with identical resonators. Then, numerical results are presented and discussed, showing

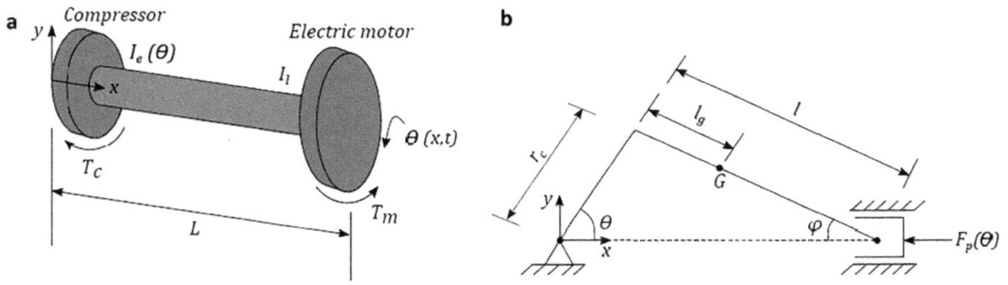

Fig. 1.1 Model of a single-cylinder reciprocating compressor: **a** simplified model and **b** crank-slider mechanism

the system response under a concentrated harmonic excitation for two cases. In the first case, the crankshaft response shows the effects of the non-constant inertia. On the other hand, the resonators attached produce a vibration attenuation located in a specific frequency range. Finally, the conclusions section summarizes the main achievements of this work.

Crankshaft Model and Response Using Modal Coordinates

The discrete system proposed by Metallidis and Natsiavas (2003) divides reciprocating machines into two components, the first represents the compressor itself, and the second is the load. In this context, the authors make sense of the use of load in the representation of the model, because it is about an internal combustion engine. However, herein this term is adapted for compressor systems. This work considers two-concentrated inertia, one representing the crank-slider mechanism inertia of the compressor that varies with rotation, $I_e(\theta)$ and one constant representing the load, I_l. These two elements are connected to a shaft represented by a continuous system. Two external torques are considered, the cylinder resistive torque is described by T_c, and the electric motor torque is T_m, as is shown in Fig. 1.1a. The compressor inertia is determined by using the equivalent crank-slider mechanism inertia, which is represented by Fig. 1.1b.

1.1.1 *Mathematical Model*

The crankshaft has a diameter d, Young's modulus of E, Poisson's ratio ν, shear's modulus $G = E/2(1+\nu)$, cross-section polar moment of inertia $J = \pi d^2/32$ and density ρ. The equation of motion is based on the problem of torsional vibration of an undamped shaft with a uniform cross-section and constant torsional stiffness GJ. The lumped inertia $I_e(\theta)$ is introduced into the equation together with an excitation torque $f(x, t)$ as follows:

$$\rho J\frac{\partial^2\theta(x,t)}{\partial t^2} + I_e(\theta)\delta(x-x_c)\frac{\partial^2\theta(x,t)}{\partial t^2} - \mathrm{GJ}\frac{\partial^2\theta(x,t)}{\partial x^2} = f(x,t) \qquad (1.1)$$

The system is free at the end where the variable inertia is anchored, and the constant inertia I_l is attached at the opposite shaft extremity, thus yielding to the boundary conditions (Inman 2007):

$$\frac{\partial\theta(0,t)}{\partial x} = 0,\ \mathrm{GJ}\frac{\partial\theta(L,t)}{\partial x} = -I_l\frac{\partial^2\theta(L,t)}{\partial t^2} \tag{1.2}$$

The variable cylinder inertia, $I_e(\theta)$, and its linearized form, I_0, are obtained by Metallidis and Natsiavas (2003):

$$I_e(\theta) = I_c + m_1 r_c^2 + I_2\kappa^2(\theta) + \left(m_2 + m_p\right) r_c^2(\theta)[1 + \kappa(\theta)]^2, \tag{1.3}$$

$$I_0 = I_c + m_1 r_c^2 + \frac{1}{2}\left(m_2 + m_p\right) r_c^2, \tag{1.4}$$

being the parameters m_1, m_2, κ and λ defined by the following expressions:

$$m_1 = \left(1 - \frac{l_g}{l}\right) m_r\ m_2 = \frac{l_g}{l} m_r\ I_2 = I_g - m_1 l_g^2 - m_2\left(l - l_g\right)^2 \tag{1.5}$$

$$\kappa(\theta) = \frac{\lambda\cos\cos(\theta)}{\sqrt{1 - \lambda^2(\theta)}}\ \text{and}\ \lambda = \frac{r_c}{l} \tag{1.6}$$

where I_2 is the additional inertia from a simplified double-mass system, I_c is the moment of inertia of the crank, I_g is the moment of inertia of the connecting rod center of mass, m_1 is the mass of the big end, m_2 is the mass of the small end, m_p is the piston mass, m_r is the mass of the connecting rod, l is the total length of the connecting rod, l_g is the length of the connecting rod center of mass, r_c is the crank length, $\delta(x)$ is the Dirac delta function, x_c is the cylinder position, x_l is the load position, and x_m is the electric motor position.

The variable inertia $I_e(\theta)$ can also be expressed as:

$$I_e(\theta) = I_0 + I_{nl}(\theta) \tag{1.7}$$

where $I_{nl}(\theta) = I_c + m_1 r_c^2 + I_2\kappa^2(\theta) + \left(m_2 + m_p\right) r_c^2(\theta)[1 + \kappa(\theta)]^2 - I_c + m_1 r_c^2 + \frac{1}{2}\left(m_2 + m_p\right) r_c^2$ corresponds to the nonlinear part.

Using the newly defined quantities, Eqs. (1.1) and (1.2) can also be expressed in the alternative form:

$$\rho J\frac{\partial^2\theta(x,t)}{\partial t^2} + I_{nl}(\theta)\delta(x - x_c)\frac{\partial^2\theta(x,t)}{\partial t^2} - \mathrm{GJ}\frac{\partial^2\theta(x,t)}{\partial x^2} = f(x,t), \tag{1.8}$$

and the boundary condition are defined by:

$$\mathrm{GJ}\frac{\partial\theta(0,t)}{\partial x}=I_0\frac{\partial^2\theta(0,t)}{\partial t^2}\ \mathrm{GJ}\ \frac{\partial\theta(L,t)}{\partial x}=-I_1\frac{\partial^2\theta(L,t)}{\partial t^2}, \tag{1.9}$$

The natural frequencies and modes are obtained using Eq. (1.8) disregarding the nonlinear inertia term $I_{\mathrm{nl}}(\theta)$ and the external excitations.

1.1.2 Solution Using Modal Expansion

The crankshaft response is determined employing a modal expansion theory as addressed by Meirovitch (1997). Despite the nonlinear nature of the system under investigation, the dynamical response is assessed using modal expansion, assuming no nonlinear modal interactions, as considered by Xia et al. (2020). The torsional displacement of the shaft is:

$$\theta(x,t)=\sum_{r=1}^{N}\eta_{\mathrm{r}}(t)\phi_{\mathrm{r}}(x), \tag{1.10}$$

where $\eta_{\mathrm{r}}(t)$ is the modal coordinate of the r-th mode, $\phi_{\mathrm{r}}(x)$ is r-th mode shape of the shaft with the concentrated inertia at the boundaries I_1 and I_0, N is the total number of modes considered in the modal expansion. The modes shapes are inertia-normalized by Eq. 1.16 yields in:

$$\phi_r(x)=A_2\left(\cos\left(\frac{\beta_{\mathrm{r}}x}{L}\right)-\frac{\beta_{\mathrm{r}}I_0}{\rho J}\sin\left(\frac{\beta_r x}{L}\right)\right),\quad r=1,2,3,\ldots,N. \tag{1.11}$$

with

$$A_2=\sqrt{\frac{2}{\rho\mathrm{JL}\left(1+\left(\frac{\beta_{\mathrm{r}}I_0}{\rho J}\right)^2\right)}}, \tag{1.12}$$

where β_r is the r-th positive and real solution of the characteristic equation given by:

$$(\beta L)=\frac{\rho\mathrm{JL}(I_0+I_1)\beta L}{I_0 I_{\mathrm{L}}(\beta L)^2-\rho^2\mathrm{J}^2\mathrm{L}^2}. \tag{1.13}$$

Introducing the Eq. 1.10 in Eq. 1.8, assuming that the system is under a concentrated harmonic excitation $f(x,t)=F\delta(x-x_{\mathrm{e}})e^{i\omega t}$ located in the excitation position x_{e}, one obtains:

$$\sum_{r=1}^{N}\rho J\,\ddot{\eta}_{\mathrm{r}}(t)\phi_{\mathrm{r}}(x)+\sum_{r=1}^{N}I_{\mathrm{nl}}(\theta)\delta(x-x_c)\ddot{\eta}_{\mathrm{r}}(t)\phi_{\mathrm{r}}(x)$$

$$-\sum_{r=1}^{N} \mathrm{GJ}\eta_{\mathrm{r}}(t)\phi_r''(x) = F\delta(x - x_{\mathrm{e}})e^{i\omega t}. \tag{1.14}$$

Multiplying the Eq. 1.14 by $\phi_s(x)$ and integrating over the domain from $x = 0$ to $x = L$, results in:

$$\sum_{r=1}^{N} \int_0^L \rho J \ddot{\eta}_{\mathrm{r}}(t)\phi_{\mathrm{r}}(x)\phi_{\mathrm{s}}(x)dx + \sum_{r=1}^{N} \int_0^L I_{\mathrm{nl}}(\theta)\delta(x - x_{\mathrm{c}})\ddot{\eta}_{\mathrm{r}}(t)\phi_{\mathrm{r}}(x)\phi_{\mathrm{s}}(x)\mathrm{d}x$$
$$-\sum_{r=1}^{N} \int_0^L \mathrm{GJ}\eta_{\mathrm{r}}(t)\phi_{\mathrm{r}}''(x)\phi_{\mathrm{s}}(x)\mathrm{d}x = \int_0^L F\phi_{\mathrm{s}}(x)\delta(x - x_{\mathrm{e}})e^{i\omega t}\mathrm{d}x. \tag{1.15}$$

The orthonormal conditions are described below:

$$\int_0^L \rho J \phi_{\mathrm{r}}(x)\phi_{\mathrm{s}}(x)\mathrm{d}x = \delta_{rs}, \quad r, s = 1, 2, \ldots \tag{1.16}$$

$$\int_0^L \mathrm{GJ}\phi_{\mathrm{r}}''(x)\phi_{\mathrm{s}}(x)\mathrm{d}x = -\omega_{\mathrm{s}}^2\delta_{rs}, \quad r, s = 1, 2, \ldots \tag{1.17}$$

where ω_{s} is the natural frequency of the s-th mode of the shaft with concentrated inertia at the boundaries I_1 and I_0, and $()'$ denotes the derivative with respect to the space variable x.

Applying the orthonormal conditions represented by Eqs.1.16 and 1.17 one obtains the undamped inertia-coupled equations of motion in modal coordinates:

$$\ddot{\eta}_{\mathrm{s}}(t) + I_{\mathrm{nl}}\left(\theta_{x_{\mathrm{c}}}\right)\phi_{\mathrm{s}}(x_{\mathrm{c}}) \sum_{r=1}^{N} \phi_{\mathrm{r}}(x_{\mathrm{c}})\ddot{\eta}_{\mathrm{r}}(t) + \omega_{\mathrm{s}}^2\eta_{\mathrm{s}}(t) = F\phi_{\mathrm{s}}(x_{\mathrm{e}})e^{i\omega t},$$
$$s = 1, 2, \ldots, N. \tag{1.18}$$

where $\theta_{x_{\mathrm{c}}} = \theta(x_{\mathrm{c}}, t)$.

Rewritten Eq. 1.18 in matrix form one obtains:

$$\begin{bmatrix} 1+I_{\mathrm{nl}}(\theta_{x_c})\phi_1^2(x_c) & I_{\mathrm{nl}}(\theta_{x_c})\phi_1(x_c)\phi_2(x_c) & \ldots & I_{\mathrm{nl}}(\theta_{x_c})\phi_1(x_c)\phi_N(x_c) \\ I_{\mathrm{nl}}(\theta_{x_c})\phi_2(x_c)\phi_1(x_c) & 1+I_{\mathrm{nl}}(\theta_{x_c})\phi_2^2(x_c) & \ldots & I_{\mathrm{nl}}(\theta_{x_c})\phi_2(x_c)\phi_N(x_c) \\ \vdots & \vdots & \ddots & \vdots \\ I_{\mathrm{nl}}(\theta_{x_c})\phi_N(x_c)\phi_1(x_c) & I_{\mathrm{nl}}(\theta_{x_c})\phi_N(x_c)\phi_2(x_c) & \ldots & 1+I_{\mathrm{nl}}(\theta_{x_c})\phi_N^2(x_c) \end{bmatrix}$$
$$\{\ddot{\eta}(t)\} + \begin{bmatrix} \omega_1^2 & 0 & 0 & \ldots & 0 \\ 0 & \omega_2^2 & 0 & \ldots & 0 \\ 0 & 0 & \omega_3^2 & \ldots & 0 \\ \vdots & \vdots & \vdots & \ddots & \vdots \\ 0 & 0 & 0 & \ldots & \omega_N^2 \end{bmatrix} \{\eta(t)\} = \begin{Bmatrix} F\phi_1(x_e)e^{i\omega t} \\ F\phi_2(x_e)e^{i\omega t} \\ F\phi_3(x_e)e^{i\omega t} \\ \vdots \\ F\phi_{\mathrm{N}}(x_e)e^{i\omega t} \end{Bmatrix}. \tag{1.19}$$

The inertia matrix is approximate as

$$\begin{bmatrix} 1+I_{\mathrm{nl}}(\theta_{x_c})\phi_1^2(x_c) & I_{\mathrm{nl}}(\theta_{x_c})\phi_1(x_c)\phi_2(x_c) & \ldots & I_{\mathrm{nl}}(\theta_{x_c})\phi_1(x_c)\phi_N(x_c) \\ I_{\mathrm{nl}}(\theta_{x_c})\phi_2(x_c)\phi_1(x_c) & 1+I_{\mathrm{nl}}(\theta_{x_c})\phi_2^2(x_c) & \ldots & I_{\mathrm{nl}}(\theta_{x_c})\phi_2(x_c)\phi_N(x_c) \\ \vdots & \vdots & \ddots & \vdots \\ I_{\mathrm{nl}}(\theta_{x_c})\phi_N(x_c)\phi_1(x_c) & I_{\mathrm{nl}}(\theta_{x_c})\phi_N(x_c)\phi_2(x_c) & \ldots & 1+I_{\mathrm{nl}}(\theta_{x_c})\phi_N^2(x_c) \end{bmatrix} \approx$$
$$\begin{bmatrix} 1+I_{\mathrm{nl}}(\theta_{x_c})\phi_1^2(x_c) & 0 & \ldots & 0 \\ 0 & 1+I_{\mathrm{nl}}(\theta_{x_c})\phi_2^2(x_c) & \ldots & 0 \\ \vdots & \vdots & \ddots & \vdots \\ 0 & 0 & \ldots & 1+I_{\mathrm{nl}}(\theta_{x_c})\phi_N^2(x_c) \end{bmatrix}. \tag{1.20}$$

The higher order mode shapes evaluated in the position $x_c = 0$ are neglecting since when $\beta_r \to \infty$, leads to $\phi_r(x_c) \to 0$. However, the lower mode shapes outside of the main diagonal are also not considered in the formulation. This simplification will introduce errors in the system response. Assuming the hypothesis described by the Eq. 1.20, since the Eq. 1.19 requires a higher computational cost solution. The undamped equations of motion are then uncoupled in modal coordinates as follows:

$$\ddot{\eta}_s(t) + I_{\mathrm{nl}}(\theta_{xc})\phi_s^2(x_c)\ddot{\eta}_s(t) + \omega_s^2\eta_s(t) = F\phi_s(x_e)e^{i\omega t},$$
$$s = 1, 2, 3, \ldots, N. \tag{1.21}$$

1.1.3 Crankshaft With Identical Resonators

This section describes the formulation when resonators are attached to the crankshaft employing the modal expansion theory. The attachments are identical and tuned at

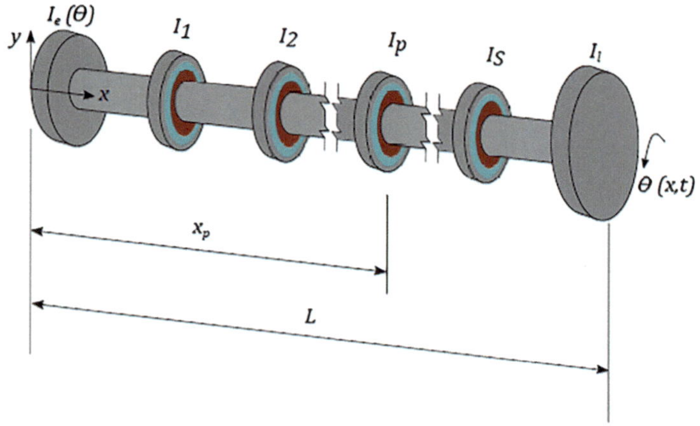

Fig. 1.2 Crankshaft with identical resonators

the same frequency, to reduce the vibration amplitude in a specific frequency range. Figure 1.2 shows the crankshaft with evenly distributed identical resonators.

According to the work proposed by Sugino et al. (2017), the equation of motion of the crankshaft with evenly distributed identical resonators attached is represented by:

$$\rho J \frac{\partial^2\theta(x,t)}{\partial t^2} + I_{\mathrm{nl}}(\theta)\delta(x-x_{\mathrm{c}})\frac{\partial^2\theta(x,t)}{\partial t^2} - \mathrm{GJ}\frac{\partial^2\theta(x,t)}{\partial x^2} - \sum_{p=1}^{S} k_{\mathrm{p}}\theta_{\mathrm{p}}(t)\delta\left(x-x_{\mathrm{p}}\right) = f(x,t), \tag{1.22}$$

and each resonator has the following equation of motion given as:

$$I_{\mathrm{p}}\ddot{\theta}_{\mathrm{p}}(t) + k_{\mathrm{p}}\theta_{\mathrm{p}}(t) + I_{\mathrm{p}}\ddot{\theta}\left(x_{\mathrm{p}},t\right) = 0,\ p = 1, 2, 3, \ldots, S \tag{1.23}$$

where k_{p} is the torsional stiffness of the p-th resonator, I_{p} is the inertia of the p-th resonator, $\theta_{\mathrm{p}}(t)$ is the torsional relative displacement of the p-th resonator, x_{p} is the position of the p-th resonator, S is the total number of resonators and $f(x,t)$ is the external torque density.

As assumed before, that the crankshaft is subjected to a concentrated harmonic excitation $f(x,t) = F\delta(x-x_{\mathrm{e}})e^{i\omega t}$. Using again the modal expansion theory with N modes as $\theta(x,t) = \sum_{r=1}^{N}\eta_{\mathrm{r}}(t)\phi_{\mathrm{r}}(x)$, neglecting the nonlinear modal interactions, and employing the orthonormalization conditions defined by Eqs. 1.16, 1.17, we obtain a set of $N+S$ coupled equation of motions:

$$\ddot{\eta}_{\mathrm{s}}(t) + I_{\mathrm{nl}}(\theta)\phi_{\mathrm{s}}^2(x_{\mathrm{c}})\ddot{\eta}_{\mathrm{s}}(t) + \omega_{\mathrm{s}}^2\eta_{\mathrm{s}}(t) - \sum_{p=1}^{S} k_{\mathrm{p}}\theta_{\mathrm{p}}(t)\phi_{\mathrm{s}}\left(x_{\mathrm{p}}\right)$$

$$= F\phi_s(x_e)e^{i\omega t}, \ s = 1, 2, 3, \ldots, N \tag{1.24}$$

$$\ddot{\theta}_p(t) + \omega_p^2\theta_p(t) + \sum_{s=1}^{N} \ddot{\eta}_s(t)\phi_s(x_p) = 0, \ p = 1, 2, 3, \ldots, S \tag{1.25}$$

where ω_p is the natural frequency of the p-th resonator, being equal for all of them.

1.1.4 Frequency Domain Sweep

This section presents a numerical procedure for determining the response of a dynamical system in the frequency domain, as there is no closed-form expression for the frequency response function (FRF) in the case of nonlinear systems. The FRF is obtained by calculating the FFT (Fast Fourier Transform) from a temporal response of a harmonic excitation sweep. The crankshaft is subject to a harmonic excitation with frequency ω, which is considered a range of frequency. From the uncouple equations in modal coordinates (Eq. 1.21), introducing modal damping ξ_s, one obtains:

$$\begin{aligned} &\ddot{\eta}_s(t) + I_{nl}(\theta_{x_c})\phi_s^2(x_c)\ddot{\eta}_s(t) + 2\xi_s\omega_s\dot{\eta}_s(t) + \omega_s^2\eta_s(t) \\ &= F\phi_s(x_e)e^{i\omega t} \ s = 1, 2, 3, \ldots, N \end{aligned} \tag{1.26}$$

where the excitation frequency ω varied linearly over time as described by

$$\omega = \frac{\omega_f - \omega_0}{t_s}t + \omega_0 \tag{1.27}$$

where ω_0 is the initial frequency, ω_f is the final excitation frequency and t_s is the total time for the numerical simulation.

The modal coordinates $\eta_s(t)$ and the harmonic excitation $f(x, t)$ are obtained from the employed fourth-order Runge–Kutta numerical method in the state equations:

$$\begin{aligned} \{s_e\} &= \begin{Bmatrix} \eta_s(t) \\ \dot{\eta}_s(t) \end{Bmatrix}, \ \{\dot{s}_e\} = \begin{Bmatrix} \dot{\eta}_s(t) \\ \ddot{\eta}_s(t) \end{Bmatrix} \\ &= \begin{Bmatrix} \dot{\eta}_s(t) \\ \frac{F\phi_s(x_e)e^{i\omega t} - 2\xi_s\omega_s\dot{\eta}_s(t) - \omega_s^2\eta_s(t)}{1 + I_{nl}(\theta_{x_c})\phi_s^2(x_c)} \end{Bmatrix}, \ \text{with } s = 1, 2, 3, \ \ldots, N. \end{aligned} \tag{1.28}$$

Therefore, the FRF can be determined in the response position x_r using the FFT from the following relation:

$$\alpha_{re}(\omega) = \frac{\text{FFT}[\theta(x_r, t)]}{\text{FFT}[f(x_e, t)]} = \frac{\text{FFT}\left[\sum_{s=1}^{N} \eta_s(t)\phi_s(x_r)\right]}{\text{FFT}\left[F\phi_s(x_e)e^{i\omega t}\right]} \tag{1.29}$$

This approach follows the same path when identical resonators are attached to the crankshaft.

1.2 Results and Discussion

This section shows the steady-state responses in the frequency domain when the crankshaft is subject to harmonic excitation, as well as evaluating the vibration reduction produced by the identical resonators. Two main configurations are presented, in the first without resonators, the nonlinear model is compared with the linearized one. In the second case, the resonators are attached to the system to attenuate torsional vibrations in a specific frequency range. The simulations are carried out for different values of the non-constant inertia $I_{nl}(\theta)$, to explore the nonlinear effects in the crankshaft response when the nonlinearity is increased. Table 1.1 presents the system's parameters used in the numerical investigations, which are based on the works of Huang et al. (2012) and Metallidis and Natsiavas (2003).

In the first case, the dynamic analysis compares the nonlinear and linear systems under harmonic excitation without resonators. The equations of motion (Eqs. 1.27) are numerically integrated employing the fourth-order Runge–Kutta method, with the

Table 1.1 System parameters

Inertia of the crank, I_c	0.01783 kg.m^2
Inertia of the connecting rod center of mass, I_g	0.0458 kg.m^2
Mass of the connecting rod, m_r	3.8 kg
Mass of the piston, m_p	4.026 kg
l_g/l	0.29
Crank length, r_c	0.0725 m
Length of the shaft, L	1 m
Diameter of the shaft, d	0.06 m
Young's modulus of the shaft, E	221 GPa
Density of the shaft, ρ	7800 kg/m^3
Poisson's ratio of the shaft, ν	0.3
Inertia of the load, I_l	0.165 kg.m^2
Connecting rod length, l	0.262 m
Modal damping of the shaft, ξ_s	0.002
Harmonic excitation amplitude, F	1×10^8 N.m
Initial harmonic excitation frequency, ω_0	1 Hz
Final harmonic excitation frequency, ω_f	18,000 Hz
Total time for the numerical simulation, t_s	1 s
Number of modes in modal expansion, N	10
Number of attached resonators, S	10

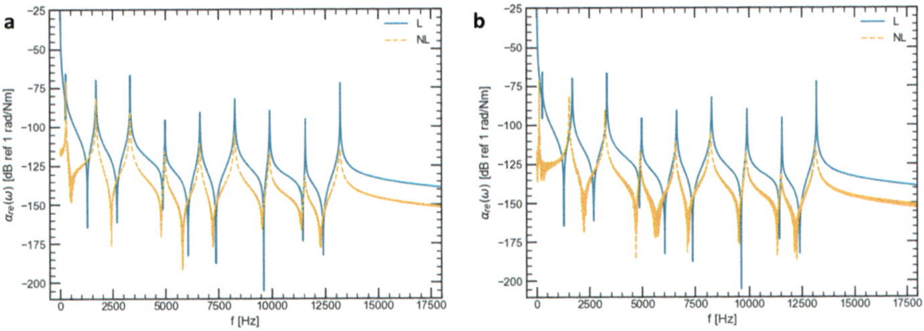

Fig. 1.3 FRFs of linear (L) and nonlinear (NL) systems for different crank lengths r_c: **a** r_c = 72.5 mm and **b** r_c = 200 smm

incremental step $h = 6.283 \times 10^{-7}$ s and the initial conditions $\eta_2(0) = \eta_3(0) = \ldots = \eta_{10}(0) = 0$ rad/m, $\dot{\eta}_2(0) = \dot{\eta}_3(0) = \ldots = \dot{\eta}_{10}(0) = 0$ rad/m.s. It is worth mentioning that the simulations consider only non-zero natural frequencies and mode shapes associated, thus, in the modal expansion presented in Eq. (1.10), the summation start from 2.

Figure 1.3 shows system responses for two crank length values. The continuous blue line represents the response of the linearized system (L), while the dashed orange line is the nonlinear (NL) system behavior. For lower crank length, r_c = 72.5 mm (Fig. 1.3a), we observe that resonance frequencies match when compared to the linear one. However, some antiresonance peaks are shifted due to the neglect of the first mode in the modal expansion. When the crank length is raised to r_c = 200 mm (Fig. 1.3b), the resonance frequencies shifted slightly backward due to the increase of the nonlinearities.

The second case presents the configuration when identical resonators are attached to the system. Figure 1.4 shows system responses without resonators represented by the continuous blue line and with attached resonators described by the dashed orange line. The shaded region is the estimated linear local bandgap for finite and continuous metastructures expressed by Sugino et al. (2017) as:

$$\omega_t < \omega < \omega_t\sqrt{1+\mu}, \tag{1.30}$$

where ω_t is the target frequency of the resonators and μ is the inertia ratio of the total inertia of the resonators to the total inertia of the crankshaft as follows:

$$\mu = \frac{\sum_{p=1}^{S} I_p}{\rho JL} = \frac{I_p S}{\rho JL}. \tag{1.31}$$

The numerical investigations assume that the inertia ratio μ = 0.3 and the resonators are tuned in ω_t = 4967.154 Hz, which corresponds to the fifth resonance peak. As in the previous case, the performance of the resonators is analysed for two crank length values. In both cases, for a low and high value of the

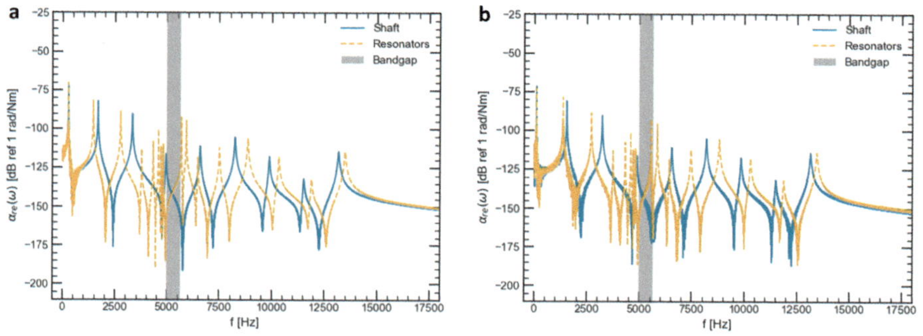

Fig. 1.4 FRFs of the nonlinear system with identical resonators for different crank lengths r_c: **a** $r_c = 72.5$ mm and **b** $r_c = 200$ mm

crank length $r_c = 72.5$ mm (Fig. 1.4a) and $r_c = 200$ mm (Fig. 1.4b), respectively, the metastructure attenuates the torsional vibrations producing a bandgap with bandwidth according to Eq. 1.30, even when the nonlinearities are increased. As mentioned before, the antiresonance peaks near of the first natural frequency are shifted. However, since the tuning frequency of the resonators is far away from this region, this will not influence in the performance of the metamaterials.

The procedure used to obtain the FRF of the nonlinear system is an alternative for a preliminary analysis with low computational cost. However, it does not capture the richness of details of the dynamic behaviour. This approach was employed as an initial assessment to identify situations where nonlinearities play a significant role. In these situations, it is necessary to evaluate the response in more detail using other suitable tools that can identify various aspects, such as the type of system response, the occurrence of dynamical jumps, and other relevant aspects.

1.3 Conclusions

This work investigates the torsional vibration attenuation in a crankshaft of a reciprocating compressor with a single-cylinder using identical resonators. The shaft is modelled by a continuous nonlinear system, considering the inertia variation of the crank-slider mechanism with the rotation $I_e(\theta)$, and subjected to a concentrated harmonic excitation. Two main configurations were considered, without and with the attached resonators. In the first case, the responses of the nonlinear and the linearized systems are compared considering two different crank length values r_c. The results show that for a low value of the crank length, the resonance frequencies match while the antiresonance peaks are shifted when compared to the linearized case, due the absence of the first natural frequency that it is null. On the other hand, when the crank length is increased, the resonance frequencies are slightly moved backward and the nonlinear effects are evident in the response. In the second case, the performance of the resonators attached to the crankshaft is evaluated for two crank length values as in the previous analysis. Results showed that the resonators successfully reduced the vibration amplitudes in the region near the target frequency, even when the nonlinearity is raised in the system response.

Acknowledgements The authors are grateful to the Brazilian National Council of Research (CNPq—Brazil) to CAPES and FAPDF, process 00193-00001139/2021-57, for their support.

References

Brusa E, Delprete C, Genta G (1997) Torsional vibration of crankshafts: effects of non-constant moments of inertia. J Sound Vib 205(2):135–150

Chandrashekar G, Raj W, Godwin C, Paul PS (2018) Study on the influence of shaft material on vibration in rotating machinery. Mater Today: Proc 5(5):12071–12076

Corbo MA, Melanoski SB (1996) Practical design against torsional vibration. In: Proceedings of the 25th turbomachinery symposium. Texas A&M University, Turbomachinery Laboratories, Dallas, pp. 189–222

Draminsky P (1965) An introduction to secondary resonance. Shipbuild Mar Eng Int 88:22–26

Gürgöze M (1984) A note on the vibrations of restrained beams and rods with point masses. J Sound Vib 96(4):461–468

Hesterman D, Stone B (1994) A systems approach to the torsional vibration of multi-cylinder reciprocating engines and pumps. Proc Inst Mech Eng C J Mech Eng Sci 208(6):395–408

Hesterman D, Stone B (1995) Secondary inertia effects in the torsional vibration of reciprocating engines—a literature review. Proc Inst Mech Eng C J Mech Eng Sci 209(1):11–15

Huang Y, Yang S, Zhang F, Zhao C, Ling Q, Wang H (2012) Non-linear torsional vibration characteristics of an internal combustion engine crankshaft assembly. Chin J Mech Eng (English Edition) 25(4):797–808

Inman DJ (2007) Engineering vibration, 3rd edn. Pearson Education, New Jersey

Koser K, Pasin F (1995) Continuous modelling of the torsional vibrations of the drive shaft of mechanisms. J Sound Vib 188(1):17–24

Koser K, Pasin F (1997) Torsional vibrations of the drive shafts of mechanisms. J Sound Vib 199(4):559–565

Li Q (2003) Torsional vibration of multi-step non-uniform rods with various concentrated elements. J Sound Vib 260(4):637–651

Li T, Chen Z, Zhang K, Wang J, Huang Z (2022) Analysis of the influence of piston–cylinder friction on the torsional vibration characteristics of compressor crankshaft system. Nonlinear Dyn 110(2):1323–1338

Liu S, Li W, Shuai Z, Chen M (2019) Vibration analysis of a single-cylinder reciprocating compressor considering the coupling effects of torsional vibration. Shock Vib 2019

Meirovitch L (1997) Principles and techniques of vibrations. Prentice Hall, New Jersey

Metallidis P, Natsiavas S (2003) Linear and nonlinear dynamics of reciprocating engines. Int J Non-Linear Mech 38(5):723–738

Pasricha MS, Carnegie WD (1979) Formulation of the equations of dynamic motion including the effects of variable inertia on the torsional vibrations in reciprocating engines, part i. J Sound Vib 66(2):181–186

Pasricha M, Carnegie W (1976) Effects of variable inertia on the damped torsional vibrations of diesel engine systems. J Sound Vib 46(3):339–345

Peng T, Yan Q (2022) Torsional vibration analysis of shaft with multi inertias. Sci Rep 12

Qiao H, Li Q, Li G (2002) Torsional vibration of non-uniform shafts carrying an arbitrary number of rigid disks. J Vib Acoust 124(4):656–659

Sugino C, Xia Y, Leadenham S, Ruzzene M, Erturk A (2017) A general theory for bandgap estimation in locally resonant metastructures. J Sound Vib 406:104–123

Wachel JC, Szenasi FR (1993). Analysis of torsional vibrations in rotating machinery. In: Proceedings of the 22nd turbomachinery symposium. Texas A&M University. Turbomachinery Laboratories, Dallas, pp 127–151

Wachel J, Tison J (1994) Vibrations in reciprocating machinery and piping systems. In: Proceedings of the 23rd turbomachinery symposium. Texas A&M University. Turbomachinery Laboratories, Dallas, pp. 243–272.

Wu JS, Lin TL (1990) Free vibration analysis of a uniform cantilever beam with point masses by an analytical-and-numerical-combined method. J Sound Vib 136(2):201–213

Wu JJ (2007) Torsional vibration analyses of a damped shafting system using tapered shaft element. J Sound Vib 306(3–5):946–954

Xia Y, Ruzzene M, Erturk A (2020) Bistable attachments for wideband nonlinear vibration attenuation in a metamaterial beam. Nonlinear Dyn 102:1285–1296

Chapter 2
How to Use OMA (Operational Modal Analysis) for Turbomachinery

Adhemar Castilho, Luiz A. Vaz Pinto, Ulisses A. V. Monteiro, and Antonio C. Troyman

Abstract Challenging applications within the high-performance centrifugal compressor industry, such as CO_2 reinjection turbo compressors on offshore platforms, have become increasingly common. These machines operate under high pressure, necessitating equipment manufacturers (OEMs) and users to develop more sophisticated manufacturing technologies and complex acceptance tests. The frontier of manufacturers' knowledge has been pushed by high-density gasses, which introduce uncertainties in machine design, particularly regarding their rotodynamic stability. The number of end users requiring stability verification has grown, highlighting the subsidiary risk of increased unbalance sensitivity in this turbomachinery. This risk is referred to as Rotodynamic Design Uncertainties. Despite a single machine performing satisfactorily on its OEM test foundation, there remains a risk that it may fail to operate in an offshore plant. Recognizing this little-acknowledged reality, we are conducting research to address this design anomaly: Rotodynamic Design Uncertainties. We propose small modifications to the ADX vibration monitoring system, developed by LEDAV/COPPE/UFRJ, to enable a dataset project to be carried out at PETROBRAS/REFAP, **followed by subsequent analysis of the results in the LEDAV environment using the ARTEMIS System.**

Keywords Operational Modal Analysis (OMA) · Turbomachinery · Compressor · Applicability · Effectiveness

2.1 Introduction

Challenging applications in the high-performance centrifugal compressor industry, particularly in oil and gas, such as those involving natural gas and CO_2 reinjection turbomachinery on offshore platforms, have become mandatory. These machines

Notice: The author(s) bear sole responsibility for the content printed within this paper.

A. Castilho (✉) · L. A. V. Pinto · U. A. V. Monteiro · A. C. Troyman
LEDAV/COPPE/UFRJ, Federal University of Rio de Janeiro, Rio de Janeiro, Brazil
e-mail: adhemar.castilho@gmail.com

P. Kurka and M. Pereira (eds.), *Advances in Structural Vibration*, Lecture Notes in Mechanical Engineering, https://doi.org/10.1007/978-3-031-71540-2_2

operate at pressures of up to 600 atmospheres, driving OEMs and oil companies to develop more sophisticated manufacturing technologies and acceptance tests to ensure smooth integration into production processes.

The use of high-density gasses, like CO_2, poses a significant challenge to manufacturers' knowledge frontiers, particularly concerning rotodynamic stability according to API 617 Standard. This has led to a breakpoint in end users requesting stability verification tests to mitigate the risks of rotordynamic instability during operational lifespans on offshore platforms.

Moreover, there exists an additional, less apparent risk of sudden emergence of new natural frequencies with low damping factors during machine operations. Such occurrences can lead to production losses due to unbalanced turbo machinery, even after thorough successful factory acceptance tests according to API 617 Running Test.

To illustrate, consider a case where a multinational oil company had to decommission its semi-submersible production platform due to an inability to produce sufficient oil. The platform relied on a gas-lift compressor, which failed to operate satisfactorily due to excessive vibration.

Despite major design changes recommended by an international consultant and subsequent successful acceptance tests, the compressor continued to vibrate excessively, rendering gas-lift operations unfeasible. The consultant later attributed the issue to inherent imbalance in the compressor, highlighting the challenges of addressing rotodynamic design uncertainties.

This risk entails a paradox: an excessive increase in rotor damping (due to sealing and bearings) aimed at preventing rotordynamic instability can inadvertently result in rotor locking, thereby reducing the system's damping capacity unexpectedly. It is crucial to understand that solely relying on internal gas sealing in these compressors cannot enhance the system's rigidity or the rotor suspension's stiffness by compressing the gas film, given the gas compressibility.

Moreover, the previously mentioned increase in rotor damping could compromise the machine's capability to dissipate unbalanced energy. This ultimately directs the unbalanced vibration energy to be entirely transmitted to the support structure, which typically possesses a low damping factor, consequently leading to significant machine vibration levels in some cases.

These low damping factors within the support structure are accountable for the elevated vibration levels associated with the first harmonic, posing a substantial risk to the operational efficacy of these machines in the field, particularly regarding unbalance issues.

Furthermore, this complex issue can be better understood through the perspective expressed by a key figure in the OEM community. In the article titled "Giuseppe Vannini et al (2017)" an experienced consultant from a renowned OEM suggests the feasibility of operating a turbomachine beyond its third critical speed. It is worth noting that manufacturers consistently strive to build machines that are smaller, lighter, with higher speeds, greater useful power, and subsequently lower manufacturing costs, all in pursuit of enhanced productivity in equipment sales processes.

While a machine like that described in this article may perform adequately on a test bench (with an inertial foundation), meeting the API 617 Standard requirements (amplification factor less than 2.5 in all critical speeds within the operating range), there is a considerable risk that the same machine might fail to perform satisfactorily in an offshore facility due to rotodynamic design uncertainties.

These uncertainties stem from the unpredictable nature of rotodynamic parameters (such as natural frequency and damping factor of the rotor) in a work cradle that cannot guarantee sufficient elastic properties, making it challenging to estimate transmitted vibration beforehand. The elastic characteristics of the support for such machines, represented by their FRF (frequency response function) and consequently their vibration performance in the field, remain undefined.

This indicates a turbomachine imbued with significant rotodynamic design uncertainty, far greater than expected from a machine operating between the first and second critical speeds, as conventionally recommended by consultants. It is essential to note that two identical platforms, constructed under the same project, may exhibit different FRFs at their anchor points.

This assertion becomes evident when considering that the rigidity of these support structures remains unknown in the initial phases of the project, only being defined upon the completion of the physical project "as built". This rigidity is profoundly influenced by assembly methods, manufacturing tolerances of metallic profiles used, construction welds, and decisions made during the assembly process, all contributing to uncertainties in rotodynamic design.

Additionally, the API 617 Standard inadequately addresses this issue, placing responsibility on the equipment buyer if the stiffness of the support structure does not exceed 3.5 times the stiffness of the bearing, an assurance impossible to provide during the turbomachinery purchasing design phase.

Despite the numerous particularities and uncertainties outlined above, there is a growing interest among turbomachinery operators in conducting tests to verify machinery rotodynamic stability. These tests serve to validate the accuracy of rotodynamic modeling and aid in mitigating the risk of "whip."

Acknowledging this reality, which often goes unrecognized by many international stakeholders, we are undertaking a project aimed at addressing a design anomaly known as rotodynamic design uncertainties.

Our initiative involves making minor modifications to the ADX vibration monitoring system, developed by the Laboratory of Tests and Analysis of Vibration at UFRJ (LEDAV/COPPE/UFRJ), to facilitate the implementation of a data storage set-up project at one of PETROBRAS' Refineries (REFAP). Subsequent analysis of the collected data will be conducted in the laboratory using the ARTEMIS code.

Upon successful completion of this project, LEDAV will be positioned to offer PETROBRAS a service/equipment package capable of enhancing rotodynamic parameters, specifically rotor natural frequency and damping factor, for both factory and field tests. Given PETROBRAS' identified need and current lack of such a "design facility", it is anticipated that they will be eager to avail themselves of this service/facility.

Negotiations for data storage at REFAP have already been initiated by Dr. Luiz Vaz, the project coordinator, and are progressing well, with implementation already in an advanced stage.

Following this introduction, it is imperative to deliberate on the optimal strategies for planning and executing acceptance tests for these machines, both at the factory and in the field.

2.2 Choosing a Test Method for Turbomachinery

Numerous tests must be conducted on turbomachinery to obtain approval before it can be deemed suitable for service and integrated into the production portfolio of an industrial plant. This study focuses on evaluating the effectiveness of rotodynamic stability tests for such turbomachinery. Given these specifics, it is imperative to identify the most suitable method for rotodynamic stability tests (modal parameters), which must be witnessed by the buyer.

Choosing the appropriate method is intricate and cannot guarantee the flawless operation of the machine in its operational environment with absolute certainty. The most recommended approach is the Experimental Modal Analysis (EMA) technique, which offers both advantages and disadvantages:

- EMA necessitates the temporary installation of an electromagnetic exciter "shaker", placed at the driver's opposite side to excite the relevant asynchronous modes. This incurs additional costs and logistical challenges, while also slightly altering the rotor dynamics.
- It portrays the machine's dynamics on a purpose-built inertial support provided by the manufacturer for factory testing (OEM testing). However, replicating this dynamic on the machine's workstation, a significantly more flexible, complex, and unpredictable structure, is unfeasible.
- Another viable option is the Operational Modal Analysis (OMA) technique, commonly employed in structural modal analysis, which also presents its own set of advantages and disadvantages:
- OMA does not require an external excitation source as the excitation is derived from the aerodynamic noise generated by the gas within the machine.
- It can be easily replicated in the field where facilities are suitably equipped, such as at REFAP.
- OMA depicts the machine's dynamics on its existing support: both in the manufacturer's facilities and in its real operational environment during in-situ field tests.

Despite its potential, OMA does not appear to be widely adopted currently, as indicated by recent literature. This research aims to investigate the reasons behind the limited interest in this method. One significant drawback is the uncertainty surrounding the resolution and efficacy of the test results, which will be elucidated in this study. This uncertainty stems from the relatively low level of internal gas noise produced within compressors that significantly impedes the method's effectiveness and applicability.

In the referenced paper by Zag et al (2019) the application of the OMA technique in a field test of rotodynamic stability for a recycle compressor is detailed.

While the test conducted by the manufacturer ensures the absence of rotordynamic instability "whip" in the machine, it is somewhat disheartening as it fails to reveal any sub-harmonic vibration manifestations in its cascade diagram, as depicted in Figure-2.1, illustrating the tests conducted at the manufacturer's facilities. The EMA test disclosed that the rotor's natural frequency is 7900 rpm. Consequently, it is expected that there will be some vibration occurring around rotor critical speed, induced by internal compressor noise, which does not happen.

This issue arose because the aerodynamic gas noise within the machine lacked sufficient energy to register on the displacement sensors in use. Some articles, such as this one, attempt to correlate the aerodynamic noise within machines with the density of compressed gas. However, this is a significant misconception, as we will discuss further in this research.

Now, let us temporarily shift our focus from the previous discussion on test strategy to delve into experimental techniques: EMA/OMA.

In mechanics, structures undergo vibrations over time, and modal analysis proves invaluable for identifying these movements by breaking them down into their natural modes of vibration. This process establishes a new coordinate system for addressing the problem.

Initially, data is acquired in the time domain with the assistance of vibration sensors and then processed numerically to state the problem in the frequency domain. Here, its "fingerprint" can be characterized, including natural frequencies, vibration modes of the system, and damping factors (modal parameters).

In Fig. 2.2, a simple structure exhibits an infinite number of vibration modes, which become increasingly complex as their natural frequencies rise. This complexity necessitates more energy for vibration at higher critical frequencies.

Experimental Modal Analysis (EMA) enables the estimation of vibration modes for simple structures within laboratory settings using external excitation sources and

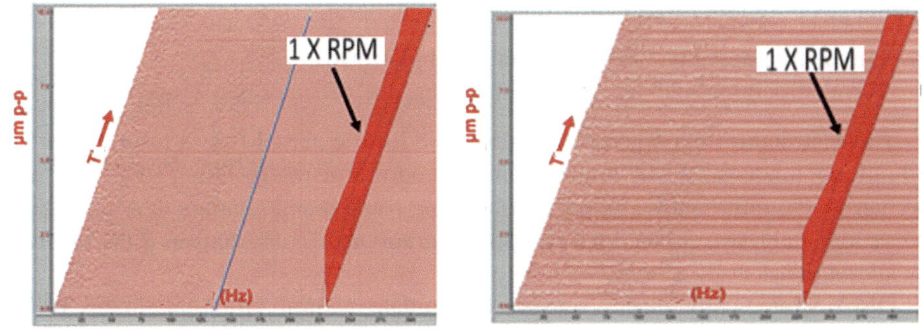

Fig. 2.1 Cascade diagram showing the absence of instability. *Source*: Zag et al (2019)

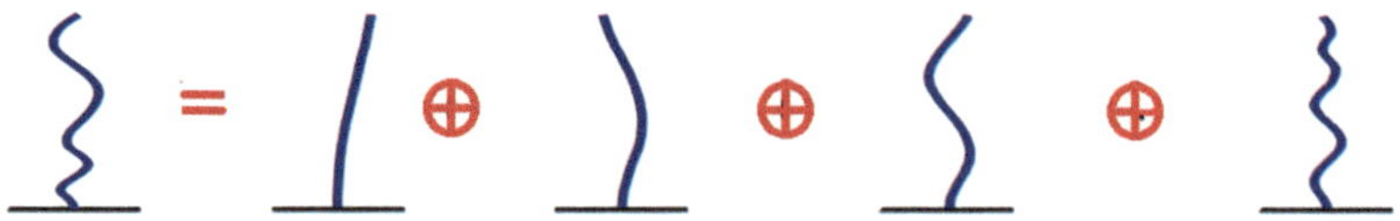

Fig. 2.2 Vibration mode decomposition of an EMA beam. *Source*: Artemis International Solutions Corporation (2024)

sensors, typically accelerometers. Various methods can be employed for external excitation, such as impact hammers or external exciters applied to freely suspended structures, with accelerometers capturing the structure's response.

In the case of large structures, variable frequency exciters can also be utilized. Signals gathered by accelerometers are transformed into the frequency domain using techniques like Fast Fourier Transform (FFT), generating input (force) and output (acceleration/displacement) frequency spectra. Dividing the output signal (displacement) by the input signal (force) yields the Frequency Response Function (FRF), crucial for accuracy. It is imperative that only the excitation force acts on the structure during measurement to avoid a contaminated FRF.

The Coherence Function serves as a tool to assess the FRF quality. Notably, EMA testing predominantly relies on a deterministic mathematical approach. However, limitations exist, necessitating consideration of complementary analysis tools. These limitations encompass:

1. Inability to apply external excitation during testing, as seen in field turbomachinery tests, compromising the FRF (natural frequencies and structural damping factors identification).
2. Boundary conditions inaccurately represent real support conditions, significantly impacting results, and altering modal parameters.
3. Improper excitation levels during testing can similarly compromise results, affecting modal parameters.
4. Difficulties may arise in exciting large and symmetric structures, hindering mode revelation.

As indicated on the ARTEMIS platform, certain mechanical structures pose challenges or render EMA nearly impossible, including:

1. Large structures like bridges, towers, and platforms.
2. Naval structures such as ships, submarines, and offshore platforms.

These structures experience uncontrollable local external forces responsible for vibrations, hindering EMA application and compromising resulting FRFs. In such scenarios, the only viable option for obtaining rotodynamic parameters is utilizing prevailing, unknown excitation forces and Operational Modal Analysis (OMA) offers an alternative.

OMA strategically places sensors to emulate natural vibration modes based on dynamic modeling studies, often employing stochastic processes to identify modal parameters. Unlike EMA's deterministic approach, OMA requires statistical

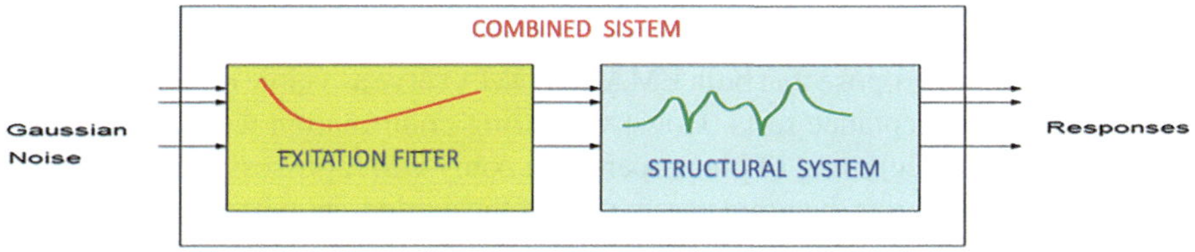

Fig. 2.3 Operational Modal Analysis Processing Strategy. *Source*: Artemis International Solutions Corporation (2024)

treatment due to unknown excitation forces inferred from stochastic processing. Stochastic treatment assumes excitation by white noise, equally exciting all modes, though in practice, certain frequency ranges may contain more energy. To address inconsistencies, stochastic processing incorporates linear filters to realistically modify white noise, representing a combination of structural dynamics and filtered excitation, as shown in Fig. 2.3.

Practical implications of these considerations include:

1. OMA assumes a broadband frequency signal covering all modes of interest, particularly the first mode. Inconsistent mode excitation leads to less representative results. Narrowband excitations are acceptable if derived from broadband excitations.
2. Some system response modes displayed in frequency peaks may originate from improperly filtered input modes rather than the structure itself.

In practical scenarios, certain peaks observed in the spectrum may not directly correlate with the structure itself but rather stem from the input signal. Particularly in turbomachinery applications, these peaks could be attributed to rotational motion and associated harmonics.

For measurement procedures, sensors can be arranged in a modal configuration, and a measurement window (time window) can be utilized, or alternatively, the sensor can be systematically repositioned to predetermined locations, facilitating multiple measurement processes. The arrangement of accelerometers for measurement purposes is denoted as the "TEST SET-UP," with measurements conducted under this configuration referred to as the "DATA-SET." Each subsequent measurement on the same set-up constitutes a new dataset, with the flexibility to span any desired duration.

In certain instances, sensors are permanently embedded within structures, such as for the continuous monitoring of turbomachinery, leading to a singular set-up configuration. Conversely, in scenarios where a set of sensors traverses a structure (for instance, monitoring a vehicle crossing a road or bridge), multiple set-up configurations may be encountered. Thus, it can be deduced that Operational Modal Analysis (OMA) should not be viewed as a replacement for Experimental Modal Analysis (EMA), but rather as an alternative technology, particularly in cases where controlling or even measuring input forces is unfeasible (as is often the case with turbomachinery).

In the context of turbomachinery, contrary to the prevailing stance in current technical literature, we propose that both EMA and OMA serve as viable modal evaluation tools in factory acceptance tests. However, during confirmation tests conducted in the field, after the beginning of plant operations, only OMA possesses the capability to discern the actual rotodynamic parameters of the equipment within its operational environment.

Returning to the initial discussion on testing strategy, we propose that:

Returning to the discourse on testing strategy, we advocate for efforts aimed at attaining more robust noise levels capable of stimulating the first vibration mode during tests. This can be accomplished by generating consistent internal noise encompassing a broad frequency band of excitation. For instance, positioning compressors near their "surge" condition can induce aerodynamic noises consistent with the physical phenomenon associated with gas boundary layer separation at the last impeller vanes of each stage (referred to as the "snowball effect"), as shown in the reference: Baldassarre et al (2015).

Another feasible test filter window is achieved by pushing the machine into a "Stonewall" condition. Within this operational condition, the induced noises can also be associated with a gas noise operation near its sonic speed within the machine, specifically associated with first rotor tip speed.

The often-requested factory test (EMA) facilitates a comparison between the measured modal parameters and those predicted by the manufacturer's inertial-based model, adding credibility and confidence in the OEM's ability to model turbomachinery's dynamic behavior under inertial-based conditions (detached from field reality).

It is imperative to bear in mind, as per API-617, that: If the dynamic stiffness of the machine foundation (support structure) falls below 3.5 times the bearing stiffness, the rotodynamic study is compromised, potentially leading to significant modifications in the machine's rotodynamic parameters.

Within specialized circles, internationally recognized consultants versed in turbomachinery rotodynamic, may agree that a substantial reduction in vibration damping coefficient can result from a support of excessive dynamic flexibility (platform structure).

2.3 Operational Modal Analysis Applications at REFAP/PETROBRAS

The proposed implementation of this test on an existing machine within a refinery, operating in its natural environment, aims to address the critical issue of the applicability of OMA in turbomachinery. It seeks to provide insights into a pertinent question posed by specialized technical literature:

Can OMA be reliably applied in turbomachinery? Under what circumstances can we confidently anticipate favorable outcomes?

This experiment endeavors to demonstrate the feasibility of measuring crucial rotodynamic parameters, specifically the natural frequency and damping factor of the first mode, within a steam turbine. Moreover, it aims to elucidate why such measurements are feasible despite prevailing doubts regarding OMA's suitability for turbomachinery, as suggested by literature associated with the ARTEMIS platform itself.

It is important to highlight that rotodynamic parameters in turbomachinery are intricately linked to essential operational aspects such as rotodynamic stability and adherence to reliable balancing. These parameters significantly influence the occurrence of failures or malfunctions in turbomachinery, particularly in offshore installations, even after passing OEM factory acceptance tests, according to API-617 Standard.

This study's second phase endeavors to demonstrate that OMA technology can be extended to the process centrifugal compressor driven by this steam turbine, notwithstanding prevalent skepticism in the literature regarding OMA's efficacy in this domain. Here, we aim to elucidate the reasons why OMA is not recommended for rotodynamic parameters test evaluation.

In the subsequent phase of this research, we intend to illustrate that positive results in determining rotodynamic parameters can be achieved for the same compressor through the implementation of pertinent measures prior to conducting measurements.

Finally, in the concluding phase of this work, we will explore the methodology for determining the rotodynamic parameters of an electric motor rotor, proposing a new measurement approach to document this final phase of the research.

First Faze Applied in Steam Turbine

The selected machine to demonstrate this feasibility was the wet gas compressor turbine of the UFCC unit at REFAP (Alberto Pasqualini Refinery of Petrobras, situated in Canoas, Rio Grande do Sul). This machine has operated reliably for many years and served as a testbed to demonstrate the proper utilization of the discussed test methodology for applying the OMA technique to turbomachinery.

The time series data was captured in June 2023, comprising approximately 8,000,000 data points collected from each of the turbine displacement sensors. These data were processed using the ARTEMIS code. The time series data from the turbine sensors are shown in Fig. 2.4.

The power spectral density of the signals obtained from turbine sensors, processed through the ARTEMIS program, is illustrated in Fig. 2.5. Here, the operational parameters of the system can be identified.

ARTEMIS processes these signals to provide a precise characterization of the steam turbine rotor's natural frequency and damping factor within its actual suspension. This involves employing statistical methods to learn the true rotodynamic parameters of the machine.

The first critical mode identified corresponds to equivalent suspension behavior in its first vibration mode. This mode results from the interaction between the rotor and its support system, revealing the stiffness characteristics of the linear system representing the rotor support and its interaction with the platform's flexible structure.

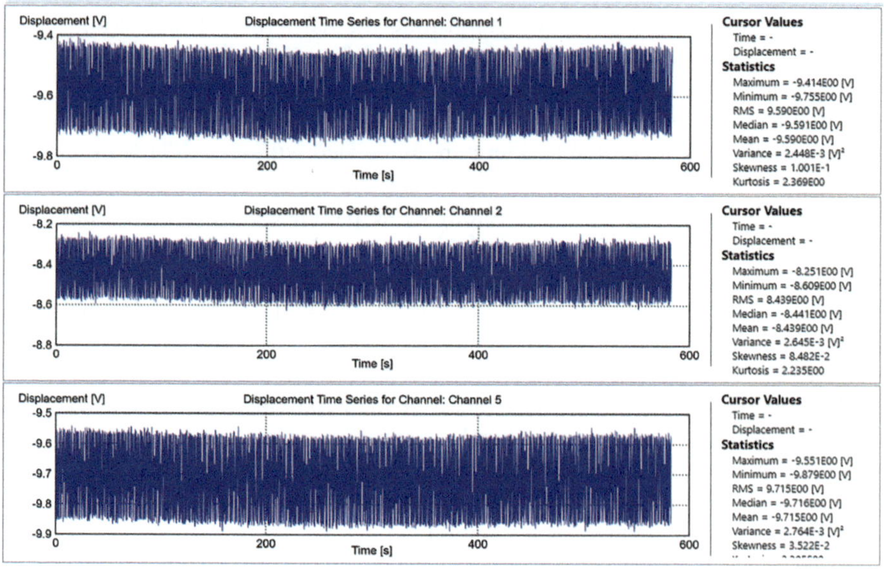

Fig. 2.4 Time series captured by the Steam Turbine sensors. *Source*: Artemis International Solutions Corporation (2024)

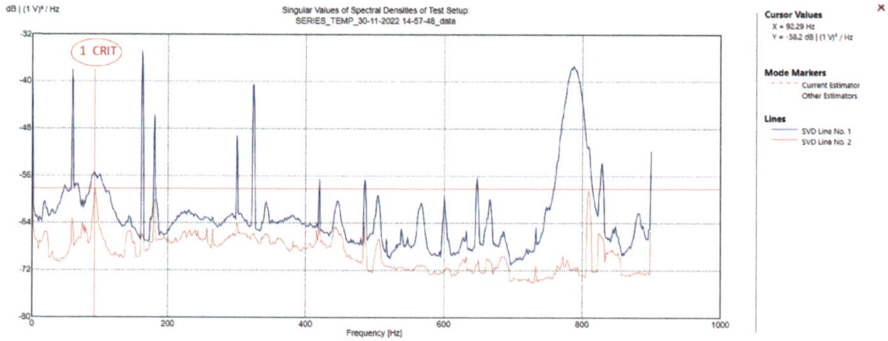

Fig. 2.5 Power spectral density of the signal from the Steam Turbine sensors after processing in ARTEMIS. *Source*: Artemis International Solutions Corporation (2024)

An Important Question Arises

Do we need to identify operational parameters for all critical rotor modes, or is it sufficient to focus on the first critical mode?

The obtained rotodynamic parameters may differ from those obtained during factory testing, especially due to the machine's support structure flexibility increases. This study concentrates solely on reliably detecting rotodynamic parameters in the field that represents an uncertainty area in the existing literature.

The results obtained here can be replicated in machine acceptance tests conducted at the manufacturer's facilities when purchasing new equipment. Moreover, the frequent use of EMA in these tests does not preclude the simultaneous utilization of OMA technology; both methods can be employed in parallel.

Figure 2.6 depicts the time signal measured by the steam turbine's vibration sensors when filtered using a band-pass filter tuned to the turbine's natural frequency.

The distinctive "bubbles" evident in the turbine vibration signals are only discernible in the filtered output. They arise as a consequence of the turbine control system's valve modulation (commonly referred to as partializing valves), under governor system control.

When these partializing valves get open, the steam flow within the turbine can escalate to, and surpass, its supersonic threshold. This engenders a broadband noise

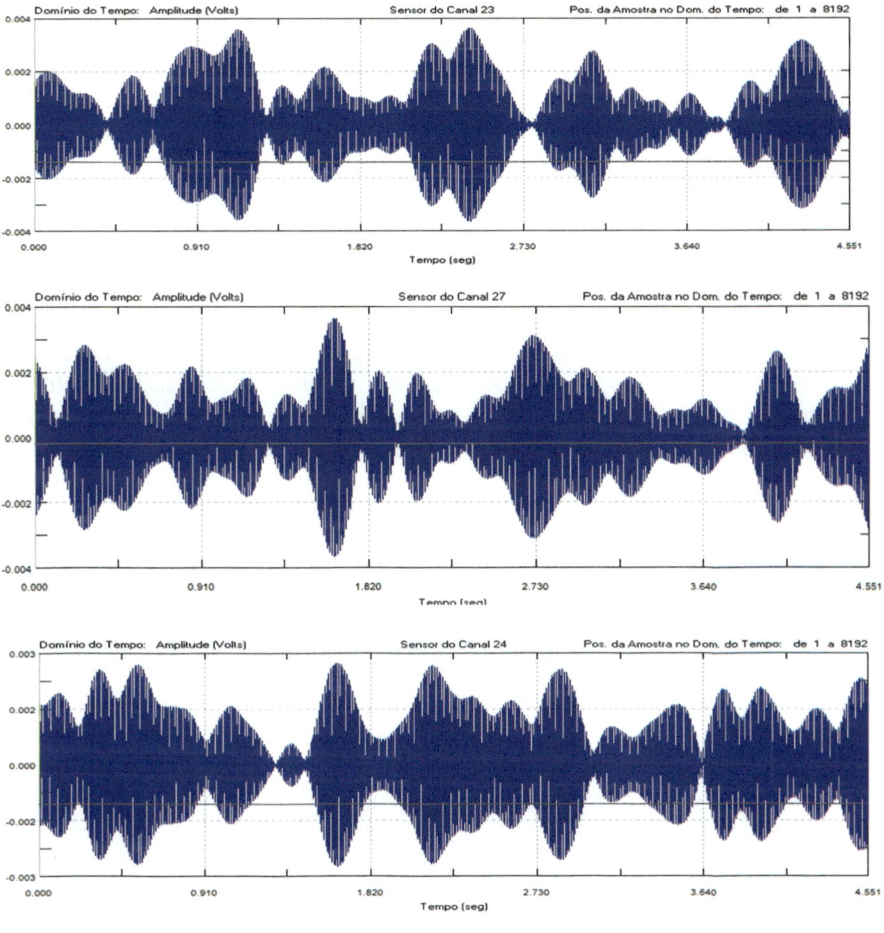

Fig. 2.6 Turbine sensors' time signals, filtered at band-pass filter tuned to natural frequency

with significant energy levels, capable of exciting the natural frequencies of the turbine rotor.

The power spectrum density derived from turbine sensors, processed through the ARTEMIS code, is capable of accurately pinpointing and validating the rotor's primary natural frequency and the damping factor of the corresponding system, as shown in Fig. 2.5.

Second Phase: Wet Gas Compressor Analysis

In the subsequent phase of this investigation, the time series data of the compressor was collected on 27/08/23, comprising four signal datasets, each containing approximately 2,000,000 data points recorded from individual compressor displacement sensors connected to the compressor. Figure 2.7 illustrates the time series data obtained from each of the compressor sensors.

The deterministic signal processing in the frequency domain, illustrated in the cascade diagram in Fig. 2.8, fails to accurately characterize the first natural frequency of the compressor rotor within its real suspension workstation, as predicted in specialized literature. The initial critical mode of the compressor remains hidden due to insufficient noise to effectively excite the rotor with the required energy and frequency range. Instead, the observed peak near the natural frequency is attributed to 60 Hz electrical noise.

During normal operation, the gas compressor maintains a flow within its designated operating range, resulting in minimal aerodynamic noise despite gas density

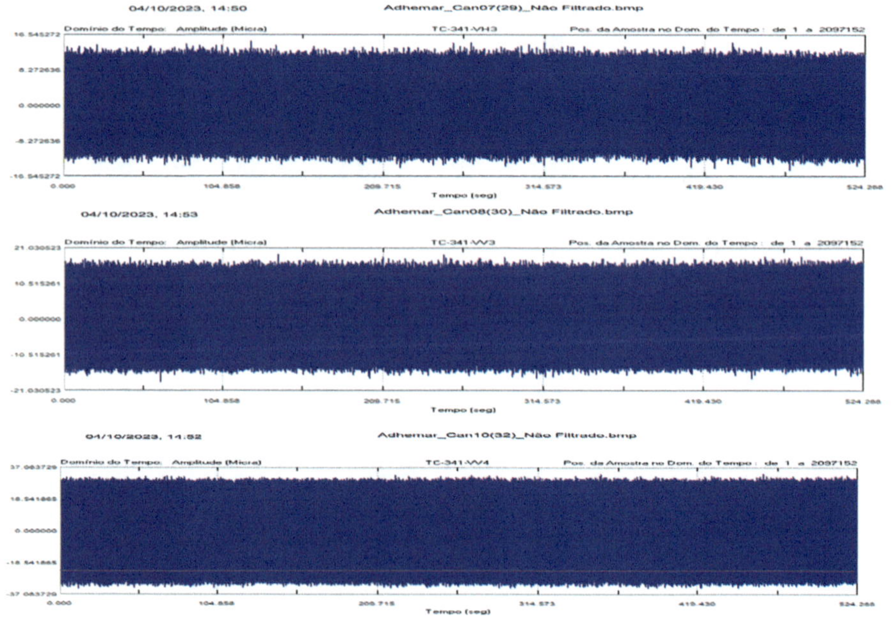

Fig. 2.7 Compressor sensors time signals

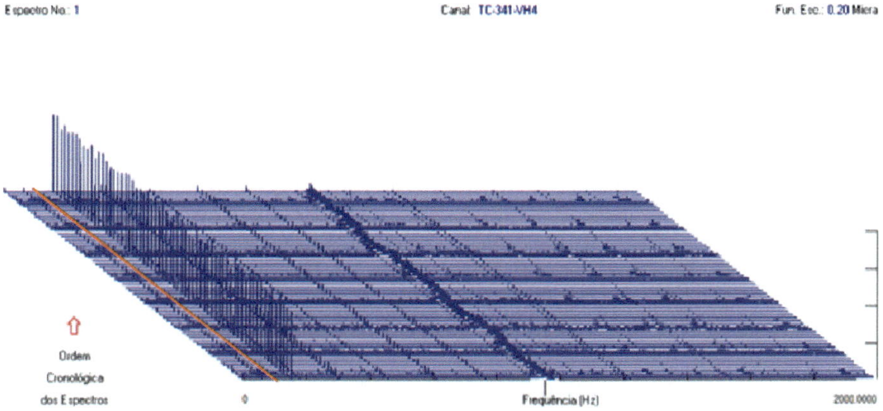

Fig. 2.8 Cascade diagram of one of the signals collected from the gas compressor

variation. This gas noise lacks the required energy to consistently excite the system with enough energy across a broadband frequency range, and consequently do not attend to (OMA) technique requirements.

In Fig. 2.9, the time series data from each compressor sensor is displayed after filtration with a band-pass filter centered around the compressor's natural frequency.

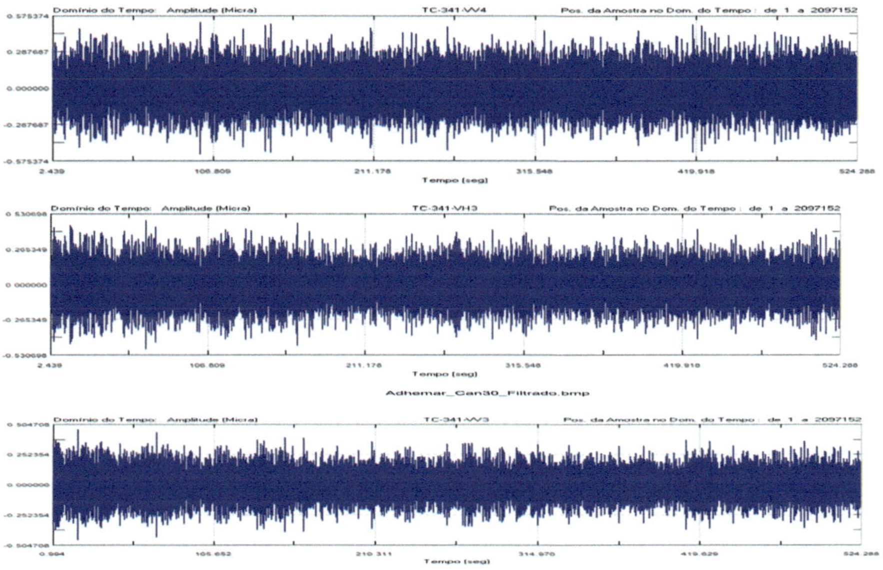

Fig. 2.9 Compressor time series filtered at the machine's natural frequency

However, identifying the natural frequency at this stage is challenging due to interference from electrical noise in the sensor signal, as previously discussed. This viewpoint is widely supported in the literature, underscoring the limitations of employing OMA for turbomachinery analysis.

To elucidate compressor suspension behavior, which corresponds to the output signal of the rotor/suspension interaction, it is imperative to induce aerodynamic noise within the compressor. This noise should possess adequate energy to consistently excite the system in a broadband, according to Operational Modal Analysis (OMA) methodology.

Such noise can be generated under two distinct abnormal operational conditions:

(a) By positioning the machine near its “surge” region, characterized by a pattern of highly turbulent gas flow at the inlet of the final compressor impeller. This condition produces broadband noise frequencies attributable to the detachment of the gas boundary layer at the leading edge of the last impeller blades (impeller eye) in the final investigated rotor stage (snowball effect).
(b) By placing the machine under the “stonewall” operational region, where a supersonic gas flow pattern occurs in the initial impeller stage, resulting in significant noise generation. This condition is compatible with the anticipated noise levels associated with OMA technology.

By subjecting the wet gas compressor of the UFCC unit at REFAP/Petrobras to a stonewall condition, it becomes feasible to discern the behavior of its equivalent suspension (rotodynamic parameters) through signal processing of data collected at the refinery, as illustrated in Fig. 2.10.

The rotodynamic parameters of the linear system, representing the compressor rotor’s support interacting with its flexible structural foundation, can now be

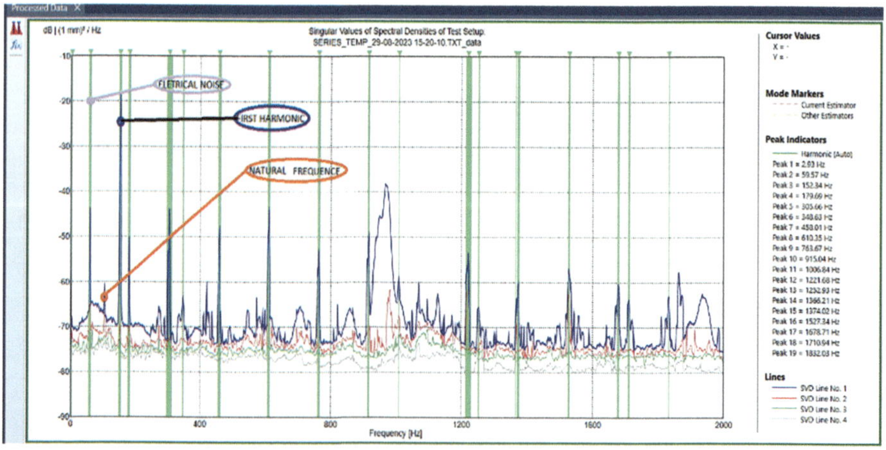

Fig. 2.10 Sensor power spectrum density (PSD—near stone wall) processed by ARTEMIS. *Source*: Artemis International Solutions Corporation (2024)

identified utilizing the ARTEMIS code, as depicted in Fig. 2.10 (rotor-structure interaction).

The obtained rotodynamic parameters are different from those acquired during the manufacturer's API 617 Running Test. This deviation is expected to amplify with an increase in the flexibility of the machine's support structure.

This study exclusively concentrates on the reliable detection capability of rotodynamic parameters in practical settings, a topic that has stirred controversy within the literature.

It is worth noting that the same results demonstrated herein could be precisely replicated during a machine acceptance test conducted at the manufacturer's premises, typically executed upon purchasing a new machine.

Despite the frequent utilization of Experimental Modal Analysis (EMA) in factory acceptance tests, it does not preclude the concurrent use of Operational Modal Analysis (OMA) during such assessments.

2.4 Conclusion

This research marks the initial phase of a broader investigation aimed at providing a tool capable of identifying the anomaly known as Rotodynamic Design Uncertainties.

At this juncture, it can be concluded that OMA technology is good enough to be applicable to both factory and field tests of high-performance turbo compressors, such as reinjection compressors utilized in offshore platforms within the pre-salt fields. The comparative analysis of results obtained from factory and field tests facilitates a potential diagnosis of the inadequacy of the support structure for the service, thereby guiding modifications in machine design to adapt the compressor's damping factors to its original condition (turning excessive damping).

2.5 Future Research Proposals

In the forthcoming phase of this work, we will delve into how the rotodynamic parameters of an electric motor, and a gas turbine rotor can be obtained, proposing a new measurement to document this final phase of the research.

The feasibility of applying OMA technology to high-power electric motors and gas turbines will be investigated, alongside discussions regarding potential necessary adaptations.

Reference (Allaire et al 2005) will also be made to a real case wherein the ammonia motor-driven compressor at FAFEN/SERGIPE encountered a severe imbalance sensitivity issue due to the flexibility of its support structure (mezzanine). Notably, in this specific instance, the problem was rectified by modifying the machine's supporting mezzanine design to detune one of its natural modes of vibration at the onshore support structure.

Implementing such a solution in an offshore structure would pose significant challenges and would likely necessitate alterations in machine design, particularly in the bearing and compressor sealing designs (both internal and external).

References

Allaire PE, Rockwell DR, Castilho A (2005) Finite element analysis of a petrobras motor/compressor and support pedestal modification. In: XI International symposium on dynamic problems of mechanic Ouro Preto, Minas Gerais Brazil

Artemis International Solutions Corporation (2024) Now Aurea Planning Solutions. https://www.aurea.com/our-acquisitions/artemis/. Accessed 6 June 2024

Baldassarre L, Guglielmo A, Catanzaro M, Zague L de O, Ishimoto L, Miranda MA, Silva RT (2015) Operational modal analysis application for the measure of logarithmic decrement in centrifugal compressor. In: 44th TurboMachinery Simpósio, Houston, Texas

Vannini G, Pelagotti A, Rizzo E, Carmicino C (2017) Rotodynamic test results from a high flexibility ratio—high pressure fully instrumented centrifugal compressor test vehicle. In: 46th Turbomachinery Symposia, Houston, Texas

Zag LO, Hunter Cloud C, Miranda MA, Campos RS (2019) Operational modal analysis of a centrifugal compressor in field operation. In: Proceedings of the 10th international conference on rotor dynamics, IFToMM 2018, MMS 61, vol 2, p 460

Chapter 3
Fluid–Structure Coupling in Intermittent Two-Phase Flows: Vibration Amplification and Its Potential for Flow Characterization

Daniely A. das Neves, Adriano Todorovic Fabro, Saon C. Vieira, Juliana R. Cenzi, and Marcelo S. Castro

Abstract Two-phase flow in ducts, especially in intermittent patterns, presents unique challenges due to the complex interaction between the liquid and gas phases. This article investigates fluid–structure coupling in horizontal ducts transporting gas–liquid two-phase flow, focusing on the amplification of duct vibration at the cut-on frequencies of circumferential wave modes. This amplification, resulting from the interaction between fluid pressure waves and the duct's structural modes, is particularly significant at specific frequencies where energy transfer occurs between the fluid and the structure. Using acceleration, pressure, and void fraction measurements, the study demonstrates that the interaction between fluid pressure waves and the duct's structural modes leads to a significant increase in vibration at specific frequencies. This amplification occurs near the cut-on frequencies of circumferential modes, where energy transfer occurs between the fluid and the structure. The research explores the localized nature of this coupling, utilizing the coherence function to show that the coupling is stronger in specific regions of the duct. Additionally, the time-domain modulation of the vibrational response caused by the passage of bubbles and liquid pistons is investigated. The results demonstrate that the bubble passage frequency can be estimated from the demodulation of the acceleration signal filtered at the cut-on frequencies. This finding suggests that the duct's vibrational response can be used as a non-intrusive indicator of the void fraction in two-phase flows.

D. A. das Neves · A. T. Fabro (✉)
Department of Mechanical Engineering, University of Brasília, Brasília-DF, Brazil
e-mail: fabro@unb.br

S. C. Vieira
PETROBRAS, Santos-SP, Brazil

J. R. Cenzi · M. S. Castro
School of Mechanical Engineering, University of Campinas, Campinas-SP, Brazil

M. S. Castro
Center for Petroleum Studies, University of Campinas, Campinas-SP, Brazil

P. Kurka and M. Pereira (eds.), *Advances in Structural Vibration*, Lecture Notes in Mechanical Engineering, https://doi.org/10.1007/978-3-031-71540-2_3

Keywords Flow-induced vibration · Fluid–structure interaction · Pipe acoustics · Two-phase flow

3.1 Introduction

Fluid–structure coupling, a phenomenon that describes interactions between moving fluids and solid structures, presents unique challenges in both single-phase and two-phase internal flows. In single-phase flows, turbulence is a significant excitation mechanism, generating vibrations in the structure (Pettigrew et al. 1998; Evans et al. 2004; Lannes & Gama 2018). These vibrations, measured by accelerometers, provide important information about flow velocity (Evans et al. 2004; Kim & Kim 1996). In two-phase flows, fluid–structure coupling becomes more complex due to the presence of two phases and their interactions. Various factors influence phase distribution in the flow, directly impacting the dynamic response of the structure. Several factors influence the phase distribution in the flow, directly impacting the structural dynamic response. Among these factors, flow pattern plays a critical role in affecting mass and momentum distribution in the flow, leading to different excitation forces on the structure (Ortiz-Vidal 2012). Void fraction, which represents the proportion of gas in the flow, influences the mixture density and, consequently, the structural vibrational response (Matos et al. 2022 part II). Additionally, added mass, representing the mass of fluid vibrating with the structure, is influenced by void fraction and flow pattern, complicating the modeling of structural response (Matos et al. 2022 part II; Charreton et al. 2015). Thus, the flow pattern in a two-phase flow significantly influences the phase distribution and, consequently, the excitation forces acting on the structure.

Intermittent flow, characterized by the periodic transport of elongated gas bubbles followed by liquid slugs (which may contain dispersed bubbles), is commonly found in industries such as petrochemical, chemical, and nuclear, can generate particularly high levels of vibration, capable of causing catastrophic failures in pipelines (Thaker & Banerjee 2016; Miwa et al. 2015; Xu et al. 2020; Fernandes et al. 2023). The representation of this flow can be seen in Fig. 3.1, it consists of alternating repetitive arrangements, called unit cells with length L_U along the pipe with a characteristic frequency. These cells are composed of liquid pistons aerated by small dispersed bubbles of length L_s followed by a Taylor bubble of length L_F. In addition, there is a turbulence zone behind the Taylor bubble passage defined as L_M. The translational velocity V_{TB} is the highest among various velocities in the system. The highly unstable nature of intermittent flow results in local fluctuations of various variables such as phase fraction, densities, velocities, and pressure, which generate periodic forces on the pipeline structure. Matos et al. (Matos et al. 2022 part I) modeled intermittent flow as a periodic unit cell, consisting of a liquid slug and an elongated gas bubble. This representation allows the use of mathematical tools such as Fourier series to describe density fluctuations along the pipeline. The intermittent two-phase flow, characterized by the movement of gas bubbles and liquid pistons within the pipe, acts as a periodic exciter of the pipe structure. Depending on the

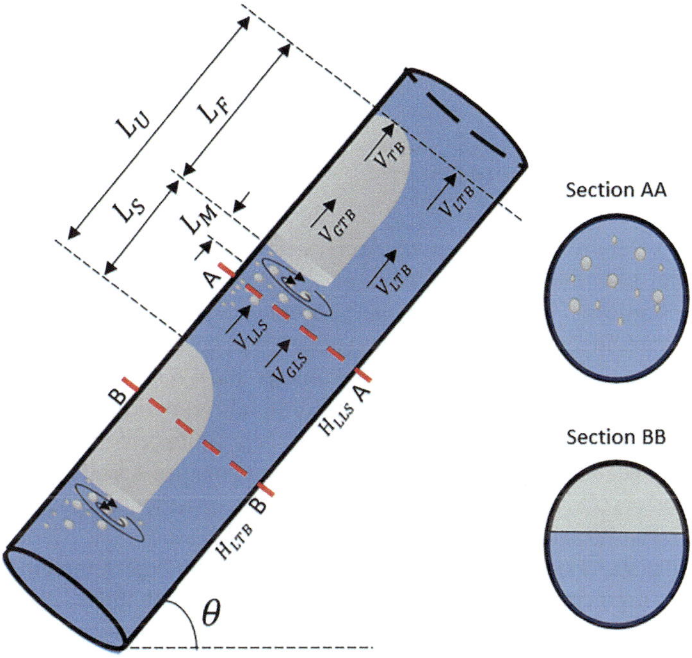

Fig. 3.1 Slug flow geometric characteristics (adapted) (Shoham 2006)

flow conditions and pipe properties, this excitation can lead to resonant vibrations. Later, Matos et al. (Matos et al. 2022 part II) investigated how the properties of two-phase flow (density and velocity) influence the propagation of flexural waves in the pipe and proposed a methodology to estimate void fraction in gas–liquid two-phase flow pipes using measurements of the wave number of flexural waves.

Das Neves et al. (Das Neves et al. 2024a, b) investigate the phenomenon of fluid–structure coupling in horizontal tubes transporting an intermittent gas–liquid two-phase flow. One of the main excitation mechanisms identified is the periodic passage of liquid slugs and elongated gas bubbles (unit cell), creating pressure fluctuations that excite the pipeline structure. The intermittent nature of the flow amplifies these fluctuations, leading to more energetic excitation compared to single-phase or homogeneous flow. The authors also observed that the amplification of vibrations is particularly significant in frequency ranges close to the natural frequencies of the circumferential modes of the pipeline. This observation suggests a strong coupling between the pressure waves generated by the two-phase flow and the vibration modes of the structure, leading to resonant behavior, thus establishing a bidirectional coupling. Thus, this study aims to experimentally investigate fluid–structure coupling in intermittent flows in horizontal ducts. To achieve this, measurements of vibration, dynamic pressure, and void fraction are used in the analysis. The paper is organized such that Sect. 3.2 presents the experimental setup used in this work.

Section 3.3 presents the experimental results and discussion and Sect. 3.4 presents the conclusion and summarizes the findings in this paper.

3.2 Experimental Setup

The findings reported in this study were derived from tests conducted at the Experimental Laboratory of Petroleum—LabPetro—of the Center for Energy and Petroleum Studies—CEPETRO—at the University of Campinas. A 20-meter-long horizontal pipeline for two-phase liquid–gas flow was developed, featuring a 6-meter steel pipe section with a 3 inches internal diameter and 4.5 mm thickness. Tap water and compressed air served as the working fluids. The experimental setup included multiple sensors for synchronous measurement, such as piezoelectric sensors 106B50 (model 2), PCB 352C33 accelerometers, and a conductance-based void meter array with four sensors (R_1, R_2, R_3, and R_4), a photo of the experimental apparatus is shown in Fig. 3.2. The pressure sensor was installed at the bottom of the pipe, while the accelerometer was externally attached just above it, at the top of the pipe. The void fraction measurement from sensor R_4 was synchronized with the accelerometer and pressure sensor using the known distances in the setup and the velocity of the Taylor bubble, following the methodology proposed by Vieira et al. (2021).

The acceleration and pressure time series were recorded at 25.6 kHz for 600 s using a National Instruments (NI) acquisition system. The void fraction measurements were performed at 400 Hz and up-sampled to match the other measurements. Measurements were carried at flow conditions given by the liquid superficial velocities J_{sl}, and gas superficial velocities J_{sg}, summarized in Table 3.1. It also shows average values for liquid piston length L_{slug}, Taylor bubble length L_{bubble}, frequency of passage of the unit cell, f_{slug}, and its measured velocity V_{TB}, obtained from the void fraction measurements by applying the binarization approach proposed by Vieira et al. (2021).

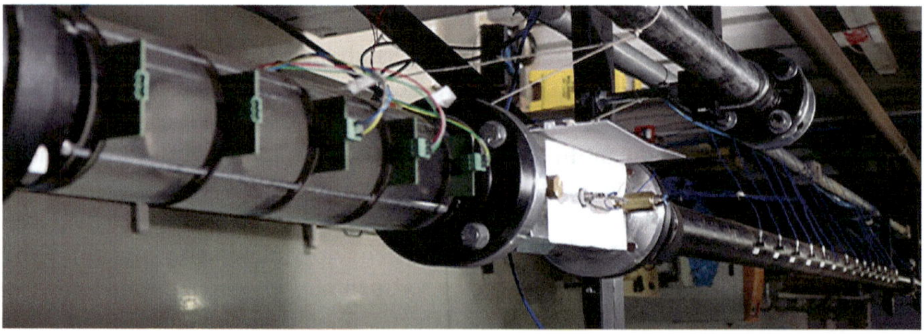

Fig. 3.2 Photograph of the experimental apparatus showing the void fraction sensor and the acceleration sensors

Table 3.1 Experimental test matrix with liquid superficial velocities (J_{sl}), gas superficial velocities (J_{sg}), liquid piston length (L_{slug}), Taylor bubble length (L_{bubble}), and flow pattern

Point	J_{sl} [m/s]	J_{sg} [m/s]	J_m [m/s]	L_{slug} [m]	L_{bubble} [m]	f_{slug} [Hz]	V_{TB} [m/s]	Flow pattern
13	1.354	1.402	2.756	1.065	2.313	0.890	3.05	Slug

3.3 Experimental Results and Discussion

In this section, the experimental results are presented and discussed. The Taylor bubble velocity, or translational velocity, is the velocity at which a Taylor bubble moves through the duct in gas–liquid intermittent flow. It is an important parameter in multiphase flow models. The Taylor bubble velocity is the sum of the Taylor bubble rise velocity, or drift velocity, which is the speed of a Taylor bubble in a stagnant liquid, plus the contribution of the mixture velocity in the preceding slug (Shoham 2006). The estimation of the Taylor bubble velocity using void fraction sensors represents an average of the unit cells (pistons) because each unit cell may have slight variations from the average velocity due to the flow hydrodynamics. Thus, the Taylor bubble velocity is estimated based on the time delay Δt_{TB} from measurements at two sensors at a known distance d as,

$$V_{TB} = \frac{d}{\Delta t_{TB}} \tag{3.1}$$

Cross-correlation-based approaches can be used for estimating the time delay of the measured signals, defined as $R_{1,2}(\tau) = E[\alpha_1(t)\alpha_2(t+\tau)]$, where τ is the time lag and $\alpha_1(t)$ and $\alpha_2(t)$ are the void fractions at each measurement position. This basic estimator can be greatly enhanced by generalized cross-correlation methods (GCC) (Hassab & Boucher 1979). In this approach, a frequency domain pre-filtering is applied on the Cross-Power Spectral Density (CPSD), prior to the inverse Fourier Transform, aiming at enhancing the signals with better signal-to-noise ratio (SNR) and pre-whitening the signals are such that the peak of the cross-correlation is sharpened Gao et al. (2006). The generalized cross-correlation $R^{g}_{1,2}(\tau)$ in this case is given as,

$$R^{g}_{1,2}(\tau) = \mathcal{F}^{-1}\left\{\Psi_g(\omega)S_{1,2}(\omega)\right\}, \tag{3.2}$$

where $\Psi_g(\omega)$ if the frequency weighing function.

Amongst the GCC methods, Vieira et al. (2021) showed that the smoothed coherence transform (SCOT) is very appealing for estimating the velocity of gas bubble because it combines a pre-whitening, which removes the dispersive effects of traveling waves, with a weighting by the coherence function, which decreases the influence of frequency bands with low SNR. For the case of traveling bubbles, or mass waves, the wave dispersion is mostly due to changes in the format of the bubble along the pipe. The SCOT weighing is given by,

$$\Psi_g(\omega) = \frac{\gamma_{1,2}(\omega)}{\left|S_{1,2}(\omega)\right|} \tag{3.3}$$

where $\gamma_{1,2}(\omega)$ is the ordinary coherence function from the estimation of the CPSD (Shin & Hammond 2008). The latter gives a frequency-dependent measure of the linear relation between both sensors. It can be shown that $\gamma_{1,2}(\omega) = 1$ for linearly related signals and $\gamma_{1,2}(\omega) = 0$ for uncorrelated signals. Values in between indicate that the signals are only partially linearly related. Typically, this is due to noise contamination, the presence of extra sources affecting one of the sensors, and/or non-linearities.

The binarization of void fraction series is performed using the Otsu method, an unsupervised and non-parametric approach that utilizes the signal from a conductance-based void fraction meter. This process simplifies the void fraction data by transforming it into a binary representation: presence of the gas phase (1) or the liquid phase (0). The resulting binarized time series are then synchronized with acceleration and pressure signals. This synchronization is based on the time delay estimated from the Taylor bubble velocity. The importance of binarization lies in its ability to determine the moments of vibration and pressure when both the elongated bubble and the liquid slug are at the sensor position, and the result can be analyzed in Fig. 3.3. Intermittent nature implies that the pressure signal will exhibit fluctuations over time, reflecting the alternating passage of liquid and gas phases at the measurement point. Therefore, the choice of piezoelectric sensors, highly sensitive and responsive to dynamic pressure fluctuations, makes them suitable for capturing the transient nature of multiphase flows. Das Neves et al. (Das Neves et al. 2024a, b) used this sensor model in their work and successfully classified the flow pattern through nonlinear time series analysis techniques.

The Frequency Response Function (FRF) is a measure that describes the behavior of a linear system as a function of frequency. In other words, the FRF indicates how the amplitude and phase of an output signal relate to the amplitude and phase of an input signal at different frequencies. The FRF is a valuable tool in vibration and acoustic analysis, often used to characterize mechanical systems subjected to vibrations. It is calculated by dividing the Fourier Transform of the response (system output) by the Fourier Transform of the excitation (system input). In the context of your question, pressure can be considered as the input signal, while acceleration can be considered as the output signal. Figure 3.4 shows the FRF between the pressure and acceleration signals, indicating an increase in FRF at certain frequencies, suggesting resonance. Resonance occurs when the excitation frequency approaches the natural vibration frequency of the structure, resulting in a significant increase in response amplitude.

The coherence function is a dimensionless statistical parameter that quantifies the linear correlation between two signals as a function of frequency. The coherence function value ranges from zero to one, indicating the strength of the relationship between the input and output signals of a system. A coherence value of one means there is a perfect linear relationship between the input and output at that specific

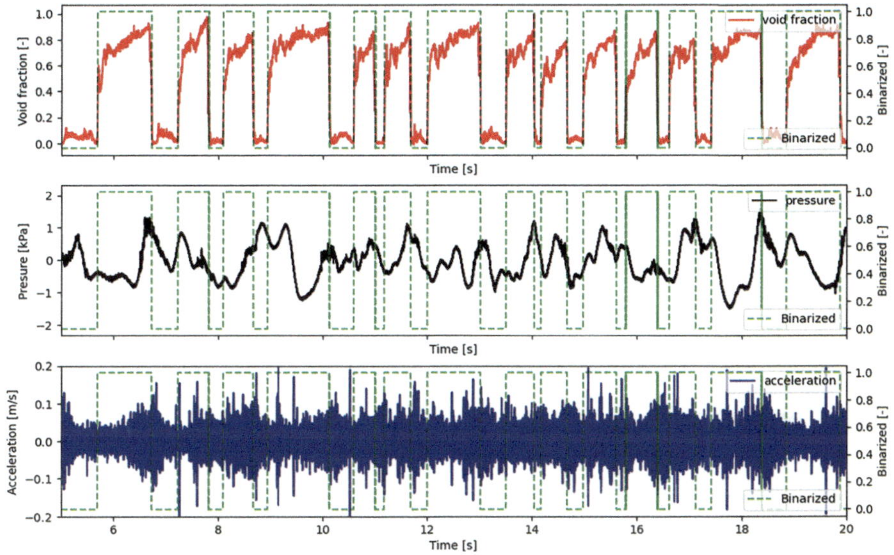

Fig. 3.3 Measured pressure and acceleration signals along with the synchronized void fraction series and its corresponding binarized time series, in seconds

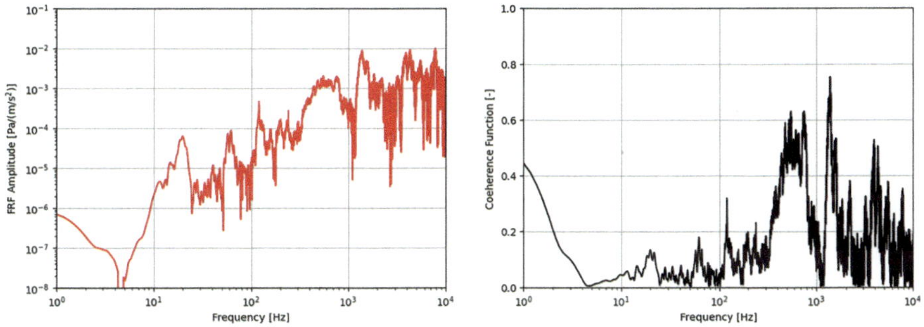

Fig. 3.4 Amplitude of the estimated FRF and the coherence as a function of frequency in Hz

frequency, while a value of zero indicates no linear relationship. In Fig. 3.4, high coherence values are observed at high frequencies where peaks in the FRF appear. The high coherence observed along with peaks in the FRF indicates that the relationship between pressure fluctuations and structural vibration is predominantly linear at these specific frequencies. This implies that the vibrational response is directly proportional to the excitation force generated by pressure fluctuations.

In a system under dynamic excitations, the presence of strong coherence at high frequencies between pressure and acceleration FRFs measured at the same point, coupled with peaks in these FRFs, suggests a robust correlation between flow-induced pressure fluctuations and the resulting structural vibrations within that specific frequency band. This correlation implies that pressure fluctuations act as a significant driving force, inducing structural vibrations. Multiple factors contribute to this phenomenon, with flow turbulence being particularly influential as it generates pressure fluctuations across a broad frequency spectrum. When these fluctuations align with the structure's natural frequencies, resonance occurs, leading to a notable increase in vibration amplitude as seen in the FRF. The observed high coherence and peaks in the FRF indicate a predominantly linear relationship between pressure fluctuations and structural vibration at this particular frequency, highlighting that the vibrational response is directly influenced by the excitation force exerted by pressure fluctuations.

3.4 Conclusions

This study has explored the complex dynamics of fluid–structure coupling in two-phase intermittent flows within horizontal pipelines. The investigation revealed that periodic passage of liquid slugs and elongated gas bubbles serves as a significant excitation mechanism, inducing pressure fluctuations that trigger structural vibrations. These vibrations are amplified particularly near the natural frequencies of the pipeline, indicating resonance phenomena. The observed coherence between pressure fluctuations and structural vibrations at high frequencies underscores a strong linear relationship, where pressure fluctuations act as a direct excitation force on the structure. This bidirectional coupling highlights the critical role of flow dynamics, including turbulence and phase distribution, in influencing structural responses. By employing advanced experimental techniques and analysis methodologies, this study contributes to a deeper understanding of the interactions between two-phase flow characteristics and structural dynamics, crucial for enhancing the reliability and safety of industrial pipelines subjected to such challenging operational conditions.

Acknowledgements We would like to acknowledge PETROBRAS (grant number 2017/00778-2) and ANP for the financial support of this research, the Federal District Research Foundation (FAPDF—Distrito Federal, Brazil) for a post-graduation scholarship of the first author, the Laboratório Experimental de Petróleo (LabPetro) "Kelsen Valente Serra", at the Center for Energy and Petroleum Studies (CEPETRO), University of Campinas for the experimental facilities, and the Post-Graduation Program in Mechanical Sciences of the University of Brasilia (UnB).

Competing Interests The authors declare that they have received financial support from PETROBRAS, which might constitute a conflict of interest.

References

Charreton C, Béguin C, Ross A, Étienne S, Pettigrew MJ (2015) Two-phase damping for internal flow: physical mechanism and effect of excitation parameters. J Fluids Struct 56:56–74

Das Neves DA, Fabro AT, Vieira SC, Cenzi JR Castro MS (2024) An experimental investigation on the fluid–structure coupling in horizontal pipes conveying two-phase intermittent flow. Int J Multiph Flow 176. https://doi.org/10.1016/j.ijmultiphaseflow.2024.104825

Das Neves DA, Vieira SC, Cenzi JR, De Paula AS, Castro MS, Fabro AT (2024) Identification of the flow pattern from the experimental pressure signal in horizontal pipes carrying two-phase flows. Exp Thermal Fluid Sci 154. https://doi.org/10.1016/j.expthermflusci.2024.111141

Evans PR, Blotter JD, Stephens AG (2004) Flow rate measurements using flow-induced pipe vibration. J Fluids Eng 106:208–285. https://doi.org/10.1115/1.1667882

Fernandes LS, Mesquita RSN, Martins FJWA, Azevedo LFA (2023) Three-component turbulent velocity fields in the liquid phase of air-water horizontal intermittent pipe flows. Int J Multiph Flow 162:104378. https://doi.org/10.1016/j.ijmultiphaseflow.2023.104378

Gao Y, Brennan MJ, Joseph PF (2006) A comparison of time delay estimators for the detection of leak noise signals in plastic water distribution pipes. J Sound Vib 292(3–5):552–570. https://doi.org/10.1016/j.jsv.2005.08.014

Hassab JC, Boucher RE (1979) Optimum estimation of time delay by a generalized correlator. IEEE Trans Acoust Speech Signal Process 27(4):373–380. ISSN 00963518. https://doi.org/10.1109/TASSP.1979.1163269

Kim YK, Kim YH (1996) A three accelerometer method for the measurement of flow rate in pipe. J Acoust Soc Am 100:717–726

Lannes DP, Gama AL (2018) On the relationship between pipe accelerance and the void fraction of internal two-phase flow. J Fluids Struct 80:350–369. https://doi.org/10.1016/j.jfluidstructs.2018.04.005

Matos PH, Das Neves DA, Vieira SC, der Geest CV, Cenzi, JR, Iceri DM, Castro MS, Fabro AT (2022) A wave-based approach for estimating void fraction in horizontal pipes conveying two-phase flow. J Pet Sci Eng 216:110766. https://doi.org/10.1016/j.petrol.2022.110766 Part II

Matos PH, Das Neves DA, Vieira SC, Der Geest CV, Iceri DM, Castro MS, Fabro AT (2022) Analytical and experimental investigation of flexural waves in horizontal pipes conveying two-phase periodic intermittent flow. Applied Acoustics 192:108714. ISSN 0003-682X. https://doi.org/10.1016/J.APACOUST.2022.108714 Part I

Miwa S, Mori M, Hibiki T (2015) Two-phase flow induced vibration in piping systems. Prog Nucl Energy 78(2015):270–284. ISSN 01491970. https://doi.org/10.1016/j.pnucene.2014.10.003

Ortiz-Vidal LH (2012) Diseno de un banco experimenta para la generacion de flujo bifasico agua-aire. Pontifica Universidad Católica Del Perú, Tesis el bachiller

Pettigrew MJ, Taylor CE, Fisher NJ, Yetisir M, Smith BAW (1998) Flow-induced vibration: recent findings and open questions. Nucl Eng Des 185:249–276

Shin K, Hammond JK (2008) Fundamentals of signal processing for sound and vibration engineers, vol 270. Wiley. ISBN ISBN 978-0-470-51188-6

Shoham O (2006) Mechanistic modeling of gas-liquid two-phase flow in pipes. ISBN 9781555631079

Thaker J, Banerjee J (2016) On intermittent flow characteristics of gas–liquid two-phase flow. Nucl Eng Des 310:363–377. https://doi.org/10.1016/j.nucengdes.2016.10.020

Vieira SC, Van der Geest C, Fabro AT, De Castro MS, Bannwart, AC (2021) Intermittent slug flow identification and characterization from pressure signature. Mech Syst Signal Process 148:107148. ISSN 08883270. https://doi.org/10.1016/j.ymssp.2020.107148

Xu W, Li Y, Ma W, Liang K, Yu Y (2020) Effects of spacing ratio on the FIV fatigue damage characteristics of a pair of tandem flexible cylinders. Appl Ocean Res 102:102299. https://doi.org/10.1016/j.apor.2020.102299

Chapter 4
Analysis of a NiTi Blade-Type Spring with Superelastic Effect According to Manufacturing Processes

Igor Jordan Guilherme de Sena, Richard Senko, Antônio Almeida Silva, and Vinicius da Silva Almeida

Abstract The application of smart materials in rotating systems as vibration attenuators is well-established due to their vibration reduction capabilities. One of the most common types of smart materials used is Shape Memory Alloys (SMA), which exhibit property changes in response to temperature (shape memory effect—SME) or mechanical stress (superelasticity—SE). The objective of this research is to compare and analyze the manufacturing of superelastic NiTi blade-type springs, considering different heat treatment methods, mechanical forming processes, and fabrication techniques (shape setting), with the aim of achieving a balance in properties. By optimizing these factors, it was possible to obtain a final austenitic temperature of 14.88 °C, an increased hysteresis damping of 78.44%, and a rupture strength of 1285.23 MPa. These results demonstrate the high potential of the springs in passive vibration reduction in rotating systems.

Keywords Shape memory alloys · Superelasticity · Manufacturing · Rotating systems

4.1 Introduction

The extensive application and studies involving smart materials are due to their ability to alter physical properties when subjected to external stimuli. Among these materials are piezoelectrics, electro and magneto-rheological fluids, and Shape Memory Alloys (SMAs) (Gandhi & Thompson 1992), with the latter being the focus of this study.

I. J. G. de Sena · A. A. Silva · V. da Silva Almeida
Academic Unit of Mechanic Engineering, Federal University of Campina Grande, Campina Grande, Brazil

R. Senko (✉)
Academic Unit of Production Engineering, Federal University of Campina Grande, Campina Grande, Brazil
e-mail: richard.senko@uaep.ufcg.edu.br

P. Kurka and M. Pereira (eds.), *Advances in Structural Vibration*, Lecture Notes in Mechanical Engineering, https://doi.org/10.1007/978-3-031-71540-2_4

SMAs are smart materials that, after being deformed, have the ability to recover their original shape when subjected to temperature or mechanical stress variations.

This ability is due to the material's crystalline structure, which exhibits different effects depending on the martensitic and austenitic phases, such as shape memory effect (SME) and superelasticity (SE). When the material is fully in the martensitic phase, changes occur in the crystalline structure when subjected to temperature and stress variations. Superelasticity occurs above the austenite finish temperature (A_f) when the mechanical stress is applied to the material, resulting in an elastic deformation of up to 8% in NiTi alloys and inducing a phase transformation to martensite.

According to the literature, these alloy behaviors are influenced by factors such as material composition, cold work, and heat treatments, among others (Lagoudas, 2008; Otsuka & Wayman, 1999; Patoor et al. 2006). Due to these characteristics, the applications of SMAs encompass various areas such as dentistry, medicine, and engineering, with the latter focusing on reducing undesirable vibrations in rotating systems.

This research aims to analyze and compare the manufacturing methods, heat treatments, and mechanical forming of superelastic NiTi blade-type springs with the study conducted by Almeida (2020), who used water quenching as a cooling method for the samples. By employing different methods, it is expected to achieve improved damping capacity and mechanical strength, as well as a final austenite temperature close to room temperature, enabling the activation of the superelastic effect without the requirement of external heating. This makes it possible to passively reduce vibrations in rotating systems.

4.2 Methodology

The methodology of this research followed the steps of the flowchart presented in Fig. 4.1.

4.2.1 Manufacturing of LMF-SE Blade-Type Spring

For the manufacturing of C-shaped LMF-SE blade-type springs as proposed by Senko et al. (2022a), Fig. 4.2, the first step involved the use of an electro erosion machine to cut the NiTi sheet with a nickel weight percentage of 55.8%, resulting in samples with a length of 64 mm, width of 5 mm, and thickness of 0.52 mm.

With the samples in these dimensions, a heat treatment (HT) was performed for homogenization and stress relief (Souza 2015), at a temperature of 450 °C for 10 min, as recommended by the manufacturer. Subsequently, natural convection cooling was carried out to achieve an austenite finish (A_f) temperature close to ambient and high mechanical strength, based on studies conducted by Heidari et al. (2016). To obtain

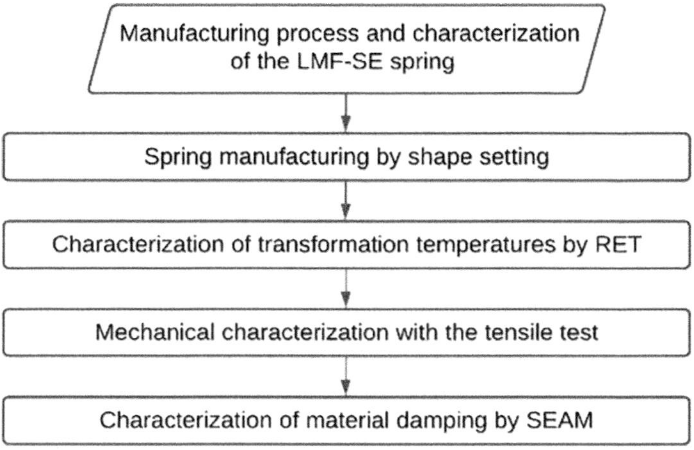

Fig. 4.1 Research flowchart. *Source* Author

Fig. 4.2 C-shaped blade spring. *Source* Adapted from Senko et al. (2022a)

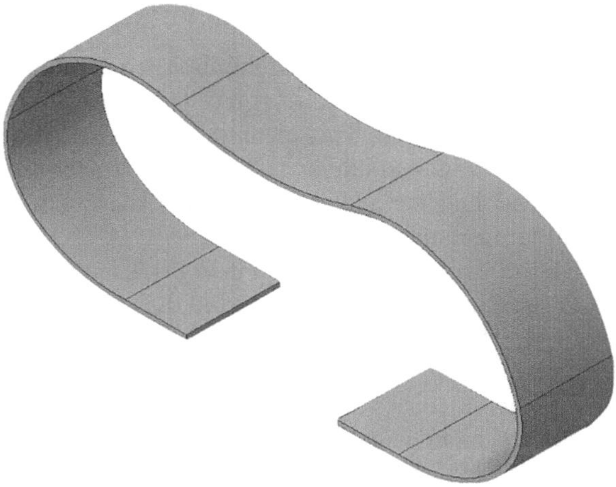

the C-shaped form (Fig. 4.2), shape setting was performed using a carbon steel mold consisting of three pieces connected by screws, as shown in Fig. 4.3(A). After inserting the sample into the mold, a HT was carried out at a temperature of 500 °C for 30 min, followed by natural convection cooling. These shape setting parameters are suggested by Rao et al. (2015). At the end of the shape-setting process, the C-shaped blade springs obtained the form shown in Fig. 4.3(B).

Fig. 4.3 **A** Carbon steel mold, **B** c-shaped blade. *Source* Author

4.2.2 *Characterization of LMF-SE Blade-Type Spring*

4.2.2.1 Characterization of Transformation Temperatures

To determine the phase transformation temperature values of the C-shaped LMF-SE blade springs, the electrical resistance variation with temperature (RET) technique was employed. This technique involves controlled temperature variation of the sample while simultaneously monitoring the temperature and electrical resistance, enabling the acquisition of the experimental curve of electrical resistance versus temperature (Dos Reis 2010).

For the test setup, as depicted in Fig. 4.4, a power source (1) is required to provide fixed current (0.574 A) and potential difference to the sample, which is fixed on a base (3). The sample is immersed in a thermal bath (2) whose temperature (T) is controlled using a device (4). Temperature measurements are obtained through a type K thermocouple attached to the sample and transmitted to the data acquisition system (5), along with the potential difference (voltage) from the power source. The acquired data is then sent to a computer (6).

4.2.2.2 Mechanical Characterization

To determine the yield strengths and verify the mechanical properties of the spring, a mechanical characterization was performed following the standard procedure for tensile testing of wires in NiTi alloys, ASTM F2516.26200 (2014). The test specimens, as shown in Fig. 4.5, were produced by electroerosion cutting of SMA alloy sheets.

This procedure was also carried out by Almeida (2020), where the samples were subjected to loading up to 6% deformation, followed by unloading to stress lower than 7 MPa, and then loaded until fracture. This was done using an Instron® 5582 tensile testing machine equipped with a video extensometer (AVE) at room temperature

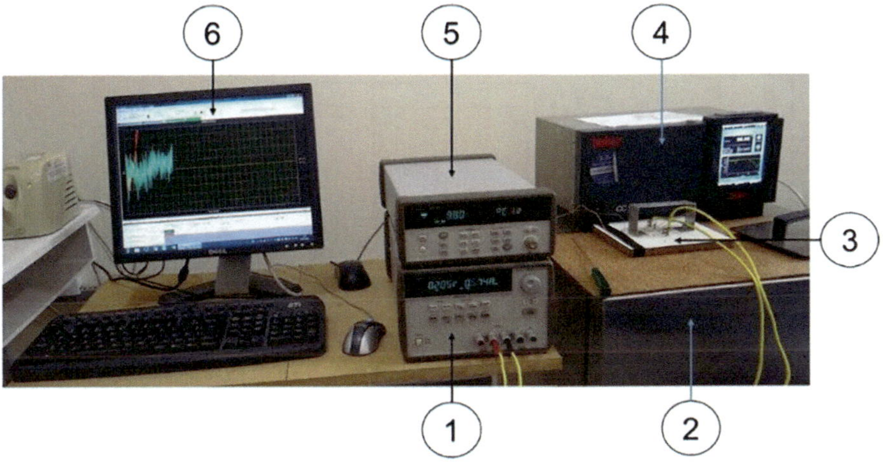

Fig. 4.4 RET. *Source* Almeida (2020)

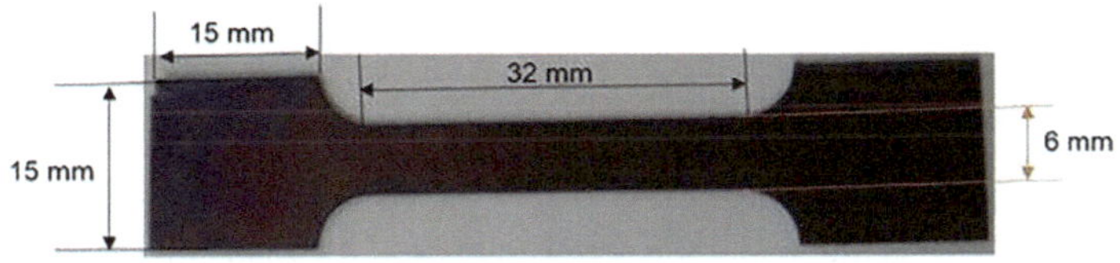

Fig. 4.5 Tensile test specimen. *Source* Almeida (2020)

(approximately 22 °C) and a deformation rate of 0.04 mm/min. According to the standard, the following results are to be observed: Upper plateau strength (UPS): Stress at 3% deformation during sample loading; Lower plateau strength (LPS): Stress at 2.5% deformation during sample unloading; Residual Elongation (EL_r): Difference between the initial deformation and the deformation at the moment the test reaches 7 MPa during unloading; Uniform Elongation: Elongation determined at the maximum force sustained by the sample before necking, cracking, or both. Additionally, the austenitic and martensitic elastic moduli and transformation stresses are also obtained.

4.2.2.3 Material Damping Characterization

Finally, to determine the damping factor of the material, experiments were conducted using the System for Material Damping Estimation (SEAM), as depicted in Fig. 4.6, as presented by Dos Reis et al. (2019). SEAM is a single-degree-of-freedom modal analysis used to determine the loss factor and stiffness of the device.

1. The blade (7) is fixed to the structure (6).

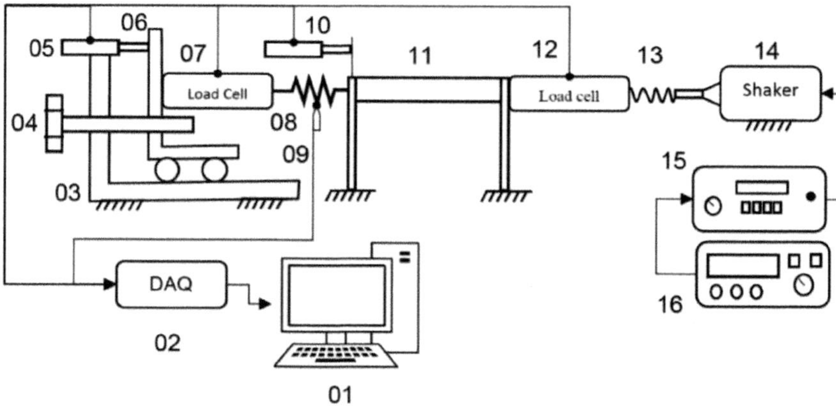

Fig. 4.6 SEAM. *Source* Adapted from Dos Reis et al. (2019)

2. The carriage (10) is moved by the screw (12) until the load cell (9) makes contact with the blade (7). This position is set as the origin of the LVDT (11).
3. The load cell values (5, 9) are zeroed.
4. The screw (12) is displaced until the LVDT (11) reaches a displacement of 2 mm, which is the pre-displacement value in the test.
5. The flexible rod (4) is compressed and coupled to the structure (6) in such a way that the load cell (5) reaches 4 N.
6. The generator (1) generates a harmonic sine signal with a linear variation from 1 to 50 Hz over 300 s, which is amplified (2) and sent to the shaker (3).
7. The load cell (5) and LVDT (8) results are acquired using the Catman Data Acquisition (DAQ) software at a rate of 2400 Hz.
8. The process is repeated four times to ensure test repeatability and generate the average of the curves.

4.3 Results

4.3.1 Characterization of Transformation Temperatures

Figure 4.7 depicts the RET curves obtained from the blade before and after forming, respectively, and Table 4.1 presents a comparison with the results obtained by Almeida (2020).

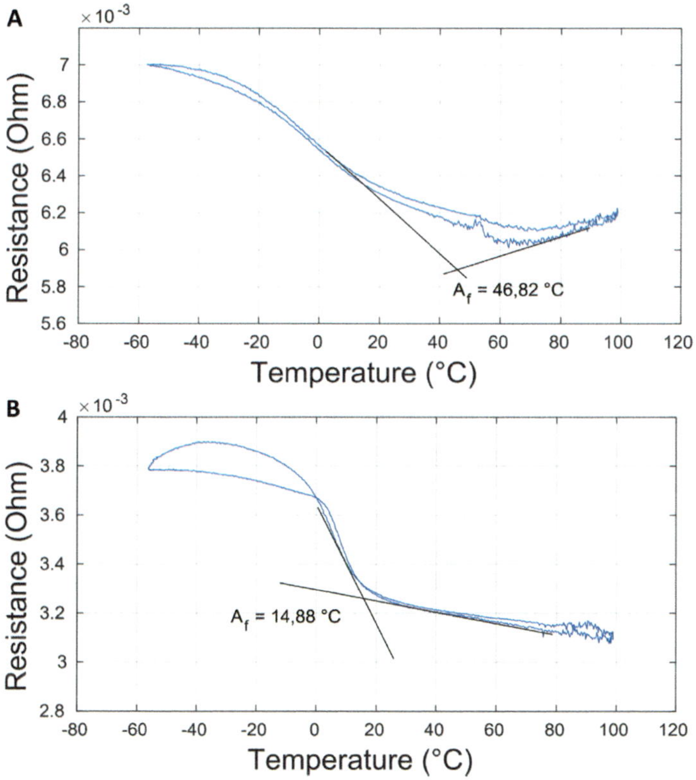

Fig. 4.7 RET of Sample 1—**A** sheet, **B** after fabrication. *Source* Author

Table 4.1 Comparison of RET results

	Manufacture spring (°C)	Almeida (2020) (°C)
$T A_f$ before fabrication	46.82	23.57
$T A_f$ after fabrication	14.88	14.10

Source Author

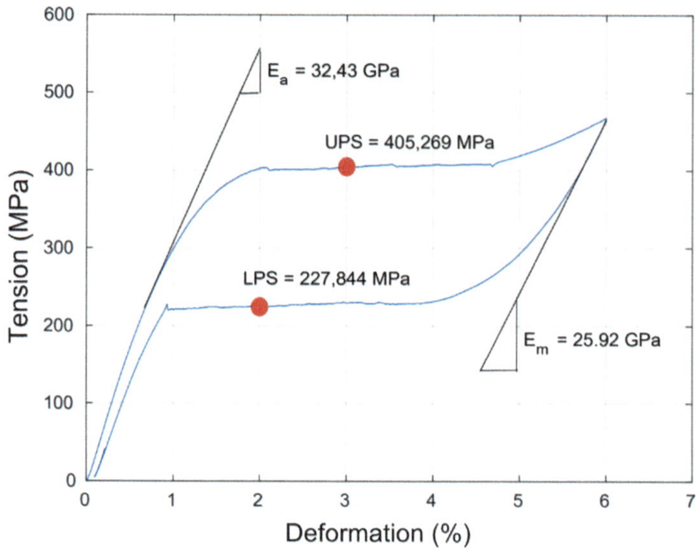

Fig. 4.8 Tensile test. *Source* Author

In this experiment, an A_f temperature of 46.82 °C was obtained before shape setting, which is above the local ambient temperature (approximately 22 °C), indicating that the material is not fully in the austenitic phase. However, after the shape-setting process, the A_f temperature was reduced to 14.88 °C (Fig. 4.8b). Therefore, the material tends to be fully in the austenitic phase at room temperature in order to take advantage of the superelasticity effect. This result was expected and is consistent with the studies by Motemani (2009), which demonstrate that the final austenitic temperature decreases as the cooling rate increases. This result was similar to those obtained by Almeida (2020) since the same type of spring was manufactured.

4.3.2 *Mechanical Characterization of NiTi LMF Sheets*

Figure 4.8 shows the stress–strain curve obtained from the tensile test (Table 4.2).

The Young's modulus obtained from the tensile test was higher in the austenitic phase than in the martensitic phase, which was expected since the stiffness of the austenitic phase is about four times higher than that of the martensitic phase. These results were relatively lower compared to those obtained by Almeida (2020), who had a higher stiffness index, making the material susceptible to failure at low load levels. The ultimate tensile strength was 1285.229 MPa, which was expected due to the natural convection cooling, similar to the method used by Senko et al. (2022b), who obtained an ultimate tensile strength of around 1200 MPa.

Table 4.2 Comparison between the results of the tensile test

	Manufactured spring	Almeida (2020)
Ea (GPa)	32.43	55.13
Em (GPa)	25.92	29.39
UPS (MPa)	405.269	434.600
LPS (MPa)	227.844	205.240
Elr (%)	0.0944	0.1790
σrup (MPa)	1285.230	638.220

Source Author

The difference between the plateau stresses (UPS and LPS) indicates approximately 44% of the UPS value, representing a high hysteresis area and therefore demonstrating a high damping potential, a value similar to that of Almeida (2020), who obtained a value of 50%. The residual elongation of the sample was 0.0944%, indicating that only a few cycles are needed to reach the stabilization of the elastic behavior.

4.3.3 *Material Damping*

Using the SEAM, the Nyquist diagram and Frequency Response Function (FRF) graphs of the spring were obtained for a pre-load of 1.5 mm to determine the hysteresis damping. Figure 4.9A shows the comparison between the system without the proposed spring and the system with the spring, while Fig. 4.9B represents the amplitude of the system. It can be observed that the Nyquist circle of the system with the spring is reduced, indicating the high damping potential of the device. The hysteresis damping of the system without the spring is 0.276, whereas with the spring, this value increased to 1.283, resulting in a 78.44% increase. In the FRF (Fig. 4.9B), a reduction of 15.092 dB in the system's amplitude can be observed when compared to the system without it.

The manufactured spring has a loss factor (η) of 14.78%, while Almeida's (2020) was 7.55%. The difference was expected, as reported by Motemani (2009), who stated that samples with higher cooling rates tend to have a reduced loss factor.

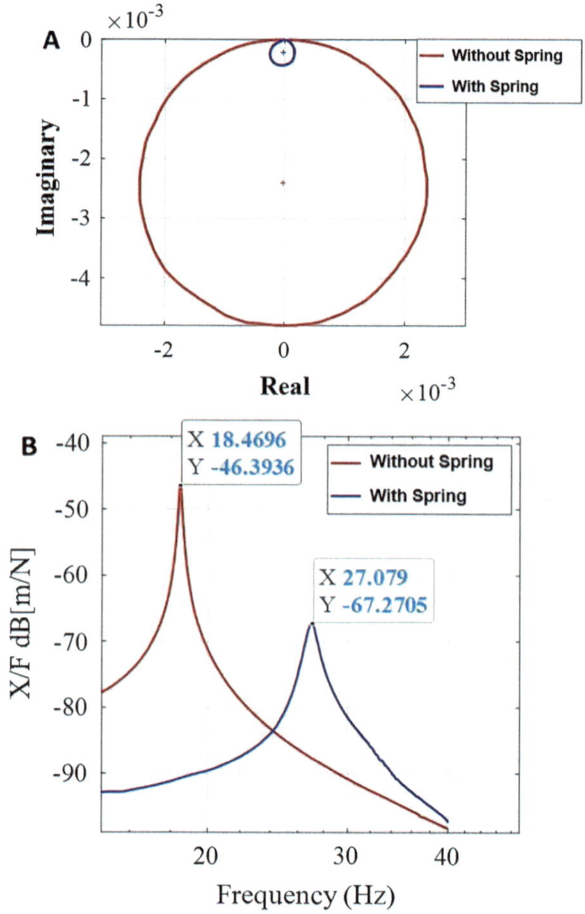

Fig. 4.9 **A** Nyquist Plot, **B** FRF. *Source* Author

4.4 Conclusion

In this research, experiments and studies were conducted on the application of heat treatment (HT) to obtain complex-shaped LMF-SE springs. Although the LMF-SE blade, after being treated, obtained a relatively high A_f temperature, after shape setting, the temperature was below ambient, which is considered ideal. This observation is consistent with Almeida (2020).

Through the tensile test, it was possible to analyze that the material exhibits stable elastic behavior. The rupture stress was also determined, and it was found to be twice as high as that of Almeida (2020). The higher mechanical strength of the spring in this research overcomes the limitation observed in Almeida's study, where the low mechanical strength hindered the application of their springs in the rotating system.

Finally, the results obtained from the SEAM indicate that the manufactured spring has a higher hysteresis damping capacity, a reduction in the system's amplitude, and a higher loss factor compared to the results obtained by Almeida (2020).

Thus, the spring presented in this research demonstrates ideal properties for its application in rotating systems and proves to be advantageous compared to Almeida (2020), as it can be applied in the rotating system without premature fracture.

References

Almeida VS (2020) Development and testing of vibration attenuation in rotating system bearings applying blade-type shape memory alloy actuators in the superelastic regime. (Master's Dissertation). Academic Unit of Mechanical Engineering, Federal University of Campina Grande

ASTM International (2014) ASTM F2516.26200: standard test methods for tensile testing of nickel-titanium superelastic materials. ASTM International

Dos Reis RPB (2010) Development of equipment for thermal characterization of shape memory alloy actuators using thermoelectric effect. Federal University of Campina Grande, p 58

dos Reis RPB, Silva PCS, Senko R, Silva AA, de Araújo CJ (2019) Methodology for the estimation of material damping as applied to superelastic shape memory alloy mini-springs. Mater Des 161:124–135. https://doi.org/10.1016/j.matdes.2018.11.012

Gandhi MV, Thompson BS (1992) Smart materials and structures. [S.l.] Chapman & Hall. ISBN 0412370107

Heidari B, Kadkhodaei M, Barati M et al (2016) Fabrication and modeling of shape memory alloy springs. Smart Mater Struct 25(12):10. https://doi.org/10.1088/0964-1726/25/12/125003

Lagoudas DC (2008) Shape memory alloys, vol 1. Springer Boston, MA, USA. ISBN 978-0-387-47684-1

Motemani Y, Nili-Ahmadabadi M, Tan MJ et al (2009) Effect of cooling rate on the phase transformation behavior and mechanical properties of Ni-rich NiTi shape memory alloy. J Alloy Compd 469(1–2):164–168. https://doi.org/10.1016/j.jallcom.2008.01.153

Otsuka K, Wayman CM (1999) Shape memory materials Cambridge University Press, Cambridge, UK. ISBN 0521663849

Patoor E, Lagoudas DC, Entchev PB et al (2006) Shape memory alloys, Part I: general properties and modeling of single crystals. Mech Mater 38(5–6):391–429. https://doi.org/10.1016/j.mechmat.2005.05.027. Available at: https://linkinghub.elsevier.com/retrieve/pii/S0167663605001195

Rao A, Srinivasa AR, Reddy JN (2015) Design of shape memory alloy (SMA) actuators. Springer International Publishing, Cham. Available at: https://doi.org/10.1007/978-3-319-03188-0 (SpringerBriefs in Applied Sciences and Technology)

Senko R, Silva AA, Rodrigues MC, Lima AGB, dos Reis RPB, Guedes CAL (2022a) Numerical and experimental analysis of superelastic sma bending springs in rotor systems. Mech Adv Mater Struct 29:579–593. https://doi.org/10.1080/15376494.2020.1780523

Senko R, Almeida VS, Dos Reis RPB, Oliveira AG, Silva AA, Rodrigues MC, De Carvalho LH, Lima AGB (2022b) Passive attenuation of mechanical vibrations with a superelastic sma bending springs: an experimental investigation. Sensors (Basel) 22(9):3195. https://doi.org/10.3390/s22093195 PMID: 35590893, PMCID: PMC9103138

Souza EF (2015) Experimental study of the thermal and dynamic behavior of a belleville spring with NiTi shape memory alloy. Federal University of Campina Grande, p 26

Chapter 5
On the Evaluation of Natural Frequencies and Mode Shapes of Beams Under Tensile Axial Loading

Kevin Mauricio Menon Ribeiro, Gabriel Ruggiero do Amaral, Jose Manoel Balthazar, Alexandre de Macêdo Wahrhaftig, and Eduardo Márcio de Oliveira Lopes

Abstract The accurate prediction of lateral vibration in beams subjected to axial loads is of great practical interest due to its wide use in describing the dynamic behavior of structures in the field of civil, mechanical, and aerospace engineering. A distinct example of its application can be found in problems involving overhead cables of transmission lines, for which the corresponding Euler–Bernoulli beam model is widely used. Although the analytical formulation for calculating natural frequencies and mode shapes for beams under tensile axial loading is well known and developed, it is worth noting that in certain applications, such as in the case of overhead cables, the results obtained may exhibit numerical instabilities for some boundary conditions. In the vast majority of studies, the mode shapes are described in terms of trigonometric and hyperbolic functions, despite the fact that hyperbolic functions tend to grow rapidly to infinity as their argument increases, which can cause numerical errors, especially in the vicinity of the beam supports. On the other hand, numerical simulations indicate that more accurate and reliable results can be obtained when the analytical solutions are expressed by linear combinations of trigonometric and exponential terms. In this context, the present work aims to compare the two aforementioned ways of describing the analytical solutions for calculating natural

K. M. M. Ribeiro (✉) · G. R. do Amaral · E. M. de Oliveira Lopes
Federal University of Paraná (UFPR), Curitiba, Brazil
e-mail: eng.kevin.mmr@gmail.com

G. R. do Amaral
Mechanical Testing Laboratory, Institute of Technology for Development—Lactec, Curitiba, Brazil

J. M. Balthazar
Department of Mechanical Engineering, Engineering School, São Paulo State University (UNESP), Bauru, Brazil

Federal University of Technology—Paraná (UTFPR), Ponta Grossa, Brazil

A. de Macêdo Wahrhaftig
Department of Construction and Structures, Polytechnic School, Federal University of Bahia (UFBA), Salvador, Brazil

P. Kurka and M. Pereira (eds.), *Advances in Structural Vibration*, Lecture Notes in Mechanical Engineering, https://doi.org/10.1007/978-3-031-71540-2_5

frequencies and their respective mode shapes of vibration with respect to overhead cables of transmission lines. In this effort solutions are developed for two classical boundary conditions (fixed–fixed and fixed-simply supported) and one non-classical boundary condition (simply supported with torsional springs). Numerical simulations are carried out in the MATLAB® environment. A significant gain is observed in the use of the preferred formulation, especially in terms of a more accurate prediction of mode shapes for the type of structure under consideration.

Keywords Analytical solution · Axial loading · Euler–Bernoulli beam · Mode shapes · Overhead cables

5.1 Introduction

Understanding the lateral vibration in beams under axial loading plays an important role in many engineering and industrial applications, including the design of overhead cables of transmission lines, which is the focus of the present work. As a result, this phenomenon is extensively discussed in classical vibration theory references. However, the majority of these references consider the problem only in the context of simply supported beams, for which the natural frequencies can be computed analytically using closed-form solutions, while the modal shapes are described by pure sine functions.

In past decades, many researchers have made considerable efforts to develop solutions for calculating natural frequencies and mode shapes as design tools for the structure of concern. Shaker (1975) studied the effect of compressive axial loads on mode shapes and natural frequencies. Subsequently, Bokaian (1990) developed analytical solutions for calculating natural frequencies and mode shapes for beams subjected to tensile and compressive axial loads, under various classical boundary conditions. Maurizi and Belles (1991) derived the characteristic equation for calculating natural frequencies of beams in compression, under non-classical boundary conditions, including translational and rotational springs at the ends. Liu et al. (1996) developed the characteristic equation and the mode shape expression for free-free beams under tension in more detail. More recent works, such as those by Ceballos and Prato (2008), Foti et al. (2021), and Geuzaine et al. (2021), present solutions related to mode shapes and natural frequencies for beams in tension subjected to the effect of torsional springs at the ends, with the aim of determining structural parameters from experimental tests.

It is a fact that the analytical formulation available in the literature on the subject is well known and developed. However, it has been observed through computational simulations that, in certain applications, such as the prediction of the dynamic behavior of overhead cables, the results can exhibit numerical instabilities for some boundary conditions of concern. That should be addressed, as pointed out below.

In general, the mode shapes found in the literature for Euler–Bernoulli beam problems are described in terms of hyperbolic and trigonometric functions. But authors such as Tang (2003), Gonçalves et al. (2007), and Mazilu (2022) have observed that describing the vibration modes using hyperbolic functions can sometimes lead to numerically unstable modes. The cause of numerical instability is often related to floating-point calculations, particularly when computing the difference between two very large numbers. Floating-point calculations can result in a zero value instead of returning a small but significant number.

Numerical simulations demonstrate that more accurate solutions for the mode shapes of Euler–Bernoulli beams under axial loading are obtained when using a linear combination of trigonometric and exponential terms. That is described in works by Franklin (1989), Main and Jones (2007), and Geuzaine et al. (2021). Nevertheless, further development of this type of solution is still needed for different boundary conditions, the results of which can be used as design tools in various applications.

In this context, the present study aims to compare the two aforementioned ways of describing the analytical solutions for calculating natural frequencies and mode shapes under classical and non-classical boundary conditions. These conditions may be encountered in real-life industrial applications and laboratory bench situations, particularly in problems involving overhead cables of transmission lines.

5.2 Mathematical Development

The differential equation of motion for the free vibration of Euler–Bernoulli beam under tensile axial load is given by (Clough and Penzien 1975):

$$\mathrm{EI}\frac{\partial^4 y(x,t)}{\partial x^4} - T\frac{\partial^2 y(x,t)}{\partial x^2} + \rho A\frac{\partial^2 y(x,t)}{\partial t^2} = 0, \tag{5.1}$$

where EI (Nm^2) is the flexural stiffness, T (N) is the axial tension force, ρA (kg/m) is the mass per unit length, and $y(x,t)$ is the lateral deflection of the beam in axial coordinate x (m) at time t (s).

Equation (5.1) can be solved by using the method of separation of variables. Therefore, the lateral deflection $y(x,t)$ can be described as a function of two variables—a spatial variable $Y(x)$ and a temporal variable $G(t)$—such that

$$y(x,t) = Y(x)G(t), \tag{5.2}$$

the expression of which allows rewriting Eq. (5.1) as

$$-\frac{\mathrm{EI}}{\rho A}\frac{1}{Y(x)}\frac{\partial^4 Y(x)}{\partial x^4} + \frac{T}{\rho A}\frac{1}{Y(x)}\frac{\partial^2 Y(x)}{\partial x^2} = \frac{1}{G(t)}\frac{\partial^2 G(t)}{\partial t^2} = -\omega^2 \tag{5.3}$$

Equation (5.3) allows dividing the problem into two parts, namely, a temporal problem and a spatial problem, represented respectively by Eqs. (5.4) and (5.5), as follows:

$$\frac{\mathrm{d}^2G(t)}{\mathrm{d}t^2}+\omega^2G(t)=0, \tag{5.4}$$

$$\mathrm{EI}\frac{\mathrm{d}^4Y(x)}{\mathrm{d}x^4}-T\frac{\mathrm{d}^2Y(x)}{\mathrm{d}x^2}-\rho A\omega^2Y(x)=0 \tag{5.5}$$

The temporal problem has a harmonic solution of the form

$$G(t)=G_{\mathrm{os}}\mathrm{sen}(\omega t)+G_{\mathrm{oc}}\cos\cos(\omega t), \tag{5.6}$$

where G_{os} and G_{oc} are arbitrary constants to be determined by the initial conditions of the problem.

On the other hand, the spatial problem defines an eigenvalue problem from which the natural frequencies ω_{n_i} and mode shapes $Y_{n_i}(x)$ are evaluated. That is discussed in the following lines.

Introducing the dimensionless coordinate $\overline{x}=\frac{x}{L}$ into Eq. (5.5), where L is the length of the beam, such that $0\leq\overline{x}\leq 1$, it follows that

$$\frac{\mathrm{d}^4Y(\underline{x})}{\mathrm{d}\overline{x}^4}-\frac{TL^2}{\mathrm{EI}}\frac{\mathrm{d}^2Y(\underline{x})}{\mathrm{d}\overline{x}^2}-\frac{\rho A\omega^2L^4}{\mathrm{EI}}Y(\overline{x})=0 \tag{5.7}$$

By also assuming that the solution of Eq. (5.7) has the form $Y(\overline{x})=Ce^{s\overline{x}}$, it yields

$$\left(s^4-\frac{TL^2}{\mathrm{EI}}s^2-\frac{\rho A\omega^2L^4}{\mathrm{EI}}\right)=0 \tag{5.8}$$

Equation 5.8 is quadratic in s^2 and hence can be immediately solved. There are four roots, two of which are pure real roots, while the other two are pure imaginary roots, such that

$$s_1=\pm M,\, s_2=\pm iN, \tag{5.9}$$

where M and N are given by

$$M=L\left\{\left(\frac{T}{2\mathrm{EI}}\right)+\left[\left(\frac{T}{2I}\right)^2+\left(\frac{\rho A}{\mathrm{EI}}\right)\omega^2\right]^{\frac{1}{2}}\right\}^{\frac{1}{2}}=\left(U+\sqrt{U^2+\Omega^4}\right)^{\frac{1}{2}},$$
$$N=L\left\{-\left(\frac{T}{2\mathrm{EI}}\right)+\left[\left(\frac{T}{2\mathrm{EI}}\right)^2+\left(\frac{\rho A}{\mathrm{EI}}\right)\omega^2\right]^{\frac{1}{2}}\right\}^{\frac{1}{2}}=\left(-U+\sqrt{U^2+\Omega^4}\right)^{\frac{1}{2}} \tag{5.10}$$

In the above equations, the dimensionless variables for flexural stiffness β, axial load U, and frequency Ω^4 are defined, respectively, as

$$\beta = \frac{\mathrm{EI}}{TL^2}, U = \frac{1}{2\beta}, \Omega^4 = \frac{\rho A\omega^2 L^4}{\mathrm{EI}} \tag{5.11}$$

The solution of the spatial problem—posed by Eq. (5.7)—is commonly provided in textbooks (Clough and Penzien 1975) as

$$Y\left(\frac{x}{L}\right) = C_1\mathrm{senh}M\overline{x} + C_2\cosh M\overline{x} + C_3\mathrm{sen}N\overline{x} + C_4\cos N\overline{x}, \tag{5.12}$$

where the trigonometric terms represent propagating waves, and the hyperbolic terms represent evanescent waves. C_1, C_2, C_3, and C_4 are integration constants that need to be determined by imposing the boundary conditions.

Bokaian (1990) provides the frequency equations and integration constants for various classical boundary conditions based on Eq. (5.12). However, it is worth noting that the use of trigonometric and hyperbolic functions to express such a solution can sometimes lead to numerical instabilities. This instability arises due to the rapid growth of hyperbolic sine and cosine terms to infinity, while the numerical difference between them remains small. For this reason, authors such as Franklin (1989), Main and Jones (2007), and Geuzaine et al. (2021) emphasize that it is preferable to express the solution through a linear combination of trigonometric and exponential terms, such as

$$Y\left(\frac{x}{L}\right) = R_1 e^{-M\overline{x}} + R_2 e^{-M(1-\overline{x})} + R_3\mathrm{sen}N\overline{x} + R_4\cos N\overline{x}, \tag{5.13}$$

where R_1, R_2, R_3, and R_4 are the new integration constants, which must be obtained by imposing the boundary conditions, as is shown in the next subsections.

5.2.1 *Simply Supported Beam*

The simply supported condition, whose boundary conditions are expressed by $Y(0) = 0$, $\frac{\mathrm{d}^2Y(0)}{\mathrm{d}x^2} = 0$, $Y(L) = 0$ and $\frac{\mathrm{d}^2Y(L)}{\mathrm{d}x^2} = 0$ is the only one that has an analytical solution in closed form. Therefore, the natural frequencies and mode shapes can be respectively calculated by the following expressions:

$$\omega_{n_j} = \frac{\pi^2}{L^2}\sqrt{\frac{\mathrm{EI}}{\rho A}}\left(j^4 + \frac{j^4 TL^2}{\pi^2\mathrm{EI}}\right)^{\frac{1}{2}} \text{ and} \tag{5.14}$$

$$Y_{n_j}(x) = \mathrm{sen}\left(\frac{j\pi x}{L}\right), \tag{5.15}$$

where $j = 1, 2, 3, \ldots$

5.2.2 Clamped—Clamped Beam

Consider now the case of a clamped–clamped beam, with boundary conditions $Y(0) = 0$, $\frac{dY(0)}{dx} = 0$, $Y(L) = 0$ and $\frac{dY(L)}{dx} = 0$. The expression for the *j*th mode shape is given by (Bokaian 1990)

$$Y_{n_j}(x) = \mathrm{senh}\left(\frac{Mx}{L}\right) + \left[\frac{M\,\mathrm{sen}N - N\,\mathrm{senh}M}{N(\cosh M - \cos N)}\right] \cosh\left(\frac{Mx}{L}\right) - \left(\frac{M}{N}\right)\mathrm{sen}\left(\frac{Nx}{L}\right) - \left[\frac{M\,\mathrm{sen}N - N\,\mathrm{senh}M}{N(\cosh M - \cos N)}\right]\cos\left(\frac{Nx}{L}\right), \tag{5.16}$$

where M and N can be obtained by solving the frequency equation

$$\Omega - U\,\mathrm{senh}(M)\mathrm{sen}(N) - \Omega\cosh\,\cosh\,(N)\cos\,\cos\,(N) = 0, \tag{5.17}$$

along with the use of the relationships provided by Eqs. (5.10) and (5.11).

The revised expression of the mode shapes for the clamped–clamped condition—now written in terms of exponential and trigonometric functions—is obtained by applying the boundary conditions in the solution given by Eq. (5.13). The resulting expression is given by

$$Y_{n_j}(x) = \left(\frac{N}{M}\right)\exp\left(\frac{-Mx}{L}\right) + \left(\frac{N}{M}\cos N - \mathrm{sen}N\right) \exp\left[\frac{M(L-x)}{L}\right] + \mathrm{sen}\left(\frac{Nx}{L}\right) - \left(\frac{N}{M}\right)\cos\left(\frac{Nx}{L}\right), \tag{5.18}$$

where M and N are now obtained by solving the following frequency equation:

$$\left(N^2 - M^2\right)\mathrm{sen}(N) + 2MN\cos(N) = 0. \tag{5.19}$$

It is worth noting that Eqs. (5.17) and (5.19), as well as the frequency equations for the other boundary conditions, take the form of a transcendental equation, which has an infinite number of roots. Therefore, these roots must generally be determined numerically.

5.2.3 Clamped—Supported Beam

Addressing the case of a clamped-supported beam, for which the boundary conditions are $Y(0) = 0$, $\frac{dY(0)}{dx} = 0$, $Y(L) = 0$ and $\frac{d^2Y(L)}{dx^2} = 0$, the expression for the *j*th mode shape is given by (Bokaian 1990)

$$
{}_{n_j}(x) = \mathrm{senh}\left(\frac{Mx}{L}\right) - \tanh M \cosh\left(\frac{Mx}{L}\right) - \frac{M}{N}\mathrm{sen}\left(\frac{Nx}{L}\right) + \frac{M}{N}\tan N \cos\left(\frac{Nx}{L}\right), \tag{5.20}
$$

where M and N can be obtained by solving the following frequency equation:

$$
M\cosh(M)\mathrm{sen}(N) - N\mathrm{senh}(M)\cos\cos(N) = 0. \tag{5.21}
$$

The revised mode shape function, now described in terms of exponential and trigonometric terms, arising from the solution provided by Eq. (5.13), is expressed as

$$
\begin{aligned}
Y_{n_j}(x) = {} & \left(\frac{N}{M}\right)\exp\left(\frac{-Mx}{L}\right) + \left(\frac{N}{M}\cos N - \mathrm{sen} N\right) \\
& \exp\left[\frac{M(L-x)}{L}\right] + \mathrm{sen}\left(\frac{Nx}{L}\right) - \left(\frac{N}{M}\right)\cos\left(\frac{Nx}{L}\right),
\end{aligned} \tag{5.22}
$$

where M and N are obtained in this case by solving the frequency equation

$$
\left(M^2 + N^2\right)[N\cos(N) - M\,\mathrm{sen}(N)]. \tag{5.23}
$$

5.2.4 Supported Beam with Torsional Springs

Despite the importance of developing solutions for classical boundary conditions, it is known that, in real situations, end constraints are rarely infinitely rigid in rotation. Thus, it is crucial to analyze boundary conditions that consider the effect of torsional springs at their ends, as illustrated in Fig. 5.1.

The boundary conditions for the structural system under consideration are as follows: at $x = 0$,

$$
Y(x = 0) = 0, \tag{5.24}
$$

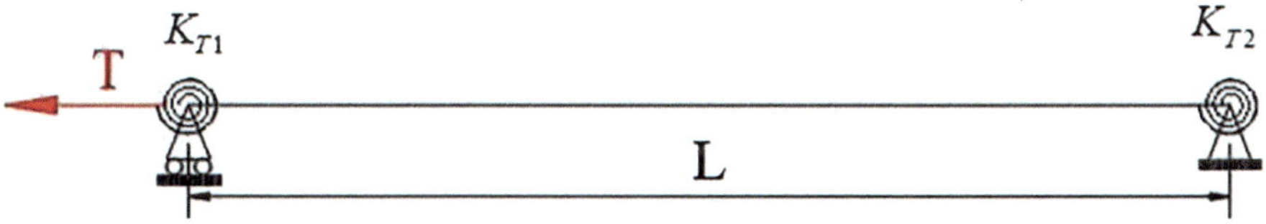

Fig. 5.1 Supported beam under tensile load with torsional springs

$$K_{T1}\frac{Y(x=0)}{dx} = EI\frac{d^2Y(x=0)}{dx^2}, \tag{5.25}$$

and at $x = L$,

$$Y(x=L) = 0, \tag{5.26}$$

$$K_{T2}\frac{dY(x=0)}{dx} = -EI\frac{d^2Y(x=0)}{dx^2}, \tag{5.27}$$

where K_{T1} and K_{T2} are torsional springs constants.

The application of these boundary conditions to Eq. (5.13) results in the following system of algebraic equations:

$$\begin{aligned}
&R_1 + ((-\exp\exp M))R_2 + R_4 = 0,\\
&(\exp\exp(-M))R_1 + R_2 + (\text{sen}N)R_3 + (\cos N)R_4 = 0,\\
&-\left(R_{T1}M^2 + M\right)R_1 + \left[\exp\exp(-M)\left(M - R_{T1}M^2\right)\right]R_2 + NR_3 + R_{T1}N^2R_4 = 0,\\
&\left[\exp(-M)\left(R_{T2}M^2 - M\right)\right]R_1 + \left(R_{T2}M^2 + M\right)R_2\\
&\quad + \left(N\cos N - R_{T2}N^2\text{sen}N\right)R_3 - \left(N\sin N + R_{T2}N^2\cos N\right)R_4 = 0,
\end{aligned} \tag{5.28}$$

where $R_{T1} = \frac{EI}{K_{T1}L}$ and $R_{T2} = \frac{EI}{K_{T2}L}$ are the dimensionless constants of the torsional springs.For the specific case of application in overhead cables, where the structural system is long, slender, and subjected to a high axial force, it is valid to consider that $\beta \ll 1$. That leads to the observation that $e^{-M} \to 0$, and therefore its effect can be neglected. Thus, Eq. (5.28) can be simplified to

$$R_1 + R_4 = 0,$$

$$R_2 + (\text{sen}N)R_3 + (\cos N)R_4 = 0,$$

$$-\left(R_{T1}M^2 + M\right)R_1 + NR_3 + R_{T1}N^2R_4 = 0,$$

$$\left(R_{T2}M^2 + M\right)R_2 + \left(N\cos N - R_{T2}N^2\text{sen}N\right)R_3 - \left(N\sin N + R_{T2}N^2\cos N\right)R_4 = 0. \tag{5.29}$$

For the solution to be nontrivial, the determinant of the coefficient matrix in Eq. (5.29) must be zero, resulting in the following frequency equation:

$$A_c\text{sen}N + B_c\cos N = 0, \tag{5.30}$$

where

Table 5.1 Limit values of the torsional springs constants

Boundary Condition	K_{T1}	K_{T2}
Simply—supported	$K_{T1} \to 0$	$K_{T2} \to 0$
Clamped—clamped	$K_{T1} \to \infty$	$K_{T2} \to \infty$
Clamped—supported	$K_{T1} \to \infty$	$K_{T2} \to 0$

$$\begin{aligned} A_c &= N^2\left(1 - R_{T1}R_{T2}N^2\right) - M^2\left(MR_{T1} + MR_{T2} + M^2R_{T1}R_{T2} + 1\right) \\ &\quad - MN^2(R_{T1} + R_{T2}) - 2R_{T1}R_{T2}M^2N^2, \\ B_c &= N^3(R_{T1} + R_{T2}) + M^2N(R_{T1} + R_{T2}) + 2MN. \end{aligned} \tag{5.31}$$

By solving the homogeneous system of algebraic equations for the unknowns R_1, R_2, R_3 and R_4, it is obtained the expression for the mode shape, given by

$$\begin{aligned} Y_{n_j}(x) &= \sin\left(\frac{Nx}{L}\right) - \sin(N)\exp\left[-M\left(\frac{L-x}{L}\right)\right] \\ &\quad + \beta_n\left\{\exp(-Mx) + \cos(N)\exp\left[-M\left(\frac{L-x}{L}\right)\right] - \cos\left(\frac{Nx}{L}\right)\right\} \end{aligned} \tag{5.32}$$

where

$$\beta_n = \frac{N}{R_{T1}N^2 + R_{T1}M^2 + M}. \tag{5.33}$$

In order to adapt the above formulation to a system with classical boundary conditions, the torsional springs constants should be modified, as shown in Table 5.1.

5.3 FEM Solution

To validate the analytical solutions presented in the previous section, a numerical model using the Finite Element Method (FEM) was also considered. This is associated to an eigenvalue problem, which can be expressed as

$$\left(K - \omega^2 M\right)\varphi = 0, \tag{5.34}$$

where M is the mass matrix of the structure, and K is the corresponding stiffness matrix. Also in Eq. (5.34), ω^2 are the eigenvalues and φ are the eigenvectors. In this case, all the eigenvalues are real and positive, apart from being the squares of the natural frequencies of the system.

The known mass and stiffness matrices for the four-degree-of-freedom linear planar beam finite elements are (Yang and Kuo 1994):

$$M^{\mathrm{e}} = \frac{\rho A L_{\mathrm{e}}}{420}\begin{bmatrix} 156 & 22L_{\mathrm{e}} & 54 & -13L_{\mathrm{e}} \\ 22L_{\mathrm{e}} & 4L_{\mathrm{e}}^2 & 13L_{\mathrm{e}} & -3L_{\mathrm{e}}^2 \\ 54 & 13L_{\mathrm{e}} & 156 & -22L_{\mathrm{e}} \\ -13L_{\mathrm{e}} & -3L_{\mathrm{e}}^2 & -22L_{\mathrm{e}} & 4L_{\mathrm{e}}^2 \end{bmatrix} \tag{5.35}$$

and

$$K^{\mathrm{e}} = K_{\mathrm{E}}^{\mathrm{e}} + K_{\mathrm{T},}^{\mathrm{e}} \tag{5.36}$$

where

$$K_{\mathrm{E}}^{\mathrm{e}} = \frac{EI}{L_e^3}\begin{bmatrix} 12 & 6L_{\mathrm{e}} & -12 & 6L_{\mathrm{e}} \\ 6L_{\mathrm{e}} & 4L_{\mathrm{e}}^2 & -6L_{\mathrm{e}} & 2L_{\mathrm{e}}^2 \\ -12 & -6L_{\mathrm{e}} & 12 & -6L_{\mathrm{e}} \\ 6L_{\mathrm{e}} & 2L_{\mathrm{e}}^2 & -6L_{\mathrm{e}} & 4L_{\mathrm{e}}^2 \end{bmatrix}$$

$$K_{\mathrm{T}}^{\mathrm{e}} = \frac{\mathrm{T}}{30L_{\mathrm{e}}}\begin{bmatrix} 36 & 3L_{\mathrm{e}} & -36 & 3L_{\mathrm{e}} \\ 3L_{\mathrm{e}} & 4L_{\mathrm{e}}^2 & -3L_{\mathrm{e}} & -L_{\mathrm{e}}^2 \\ -36 & 3L_{\mathrm{e}} & 36 & -3L_{\mathrm{e}} \\ 3L_{\mathrm{e}} & -L_{\mathrm{e}}^2 & -3L_{\mathrm{e}} & 4L_{\mathrm{e}}^2 \end{bmatrix} \tag{5.37}$$

and L_e is the element length.

The FEM model is implemented in the MATLAB® environment with 150 elements.

5.4 Numerical Simulations

The formulations presented in the preceding sections were compared by means of numerical simulations. In those, an aluminum alloy conductor (CAL), used as an overhead cable in power lines, was considered, with the following parameters: flexural stiffness EI $= 271.3$ Nm^2, mass per unit length $\rho A = 0.6870$ kg/m, span length $L = 30.2$ m, and tensile force $T = 13091$ N.

All the analytical solutions developed in Sect. 5.2 were implemented in the MATLAB® environment. In order to solve the transcendental frequency equations, the bisection method was applied using the existing fzeros function in Matlab. Consequently, once the jth natural frequency ω_{n_j} is obtained, it is possible to calculate the corresponding jth mode shape $Y_{n_j}(x)$.

The following abbreviations are also used below: E—T denotes the solution described in exponential and trigonometric terms, while H—T refers to the solution described in hyperbolic and trigonometric terms.

5.4.1 Clamped—Clamped Condition

For the clamped–clamped boundary condition, the first eight natural frequencies obtained from the frequency equations given by Eqs. (5.17), (5.19) and by the Eq. (5.30) with $K_{T1} \rightarrow \infty$ and $K_{T2} \rightarrow \infty$, as well as the corresponding natural frequencies obtained by FEM are shown in Table 5.2.

Figures 5.2 and 5.4 show the 7th and 8th vibration mode shapes, respectively, obtained from the numerical simulations for the clamped–clamped condition. Figures 5.3 and 5.5 provide an enlarged view of the numerical results, focusing on the regions surrounding the right and left ends of the cable. In addition, a zoomed-in view of the numerically unstable region is presented for the 7th and 8th mode shapes.

From Figs. 5.2, 5.3, 5.4 and 5.5, it is evident that the exponential and trigonometric-based solution shows strong agreement with FEM results for the clamped–clamped condition. Moreover, this solution provides a more accurate depiction, particularly

Table 5.2 Values of natural frequencies in Hz for the clamped–clamped boundary condition

Natural frequencies	$j=1$	$j=2$	$j=3$	$j=4$	$j=5$	$j=6$	$j=7$	$j=8$
Exponential and trigonometric terms[a]	2.306	4.614	6.239	9.239	11.560	13.887	16.222	18.567
Hyperbolic and trigonometric terms[b]	2.306	4.614	6.239	9.239	11.560	13.887	16.222	18.567
Rotational Springs[c]	2.306	4.614	6.239	9.239	11.560	13.887	16.222	18.567
FEM	2.306	4.614	6.239	9.239	11.560	13.887	16.223	18.568

[a] Eq. (5.17); [b] Eq. (5.19); [c] Eq. (5.30) with $K_{T1} \rightarrow \infty$ and $K_{T2} \rightarrow \infty$

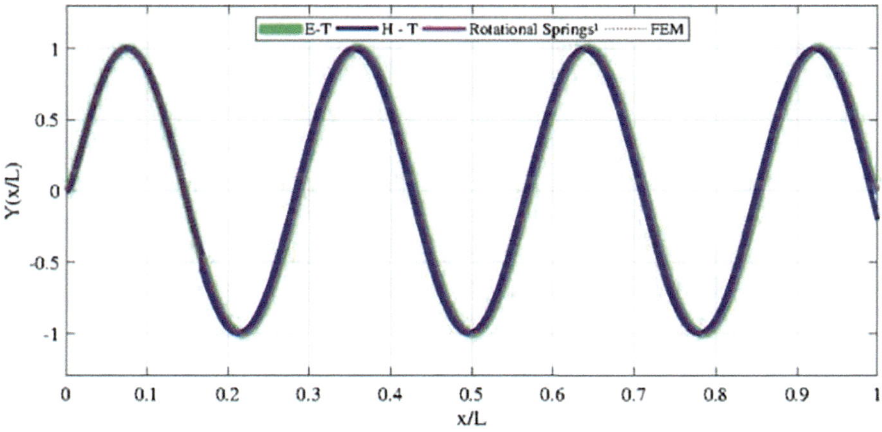

Fig. 5.2 Clamped—clamped boundary condition—7th mode shape. Equation (5.32) with $K_{T1} \rightarrow \infty$ and $K_{T2} \rightarrow \infty$

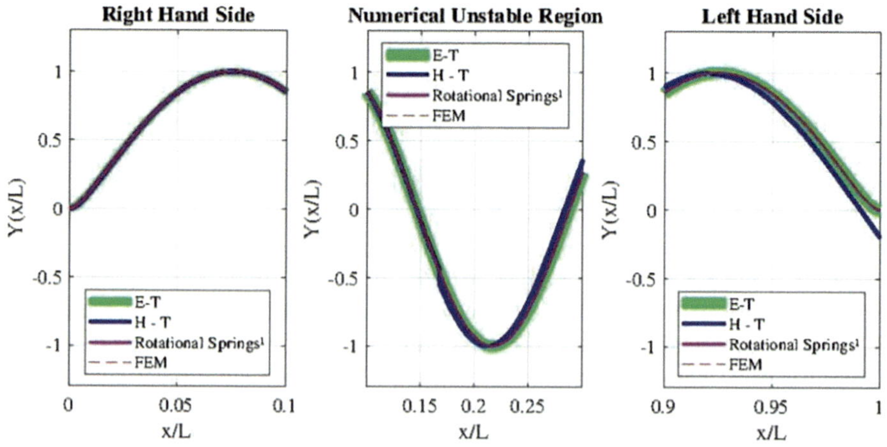

Fig. 5.3 Zoomed-in view of the right and left-hand side ends—clamped—clamped boundary condition—7th mode shape. a Eq. (5.32) with $K_{T1} \to \infty$ and $K_{T2} \to \infty$

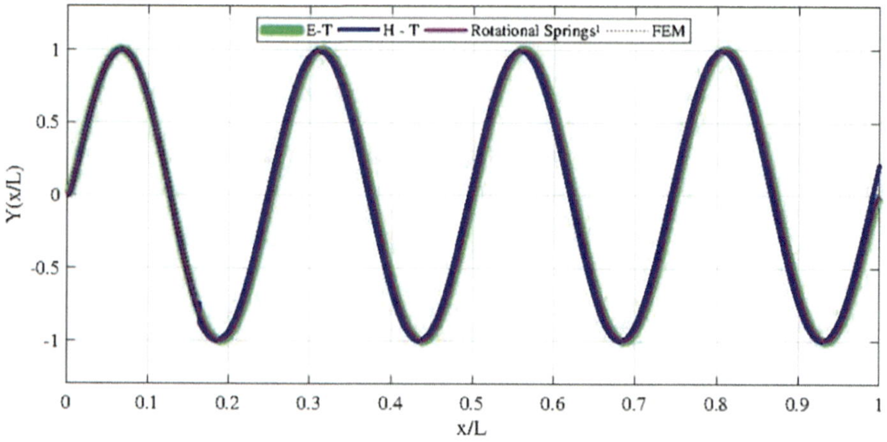

Fig. 5.4 Clamped—clamped boundary condition—8th mode shape. a Eq. (5.32) with $K_{T1} \to \infty$ and $K_{T2} \to \infty$

in the boundary regions, when compared to the hyperbolic and trigonometric-based solution.

5.4.2 Clamped—Supported Condition

For the clamped-supported boundary condition, the first eight natural frequencies obtained from the frequency equations given by the Eqs. (5.21), (5.23) and by the

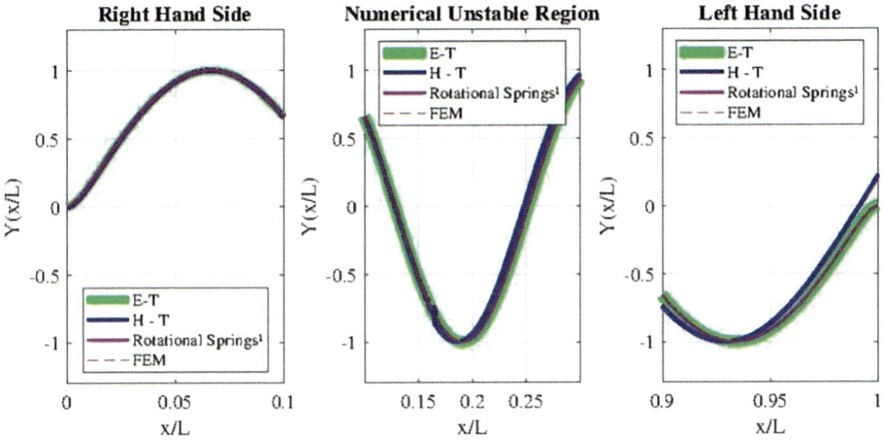

Fig. 5.5 Zoomed-in view of the right and left-hand side ends—clamped—clamped boundary condition—8th mode shape. a Eq. (5.32) with $K_{T1} \rightarrow \infty$ and $K_{T2} \rightarrow \infty$

Eq. (5.30) with $K_{T1} \rightarrow \infty$ and $K_{T2} \rightarrow 0$, as well as the corresponding natural frequencies obtained by FEM are shown in Table 5.3.

Figures 5.6 and 5.8 show the 7th and 8th vibration mode shapes, respectively, obtained from the numerical simulations for the clamped-supported condition. Figures 5.7 and 5.9 show an enlarged view of the numerical results, focusing on the regions around the right and left ends of the cable. These figures also provide a zoomed-in view of the numerically unstable region for the 7th and 8th mode shapes.

Figures 5.6, 5.7, 5.8 and 5.9 clearly show the strong agreement between the exponential and trigonometric-based solution and the FEM results for the clamped-supported condition.

Therefore, it can be stated that for both the clamped–clamped condition and the clamped-supported condition, there is a strong agreement between the results obtained by the Finite Element Method (FEM) and those obtained from Eqs. (5.18),

Table 5.3 Values of natural frequencies in Hz for the clamped-supported boundary condition

Natural Frequencies	$j=1$	$j=2$	$j=3$	$j=4$	$j=5$	$j=6$	$j=7$	$j=8$
Exponential and trigonometric terms[a]	2.296	4.593	6.893	9.197	11.507	13.824	16.149	18.483
Hyperbolic and trigonometric terms[b]	2.296	4.593	6.893	9.197	11.507	13.824	16.149	18.483
Rotational Springs[c]	2.296	4.593	6.893	9.197	11.507	13.824	16.149	18.483
FEM	2.296	4.593	6.894	9.198	11.508	13.825	16.150	18.484

[a] Eq. (5.21); [b] Eq. (5.23); [c] Eq. (5.30) with $K_{T1} \rightarrow \infty$ and $K_{T2} \rightarrow 0$

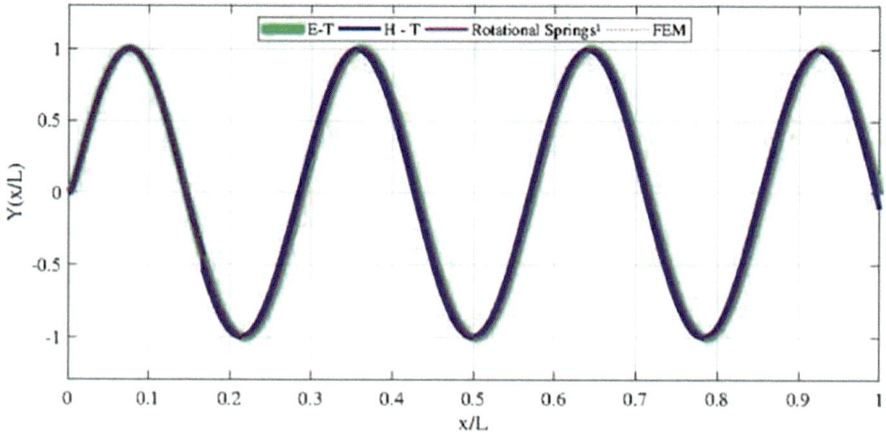

Fig. 5.6 Clamped—supported boundary condition—7th mode shape. a Eq. (5.32) with $K_{T1} \rightarrow \infty$ and $K_{T2} \rightarrow 0$

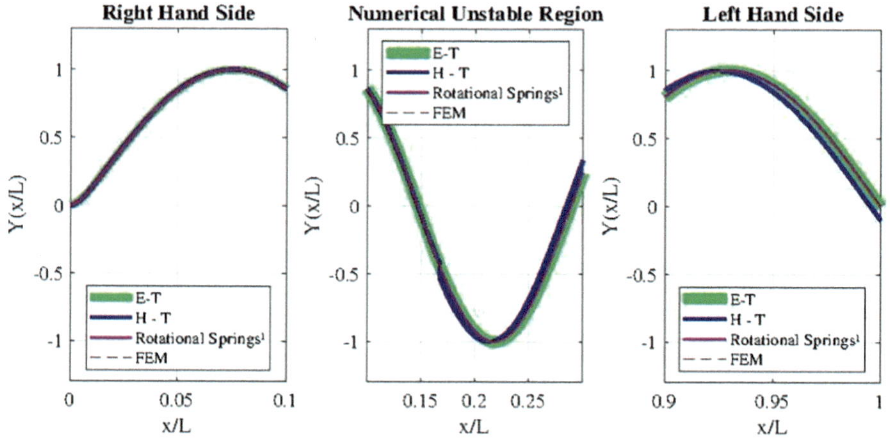

Fig. 5.7 Zoomed-in view of the right and left-hand side ends—clamped—supported boundary condition—7th mode shape. a Eq. (5.32) with $K_{T1} \rightarrow \infty$ and $K_{T2} \rightarrow 0$

(5.22) and (5.32), where the solution is expressed in exponential and trigonometric terms for the corresponding boundary conditions.

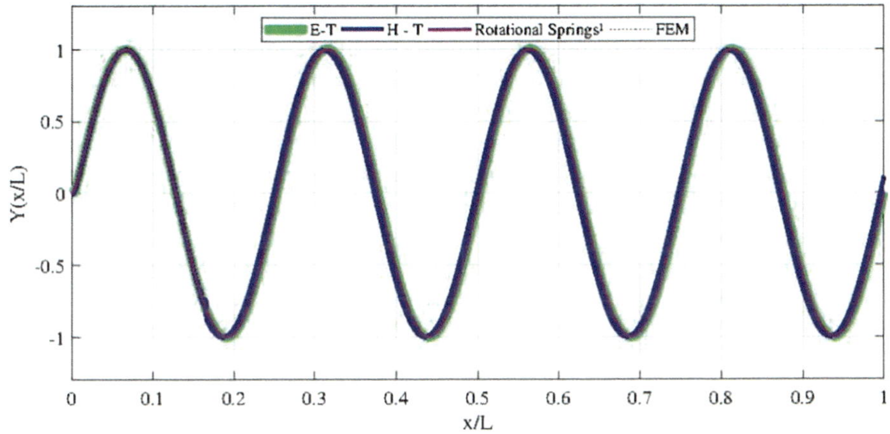

Fig. 5.8 Clamped—supported boundary condition—8th mode shape. a Eq. (5.32) with $K_{T1} \to \infty$ and $K_{T2} \to 0$

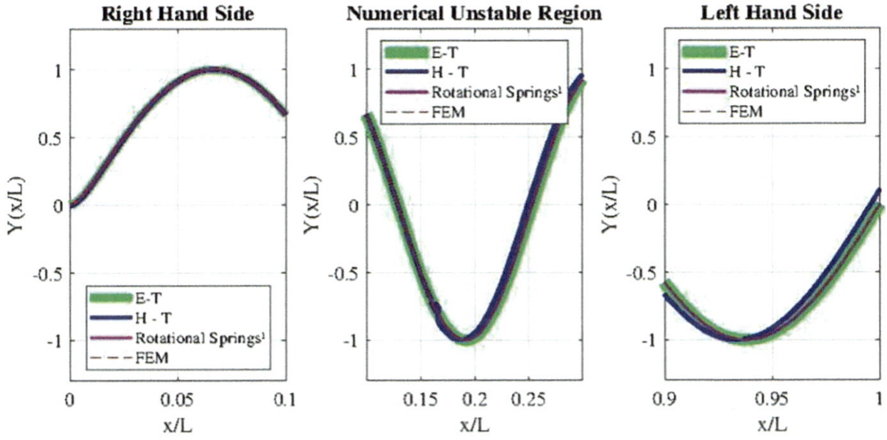

Fig. 5.9 Zoomed-in view of the right and left-hand side ends—clamped—supported boundary condition—8th mode shape. a Eq. (5.32) with $K_{T1} \to \infty$ and $K_{T2} \to 0$

5.5 Conclusions

It has been shown that the solution expressed in exponential and trigonometric terms gives a remarkably accurate prediction, especially near the ends of the considered cable. Therefore, it is understood that the formulation presented here constitutes a significant improvement in the prediction of vibration mode shapes for problems involving overhead cables of transmission lines.

Moreover, it is essential to emphasize that achieving a more accurate representation of mode shapes and natural frequencies facilitate a more detailed characterization of Frequency Response Functions (FRFs) in studies of concern. These FRFs hold significant potential for equivalent parameter identification in overhead cables, in which there are some typically unknown dynamic pieces of information, such as those related to damping. This identification process can be carried out through comparative analysis with experimental data, aiming the better understanding of the mechanical properties of those cables.

Acknowledgements José M. Balthazar, Alexandre M. Wahrhaftig and Eduardo M. O. Lopes acknowledge the support of CNPq—National Council for Scientific and Technological Development for their scholarships

References

Bokaian A (1990) Natural frequencies of beams under tensile axial loads. J Sound Vib 142(3):481–498

Ceballos MA, Prato CA (2008) Determination of the axial force on stay cables accounting for their bending stiffness and rotational end restraints by free vibration tests. J Sound Vib 317(1–2):127–141

Clough R, Penzien J (1975) Dynamics of Structures. McGraw-Hill NY, USA

Foti F, Geuzaine M, Denoël V (2021) On the identification of the axial force and bending stiffness of stay cables anchored to flexible supports. Appl Math Model 92:798–828

Franklin PA (1989) Modal analysis of long systems: a transfer matrix approach. Ph.D. thesis, Johns Hopkins University

Geuzaine M, Foti F, Denoël V (2021) Minimal requirements for the vibration-based identification of the axial force, the bending stiffness and the flexural boundary conditions in cables. J Sound Vib 511:116326

Gonçalves P, Brennan M, Elliott S (2007) Numerical evaluation of high-order modes of vibration in uniform Euler-Bernoulli beams. J Sound Vib 301(3–5):1035–1039

Liu X, Ertekin R, Riggs H (1996) Vibration of a free-free beam under tensile axial loads. J Sound Vib 190(2):273–282

Main JA, Jones NP (2007) Vibration of tensioned beams with intermediate damper: formulation, influence of damper location. J Eng Mech 133(4):369–378

Maurizi M, Belles P (1991) General equation of frequencies for vibrating uniform one-span beams under compressive axial loads. J Sound Vibration 145(2):345–347

Mazilu T (2022) Numerically stable form of Green's function for a free-free uniform Timoshenko beam. Mathematics 11(1):86

Shaker FJ (1975) Effect of axial load on mode shapes and frequencies of beams. Nasa technical note (nasa-tn-d-8109)

Tang Y (2003) Numerical evaluation of uniform beam modes. J Eng Mech 129(12):1475–1477

Yang YB, Kuo SR (1994) Theory & analysis of nonlinear framed structures. Pearson Ed

Chapter 6
A Web-Based Simulator for Leak Detection Problems in Water Distribution Pipes Using Vibro-Acoustic Techniques

Fabrício César Lobato de Almeida, Michael John Brennan, Maurício Kiotsune Iwanaga, Sillas de Oliveira Cézar, Luís Armando Ferreira Bispo, and Oscar Scussel

Abstract Reduction of non-revenue water is one of the main goals of the water industry. It is estimated that 40% of water pumped via distribution systems in Brazil is lost, but there are states which have losses of 70%. One way of minimizing this problem is by massive investment in replacing old metallic pipes by plastic pipes made from PVC (Polyvinyl Chloride) and HDPE (High Density Polyethylene) material. Furthermore, early detection and location of leaks plays a major role in reducing losses. This can be conducted using vibro-acoustics techniques, but there are problems with plastic pipes compared to metallic pipes, especially when using leak noise correlators. This is because the leak noise that propagates through the pipe is heavily attenuated in plastic pipes reducing the amplitude and hence distance that this noise can propagate. Leak noise correlators use knowledge of the leak noise velocity and an estimate of the difference between the arrival times of the leak noise at two sensors placed on the pipe at available access points. Generally, the pipe geometry (wall thickness and pipe nominal radius) and its Young's modulus have a profound effect on leak noise attenuation and velocity. Moreover, the surrounding medium also has an additional effect. In this paper some web-based software is described that simulates different scenarios to help users of leak noise correlators to understand the physics behind the problem and to help them select the properties of some filters that need to be set to obtain reliable results. The software is based on a wave-type model which takes into account the pipe vibration response due to excitation by a leak. Inputs to the model are the pipe geometry, material properties of the pipe and surrounding medium, distance between the sensors and sensor type (accelerometer, geophone and hydrophone). These parameters are then used to estimate the leak velocity, time delay, frequency bandwidth (CPSD-Cross-Power Spectral Density) and leak location, which are given in a user-friendly graphical interface.

F. C. L. de Almeida (✉) · M. J. Brennan · S. de Oliveira Cézar · L. A. F. Bispo
Department of Mechanical Engineering, School of Engineering, UNESP, Bauru, Brazil
e-mail: fabricio.lobato@unesp.br

M. K. Iwanaga · O. Scussel
Department of Mechanical Engineering, Faculty of Engineering, UNESP, Ilha Solteira, Brazil

P. Kurka and M. Pereira (eds.), *Advances in Structural Vibration*, Lecture Notes in Mechanical Engineering, https://doi.org/10.1007/978-3-031-71540-2_6

6.1 Introduction

Climate change and high demand of potable water in big cities have increased the need to manage water loss in water distribution systems (Eliasson 2015; Hamilton and McKenzie 2014). Recently, long draught seasons have stressed cities around Brazil such as São Paulo city in 2014/2015 (Millington 2018), which was very close to collapse due to the lack of potable water, and more recently in 2021 (Getirana et al. 2021). This problem, however, is a worldwide issue that affects many countries around the globe (Greve et al. 2018; Liemberg and Wyatt 2019). Hence, strategies to reduce non-revenue water have been put in place by water distribution companies (Zaman et al. 2020). These strategies use different technologies to locate leaks in buried water pipes (El-Zahab and Zayed 2019; Islam et al. 2022). The simplest one is the balance between the pressure versus flow to locate regions where leaks are more likely to be present (Al Qahtani et al. 2020).

Many techniques can be applied to refine the search for the leak position. Some of these are invasive, such as smart ball (Fletcher and Chandrasekaran 2008), and others not invasive like the ground penetrating radar (Cataldo et al. 2014; Abouhamad et al. 2016), thermal analysis (Fahmy and Moselhi 2009) and more recently satellite imaging (Martins et al. 2019). Furthermore, vibro-acoustic techniques are generally more popular in leak detection surveys. The simplest vibro-acoustic technique involves a listening stick, which is a metallic rod placed on the pipe, so that the pipe vibration is transmitted through the rod to the "user's ear" (Hamilton and McKenzie 2014). Similar to the listening stick there are geophones that are placed on the surface right above the buried pipe to detect the leak and estimate its position (Hamilton and McKenzie 2014). These two techniques rely on user experience and provide a single measurement only, so that leak detection is subjective and the signals are easy to misinterpret. Leak noise correlators employ a vibro-acoustic technique where at least two sensors are used to pinpoint the position of a leak position (Fuchs and Riehle 1991; Hamilton and McKenzie 2014). This technique can be invasive when hydrophones are used, but generally accelerometers are applied, which are placed either directly on the pipe wall or to any access point where the pipe vibration can be measured, i.e. hydrants (Gao et al. 2004; Almeida et al. 2014). Leak noise correlators use knowledge of the leak noise velocity, and an estimate of the difference between the arrival times of the leak noise at two sensors placed on the pipe at available access points. It needs a good estimation of the time delay and leak velocity to successfully pinpoint the leak position (ADS 2010; Aquasave 2015; FUJI TECOM 2022). Although it seems straight forward, the user has to set up filters prior to calculating the estimated time delay together with defining the pipe properties to estimate the leak velocity (Gao et al. 2004). In most cases, commercial leak noise correlators do not take into account the effects of the surrounding medium on the leak noise, which has a direct effect on the leak noise attenuation (filter bandwidth) and velocity (Brennan et al. 2018).

This book chapter describes some web-based software developed to aid professionals that use leak noise correlators. The software is based on a wave model that accounts for the pipe properties (material and geometry), surrounding medium (soil type) and sensor (hydrophone, geophone or accelerometer) used to measure pipe vibration or acoustic pressure due to a leak. These inputs are provided by the user, so that a normalized modulus of the cross-power spectral density (CPSD), the phase, normalized cross-correlation and leak velocity estimation are provided as the software outputs. The first is used to estimate the frequency bandwidth over which the leak energy is mostly located. The phase and normalized modulus of the cross-correlation function are used to estimate the time delay in the frequency and time domains, respectively. Finally, the leak noise velocity is estimated via the wave that carries most of the leak energy (Pinnington and Briscoe 1994; Brennan et al. 2018; Scussel et al. 2023). These simulations are performed using two different user-friendly interfaces. The results provided by the web-based software are compared to actual leak noise data to show this software is representative of actual situations and can be used to aid leak detection teams.

6.2 Classic Signal Processing for Time Delay Estimation

As mentioned previously, a leak noise correlator uses the estimation of the time of arrival between the leak excitation signal collected via at least two sensors placed at available access points. Moreover, the speed at which the leak noise propagates along the pipe is also used to pinpoint the leak position. Figure 6.1 shows a schematic depicting how the leak noise energy is dissipated along the pipe and into the surrounding medium. The model is discussed in the next section. This section focuses on the signal processing used to estimate the time delay and leak velocity.

The leak position with reference to sensor 1 is given by (Gao et al. 2004):

$$d_1 = \frac{d - cT_o}{2}, \tag{6.1}$$

where d is the total distance between the sensors in metres, c is the leak noise velocity in m/s and T_o is the time of arrival or the time delay in seconds between $x_1(t)$ and $x_2(t)$, which are the leak noise signals at Sensors 1 and 2 respectively

The process of time delay estimation has a few steps. First it is important to calculate the Cross Power Spectral Density (CPSD) function, which can be used to observe the power distribution of the signal over a frequency range. This distribution can be used to provide an initial estimation of the frequency range where the leak signal is likely to be located in practice. This estimation is used to set up a bandpass filter prior to the calculation of the time delay via the cross-correlation function. The CPSD function $S_{x_1x_2}(\omega)$ is defined as (Shin and Hammond 2008):

$$S_{x_1x_2}(\omega) = \lim_{T\to\infty} \frac{E[X_1^*(\omega)X_2(\omega)]}{T} \tag{6.2}$$

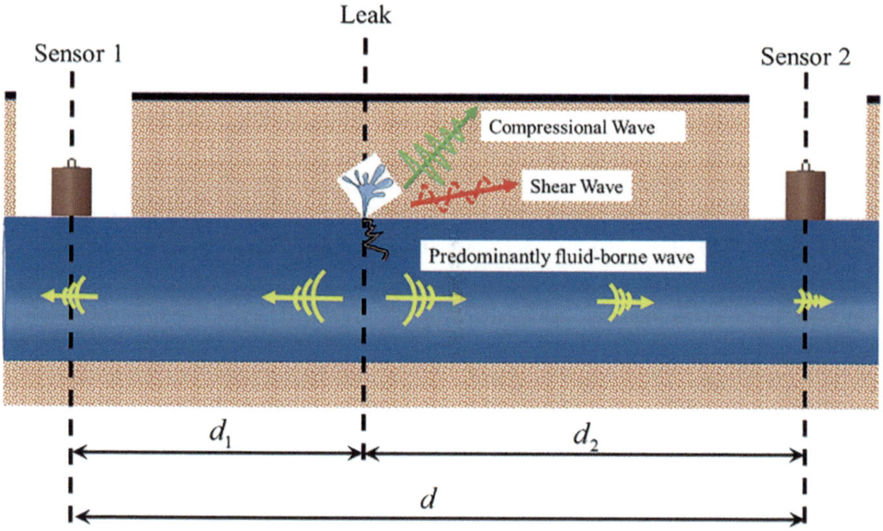

Fig. 6.1 Schematic diagram depicting how the leak energy propagates along the pipe and radiates throughout the soil via different wave types

where $E[]$ is the expectation operator, $X_1(\omega)$ and $X_2(\omega)$ are the Fourier transforms of $x_1(t)$ and $x_2(t)$ respectively, $*$ denotes complex conjugate, ω is frequency in rad/s, and T is the time duration of a signal segment. Note that $S_{x_1x_2}(\omega)$ can be written as $\left|S_{x_1x_2}(\omega)\right|e^{j\phi(\omega)}$ in which $\left|S_{x_1x_2}(\omega)\right|$ is the modulus of the CPSD, $\phi(\omega)$ is the phase between the two signals and $j = \sqrt{-1}$. . If there is a pure delay in the system (buried pipe excited by a leak) given by T_0, then $\phi(\omega) = \omega T_0$. It is important to note that the CPSD contains modulus and phase information, where the modulus can be used to estimate the frequency bandwidth as mentioned previously, and the phase is used to estimate the time delay (time of arrival) between the signals $x_1(t)$ and $x_2(t)$. These signals are collected using sensors, generally placed at pipe access points (Gao et al. 2004).

The cross-correlation function between the measured signals $x_1(t)$ and $x_2(t)$ whose means are set to zero, is given by (Shin and Hammond 2008):

$$R_{x_1x_2}(\tau) = E[x_1(t)x_2(t+\tau)], \tag{6.3}$$

where τ is the delay. The delay that results in the highest similarity between the two signals $x_1(t)$ and $x_2(t)$ (indicated by a peak in the cross-correlation function) is considered to be an estimate T_0. The cross-correlation function is also related to the Fourier Transform of the CPSD function by (Shin and Hammond 2008):

$$R_{x_1x_2}(\tau) = \int_{-\infty}^{+\infty} S_{x_1x_2}(\omega)e^{(j\omega\tau)}d\omega \tag{6.4}$$

The velocity of leak noise in the pipe can be calculated via standard tables which come with commercial correlators or via analytical equations. The latter is based on wave models being one of which described in the next section. This is the same model used in the web-based software introduced in this work.

6.3 Wave Model Overview

The CPSD between the two signals $x_1(t)$ and $x_2(t)$ for a buried water pipe can also be estimated using the system's Frequency Response Function (FRF) and the leak spectrum characteristics $S_{ll}(\omega)$ (Scussel et al. 2021) by

$$S_{x_1x_2}(\omega) = S_{ll}(\omega)H_1^*(\omega, d_1)H_2(\omega, d_2), \tag{6.5}$$

where $H_1(\omega, d_1)$ and $H_2(\omega, d_2)$ are the FRFs of the pipe sections with length d_1 and d_2, respectively. Furthermore, to simplify the model the leak spectrum is assumed to have white noise characteristics (Gao et al. 2004), hence $S_{ll}(\omega) = S_o$. The pipe model FRF used in this paper assumes that the leak noise energy is transmitted along the pipe via a predominantly fluid-borne wave, in which there is a strong coupling between the water and the pipe wall (Pinnington and Briscoe 1994; Muggleton et al. 2002). This wave is shown in Fig. 6.1 and its main characteristic is such that either acoustic (i.e. hydrophones) or vibration (i.e. accelerometers/geophones) sensors can be used with leak noise correlators. Furthermore, the leak energy can also radiate through the soil via shear and compressional waves, also depicted in Fig. 6.1. These two mechanisms can occur simultaneously in the system when the soil presents a sandy-like characteristic, but only compressional waves are excited in stiff clay-like soils as discussed by Brennan et al. (2018). Thus, the surrounding medium and the pipe material have a direct effect on the leak energy attenuation and velocity in the pipe. This is taken into account in the model described herein. Reflections in the model are also neglected and the frequency content where the model is valid is well below the pipe ring frequency. Moreover, the acoustic wavelength of water is assumed to be much greater than the diameter of the pipe). Hence, the FRF for the pipe is given by (Gao et al. 2005).

$$H(\omega, x) = (j\omega)^n e^{-j\omega x/c} e^{-\omega\beta x}, \tag{6.6}$$

where x is the distance from the leak position to any measurement position, i.e. d_1 or d_2, $c = \omega \big/ \mathrm{Re}\{k\}$, $\beta = -\mathrm{Im}\{k\} \big/ \omega$ is related to the wave attenuation, and k is the wavenumber of the predominantly fluid-borne wave. The term $(j\omega)^n$ is related to the type of sensor used for collecting the pipe response, so n can be set to 0, 1 or 2 for pressure/displacement, velocity and acceleration measurements, respectively. It is important to note that the attenuation is a function of d. The wavenumber is given by (Scussel et al. 2023):

$$k = k_{\mathrm{w}}\left(1 + \frac{K_{\mathrm{w}}}{K_{\mathrm{p}} + K_{\mathrm{p\_m}} + K_{\mathrm{m}}}\right)^{1/2} \tag{6.7}$$

where $k_{\mathrm{w}} = \omega / c_{\mathrm{w}}$, is the free water wavenumber in which $c_{\mathrm{w}} = 1500$ m/s is the speed of sound in water, and $K_{\mathrm{w}} = (2B_{\mathrm{w}}/a) / \left[\left(k^2 / k_{\mathrm{w}}^2\right) - 1\right]$, $K_{\mathrm{p}} = E_{\mathrm{p}}^{*} b / [a^2(1 - \nu_{\mathrm{p}}^2)] - \omega^2 \rho_{\mathrm{p}} b$, $K_{\mathrm{p\_m}} = (\overline{K}_{\mathrm{p}} + \overline{K}_{\mathrm{m}})^2 / (\tilde{K}_{\mathrm{p}} + \tilde{K}_{\mathrm{m}})$ and K_{m} are, respectively, the wave dynamic stiffnesses of the water, pipe, interaction between pipe and surrounding medium and the dynamic stiffness of the surrounding medium (soil); B_{w} is the bulk modulus of water, $B_{\mathrm{m}}^{*} = B_{\mathrm{m}}(1 + j\eta_{\mathrm{d}})$ and $G_{\mathrm{m}}^{*} = G_{\mathrm{m}}(1 + j\eta_{\mathrm{s}})$ are the complex bulk and shear moduli of the surrounding medium (soil) respectively, in which B_{m} is the Bulk modulus and η_{d} is the compressional loss factor, and G_{m} is the shear storage modulus, η_{s} is the shear loss factor; the pipe geometry and material properties are given by the mean radius a, the wall thickness b, the density ρ_{p}, the Poisson's ratio ν_{p}, and the complex Young's modulus $E_{\mathrm{p}}^{*} = E_{\mathrm{p}}(1 + j\eta_{\mathrm{p}})$ in which E_{p} is the storage modulus and η_{p} the loss factor. Furthermore, it can be seen that $K_{\mathrm{p\_m}}$ involves several terms, which are given by (Scussel et al. 2023):

$$\overline{K}_{\mathrm{p}} = j\nu_{\mathrm{p}} k E_{\mathrm{p}}^{*} b / [a(1 - \nu_{\mathrm{p}}^2)] \tag{6.8}$$

$$\overline{K}_{\mathrm{m}} = G_{\mathrm{m}}\big[j(2 - \alpha \overline{H}_{\mathrm{d}} k)\big] \tag{6.9}$$

$$\tilde{K}_{\mathrm{p}} = k^2 E_{\mathrm{p}}^{*} b / (1 - \nu_{\mathrm{p}}^2) - \omega^2 \rho_{\mathrm{p}} b \tag{6.10}$$

$$\tilde{K}_{\mathrm{m}} = -G_{\mathrm{m}} \alpha k_{\mathrm{d}}^{\mathrm{R}} \tag{6.11}$$

where $\alpha = k_{\mathrm{s}}^2 / (k_{\mathrm{d}}^{\mathrm{R}} k_{\mathrm{s}}^{\mathrm{R}} \overline{H}_{\mathrm{s}} + k^2 \overline{H}_{\mathrm{d}})$, $\overline{H}_{\mathrm{s}} = H_0(k_{\mathrm{s}}^{\mathrm{R}} r) / H_0'(k_{\mathrm{s}}^{\mathrm{R}} r)$, $\overline{H}_{\mathrm{d}} = H_0(k_{\mathrm{d}}^{\mathrm{R}} r) / H_0'(k_{\mathrm{d}}^{\mathrm{R}} r)$ in which $H_0(\bullet)$ is a Hankel function of the second kind of zero order related to conical waves that radiate throughout the soil, $'$ denotes a spatial derivative, r is the radial distance from the pipe center, and the superscript R denotes the radial direction; $k_{\mathrm{s}}^{\mathrm{R}} = \sqrt{k_{\mathrm{s}}^2 - k^2}$, where $k_{\mathrm{s}} = \omega / c_{\mathrm{s}}$ is the shear wavenumber for the soil and $c_{\mathrm{s}} = \sqrt{G_{\mathrm{m}}^{*} / \rho_{\mathrm{m}}}$ is the related wave speed; $k_{\mathrm{d}}^{\mathrm{R}} = \sqrt{k_{\mathrm{d}}^2 - k^2}$, in which $k_{\mathrm{d}} = \omega / c_{\mathrm{d}}$ is the compressional/dilatational wavenumber for the soil and $c_d = \sqrt{(B_{\mathrm{m}}^{*} + 4G_{\mathrm{m}}^{*} / 3) / \rho_{\mathrm{m}}}$ is the related wave speed. The medium stiffness is given by

$$K_{\mathrm{m}} = \frac{2}{a} + \alpha k_{\mathrm{s}}^{\mathrm{R}} \overline{H}_{\mathrm{s}} \overline{H}_{\mathrm{d}} \tag{6.12}$$

Hence, the cross-correlation function for the model is computed as,

$$R_{x_1x_2}(\tau) = S_o \int_{-\infty}^{+\infty} W(\omega)H_1^*(\omega, d_1)H_2(\omega, d_2)e^{(j\omega\tau)}d\omega \tag{6.13}$$

where $W(\omega)$ is the FRF of the filter used prior to perform the correlation. The frequency limits of the filter are determined by analyzing the modulus of the CPSD. This is covered in the next section when the web-based software interface is introduced.

6.4 The Web-Software Interface

The web-software can be accessed at http://correlator.feb.unesp.br/ and has two interfaces. One is named *frequency bandwidth estimation for leak noise correlators*, which is shown in Fig. 6.2a. The other is named *cross-correlation and phase estimation for correlators* which is depicted in Fig. 6.2b. The inputs for both interfaces are:

- Pipe material properties [E_p^* and ρ_p]: This can be PVC (Polyvinyl Chloride), MDPE (High Density Polyethylene), Cast Iron and Concrete. The default values for each material used in the simulator are given in Table 6.1.
- Sensor Type [$(j\omega)^n$]: hydrophone ($n = 0$, pressure/displacement measurements), geophone ($n = 1$, , velocity measurements) and accelerometers ($n = 2$, , acceleration measurements).
- Soil Type [B_m^* and G_m^*]: sandy, clay and air (no surrounding medium). The default values for each soil type used in the simulator are given in Table 6.2.
- Pipe geometry properties [a and b].
- Distance between the sensors [d].

The water in the pipe has bulk modulus B_w which is 2 10^9 N/m^2. The output for the frequency bandwidth interface is the modulus of the CPSD normalized by its largest value, so that $|\overline{S}_{x_1x_2}(\omega)| = |S_{x_1x_2}(\omega)| / \max[|S_{x_1x_2}(\omega)|]$. The lower and upper limits of the filter are determined by the condition $|\overline{S}_{x_1x_2}(\omega)| \geq 0.1$. This is depicted in Fig. 6.2a as the software output. The frequency limits are also highlighted and the frequency range over which the CPSD is useful is shaded in the graph. The frequency where $|\overline{S}_{x_1x_2}(\omega)|$ is maximum is also highlighted for convenience. It is important to note that this is an estimation of the frequency range where the leak noise is likely to be located, so that the frequency range over which the time delay information can be found.

The same inputs are used for the second interface where the outputs are the normalized cross-correlation, phase and wave speed. The cross-correlation is also normalized by its maximum value (peak), so that $\overline{R}_{x_1x_2}(\tau) = R_{x_1x_2}(\tau) / \max[R_{x_1x_2}(\tau)]$. The distance between the sensors, however, is updated via a sliding bar, so that the phase and the normalized cross-correlation coefficient can be performed instantly according to this input. Such a tool can be useful to show how the correlation and phase works

(a)

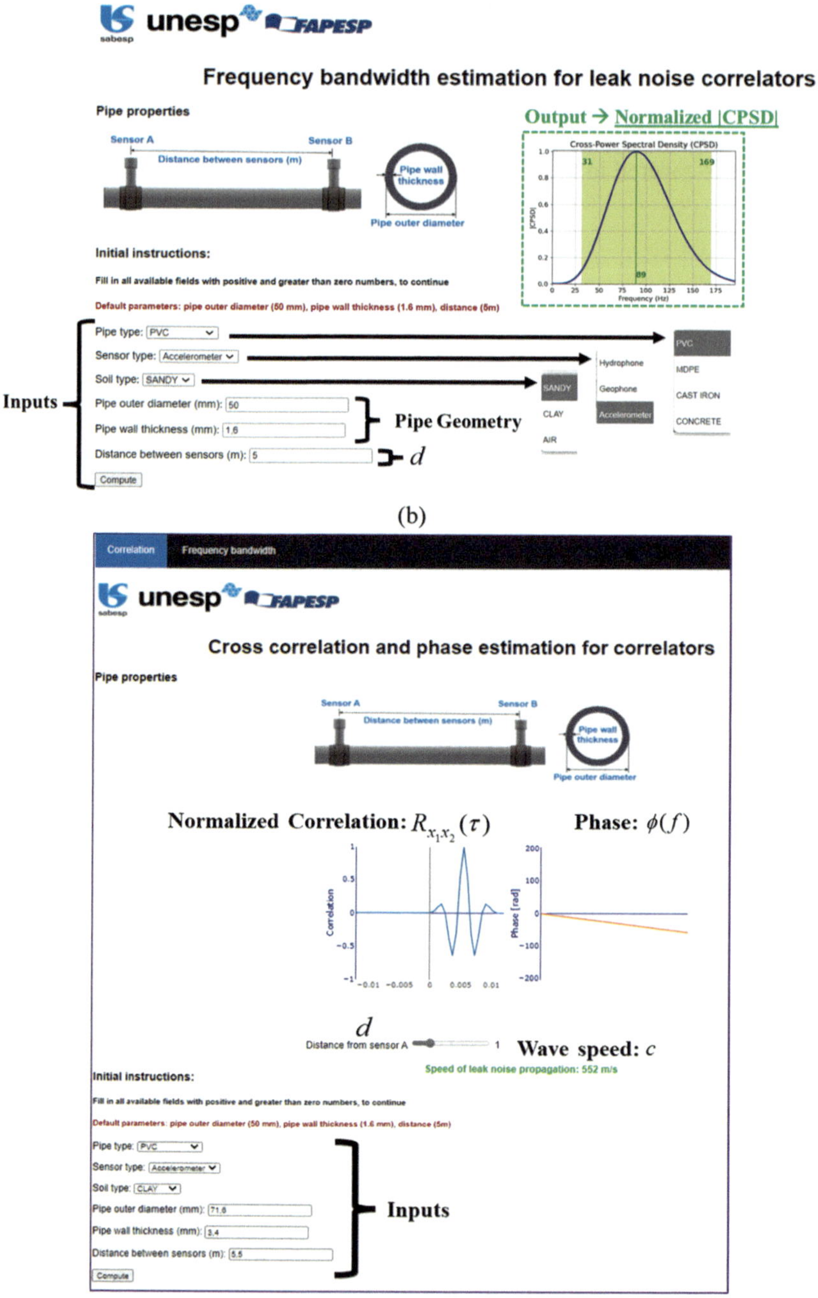

Fig. 6.2 User-friendly interfaces of the web-based software **a** Interface is named as frequency bandwidth estimation for leak noise correlators. **b** Interface is named as cross-correlation and phase estimation for correlators. (http://correlator.feb.unesp.br/)

Table 6.1 Pipe properties used in the web-based software

Soil Type Properties	Young's Modulus (N/m^2)	Density (kg/m^3)	Loss Factor	Poisson's ratio
PVC	4 10^9	1450	0.06	0.4
MDPE	2 10^9	930	0.06	0.4
Cast Iron	170 10^9	7800	0.001	0.25
Concrete	20 10^9	2240	0.06	0.205

Table 6.2 Soil properties used in the web-based software

Soil type properties	Bulk modulus (N/m^2)	Density (kg/m^3)	Shear modulus (N/m^2)
Sandy	4.7 10^7	2 10^3	1.5 10^7
Clay	4.5 10^9	2 10^3	2.4 10^8

according to the leak position (closer or further from one sensor, i.e. sensor 1). It is important to note here that the wave-speed calculated in this interface is the predominantly fluid-borne wave affected by the surround medium and pipe properties. This is related to the leak velocity which is usually calculated via analytical equations or given by the standard tables.

6.5 Simulations

The simulations presented in this section are compared to actual leak data collected in two different test rigs to show the good agreement between the simulated data provided by the web-based software and the data collected from those test rigs. Figures 6.3a and b respectively show the test rig located in Brazil and the test rig located in the UK. More information about the test rig in Brazil can be found in Brennan et al. (2018), and more information of the test rig in the UK can be found in Almeida et al. (2014). The two main differences between these test rigs are the surrounding medium, and the pipe material and geometry. Table 6.3 summarizes such characteristics, so that it is possible to conduct two different analyses via the web-based software for these two scenarios. Accelerometers were used to collect the pipe vibration in both test rigs.

Figures 6.4a, b and c show the normalized modulus of the CPSD, the phase and the normalized cross-correlation for the test rig in Brazil and the UK, respectively. The blue thick lines are related to the simulated data provided by the web-based software and the red thin lines are related to the actual leak data. The simulated phase given by the straight line $\phi(\omega) = \omega T_0$ is depicted in dashed blue line for the simulated data to highlight the very good agreement to the actual data. The time delay and leak noise velocity for the Brazil and UK test rigs are approximately 0.01 s and 552 m/s, and 0.079 s and 376 m/s, respectively. The threshold value of 0.1 for the normalized

Fig. 6.3 Data from the test rig **a** located in Brazil and the test rig **b** located in the UK to show how representative is the data generated by the web-based software

Table 6.3 Test rig characteristics

Test rig	Soil type	Pipe material	Pipe radius (mm)	Pipe wall thickness (mm)	Distance between the sensors (m)
Brazil	Clay	PVC	35.8	3.4	5.5
UK	Sandy	HDPE	84.5	11	30

modulus of the CPSD used to estimate the frequency bandwidth (shaded area) is also depicted in the figure for convenience.

Although the actual normalized modulus of the CPSD is a little erratic, it is observed that the actual data is contained within the frequency range given by the simulations. This can be useful for leak noise correlator users to check if the selection of the filter bandwidth is in accordance to the one in actual situations. Moreover, this can be used to check if the bandwidth is too narrow for a specific situation (for a given input), such that leak noise correlators may not be effective in pinpointing the leak. Although the actual normalized cross-correlation is not smooth as with the simulation, there is good agreement concerning the peak (time delay), which corresponds to the phase gradient (the analytical one overlays very well to the actual phase).

6.6 Final Remarks

This book chapter has described some web-based software that can aid leak detection professionals which are using commercial leak noise correlators. The model behind the software is based on the wave approach taking into account the buried pipe properties (pipe material and pipe geometry) and the surrounding medium (soil type). Furthermore, this model is related to the predominantly fluid-borne wave which is

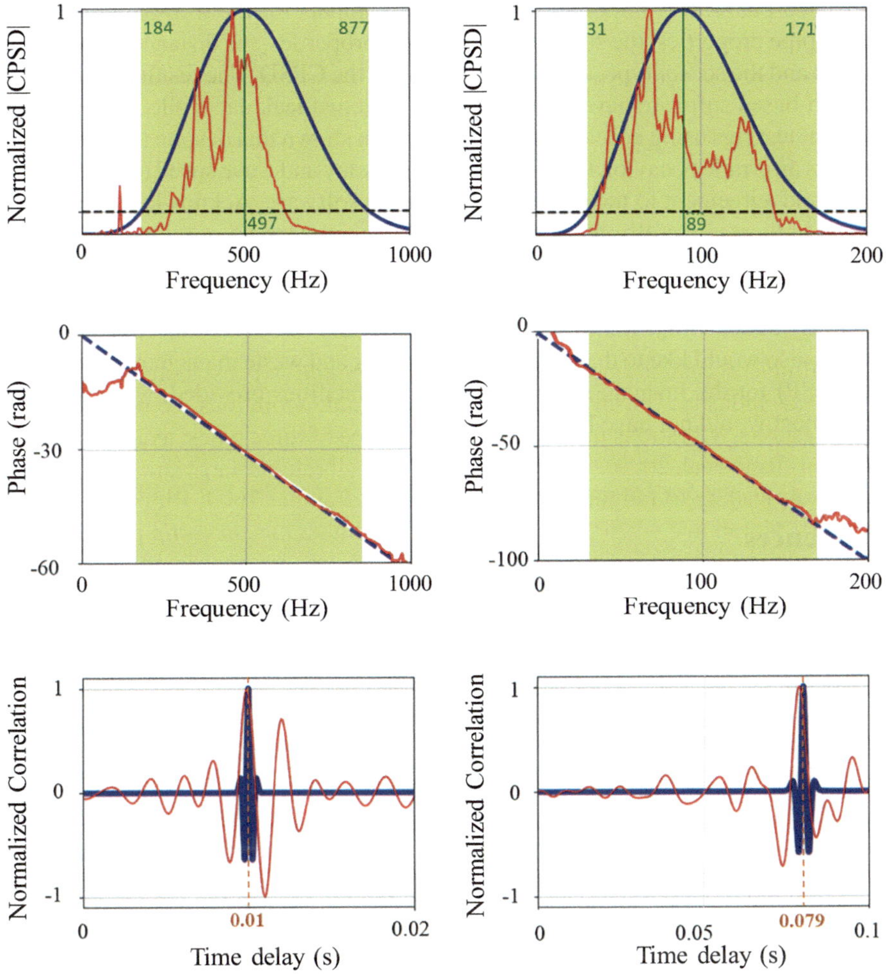

Fig. 6.4 Comparison between the actual data collected from "i" the test rig in Brazil and "ii" the test rig in the UK and the results provided by the web-based software for the **a** Normalized modulus of the CPSD, **b** the unwrapped phase and **c** the normalized cross-correlation. The thick blue lines are related to the simulated data provided by the web-based software and the thin red lines are related to the actual leak data. The simulated phases are depicted as dashed blue lines for convenience

responsible for propagating leak noise energy in the pipe. Hence, measurements in the water (acoustic pressure) or on the pipe wall (displacement, velocity and acceleration) can be simulated. The speed of such a wave is the one calculated in commercial leak noise correlators. The surrounding medium affects this velocity and the frequency range over which the leak noise information is located. This frequency range is important in the calculation time delay and the wave-speed and is estimated from the normalized modulus of the cross-power spectral density (CPSD) and the associated phase. It is important to note that the inputs for the web-based software are the pipe properties, the surrounding medium properties, the distance between the sensors and the sensor type, and one key output is the CPSD. The results provided by the web-based software have been compared to actual leak data collected from two different test rigs using accelerometers. It has been shown that despite the actual data presents an erratic behavior, the bandwidth, time delay and wave speed (leak velocity) have good agreement to the ones provided by the software. Acknowledgments.

The authors gratefully acknowledge the financial support provided by the São Paulo Research Foundation (FAPESP) under Grant numbers 2018/25360-3 and 2020/12251-1, and the Coordination for the Improvement of Higher Education Personnel (CAPES) under Grant numbers 432272/2018-6 and 88887.374001/2019-00. The authors also would like to thank the Brazilian water and waste management company (SABESP) for the financial support and test rig facilities provided, without which this project would not have been possible.

References

Abouhamad M, Zayed T, Moselhi O (2016) Leak detection in buried pipes using ground penetrating radar—a comparative study. In: Pipelines 2016: out of sight, out of mind, not out of risk. ASCE, p 417–424

ADS (2010) Eureka2R leak noise correlator user manual. https://www.adsenv.com/sites/default/files/manuals/adseureka2rusermanuala1.pdf. Accessed 10 May 2024

Al Qahtani T, Yaakob MS, Yidris N, Sulaiman S, Ahmad KA (2020) A review on water leakage detection method in the water distribution network. J Adv Res Fluid Mech Thermal Sci 68(2):152–163

Almeida F, Brennan M, Joseph P, Whitfield S, Dray S, Paschoalini A (2014) On the acoustic filtering of the pipe and sensor in a buried plastic water pipe and its effect on leak detection: an experimental investigation. Sensors 14:5595–5610

Aquasave (2015) Leak detection. http://www.aquasave.mk/en-GB/Home/ServiceDetails/15. Accessed 10 May 2024

Brennan MJ, Karimi M, Muggleton JM, Almeida FCL, Lima FK, Ayala PC, Obata D, Paschoalini AT, Kessissoglou N (2018) On the effects of soil properties on leak noise propagation in plastic water distribution pipes. J Sound Vib 427:120–133

Cataldo A, Persico R, Leucci G, De Benedetto E, Cannazza G, Matera L, De Giorgi L (2014) Time domain reflectometry, ground penetrating radar and electrical resistivity tomography: a comparative analysis of alternative approaches for leak detection in underground pipes. NDT and E Int 62:14–28

Eliasson J (2015) The rising pressure of global water shortages. Nature 517(6)

El-Zahab S, Zayed T (2019) Leak detection in water distribution networks: an introductory overview. Smart Water 4(5)

Fahmy M, Moselhi O (2009) Detecting and locating leaks in underground water mains using thermography. In: International symposium on automation and robotics in construction—ISARC 2009. Austin, United States of America

Fletcher R, Chandrasekaran M (2008) SmartBall™: a new approach in pipeline leak detection. In: Proceedings of the 7th international pipeline conference 2008. ASME, Calgary, Alberta, Canada

Fuchs HV, Riehle R (1991) Ten years of experience with leak detection by acoustic signal analysis. Appl Acoust 33(1):1–19

FUJI TECOM (2022) Water Leak Detection LC-5000. https://www.fujitecom.com/products/pcl.html. Accessed 10 May 2024

Gao Y, Brennan MJ, Joseph PF, Muggleton JM, Hunaidi O (2004) A model of the correlation function of leak noise in buried plastic pipes. J Sound Vib 277(1–2):133–148

Gao Y, Brennan MJ, Joseph PF, Muggleton JM, Hunaidi O (2005) On the selection of acoustic/vibration sensors for leak detection in plastic water pipes. J Sound Vib 283(3–5):927–941

Getirana A, Libonati R, Cataldi M (2021) Brazil is in water crisis—it needs a drought plan. Nature 600:218–220

Greve P, Kahil T, Mochizuki J, Schinko T, Satoh Y, Burek P, Fischer G, Tramberend S, Burtscher R, Langan S, Wada Y (2018) Global assessment of water challenges under uncertainty in water scarcity projections. Nature Sustainability 1:486–494

Hamilton S, McKenzie R (2014) Water management and water loss. IWA Publishing (13). https://doi.org/10.2166/9781780406367

Islam MR, Azam S, Shanmugan B, Mathur D (2022) A review on current technologies and future direction of water leakage detection in water distribution network. IEEE Access 10:107177–107201

Liemberg R, Wyatt A (2019) Quantifying the global non-revenue water problem. Water Supply 19(3):831–837

Martins R, Azevedo A, Fortunato AB, Alves E, Oliveira A, Carvalho A (2019) An innovative and reliable water leak detection service supported by data-intensive remote sensing processing. In: International conference on computational scienc—ICCS 2019 Lecture notes in computer science, vol 11539. Springer

Muggleton JM, Brennan MJ, Pinnington RJ (2002) Wavenumber prediction of waves in buried pipes for water leak detection. J Sound Vib 249(5):939–954

Millington N (2018) Producing water scarcity in São Paulo, Brazil: The 2014–2015 water crisis and the binding politics of infrastructure. Polit Geogr 65:26–34

Pinnington RJ, Briscoe AR (1994) Externally applied sensor for axisymmetric waves in a fluid filled pipe. J Sound Vib 173(4):503–516

Scussel O, Brennan MJ, Almeida FCL, Muggleton JM, Rustighi E, Joseph PF (2021) Estimating the spectrum of leak noise in buried plastic water distribution pipes using acoustic or vibration measurements remote from the leak. Mech Syst Signal Process 147:107059

Scussel O, Brennan MJ, Almeida FCL, Iwanaga MK, Muggleton JM, Joseph PF, Gao Y (2023) Key factors that influence the frequency range of measured leak noise in buried plastic water pipes: theory and experiment. Acoustics 5:490–508

Shin K, Hammond JK (2008) Fundamentals of signal processing for sound and vibration engineers. Wiley, Chichester

Zaman D, Tiwari MJ, Gupta AK, Sen D (2020) A review of leakage detection strategies for pressurized pipeline in steady-state. Eng Fail Anal 109:104264

Chapter 7
Use of Natural Fibers for Sound Absorption: A Chronological Literature Review

Gabriel Weiss Mattioli, Flávio Issao Kubota, Vinicius Pires de Oliveira Doca, and Suzana Regina Moro

Abstract The rise in pollution has prompted an increased demand for sustainable materials. In this review, we explore studies that investigate natural materials for acoustic absorption panels. This chronological literature review surveys studies that consider natural fibers for sound absorption, considering their technical feasibility by tabulating the Noise Reduction Coefficient. We collected and analyzed articles from Scopus, Engineering Village, Google Scholar, and ScienceDirect databases. For the selection of fibers, the following criteria were applied: the sound absorption coefficient must be obtained experimentally for greater accuracy, be single-layered, not present any added binder, have tables or graphs of the sound absorption coefficient by frequencies of at least 250–2000 Hz, and the difference in thickness between the thinnest and thickest layer tabulated should be a maximum of 0.05 m. NRC revealed low values for banana pseudostem (0.15), cotton (0.30), corn fiber (0.30), and cork (0.30). On the other hand, other fibers such as sheep wool (0.70), kenaf (0.70), coconut fiber (0.50), and betung bamboo (0.50) demonstrated exceptional results, confirming that natural fibers are viable options for sound absorption purposes.

Keywords Natural fibers · Acoustic absorption · Sound absorption · Chronological literature review

G. Weiss Mattioli (✉) · F. I. Kubota · V. P. de Oliveira Doca
Universidade Federal Do Paraná, Curitiba, Brazil
e-mail: gabrielwmattioli@ufpr.br

F. I. Kubota
e-mail: flavio.kubota@ufpr.br

V. P. de Oliveira Doca
e-mail: vinicius.doca@ufpr.br

S. R. Moro
Universidade Estadual de Campinas, São Paulo, Brazil
e-mail: smoro@unicamp.br

P. Kurka and M. Pereira (eds.), *Advances in Structural Vibration*, Lecture Notes in Mechanical Engineering, https://doi.org/10.1007/978-3-031-71540-2_7

7.1 Introduction

Recent studies have taken an interest in the relationship between human development and the limits of planetary growth, which is connected with exploring environmental sustainability. As such, research and discussions on this topic have increased. Brazil's harvest of natural fibers provides a variety of materials. The 21st Ordinary Meeting of the Natural Fibers Sector Chamber CSFN/MAPA (2019) reported that the 2018 harvest yielded 517 tons of silk, 400,000 tons of bamboo, 58,000 tons of sisal, 7264 tons of jute, and 4500 tons of coconut. However, this production process generates waste, presenting disposal challenges.

Additionally, latest research has considered the potential applications of natural-origin materials, with sound-absorbing panels as one example. Sound absorption' refers to a material's capacity to absorb sound wave energy and convert it into heat, which may prevent sound propagation (Zhu et al. 2014). However, some synthetic acoustic materials have been associated with energy consumption and greenhouse gas emissions (Dissanayake et al. 2021).

Given these observations, the current review explores publications discussing the use of natural fiber in sound absorption, including analyzing their sound absorption coefficients. This review is conducted chronologically and aims to collate studies on the potential applications of natural fibers in sound absorption and provide an overview of existing work in this field. This study compiled and reviewed literature from Scopus, Engineering Village, Google Scholar, and ScienceDirect databases. The following keywords were searched for in the title, abstract, and keywords fields: natural fibers, acoustics, absorption panel, and sound absorption.

7.2 Chronological Literature Review

Biot (1956a; b) proposed one of the early mathematical models to predict sound absorption mechanisms. This work offered a conceptual and mathematical framework for analyzing elastic wave propagation in saturated porous media and mathematical equations to help describe this phenomenon. Delany and Bazley (1970) developed mathematical models to describe sound properties such as impedance and wave propagation in fibrous media. In order to utilize their model for predicting acoustic properties, a comprehensive understanding of the material's intrinsic properties, including porosity, tortuosity, and airflow resistance, is helpful. Dunn and Davern (1986) were inspired by the mathematical model of Delany and Bazley (1970) to devise a method for calculating the acoustic impedance of foamed multilayer sound absorbers. They then compared the results generated by their mathematical model to those obtained from the impedance tube, and the outcomes were satisfactory. Interestingly, they also evaluated the applicability of Delany and Bazley's (1970) equation forms for accurately determining parameters for the foamed materials.

Johnson et al. (1987) introduced significant improvements in describing viscous forces, defining Tortuosity and Viscous Dynamic Permeability. They described these as a structural factor dependent on the porous frame and a complex parameter relating the pressure gradient to fluid velocity in an isotropic porous medium, respectively. Drawing upon a large amount of experimental data, Qunli (1988) proposed empirical relationships between the characteristic impedance and propagation constant of porous plastic open-cell foams and the unit-depth flow resistivity. This same study compared the resulting data to those obtained from the Dunn and Davern (1986) and Delany and Bazley (1970) models, noting some findings.

Miki (1990) evaluated the Delany-Bazley mathematical model for the case of multiple layers and observed that the genuine part of the surface impedance tended towards negative values at low frequencies. He subsequently adapted this model to achieve positive values for higher frequency ranges. In response to Delany and Bazley's (1970) work, Allard and Champoux (1992) introduced new equations for sound propagation in fibrous materials. Drawing on Johnson et al. (1987) publication, they presented a subset of equations to predict fibrous materials' acoustic absorption properties. These were compared to Delany-Bazley's (1970) equations and indicated a higher accuracy for low frequencies.

Next, Allard (1993) delved into the critical theories underpinning sound absorption in porous materials, providing an in-depth discussion of mathematical models, including those of Johnson et al. (1987) and Biot (1956a; b). Around the turn of the twenty-first century, paralleling the growing societal awareness of environmental degradation, interest in studying the acoustic insulation behavior of natural fibers began to increase. An early example is the work of Koizumi et al. (2002), which explored the acoustic insulation properties of bamboo fiber using an impedance tube. They observed correlations between the sound absorption coefficient and bamboo fibers' thickness, bulk density, and diameter. The study utilized samples with thicknesses of 25, 50, and 75 mm; fiber diameters ranging from 90 to 425 μm; and bamboo fibers with a bulk density of 80, 120, and 160 kg/m^3. Through this experiment, they noted similarities between bamboo fiber's and glass wool's acoustic insulation properties, which had relative densities of 120 kg/m^3 and 32 kg/m^3, respectively (Koizumi et al. 2002), confirming the potential of natural fiber as an effective acoustic insulation material.

Yang et al. (2003) conducted a study on rice straw's physical, mechanical, and acoustic properties intending to replace wood-based panels as building materials. Their acoustic behavior evaluation involved samples mixed with an adhesive resin hardener binder, which used an impedance tube to measure the sound absorption coefficient within a range of 0–8000 Hz. From the results they obtained, they concluded that rice straw makes a suitable sound-absorbing alternative to wood-based panels. Similarly, Nor et al. (2004) investigated the acoustic properties of coconut fiber theoretically, assessing the effectiveness of the air layers, known as airspace layers, between layers of insulating material and the influence of microperforated aluminum plates. They used the WinFLAG simulation program to calculate the sound absorption coefficient for different layers of materials. Their observations revealed that the air layer between the insulating material enhanced sound absorption effectiveness for low frequencies. While the microperforated aluminum plate also improved sound absorption for low frequencies, it had the opposite effect at high frequencies (Nor et al. 2004).

Asdrubali (2007) emphasized the need to use sustainable materials and techniques for noise control in buildings. The environmental repercussions of applying conventional sound insulation materials were addressed, comparing Non-Renewable Energy, Global Warming Potential, and Acidification Power data between traditional and natural materials. An evaluation was carried out on several synthetic and natural materials categories by comparing their Noise Reduction.

Coefficient (NRC) values, which are averages of the sound absorption coefficients at frequencies of 250, 500, 1000, and 2000 Hz. Based on the findings, Asdrubali (2007) suggested that natural materials offer numerous benefits, with a lower environmental impact and satisfactory sound absorption results among the key advantages.

Zulkifli et al. (2008) collected sound absorption data from two insulating panels made from coconut fiber; one was a regular panel, and the other was a coconut fiber panel sheathed in a microperforated outer layer of composite material. They gathered this information through experimental methods using a reverberation room compliant with ISO 354 and ISO 717-1 standards and theoretical methods using simulations based on the WinFLAG application, capable of calculating the sound absorption coefficient for different layers of materials. The similarity between simulation and experimental results indicated the potential for coconut fiber as a cheaper, lighter, and ecologically superior alternative to synthetic acoustic insulation (Zulkifli et al. 2008).

In a separate study, Ayub et al. (2009) examined the sound absorption of natural coconut fibers using the Delany-Bazley theoretical model and an experimental method involving an impedance tube. This study investigated the sound absorption capacity of coconut fibers under various conditions, such as varying thicknesses between samples and the effects of the air gap on the sound absorption coefficient. The analysis found favorable sound absorption properties in coconut fibers, with the Delany-Bazley model demonstrating greater accuracy at low frequencies, except in cases of multiple layers, as previously highlighted by Miki (1990).

Fouladi et al. (2011) examined the sound absorption properties of coconut fiber, partitioned into two sets of 15 samples each. One set comprised ordinary coconut fiber, while the other consisted of coconut fiber treated with a binder to enhance fiber hardness, inhibit fungal growth, and improve flammability. To obtain results, the researchers applied three scientific methods—two theoretical (Delany-Bazley and Biot-Allard) and one practical (using an impedance tube). The binder was considered part of the fibers for the Biot-Allard method on samples with additives, and compensation was made. The study found that samples with a binder exhibited lower sound absorption as their peaks were flatter, and they demonstrated a lower sound absorption rate for low frequencies. The authors suggested techniques like airspace layers and microperforated plates to ease these issues.

Abdullah et al. (2011) investigated the potential of dried rice straw fibers as an acoustic-absorbing material, focusing on assessing these fibers' acoustic properties and sound absorption capacity. Their findings pointed towards the promising potential of dried rice straw fibers as acoustic absorptive materials. However, they also identified that using a methylcellulose binder impacted the sound absorption coefficient at lower frequencies. The authors proposed that this obstacle could be mitigated by increasing the sample's thickness or coupling it with a perforated panel.

Fatima and Mohanty (2011) explored jute composite materials' acoustic and fire-retardant properties. They set out to investigate these materials' acoustic performance and resistance to fire, conducting tests to measure the sound absorption coefficient of untreated and alkali-treated jute samples using an impedance tube. Their results indicated that jute composites displayed strong acoustic performance across various frequency ranges. The authors concluded that jute composite materials are promising candidates for acoustic insulation and fire protection with their favorable acoustic properties and fire resistance.

Karlinasari et al. (2012) investigated the acoustic properties of particle panels made from Betung bamboo (*Dendrocalamus asper*) as building materials. They aimed to gauge these bamboo panels' sound absorption performance. To this end, they conducted tests on Betung bamboo panels with particles of three different sizes (fine, medium, and wool or excelsior) and two different densities (0.5 and 0.8 g/cm^3) to measure the sound absorption coefficient using an impedance tube. According to Karlinasari et al. (2012), particleboards crafted from Betung bamboo hold potential as building materials, particularly from the standpoint of acoustic absorption.

Fouladi et al. (2013) use an impedance tube to study the acoustic behavior of natural fibers found in Malaysia, such as Corn, Grass, Coir, and Sugarcane, all without added binders, and compare them with synthetic materials such as Carpet, Plywood, Drapery, and Fiberglass. The values found point to Fiberglass as an exception among the fibers, standing out among the others. Otherwise, the values found among the other fibers are close, making natural fibers great alternatives to replace existing synthetic materials.

In their study, Rahman et al. (2013) carried out an experimental investigation on natural fibers as potential sustainable materials for sound absorption. Their examination centers on two types of fibers: Date Palm Fiber (DPF) and Coconut Coir Fiber (CCF). Using an impedance tube, the researchers assess these fibers for their sound absorption capabilities, considering characteristics such as thickness and density. The authors position these fibers as possible replacements for conventional synthetic fibrous materials, including glass wool, rock wool, and asbestos. Their findings indicate that fiber size significantly influences sound absorption efficiency, with smaller fiber diameters leading to more effective sound absorption.

Wang and Choy (2015) focus on improving the sound absorption efficiency of a perforated panel absorber composed of support cavities partially filled with polymeric materials. Three polymeric materials are considered: expandable polystyrene foam, polymethacrylimide foam, and polyester fiber. The paper simulates the acoustic behavior of a perforated panel absorber. Simulating the acoustic behavior of the composite sound absorber, the article uses numerical models of finite element simulations and experimental means using the impedance tube. A strong agreement was observed between the numerical predictions and the experimental results, which validates the use of the finite element modeling method for the polymer composite absorber. The author finally concludes that the addition of polymer foam can improve the sound absorption performance of the composite absorber, but the improvement is superficial.

Using an impedance tube, Berardi and Iannace (2015) investigate the sound insulation properties of various natural fibers, including kenaf, wood, hemp, coconut fiber, cork, reed, cardboard, and wool. These practical results are compared with values derived from theoretical models, specifically the model proposed by Delany and Bazley (1970). The authors notice that Delany and Bazley's model needs to be revised to predict sound absorption for high-thickness fibers accurately. This discrepancy arises as the original model is calibrated using fiber samples with diameters ranging from 1 to 10 μm. For example, hemp, which exhibits an average fiber diameter of 22 μm, exhibits many errors: 67, 56, 96, and 59% at frequencies of 250 Hz, 500 Hz, 1000 Hz, and 2000 Hz, respectively. The researchers also assess each material's Noise Reduction Coefficient (NRC). The NRC values generally prove high for the evaluated materials, affirming the viability of natural fibers as acoustic absorbers. Exceptionally high sound absorption efficiency emerges from the results for sheep's wool and coconut fiber (Berardi and Iannace 2015).

Sambu et al. (2016) investigate the impact of physical properties on the acoustic performance of natural fibers derived from palm oil fronds. The research examines these fibers' physical properties and their relation to acoustic performance, particularly sound absorption. The study conducts experimental analyses of the density, porosity, tortuosity, and airflow resistivity of the material with binder compositions of 40, 30, 20, and 0%. Sound absorption measurements are also performed at varying frequencies using an impedance tube. The findings suggest a significant influence of the fibers' physical properties on acoustic performance. The authors note that as the binder content in the material's composition increases, there tends to be a higher density and tortuosity and lower porosity. The palm oil frond fibers demonstrate effective sound absorption at specific frequencies, suggesting their potential use as acoustic absorption materials (Sambu et al. 2016).

Berardi et al. (2017) investigate the acoustic insulating properties of Broom fiber, a plant native to Europe and Asia. The study uses an experimental method with an impedance tube on samples with diameters ranging from 1.5 to 4 mm and thicknesses of 60, 80, and 120 mm. According to the authors, this plant, widely available in Mediterranean countries, demonstrated promising acoustic absorption results, thus encouraging the use of such sustainable materials (Berardi et al. 2017).

Lim et al. (2018) examine the sound absorption performance of natural kenaf fibers. Experiments were carried out to evaluate the sound absorption coefficient using an impedance tube and to measure kenaf fibers' random-incidence sound absorption coefficient at various frequencies in a reverberation chamber. The findings suggest that kenaf fibers excel in sound absorption. The study also investigates the effects of air gaps and bulk density on the sound absorption coefficient, revealing that adding air gaps significantly enhances absorption for frequencies below 1500 Hz while increasing the bulk density improves sound absorption overall. As per the authors, kenaf fiber's performance in sound absorption is comparable to that of synthetic fiber Rock Wool (Lim et al. 2018).

Zhu et al. (2018) explore innovations in sound absorption technology, which is mainly focused on broadening the sound absorption bandwidth of a perforated panel using a membrane-type resonator. The authors explore how incorporating a membrane resonator can improve the perforated panel's performance as an acoustic absorber. The author uses a numerical model in which the performance of the perforated panel as a sound absorber is obtained analytically by combining the vibration equation of the membrane-type resonator with the acoustic impedance equation of the perforated panel and then validates the results found through the impedance tube. Finally, the author concludes that the membrane-type resonator can widen the perforated panel's sound absorption band.

Silva et al. (2019) investigate the sound insulation characteristics of sisal, coconut, and sugarcane fibers at low frequencies, utilizing the Delany and Bazley (1970) and Allard and Champoux (1992) models and experimentally through an impedance tube. The transfer function technique assessed the sound absorption coefficient per ASTM 1050-12 standard. With ten samples of each material having thicknesses of 20, 30, and 40 mm, they evaluated sound absorption coefficients within a frequency range of 200–1600 Hz. The authors found that the Allard and Champoux (1992) equations yielded more accurate values for low and medium frequencies, particularly notable when the samples had small thicknesses and airflow resistivity. The theoretical methods exhibited greater accuracy with high airflow resistivity values. Sugarcane fiber displays a higher sound absorption quality when compared to coconut fiber and sisal. The article also proposes an efficient inverse methodology for sample characterization, offering a viable alternative for adjusting Delany-Bazley's coefficients for the natural fibers studied. (Silva et al. 2019).

Liu et al. (2021) researched the airflow resistivity measurement and sound absorption performance analysis of sound-absorbing cotton. The study aimed to assess cotton's acoustic properties using airflow resistivity measurement as a significant parameter. Three experiments were performed: two theoretical and one experimental to measure the sound absorption coefficient for different thicknesses of cotton samples. The theoretical methods employed were the Delany and Bazley (1970)—Miki (1990) and Johnson et al. (1987)—Allard and Champoux (1992) mathematical models, whereas the experimental method utilized the impedance tube. The authors concluded that the Johnson-Champoux-Allard model aligns more accurately with the measured sound absorption coefficient, thus showcasing its capability to predict and design the sound absorption performance of new fibrous materials (Liu et al. 2021).

Mendes and Nunes (2022) examined the acoustic properties of banana pseudostem fibers using a sound impedance tube, assessing frequencies from 100 to 6300 Hz. They employed the transfer function technique described in ISO 10534:2 to verify the sound absorption coefficient. The researchers divided the samples into groups A, B, and C, each consisting of eight samples. Group A featured fibers grouped without additional substances; Group B had fibers bound by a natural glue derived from cassava starch and water; and Group C contained fibers bound with the same natural glue but sandwiched between two plaster plates. They observed that Group C showed the highest Noise Reduction Coefficient values, reaching a sound absorption coefficient of 0.89 at 6300 Hz. However, Mendes and Nunes (2022) noted that adding natural glue did not influence sound absorption, as Groups A and B exhibited similar NRCs.

7.3 Experimental Procedures

In this study, the aim is to determine which natural fibers exhibit high performance in sound absorption. To this end, a tabulation was made based on their sound absorption coefficient. This process is enhanced by using the Noise Reduction Coefficient which Berardi and Iannace (2015) define as the calculation of the arithmetic mean of the absorption coefficients in the octave band of 250, 500, 1000 and 2000 Hz, rounded to the nearest multiple of 0.05. This coefficient, known as NRC, facilitates comparison between sound-absorbing materials (Berardi and Iannace 2015). The equation for this coefficient is presented in Eq. 7.1,

$$\mathrm{NRC} = \frac{\alpha 250\,\mathrm{Hz} + \alpha 500\,\mathrm{Hz} + \alpha 1000\,\mathrm{Hz} + \alpha 2000\,\mathrm{Hz}}{4} \tag{7.1}$$

where α250, α500, α1000, and α2000 Hz denote the acoustic absorption coefficients at their respective frequencies.

The selection criteria for the natural fibers included exclusively in the tabulation were as follows:

1. The sound absorption coefficient should be derived from experimental results, specifically the impedance tube.
2. The material should be single-layered.
3. No added binder should be present.
4. Tables or graphs of the sound absorption coefficient by frequencies of at least 250–2000 Hz should be provided.
5. The difference in thickness between the thinnest and thickest layer tabulated should be a maximum of 0.05 m.

All Noise Reduction Coefficients were calculated despite potential variations in density, porosity, fiber diameter, and airflow resistivity between samples within the same article. The best performers from each article were then presented.

The Noise Reduction Coefficient was extracted from the articles. It was gathered when directly quoted; if not directly presented, a proportionality method was applied using pixel counts taken from the graphs of the sound absorption coefficient by frequency using ImageJ software. This approach measures the pixel count in the vertical distance between a specific point on the curve and the abscissa axis at 250, 500, 1000, and 2000 Hz. This measurement was then used to calculate a value proportional to the pixel count on the ordinate axis of the sound absorption coefficient, α.

7.4 Results and Discussion

Table 7.1 was based on the articles mentioned in the chronological review, according to the five criteria previously established, which provides data for each type of natural fiber selected and their respective Noise Reduction Coefficient (NRC).

The NRC results found respectively by Mendes and Nunes (2022), Liu et al. (2021), Fouladi et al. (2013), and Berardi and Iannace (2015) reported low values for banana pseudostem (0.15), cotton (0.30), corn fiber (0.30), and cork (0.30), suggesting these materials may not be ideal for sound absorption if used individually. However, increasing thickness and utilizing techniques like microperforated plating, increasing bulk density, multi-layering, and airspace layering may enhance their performance, as already researched by Lim et al. (2018), Nor et al. (2004), Zulkifli et al. (2008), Mendes and Nunes (2022), and suggested by Fouladi et al. (2011) and Abdullah et al. (2011).

Table 7.1 Noise reduction coefficient of each sample analyzed among filtered articles

Material	Thickness (m)	∝ 250 Hz	∝ 500 Hz	∝ 1000 Hz	∝ 2000 Hz	NRC
Betung Bamboo	0.01	0.2354	0.2483	0.5903	0.8483	0.50
Shredded Broom	0.06	0.081	0.159	0.647	0.929	0.45
Banana Pseudostem	0.01	0.07	0.1033	0.1566	0.2533	0.15
Kenaf	0.06	0.3	0.61	0.99	0.95	0.70
Wood	0.06	0.4	0.5	0.65	0.91	0.60
Hemp	0.03	0.15	0.25	0.51	0.7	0.40
Coconut	0.05	0.20	0.34	0.67	0.79	0.50
Cork	0.03	0.02	0.1	0.3	0.86	0.30
Sheep wool	0.06	0.28	0.66	0.95	0.94	0.70
Reed	0.04	0.12	0.38	0.64	0.62	0.45
Cotton	0.0151	0.038	0.0952	0.2761	0.7666	0.30
Oil palm frond	0.05	0.1157	0.2912	0.7052	0.7473	0.45
Jute fiber	0.0254	0.1587	0.438	0.8825	0.8031	0.60
Sugar cane	0.02	0.05	0.13	0.88	0.63	0.40
Corn fiber	0.02	0.06	0.16	0.28	0.81	0.30
Grass	0.02	0.08	0.14	0.45	0.98	0.40

Contrastingly, other absorbent fibers yielded high-performance NRC results, as reported by the authors Berardi and Iannace (2015), Fatima and Mohanty (2011), Karlinasari et al. (2012), notably sheep wool (0.70), kenaf (0.70), jute fiber (0.60), coconut fiber (0.50), and betung bamboo (0.50). Among these, kenaf and sheep wool (both 0.6 m) were the thickest, followed by coconut fiber (0.05 m), jute fiber (0.0254 m), and betung bamboo (0.01 m), despite being the thinnest, betung bamboo still exhibited a relatively high Noise Reduction Coefficient.

Natural fiber's shorter decomposition time than synthetic fibers contributes to their environmental sustainability. However, this benefit also implies less durability, requiring consideration of material exchange or maintenance ease. For instance, using natural fibers in building construction for soundproofing may not be advantageous if the insulating panels are located between walls with difficult access.

7.5 Conclusion

This study aimed to investigate the viability of natural fibers for sound absorption and to summarize significant research in this field chronologically. Researchers may consider incorporating binders or additives to improve their shape retention and flammability resistance when using natural fibers as absorbent materials. Such modifications could enable the placement of absorbent panels in locations with easy access, simplify their exchange and maintenance, and make them suitable for framed, exposed, and lined applications.

Sustainable development remains a prominent focus in today's society. Concurrently, scientific research considers natural fibers as potential materials for sound wave absorption. The hypothesis that natural fibers could substitute synthetic fibers in sound absorption applications originates from mathematical models and the inherent properties of natural fibers. These fibers exhibit promising potential for sound wave absorption. Additionally, they could provide environmental benefits, such as repurposing agricultural waste, offering biodegradability, boosting widespread availability, and enhancing accessibility and cost-effectiveness. Further research may consider the assessment of those natural fibers in more industrial applications, for instance, civil construction, household appliances, and the automotive industry.

References

Abdullah Y, Putra A, Effendy H, Farid W, Ayob M (2011) Dried paddy straw fibers as an acoustic absorber: a preliminary study. Int J Renew Energy Resources 3(2013):52–56

Allard JF (1993) Propagation of sound in porous media, modelling sound absorbing materials. Elsevier Applied Science, London

Allard JF, Champoux Y (1992) New empirical equations for sound propagation in rigid flame fibrous materials. J Acoustical Soc Am 91:3346–3353

Asdrubali F (2007) Green and sustainable materials for noise control in buildings. In: Proceedings of the 19th international congress on acoustics—ICA 2007. Madrid, Spain

Ayub M, Nor JM, Amin N, Zulkifli R, Fouladi MH (2009) Analysis on sound absorption of natural coir fiber using delany-bazley model. In: Proceedings of the international conference on mechanical engineering, pp 1–6. Dhaka, Bangladesh

Berardi U, Iannace G (2015) Acoustic characterization of natural fibers for sound absorption. Build Environ 94(Part 2):840–852

Berardi U, Iannace G, Di Gabriele M (2017) The acoustic characterization of broom fibers. J Nat Fibers 14(6):858–863

Biot MA (1956a) Theory of propagation of elastic waves in a fluid-saturated porous solid. I. Low-frequency range. J Acoustical Soc Am 28(2):168–178

Biot MA (1956b) Theory of propagation of elastic waves in a fluid-saturated porous solid. II. Higher frequency range. J Acoustical Soc Am 28(2):179–191

CFSN/MAPA (2019) Fibras do Brasil: Breve panorama do setor de Fibras Naturais para orientar uma proposta de governança. Available (In Portuguese) at: https://www.gov.br/agricultura/pt-br/assuntos/camaras-setoriais-tematicas/documentos/camaras-setoriais/fibras-naturais/2019/23a-ro/csfn-23a-ro_diagnostico-setor-fibras-naturais_nov-2019.pdf. Accessed on 13 Jun 2023

Delany ME, Bazley EN (1970) Acoustical properties of fibrous absorbent materials. Appl Acoust 3(2):105–116

Dissanayake DGK, Weerasinghe DU, Thebuwanage LM, Bandara UAAN (2021) An environmentally friendly sound insulation material from post-industrial textile waste and natural rubber. J Buid Eng 33:101606

Dunn IP, Davern WA (1986) Calculation of acoustic impedance of multi-layer absorbers. Appl Acoust 19:321–334

Fatima S, Mohanty AR (2011) Acoustical and fire-retardant properties of jute composite materials. Appl Acoust 72(2–3):108–114

Fouladi MH, Ayub M, Nor MJM (2011) Analysis of coir fiber acoustical characteristics. Appl Acoust 72:35–42

Fouladi MH, Nassir MH, Ghassem M, Shamel M, Peng SY, Wen SY, Xin PZ, Nor MJM (2013) Utilizing Malaysian natural fibers as sound absorber. Model Measur Meth Acoustic Waves Acoustic Microdev 161–170

Johnson DL, Koplik J, Dashen R (1987) Theory of dynamic permeability and tortuosity in fluid-saturated porous media. J Fluid Mech 176:379–402

Karlinasari L, Hermawan D, Maddu A, Martianto B, Lucky IK, Nugroho N, Hadi YS (2012) Acoustical properties of particleboards made from Betung bamboo (Dendrocalamus asper) as building construction material. BioResources 7(4):5700–5709

Koizumi T, Tsujiuchi N, Adachi A (2002) The Development of sound absorbing materials using natural bamboo fibers. WIT Trans Built Environ 59:157–166

Lim ZY, Putra A, Nor MJM, Yaakob MY (2018) Sound absorption performance of natural kenaf fibres. Appl Acoust 130:107–114

Liu X, Xiong X, Pang J, Wu L, Zhang H (2021) Airflow resistivity measurement and sound absorption performance analysis of sound-absorb cotton. Appl Acoust 179:108060

Mendes COB, Nunes MAA (2022) Acoustic performance of the banana pseudostem fiber. Appl Acoust 191:108657

Miki Y (1990) Acoustical properties of porous materials—modifications of Delany-Bazley models. J Acoust Soc Jpn (E) 11(1):19–24

Nor MJM, Jamaludin N, Tamiri FM (2004) A preliminary study of sound absorption using multi-layer coconut coir fibers. Electron J "Tech Acoust" 3:1–8

Qunli W (1988) Empirical relations between acoustical properties and flow resistivity of porous plastic open-cell foam. Appl Acoust 25:141–148

Rahman LA, Raja RI, Rahman RA (2013) Experimental study on natural fibers for green acoustic absorption materials. Am J Appl Sci 10(10):1307–1314

Sambu M, Yahya MN, Latif HA, Roslan MAB, Ghazali MIB (2016) Influence of physical properties on the acoustical performance of the oil palm frond natural fibre. ARPN J Eng Appl Sci 11(10)
Silva CCB, Terashima FJH, Barbieri N, Lima KF (2019) Sound absorption coefficient assessment of sisal, coconut husk and sugar cane fibers for low frequencies based on three different methods. Appl Acoust 156:92–100
Wang CQ, Choy YS (2015) Investigation of a compound perforated panel absorber with backing cavities partially filled with polymer materials. ASME. J Vibr Acoust 137(4):044501. https://doi.org/10.1115/1.4029771
Yang HS, Kim DJ, Kim HJ (2003) Rice straw–wood particle composite for sound absorbing wooden construction materials. Biores Technol 86(2):117–121
Zhu X, Kim BJ, Wang Q, Wu Q (2014) Recent advances in the sound insulation properties of bio-based materials. BioResources 9(1):1764–1786
Zhu X, Chen Z, Jiao Y, Wang Y (2018) Broadening of the sound absorption bandwidth of the perforated panel using a membrane-type resonator. ASME J Vibr Acoust 140(3):031014. https://doi.org/10.1115/1.4038942
Zulkifli R, Nor MJM, Mat Tahir MF, Ismail AR, Nuawi MZ (2008) Acoustic properties of multi-layer coir fibres sound absorption panel. J Appl Sci 8(20):3709–3714

Chapter 8
Identification of Modal Parameters of Coupled Rotor Foundation System via Automatic Operational Modal Analysis

Nathali Dreher, Tiago Machado, Thomas Paulsen, and Ilmar Santos

Abstract Operational Modal Analysis (OMA) extracts modal parameters of systems that are excited by unknown ambiental excitation, being broadly used in the monitoring of civil structures. In the past decades, research has been made to enable the application of OMA in rotating machines, dealing with challenges such as harmonic excitation, nonlinearities, and the lack of proper excitation. With the popularization of machine learning techniques, many researchers have been using these tools to overcome some challenges in this research field. Clustering techniques, that can group information about datasets without prior knowledge of their characteristics, has been used along with statistical methods to automate OMA so that it can be used for condition monitoring. Recently, automatic OMA (AOMA) was applied to rotating machinery data. This paper evaluates one of these AOMA algorithms, successfully tested with data from a test rig with a rotor supported by hydrodynamic bearings, in a more complex dataset, with data from a rotor supported by magnetic bearings and influenced by gas seal. The results show that the proposed algorithm can extract modal parameters close to the ones extracted by EMA and by the mathematical modeling of the test rig, being robust even when a more complex system is analyzed.

Keywords Operational modal analysis · Rotating machinery · Machine learning · System identification

N. Dreher (✉) · T. Machado
School of Mechanical Engineering, University of Campinas (UNICAMP), 200 Mendeleyev Street, Campinas, SP 13083-860, Brazil
e-mail: n265296@dac.unicamp.br

T. Paulsen · I. Santos
Department of Civil and Mechanical Engineering, Technical University of Denmark (DTU), Koppels Allé 404, 2800 Kgs Lyngby, Denmark

P. Kurka and M. Pereira (eds.), *Advances in Structural Vibration*, Lecture Notes in Mechanical Engineering, https://doi.org/10.1007/978-3-031-71540-2_8

8.1 Introduction

Vibration patterns of structures, machines, and vehicles are described by the modal parameters of the system. Changes in these parameters indicate that the system has been altered in some way, a possible indication of faults or failures. Therefore, extracting and monitoring the system's modal parameters is a matter of interest in the context of condition and health monitoring. Generally, the modal parameters are computed through mathematical modeling and extracted through Experimental Modal Analysis (EMA) tests in the existing system, which requires knowledge of the dynamic behavior of the system, and known and controllable excitation sources, respectively. Operational Modal Analysis (OMA) has been an important topic of research in the past decades because it enables modal parameter extraction without the need of known excitation sources and keeping the system in operation for testing. OMA considers that the excitation found during the system's operation is white noise with a broad frequency range that is capable of exciting all modes of interest. Moreover, it assumes that the system is linear and time invariant. OMA has been widely used for the modal analysis of civil structures because they usually can be considered linear and time invariant and they are subjected to similar white noise excitation.

The application of OMA in rotating machines is a topic of growing interest especially because of its advantages. However, rotating machines are subjected to complex load conditions and are themselves more complex than civil structures, requiring additional data processing to accurately extract the modal parameters of interest. Some solutions to these challenges have already been studied, such as the application of OMA in non-linear or time-varying systems (Kang and Zeng 2023; Vesterholm et al. 2019), in systems that are subjected to harmonic excitation (Brandt 2019; Gioia et al. 2019a, b; Gres et al. 2021; Xu et al. 2023), and in rotating systems with closely spaced forward and backward modes (Dreher et al. 2022).

Another topic of interest in the context of OMA and condition and health monitoring is the Automatic OMA (AOMA). Although OMA is more versatile than traditional EMA, the results' analysis is still highly dependent on a qualified professional. Therefore, several algorithms that combine statistical and machine learning techniques with OMA techniques have been proposed in the last few decades to fully automate OMA. Some classical research on the subject are the works of Magalhães et al. (2009) and Reynders et al. (2012). More recently, Priou et al. (2022) proposed an algorithm to automatically extract modal parameters considering the modes' uncertainties, validating the algorithm on vibration data from two bridges. Mugnaini et al. (2022) proposed a fully automated approach, successfully applying it to a numerical case study and to data from an Airbus H135 helicopter blade. Shu et al. (2023) developed a multi-task deep neural network that can automatically extract independent modes from multi-mode vibration responses of structures, further determining the modal parameters with conventional random decrement technique (RDT) and curve fitting approach. Once the network is trained, there is no need to manually set thresholds or indexes. When applied to data from a spring-mass-damper model and a long-span cable-stayed bridge, the proposed neural network achieved high accuracy and computational efficiency. Volkmar et al. (2023) presented a fully automated modal analysis approach that learns optimal parameters in a supervised manner based on a single manual modal analysis and choses new parameters with statistical methods. The approach was evaluated in ground vibration test data (EMA) and flight vibration data (OMA), showing good accuracy and computational efficiency. Zhong et al. (2023) proposed an AOMA method that uses an adaptive clustering method based on DBSCAN clustering with automatic selection of parameters. The method's performance was investigated via numerical analyses and the measured data from an actual bridge, being able to identify the modal parameters under various scenarios, with low bias and variance.

All research mentioned so far were developed for and tested with bridges and other simple models and structures. However, AOMA algorithms that focus on extracting modal parameters of coupled rotor foundation systems were already proposed by Paulsen et al. (2023) and Dreher et al. (2023). Paulsen et al. (2023) developed an AOMA algorithm to extract modal parameters for a bed plate of a rotor-foundation system when the rotor is affected by active magnetic bearings and gas seals. Dreher et al. (2023) proposed an AOMA method to extract the rotor's modal parameters, validating it in vibration signals from a test rig supported by hydrodynamic bearings and rolling bearings. It is important to highlight that both papers worked with rotating machinery parts, which is not common in the context of OMA because of the above-mentioned challenges. To investigate the robustness of the algorithm proposed by Dreher et al. (2023), this paper applies the algorithm on the data collected and analyzed by Paulsen et al. (2023), that comes from a more complex system and has mainly modal parameters from the foundation. The algorithm from Dreher et al. (2023) is briefly presented in Sect. 8.2.1 and the dataset from Paulsen et al. (2023) is presented in Sect. 8.2.2. The results are analyzed in terms of the stabilization diagrams obtained with the Stochastic Subspace Identification (SSI) method and the extracted modal parameters, which are compared with EMA modal parameters from the foundation, that supports the rotor and is influenced by it, and OMA results from Paulsen et al. (2023), as presented in Sect. 8.3. Focus is given to the modal analysis of a non-rotating structure that supports the rotor rather than the rotor itself.

8.2 Methodology

8.2.1 AOMA Algorithm

The algorithm proposed by Dreher et al. (2023) can be summarized into the following steps:

1. Create the stabilization diagram using the SSI algorithm and classify each pole based on stabilization criteria,
2. Clear the stabilization diagram using the Hard Validation Criteria (HVC),

3. Group stable poles that represent the same mode using agglomerative hierarchical clustering,
4. Remove from each cluster poles of repeated orders, so that only one pole of this order remains,
5. Eliminate small clusters that probably represent clusters of spurious or mathematical poles,
6. Perform an outlier detection based on the boxplot method,
7. Describe the global modes by the clusters mean frequency, mean damping, and mean mode shape,
8. Group poles with mode shapes of high correlation using agglomerative hierarchical clustering.

In order to support the results' analysis presented in Sect. 8.3, it is important to further explain the characteristics of the stabilization diagram and the final result of the algorithm.

The stabilization diagram is built performing the SSI with increasing model orders and plotting the natural frequency of the resulting poles in the x axis and the order in which they were extracted in the y axis. All poles are inspected and classified according to the following evaluation. A k-order pole is stable if there is at least one pole in the (k-1)-order that satisfies the following stabilization criteria from one order to the other: frequency variation limit, damping ratio variation limit, and Modal Assurance Criterion (MAC) limit. All poles that fulfill the stabilization criteria are classified as stable, and all that do not are classified as not stable, in accordance with the algorithm's first step. This is important because physical modes of the system can be identified by the alignment of stable poles along different model orders.

In sequence, each pole (stable and not stable) is analyzed to verify if it respects the Hard Validation Criteria (HVC). The idea behind the HVC is to remove all certainly spurious poles from the following steps, that is, poles with negative damping ratio and no complex conjugated pair. Moreover, performing initial tests allow the analyst to know the normal behavior of the system, including the normal range of damping ratios. Thus, poles with damping ratios out of this range are also considered spurious. In that way, the stabilization diagram presents stable, not stable, and spurious poles. Only stable poles are considered for the following steps of the algorithm.

The hierarchical clustering is then employed to group poles that characterize the same physical modes. The following steps are used to remove spurious or inaccurate poles that were not excluded by the stabilization criteria or the HVC, making the results more accurate. The algorithm concludes by taking mean frequencies, mean damping ratios, and mean mode shapes of each group of poles to represent each global mode.

8.2.2 Dataset Description

The test rig employed by Paulsen et al. (2023) is displayed on Fig. 8.1, being composed of an electric motor connected to a drive shaft through a belt drive. The drive shaft controls the main rotor through a flexible coupling. The main rotor is levitated in the lateral directions by active magnetic bearings and influenced by two smooth annular seals mounted in a back-to-back configuration. The modules carrying the motor, drive shaft, active magnetic bearings, and seal geometries are placed on a bed plate. The authors focused the investigation on the identification of vibration patterns present in the bed plate when the seal inlet pressure changes.

EMA tests were performed using an excitation hammer to obtain estimates of the modal parameters of the system during three operating conditions, seal inlet pressures of 0 bar, 1.3 bar and 2.7 bar. The modal parameters for each operating condition were also estimated with a mathematical model of the test rig. Further details about the EMA tests and the mathematical model can be found in Paulsen et al. (2023).

Data for OMA tests, on the other hand, were collected for five operating conditions, pressures of 0 bar and white noise excitation, 1.3 bar and no excitation, 1.3 bar and white noise excitation, 2.7 bar and no excitation, and 2.7 bar and white noise excitation. The authors' idea was to evaluate whether the seal forces could be used as white noise excitation for OMA, which was proved possible. For OMA, data was collected with one stationary accelerometer and one roving accelerometer that was placed in 13 different positions of the bed plate, as presented in Fig. 8.2. For all 5 conditions of OMA tests, 13 measurements were collected with both accelerometers, the stationary and roving ones, during 128s and with sampling frequency of 3,012 Hz.

Summarizing the available data for OMA, tests with the following conditions are available for each of the 13 configurations:

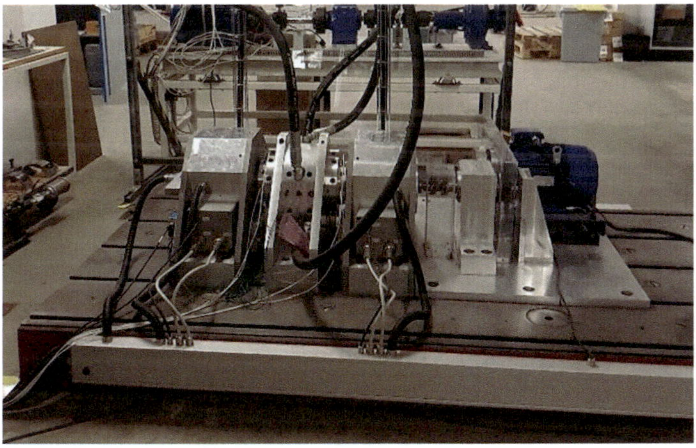

Fig. 8.1 Test rig of a rotor supported by active magnetic bearings and influenced by gas seal

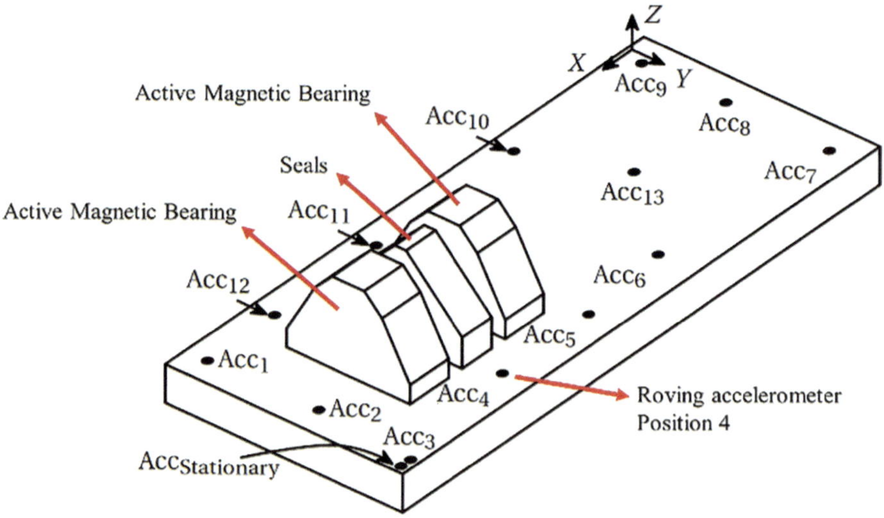

Fig. 8.2 Test rig with the 13 positions of the roving accelerometer, adapted from Paulsen et al. (2023)

1. 0 bar pressure and white noise excitation,
2. 1.3 bar pressure and no excitation,
3. 1.3 bar pressure and white noise excitation,
4. 2.7 bar pressure and no excitation,
5. 2.7 bar pressure and white noise excitation.

8.3 Results

In this section, the results of the application of the algorithm presented by Dreher et al. (2023) in the data collected by Paulsen et al. (2023) are discussed. Since the AOMA algorithm depends on the results of the SSI algorithm, the parameters of this algorithm need to be selected, which is presented in Sect. 8.3.1. Then, the AOMA algorithm's robustness is proved in Sect. 8.3.2, varying the AOMA's parameters and comparing the results with the ones obtained in Paulsen et al. (2023) from EMA and OMA analysis.

8.3.1 Stabilization Diagram Analysis

The quality of the stabilization diagram generated by the SSI algorithm is based on the right selection of the number of block rows in the Hankel matrix (i), the method (Unweighted Principal Component – UPC, Principal Component – PC, and

Canonical Variate Algorithm – CVA) of the data driven SSI (SSI-DATA), and the maximum order of the model. These parameters are dependent on the systems' characteristics and are selected by experienced users from case to case.

For the dataset employed in Dreher et al. (2023), the selected number of block rows was 100, the SSI-DATA method was the UPC, and the maximum order was 100. In Paulsen et al. (2023), the authors found the ideal number of block rows as 256 (28) and employed the covariance driven SSI (SSI-COV) with maximum order of 128 (27). In order to define the best parameters for this dataset using the algorithm from Dreher et al. (2023), the stabilization diagrams generated with different parameters were visually analyzed. The stabilization diagrams are used because the choice of these parameters directly affect the diagrams' quality.

The stabilization diagrams in Fig. 8.3 were generated with measurements from condition 1 (0 bar and white noise excitation) and roving sensor 6, using 100 (a) and 256 (b) block rows. The SSI results are presented as circles (poles) that indicate the natural frequency in Hz (x axis) and the model order (y axis) of each possibly physical mode of the system. Each pole is identified as stable (red), not stable (green), and spurious (blue) in accordance with the natural frequency, damping ratio, and mode shape. The presence of a physical mode is indicated by the alignment of stable poles, and the presence of a mathematical mode or a harmonic frequency exciting the system is indicated by not stable or spurious poles. The curves in the background are the results from the Frequency Domain Decomposition (FDD) method. From the results, one can see that the diagram generated with 256 block rows has a higher number of stable poles and better-defined alignments of stable poles, being better for modal parameter identification. The stabilization diagrams of different conditions and sensors were also analyzed, and the results were similar.

The higher the number of block rows, the higher is the computational cost. Considering that the results with 256 block rows were good for modal identification and the number is the same found as ideal in Paulsen et al. (2023), it will be selected as the ideal number of block rows for this case. Still, an increase of 17% in the number of block rows was tested, from 256 (Fig. 8.4a) to 300 (Fig. 8.4b) for the measurement from condition 5 and roving sensor 6. From the results of the stable poles, displayed in Fig. 8.4, one can see that the increase in the number of block rows

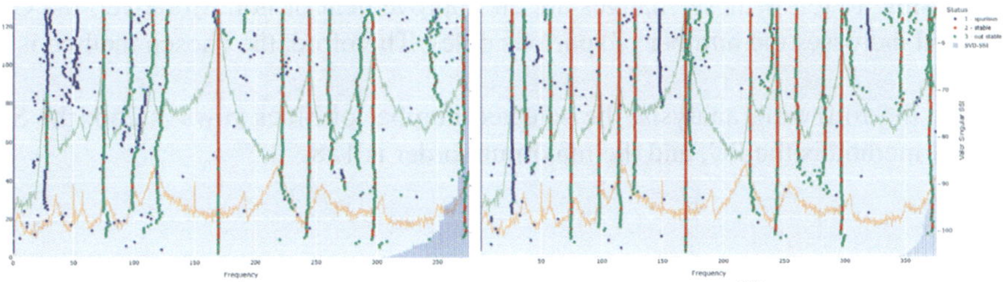

(a) (b)

Fig. 8.3 Stabilization diagrams from condition 1 and roving sensor 6 with $i = 100$ (**a**) and $i = 256$ (**b**)

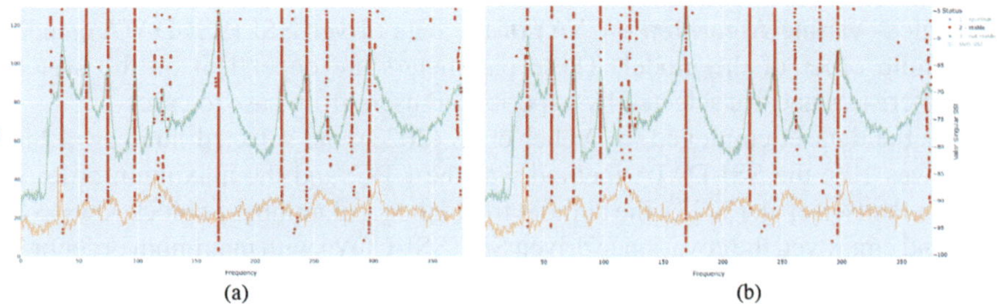

Fig. 8.4 Stabilization diagrams from condition 5 and roving sensor 6 with $i = 256$ (**a**) and $i = 300$ (**b**)

slightly increased the number of stable poles near the frequency of 120 Hz, which has a physical mode of interest. However, the stability of the other frequency ranges of interest didn't change significantly. Moreover, the FDD results indicate that the main reason for the weak visualization of the frequency of 120 Hz is the small number of sensors or the weak excitation of this mode, and not the SSI method or parameters.

The influence of the maximum model order was also evaluated. Stabilization diagrams with maximum orders of 128 and 256 were compared for both $i = 128$ and $i = 256$. From the results, it was concluded that the increase in the number of block rows still provides more stable poles than the increase in the order. And the increase in both parameters (order and number of block rows) leads to an enormous number of probably spurious poles that were not identified as such with HVC. This happens because the SSI algorithm attempts to fit the data precisely. The higher the model order, the more modes are used for fitting. Therefore, the algorithm results in fitting noise and other characteristics of the vibration signal that are not related to the dynamics of the system.

The stabilization diagrams were generated with all SSI-DATA methods, UPC, PC, and CVA. From a visual inspection of all diagrams, the results with UPC and PC methods were very similar, but there is a slight improvement in the poles' stability using the PC method. Comparing the PC and the CVA methods, the poles stability is improved only on the higher orders of the CVA method. The damping ratios are slightly affected by the change, however, even with the EMA's expected results, it is impossible to tell if these changes are an improvement or not. Moreover, the CVA method increases the number of spurious poles. Therefore, the chosen method is the PC.

For the following analysis, the selected number of block rows is 256, the SSI-DATA method is the PC, and the maximum order is 128.

8.3.2 Modal Parameter Analysis

All results presented so far were compared in terms of the stabilization diagram. To verify how the limits of stabilization and damping ratio and the hierarchical clustering threshold influence the AOMA algorithm, the results will be compared in terms of the extracted global modes. The global modes are analyzed because the choice of these parameters doesn't affect the poles alignment in the stabilization diagram, only their classification. Since there are 13 measurements (13 positions for the roving sensor) for each operating condition, 13 modal analysis results are obtained for each condition. In order to compare these results with the ones presented in Paulsen et al. (2023), similar global modes within a single operating condition were grouped together in terms of mean value, which provided a single group of modal parameters for each condition. Therefore, each change in the SSI or AOMA parameters will be evaluated in terms of the mean modal parameters extracted with these changes.

In Dreher et al. (2023), the damping ratio limit was chosen as 0.3–10%, the frequency variation limit as 0.2%, the damping ratio variation limit as 2%, the minimum MAC value as 95%, and the threshold value for the hierarchical clustering as 0.01. The same parameters were used to process the dataset from Paulsen et al. (2023). The results are displayed in Tables 8.1, 8.2 and 8.3, along with the EMA and OMA results from Paulsen et al. (2023). From the overview, one can see that the results obtained with my AOMA algorithm from Dreher et al. (2023) are similar to the EMA reference and to the results obtained by Paulsen et al. (2023), that were validated through the EMA reference and the mathematical model of the test rig. There are only a few modes that are identified by one algorithm and not the other, or that are identified with different damping ratios when compared to the EMA reference. For most modes that the present algorithm is unable to identify or identifies with significant deviation from the reference, the problem can be traced back to the stabilization diagram, not the algorithm's ability to automatically interpret the stabilization diagram. It is important to highlight that the AOMA parameters are the same configured for the test rig with a rotor supported by a hydrodynamic bearing, that is simpler than the one from Paulsen et al. (2023). This means that even for the more complex case, the AOMA algorithm from Dreher et al. (2023) was able to identify most physical modes of interest with high accuracy compared to EMA and to the mathematical model. This indicates the robustness of the AOMA algorithm.

One can see that the damping ratio of the first mode (around 25 Hz and 9% damping) is difficult to estimate via OMA, and the estimations are usually lower than the reference. Taking a closer look to Tables 8.1, 8.2 and 8.3, there are also physical modes extracted by EMA with damping ratio above 10% (around 45 Hz and 13% damping), values above the damping ratio limit employed for AOMA. Therefore, another set of AOMA results was generated using the damping ratio limit of 15%. The results for the modes of interest are displayed in Table 8.4, from which one can see that the change in the damping ratio limit improved some of the extracted damping ratios (see conditions 1 and 2). But OMA is still unable to identify some modes of interest around 42 Hz.

Table 8.1 Results with original parameters from Dreher et al. (2023) for condition with 0 bar pressure and white noise

EMA		OMA (Paulsen et al. 2023)		OMA (Dreher et al. 2023)	
ω_n[Hz]	ζ[%]	ω_n[Hz]	ζ[%]	ω_n[Hz]	ζ[%]
25.64	8.67	25.84	5.50	26.27	8.84
34.90	3.59	35.62	3.73	35.46	3.15
44.24	14.81	–	–	–	–
56.34	2.31	55.32	3.00	54.58	2.59
74.09	2.05	74.61	1.55	74.51	1.37
95.93	0.98	98.10	0.97	98.07	0.97
114.30	1.96	114.02	1.63	114.30	1.67
120.16	3.04	127.02	1.58	127.33	2.39
169.43	0.72	169.01	1.20	169.41	1.09
190.46	0.90	186.01	1.18	–	–

Table 8.2 Results with original parameters from Dreher et al. (2023) for conditions with 1.3 bar pressure

EMA		OMA (Paulsen et al. 2023) No excitation		OMA (Dreher et al. 2023) No excitation		OMA (Paulsen et al. 2023) White noise		OMA (Dreher et al. 2023) White noise	
ω_n[Hz]	ζ[%]	ω_n[Hz]	ζ[%]	ω_n[Hz]	ζ[%]	ω_n[Hz]	ζ[%]	ω_n[Hz]	ζ[%]
24.73	8.69	25.81	2.27	26.46	4.31	26.03	4.69	26.37	8.24
34.71	4.00	35.42	3.19	34.07	1.48	32.44	4.96	36.00	3.79
42.84	13.02	–	–	–	–	36.13	3.93	–	–
55.94	2.08	57.10	1.04	56.93	1.86	56.39	5.67	56.40	3.04
73.99	1.91	74.41	1.32	74.55	1.57	74.51	1.38	74.53	1.38
95.10	1.55	98.60	0.85	98.65	1.06	98.20	0.91	98.13	0.90
114.18	1.56	114.02	1.63	114.61	1.83	115.01	1.52	114.47	1.49
121.34	2.83	–	–	–	–	127.02	1.85	127.58	2.23
–	–	157.00	0.60	–	–	–	–	–	–
169.40	0.82	169.01	1.19	169.43	1.20	169.01	1.13	169.37	1.10
–	–	177.01	1.18	–	–	–	–	–	–
190.29	1.02	–	–	190.98	0.88	183.01	0.81	–	–

To verify if the results can be further improved, identifying more physical modes of interest or modal parameters with higher accuracy, the following changes will also be adopted: frequency variation limit of 0.5% (2.5 $\times$ the original), damping variation limit of 4% (2 $\times$ the original), MAC limit of 90% (5% lower than the original). It is important to highlight that the changes are all in the sense of making the stabilization criteria wider, enabling the classification of more poles as stable

Table 8.3 Results with original parameters from Dreher et al. (2023) for conditions with 2.7 bar pressure

EMA		OMA (Paulsen et al. 2023) No excitation		OMA (Dreher et al. 2023) No excitation		OMA (Paulsen et al. 2023) White noise		OMA (Dreher et al. 2023) White noise	
ω_n[Hz]	ζ[%]	ω_n[Hz]	ζ[%]	ω_n[Hz]	ζ[%]	ω_n[Hz]	ζ[%]	ω_n[Hz]	ζ[%]
25.25	9.53	26.63	4.39	25.79	4.96	26.15	5.93	26.47	7.86
34.91	2.88	35.35	5.54	35.46	2.29	36.21	2.87	35.87	3.35
48.46	9.30	–	–	40.27	1.53	40.30	6.92	42.63	8.10
55.79	1.92	56.71	1.92	56.84	2.55	56.54	3.73	57.04	3.44
74.07	1.77	75.01	1.92	74.78	2.37	73.91	1.69	74.90	1.95
95.61	2.52	98.20	0.82	98.39	0.86	97.70	0.97	97.63	0.90
114.44	2.42	115.02	1.89	115.19	1.59	115.01	1.50	115.00	1.30
120.91	4.36	–	–	–	–	–	–	121.63	1.32
–	–	–	–	–	–	139.04	2.34	129.11	2.33
169.16	0.87	169.01	1.25	169.49	1.26	169.01	1.15	169.32	1.18
191.63	1.42	–	–	191.04	0.81	–	–	–	–

Table 8.4 Results with damping ratio limit of 15% for all conditions

Condition	EMA		OMA—original parameters		OMA—damping ratio limit of 15%	
	ω_n[Hz]	ζ[%]	ω_n[Hz]	ζ[%]	ω_n[Hz]	ζ[%]
1	25.64	8.67	25.80	5.50	26.36	10
	44.24	14.81	–	–	–	–
2	24.73	8.69	26.46	4.31	27.05	5.91
	42.84	13.02	–	–	–	–
3	24.73	8.69	26.37	8.24	26.46	9.29
	42.84	13.02	–	–	–	–
4	25.25	9.53	25.79	4.96	25.79	4.96
	48.46	9.30	40.27	1.53	40.27	1.53
5	25.25	9.53	26.47	7.86	26.52	8.10
	48.46	9.30	42.63	8.10	42.55	8.79

and, therefore, providing more poles for the AOMA method to process. The results of these changes are displayed on Tables 8.5, 8.6 and 8.7. For condition 0 (Table 8.5), one can see that the changes enabled the identification of one more physical mode of interest, around 190 Hz, and didn't alter the other results, that were already close to the EMA reference. For condition 2 (Table 8.6), the mode of 34.68 Hz was identified with a damping ratio closer to the reference. For conditions 3 (Table 8.5), the mode of 56 Hz was split into two. Evaluating the stabilization diagrams, one can verify that a single strong alignment of stable poles around that frequency is split into two at higher orders. Since using greater limits for the stabilization diagram enables a wider frequency range of poles around this mode to be identified as stable, these two alignments were identified by the AOMA algorithm. Further investigations are needed to verify if there are two closely spaced modes around this frequency or if these two alignments appear as an attempt of the SSI method to fit the data. All other modes of these conditions were similar to the ones extracted using the original stabilization parameters. For conditions 4 (Table 8.7), the results were similar to the ones extracted using the original stabilization parameters and to the EMA reference. Finally, for condition 5 (Table 8.7) there are several physical modes of interest that were split into two, results that can also be traced back to the stabilization diagram's result. Apart from the two modes near 56 Hz, the duplications are caused because there is not a strong stabilization around the frequencies of interest.

The threshold limit of the hierarchical clustering was set to 0.013 (the original value was 0.01), so that more poles could be clustered in the same group. With this change, most modes that were divided into two groups in Table 8.7 were merged back to a single group. Moreover, further steps can be implemented to evaluate if these groups with close mean frequencies are groups of the same physical mode, as done by Paulsen et al. (2023), and, if so, either group them together or choose one of them as more accurate.

Even with changes in the AOMA parameters, it was not possible to identify all modes of interest for all conditions. Physical modes of 40, 127, and 190 Hz presented weak stabilization in the stabilization diagrams of most conditions, which explain their absence in the results. Thus, they would likely not be identified as physical modes of the system by an experienced analyst and could not be identified by the AOMA algorithm. Some extra modes, that were not identified in EMA tests, were identified by the AOMA algorithms from Paulsen et al. (2023) and Dreher et al. (2023). Evaluating the stabilization diagrams, one can see a strong alignment of

Table 8.5 Results with new parameters for condition with 0 bar pressure and white noise

EMA		OMA—original parameters		OMA—new parameters	
ω_n[Hz]	ζ[%]	ω_n[Hz]	ζ[%]	ω_n[Hz]	ζ[%]
25.64	8.67	26.27	8.84	26.32	9.07
34.90	3.59	35.46	3.15	35.55	3.40
44.24	14.81	–	–	–	–
56.34	2.31	54.58	2.59	56.15	3.35
74.09	2.05	74.51	1.37	74.51	1.37
95.93	0.98	98.07	0.97	98.06	0.97
114.30	1.96	114.30	1.67	114.29	1.66
120.16	3.04	127.33	2.39	127.35	2.45
169.43	0.72	169.41	1.09	169.40	1.09
190.46	0.90	–	–	190.18	0.76

Table 8.6 Results with new parameters for conditions with 1.3 bar pressure

EMA		OMA—no excitation Original parameters		OMA—no excitation New parameters		OMA—white noise Original parameters		OMA—white noise New parameters	
ω_n[Hz]	ζ[%]	ω_n[Hz]	ζ[%]	ω_n[Hz]	ζ[%]	ω_n[Hz]	ζ[%]	ω_n[Hz]	ζ[%]
24.73	8.69	26.46	4.31	26.57	4.59	26.37	8.24	26.35	8.29
34.71	4.00	34.07	1.48	34.37	3.54	36.00	3.79	35.90	3.90
42.84	13.02	–	–	–	–	–	–	–	–
–	–	–	–	–	–	–	–	55.54	3.41
55.94	2.08	56.93	1.86	56.94	1.96	56.40	3.04	56.54	3.02
73.99	1.91	74.55	1.57	74.54	1.59	74.53	1.38	74.53	1.38
95.10	1.55	98.65	1.06	98.66	1.07	98.13	0.90	98.13	0.89
114.18	1.56	114.61	1.83	114.61	1.84	114.47	1.49	114.47	1.50
121.34	2.83	–	–	–	–	127.58	2.23	127.58	2.21
169.40	0.82	169.43	1.20	169.43	1.20	169.37	1.10	169.37	1.10
190.29	1.02	190.98	0.88	190.52	0.75	–	–	–	–

Table 8.7 Results with new parameters for conditions with 2.7 bar pressure

EMA		OMA—no excitation Original parameters		OMA—no excitation New parameters		OMA—white noise Original parameters		OMA—white noise New parameters	
$\omega_n[Hz]$	ζ[%]	$\omega_n[Hz]$	ζ[%]	$\omega_n[Hz]$	ζ[%]	$\omega_n[Hz]$	ζ[%]	$\omega_n[Hz]$	ζ[%]
25.25	9.53	25.79	4.96	25.45	4.82	26.47	7.86	26.62	8.04
34.91	2.88	35.46	2.29	35.56	2.25	35.87	3.35	34.60	4.15
–	–	–	–	–	–	–	–	36.08	3.25
48.46	9.30	40.27	1.53	40.17	1.60	42.63	8.10	42.61	8.04
55.79	1.92	56.84	2.55	56.87	2.77	57.04	3.44	56.75	3.37
–	–	–	–	–	–	–	–	57.71	3.63
–	–	–	–	–	–	–	–	73.84	2.56
74.07	1.77	74.78	2.37	74.78	2.38	74.90	1.95	74.91	1.95
95.61	2.52	98.39	0.86	98.29	0.82	97.63	0.90	97.63	0.90
114.44	2.42	115.19	1.59	115.19	1.59	115.00	1.30	115.08	1.35
120.91	4.36	–	–	–	–	121.63	1.32	121.65	1.15
–	–	–	–	–	–	129.11	2.33	128.42	2.32
–	–	–	–	–	–	–	–	163.87	0.88
169.16	0.87	169.49	1.26	169.49	1.27	169.32	1.18	169.32	1.18
191.63	1.42	191.04	0.81	190.85	1.02	–	–	–	–

these modes (129 and 164 Hz), that look like true physical modes present in the measurements. These findings indicate that the AOMA algorithm could properly interpret the stabilization diagrams.

8.4 Conclusions

In this paper the Automatic Operational Modal Analysis (AOMA) algorithm proposed by Dreher et al. (2023) and tested in a test rig with a rotor supported by hydrodynamic bearings was evaluated. The algorithm was applied to a more complex dataset, with a rotor supported by active magnetic bearings and influenced by gas seal pressure. The SSI parameters were adjusted for the new dataset, which is a common practice since their values vary from system to system. The parameters of the AOMA algorithm were also adjusted, and the results demonstrated that even for a more complex rotating system, the original parameters could accurately extract the majority of the physical modes of interest, that are related to the foundation that supports the rotor and is influenced by it. Only small changes in the parameters were made to slightly improve the results. The modes that were not identified by the AOMA method were weakly excited modes, with weak stabilization in the diagrams, and it is likely that an experienced analyst wouldn't classify these modes as physical modes of the system. Finally, from these results, one can see that the AOMA algorithm from Dreher et al. (2023) is good for extracting modal parameters from systems that are composed of a foundation supporting a rotating machine, which validates its robustness.

Acknowledgements The authors would like to thank CAPES, Coordination for the Improvement of Higher Education Personnel, and PRPG, Pro-Rectory of Graduate Studies at UNICAMP, for the financial support, and the Department of Civil and Mechanical Engineering from Technical University of Denmark for the support that made this work possible.

Responsibility Notice The authors are the only responsible for the printed material included in this paper.

References

Brandt A (2019) A signal processing framework for operational modal analysis in time and frequency domain. Mech Syst Sig Proc 115:380–393. https://doi.org/10.1016/j.ymssp.2018.06.009

Dreher NR, Storti GC, Machado TH (2023) Automated operational modal analysis for rotating machinery based on clustering techniques. Sensors. https://doi.org/10.3390/s23031665

Dreher NR, Storti GC, Machado TH (2022) Directional coordinates for the identification of backward and forward frequencies of rotating machines via OMA. In: Proceedings of the 9th international operational modal analysis conference—IOMAC 2022, Vancouver, Canada, 3–6 July 2022

Gioia N, Daems PJ, Peeters C, El-Kafafy M, Guillaume P, Helsen J (2019a) Influence of the harmonics on the modal behavior of wind turbine drivetrains. In: Di Maio D (eds) Rotating machinery, vibro-acoustics laser vibrometry, Volume 7. Conference proceedings of the society for experimental mechanics series. Springer, Cham. https://doi.org/10.1007/978-3-319-74693-7_22

Gioia N, Peeters C, Guillaume P, Helsen J (2019b) Identification of noise, vibration and harshness behavior of wind turbine drivetrain under different operating conditions. Energies. https://doi.org/10.3390/en12173401

Gres S, Döhler M, Andersen P, Mevel L (2021) Kalman filter-based subspace identification for operational modal analysis under unmeasured periodic excitation. Mech Syst Sig Proc. https://doi.org/10.1016/j.ymssp.2020.106996

Kang J, Zeng S (2023) Uncertainty quantification in operational modal analysis of time-varying structures based on time-dependent autoregressive moving average model. J Sound Vibr. https://doi.org/10.1016/j.jsv.2022.117549

Magalhães F, Cunha A, Caetano E (2009) Online automatic identification of the modal parameters of a long span arch bridge. Mech Syst Sig Proc. https://doi.org/10.1016/j.ymssp.2008.05.003

Mugnaini V, Fragonara LZ, Civera M (2022) A machine learning approach for automatic operational modal analysis. Mech Syst Sig Proc. https://doi.org/10.1016/j.ymssp.2022.108813

Paulsen TT, Damsgaard SV, Santos IF (2023) Automated modal parameter estimation for coupled rotor-foundation systems using seal forces as excitation source. Mech Syst Sig Proc. https://doi.org/10.1016/j.ymssp.2024.111293

Priou J, Gres S, Perrault M, Guerineau L, Döhler M (2022) Automated uncertainty-based extraction of modal parameters from stabilization diagrams. In: Proceedings of the 9th international operational modal analysis conference—IOMAC 2022, Vancouver, Canada, 3–6 July 2022

Reynders E, Houbrechts J, De Roeck G (2012) Fully automated (operational) modal analysis. Mech Syst Sig Proc 29:228–250. https://doi.org/10.1016/j.ymssp.2012.01.007

Shu J, Zhang C, Gao Y, Niu Y (2023) A multi-task learning-based automatic blind identification procedure for operational modal analysis. https://doi.org/10.1016/j.ymssp.2022.109959

Vesterholm KK, Brincker R, Brandt AB (2019) Detection of nonlinear behavior using the Random Decrement Technique. In: Proceedings of the 8th international operational modal analysis conference—IOMAC 2019, Copenhagen, Denmark, 13–15 May 2019

Volkmar R, Soal K, Govers Y, Böswald M (2023) Experimental and operational modal analysis: automated system identification for safety-critical applications. Mech Syst Sig Proc. https://doi.org/10.1016/j.ymssp.2022.109658

Xu M, Au FTK, Wang S, Tian H (2023) Operational modal analysis under harmonic excitation using Ramanujan subspace projection and stochastic subspace identification. J Sound Vib. https://doi.org/10.1016/j.jsv.2022.117436

Zhong QM, Chen SZ, Sun Z, Tian LC (2023) Fully automatic operational modal analysis method based on statistical rule enhanced adaptive clustering method. Eng Struct. https://doi.org/10.1016/j.engstruct.2022.115216

Chapter 9
Investigation of a Novel Metastructure with Trapped, Fluid-Filled Unit Cells

Vinícius Mauro de Souza Santos, Thiago de Paula Sales, and Morvan Ouisse

Abstract This work investigates a novel metamaterial concept using the Wave-based Finite Element Method. The metamaterial comprises a periodic-like structure manufactured through fused filament deposition, featuring internal cavities filled with water. Experimental characterization of the dynamics of the periodic system without internal fluid confirms good agreement with numerical predictions obtained through frequency response function measurements. Furthermore, the dynamic behavior of the two-phase periodic metastructure is experimentally examined, where waves interact within the heterogeneous medium consisting of both fluid and solid phases. In this case, the resulting wave characteristics depend on the properties of both phases. It was shown that the fluid-filled metastructure exhibits vibration reduction through the whole frequency range compared to the case lacking internal fluid. Additionally, it was seen that the frequency range near the second attenuation band of the periodic metastructure without fluid can be enlarged after the fluid inclusion within the cavities of its unit cells, as a consequence of mass increase and damping effects. Consequently, this work presents a promising avenue for metastructure design, with potential applications in structural dynamics and acoustics.

Keywords Metamaterial · Bandgap · Periodic structures · Fluid–structure interaction · Wave-based finite element method

V. Mauro de Souza Santos (✉) · T. de Paula Sales
Mechanical Engineering Division, ITA—Aeronautics Institute of Technology, Praça Marechal Eduardo Gomes, 50, Vila das Acácias, São José Dos Campos, SP 12228-900, Brazil
e-mail: vinicius.santos@ga.ita.br

T. de Paula Sales
e-mail: tpsales@ita.br

V. Mauro de Souza Santos · M. Ouisse
SUPMICROTECH, Institut FEMTO-ST, Université de Franche-Comté, CNRS, Besançon, France
e-mail: morvan.ouisse@femto-st.fr

P. Kurka and M. Pereira (eds.), *Advances in Structural Vibration*, Lecture Notes in Mechanical Engineering, https://doi.org/10.1007/978-3-031-71540-2_9

9.1 Introduction

The study of periodic and quasi-periodic structures opened possibilities to explore several properties that are not naturally encountered in nature, such as waveguiding (Pal and Ruzzene 2017), bandgap formation (Wang et al. 2019), confinement and mode localization phenomena (Wu et al. 2009), invisibility cloaks (Ning et al. 2020), mode conversion (Wang et al. 2016), and some negative equivalent properties such as refractive index, elastic modulus, and Poisson's ratio (Gao et al. 2022). Such structures are usually referred to as metastructures just because they can exhibit apparent properties, as listed earlier, which are not typically observed in conventional systems. Probably, in the context of structural dynamics, the most studied phenomenon related to periodic structures is what one knows as bandgaps, which comprise frequency ranges where only evanescent wavemodes are allowed to occur (Miranda et al. 2019). This phenomenon can arise from Bragg-scattering (Nobrega et al. 2016), periodic arrangements of resonators within a host structure (Gao et al. 2022), or specially designed unit cells incorporating inertial amplification mechanisms (IAMs) (Mizukami et al. 2021).

In certain scenarios, both metastructures and conventional systems can exhibit complex dynamic behavior due to multiphysics problems. These may involve fluid–structure interaction (FSI), thermoelasticity, piezoelectric materials, magnetic forces, acoustic-structure interaction (ASI), viscoelasticity, and more. In this work, particular emphasis is placed on the dynamics of systems with enclosed fluid. The coupling of fluid dynamics with structural systems has been a longstanding concern in various fields, including structural engineering, aeroelasticity, hydroelasticity, vibroacoustics, wind engineering, ocean engineering, biomechanics, offshore engineering, aerospace engineering, and civil engineering. For instance, in previous research by Jamshidiat and SenGupta (1991), a method based on traditional finite element (FE) procedures was developed to predict the natural frequencies and modes of an aircraft fuselage considering the enclosed fluid medium.

Although extensive research on FSI has been conducted in the aforementioned fields, it is crucial to consider the dynamics of metamaterials interacting with acoustic fluids. One notable investigation by Spadoni et al. (2014) focused on studying the dynamic behavior of closed-cell crystalline foams to explore the potential of these relatively uncharted metamaterials. The authors highlighted that the complex microstructure of closed-cell crystalline foams inherently possesses internal resonances, facilitating local resonance phenomena. Through a classical vibroacoustic model, Spadoni et al. (2014) demonstrated that truncated-octahedron (or Kelvin foam), face-centered cubic rhombic dodecahedron, and Weaire-Phelan foams exhibit superanisotropic properties, capable of behaving selectively as either a fluid or a solid. These foams display pentamode characteristics and feature regimes characterized by film resonances and a high density of states. These extraordinary properties can find effective utilization in various engineering and scientific disciplines, such as aerospace engineering, civil engineering, mechanical engineering, and acoustics, particularly in the design of metamaterials for specific purposes. The study of

Spadoni et al. (2014) also revealed the existence of persistent modes in the foams, i.e., those modes that remain unaffected by the presence or absence of an entrained acoustic fluid. It was observed that the propagating shear modes, away from film resonances, are persistent modes, suggesting the hydrostatic behavior of the fluid-filled unit cell—in this case, the fluid does not influence the volumetric deformations of the foam.

Dorodnitsyn and Damme (2016) introduced and experimentally investigated a novel acoustic metamaterial incorporating an entrained fluid. The authors tackled the interaction between waves traveling through combined media, consisting of both fluid and solid components, which leads to coupled elastic waves with dynamics influenced by both phases. Notably, the acoustic metamaterial studied by Dorodnitsyn and Damme (2016) exhibits a negative refractive index behavior, owing to the opposite signs of the system's group and phase velocity. This research has opened up a new and straightforward pathway for designing acoustic metamaterials with exceptional properties not readily found in nature. These findings hold promise for various practical engineering applications and contribute to the vibroacoustic community.

Most of the time, analytical expressions are employed with the aim of modeling the aforementioned FSI problems. However, it is well known that these expressions are limited to approximate theories, and as a result, the established models are valid only under certain specific conditions. One cannot leave mentioning that the majority of analytical models in FSI typically represent simple systems, such as fluid-filled shells. In addition, mainly seminal works published in the fifties and sixties have been focused on wave propagation behavior in empty and fluid-filled cylindrical shells with real wavenumbers, which may not accurately reflect various practical conditions (McNiven et al. 1966; Mirsky and Herrmann 1957).

In an attempt to address some of these limitations, Bao et al. (2000) presented closed-form expressions and conducted a theoretical study on dispersion curves of empty and fluid-filled cylindrical shells under axially symmetric waves, relying on Bessel functions, based on expressions derived by Kumar and Stephens (1972). Parametric analyses considering the wall thickness of both empty and fluid-filled systems were performed, yielding results similar to those reported by Kumar and Stephens (1972). In summary, Bao et al. (2000) demonstrated that dispersion curves of such cylindrical shells can exhibit purely real, purely imaginary, and complex branches. The degree of coupling between the internal fluid and the shell was found to be strongly influenced by the shell thickness. Furthermore, it was observed that the internal fluid could considerably alter the wave propagation characteristics of the cylindrical shell. It has the ability to shift the cut-off frequencies of the wavemodes, providing a valuable mean to control the dynamic behavior of a metamaterial, which was not discussed at that time. Moreover, Bao et al. (2000) emphasized that certain branches seen in the dispersion curve of the fluid-filled cylindrical shell arise exclusively due to the presence of the fluid and may be purely evanescent. This realization suggests that fluid-filled systems can be exploited for designing specially engineered structures with outstanding vibration attenuation performance for those pursuing such goals.

Fuller and Fahy (1982) derived closed-form expressions to predict the characteristics of elastic and acoustic wave propagation in a cylindrical elastic shell filled with fluid. The authors showed that when two fields, such as structural and acoustic, are coupled, the resulting wave behavior becomes complex. They demonstrated that when a cylindrical shell is filled with fluid, a fluid loading term arising from the presence of the fluid acoustic field must be considered in the system's free vibration. Naturally, the magnitude of this term can vary significantly, determining the level of coupling between the shell and the fluid. Fuller and Fahy (1982) also conducted several parametric analyses considering factors such as the shell wall thickness and the density ratio between the shell material and the internal fluid that provide valuable insights into the dynamics of more complex fluid-filled systems.

As seen before, earlier studies have effectively formulated analytical expressions to comprehend wave propagation in fluid-filled elastic pipes despite their inherent simplifications. These expressions often provide a clearer and easier interpretation of wave propagation phenomena compared to numerical methods. However, as discussed previously, it is widely recognized that analytical solutions have limitations, as they are constrained to simple geometries and classical wave theories. They are also constrained to a narrow range of vibroacoustic problems due to complex interactions that may occur between the acoustic fluid and the structure. To address these limitations, Mencik and Ichchou (2007) resorted to a numerical method known as the Wave-based Finite Element Method (WFEM), in which the governing elasto-acoustic equations of motion are discretized using the FEM, and the problem is formulated and solved in the space spanned by wave-modes. The authors showed two mathematical strategies for handling fluid-filled structure problems: the $(\mathbf{U}, \mathbf{p})$ and $(\mathbf{U}, \Psi)$ formulations. In the first approach, the equations of motion for the acoustic fluid and the structure depend on the acoustic pressures of the fluid $(\mathbf{p})$ and the structural $(\mathbf{U})$ degrees of freedom (DoFs), resulting in an asymmetrical set of equations. On the other hand, the second approach is a modified version of the $(\mathbf{U}, \mathbf{p})$ formulation that introduces the acoustic velocity potential Ψ to formulate the elasto-acoustic problem symmetrically. Mencik and Ichchou (2007) demonstrated the robustness of their numerical approach compared to analytical formulations, such as the one presented in Fuller and Fahy (1982). It was confirmed that the simplifying assumptions used in the analytical formulation are not entirely adequate and may hinder the accurate assessment of the dynamic behavior of fluid-filled systems.

Numerical methods, including the WFEM, have strengths and weaknesses in engineering applications. Challenges can arise when applying the WFEM to composite structures and FSI problems. In such cases, a fine unit cell mesh is required to accurately capture the structural dynamics. A coarse mesh can fail to predict the eigenvalues and eigenvectors related to the wave propagation characteristics throughout the system. Droz et al. (2014) addressed this issue and proposed a model reduction strategy that utilizes a reduced basis to model the structural dynamic behavior. Cut-on frequencies were identified within the analyzed frequency band, which allowed for the solution of a low-order modified eigenvalue problem for these frequencies. Eigenvectors associated with propagating waves (for which the real part of the wavenumber is much greater than its imaginary part) are chosen to form the reduced basis, using

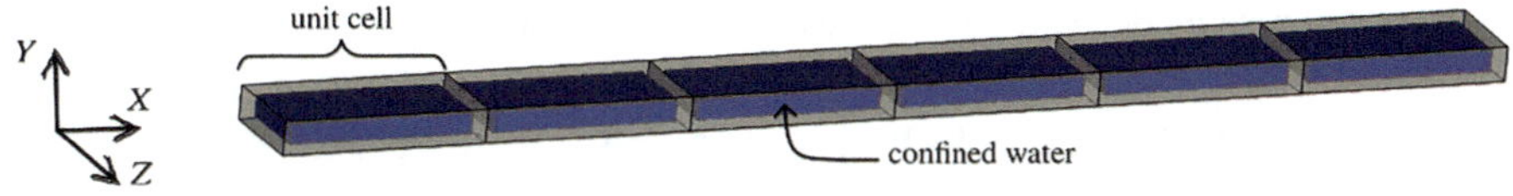

Fig. 9.1 A periodic metastructure filled with water fluid consisting of six identical unit cells

an iterative scheme to avoid redundant wave shapes, and then, a reduced eigenvalue problem, with size dependent on the number of propagating waves in the analyzed frequency range, is solved. This methodology was validated through a three-layered sandwich beam, and an elastic pipe filled with acoustic fluid in ref. (Droz et al. 2014). Additionally, Mencik (2018) also presented a valuable way to increase the computational efficiency of the WFEM, based on the Craig-Bampton (CB) model order reduction technique.

From this literature overview, which addressed various modeling strategies for tackling FSI problems, and also highlighted promising applications of FSI in the dynamics of metamaterials, in this study one aims to propose a novel concept of metamaterial with completely closed unit cells filled with water, as depicted in Fig. 9.1. In this illustration, the metastructure consists of six identical unit cells, with the internal cavity represented by the blue color, indicating the confined water. Firstly, a numerical model of the system without internal fluid is developed using the WFEM. Based on the Bloch-Floquet theorem and the matrices provided by the FEM, dispersion curves for the underlying unit cell are computed. Experimental frequency response functions (FRFs) for longitudinal and bending dynamics, obtained through hammer tests, are also made available and compared with numerical predictions. Subsequently, the dynamic behavior of the proposed fluid-filled metastructure is experimentally evaluated in order to understand its dynamic characteristics.

Beyond this introduction, this work has been divided into five sections. First, in Sect. 9.2, we review the WFEM, which can be used to model the fluid-filled metastructure shown in Fig. 9.1. Dispersion curves and numerical and experimental FRFs are shown in Sect. 9.3. Discussions about the dynamics of the system with and without fluid are supplied, as well. A summary of our findings and conclusions are given in Sect. 9.4. Acknowledgments and references are provided at last.

9.2 Modeling

Assume that one unit cell of the periodic structure portrayed in Fig. 9.1 is meshed employing three-dimensional FEs, considering the acoustic fluid and solid behaviors. Their corresponding mass and stiffness matrices are then extracted from conventional FE software, and by considering fluid and structural damping effects, one can write (Mencik and Ichchou 2007):

$$\hat{\mathbf{M}}\,\ddot{\hat{\mathbf{q}}} + \hat{\mathbf{C}}\,\dot{\hat{\mathbf{q}}} + \hat{\mathbf{K}}\hat{\mathbf{q}} = \hat{\mathbf{F}}, \qquad (9.1)$$

where $(\cdot)$ represents derivatives with respect to time and:

$$\hat{\mathbf{M}} = \begin{bmatrix} \mathbf{M}_\mathrm{S} & \mathbf{0} \\ \rho\mathbf{R}^\mathrm{T} & \mathbf{M}_\mathrm{A} \end{bmatrix},\ \hat{\mathbf{C}} = \begin{bmatrix} \mathbf{C}_\mathrm{S} & \mathbf{0} \\ \mathbf{0} & \mathbf{C}_\mathrm{A} \end{bmatrix},\ \hat{\mathbf{K}} = \begin{bmatrix} \mathbf{K}_\mathrm{S} & -\mathbf{R} \\ \mathbf{0} & \mathbf{K}_\mathrm{A} \end{bmatrix},\ \hat{\mathbf{q}} = \begin{Bmatrix} \hat{\mathbf{U}} \\ \hat{\mathbf{p}} \end{Bmatrix},\ \hat{\mathbf{F}} = \begin{Bmatrix} \hat{\mathbf{F}}_\mathrm{S} \\ \hat{\mathbf{F}}_\mathrm{A} \end{Bmatrix}. \tag{9.2}$$

In the two previous equations, $\mathbf{M}_j$, $\mathbf{C}_j$, $\mathbf{K}_j$ ($j \in \{\mathrm{S}, \mathrm{A}\}$) stand for mass, viscous damping and stiffness matrices of structure (S) and acoustic-fluid (A) parts; $\mathbf{R}$ stands for the fluid–structure coupling matrix; $\ddot{\hat{\mathbf{U}}}$, $\dot{\hat{\mathbf{U}}}$, $\hat{\mathbf{U}}$, and $\hat{\mathbf{F}}_\mathrm{S}$ are generalized acceleration, velocity, displacement, and load vectors of the structure domain; $\ddot{\hat{\mathbf{p}}}$, $\dot{\hat{\mathbf{p}}}$, $\hat{\mathbf{p}}$, and $\hat{\mathbf{F}}_\mathrm{A}$ are generalized acceleration, velocity, displacement, and load vectors related to the internal fluid ($\hat{\mathbf{p}}$ referring to pressure); ρ is the fluid density; and $(\,)^\mathrm{T}$ is the transpose operator.

As discussed in Sect. 9.1, the use of WFEM may become challenging in certain conditions where a high number of internal DoFs is found. To overcome this computational trouble, DoFs that are not located in the unit cell's left (L) and right (R) interfaces, namely internal (I), can be reduced by employing the CB model order reduction method. The procedure, basically, consists of establishing the following transformation matrix:

$$\boldsymbol{\alpha} = \begin{bmatrix} \mathbf{I} & \mathbf{0} \\ \Phi_\mathrm{C} & \Phi_\mathrm{N} \end{bmatrix}, \tag{9.3}$$

where $\mathbf{I}$ is an identity matrix, $\Phi_\mathrm{C} = -\hat{\mathbf{K}}_\mathrm{II}^{-1}\hat{\mathbf{K}}_\mathrm{IB}$ (B accounts for boundary DoFs comprising those located in both left and right interfaces), and Φ_N is a matrix of eigenvectors obtained from an eigenproblem that represents the dynamics of the considered unit cell with boundary DoFs constrained (Mencik 2018). Since $\hat{\mathbf{q}} = \boldsymbol{\alpha}\mathbf{q}$, the reduced set of equations of motion can be written as:

$$\mathbf{M}\ddot{\mathbf{q}} + \mathbf{C}\dot{\mathbf{q}} + \mathbf{K}\mathbf{q} = \mathbf{F}, \tag{9.4}$$

with $\mathbf{M} = \boldsymbol{\alpha}^\mathrm{T}\hat{\mathbf{M}}\boldsymbol{\alpha}$, $\mathbf{C} = \boldsymbol{\alpha}^\mathrm{T}\hat{\mathbf{C}}\boldsymbol{\alpha}$, $\mathbf{K} = \boldsymbol{\alpha}^\mathrm{T}\hat{\mathbf{K}}\boldsymbol{\alpha}$, and $\mathbf{F} = \boldsymbol{\alpha}^\mathrm{T}\hat{\mathbf{F}}$. It is worth noting that a modified version of the previously mentioned model reduction methodology might be employed to address structural systems interacting with enclosed acoustic fluids, as discussed elsewhere (Maess and Gaul 2006).

The reduced coupled fluid–structure equations seen before in Eq. (9.4) can be rewritten in the frequency domain as $\mathbf{D}\mathbf{q} = \mathbf{F}$, where $\mathbf{D} = -\omega^2\mathbf{M} + \mathrm{i}\omega\mathbf{C} + \mathbf{K}$, i is the imaginary unit, and ω is the circular frequency. Following the basic procedure of the WFEM, the reduced DoFs and load vectors can be explicitly organized with respect to their position in the unit cell:

$$\begin{bmatrix} \mathbf{D}_\mathrm{LL} & \mathbf{D}_\mathrm{LR} & \mathbf{D}_\mathrm{LI} \\ \mathbf{D}_\mathrm{RL} & \mathbf{D}_\mathrm{RR} & \mathbf{D}_\mathrm{RI} \\ \mathbf{D}_\mathrm{IL} & \mathbf{D}_\mathrm{IR} & \mathbf{D}_\mathrm{II} \end{bmatrix} \begin{Bmatrix} \mathbf{q}_\mathrm{L} \\ \mathbf{q}_\mathrm{R} \\ \mathbf{q}_\mathrm{I} \end{Bmatrix} = \begin{Bmatrix} \mathbf{F}_\mathrm{L} \\ \mathbf{F}_\mathrm{R} \\ \mathbf{F}_\mathrm{I} \end{Bmatrix}, \tag{9.5}$$

which may be used to obtain an expression that relates DoFs and load vectors of the left interface between two adjacent unit cells of a periodic structure, the $(n+1)$ and (n), i.e., $\mathbf{u}_{\mathrm{L}}^{(n+1)} = \mathbf{S}\mathbf{u}_{\mathrm{L}}^{(n)} + \mathbf{b}^{(n)}$. This recurrence relation incorporates the unit cell state vector $\mathbf{u}_{\mathrm{L}} = \{\mathbf{q}_{\mathrm{L}}^{\mathrm{T}}\ \mathbf{F}_{\mathrm{L}}^{\mathrm{T}}\}^{\mathrm{T}}$, a transfer matrix $\mathbf{S}$, and a forcing vector related to external loads $\mathbf{b}$ (Hoang et al. 2020).

Bloch-Floquet theorem can be resorted to, so that the recurrence relation stated before leads to $(\mathbf{S} - \mu_j\mathbf{I})\boldsymbol{\varphi}_j = \mathbf{0}$. Here, μ_j and $\boldsymbol{\varphi}_j (j \in \{1, ..., 2 \times n_b\})$, being n_b the number of DoFs in the left (or right) interface of the unit cell, are eigen-solutions (wave-modes) of the transfer matrix eigenproblem that physically represent the propagation constant and wave shapes, respectively. Wavenumbers can be computed from propagation constants as $k_j = (\ln \mu_j)/(-i\Delta)$, where Δ is the length of the unit cell along the direction of wave propagation—assumed to be in the X-direction, cf. Fig. 9.1. As discussed by Waki et al. (2009a) and Mencik (2014), the standard eigenvalue problem described earlier may be prone to numerical issues due to ill-conditioning problems. Thus, to alleviate potential computational challenges, one can resort to the alternative formulation $(\mathbf{NJL}^{\mathrm{T}} + \mathbf{LJN}^{\mathrm{T}} - \lambda_j\mathbf{LJL}^{\mathrm{T}})\mathbf{z}_j = \mathbf{0}$ (Zhong and Williams 1995), where λ_j and $\mathbf{z}_j$ are eigen-solutions of Zhong's eigenvalue problem—expressions for $\mathbf{N}$, $\mathbf{J}$, and $\mathbf{L}$ are provided by Mencik and Duhamel (2015). The set of wave-modes associated with the transfer matrix eigenvalue problem $(\mu_j, \boldsymbol{\varphi}_j)$ can be retrieved from λ_j and $\mathbf{z}_j$, following the procedure outlined by the same authors.

Wave-modes computed as discussed previously can be grouped with respect to the direction of wave propagation, i.e., in positive- $(\mu_j, \boldsymbol{\varphi}_j)$ and negative-going waves $(\mu_j^*, \boldsymbol{\varphi}_j^*)$ as follows (Duhamel et al. 2006; Waki et al. 2009b):

$$\boldsymbol{\mu} = \mathrm{diag}(\mu_1, \ldots, \mu_{n_b}), \quad \boldsymbol{\mu}^* = \mathrm{diag}(\mu_1^*, \ldots, \mu_{n_b}^*),$$
$$\boldsymbol{\Phi} = [\boldsymbol{\varphi}_1 \ldots \boldsymbol{\varphi}_{n_b}] = \begin{bmatrix} \boldsymbol{\Phi}_q \\ \boldsymbol{\Phi}_F \end{bmatrix}, \quad \boldsymbol{\Phi}^* = [\boldsymbol{\varphi}_1^* \ldots \boldsymbol{\varphi}_{n_b}^*] = \begin{bmatrix} \boldsymbol{\Phi}_q^* \\ \boldsymbol{\Phi}_F^* \end{bmatrix}, \tag{9.6}$$

with subscripts q and F used to denote partitions of the wave shapes related to DoFs and loads, respectively. To develop expressions presented afterward, one highlights that eigenvectors must be normalized as described by Mencik (2014). In addition, since the wave-modes are computed for each desired frequency independently, some frequency tracking procedure might be used to better visualize dispersion curves, i.e., the relation between wavenumbers and frequency. Given a pair of wave-modes related to eigenvalues μ_a and μ_a^*, with $\mu_a = 1/\mu_a^*$, the wave-mode corresponding to μ_a at the consecutive frequency $\omega + \delta_\omega$ is chosen so that (Mencik 2010):

$$\left| \frac{\boldsymbol{\varphi}_a^{*\mathrm{T}}(\omega)}{\|\boldsymbol{\varphi}_a^*(\omega)\|} \mathbf{J} \frac{\boldsymbol{\varphi}_a(\omega+\delta_\omega)}{\|\boldsymbol{\varphi}_a(\omega+\delta_\omega)\|} \right| = \max_{b\in\{1,\ldots,n_b\}} \left\{ \left| \frac{\boldsymbol{\varphi}_a^{*\mathrm{T}}(\omega)}{\|\boldsymbol{\varphi}_a^*(\omega)\|} \mathbf{J} \frac{\boldsymbol{\varphi}_b(\omega+\delta_\omega)}{\|\boldsymbol{\varphi}_b(\omega+\delta_\omega)\|} \right| \right\}, \tag{9.7}$$

where $\|\ \|$ denotes the norm of a vector and $|\ |$ denotes the absolute value of a scalar.

Lastly, the forced response of a finite structure with N unit cells can be calculated using the following expressions (Hoang et al. 2020):

$$\mathbf{q}_{\mathrm{L}}^{(n)} = \Phi_q \mu^{n-1}\mathbf{Q} + \Phi_q^* \mu^{N+1-n}\mathbf{Q}^* + \Phi_q \sum_{k=1}^{n-1} \mu^{n-k-1}\mathbf{Q}_B^{(k)} - \Phi_q^* \sum_{k=n}^{N} \mu^{k+1-n}\mathbf{Q}_B^{*(k)}, \tag{9.8}$$

$$-\mathbf{F}_{\mathrm{L}}^{(n)} = \Phi_F \mu^{n-1}\mathbf{Q} + \Phi_F^* \mu^{N+1-n}\mathbf{Q}^* + \Phi_F \sum_{k=1}^{n-1} \mu^{n-k-1}\mathbf{Q}_B^{(k)} - \Phi_F^* \sum_{k=n}^{N} \mu^{k+1-n}\mathbf{Q}_B^{*(k)}, \tag{9.9}$$

being $\mathbf{Q}_B^{(k)}$ and $\mathbf{Q}_B^{*(k)}$ amplitudes of externally applied loads in the wave-basis space; $\mathbf{Q} \equiv \mathbf{Q}^{(1)}$ and $\mathbf{Q}^* \equiv \mathbf{Q}^{*(N+1)}$ wave-mode amplitudes for the first (1) and $(N+1)$ unit cells. Equations (9.8) and (9.9) can be used to enforce boundary conditions and establish a linear system of equations for the unknowns $\mathbf{Q}$ and $\mathbf{Q}^*$. Solving this system enables the determination of any response of the periodic structure. For the sake of clarity, in our case, aiming to simulate a free-free condition, $\mathbf{F}_{\mathrm{L}}^{(1)} = \mathbf{0}$ and $\mathbf{F}_{\mathrm{L}}^{(N+1)} = \mathbf{0}$ must be written, so that one obtains (Santos 2022):

$$\begin{bmatrix} \mathbf{I} & \mathbf{\Phi}_F^{-1}\mathbf{\Phi}_F^*\mu^N \\ \mathbf{\Phi}_F^{*-1}\mathbf{\Phi}_F\mu^N & \mathbf{I} \end{bmatrix} \begin{Bmatrix} \mathbf{Q} \\ \mathbf{Q}^* \end{Bmatrix} = \begin{Bmatrix} \mathbf{\Phi}_F^{-1}\mathbf{\Phi}_F^* \sum_{k=1}^{N} \mu^k \mathbf{Q}_B^{*(k)} \\ -\mathbf{\Phi}_F^{*-1}\mathbf{\Phi}_F \sum_{k=1}^{N} \mu^{N-k}\mathbf{Q}_B^{(k)} \end{Bmatrix}. \tag{9.10}$$

9.3 Experimental and Numerical Results

In the following, experimental and numerical results will be presented, as well as the experimental procedures we employed. Section 9.3.1 first presents manufacturing considerations and details about the FE model of a unit cell of the periodic structure without confined fluid. Related numerical and experimental findings are compared in Sect. 9.3.2, where FRFs of the fluid-filled metastructure are made available to assess its wave propagation characteristics.

9.3.1 Manufacturing Considerations and Finite Element Model

The three-dimensional systems examined in this study comprise rectangular parallelepiped unit cells with external dimensions equal to 60 mm × 10 mm × 30 mm ($\Delta = 60$ mm), containing a perfectly centered internal rectangular parallelepiped void, measuring 53.6 mm × 6 mm × 23.6 mm. The numerical and experimental analyses presented here initially focus on the dynamic behavior of the periodic metastructure illustrated in Fig. 9.1 without internal fluid. A three-dimensional model of the system was created using computer-aided engineering (CAD) software and was subsequently made in separate parts using additive manufacturing, as depicted in Fig. 9.2a. Due to manufacturing limitations, the base of the periodic structure was printed first, resulting in the configuration depicted in Fig. 9.2b. Subsequently, the cap was produced with a nominal thickness of 2 mm, leading to the geometry displayed in Fig. 9.2c. Finally, the cap was bonded to the base of the metastructure using cyanoacrylate superglue, completing the system shown in Fig. 9.2d.

To conduct numerical investigations on the previously depicted system, the terms appearing in the equations of motion for the unit cell, as shown in Eq. (9.1), are reduced to $\hat{\mathbf{M}} = \mathbf{M}_S$, $\hat{\mathbf{C}} = \mathbf{C}_S$, $\hat{\mathbf{K}} = \mathbf{K}_S$, $\hat{\mathbf{q}} = \hat{\mathbf{U}}$, and $\hat{\mathbf{F}} = \hat{\mathbf{F}}_S$. The mass and stiffness matrices mentioned earlier are obtained from conventional FE software employing linear interpolation functions and material with linear behavior, and utilizing the FE mesh illustrated in Fig. 9.3. Specifically, 4-noded tetrahedral elements with three DoFs per node are used, i.e., encompassing translations in the X, Y, and Z-directions cf. Figure 9.1. Furthermore, the modelling of the unit cell incorporates a total of 75242 FEs, which satisfactorily ensures the model convergence up to the maximum analyzed frequency of 5000 Hz. The structure is made of polylactic acid (PLA), for which one assumed an elastic modulus of 2.24 GPa, density of 1052.6 kg m^{-3}, and Poisson's ratio of 0.3. Viscous damping was not considered in this study ($\hat{\mathbf{C}} = \mathbf{0}$). However, dissipation effects in the system's forced responses were accounted for through hysteretic damping, i.e., using $\hat{\mathbf{K}} = (1 + \mathrm{i}\eta)\mathbf{K}_S$, with η being the structural damping coefficient, set to 0.02. Moreover, a complete wave basis comprising 714 left- and right-going wave-modes (i.e., $n_b = 357$) was employed to characterize the dynamics of the unit cell under investigation. This choice was based on the demonstrated relevance of some evanescent wave-modes in the dynamics of the examined system, during numerical validations of the wave-based approach against FE analyses. These evanescent wave-modes may be reflected by the internal boundaries near the excitation location and, therefore, should not be neglected when computing responses for a finite system with relatively small length. It is noteworthy that even when considering the complete wave basis to describe the behavior of the metastructure, one solves a significantly smaller linear system of equations compared to traditional FEM.

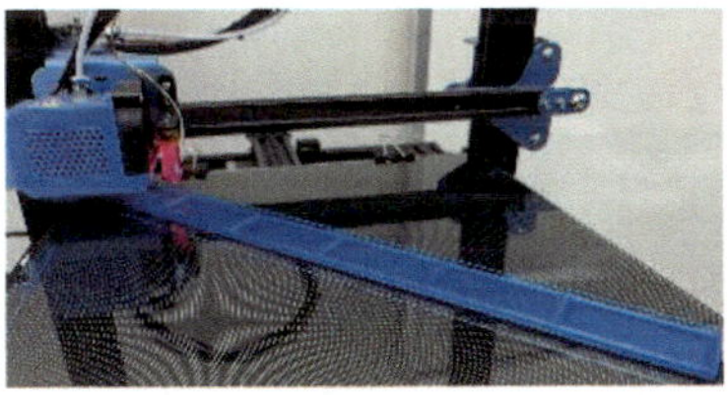

(a) Fused-filament deposition manufacturing of the periodic metastructure.

(b) Base.

(c) Cap.

(d) Assembled system.

Fig. 9.2 Manufacture details of the periodic metastructure with voids (without fluid)

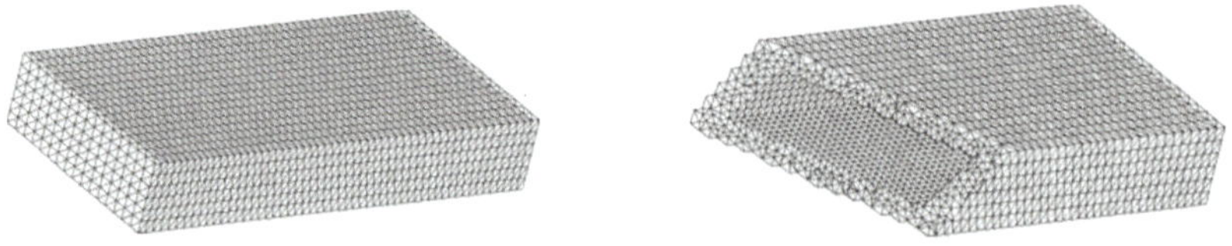

(a) Unit cell mesh.

(b) Cut revealing the internal void and wall mesh.

Fig. 9.3 Finite element mesh used to model the unit cells of the periodic metastructure shown in Fig. 9.1, without internal fluid, using the wave-based finite element method

9.3.2 Comparison Between Numerical and Experimental Results

First and foremost, the theory briefly discussed in Sect. 9.2 can be employed to calculate the dispersion curves for an infinite and undamped ($\eta = 0$) metastructure consisting of unit cells without internal fluid. Since the analyzed system does not exhibit any non-reciprocal properties in terms of wave propagation characteristics, the dispersion curves show symmetry with respect to the frequency axis. Therefore, Fig. 9.4 displays only the positive values of $\text{Re}\{k_j\}$ and the corresponding negative values of $\text{Im}\{k_j\}$ to visualize the dispersion branches. In Fig. 9.4, only the first four pairs of wavenumbers, related to wave-modes that propagate (or spatially decay) to left and right directions, were depicted instead of all those 714 computed for convenience. It is important to note that these wave-modes are related to propagating and evanescent modes and may be associated with longitudinal, bending, torsional, and shear wave types. The focus here is on the wave-modes that most influence the longitudinal and bending dynamics in the X- and Y-directions, respectively, cf. Figure 9.1, whose corresponding dispersion branches are duly identified in Fig. 9.4.

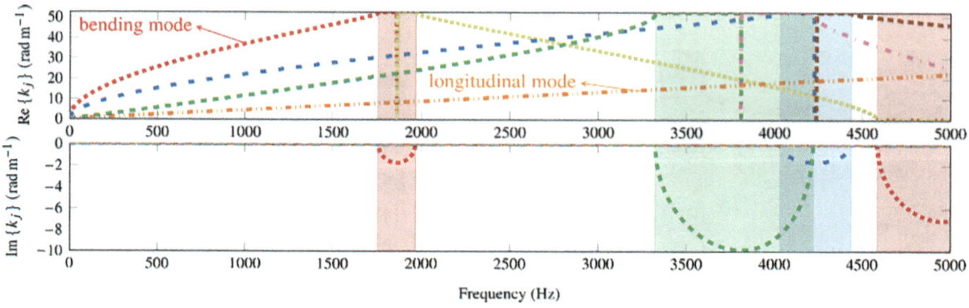

Fig. 9.4 Dispersion curves, positive values of $\text{Re}\{k_j\}$ and negative values of $\text{Im}\{k_j\}$, and bandgap frequency ranges ($\text{Im}\{k_j\} \neq 0$) computed for the undamped ($\eta = 0$) unit cell of the periodic structure shown in Fig. 9.1, without confined fluid. The correspondence between wavenumbers depicted in $\text{Re}\{k_j\}$ and $\text{Im}\{k_j\}$ is illustrated by the dispersion curves with matching colors and line styles

In addition to the dispersion curves shown in Fig. 9.4, bandgap frequency ranges are also illustrated, corresponding to those frequency ranges for which $\text{Im}\{k_j\} \neq 0$. To aid comprehension, these frequency bands are highlighted using shaded areas in colors that correspond to those used to plot the negative values of $\text{Im}\{k_j\}$. Furthermore, to enhance clarity, a summary of these bandgap frequency ranges is provided in Table 9.1. Thus, based on the curves shown in Fig. 9.4 and the results provided in Table 9.1, it is evident that the investigated system exhibits four distinct forbidden zones up to 5 kHz.

To validate the previously observed numerical predictions regarding the bandgap frequency ranges calculated for an infinite system, the wave propagation behavior of a finite system with $N = 6$ unit cells is considered, cf. Fig. 9.2d. Accelerance-type FRFs are measured in the longitudinal and transversal directions (X- and Y-directions, respectively), with the experimental setups for these measurements shown in Fig. 9.5a and b. For the investigation of the longitudinal behavior of the periodic system, the experimental scheme shown in Fig. 9.5a was utilized. An accelerometer PCB 352C22 was positioned at the leftmost face of the structure, while an impulsive force was applied at its right end. Both the input and output positions were located precisely at the center of the relevant cross-sections. On the other hand, the bending behavior of the system was examined using the experimental setup depicted in Fig. 9.5b. In this case, the structure was spatially reoriented to minimize the influence of nylon strings on the free vibration of the system. For this scenario, the accelerometer was again placed at the midpoint of the leftmost boundary, but in the XZ-plane. The hammer excitation force was directed along the Y direction, being applied at the rightmost edge of the structure. Comparisons between numerical and experimental findings are presented in Fig. 9.5c and d for longitudinal and bending dynamics, respectively. It is important to note that the numerical curves were computed using the theory discussed in Sect. 9.2, specifically by applying Eq. (9.8).

As one may notice from Fig. 9.5c, the numerical prediction matches well the experimental result obtained through the hammer test across the entire frequency range considered. Both resonance frequencies related to the longitudinal motion of the system are accurately predicted, corresponding to approximately 1960 Hz and 3890 Hz. In addition, the experimental FRF also exhibit strong agreement with the wave-based model far from the resonance conditions. Moreover, the numerical

Table 9.1 A summary of the bandgap frequency ranges ($\text{Im}\{k_j\} \neq 0$) seen in Fig. 9.4, computed for an undamped ($\eta = 0$) and infinite metastructure made with the unit cell shown in Fig. 9.1, without internal fluid

Line styles used in Fig. 9.4	Bandgap frequency ranges (Hz)
······ and ······	(1756–1971), (4586–5000)
- - - and - · -	(3321–4221)
- - and ····	(4031–4436)

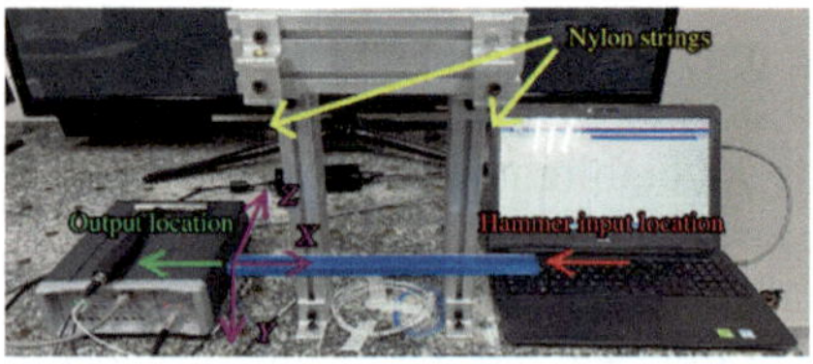

(a) Setup used to investigate the longitudinal behavior of the periodic metastructure.

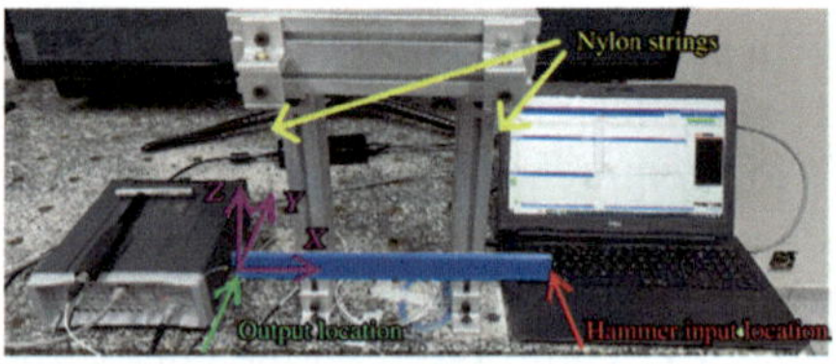

(b) Setup used to investigate the bending behavior of the periodic metastructure.

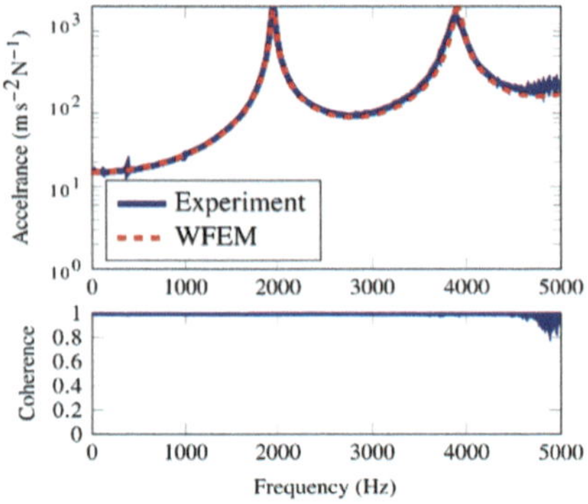

(c) A comparison between numerical and experimental FRFs corresponding to the input and output locations seen in Fig. 9.5a.

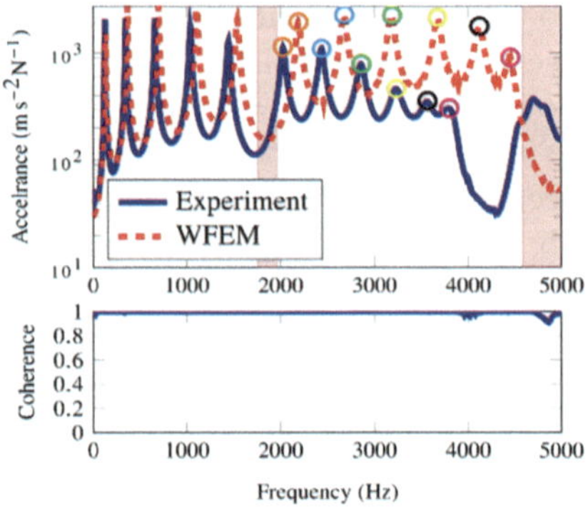

(d) A comparison between numerical and experimental FRFs corresponding to the input and output locations seen in Fig. 9.5b.

Fig. 9.5 Experimental setups and comparisons between numerical and measured FRFs used to investigate the longitudinal and bending behavior of the periodic metastructure portrayed in Figs. 9.1 and 9.2, without internal fluid

predictions regarding the wave propagation behavior of longitudinal modes, provided by the Bloch-Floquet theorem, as depicted in Fig. 9.4, are further supported by these FRFs. This means that the finite system did not experience attenuation zones in the considered frequency range regarding the propagation of longitudinal waves, since the dispersion curve seen in Fig. 9.4, related to this kind of motion, is purely real. Lastly, the coherence function depicted in Fig. 9.5c confirms the good quality of the experimental data, indicating a well-conducted experiment.

On the other hand, a comparison between numerical and experimental results depicted in Fig. 9.5d reveals that the wave-based model does not accurately describe the bending motion of the periodic structure without fluid, even with its longitudinal dynamics being properly predicted, as shown by Fig. 9.5c. In fact, the natural frequencies computed using the WFEM are larger than those identified from experimental data, indicating that the model exhibits a stiffer behavior. However, it is noteworthy that the number of resonance peaks observed in the numerical results perfectly matches those related to experiments—cf. indicated in Fig. 9.5d. Furthermore, the predictions derived from the Bloch-Floquet theorem regarding the attenuation zone of bending modes are confirmed by the numerical FRF shown in Fig. 9.5d.

This shows that the finite system exhibits two forbidden zones up to 5 kHz, as illustrated by the shaded areas in red. These attenuation zones are close to the steep valleys related to the blue curve, potentially corresponding to the actual attenuation zones of the tested finite system. Nevertheless, the forbidden zones identified through experiments are slightly shifted to the left, as previously discussed for the natural frequencies. This frequency shift can be attributed mainly to our assumption of a linear, isotropic constitutive relationship for the material, which does not fully reflect the system's actual behavior, as well as uncertainties in the PLA material properties assumed by us. For instance, the material is known to exhibit viscoelastic behavior (Müller et al. 2022), and the additive manufacturing process we employed favors the resulting media to possess orthotropic symmetry (Torre and Brischetto 2022). Notwithstanding, we chose to consider a linear model for our analysis, which still yields a fair match with the experimental data. It is also worth mentioning that the coherence function associated with the bending-type motion test is good, similarly to the results shown in Fig. 9.5c, also indicating well-conducted experiments.

To experimentally assess the dynamic behavior of the system depicted in Fig. 9.2d, with unit cells completely filled with water, new experiments were conducted. Firstly, another periodic structure was manufactured, following the procedures described in Sect. 3.1. The surfaces of the six-cell periodic system that could potentially come into contact with water were sealed using varnish, to waterproof the prototype model. Subsequently, the internal cavities were filled with fluid, and the cap was glued to the superior part of the base of the periodic structure. By adopting the same experimental setups discussed earlier and shown in Fig. 9.5a and b, longitudinal and bending accelerance-type FRFs were measured employing hammer tests for the water-filled system. Associated results are provided in Fig. 9.6. The same figure also includes the experimental FRFs obtained for the metastructure without fluid, previously shown in Fig. 9.2c and d, for convenience and ease of comparison.

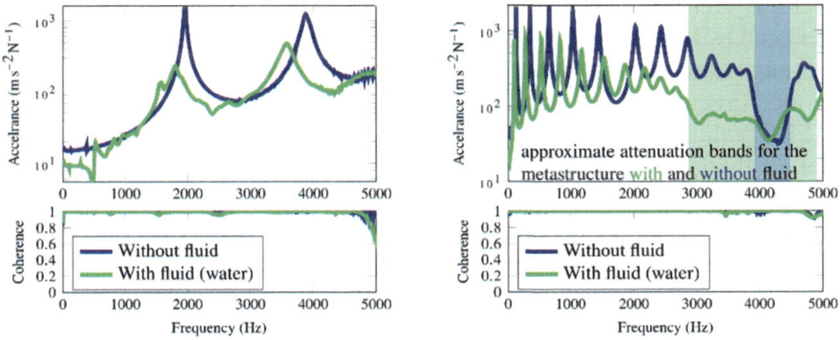

(a) Longitudinal FRFs corresponding to the input and output locations seen in Fig. 9.5c.

(b) Bending FRFs corresponding to the input and output locations seen in Fig. 9.5d.

Fig. 9.6 Comparisons between experimental FRFs measured for the periodic system portrayed in Fig. 9.1, with and without internal fluid, to investigate its longitudinal and bending dynamic behavior

First, Fig. 9.6a and b show that most resonance peaks exist in both conditions when the system is without or completely filled with fluid. This observation was previously made by Spadoni et al. (2014), where these modes were referred to as persistent modes. In addition, a comparison between the blue and green curves depicted in Fig. 9.6a and b reveals that the internal fluid significantly contributes to damping effects in the finite periodic system. This is evident from the noticeable attenuation of amplitudes at the resonance frequencies associated to the green curve in Fig. 9.6b. Turning attention to Fig. 9.6a, additional resonances emerge when the periodic system is filled with fluid, which may be associated with acoustic modes. Confirmation of this observation would require an examination of the system's deformed shapes at the corresponding frequencies. Additionally, the results shown in Fig. 9.6b, related to bending dynamics, demonstrate that the region corresponding to the second experimental attenuation zone is significantly wider when the cavities are completely filled with water, in comparison to the condition without internal fluid, as illustrated schematically using shading. This finding suggests a novel way to broaden attenuation zones in metamaterials by leveraging the dynamic behavior of fluid-filled cells. Moreover, this realization confirms the outstanding performance of fluid-filled metamaterials in vibroacoustics in terms of attenuating elasto-acoustic waves.

9.4 Concluding Remarks

This study investigated the dynamic behavior of periodic structures with internal cavities completely filled with water and in the absence of it. We reviewed a general formulation based on the WFEM to analyze wave propagation phenomena in fluid-filled systems. Utilizing this wave-based modeling approach, simulations were conducted to investigate the dynamics of the periodic system without internal fluid. Additionally, dispersion curves were computed using the Bloch-Floquet theorem for the unit cell lacking fluid, providing insights into the wave propagation characteristics through the considered finite system. Experimental tests were conducted on a finite metastructure comprising six unit cells without fluid, and measured accelerance FRFs exhibited good agreement with predictions provided by the wave-based model. Furthermore, attenuation zones predicted by the Bloch-Floquet theorem were confirmed in both numerical and experimental accelerance responses. Moreover, the dynamic behavior of the metastructure made with fluid-filled unit cells was experimentally assessed through hammer tests. A comparison between the experimental FRFs of the regarded periodic structure without and with internal fluid revealed that the frequency range near the second attenuation band observed in the former case can be significantly enlarged in the presence of fluid within the unit cells' cavities. This indicates the potential of using such metamaterials as a novel passive vibration absorber mechanism in various vibroacoustic engineering applications.

Acknowledgements V. M. S. Santos acknowledges the São Paulo Research Foundation (FAPESP) for the grants #2023/11207-7 and #2024/07549-2, and the Brazilian Coordination for the Improvement of Higher Education Personnel (CAPES). V. M. S. Santos and T. P. Sales are grateful to the FAPESP for its support to the thematic grant #2018/15894-0, related to the "Periodic structure design and optimization for enhanced vibroacoustic performance: ENVIBRO" research project. M. Ouisse acknowledges the graduate school EIPHI (project ANR-17-EURE-0002).

References

Bao XL, Überall H, Raju PK, Ahyi AC, Bjørnø IK, Bjørnø L (2000) Dispersion of axially symmetric waves in fluid-filled cylindrical shells. J Acoust Soc Am 107:2847–2848. https://doi.org/10.1121/1.429206

Dorodnitsyn V, Damme BV (2016) Two-dimensional fluid-filled closed-cell cellular solid as an acoustic metamaterial with negative index. Phys Rev B 93:134302. https://doi.org/10.1103/physrevb.93.134302

Droz C, Lainé J-P, Ichchou MN, Inquiété G (2014) A reduced formulation for the free-wave propagation analysis in composite structures. Compos Struct 113:134–144. https://doi.org/10.1016/j.compstruct.2014.03.017

Duhamel D, Mace BR, Brennan MJ (2006) Finite element analysis of the vibrations of waveguides and periodic structures. J Sound Vib 294:205–220. https://doi.org/10.1016/j.jsv.2005.11.014

Fuller CR, Fahy FJ (1982) Characteristics of wave propagation and energy distributions in cylindrical elastic shells filled with fluid. J Sound Vib 81:501–518. https://doi.org/10.1016/0022-460x(82)90293-0

Gao N, Zhang Z, Deng J, Guo X, Cheng B, Hou H (2022) Acoustic metamaterials for noise reduction: a review. Adv Mater Technol 7:2100698. https://doi.org/10.1002/admt.202100698

Hoang T, Duhamel D, Foret G (2020) Wave finite element method for waveguides and periodic structures subjected to arbitrary loads. Finite Elem Anal Des 179:103437. https://doi.org/10.1016/j.finel.2020.103437

Jamshidiat H, SenGupta G (1991) Dynamics of periodic structures interacting with an enclosed fluid medium. J Sound Vib 148:103–115. https://doi.org/10.1016/0022-460x(91)90822-2

Kumar R, Stephens RWB (1972) Dispersion of flexural waves in circular cylindrical shells. Proc R Soc A 329:283–297. https://doi.org/10.1098/rspa.1972.0114

Maess M, Gaul L (2006) Substructuring and model reduction of pipe components interacting with acoustic fluids. Mech Syst Sig Pr 20:45–64. https://doi.org/10.1016/j.ymssp.2005.02.008

McNiven HD, Sackman JL, Shah AH (1966) Axially symmetric waves in hollow, elastic rods. Part II J Acoust Soc Am 40:1073–1076. https://doi.org/10.1121/1.1910190

Mencik J-M (2010) On the low- and mid-frequency forced response of elastic structures using wave finite elements with one-dimensional propagation. Comput Struct 88:674–689. https://doi.org/10.1016/j.compstruc.2010.02.006

Mencik J-M (2014) New advances in the forced response computation of periodic structures using the wave finite element (WFE) method. Comput Mech 54:789–801. https://doi.org/10.1007/s00466-014-1033-1

Mencik J-M (2018) A wave finite element approach for the analysis of periodic structures with cyclic symmetry in dynamic substructuring. J Sound Vib 431:441–457. https://doi.org/10.1016/j.jsv.2018.05.027

Mencik J-M, Duhamel D (2015) A wave-based model reduction technique for the description of the dynamic behavior of periodic structures involving arbitrary-shaped substructures and large-sized finite element models. Finite Elem Anal Des 101:1–14. https://doi.org/10.1016/j.finel.2015.03.003

Mencik J-M, Ichchou MN (2007) Wave finite elements in guided elastodynamics with internal fluid. Int J Solids Struct 44:2148–2167. https://doi.org/10.1016/j.ijsolstr.2006.06.048
Miranda EJP, Nobrega ED, Ferreira AHR, Santos JMCD (2019) Flexural wave band gaps in a multi-resonator elastic metamaterial plate using Kirchhoff-Love theory. Mech Syst Sig Pr 116:480–504. https://doi.org/10.1016/j.ymssp.2018.06.059
Mirsky I, Herrmann G (1957) Nonaxially symmetric motions of cylindrical shells. J Acoust Soc Am 29:1116–1123. https://doi.org/10.1121/1.1908716
Mizukami K, Funaba K, Ogi K (2021) Design and three-dimensional printing of carbon-fiber-composite elastic metamaterials with inertial amplification mechanisms. J Sound Vib 513:116412. https://doi.org/10.1016/j.jsv.2021.116412
Müller M, Šleger V, Kolář V, Hromasová M, Piš D, Mishra RK (2022) Low-cycle fatigue behavior of 3D-printed PLA reinforced with natural filler. Polymers 14:1301. https://doi.org/10.3390/polym14071301
Ning L, Wang Y-Z, Wang Y-S (2020) Active control cloak of the elastic wave metamaterial. Int J Solids Struct 202:126–135. https://doi.org/10.1016/j.ijsolstr.2020.06.009
Nobrega ED, Gautier F, Pelat A, Santos JMCD (2016) Vibration band gaps for elastic metamaterial rods using wave finite element method. Mech Syst Signal Pr 79:192–202. https://doi.org/10.1016/j.ymssp.2016.02.059
Pal RK, Ruzzene M (2017) Edge waves in plates with resonators: an elastic analogue of the quantum valley Hall effect. New J Phys 19:025001. https://doi.org/10.1088/1367-2630/aa56a2
Santos VMS (2022) Numerical investigation of wave propagation in beams coupled to metastructures combining spectral and wave-finite element methods
Spadoni A, Höhler R, Cohen-Addad S, Dorodnitsyn V (2014) Closed-cell crystalline foams: self-assembling, resonant metamaterials. J Acoust Soc Am 135:1692–1699. https://doi.org/10.1121/1.4867375
Torre R, Brischetto S (2022) Experimental characterization and finite element validation of orthotropic 3D-printed polymeric parts. Int J Mech Sci 219:107095. https://doi.org/10.1016/j.ijmecsci.2022.107095
Waki Y, Mace BR, Brennan MJ (2009a) Numerical issues concerning the wave and finite element method for free and forced vibrations of waveguides. J Sound Vib 327:92–108. https://doi.org/10.1016/j.jsv.2009.06.005
Waki Y, Mace BR, Brennan MJ (2009b) Free and forced vibrations of a tyre using a wave/finite element approach. J Sound Vib 323:737–756. https://doi.org/10.1016/j.jsv.2009.01.006
Wang H, Xu Y, Genevet P, Jiang J-H, Chen H (2016) Broadband mode conversion via gradient index metamaterials. Sci Rep 6:24529. https://doi.org/10.1038/srep24529
Wang K, Zhou J, Wang Q, Ouyang H, Xu D (2019) Low-frequency band gaps in a metamaterial rod by negative-stiffness mechanisms: design and experimental validation. Appl Phys Lett 114(25). https://doi.org/10.1063/1.5099425
Wu T-C, Wu T-T, Hsu J-C (2009) Waveguiding and frequency selection of Lamb waves in a plate with a periodic stubbed surface. Phys Rev B 79:104306. https://doi.org/10.1103/physrevb.79.104306
Zhong WX, Williams FW (1995) On the direct solution of wave propagation for repetitive structures. J Sound Vib 181:485–501. https://doi.org/10.1006/jsvi.1995.0153

Chapter 10
Modeling and Analysis of Floating Solar Panels with a Tracking System for Hydroelectric Plants in the Amazon Region of Brazil

Cláudio Henrique Cerqueira Costa Basquerotto, Jhonny Feldhaus, Antônio Eduardo Turra, and João Antonio Pereira

Abstract Renewable energy sources are gaining global attention as solutions to the challenges of fossil fuels. Solar, wind, hydro, geothermal, and biomass energy harnesses continuously replenished resources, reducing environmental impacts and lessening reliance on non-renewables. An emerging concept in solar energy is the installation of floating photovoltaic plants on water bodies using platforms to mount solar panels. However, variable solar radiation intensity due to location requires efficient utilization, and intermittency can be mitigated by energy storage. Adaptable solar panel positioning through tracking systems is crucial for optimizing sunlight exposure. While these systems are common on land, their application to floating plants is complex. This study aims to evaluate the potential benefits of incorporating solar tracking systems in floating photovoltaic plants, specifically in the Amazon region of Brazil. Mathematical models and simulations will analyze panel dynamics under equatorial conditions to improve the energy production efficiency and sustainability of Amazon floating solar installations. This research contributes to the advancement of renewable technology and sustainable energy practices in the region.

Keywords Renewable energy sources · Photovoltaic plants · Dynamical analysis · Solar tracking systems · Floating photovoltaic plants

C. H. C. C. Basquerotto (✉)
Instituto de Geociências e Engenharias, Faculdade de Engenharia Mecânica, Universidade Federal do Sul e Sudeste do Pará (Unifesspa), Marabá, Pará, Brazil
e-mail: cbasquerotto@unifesspa.edu.br

J. Feldhaus
Norte Energia S.A.—NESA, Vitória do Xingu, Pará, Brazil

A. E. Turra · J. A. Pereira
Faculdade de Engenharia, Universidade Estadual Paulista (UNESP), Ilha Solteira, São Paulo, Brazil

P. Kurka and M. Pereira (eds.), *Advances in Structural Vibration*, Lecture Notes in Mechanical Engineering, https://doi.org/10.1007/978-3-031-71540-2_10

10.1 Introduction

As the world faces a pressing global energy crisis and strives to combat the detrimental impacts of climate change, the adoption of renewable energy sources has emerged as a key solution (Alberti et al. 2023; Hammoumi et al. 2022; Pawar et al. 2021). Renewable energy sources are gaining global attention as solutions to the challenges of fossil fuels. Solar, wind, hydro, geothermal, and biomass energy harness continuously replenished resources, curbing environmental impacts and lessening reliance on non-renewables. An emerging concept in solar energy is the installation of floating photovoltaic plants on water bodies, using floating platforms to mount solar panels. Within this context, the Amazon region of Brazil stands out as an extraordinary opportunity for sustainable power generation by capitalizing on solar energy (de Freitas 2022). This can be achieved through the strategic integration of advanced floating solar plants integrated within existing hydroelectric plants (Piancó et al. 2022; Rauf et al. 2020; Vidovic et al. 2023). This innovative fusion of solar photovoltaic technology and hydroelectric infrastructure has tremendous potential to revolutionize energy production in the Amazon, taking advantage of its abundant sunlight resources and using already established hydroelectric facilities (Claus and López 2023; Rashid et al. 2022).

This study focuses primarily on performing a thorough modeling and dynamic analysis of floating modules of solar panels equipped with specialized tracking systems (Ferrer-Gisbert et al. 2013; Yasmeena and Das 2015). These systems are purposefully designed for integration with hydroelectric plants located within the Amazon region. The main goal is to evaluate the behavior and performance of this integrated system, with the aim of improving its overall efficiency and maximizing its power generation capabilities (Trapani and Millar 2016).

Using the techniques of Choi et al. (2021, 2022), Abbasnia et al. (2022), and Exley et al. (2022), these studies aimed to investigate the dynamic characteristics of floating solar panels with tracking systems. Factors such as solar irradiation patterns, water flow dynamics, and the distinct environmental conditions found in the Amazon region will be considered (Choi et al. 2023; Prinsloo et al. 2023). By comprehensively understanding the interaction between variables such as angles of inclination, angular velocities, and solar irradiation position, it is possible to pave the way for sustainable and efficient energy solutions in the Amazon. This work aims to focus on the modeling and dynamic analysis of floating solar panels with tracking systems aimed at integration with a hydroelectric plant in the Amazon region of Brazil to determine how waves affect azimuth and elevation angles and thus the overall efficiency of solar panels. Through this analysis, the potential of this innovative technology for its contribution to sustainable energy generation in the Amazon is exploited.

This work is organized into four sections as follows. First, in Sect. 10.2, the solar panel model with the tracking system and corresponding equations are presented. In Sect. 10.3, a numerical simulation of the motion of a floating solar panel with a tracking system was conducted, obtaining information regarding the angular velocities of the system, tracking angles and other parameters. Finally, the final remarks and future work are discussed.

10.2 A Floating Solar Panel with a Tracker System

A floating photovoltaic plant is composed of a series of connected floating modules that form a flexible platform, as shown in Fig. 10.1, in which the walking module and the main flouting module move in relation to each other when subjected to the effects of waves. The float modules provide support to ensure that the structure remains on the water surface. The photovoltaic (PV) panels are mounted under the main flouting modules separated by a row of walking modules (Fig. 10.1b). The flouting modules are constructed from high-density polyethylene (HDPE), plastic materials that resist UV light, are safe for the environment, require minimal maintenance, and have strong durability. The anchoring mechanism ensures the stability of the floating photovoltaic (FPV) plant, restricting excessive motion to avoid harm to the plant or nearby floating entities. In freshwater initiatives, materials such as artificial fiber cords, stretchable rubber cables, or a mix of these materials are employed for the mooring system of the floating plant. In the analysis of the dynamics of the floating system, only one of the components comprising the maid floating and the PV panel was considered. Figure 10.1b shows a solar panel with a tracker system under a main floating module.

In the analysis of the dynamics of the floating system, only one of the components comprising the maid floating and the PV panel was considered. Figure 10.1c shows a solar panel with a tracker system under a main floating module. First, the point O is taken as a reference, $\mathcal{I}$, with the axes (x, y, z) and $\{\hat{\imath}, \hat{\jmath}, \hat{k}\}$ centered on O. The structure rotates around the axis y with the angular velocity ω constant. The solar panel rotates[1] around $x \equiv x_1$ in a positive way, with angular velocity $\dot{\theta}$ and angle θ in the reference system $\mathcal{B}_1$ with (x_1, y_1, z_1) and $\{\hat{\imath}_1, \hat{\jmath}_1, \hat{k}_1\}$. The center of mass of the panel is represented by point O, the distance from point A to this point along axis x is ℓ and the distance from point B to point A along axis y_1 is h.

By defining the transformation matrices between the coordinate systems $\mathcal{I}$ and $\mathcal{B}_1$:

$$\underset{(x,y,z)}{\mathcal{I}} \longrightarrow \underset{(x_1,y_1,z_1)}{\mathcal{B}_1}$$

$$^{\mathcal{I}}\mathbf{T}^{\mathcal{B}_1} = \mathbf{T}_\theta = \begin{bmatrix} 1 & 0 & 0 \\ 0 & \cos\theta & \sin\theta \\ 0 & -\sin\theta & \cos\theta \end{bmatrix}$$

The vectors of angular velocities are given by:

[1] According to the right-hand rule of axis rotation.

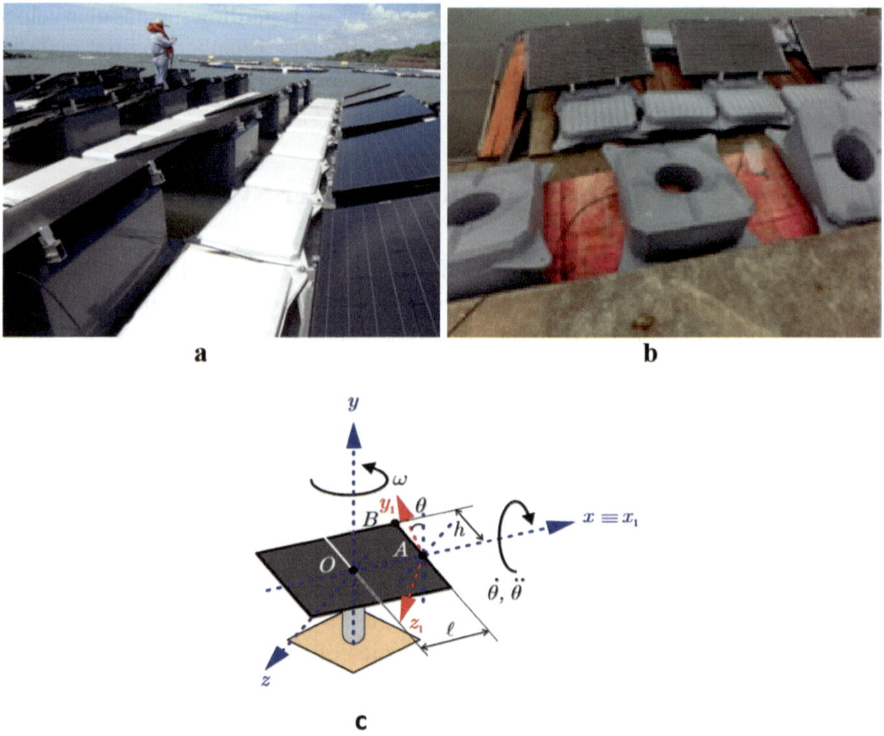

Fig. 10.1 Floating solar panel with a tracker system

$$
{}_{\mathcal{I}}\boldsymbol{\omega}_{\mathcal{B}_1} = \begin{bmatrix} 0 \\ \omega \\ 0 \end{bmatrix}, \quad {}_{\mathcal{I}}\boldsymbol{\Theta}_{\mathcal{B}_1} = \begin{bmatrix} \dot{\theta} \\ 0 \\ 0 \end{bmatrix}, \quad {}_{\mathcal{I}}\boldsymbol{\Omega}_{\mathcal{B}_1} = \begin{bmatrix} \dot{\theta} \\ \omega \\ 0 \end{bmatrix}
$$

where $\boldsymbol{\Omega}$ is the vector of angular velocities resulting from the sum of $\boldsymbol{\Theta}$ and $\boldsymbol{\omega}$.

The model of the floating module used in this study is similar to that presented by Jambersi et al. (2017). The floating system oscillates around a line of static equilibrium and is subjected to forces resulting from the action of lake waves. These efforts have been modeled as concentrated forces acting at two points, F (frontal) and R (rear), of the float. Despite the nomenclature used here, waves "impact" first at the R point and "exit" through the F point. Figure 10.2 shows the FBD (free-body diagram) of floating in this situation.

The terms F_{kF} and F_{kR} represent the hydrostatic stiffness forces, or vertical lift/restore (buoyancy) forces, caused by the action of lake waves acting at points F and R on the float. F_{bF} and F_{bR} are the hydrodynamic damping forces, which are also caused by the incidence of lake waves acting at points F and R and are of a dissipative nature. Due to the restoring nature of these efforts, these forces always act in the opposite direction of motion and in the vertical direction (axis y of the

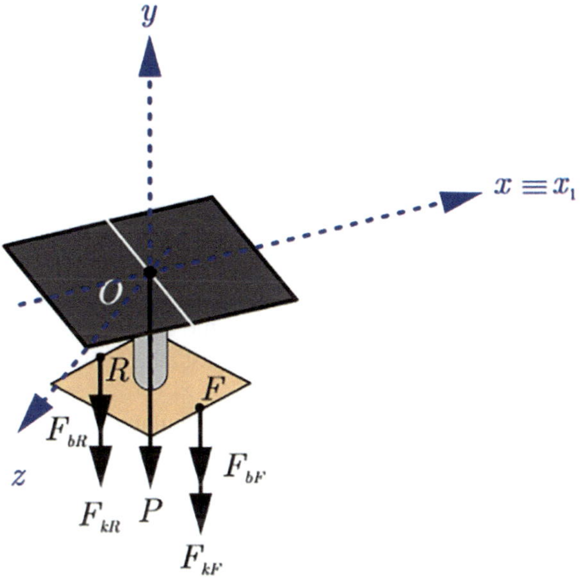

Fig. 10.2 Free body diagram of the floating solar panel (represented here with a simple geometry) subjected to restoring forces applied at *F* and *R*

inertial reference system), and their magnitudes vary with time. These magnitudes are given by:

$$F_{kF} = k_F(y_F - w_F) \tag{10.1}$$

$$F_{kR} = k_R(y_R - w_F) \tag{10.2}$$

$$F_{bF} = b_F(\dot{y}_F - \dot{w}_F) \tag{10.3}$$

$$F_{bR} = b_R(\dot{y}_R - \dot{w}_R) \tag{10.4}$$

where k_F and k_R are the hydrostatic stiffness constants; [N/m], b_F and b_R are the hydrodynamic damping constants; [Ns/m], y_F and y_R are the displacements; $\dot{y}_F$ and $\dot{y}_R$ are the instantaneous velocities; w_F and w_R are the amplitudes; and $\dot{w}_F$ and $\dot{w}_R$ are the velocities of the waves, with respect to *F* and *R*.

To determine wave forces in an aquatic FPV framework, it is crucial to consider wave motion principles. Among them, the Airy wave model is particularly prevalent. Equation (1) illustrates the variations in the water level in real time due to a wave moving in the *x* direction:

$$\eta(x, t) = \frac{H}{2}\cos\left(\frac{2\pi}{\lambda} - \frac{2\pi}{T}t\right) \tag{10.5}$$

where t is the time, H is the wave height, λ is the wavelength, and T is the wave period.

The angular momentum, $\mathbf{H}_O$, obtained by considering the center of mass, O, is given by:

$$ {}_{\mathcal{I}}\mathbf{H}_O = {}_{\mathcal{I}}\mathbb{I}_O\, {}_{\mathcal{I}}\boldsymbol{\Omega} + \underbrace{m_{\mathcal{I}}r_{O-O} \times_{\mathcal{I}} v_O}_{=0} = \begin{bmatrix} 6.8736 & -0.0053 & 2.9999 \\ -0.0053 & 5.8159 & 0.0009 \\ 2.9999 & 0.0009 & 14.5383 \end{bmatrix} \begin{Bmatrix} \dot{\theta} \\ \omega \\ 0 \end{Bmatrix} = \begin{Bmatrix} 6.8736\dot{\theta} - 0.0053\omega \\ -0.0053\dot{\theta} + 5.8159\omega \\ 2.9999\dot{\theta} + 0.0009\omega \end{Bmatrix} \quad (10.6) $$

where ${}_{\mathcal{I}}\mathbb{I}_O$ is the tensor of inertia of the solar panel, whose values can be extracted within the CAD software where the floating solar panel was modeled. Since point O was considered the center of mass, the position vector from the center of mass to point O is zero.

The gyroscopic torque can be calculated using:

$$ {}_{\mathcal{I}}\boldsymbol{\Omega} \times_{\mathcal{I}} \boldsymbol{H}_O = \begin{Bmatrix} 2.9999\dot{\theta}\omega + 0.0009\omega^2 \\ -2.9999\dot{\theta}^2 - 0.0009\dot{\theta}\omega \\ -0.0053\dot{\theta}^2 - 1.0577\dot{\theta}\omega + 0.0053\omega^2 \end{Bmatrix} \quad (10.7) $$

The Euler equation can be easily calculated. This equation relates the sum of the external torques applied in relation to the center of mass to the rate of change in angular momentum, which is also calculated in relation to the center of mass of the float. So:

$$ {}_{\mathcal{I}}\dot{\mathbf{H}}_O = {}_{\mathcal{I}}\mathbb{I}_O \frac{d}{dt}({}_{\mathcal{I}}\boldsymbol{\omega}) + {}_{\mathcal{I}}\Omega \times ({}_{\mathcal{I}}\mathbb{I}_O {}_{\mathcal{I}}\boldsymbol{\omega}) + m_{\mathcal{B}}\mathbf{r}_{O-O} \times_{\mathcal{I}} \mathbf{a}_O = {}_{\mathcal{I}}\mathbf{T}_O \quad (10.8) $$

Therefore, the motion equations can be obtained in matricial form as:

$$ \begin{bmatrix} m & 0 \\ 0 & (I_{xx} + I_{yx} + I_{zx}) \end{bmatrix} \begin{Bmatrix} \ddot{y} \\ \ddot{\theta} \end{Bmatrix} = \begin{Bmatrix} -(F_{kF} + F_{kR} + F_{bF} + F_{bR}) \\ [(F_{kF} + F_{bF}) - (F_{kR} + F_{bR})]\frac{\ell}{2}\cos\theta \end{Bmatrix} \quad (10.9) $$

where I_{xx} is the moment of inertia and I_{yx} and I_{zx} are the products of inertia.

10.3 Floating Solar Panel Simulation with a Two-Axis Tracking System

This section presents numerical simulations of the equations that govern the motion of floating photovoltaic solar panels. The simulations aim to demonstrate how the horizontal and vertical angles of the panel behave as the sun occupies different positions, as well as to assess the behavior of the angular velocities of this system. Finally, the simulations also analyze the energy generated by the solar panel during the operational regime.

For the integration of the equations, the Runge–Kutta method was used, along with null initial conditions for $\theta(0)$ and $\dot{\theta}(0)$. The integration step used was $\Delta_t = 0.005$ s. The weight was $m = 33.3$ kg, the dimensions of the module were $2.384 \times 1.303 \times 0.033$ m, the water density was 1000 kg/m^3, the solar intensity was 1000 W/m^2 (Vitória do Xingu—Pará, latitude $-3.1984°$ and longitude $-52.5688°$), the wave height considered was 0.5 m, the wind speed was 5 m/s, the air density was 1.225 kg/m^3, and the angle of the incidence of the wave was 5°. Definition of minimum and maximum frequencies: The domain must be defined to characterize the largest number of waves possible since the relationship between frequency and period is inverse and the incidence of swell waves in Belo Monte Lake is not considerable. In the case of this work, the domain used was between 0.3 and 1 Hz (Sobrinho et al. 2011). The values of the waves used in this work can be found in de Oliveira et al. (2021).

Figure 10.3a shows the behavior of the azimuth angle of the tracker. The azimuth angle of a floating solar panel with a tracking system is an important parameter that determines the panel's orientation in the horizontal plane. It is measured in radians and represents the angle between the direction of the panel's face and the true north. The optimal azimuth angle for a floating solar panel would typically be toward the geographical south, as this orientation allows the panel to receive the maximum amount of sunlight throughout the day. However, with a tracking system, the panel can dynamically adjust its azimuth angle to follow the movement of the sun across the sky. By continuously tracking the sun's position, the floating solar panel with a tracking system can maintain an optimal azimuth angle, ensuring that the panel's surface is directly facing the sun for the highest possible solar energy capture. This tracking capability allows the panel to maximize its energy output by effectively utilizing the available sunlight, even as the sun moves from east to west during the day. Figure 10.3b shows the elevation angle of the solar panel. The elevation angle refers to the vertical angle at which the panel is tilted or inclined relative to the horizontal plane. Maximizing direct exposure to sunlight throughout the year plays a crucial role in optimizing solar panel performance. This adjustment allows the solar panel to capture the maximum amount of solar energy throughout the year. By optimizing this angle, the system can achieve higher energy efficiency and contribute to the overall sustainability of solar power generation.

Figure 10.4 shows the azimuth and elevation angular velocities ω and $\dot{\theta}$, respectively. Although long, the velocities are compatible with the motion of the sun. Waves introduce disturbances in the azimuth and elevation velocities, which, in turn, can introduce spikes or irregularities in the azimuth velocity as the tracking system compensates. If these disturbances are significant and frequent, it is possible to observe more jagged or oscillatory behavior in the velocity profiles.

Figure 10.5 shows the cumulative energy generated in one day and it was computed by using the equation founded in Hayat et al. (2019). At the start of the day, the cumulative energy generated will be zero or very low. As the Sun begins to rise, energy generation will start, and energy will continue to accumulate. There will be a gradual increase as the sun moves toward its zenith. Around midday (solar noon), when the Sun reaches its highest point in the sky (its zenith), the potential

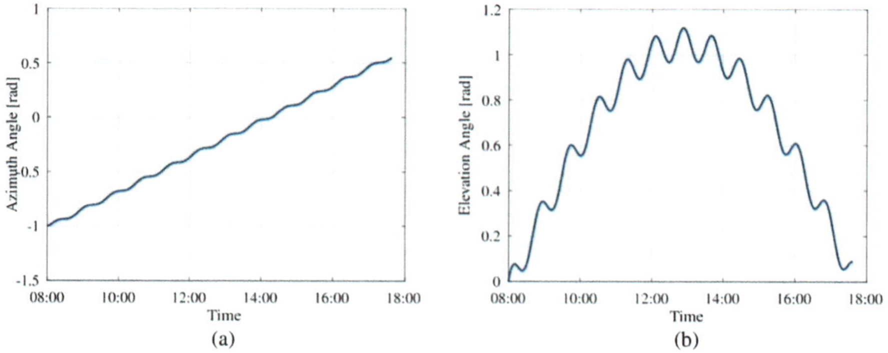

Fig. 10.3 Azimuth and elevation angles of the solar panel. **a** Tracker azimuth angle. **b** Tracker elevation angle

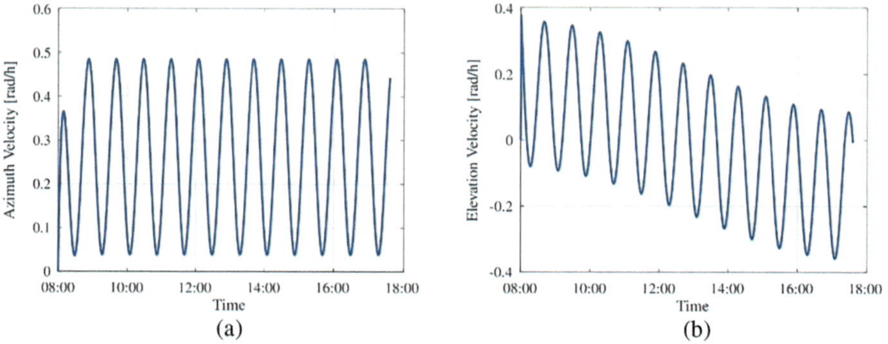

Fig. 10.4 Angular velocities of the solar panel. **a** Azimuth velocity. **b** Elevation velocity

for energy generation is at its maximum. If the tracking system works perfectly and wave disturbances are minimal, the photovoltaic system will capture the maximum amount of sunlight, leading to the steepest slope on the cumulative energy graph.

10.4 Final Remarks

The modeling and analysis of floating solar panel systems integrated in a hydroelectric plant in the Amazon region of Brazil have significant potential for improving renewable energy generation. By integrating solar panels on the water surface of hydroelectric reservoirs, this approach takes advantage of underutilized resources and offers several advantages. The implementation of floating solar panels with tracking systems should improve sunlight capture throughout the day, maximizing energy production.

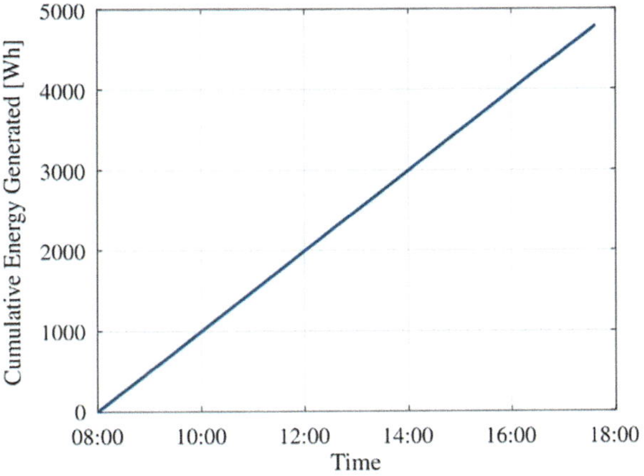

Fig. 10.5 Energy generated by the solar panel

This work presented a dynamic model of a floating solar panel with a tracking system. Through numerical simulations, the behaviors of the elevation and azimuth angles of the floating solar panel, as well as their respective velocities, could be studied. This allowed an understanding of the dynamics of this mechanical system and its potential implementation at the Belo Monte hydroelectric power plant. The tracking system successfully detected and tracked the apparent position of the Sun during periods of high solar irradiation. Additionally, the system incorporated a mechanism to handle exceptions during low irradiation intervals, aiming to minimize unnecessary movements in the tracker by either seeking or avoiding them.

It is possible to observe that the trajectory of the Sun across the sky sets the baseline for potential energy collection. When the Sun is at its zenith (at its maximum elevation), the potential solar irradiance on a perfectly aligned photovoltaic panel is at its peak. This time is often around solar noon but not always exactly at 12 PM due to the equations of time, latitude, and other factors. As the day progresses, the solar irradiance potential decreases as the sun approaches sunset. The tracking system attempts to keep the photovoltaic panel oriented toward the sun, maximizing energy collection. In the absence of any disturbances (such as waves), a perfect tracking system ensures that the photovoltaic panel is always normal to the sun's rays, ensuring maximum energy capture. The difference between a fixed system and a tracking system can be significant in terms of energy capture over a day. Waves introduce disturbances that cause the photovoltaic panel to deviate from its ideal orientation. This reduces the energy collection as the panel becomes misaligned with the sun. The magnitude of this effect depends on the amplitude and frequency of the wave disturbances. Larger waves or faster wave patterns can cause more significant misalignments.

To consider low irradiation intervals, it is necessary to address moments when the Sun is low on the horizon or when other conditions (e.g., cloud cover) significantly reduce the solar irradiance. These conditions can have an impact on the effectiveness of a tracking system and consequently on the energy generated by the solar panel. To this end, a more detailed and in-depth analysis of the tracking system control system needs to be carried out. The idea is to create a control system combined with the use of solarimetric charts. To do this, it is necessary to survey the types of sensors that can be used and how to combine them.

Acknowledgements This study was carried out under the aegis of the Research and Technological Development Program for the Electricity Sector subject to the Brazilian Regulator (ANEEL) as part of the R&D project PD-07427-0122/2022, funded by Norte Energia S.A.

References

Abbasnia A, Karimirad M, Friel D, Whittaker T (2022) Fully nonlinear dynamics of floating solar platform with twin hull by tubular floaters in ocean waves. Oc Eng 257:111320

Alberti EL, Paludo R, Portella KF et al (2023) New concept on 100.74 kWp floating solar photovoltaic plant and a real mechanical failures assessment: Case of study at Santa Clara hydroelectric power plant reservoir in southern Brazil. Sus E Tec and Ass 60:103455

Choi SM, Lee GR, Park CD et al (2021) Wind load on the solar panel array of a floating photovoltaic system under extreme hurricane conditions. Sus E Tec and Ass 48:101616

Choi SM, Park CD, Cho SH et al (2022) Effects of wind loads on the solar panel array of a floating photovoltaic system—experimental study and economic analysis. E 256:124649

Choi SM, Park CD, Cho SH et al (2023) Effects of various inlet angle of wind and wave loads on floating photovoltaic system considering stress distributions. J of Cl Prod 387:135876

Claus R, López M (2023) A methodology to assess the dynamic response and the structural performance of floating photovoltaic systems. S E 262:111826

de Freitas BMR (2022) What's driving solar energy adoption in Brazil? Exploring settlement patterns of place and space. E Res and so Sci 89:102660

de Oliveira MG, de Freitas MG, de Oliveira FF et al (2021) Estimativa de altura de ondas geradas por ventos no reservatório de Belo Monte—PA/Predicting wind waves's height in Belo Monte reservoir—PA. Br J Dev 7(8):80255–80271

Exley G, Page T, Thackeray SJ et al (2022) Floating solar panels on reservoirs impact phytoplankton populations: a modeling experiment. J of Env Man 324:116410

Ferrer-Gisbert C, Ferrán-Gozálvez JJ, Redón-Santafé M et al (2013) A new photovoltaic floating cover system for water reservoirs. Ren E 60:63–70

Hammoumi AE, Chtita S, Motahhir S et al (2022) Solar PV energy: from material to use, and the most commonly used techniques to maximize the power output of PV systems: a focus on solar trackers and floating solar panels. E R 8:11992–12010

Hayat MB et al (2019) Solar energy—a look into power generation, challenges, and a solar-powered future. Int J of En Res 43(3):1049–1067

Jambersi AB, Pereira JA, Turra AE et al (2017) Modelo dinâmico de um módulo da planta solar fotovoltaica flutuante instalada no lago da usina hidrelétrica de Porto Primavera-SP. In: Conferência Brasileira de Dinâmica, Controle e Aplicações, São José do Rio Preto

Pawar P, Pawale P, Nagthane T et al (2021) Performance enhancement of dual axis solar tracker system for solar panels using proteus ISIS 7.6 software package. Gl Trans Pro 2(2):455–460

Piancó F, Moraes L, dos Prazeres I et al (2022) Hydroelectric operation for hybridization with a floating photovoltaic plant: a case of study. Ren E 201:85–95

Prinsloo F, Schmitz P, Lombard A (2023) System dynamics characterization and synthesis of floating photovoltaics in terms of energy, environmental and economic parameters with WELF nexus sustainability features. Sus E Tec and Ass 55:102901

Rashid FL, Al-Obaidi MA, Hussein AK et al (2022) Floating solar stills and floating solar-driven membranes: recent advances and overview of designs, performance and modern combinations. S E 247:355–372

Rauf H, Gull MS, Arshad N (2020) Complementing hydroelectric power with floating solar PV for daytime peak electricity demand. Ren E 162:1227–1242

Sobrinho MD, Trovati LR, de Freitas Maciel G et al (2011) Monitoramento de ondas em reservatórios com sensor de pressão e comparação com dados de ADCP-waves. In: Anais do XIX Simpósio Brasileiro de Recursos Hídricos, Maceió

Trapani K, Millar D (2016) Floating photovoltaic arrays to power the mining industry: a case study for the McFaulds lake (ring of fire). Env P & Sus E 35

Vidovic V, Krajacic G, Matak N et al (2023) Review of the potentials for implementation of floating solar panels on lakes and water reservoirs. Ren and Sus E R 178:113237

Yasmeena S, Das GTR (2015) A review on new era of solar power systems: floatovoltaic systems or floating solar power plants. J on Instr and Con Eng 3(1):1–8

Chapter 11
Spectral Analysis of Fatigue in Drill Strings Under Torsional Vibrations

Sandro C. Valente, Daniel A. Castello, and Thiago G. Ritto

Abstract In oil exploration and production, drilling is one of the most important processes. Undesirable vibrations, intrinsic to the drilling process, can lead to fatigue failure, resulting in loss of efficiency due to downtime and, consequently, loss of productivity and high costs. The present work proposes the analysis of fatigue in drill strings under torsional vibrations with the stick–slip phenomenon, considering the frequency domain (or spectral) approach as an alternative to the time domain approach. The stick–slip occurs when the torque imposed by the top drive is not enough to overcome the resistance offered by the environment and, in the process of rock-bit interaction, the bit becomes stationary for a period, causing large fluctuations in the angular speed and, in turn, in shear stress, characterizing failure by fatigue. Spectral Methods for fatigue life analysis, although recent, have shown great potential in several applications and therefore, there is much to be explored given their viability in engineering problems due to the lower computational cost and time processing, being advantageous for complex problems with large amounts of data. In this approach, concepts such as Power Spectral Density (PSD) are used to calculate the Spectral Moments, employed in the various methods of this nature for determining fatigue life. The Dirlik and Tovo-Benasciutti Methods, being widely used not only in academia, have proved to be advantageous. These methods are compared with the rainflow method, used as a reference.

Keywords Drill strings dynamics · Torsional vibration · Stick–Slip · Fatigue · Spectral analysis

---

S. C. Valente (✉) · D. A. Castello · T. G. Ritto
Department of Mechanical Engineering, Universidade Federal Do Rio de Janeiro, Cidade Universitária—Ilha Do Fundão, Rio de Janeiro (RJ) 21945-970, Brazil
e-mail: scarvalho@mecanica.coppe.ufrj.br

P. Kurka and M. Pereira (eds.), *Advances in Structural Vibration*, Lecture Notes in Mechanical Engineering, https://doi.org/10.1007/978-3-031-71540-2_11

11.1 Introduction

The drilling process is one of the most complex and expensive activities in the oil and gas industry. During this process, undesirable vibrations are present due to several factors. Vibrations in these structures can be of three types: axial, lateral and torsional (Ritto et al. 2009). Associated with these vibrations, large stress amplitudes can occur, leading to fatigue failure (Moradi and Ranjbar 2009; Ritto et al. 2010). Torsional vibrations, when critical, are associated with the *stick–slip* phenomenon, characterized by the bit becoming stationary for a period due to the inability to win the resistance offered by the environment, leading to large variations in angular velocity and therefore in shear stress.

The objective of this work is the analysis of fatigue in drill strings subject to torsional vibrations and the *stick–slip* phenomenon considering the classical approach (*rainflow* method, in the time domain) and the frequency domain (or *spectral*) approach as an alternative. These methods, although recent, have shown great potential in many applications in engineering. The *Dirlik* (Dirlik 1985) and *Tovo-Benasciutti* (Benasciutti and Tovo 2006) methods were considered due to their good approximation in several works, for example, to compare the different methods or in applications to specific problems (see for example Mršnik et al. 2013; Zorman et al. 2023).

This chapter is organized as follows. Section 11.2 presents the theoretical background, with a mathematical model to describe the dynamics of drill strings subjected to torsion and the fatigue analysis in the time and frequency domains. For this last, are present the concepts used for random processes and the *Dirlik* and *Tovo-Benasciutti* methods. Section 11.3 presents some results considering different values for the angular velocity imposed by the *top drive* and the *weight on bit (WOB)*, important parameters to be controlled in the drilling process. The fatigue life is calculated, and the frequency domain methods are compared with the *rainflow* method. Section 11.4 presents the conclusions.

11.2 Theoretical Background

11.2.1 Mathematical Model for Torsional Vibrations in Drill Strings

To simulate the *stick–slip* phenomenon, the drill string was modeled as a simple torsional pendulum with one degree of freedom (1-DOF) with torsional stiffness k_t, damping c_t, inertia J and angular position θ. It is considered that there is no fluid between the bit and the bottom of the well and the speed Ω imposed by the *top drive* is constant. For the bit-rock interaction, the model proposed by Ritto and Ghandchi-Tehrani (2019) was adopted. This model is illustrated in Fig. 11.1.

The governing equation, therefore, is expressed by

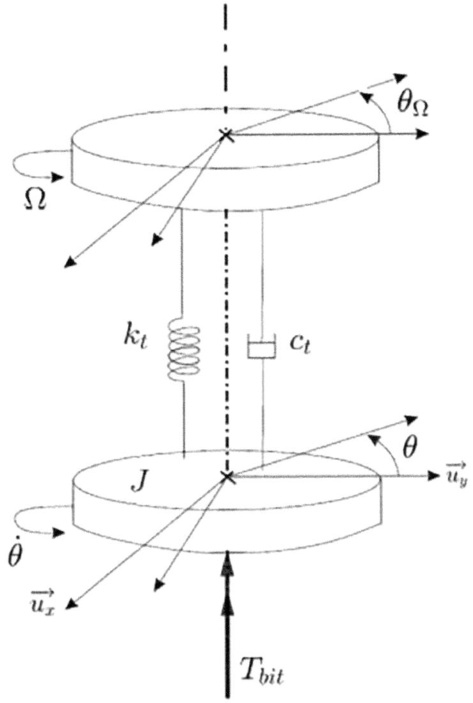

Fig. 11.1 Torsional pendulum model representation. Adapted from Navarro-Lopez and Suarez (2004)

$$J\ddot{\theta}_{\text{bit}}(t) + c_t\dot{\theta}_{\text{bit}}(t) + k_t\theta_{\text{bit}}(t) = c_t\Omega + k_t\Omega t + T_{\text{bit}}, \tag{11.1}$$

where T_{bit} is the torque on bit given by Tucker and Wang (1997),

$$T_{\text{bit}} = -b_0\text{WOB}\left(\tan h\big(b_1\dot{\theta}_{\text{bit}}\big) + \frac{b_2\dot{\theta}_{\text{bit}}}{1 + b_3\dot{\theta}^2_{\text{bit}}}\right), \tag{11.2}$$

being WOB the weight on bit and b_0, b_1, b_2 and b_3 positive constants that depend on the rock properties and characteristics of the bit. Once the angular position θ is determined, we can evaluate the shear stress.

$$\tau = G\frac{(\theta_\Omega - \theta)}{L}r^{\text{ext}}, \tag{11.3}$$

where G is the shear modulus, θ_Ω the *top drive* angular position with L and r^{ext} being the length and external radius of the *drill pipe* respectively. From the signal generated by the stress and its amplitudes, the fatigue life is estimated.

11.2.2 The Time Domain Approach to a Fatigue Life Estimation

The fatigue life is estimated using the *Whöler S–N curve* (Wöhler 1970), given by

$$\sigma = \mathrm{CN}^k, \tag{11.4}$$

with C and k being material parameters, σ the stress amplitude and N the number of cycles until the crack initiation. The *rainflow* method (Matsuishi and Endo 1968) is a counting method that transforms load history data into a series of reverse loads (Ogrinec et al. 2019), used in loads with variable amplitude.

For the damage accumulation is commonly adopted the *Palmgren–Miner* linear damage accumulation rule, expressed by

$$D = \sum_{i=1}^{N} D_i = \sum_{i=1}^{N} \frac{n_i}{N_i}, \tag{11.5}$$

where for each class of stress i, D_i is the accumulated damage, n_i the number of cycles accounted and N_i the number of cycles until nucleation occurs.

11.2.3 The Frequency Domain Approach to a Fatigue Life Estimation

Fatigue life analysis in the frequency domain is based on the *Power Spectral Density* (*PSD*) which is obtained from the loading using the *Fourier Transform.* From the PSD, the spectral moments applied in the methods are calculated and then estimated fatigue life. Several methods have been proposed and, in the present work, we will use the methods of *Dirlik* and *Tovo-Benasciutti* because they are generally most used and characterized by the approximation of the stress cycles through a *Probability Density Function* (*PDF*). Next, we briefly present the properties of a random process relevant to this new methodology presented in this work, as well as the methods used in it.

11.2.3.1 Random Process Properties

Random loads in the frequency domain are defined from the *Power Spectral Density* (*PSD*), represented by $S_{\mathrm{XX}}(\omega)$, where ω is the frequency in rad/s. In general, the unilateral *PSD* is used, characterized by the positive part and expressed by $G_{\mathrm{XX}}(\omega) = 2S_{\mathrm{XX}}(\omega)$ (Zorman et al. 2023). The statistical properties of a stationary random process are expressed by the *spectral moments*, defined by

$$m_i = \int_0^{+\infty} \omega^i G_{XX}(\omega) d\omega. \tag{11.6}$$

From these moments, we can calculate the positive slope zero crossing v_0, the expected peak frequency v_p, and the bandwidth parameter α_i, , that depend on the spectral moments (Zorman et al. 2023). These parameters are fundamental for the estimation of fatigue life in the frequency domain, whose methods discussed here will be presented below.

11.2.3.2 The Narrowband Method

1. If the stress amplitude σ can be treated as a continuous random variable, the accumulated damage for *narrowband process* is defined as (Dirlik and Benasciutti 2021)

$$D = \frac{v_p T}{C} \int_0^{+\infty} \sigma^k \rho(\sigma) d\sigma, \tag{11.7}$$

2. or, per time unit T according to

$$D = \frac{v_p}{C} \int_0^{+\infty} \sigma^k \rho(\sigma) d\sigma, \tag{11.8}$$

3. where $D = D/T$, being T the observation time, v_p the expected peaks, σ the stress amplitude with C and k being fatigue parameters of the materials.

The Eq. (11.8), once the analytical expression of $\rho(\sigma)$ is known, can be obtained in the closed form (Dirlik and Benasciutti 2021). Many concepts used to estimate fatigue in the frequency domain had their origins in Rice's research. One is the probability density function (PDF) of the peak amplitude for a general broadband (Rice 1945). For narrow band processes, peaks and valleys present symmetry about the value average, and each peak coincides with a cycle (in this process also v_p coincides with v_0). From (11.8) the accumulated damage is expressed by

$$D_{\text{NB}} = \frac{v_0 T}{C} \left(\sqrt{2m_0}\right)^k \Gamma\left(\frac{k+2}{2}\right), \tag{11.9}$$

being $\Gamma(\cdot)$ is *Euler's Gamma function* where

$$\Gamma(z) = \int_0^{\infty} t^{z-1} e^{-z} \mathrm{d}t. \tag{11.10}$$

Note, that according to according to Bendat (1964), the failure will occur for $D = 1$, therefore, making $T = T_{\text{life}}$ the life is expressed by

$$T_{\text{life}} = \frac{C}{\nu_0 \left(\sqrt{2m_0}\right)^k \Gamma\left(\frac{k+2}{2}\right)}. \tag{11.11}$$

The methods presented below, as well as others found in the literature, were developed to extend this theory to problems characterized by *broadband* processes.

11.2.3.3 The Dirlik Method

The Dirlik Method (Dirlik 1985) approximates the stress cycle distribution by means of the combination of an exponential distribution and two Rayleigh distributions, being an empirical method based on several simulations and, in addition to being able to be applied both to *narrowband* and *broadband* process, is widely used. Dirlik's *PDF* is given by

$$\rho_{\text{DK}} = \frac{1}{\sqrt{m_0}} \left[\frac{C_1}{Q} e^{-\frac{Z}{Q}} + \frac{C_2 Z}{R^2} e^{-\frac{2Z^2}{2R^2}} + C_3 Z e^{-\frac{Z^2}{2}} \right], \tag{11.12}$$

where Z is the normalized amplitude, x_m is the average frequency and C_1, C_2, C_3, Q and R are model adjustment parameters. The relation between the parameters are found in Dirlik (1985). The accumulated damage in the Dirlik model is

$$D_{\text{DK}} = \frac{\nu_p T}{C} \left(\sqrt{m_0}\right)^k \left[C_1 Q^k \Gamma(1+k) + \left(\sqrt{2}\right)^k \Gamma\left(\frac{k+2}{2}\right) \left(C_2 |R|^k + C_3\right) \right]. \tag{11.13}$$

11.2.3.4 The Tovo-Benasciutti Method

Benasciutti (2004) proposed an approximation based on the linear combination of the lower and upper limits of damage intensity, which is expressed by

$$D_{\text{TB}} = \left[b + (1-b)\alpha_2^{k-1} \right] \alpha_2 D_{\text{NB}}, \tag{11.14}$$

where b is a factor obtained through numerical simulations and proposed in two ways:

$$b_{\mathrm{TB}_1} = \min\left\{\frac{\alpha_1 - \alpha_2}{1 - \alpha_1}, 1\right\} \tag{11.15}$$

and

$$b_{\mathrm{TB}_2} = \frac{(\alpha_1 - \alpha_2)\left[1.112(1 + \alpha_1\alpha_2 - (\alpha_1 + \alpha_2))e^{2.11\alpha_2} + (\alpha_2 - \alpha_1)\right]}{(\alpha_2 - \alpha_1)^2}. \tag{11.16}$$

11.3 Simulations, Results and Discussions

Aiming at simulating the dynamics of a drill string and the *stick–slip* phenomenon, the parameters obtained from Ritto and Ghandchi-Tehrani (2019) were used, which are presented in Table 11.1. For the evaluation of the fatigue life, the material parameters C and k were obtained from Netto et al. (2008) and are presented in Table 11.2. For the simulations for the calculation of fatigue life, the steady state was considered and they were carried out with the support of the *FLife (Vibration Fatigue by Spectral Methods)* library of *open source research in structural dynamics* (Zorman et al. 2023).

Figure 11.2 presents results for four different configurations, where for the speed $\Omega = 5.0$ rad/s the weight on bit was varied using 50.0, 60.0, 80.0 and 100.0 kN. It can be observed that the angular position θ (*a*) presents a linear behavior in stationary regime and, about the relative position (*b*) there is a greater difference $\theta_\Omega\ \theta$ with

Table 11.1 Parameters used in the simulations for the torsional model

L	4733.6	Pipe length, m
r_i	0.0595	Inside radius of the drill pipe, m
r_e	0.07	Outside radius of the drill pipe, m
E	220.00	Modulus of elasticity, GPa
Sut	690.0	Tensile strength, MPa
S_y	580.0	Yield strength, MPa
ν	0.29	Poisson's ratio
ξ	0.25	Damping rate
b_0	5671.0	Model parameter, m
b_1	0.4775	Model parameter, s
b_2	8.8754	Model parameter, s
b_3	4.5595	Model parameter, s^2

Table 11.2 Material parameters used to calculate fatigue life

k	3.0	S–N curve inverse Slope
C	$4.16 \cdot 10^{12}$	S–N curve intercept, mpa^k

the increase of the weight on bit after a short time interval and the angular speed converges to the value imposed by the rotary table Ω as per (*c*). Since the shear stress (*d*) is a function of the variation in angular position, as seen in (3), greater amplitudes are associated with greater values of *WOB*. These configurations present scenarios without cyclic stressed after the transient, with the stress amplitude around the mean tending to zero.

Figure 11.3 presents the phenomenon of *stick–slip* for different values of weight on bit (120.0, 200.0, 300.0 and 500.0 kN) considering $\Omega = 4.0$ rad/s being its occurrence due to high weights on bit and low rotations. According to (a) and (c), respectively, it can be seen the period when the bit does not move and then speeds up. From (b) and (c) there are amplitudes increasing with the weight on bit, increasing the shear stress (d) and therefore, affecting fatigue life.

Figure 11.4 presents the *PDF's* obtained by the methods of *Dirlik* and *Tovo-Benasciutti* corresponding to the scenarios presented in Fig. 11.3. It is observed, with the increase of the weight on bit, the decrease of *PDF* values associated with a greater range of stress amplitude.

Figures 11.5, 11.6, 11.7 and 11.8 present the fatigue life (in seconds) respectively, for the *rainflow*, *Dirlik*, *Tovo-Benasciutti 1* and *Tovo-Benasciutti 2* methods while

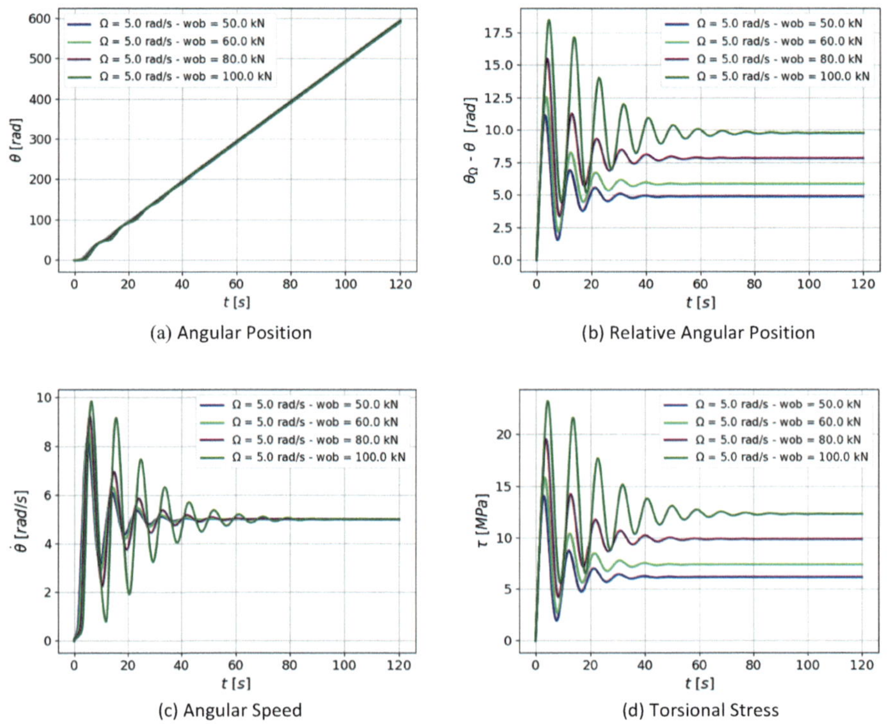

Fig. 11.2 Operating conditions without *Stick–Slip*

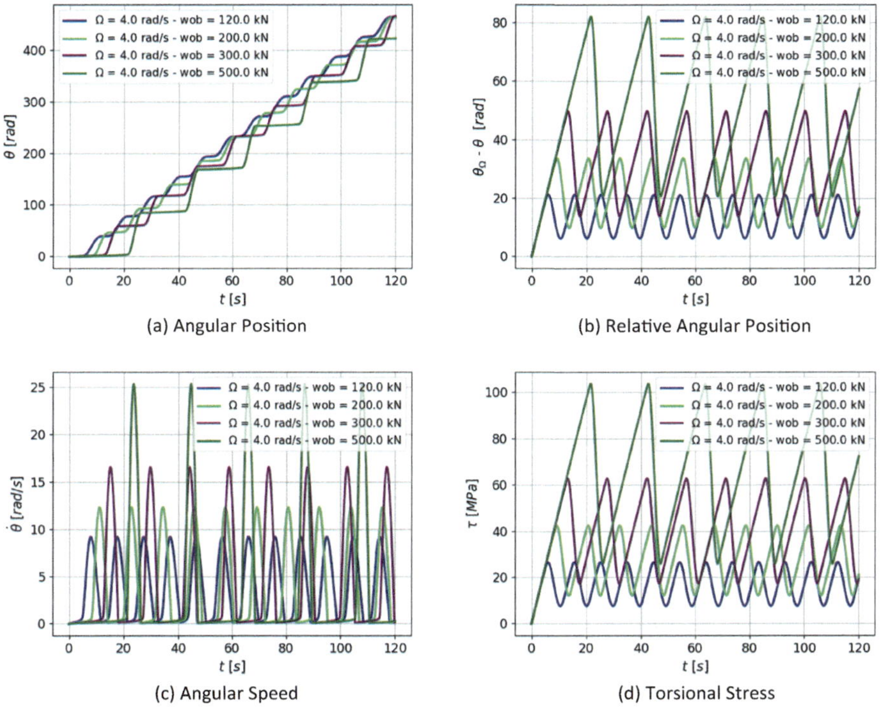

Fig. 11.3 Operating conditions with *Stick–Slip*

Figs. 11.9, 11.10 and 11.11 presents the relative errors considering the *rainflow* method as a reference. Values of Ω from 4.0 rad/s (38.19 rpm) to 16.0 rad/s (152.78 rpm) were considered here, and for the weight on bit *WOB* an interval of 50.0–500.0 kN. For infinite life, the limit value of 1.0 10^{10} s was established. It is observed that, for a given value of Ω, the life decreases by increasing the weight on bit. The same occurs with the increase of Ω for a given weight on bit until a scenario in which the estimate reaches the configuration for infinite life, which can be justified by the fact that the imposed rotation wins the resistance offered by the medium. Figure 11.12 presents some of these results on a logarithmic scale for a better view of the methods treated here.

Although the processing time depends on the characteristics and configurations of each computer and other factors, Table 11.3 presents, as an example, some values for some cases regarding the methods used.

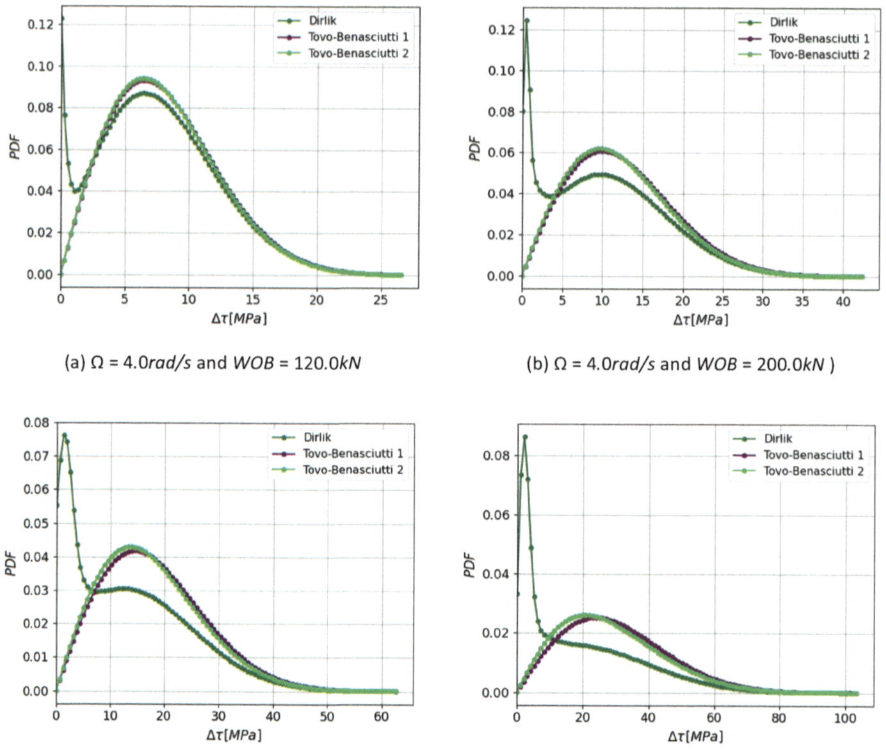

(a) Ω = 4.0*rad/s* and *WOB* = 120.0*kN*

(b) Ω = 4.0*rad/s* and *WOB* = 200.0*kN*)

(c) Ω = 4.0*rad/s* and *WOB* = 300.0*kN*

(d) Ω = 4.0*rad/s* and *WOB* = 500.0*kN*

Fig. 11.4 *PDF's* associated with *Dirlik* e *Tovo-Benasciutti* methods for different operation settings

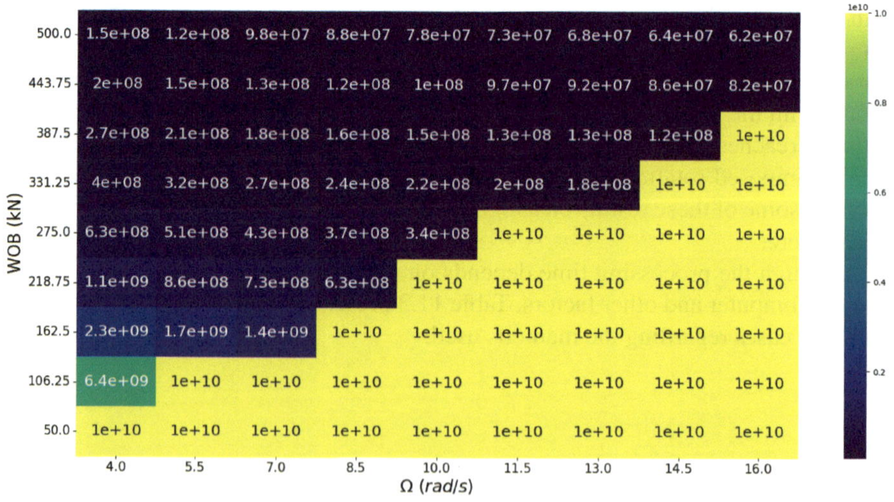

Fig. 11.5 Rainflow—Life (s)

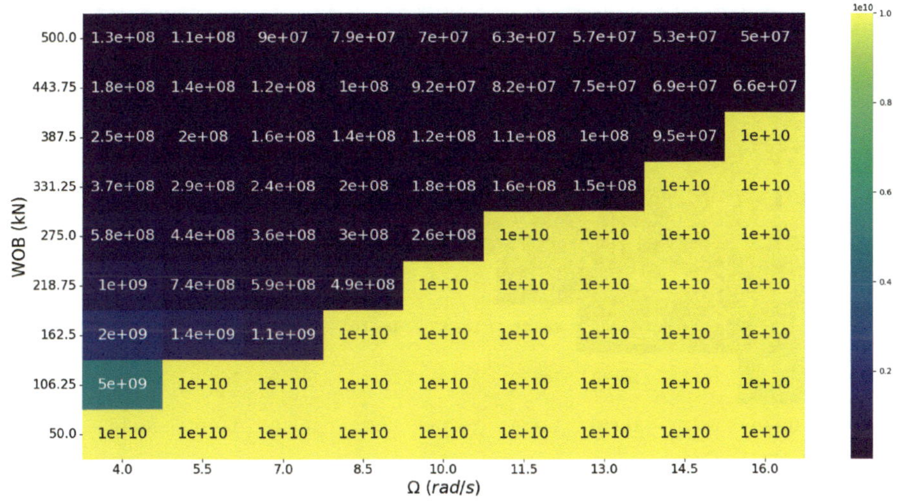

Fig. 11.6 Dirlik—Life (*s*)

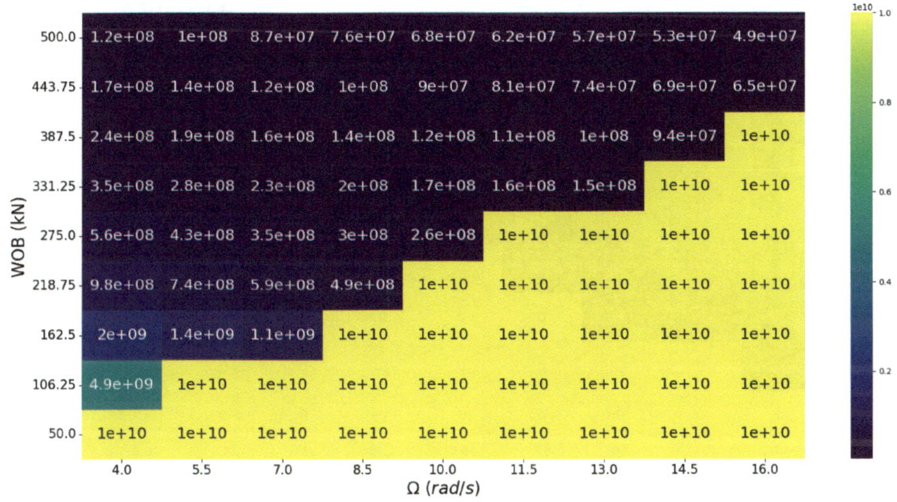

Fig. 11.7 Tovo-Benasciutti 1—Life (*s*)

11.4 Conclusions

The present work proposed to analyze fatigue in drill strings subjected to torsional vibrations with the stick–slip phenomenon. The fatigue life was computed in the time domain by the rainflow method and in the frequency **domain using** the *Dirlik* and *Tovo-Benasciutti* methods. The results for the spectral approach were conservative with the second method *Tovo-Benasciutti* presenting the best results in addition to

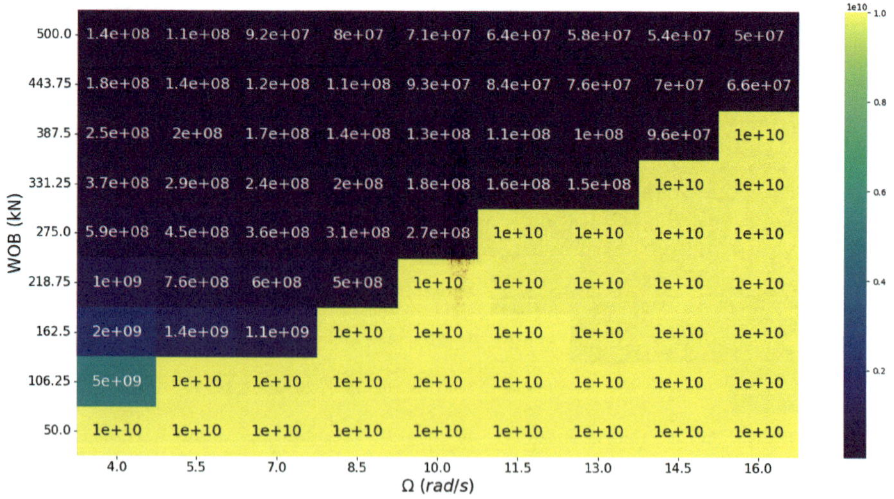

Fig. 11.8 Tovo-Benasciutti 2—Life (s)

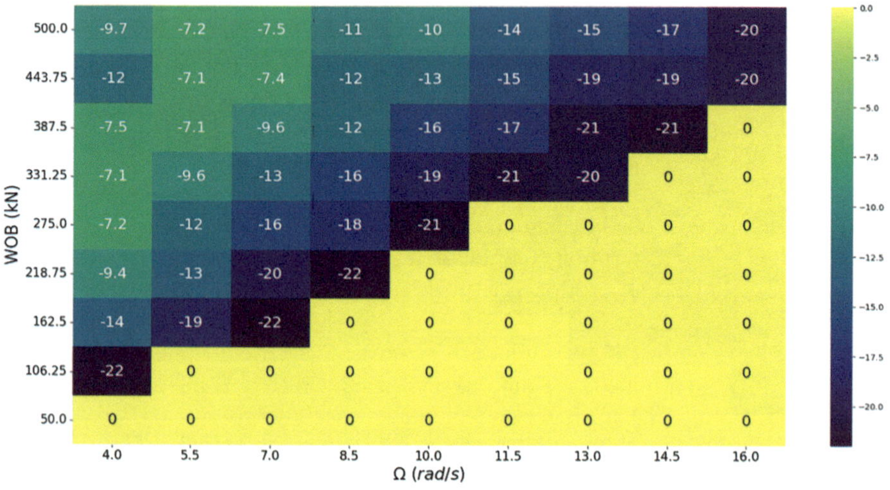

Fig. 11.9 Dirlik—Rel.Error (%)

these methods presenting a shorter simulation time. Therefore, the spectral approach to deal with fatigue life has shown viable to the problem analyzed. Other investigations must be carried out, especially concerning multi-axial fatigue where lateral or torsional vibrations must be included.

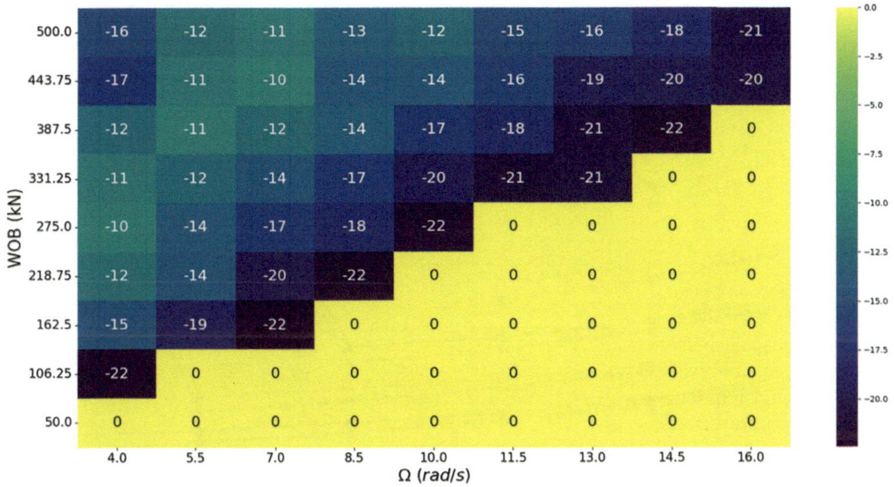

Fig. 11.10 Tovo-Benasciutti 1—Rel.Error (%)

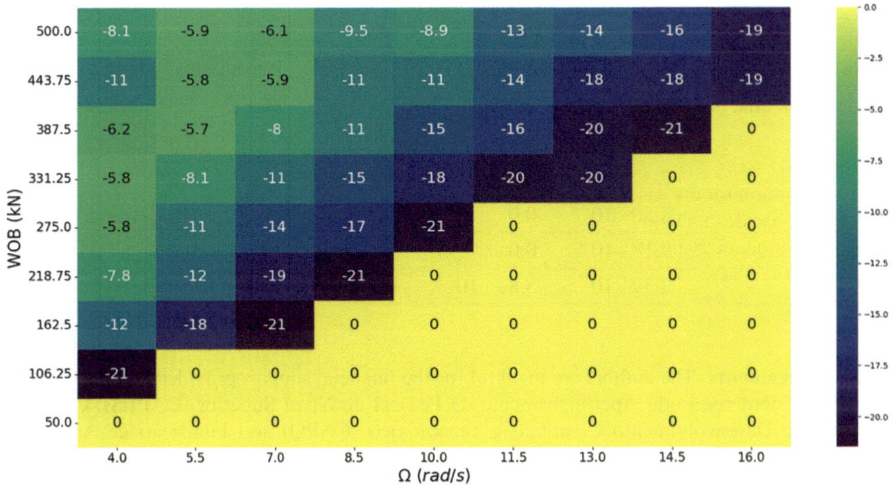

Fig. 11.11 Tovo-Benasciutti 2—Rel.Error (%)

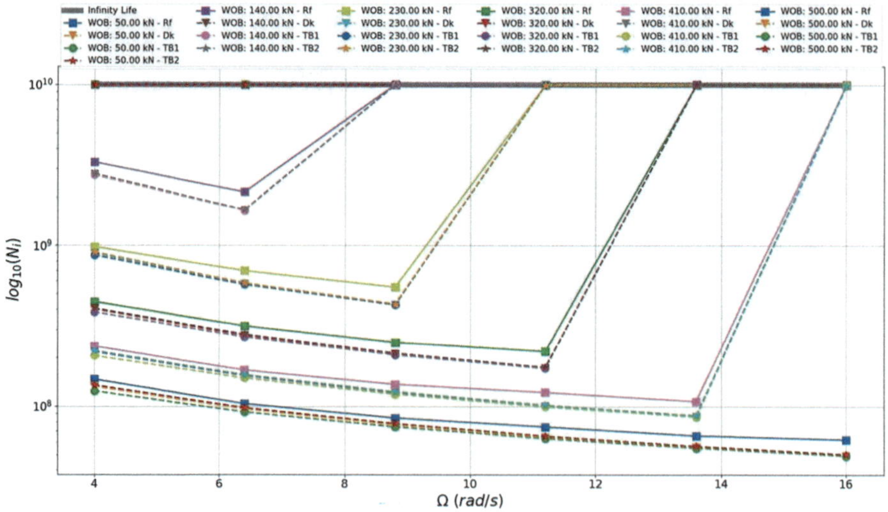

Fig. 11.12 Fatigue life as function of Ω and *WOB* for *rainflow* (Rf), *Dilik* (Dk), *Tovo-Benasciutti 1* (TB1) and *Tovo Benasciutti 2* (TB2) methods

Table 11.3 Calculation time used in methods (*s*) (Ω = 4.0 rad/s—WOB = 500.0 kN)

Observation time (*s*)	Rainflow	Dirlik	Tovo-Benasciutti 1	Tovo-Benasciutti 2
$1.20 \cdot 10^3$	$1.04 \cdot 10^{-3}$	0.0	0.0	0.0
$1.20 \cdot 10^4$	$7.32 \cdot 10^{-3}$	0.0	0.0	0.0
$1.20 \cdot 10^5$	$8.50 \cdot 10^{-2}$	0.0	0.0	0.0
$1.20 \cdot 10^6$	$9.18 \cdot 10^{-1}$	0.0	0.0	0.0
$1.20 \cdot 10^7$	$2.39 \cdot 10^{1}$	$3.87 \cdot 10^{-3}$	$1.06 \cdot 10^{-3}$	$2.67 \cdot 10^{-3}$

Acknowledgements The authors are thankful for the financial support provided by the Brazilian Agencies: Coordenação de Aperfeiçoamento de Pessoal de Nível Superior (CAPES), Conselho Nacional de Desenvolvimento Científico e Tecnológico (CNPQ) and Fundação de Amparo à Pesquisa do Estado do Rio de Janeiro (FAPERJ).

Responsibility Notice The authors are solely responsible for the printed material included in this chapter.

References

Benasciutti D, Tovo R (2006) Fatigue life assessment in non-gaussian random loadings. Int J Fatigue 28(7):733–746

Benasciutti D (2004) Fatigue analysis of random loadings. Thesis, University of Ferrara, Ferrara, Italy

Bendat J (1964) Probability functions for random responses-prediction of peaks, fatigue damage, and catastrophic failures
Dirlik T, Benasciutti D (2021) Dirlik and Tovo-Benasciutti spectral methods in vibration fatigue: a review with a historical perspective. Metals
Dirlik T (1985) Application of computer in fatigue analysis. Thesis, University of Warwick, Coventry, England
Matsuishi M, Endo T (1968) Fatigue of metals subjected to varying stress. Jpn Soc Mech Eng, Fukuoka, Jpn 68(2):37–40
Moradi S, Ranjbar K (2009) Experimental and computational failure analysis of drillstrings. Eng Fail Anal 16(3):923–933. https://doi.org/10.1016/j.engfailanal.2008.08.019
Mršnik M, Slavič J, Boltežar M (2013) Frequency-domain methods for a vibration-fatigue-life estimation—application to real data. Int J Fatigue 47:8–17
Navarro-Lopez E, Suarez R (2004) Practical approach to modelling and controlling stick-slip oscillations in oilwell drillstrings. In: Proceedings of the 2004 IEEE international conference on control applications, vol 2, pp 1454–1460
Netto T, Lourenço M, Botto A (2008) Fatigue performance of pre-strained pipes with girth weld defects: full-scale experiments and analyses. Int J Fatigue 30(5):767–778
Ogrinec P, Slavič J, Česnik M, Boltežar M (2019) Vibration fatigue at half-sine impulse excitation in the time and frequency domains. Int J Fatigue 123:308–317. ISSN 0142-1123
Rice SO (1945) Mathematical analysis of random noise. Bell Syst Tech J 24(1):46–156
Ritto TG, Ghandchi-Tehrani M (2019) Active control of stick-slip torsional vibrations in drill-strings. J Vib Control 25(1):194–202
Ritto T, Soize C, Sampaio R (2009) Non-linear dynamics of a drill-string with uncertain model of the bit-rock interaction. Int J Non-Linear Mech 44(8):865–876. https://doi.org/10.1016/j.ijnonlinmec.2009.06.003
Ritto T, Soize C, Sampaio R (2010) Robust optimization of the rate of penetration of a drill-string using a stochastic nonlinear dynamical model. Comput Mech 45(5):415–427. https://doi.org/10.1007/s00466-009-0462-8
Tucker R, Wang C (1997) The excitation and control of torsional slip-stick in the presence of axial-vibrations. pp 1–5. http://www.lancs.ac.uk/users/SPC/Physics.htm
Wöhler A (1970) Über die Festigkeitsversuche mit Eisen und Stahl
Zorman A, Slavič J, Boltežar M (2023) Vibration fatigue by spectral methods—a review with open-source support. Mech Syst Signal Process 190:110149

Chapter 12
Aeroacoustics Effects of the Wavy Leading Edge on a Fan-Rig SDT Stators

Andressa Caroline Rodrigues Garcia, Juan Antonio Flores Mezarina, and Hernán Darío Cerón-Muñoz

Abstract The interaction between the turbulent wake of the rotor and the stator vanes has been identified as one of the broadband noise main sources in a turbofan engine, in addition to the tonal noise. Studies inspired by the owl's silent flight have shown that the application of certain natural properties of its wings, such as the serrated pattern on the leading edge (Wavy Leading Edge), are effective for reducing aircraft noise. The objective of this work is to analyze it on the stator vanes source diagnostic test (SDT) low-noise model of the EESC/USP rig-fan bench. The geometry of the new vanes was defined and printed for the experimental tests and the analysis of noise emissions. The bench has acoustic instrumentation for data acquisition, and for its spectral analysis, the Welch method is used in signal processing. Tests were carried out with fan rotation speeds from 800 to 1800 rpm for both stators. The acoustic measurements of the WLE stator showed a reduction of 1.5–4 dB in low and medium frequencies in broadband noise, while there was an increase at high frequencies of up to 1.2 dB. From this study, it is understood that there is no ideal serration that reduces noise for all types of conditions that are imposed; nevertheless, it is more efficient at low rotation speeds, which could be used to improve the aircraft landing or takeoff conditions.

Keywords Wavy leading edge · SDT · Reduction noise · Rig fan

12.1 Introduction

One of the noise main sources in an aircraft landing and takeoff is the engine, which produces tonal and broadband noise. Between 1940 and 1990, the jet engine was the most common engine among aircrafts. The turbofan engines were introduced in the market in 1970, proposing better fuel efficiency and lower noise emission. With the high bypass technology—high the ratio between the airflow through the

A. C. R. Garcia (✉) · J. A. F. Mezarina · H. D. Cerón-Muñoz
São Carlos School of Engineering, University of São Paulo, São Carlos, Brazil
e-mail: agarcia@usp.br

P. Kurka and M. Pereira (eds.), *Advances in Structural Vibration*, Lecture Notes in Mechanical Engineering, https://doi.org/10.1007/978-3-031-71540-2_12

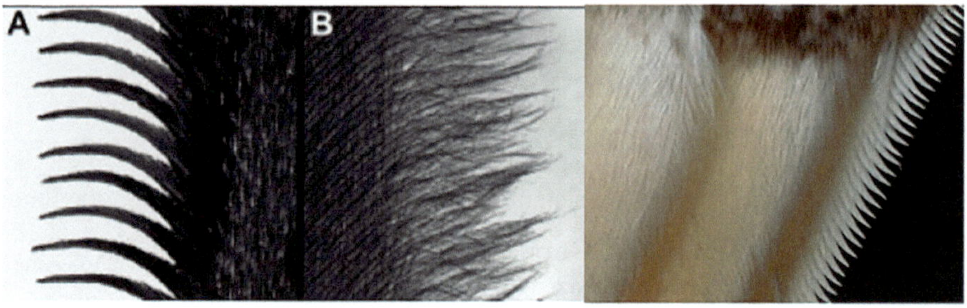

Fig. 12.1 Physical characteristics of the owl's wing

duct and center of the engine—the aircraft became more silent. As a result, fan noise predominated as one of the new sources of noise. Which is mainly generated by the rotor blades turbulent wake that incidents the stator vanes (Glegg and Devenport 2017).

Leading edge serrations, named as Wavy Leading Edge, are devices originated from the physical characteristics of the owl wings (Fig. 12.1), being identified as an effective collaborator in the reduction of sound sources and for better aerodynamic performance (Wang et al. 2019). These devices are applied to reduce noise radiation in aerodynamic profiles. Vortices are generated in the valleys that delay the separation of the boundary layer, and thus, the total turbulent wake is reduced.

The performance of the WLE is governed by parameters such as its wavelength (λ) and amplitude (A) (Fig. 12.2a). Broadband noise reduction is more sensitive to amplitude variations than to wavelength variations (Chong et al. 2015). Also, when (λ) decreases, noise reduction is enhanced at low and mid frequencies, while it is degraded at higher frequencies. If this parameter (λ) is large enough, an increase in noise at high frequencies may occur (Wang et al. 2019), as shown in Fig. 12.2b.

Claire et al. (2013) also studied the acoustic effects of three WLE models on the NACA 65 airfoil to reduce turbofan interaction noise. At high flow velocities, the WLE behaved like a low-pass filter, as they were more efficient above 1000 Hz. They demonstrated a reduction of about 3–4 dB without significant effects on the aerodynamic performance of the tested configurations with larger WLE amplitude values. At 3 kHz, noise reduction was nearly 6 dB.

Turbofan engine's broadband noise is generated by the turbulent flow, associated with the high Reynolds number, mainly produced by the wake of the rotor blades (Smith 1989). When this disturbance interacts with the stator vanes, pressure fluctuations are generated in the vanes that produce tonal and broadband noise, known as rotor/stator interaction noise (Olausson 2011). In the fan stage, tonal noise is produced by the periodicity of the rotor wake, which is noise concentrated in a narrow part of the spectrum and contains a high proportion of energy at a single frequency.

The experimental study of aircraft engine noise can be performed using aeroacoustic tunnels. Reference models such as the source diagnostic test (SDT) from NASA have a rotor/stator set with distinct geometries for simulations: baseline, low

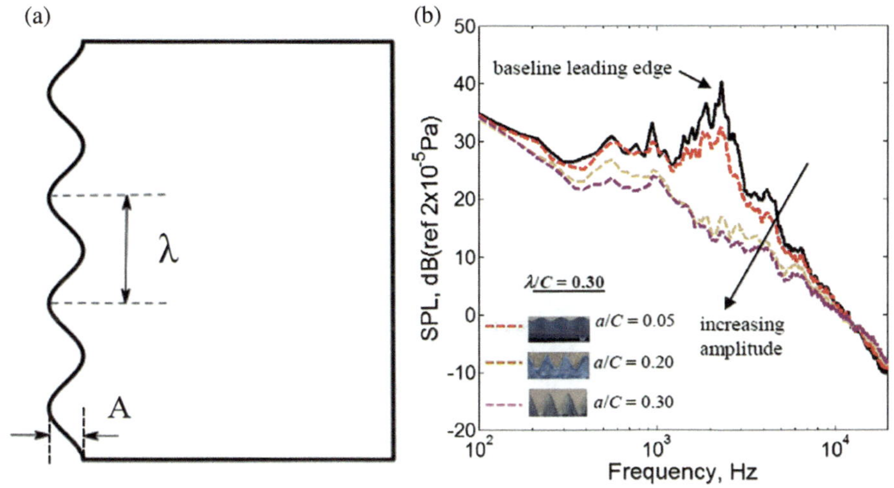

Fig. 12.2 **a** WLE—serrations on leading edge; **b** Noise levels with amplitude variation, Chong et al. (2015)

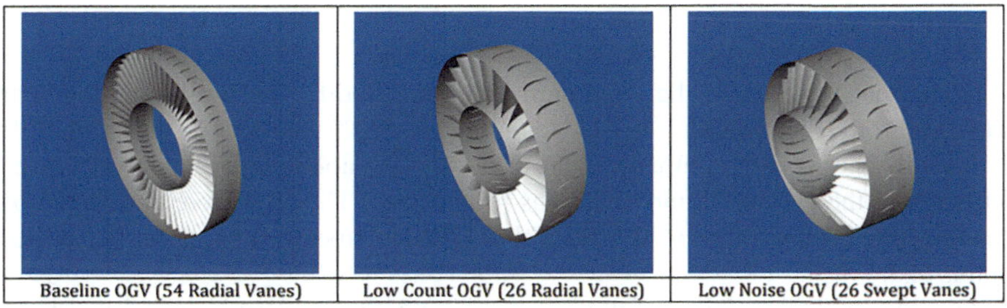

Fig. 12.3 Stator geometries, NASA Glenn (2016)

count, and low noise (Fig. 12.3). The main function of the stators is to direct the flow to the compression stage, maximizing the thrust.

In this study, the WLE aeroacoustic effects in a turbofan were assessed by wind tunnel tests at the EESC/USP Rig-Fan bench.

12.2 Experimental Setup

The tests were carried out at the Aircraft Laboratory of the Department of Aeronautical Engineering at EESC-USP, using the Rig-Fan bench. The tunnel rotor is composed by a set of 22 blades, while the reference stator by 26 vanes of the low-noise SDT configuration, which have a sweep angle of 30° at the leading edge. Due

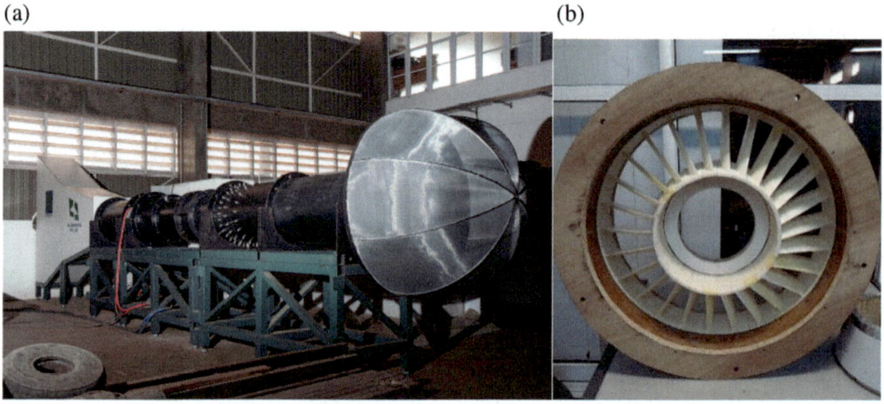

Fig. 12.4 **a** Rig-Fan bench; **b** Low-noise model stator

Table 12.1 Test operations—low-noise stator

Parameter	Specification
Fan speed	800: 200: 1800
Throttle	0, 25, 40, 55%

to equipment limitations, the maximum speed reached by the fan during the tests was 1800 rpm (Fig. 12.4).

Initially, the acoustic campaign of the reference low-noise model stator was carried out without the geometric modifications in its vanes. The tests were performed by changing the fan speed every 200 rpm. Table 12.1 describes the operations performed in this test:

For flow control, a throttle following the ISO-5136 standard is used for exhaustion, which can restrict the outlet area by up to 70%. There is a tendency to reduce broadband noise levels by decreasing the flow rate (Cuenca 2017).

In aeroacoustic instrumentation, the Rig-Fan has G.R.A.S 40PH-S2 φ7 mm model condenser microphones of high precision in the frequency range from 20 Hz to 20 kHz, which can be divided into three rings on the wall of the fan module. The acquisition system is composed of an NI PXI-1042Q board and data, acquisition is performed by cDAQ on the bench computer. For aerodynamic measurements, a pitot tube and a TSI model 8705 DP-Calc micromanometer are used. In addition, there is a meteorological station for data on pressure, density, and flow velocity.

12.2.1 Modified Stator

For the geometry of the vanes, the leading edge of the low-noise model was modified by adding WLE (serrations, see Fig. 12.5a), following the literature by Chong et al.

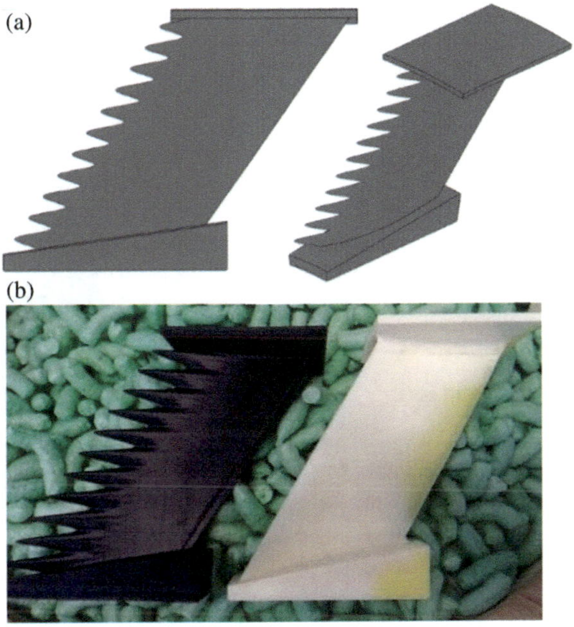

Fig. 12.5 **a** WLE vane model; **b** Vane modified × low noise

Table 12.2 Geometry WLE vanes (in % of chord)

Parameter	%C	
λ	15	
A	20	
A/λ		1.33

(2015), Clair et al. (2013), and Wang et al. (2019). The vanes were designed and printed in PLA material (Fig. 12.5b), following the geometrical parameters shown in Table 12.2. The λ value was defined to be small enough to not increase noise at high frequencies (above 6 kHz) and large enough to be greater than amplitude (*A*). According to Howe (1991), the ratio of the parameters of the serrated devices must be $A/\lambda > 1$.

All 26 vanes were printed and mounted on the duct's fan module, which is 600 mm in diameter (Fig. 12.6). The same test operations carried out with the low-noise stator are also used for the modified stator model.

12.2.2 Acoustic Processing and Testing

The acoustic measurements performed by the Rig-Fan instrumentation refer to the noise power, being necessary to carry out conversions and data processing to obtain

Fig. 12.6 Stator WLE

the desired type of signal. Microphones respond to pressure fluctuations by converting them into electronic signals. These signals arrive at the acquisition system and a time series is recorded and discretized at an acquisition rate of 52.1 k samples per second. The analysis is carried out by converting the signal from the microphones to the frequency domain, in order to obtain its spectrum in sound pressure levels (SPL) (Eq. 12.1).

$$\mathrm{SPL} = 10\log_{10}\left(\frac{p^2}{p_{\mathrm{ref}}^2}\right) \tag{12.1}$$

where p^2 is the estimated PSD and the reference pressure pref $= 20\mu\mathrm{Pa}$.

For this, it is necessary to obtain the power spectral density (PSD), which shows how the power is distributed along the frequencies. The spectra consist of methods for obtaining the PSD estimate (periodograms) without signal parameterization, and one of them is the Welch method, which is used for data processing:

$$[\mathrm{Pxx}, f] = \mathrm{pwelch}(X, \mathrm{Window}, \mathrm{Noverlap}, F, \mathrm{fs})$$

The input data (x) are exported from the bench acquisition system, where pxx is the PSD. Signal (x) is segmented into eight sections of equal length, each with 50% overlap, using the Hanning window (determines the estimator resolution and minimizes Lackage effects). The frequency is the Nyquist frequency which is half of the sampling rate used.

For the acoustic measurements, nine microphones were used, positioned in circles on the duct wall (Fig. 12.7) and calibrated for data processing. The first ring, with eight microphones, is at a distance of 0.70 m from the fan module and the second with

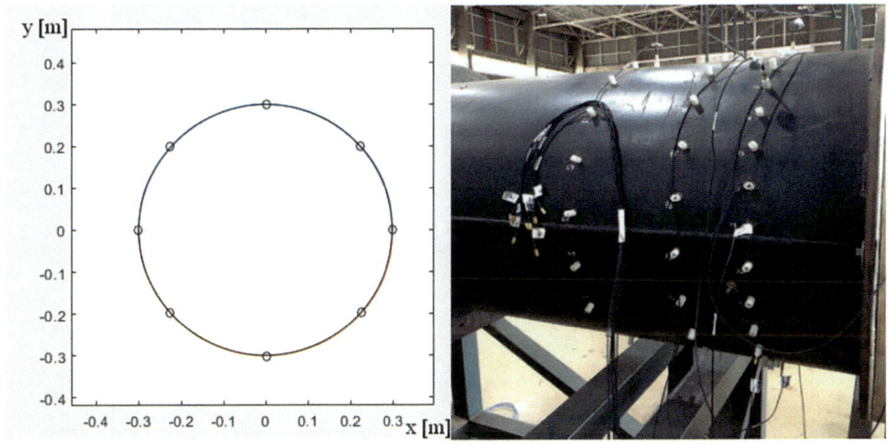

Fig. 12.7 Microphones array

one microphone at 0.10 m from the first. Acquisitions were performed at a sampling rate of 51.2 kHz with a duration of 10 s for each rotor speed.

12.3 Acoustics tests

The first tests were carried out with the low-noise stator, which is used as a baseline for comparing the noise measurements with the new geometric model. In Fig. 12.8, it is possible to verify that with the increase in fan speed, there is also a considerable increase in noise levels, which was already expected. Through these spectra, it is possible to observe both tonal and broadband noise.

Another parameter that changed during the tests was the exhaust area of the duct. Three throttles (plates) were used for flow control. This area restriction reduces broadband noise compared to open duct (0% throttle). The first plate causes about 25% restriction in the duct output, which causes a reduction in noise levels at low frequencies—100–500 Hz, when tested at a speed of 800 rpm (Fig. 12.9).

Increasing the throttle (40 and 55%) generates an increase in noise at low frequencies, but a reduction in broadband noise, in the range of medium to high frequencies—1.5–6 kHz, because the turbulence in the wake is smaller due to the reduction of the axial velocity of the flow for the same rotation (Cuenca 2017).

Results of the SPLs emitted by the WLE model stator can be seen in Fig. 12.10, with tested speeds from 800 to 1800 rpm, and without area restriction (throttle). It was found that serrations are more effective at lower rotational speeds, as well as studied by Soderman (1973) having better noise reductions at low and medium frequencies.

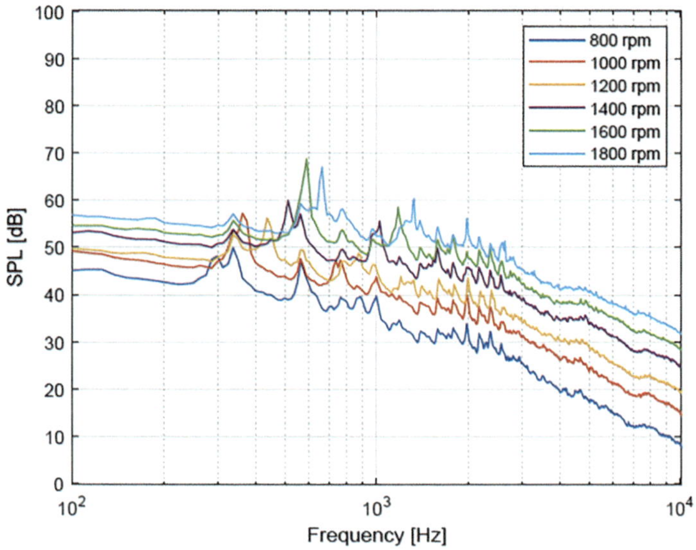

Fig. 12.8 Rotor noise levels

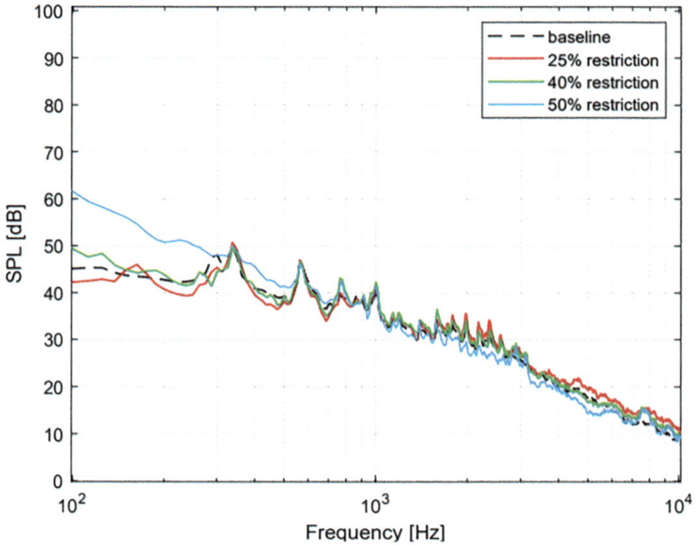

Fig. 12.9 Sound pressure levels at 800 rpm in all area restrictions

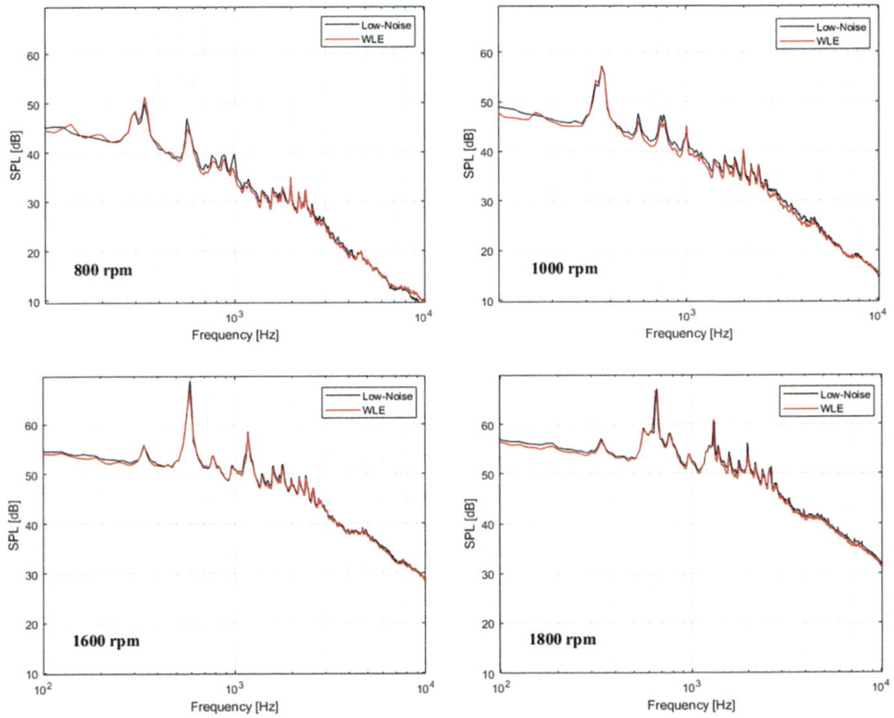

Fig. 12.10 Comparisons—low noise and WLE (800–1800 rpm)

Given the conditions imposed in these tests, serrations were more effective in reducing sound radiation at low rotation speeds (800–1000 rpm), because at high speeds, SPL levels are also higher, with more energy in the flow. Noise reduction decreases as the flow velocity increases. Because of this, the vortices that are generated in the serrated valleys (where the pressure is lower) were insufficient to reduce the periodicity of the turbulent wake.

The frequencies that had the highest |ΔSPL| calculated from 100 Hz to 10 kHz, which indicates a reduction or increase in SPLs (Fig. 12.11), the noise reduction reached 1.5–4 dB in low and medium frequencies, while an increase was verified in high frequencies in the order of up to 1.2 dB. Noise variation was negligible in all frequency bands when tests were performed using throttle.

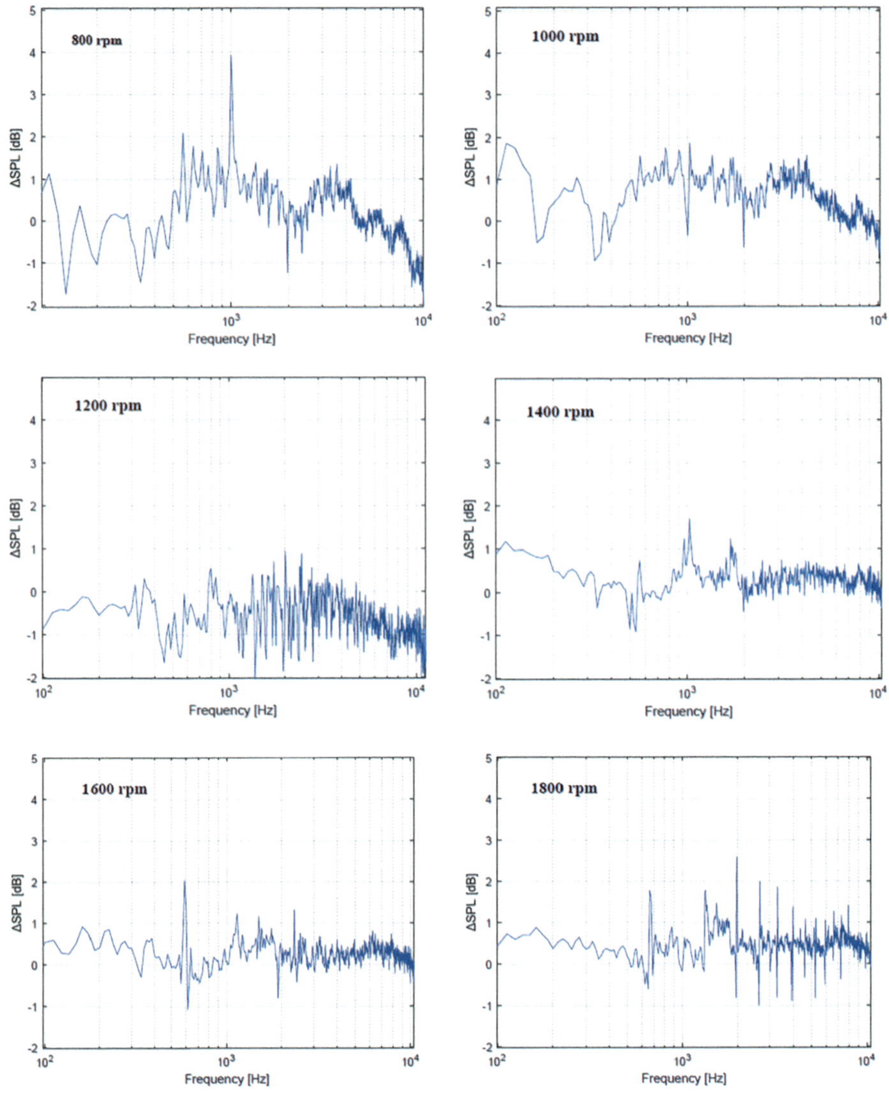

Fig. 12.11 |ΔSPL| (800–1800 rpm)

12.4 Conclusion

In this study, it was assessed the effects of the Wavy Leading Edge, a bio-inspired device, on sound radiation and noise reduction of the rotor/stator interaction, when applied to the stator vanes. Referenced research reported that the WLE is an effective device for noise reduction, up to about 10 dB depending on its configuration and rotor frequency.

The tests were carried out at an aeroacustic Rig-Fan bench, where it is possible to change test conditions such as rotor speed, stator geometry, turbulence levels, etc. At the bench, the SPLs of the base low-noise geometry SDT stator and the same stator with WLE were measured. The Welch method, which uses Fourier transforms to decompose temporal functions and process the spectrum of interest over frequency, was used to process the data obtained by the microphones.

It was found that the vanes with WLE reduced SPLs at low and medium frequencies, although it was observed a small increase in noise at high frequencies. It is considered that it is necessary a deeper investigation of the geometry of the serrations that caused this increase in noise at high frequencies. It is also concluded that there is no ideal WLE configuration that reduces noise at all tested conditions, since it showed better performance at lower rotational speeds; therefore, the WLE could be used to reduce noise emissions at the landing or takeoff conditions of a turbofan engine aircraft.

Responsibility notice The author(s) is (are) the only responsible for the printed material included in this paper.

References

Chong TP et al (2015) Aeroacoustic and aerodynamic performances of an aerofoil subjected to sinusoidal leading edges. AIAA J n(June):1–19. https://doi.org/10.2514/6.2015-2200

Clair V et al (2013) Experimental and numerical investigation of turbulence-airfoil noise reduction using wavy edges. AIAA J 51(11):2695–2713. ISSN 00011452. https://doi.org/10.2514/1.J052394

Cuenca RG (2017) Aeroacústica de Motores Aeronáuticos: Uma Abordagem por Meta-Modelo. 2017. 175 f. PhD Thesis (Aeronautical Engineering)—Escola de Engenharia de São Carlos, Universidade de São Paulo, São Carlos

Glegg S, Devenport W (2017) Aeroacoustics of low mach number flows. Elsevier, S.l., 368 p. ISBN 9780128096512

Howe MS (1991) Aerodynamic noise of serrated trailing edge. J Fluids Struct 5(1):33–45. ISSN 10958622

Olausson M (2011) Turbomachinery aeroacoustic calculations using nonlinear methods. PhD Thesis—Chalmers University of Technology

Smith MJT (1989) Aircraft noise, vol 34. Cambridge aerospace series edition. https://doi.org/10.1260/095745603321832471

Soderman PARC (1973) Leading edge serrations which reduce the noise of low-speed rotors. NASA Technical Notes

Wang Y et al (2019) Bio-inspired aerodynamic noise control: a bibliographic review. Appl Sci (Switzerland) 9(11):1–43. ISSN 20763417

Chapter 13
Sliding Mode Control for Suppression of Stick-Slip Vibrations in Oilwell Drillstrings

André Victor da Silva Castilho, Paulo Victor Gomes dos Santos, and Pedro Henrique Gomes dos Santos

Abstract In drill strings, stick-slip vibrations are a particular case of torsional vibrations, which cause large fluctuations in the angular velocity at the bottom of the well, resulting in premature wear of the bit and other components of the drilling system. This work introduces a sliding mode controller designed to regulate the torque applied to the top of the column in order to mitigate or eliminate stick-slip vibrations in the dynamic behavior of the drillstring. The drillstring is modeled as a two-degree-of-freedom lumped parameter model, representing its torsional behavior at both the top of the column and the bottom hole. The controller stability analysis is performed from the Lyapunov stability theory, in order to guarantee the system convergence to the desired state. Initially, the behavior of the system without the action of the controller is investigated and later, its influence on the dynamic behavior of the system is analyzed. For the studied cases, the controller was able to eliminate stick-slip vibrations, keeping the column angular velocity at the desired value, presenting results in agreement with experimental and field data. The chattering behavior is also verified and its elimination through a signal saturation function, guaranteeing an adequate behavior to the developed controller.

Keywords Drill string · Stick-slip · Sliding mode control · Non-linear dynamics

A. V. da Silva Castilho · P. V. G. dos Santos (✉)
CEFET/RJ, Centro Federal de Educação Tecnológica Celso Suckow da Fonseca, Rio de Janeiro, Brazil
e-mail: paulo.gomes@cefet-rj.br

A. V. da Silva Castilho
UERJ, Universidade do Estado do Rio de Janeiro, Rio de Janeiro, Brazil

P. H. G. dos Santos
Rheimmetall, Rome, Italy

P. Kurka and M. Pereira (eds.), *Advances in Structural Vibration*, Lecture Notes in Mechanical Engineering, https://doi.org/10.1007/978-3-031-71540-2_13

13.1 Introduction

The drilling phase represents a significant portion of the total cost of oil and gas exploration, accounting for approximately 40% of the overall exploration and production cost (Kriesels et al. 1999). Due to its high cost, the drilling process has become one of the major challenges in oil and gas exploration, attracting the attention of researchers worldwide.

The drilling column is a long, slender tube with a variable wall thickness that comprises drill pipes, tools, the drill bit, and additional downhole equipment. Its main function is to transmit the torque generated at the top of the column by the rotary system, while also allowing for the circulation of drilling fluid, removal of cuttings, and application of weight on the drill bit to advance through the rock formation during the drilling process.

From a dynamic perspective, during the drilling process, the column is subjected to three types of vibration: axial, lateral, and torsional. These vibrations can have several causes, such as mass imbalances, misalignments and bending of the column, irregularities in the wellbore, friction between the drill bit and column with the wellbore and well wall, as well as the dynamics involved in the interaction of the drill bit with the formation, among others (Besaisow and Payne 1988).

Stick-slip vibrations are defined by Kamel and Yigit (2014) as torsional vibrations where the drill bit periodically undergoes two phases: stick, where the drill bit completely ceases its rotational motion, and slip, where the drill bit is released from the rock formation and rotates at angular velocities considerably higher than its nominal angular speed. Studies focused on torsional vibration analysis indicate that stick-slip behavior occurs during 50% of the drilling time in a well, which can excite lateral and axial vibrations. These vibrations are responsible for approximately 80% of premature damages to drilling columns, representing a loss of approximately 5% of the total cost of well exploration (Spanos et al. 2003). Therefore, finding effective ways to reduce torsional vibrations and eliminate stick-slip is highly valuable in terms of reducing process time and preventing premature drilling equipment failures.

The objective of this work is to numerically reproduce the torsional behavior of the column in the presence and absence of stick-slip vibrations using a well-established mathematical model simulation, considering the drill-bit-rock interaction, in order to identify operational conditions where such vibrations occur, leading to a better understanding of this phenomenon. Furthermore, based on this initial analysis, the aim of this work is to propose a sliding mode torque controller at the top of the column to prevent the column from operating in the presence of these vibrations.

13.2 Literature Review

The studies on drilling string dynamics began in the late 1950s. Bailey and Finnie (1960)conducted an analytical approach that starts with the wave equation to analyze the axial and torsional dynamics of the drilling string, aiming to calculate its natural frequencies. Deily et al. (1968) measured force and displacement at the bottom of the well, observing axial, lateral, torsional loads, and bending moments. Analyzing the data obtained by Deily et al. (1968), Cunningham(1968) observed an alternation between zero rotation values and rotations higher than the column's rotation, making the first observation of the stick-slip phenomenon in drilling strings.

Halsey et al. (1988) conducted an analytical study on the torsional dynamics of the drilling string using a mathematical model based on a simple torsional pendulum, considering friction between the string and the rock formation. This model allowed for the formulation of analytical expressions to represent the stick-slip phenomenon. In this study, a speed controller was also proposed with the aim of maintaining the rotation speed of the string as constant as possible. This was the first controller developed to prevent torsional vibrations in drilling strings.

Starting in the early 1990s, with significant advancements in computing power, investigations into the dynamics of drilling strings began to be conducted numerically. The pioneering numerical analysis on this subject was presented by Lin and Wang (1990) in the study of torsional vibrations, where it was found that the fluctuation of the angular velocity of the drill bit increased as the rotation at the top of the string was also increased. However, above a critical value, such fluctuation ceased to occur. In other words, it was observed that stick-slip vibrations were eliminated when the string rotated above a critical angular velocity.

In Javanmardi and Gaspard (1992), using a system developed by the Shell Research Institute, an enhancement to the controller developed in Halsey et al. (1988) was proposed. Based on angular velocity and torque information obtained from current and voltage measurements of drill rig motors, the authors present a system for torque control through modifications to the speed controller of the rotary table.

The authors Christoforou and Yigit initiated studies on the dynamics of a drilling string in Christoforou and Yigit (1996), using a model with and without coupling between axial and lateral vibrations. They observed that when the coupling between modes was considered, the system exhibited unstable behavior. In Yigit and Christoforou (2006) the authors presented a model where axial, torsional, and lateral vibrations are fully coupled. This model has become one of the most well-established models on the subject and has served as a basis for numerous subsequent works. Additionally, in this study the authors proposed a linear controller for the elimination of stick-slip vibrations.

In Navarro-López and Licéaga-Castro(2009), the authors investigate the behavior of stick-slip torsional vibrations using dynamic analysis tools to obtain parameters for stable drilling operations. This analysis is performed through numerical simulation using a more generalized model than those previously presented in the literature.

However, the model proposed by the authors takes into account the variation of the string length as the drilling process progresses. In the study, the authors defined operating ranges for key parameters such as weight on the bit, motor torque and rotation speed to reduce oscillatory phenomena.

In Navarro-López and Licéaga-Castro (2009), the author presents a four degrees of freedom model for the torsional dynamics of the drilling string. The aim of this work was to conduct a more detailed investigation of the sliding mode controller developed in Navarro-López and Cortés (2007b). Based on the obtained results, the authors concluded that the studied sliding mode controller exhibits more attractive characteristics compared to proportional integral (PI) controllers.

In Bayliss et al. (2012), the author develops an adaptive controller using a recursive least squares algorithm. This controller continuously monitors the input value for the rotation speed of the drilling string, aiming to minimize the fluctuation of the drill bit's angular velocity and eliminate stick-slip vibrations during the drilling process. Under the action of the developed controller, it was observed that after 50 seconds, the stick-slip vibrations are completely eliminated.

In Vromen et al. (2015), the author proposes a robust controller for the elimination of stick-slip torsional vibrations, resulting in a four degrees of freedom model similar to the one presented in Fig. 13.1. The controller takes into account the nonlinearity of the interaction between the drill bit and the rock formation. The aim of the developed controller in this work is to maintain a constant velocity throughout the length of the drilling string. The input variables of the controller are the rotation speed and the torque at the top of the string. One of the main advantages of the controller developed by the author is that only measured data at the top of the string are required.

In Monteiro and Trindade(2017), the authors analyzed the performance of a proportional integral (PI) controller to assess its efficiency in eliminating stick-slip vibrations in the torsional dynamics of a drilling string. The study investigated the combination of proportional and integral gains to generate contour plots that depict the standard deviation of the drill bit's angular velocity for different values of weight

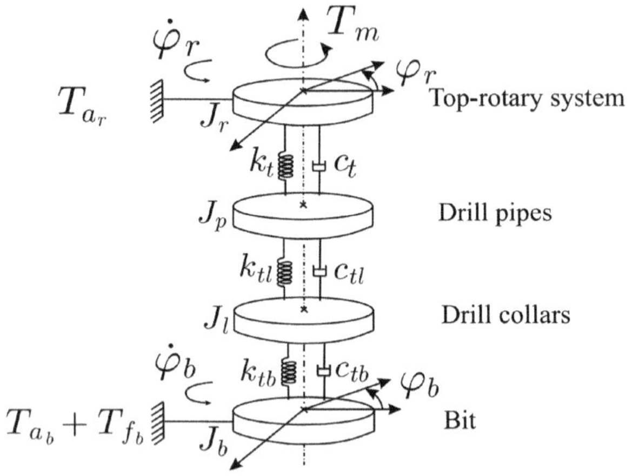

Fig. 13.1 Mechanical model describing the torsional behaviour of a simplified drillstring (Navarro-López and Licéaga-Castro 2009)

on the bit. Based on this initial analysis, the authors examined the influence of the controller for combinations of gains that resulted in different standard deviation values for the drill bit's velocity. It was observed that the controller was able to eliminate stick-slip vibrations for standard deviation values below 69%.

13.3 Problem Formulation

The mathematical model chosen to represent the rotary system and the drill-bit-rock interaction is based on Navarro-López and Suárez (2004). This model is widely used in the field due to its ease of numerical implementation and its ability to accurately reproduce experimental and field results. The motion equations are derived based on the following assumptions: perfectly vertical well, no lateral movement of the drill bit, non-zero velocity at the top of the string, neglecting friction between the pipe connections and between the pipes and the walls, considering a viscous friction coefficient to represent drilling mud, assuming laminar drilling fluid flow, and that the drill string operates under the assumption of the absence of bit-bounce, which refers to the loss of contact between the drill bit and the rock formation. These considerations form the basis of the mathematical model used in various works in the field of drilling dynamics.

The Fig. 13.2 presents the concentrated parameter model with two torsional degrees of freedom used in this study and proposed in Navarro-López and Suárez (2004).

Regarding the rotary system, at the top of the string, there is a torque T_m exerted by the electric motor. In addition, there are restorative and dissipative torques resulting from the stiffness and damping coefficient of the string, denoted by T_{res} and T_d respectively. The restorative torque at the top of the string is calculated based on the

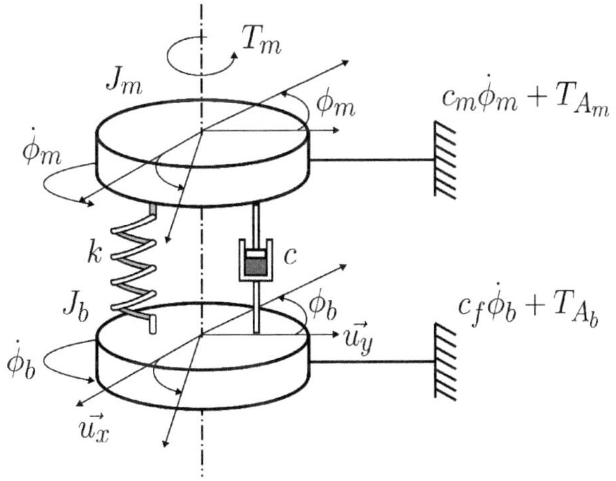

Fig. 13.2 Concentrated parameter model for drilling column. Adapted from Navarro-López and Suárez (2004)

torsional stiffness coefficient k of the string and the difference between the angular displacement of the drill bit ϕ_b and the angular displacement at the top of the string ϕ_m, as follows:

$$T_{\text{res}} = k(\phi_m - \phi_b) \tag{13.1}$$

The dissipative torque is calculated considering a damping coefficient associated with the drilling string c and the difference between the angular velocities of the degrees of freedom of the system, as follows:

$$T_d = c\left(\dot{\phi}_m - \dot{\phi}_b\right) \tag{13.2}$$

where $\dot{\phi}_b$ is the angular velocity of the drill bit, and $\dot{\phi}_m$ is the angular velocity at the top of the string. Additionally, there is a variable torque acting on the rotary system resulting from dry friction T_{A_m} and viscous damping at the top of the string, given by:

$$T_r\left(\dot{\phi}_m\right) = c_m\dot{\phi}_m + T_{A_m}\text{sign}\left(\dot{\phi}_m\right) \tag{13.3}$$

Thus, the equation of motion for the rotary system, considering J_m as the moment of inertia associated with the top of the string, is given by:

$$J_m\ddot{\phi}_m + c\left(\dot{\phi}_m - \dot{\phi}_b\right) + k(\phi_m - \phi_b) = T_m - T_r\left(\dot{\phi}_m\right) \tag{13.4}$$

The bit-rock interaction model adopted predicts that the external torque J_m exerted on the drill bit is composed by a term related to the viscous damping caused by the drilling fluid c_f, and by another term related to the friction T_{A_b} between the drill bit and the rock formation, as follows:

$$T_b = c_f\dot{\phi}_b + T_{A_b} \tag{13.5}$$

The friction torque exerted by the rock formation on the drill bit T_{A_b} is defined by a discontinuous system of equations characterized by three different states: stick, when the drill bit is stuck in the rock formation; slip, when there is relative angular velocity between them; and a transition phase that defines the onset of drill bit detachment. To differentiate between the three phases of the torque model, let's consider D_v as a reference value for the drill bit's angular velocity, representing a small region near $\dot{\phi}_b = 0$, and the magnitude of the static friction torque T_{e_b} exerted by the rock formation defined as follows:

$$T_{e_b} = R_b P \mu_{e_b} \tag{13.6}$$

where R_b is the radius of the drill bit, P is the applied load at the top of the string to advance the drill bit through the rock formation, and μ_{e_b} is the coefficient of static friction between the drill bit and the rock formation. When the angular velocity of the drill bit is lower than the reference value D_ν, the system can be in the stick phase or the transition phase. To distinguish between these phases, a comparison is made between the values of the friction torque T_{A_b} and the magnitude of the static friction. If the variable friction torque is smaller than the static friction torque, i.e. $T_{e_b} < T_{\text{stick}}$, the system will be in the stick phase. In this case, the torque is given by the variable static friction torque expression proposed by Leine et al. (1998):

$$T_{\text{stick}} = c(\dot{\phi}_m - \dot{\phi}_b) + k(\phi_m - \phi_b) - c_b\dot{\phi}_b \tag{13.7}$$

When the angular velocity of the drill bit is below D_ν and the value of the static friction torque is lower than the torque calculated by Eq. 13.7, the system will be in the transition from the stick phase to the slip phase. In this situation, the torque on the drill bit will be given by:

$$T_{tr} = T_{e_b}\text{sign}(T_{\text{stick}}) \tag{13.8}$$

If the magnitude of the drill bit's angular velocity is above the reference value D_ν, the system enters the slip phase. In this condition, the torque on the drill bit is calculated as follows:

$$T_{\text{slip}} = R_b P \mu_b(\dot{\phi}_b)\text{sign}(\dot{\phi}_b) \tag{13.9}$$

where $\mu_b(\dot{\phi}_b)$ is the variable friction coefficient as a function of velocity, described as:

$$\mu_{\dot{\phi}_b} = \mu_{c_b} + (\mu_{e_b} + \mu_{c_b})e^{-\gamma|\dot{\phi}_b|} \tag{13.10}$$

where μ_{c_b} is the Coulomb friction coefficient and γ is a positive constant associated with the rate of friction coefficient decay with velocity. With the information provided, the friction torque exerted by the rock formation on the drill bit can be described as a discontinuous system of equations, expressed as follows:

$$T_{A_b} = T_{\text{stick}}\ \mathit{if}\ |\dot{\phi}_b| < D_\nu \text{and} |T_{\text{stick}}| \le T_{e_b}$$

$$T_{A_b} = T_{e_b}\text{sign}(T_{\text{stick}})\ \text{if}\ |\dot{\phi}_b| \langle D_\nu \text{and} |T_{\text{stick}}| \rangle T_{e_b}$$

$$T_{A_b} = R_b P_B(\dot{\phi}_b)\text{sign}(\dot{\phi}_b)\ \text{if}\ |\dot{\phi}_b| > D_\nu \tag{13.11}$$

Considering J_b as the moment of inertia of the drill bit, the equation of motion for the drill bit can be defined as:

$$J_b\ddot{\phi}_b + c(\dot{\phi}_b - \dot{\phi}_m) + c_b\dot{\phi}_b + k(\phi_b - \phi_m) = -T_b(\dot{\phi}_m) \tag{13.12}$$

The system of equations obtained, characterized by Eqs. 13.4 and 13.12, consists of a set of second-order coupled differential equations with a discontinuity arising from the adopted model for describing the bit-rock interaction.

13.4 Sliding Mode Control Implementation

The change of variables presented in Eq. 13.13 reduces the number of state variables in the system. However, to understand its dynamics and for the development of the controller, it is sufficient to know the difference between the angular displacements of the table and the drill bit.

$$X = \begin{pmatrix} x_1 \\ x_2 \\ x_3 \end{pmatrix} = \begin{pmatrix} \dot{\phi}_m \\ \phi_m - \phi_b \\ \dot{\phi}_b \end{pmatrix} \quad (13.13)$$

Re-writing the system in its explicit form, considering the change of variables, we obtain:

$$\dot{X} = \begin{pmatrix} \dot{x}_1 \\ \dot{x}_2 \\ \dot{x}_3 \end{pmatrix} = \begin{pmatrix} \frac{1}{J_m}[T_m - T_r(x_1) - c(x_1 - x_3) - kx_2] \\ x_1 - x_3 \\ \frac{1}{J_b}[-T_b(x_3) - c(x_3 - x_1) + kx_2] \end{pmatrix} \quad (13.14)$$

In the present work, the goal of the controller is to eliminate stick-slip vibrations at the bottom of the well, keeping the drill bit rotating at a desired angular velocity Ω_d. For this purpose, a surface must be proposed in which the trajectories of the system converge to a state where the angular velocity of the drill bit tends to the angular velocity of the rotating system, which in turn tends to Ω_d. Thus, the vector of the tracking errors of the variables of interest can be written as:

$$\tilde{x} = \begin{pmatrix} x_1 - \Omega_d \\ x_1 - x_3 \end{pmatrix} \quad (13.15)$$

Let $S(t) \in R^2$ be the sliding surface defined by the equation $s(\tilde{x}, t) = 0$ and given, for the presented second order system, by Eq. 13.16:

$$s = (x_1 - \Omega_d) + \lambda \int_0^t x_1 - \Omega_d d\tau + \lambda \int_0^t x_1 - x_3 d\tau \quad (13.16)$$

The foundation of the design of a sliding mode controller is to convert an n in x th—order path-following problem into a first-order stabilization problem in s. However, the sliding surface $s(x, t)$ defined by the designer, as well as the values of the tracking errors, must tend to zero in a finite time interval. That is, the sliding mode controller forces the states of the system to converge to the region defined by the sliding surface and after reaching it, the system error should converge to zero (Slotine et al. 1991). The control law is obtained from the Fillipov construction for the equivalent dynamics of the system obtained in $\dot{s}(x, t) = 0$.

The time derivative of the sliding surface shown in Eq. 13.16, substituting x_1 by its equivalent expression in Eq. 13.14, is:

$$\begin{aligned}\dot{s} &= \frac{1}{J_m}[T_m - T_r(x_1) - c(x_1 - x3) - kx_2] \\ &\quad + \lambda(x_1 - \Omega_d) + \lambda(x_1 - x_3)\end{aligned} \tag{13.17}$$

The expression for the torque at the top of the column is given by the sum of the equivalent torque T_{eq} obtained by the control law and a switching torque T_g for the system to reach the slip surface from an initial state. Therefore, the torque obtained by the control law considering $\dot{s} = 0$ can be written as:

$$T_{eq} = T_r(x_1) + c(x_1 - x_3) + kx_2 - J_m\lambda(x_1 - \Omega_d) - J_m\lambda(x_1 - x_3) \tag{13.18}$$

With switching term defined as $T_g = -k\,\text{sign}(s)$ or, to avoid chattering issues, $T_g = -\text{ksat}\left(\frac{s}{\Phi}\right)$ where Φ is the boundary layer width and describes the distance of the system response to the sliding surface $S(t)$.

When applying the sliding mode controller, the entire trajectory of the system reaches and remains in the region $s = 0$ in a finite time. This can be proved from the selection of a Lyapunov Function V and its derivative with respect to time given as per Liu (2015).

From the moment the system reaches the state $s = 0$, the system converges asymptotically to an equilibrium state, called the desired equilibrium state, defined here by:

$$\overline{X}_d = \begin{pmatrix} \bar{x}_{1_d} \\ \bar{x}_{2_d} \\ \bar{x}_{3_d} \end{pmatrix} = \begin{pmatrix} \Omega_d \\ \frac{c_m\Omega_d + T_b}{k} \\ \Omega_d \end{pmatrix} \tag{13.19}$$

This statement can be verified by the stability analysis of the controller. To do so, the new Lyapunov Function proposed in Vaziri et al. (2018) is considered:

$$\begin{aligned}\overline{V} &= \frac{1}{2}\left[J_m\left(x_1 - \bar{x_1}\right) + k\left(x_2 - \bar{x_2}\right) + J_m\left(x_3 - \bar{x_3}\right)\right] \\ \dot{\overline{V}} &= J_m\dot{x}_1(x_1 - \bar{x}_1) + k\dot{x}_2(x_2 - \bar{x}_2) + J_m\dot{x}_3(x_3 - \bar{x}_3)\end{aligned} \tag{13.20}$$

When analyzing Equation 13.20, it is noted that for the equilibrium state, $\dot{\overline{V}} = 0$. If the desired equilibrium state presented in Eq. 13.19 is now considered, Eq. 13.20

can then be defined as:

$$\dot{\overline{V}}\left(\overline{X}_d\right) = (-T_d - K) \tag{13.21}$$

With Ω_d and K being strictly positive constants, it can be said that $\dot{\overline{V}} \leq 0$. Thus, it can be stated that the system is asymptotically stable, therefore, when reaching the slip surface, the system will remain on it or in a small neighborhood.

13.5 Results and Discussions

In this section, the results of the numerical simulation for the mathematical model of the drilling string obtained in the previous section are presented and discussed. The numerical integration of the system's equations of motion is performed using the fourth-order Runge-Kutta method. In all the results, initial conditions are assumed to be zero, except for the rotary system at the top of the string, which has an initial velocity equivalent to the desired angular velocity value Ω_d.

The parameters adopted for the drilling string characteristics and the controller parameters are taken from Vaziri *et al.*, considering a reduced model for the drilling string with a length of 1 meter. The parameters for the torque model are taken from Navarro-López and Suárez (2004). These values are presented in Table 13.1.

The results will be presented in three subsections. The first subsection aims to verify the torsional dynamics of the drilling string under stick-slip regime by comparing the obtained results with well-established results in the literature, in order to ensure the validity of the motion equations and the numerical integration code. The second subsection verifies the operation of the ideal controller obtained from

Table 13.1 Parameters used in simulations

Parameters	Value
J_m (kg/m^2)	13.92
J_b (kg/m^2)	0.5
c_m (Nms/rad)	11.38
c_b (Nms/rad)	0.03
c (Nms/rad)	0.005
k (Nm)	10
T_{A_m} (Nm)	0.5
P (N)	1790
D_ρ	10^{-6}
γ	0.9
μ_{e_b}	0.3
μ_{c_b}	0.2

the control law, disregarding the effect of the switching term on its operation. In the last subsection of this chapter, the influence of the switching term on the controller's operation is examined. In this last subsection, the influence of the saturation function on mitigating or eliminating excessive switching phenomenon or chattering is also investigated.

13.5.1 Analysis of Drill String Dynamics in Stick-Slip Regime

Initially, the dynamic behavior of the system is examined without considering any action from the controller. For this purpose, it is assumed that the torque at the top of the string T_m has a constant value of 39.57 Nm. Figure 13.3a shows the angular velocities at the top of the string, the rotary system, and the drill bit, represented in black and red, respectively. It can be observed that while the rotary system exhibits an angular velocity that oscillates around the desired value of 3.1 rad/s, the drill bit alternates its angular velocity between periods of zero value, indicating the stick phase, and periods where it reaches magnitudes of more than twice the reference value of the rotary system at the top of the string, indicating the slip phase. The stick-slip behavior can also be easily observed in the phase space of the system, as shown in Fig. 13.3b. While the drill bit is in the stick phase, indicated by the horizontal line at $\dot{\phi}_b$, the angular distortion increases until the energy accumulation is large enough to release the drill bit from the rock formation, causing it to enter the slip phase.

The behavior of the torque exerted by the rock formation on the drill bit is illustrated in Fig. 13.3c. Maximum and minimum values can be observed corresponding to the stick and slip phases, as indicated by the periods shown in Fig. 13.3a. Regions of instability can also be noticed in relation to the stick phases, which are related to the transition range in the adopted torque model.

It is worth noting that the results presented in Fig. 13.3a and b are consistent with those shown in Vaziri et al. (2018), while the behavior illustrated in Fig. 13.3c aligns with the torque behavior shown in Navarro-López and Suárez (2004), validating the obtained results.

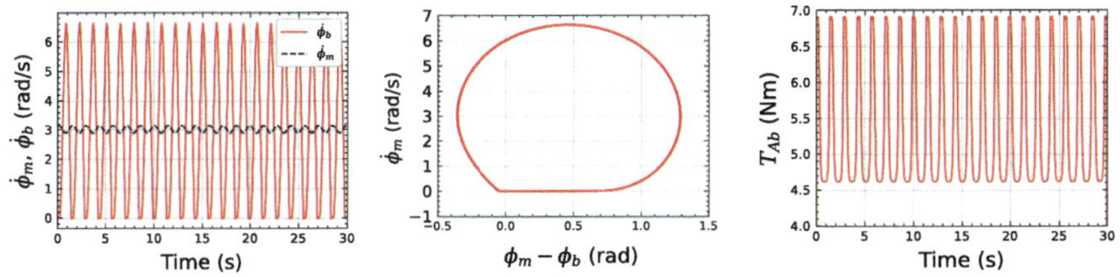

(a) Angular velocity at top of string and drill for constant torque

(b) Trajectory in phase subspace for constant motor torque

(c) Torque on drill for constant motor torque.

Fig. 13.3 Drill string dynamics in stick-slip regime

13.5.2 Ideal Controller

At this point, the behavior of the system under the effect of the ideal controller is analyzed. It is assumed that the control torque at the top of the string is given for a switching term gain of zero, i.e., $K = 0$. For an initial analysis, the behavior under the same conditions as in the previous section is studied, with a motor torque of 39.57 Nm and a desired angular velocity of 3.1 rad/s. However, after the initial twenty seconds, the motor torque at the top of the string is defined by the equivalent control law described in Eq. 13.18, with $\lambda = 0.8$. The value of the torque at the top of the string over time is illustrated in Fig. 13.4a. At the beginning of the controller's operation, a significant oscillation with large amplitude can be observed. As time progresses, the oscillation decays significantly until it converges to a constant torque value slightly higher than the initial torque value.

When the control torque value converges to an approximate value of the initial torque at the top of the string, the system converges to the desired angular velocity of 3.1 rad/s. This can be observed by analyzing Fig. 13.4b, where initially, when subjected to a constant torque of 39.57 Nm at the top of the string, the system operates under stick-slip regime. However, when the sliding mode controller starts to act, both the angular velocity at the top of the string and the angular velocity of the drill bit converge to the desired angular velocity value of 3.1 rad/s.

By analyzing Fig. 13.5, where the controller is activated only during the time intervals of 30–60 and 130–160 s, it is possible to verify its effectiveness in eliminating stick-slip vibrations. When the controller is activated, the angular velocity of the system converges to the desired value, while during the moments when the torque at the top of the string is given by a constant value, the system returns to the stick-slip regime, highlighting the significant effect of the controller on the torsional dynamics of the string.

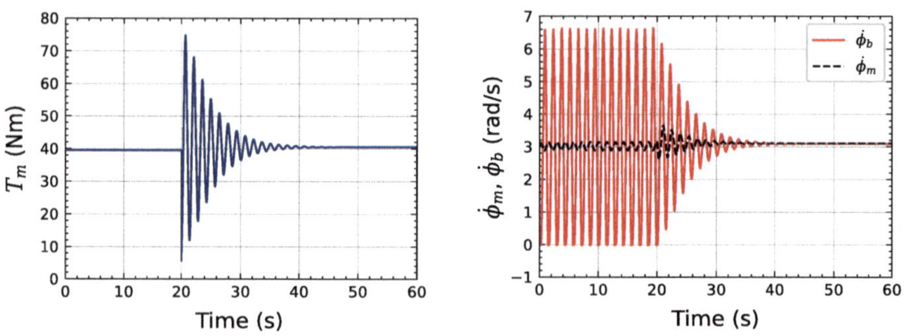

(a) Torque at top of column with equivalent controller for $\lambda = 0.8$.

(b) Angular velocity at the top of the column and at the drill under the effect of the equivalent controller for $\lambda = 0.8$.

Fig. 13.4 Equivalent controller for $\lambda = 0.8$

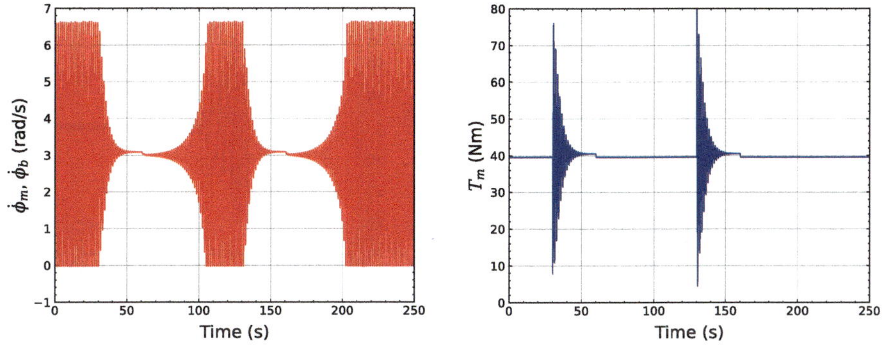

(a) Angular velocity at the top oh the column with equivalent controller at different time intervals to $\Omega_d = 3.1 rad/s$

(b) Torque at top of column with equivalent controller at different time intervals to to $\Omega_d = 3.1 rad/s$

Fig. 13.5 Controller effect when activated at different time intervals to to $\Omega_d = 3.1$rad/s

13.5.3 Controller with Switching Term

As seen in the previous section, the torque controller obtained from the control law, with $K = 0$, proved to be capable of eliminating stick-slip in the system. However, for real cases, the influence of uncertainties in parameter calculation and the influence of unknown phenomena can negatively affect the behavior of the system and the torque controller, which, in turn, may result in the system not converging to the desired operating condition. Therefore, the addition of the switching term in the controller, for real-world implementation, becomes necessary for the system to behave as desired.

Figure 13.6 illustrates the temporal behavior of the system's angular velocity and the control torque when considering the switching term with a unity gain. It can be observed that the angular velocity at the top of the string and the drill bit exhibit behavior identical to that shown in Fig. 13.4b for the same conditions. However, when comparing the control torque with that shown in Fig. 13.4a, it can be seen that for the case analyzed in this moment, the phenomenon of chattering appears in its behavior, characterized by high-frequency oscillations after the convergence of the system's angular velocities to the desired state.

The amplitude of chattering is directly related to the intensity of the gain of the switching term in the control law. The higher the gain value, the larger the amplitude of this phenomenon. It is worth noting that for the gain values analyzed in this image, all of them were able to bring the system to the desired operating conditions. However, the improper selection of the gain can lead to premature wear of the components of the actual controller.

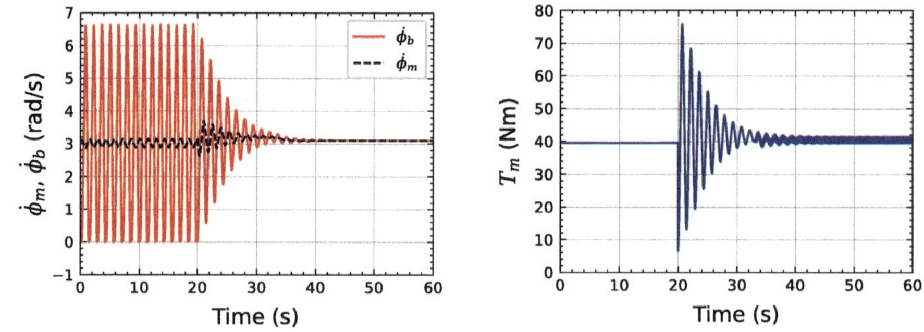

(a) Torque at top of column with equivalent controller for $K = 1$.

(b) Angular velocity at top of the column and at the drill under the effect of the equivalent controller for $K = 1$.

Fig. 13.6 Control considering switching term for $K = 1$

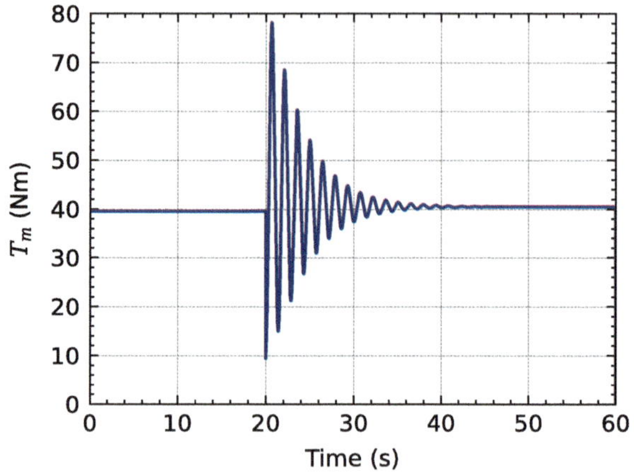

Fig. 13.7 Control torque considering saturation function with $\Phi = 10^{-4}$

If we consider the switching term, according to $T_g = -k \text{sat}\left(\frac{s}{\Phi}\right)$, the chattering phenomenon is completely eliminated for the analyzed gain values, granting the controller with the switching term a behavior similar to the ideal controller, as shown in Fig. 13.7.

13.6 Conclusions

In this work, the development and functionality verification of a sliding mode torque controller for the elimination of stick-slip vibrations in drilling strings have been studied. Based on the presented results, it was possible to verify that the mathematical model satisfactorily reproduced the angular behavior of the system under stick-slip regime, while the sliding mode torque controller proved to be fully capable of eliminating such vibrations in drilling strings, yielding results consistent with those found in the literature. It was observed that the consideration of non-zero gain values for the control maintains the ability to drive the system to the desired state. However, the phenomenon of excessive switching or chattering arises in the behavior of the controller, which can lead to premature wear of control device components. Finally, it was found that considering a saturation function in the switching term ensures that the controller exhibits a behavior similar to the ideal controller, mitigating the chattering phenomenon.

References

Bailey J, Finnie I (1960) An analytical study of drill-string vibration

Bayliss MT, Panchal N, Whidborne JF (2012) Rotary steerable directional drilling stick/slip mitigation control. IFAC Proc Vol 45(8):66–71

Besaisow AA, Payne ML (1988) A study of excitation mechanisms and resonances inducing bottomhole-assembly vibrations. SPE Drill Eng 3(01):93–101

Christoforou A, Yigit A (1996) Impact of composite structures—the momentum balance method. J Comp Mater 30(10):1068–1087

Cunningham R (1968) Analysis of downhole measurements of drill string forces and motions

Deily F, Dareing D, Paff G, Ortloff J, Lynn R (1968) Downhole measurements of drill string forces and motions

Halsey G, Kyllingstad A, Kylling A (1988) Torque feedback used to cure slip-stick motion. In: SPE annual technical conference and exhibition. OnePetro

Javanmardi K, Gaspard D (1992) Application of soft-torque rotary table in mobile bay. In: IADC/SPE drilling conference. OnePetro

Kamel JM, Yigit AS (2014) Modeling and analysis of stick-slip and bit- bounce in oil well drillstrings equipped with drag bits. J Sound Vibr 333(25):6885–6899

Kriesels P, Keultjes W, Dumont P, Huneidi I, Owoeye O, Hartmann R (1999) Cost savings through an integrated approach to drillstring vibration control. In: SPE/IADC middle east drilling technology conference. OnePetro

Leine R, Van Campen D, De Kraker A, Van Den Steen L (1998) Stick-slip vibrations induced by alternate friction models. Nonlinear Dyn 16:41–54

Lin YQ, Wang YH (1990) New mechanism in drillstring vibration. In: Offshore technology conference. OnePetro

Liu Y (2015) Suppressing stick-slip oscillations in underactuated multibody drill-strings with parametric uncertainties using sliding-mode control. IET Control Theory Appl 9(1):91–102

Monteiro HL, Trindade MA (2017) Performance analysis of proportional-integral feedback control for the reduction of stick-slip-induced torsional vibrations in oil well drillstrings. J Sound Vibr 398:28–38

Navarro-López EM, Cortés D (2007a) Avoiding harmful oscillations in a drillstring through dynamical analysis. J Sound Vibr 307(1–2):152–171

Navarro-López EM, Cortés D (2007b) Sliding-mode control of a multi-dof oilwell drillstring with stick-slip oscillations. In: 2007 American control conference. IEEE, pp 3837–3842

Navarro-López EM, Suárez R (2004) Practical approach to modelling and controlling stick-slip oscillations in oilwell drillstrings. In: Proceedings of the 2004 IEEE international conference on control applications, 2004, vol 2. IEEE, pp 1454–1460

Navarro-López EM, Licéaga-Castro E (2009) Non-desired transitions and sliding-mode control of a multi-dof mechanical system with stick-slip oscillations. Chaos, Solitons Fractals 41(4):2035–2044

Slotine JJE, Li W et al (1991) Applied nonlinear control, vol 199. Prentice hall Englewood Cliffs, NJ

Spanos PD, Chevallier A, Politis N, Payne M (2003) Oil and gas well drilling: a vibrations perspective. In: The shock and vibration digest, vol 35, no 2, pp 85–103

Vaziri V, Kapitaniak M, Wiercigroch M (2018) Suppression of drill-string stick–slip vibration by sliding mode control: Numerical and experimental studies. Euro J Appl Math 29(5):805–825

Vromen T, Dai CH, van de Wouw N, Oomen T, Astrid P, Nijmeijer H (2015) Robust output-feedback control to eliminate stick-slip oscillations in drill-string systems. IFAC-papersonline 48(6):266–271

Yigit AS, Christoforou AP (2006) Stick-slip and bit-bounce interaction in oil-well drillstrings. Trans ASME 128:268–274

Chapter 14
Experimental Methods for Identifying Physical Parameters of Four-Wheeled Autonomous Vehicles

João Bezerra, Felipe W. Varga, Ely de Paiva, Niederauer Mastelari, Rafael A. Cordeiro, and Mauro F. Koyama

Abstract Four-Wheeled Autonomous Vehicles have emerged as an innovative technology with significant potential to revolutionize the transportation industry. In the past decade, research efforts in this field aim to develop strategies ranging from the development of Advanced Driver Assistance Systems (ADAS) to the creation of fully autonomous navigation strategies. In this context, mathematical models of the four-wheeled robots are an essential part of developing autonomous navigation techniques. Whether using complete or simplified mathematical models of the vehicle, the control synthesis requires accurate model parameters. This work presents experimental methods to obtain some of the main physical parameters that determine the dynamic model of four-wheeled vehicles, including its mass, center of gravity, and principal moments of inertia. The uncertainties of these measurements are presented aiming to estimate the reliability and precision of the results obtained. Furthermore, a cornering stiffness estimation method based on the vehicle lateral dynamics is proposed. Experimental results obtained from a four-wheeled 1:5 scaled electrical vehicle, with real electronic differential distribution are presented.

Keywords Dynamical models · Vehicle dynamics · Cornering stiffness · Experimental methods · Parameters determination

14.1 Introduction

In recent years, there has been significant research attention dedicated to autonomous vehicle driving technology, both in academy as well as in industry. Many companies and research laboratories have been looking for improvements in performance, safety,

---

J. Bezerra (✉) · F. W. Varga · E. de Paiva · N. Mastelari · M. F. Koyama
School of Mechanical Engineering, University of Campinas (UNICAMP), Campinas, Brazil
e-mail: joaovictorapb@gmail.com

R. A. Cordeiro
Electrical Engineering Department, Federal University of Espirito Santo (UFES), Vitória, Brazil

P. Kurka and M. Pereira (eds.), *Advances in Structural Vibration*, Lecture Notes in Mechanical Engineering, https://doi.org/10.1007/978-3-031-71540-2_14

and comfort of self-driving cars (Lutin 2018). The use of scaled-robotic vehicles is a way to demonstrate and facilitate the investigations in autonomous vehicle control and sensing systems. Researchers often resort to using car-like robots that shorten the path of building, testing, validating, and scaling algorithms for large cars (Goldfain et al. 2019; Polley and Alleyne 2004).

This is the case proposed in the VERDE (Robotic Electric Vehicle with Electronic Differential) Project (Nogueira et al. 2019). The VERDE prototype is a four-wheeled 1:5-scaled vehicle, which has an electronic differential for optimal distribution of torque and/or spin speed, aiming at autonomous driving applications in aggressive environments, like in agricultural and mining applications, under slippery and highly uneven terrains (Bonacini et al. 2023; Lemos et al. 2017; Ribeiro et al. 2022). The VERDE vehicle, shown in Fig. 14.1, is a fully instrumented platform with inertial navigation system INS/GPS, on-board computer, CAN bus, long-distance laser scan (30 m), camera, and sensors for measuring wheel speed and steering.

Most developments in autonomous navigation techniques require an accurate knowledge of the vehicle model (Cordeiro 2013). Providing accurate physical parameters is crucial for vehicle models to ensure reliable, high-performance, and safe driving.

Numerous studies have been conducted in searching for techniques to accurately estimate vehicle parameters in order to enhance their reliability and performance (Cordeiro et al. 2014; Ribeiro et al. 2021).

This work presents experimental methods to identify fundamental vehicle parameters such as vehicle mass, center of gravity (CG), and principal moments of inertia in the x, y, and z directions, using standard weighing scales and pendulums. The measurements of these parameters and their respective uncertainties were obtained by applying the uncertainty analysis presented by Taylor (1997), to evaluate the quality of measurements and methods used. We also propose a cornering stiffness estimation method based on the lateral dynamics of the vehicle and an experimental dataset, taking into account the different characteristics of the front and rear tires.

Fig. 14.1 VERDE robotic platform

Autonomous vehicles in aggressive environments, such as the VERDE prototype, have been sought in a variety of economic activities, for example, in the agricultural sector, there is a demand for automated harvesters; in mining and construction camps, autonomous vehicles are desired for transporting heavy cargo; in aerospace exploration of planets.

The work proposed herein addresses this demand head-on by providing practical solutions for allowing navigation in challenging terrains and environments. In the current landscape of engineering and industry, the demand for advancements in autonomous vehicle technology is palpable. By applying experimental methods to identify fundamental vehicle parameters and proposing estimation techniques, we streamline parameter estimation using standard sensors, eliminating the need for costly equipment. This approach significantly enhances accessibility and affordability, making our methods readily applicable in real-world scenarios. Through these efforts, this work contributes to advancing the field of autonomous vehicle technology, making it more accessible and effective for real-world applications.

This Chapter is organized as follows: Sect. 14.2 presents the methods and techniques employed in this work. Section 14.3 shows the experimental setups and results and Sect. 14.4 presents the final conclusions.

14.2 Approach for Parameter Identification: Methods and Techniques

In order to obtain the main parameters required for the dynamic modelling of a terrestrial robotic vehicle, we describe the techniques and methods for identifying these parameters, which are essential for analyzing the dynamics of vehicles.

This section presents the methodology employed for obtaining the main parameters required for the dynamic modelling of a terrestrial robotic vehicle and describes the techniques and methods for identifying these parameters, which are essential for analyzing vehicle dynamics.

14.2.1 Total Mass

The main parameter for modelling vehicle dynamics is its total mass. The total mass of the vehicle can be obtained directly through the use of scales. To carry out the measurements, we place four scales under each wheel over a flat surface. The static equilibrium analysis of the free-body diagram shown in Fig. 14.2 leads to:

$$\frac{F_z^{\mathrm{FL}}}{g} + \frac{F_z^{\mathrm{FR}}}{g} + \frac{F_z^{\mathrm{RL}}}{g} + \frac{F_z^{\mathrm{RR}}}{g} = m \tag{14.1}$$

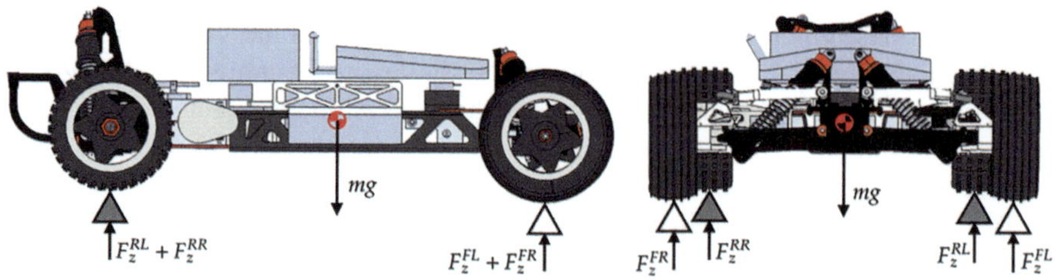

Fig. 14.2 Free-body diagram of the vehicle

The indices FL, FR, RL, and RR correspond to the front-left, front-right, rear-left, and rear-right corners, respectively, and g is the gravitational acceleration.

14.2.2 Center of Gravity

The position of the Center of Gravity (CG) is required for modelling the vehicle dynamics. Figure 14.3 shows the parameters that provide the position of the vehicle's center of gravity.

The dimensions a, b, c, d, c', and d' are defined from fixed points of the vehicle body, and h is defined with respect to the lower end of the vehicle.

To determine these parameters, it is necessary to measure the normal forces at each of the vehicle extremities in two situations: with the vehicle on a horizontal plane, to ensure a uniform load distribution, and with the vehicle on an inclined plane (Jazar 2008). In the case of the horizontal plane, the static equilibrium analysis in the x-axis rotation results in:

$$(F_z^{\mathrm{FL}} + F_z^{\mathrm{FR}})a = (F_z^{\mathrm{RL}} + F_z^{\mathrm{RR}})b \tag{14.2}$$

This leads us to a scenario with one equation and two unknowns. The second equation can be derived by considering that the parameter $(a + b)$ represents the distance between the front and rear wheel axles of the vehicle, which is a known value.

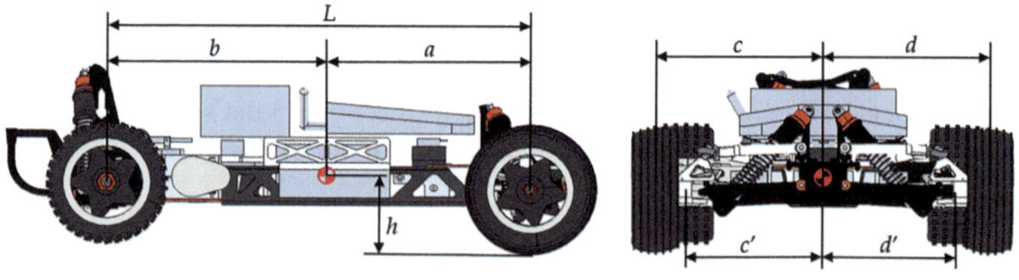

Fig. 14.3 Location of center of gravity

Similarly, c and d dimensions can be determined by the normal forces measurements and the static equilibrium analysis in the y-axis rotation. Since the rear track width is smaller than the front track width, the torques generated by the rear wheels are functions of c' and d'. Therefore, the static analysis provides:

$$F_z^{\mathrm{FR}} c + F_z^{\mathrm{RR}} c' = F_z^{\mathrm{RL}} d' + F_z^{\mathrm{FL}} d \tag{14.3}$$

The remaining equations to solve for c can be derived by the sum $(c + d)$ and $(c' + d')$, which are equal to the front and rear track distances, respectively, and the linear relation between the distances given by $c - c' = d - d'$. To obtain the height h of the CG, we must measure the support forces of the vehicle on an inclined plane, at a known angle theta with respect to the ground plane. By balancing the sum of torques in the y-axis direction on point O (center of front wheel), see Fig. 14.4, the height h can be obtained from Eq. 14.4.

$$h = \frac{-(F_z^{\mathrm{RL}} + F_z^{\mathrm{RR}}) L \cos\theta + \mathrm{mg}(a \cos\theta + r \sin\theta)}{\mathrm{mg} \sin\theta} \tag{14.4}$$

According to Uys (2006) the accuracy of the value of h increases as the slope increases and then obtains results that suggest that the angle θ must be above 10°.

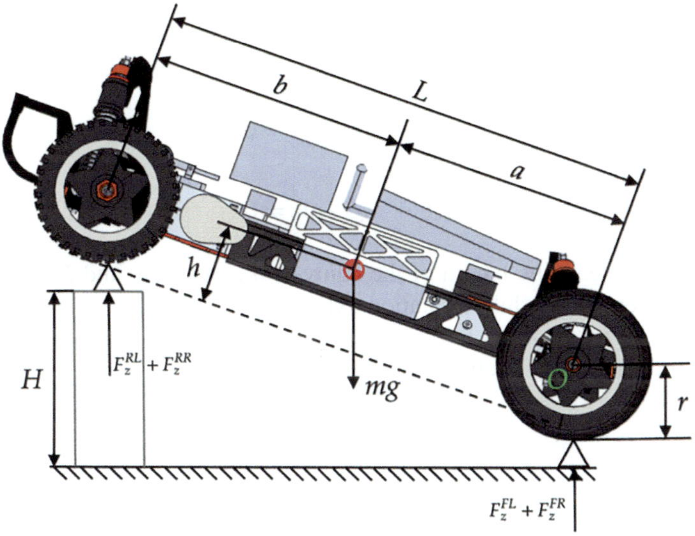

Fig. 14.4 Forces diagram of inclined vehicle experiment

14.2.3 Main Moments of Inertia

The moments of inertia play a crucial role in vehicle dynamics as they are directly linked to angular motions. The principal moments of inertia capture the mass distribution around the principal axes, with the origin located at the center of gravity. The moment of inertia provides valuable insights into the vehicular dynamics.

In (Schedlinski and Link 2001) several methods for measuring moments of inertia are described, classifying them as: (a) methods without movement limits, (b) methods with low angular movement, and (c) methods with very low angular movement.

A widely used method for measuring moments of inertia involves pendulums (Dunmore 1961; Schedlinski and Link 2001). This approach, although simple and accurate, requires performing six separate tests to obtain the parameters of the inertia tensor.

By assuming a symmetric mass distribution of the vehicle across the x, y, and z axes, the coordinate system aligns with the principal axes of rotation. Consequently, the cross-parameters can be disregarded, resulting in a reduction to three required tests (Cordeiro et al. 2014).

The first method applied is the simple pendulum, commonly employed to determine the moments of inertia around the x and y axes. This method allows for experimentation with the vehicle in a horizontal position, ensuring enhanced safety during the procedure. The measurement using a simple pendulum relies on the direct application of the Newton–Euler motion equation. As shown in Fig. 14.5, the body is suspended by strings, and a controlled rotation around the axis (in this case, x or y) is induced. Consequently, for small angles:

$$I_x'\ddot{\emptyset} + \mathrm{mg}L_{\mathrm{CG}}\emptyset = 0 \tag{14.5}$$

$$I_y'\ddot{\theta} + \mathrm{mg}L_{\mathrm{CG}}\theta = 0 \tag{14.6}$$

Here, L' represents the length of the equivalent pendulum, and L_{CG} denotes the distance between the pendulum rotation axis and the CG of the vehicle, specifically given by $L_{\mathrm{CG}} = L' - h$. It is important to note that this analysis assumes an ideal pendulum, devoid of any friction or air resistance, thus:

$$\emptyset = \sin(2\pi f_{\emptyset} t) \tag{14.7}$$

$$\theta = \sin\sin(2\pi f_{\theta} t) \tag{14.8}$$

Let $f_{\emptyset}$ and f_{θ} denote the oscillation frequencies relative to the x and y axes, respectively. By performing a double time derivative on Eqs. 14.7 and 14.8 with respect to time and replacing them into Eqs. 14.5 and 14.6, respectively, we derive the following expressions:

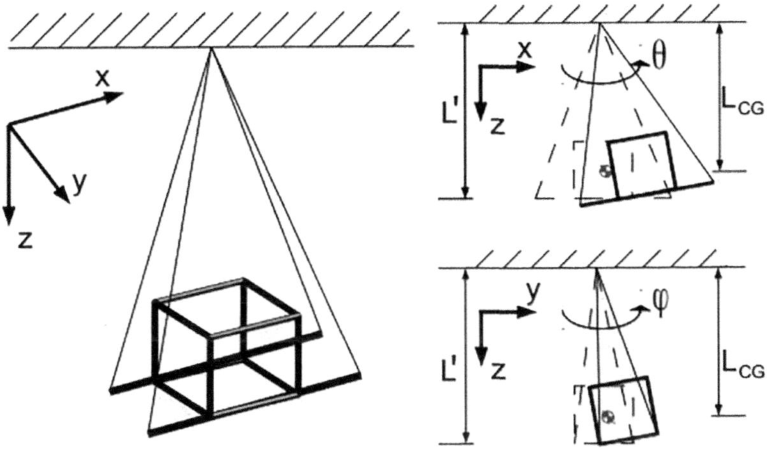

Fig. 14.5 Simple gravitational pendulum

$$I'_x = \frac{mgL_{CG}}{(2\pi f_{\emptyset})^2} \tag{14.9}$$

$$I'_y = \frac{mgL_{CG}}{(2\pi f_{\theta})^2} \tag{14.10}$$

The inertia values calculated by Eqs. 14.9 and 14.10 are determined with the axes passing through the point of attachment of the pendulum strings. To obtain the moment of inertia about the CG, it is imperative to employ the Huygens-Steiner theorem (Gobbi et al. 2011), commonly referred to as the Parallel Axis Theorem. Therefore, the moments of inertia with respect to the x and y axes are determined by the following equations:

$$I_x = \frac{mgL_{CG}}{(2\pi f_{\emptyset})^2} - mL_{CG}^2 \tag{14.11}$$

$$I_y = \frac{mgL_{CG}}{(2\pi f_{\theta})^2} - mL_{CG}^2 \tag{14.12}$$

For the determination of I_z, the use of torsional pendulums is a simpler and safer alternative compared to the simple pendulum method (Dunmore 1961; Schedlinski and Link 2001). In the implementation of the torsional pendulum, two or more wires are employed to support the vehicle, and a torsional force is applied to the pendulum. As a result of the rotation, the vehicle elevates, converting the kinetic energy of motion into potential energy (Nestorides 1958). Figure 14.6 illustrates a representation of the torsional pendulum.

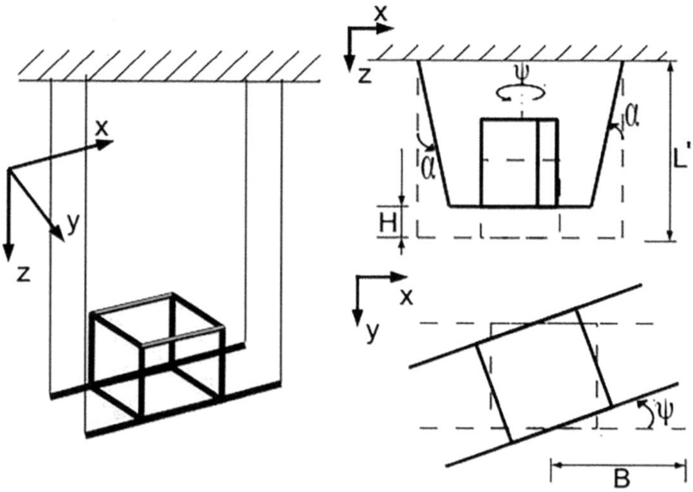

Fig. 14.6 Multifilar torsional pendulum

Considering the depicted configuration in Fig. 14.6, where the cables are arranged orthogonally to the vehicle frame, an analysis of the potential energy and geometry allows for the determination of the inertia along the z axis. By employing this approach, the equation for the I_z, can be expressed by Eq. 14.13.

$$I_z = \frac{\mathrm{mg}B^2}{L'(2\pi f_\psi)^2} \tag{14.13}$$

where B is the radius achieved by the torsional pendulum and f_ψ is the frequency of oscillation around the z axis.

14.2.4 Tire Cornering Stiffness

Cornering stiffness plays a crucial role in the lateral vehicle dynamics model as it directly influences the vehicle behaviour and stability. The accurate value of cornering stiffness in the model is vital for predicting and analyzing the vehicle's lateral response.

In order to estimate the cornering stiffness, we employed a vehicle bicycle model as the basis, due to it provides a description of the lateral and yaw dynamics of the vehicle (Snider 2009). A representation of the bicycle model is presented in Fig. 14.7.

Considering the forces in the free-body diagram, and applying Newton's second law, the mathematical model is achieved (Rajamani 2011).

$$\mathrm{m}a_y = F_{y,R} + F_{y,R}\cos\delta \tag{14.14}$$

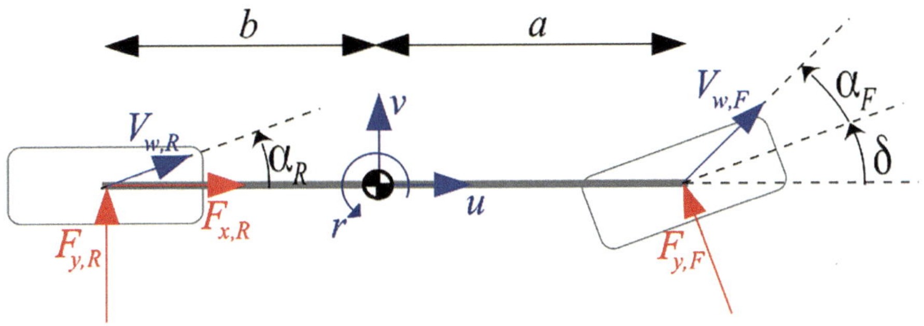

Fig. 14.7. 2D-yaw bicycle model

where a_y is the lateral acceleration, u is the longitudinal speed, r is the angular velocity around the z axis, and δ is the front-wheel steering angle. For the front wheel the longitudinal slip will be considered negligible, i.e., $F_{x,\,F} \approx 0$.

The lateral force is modelled as linearly dependent on the slip angle in the corresponding wheel (considering low speeds and good grip):

$$F_{y,W} = -2C_{\alpha,W}\alpha^{W} \tag{14.15}$$

where the sub-index W can be R (rear wheel) or F (front wheel), and α^{W} is the angle between the wheel velocity and the longitudinal axis of the vehicle, defined as follows:

$$\alpha^{F} = \text{arctan arctan}\left(\frac{v + \text{ar}}{u}\right) - \delta \tag{14.16}$$

$$\alpha^{R} = \text{arctan arctan}\left(\frac{v + \text{br}}{u}\right) \tag{14.17}$$

By substituting Eqs. 14.15, 14.16, and 14.17 into Eq. 14.14 and linearizing, we obtain (Sierra et al. 2006):

$$\text{ma}_y = \left(\frac{v + \text{ar}}{u} - \delta\right)C_{\alpha,F} + \tag{14.18}$$

where v is the lateral velocity of the vehicle. The signals a_y, v, u, and r can be obtained (directly or indirectly) by sensors, thus transforming the problem into a system of one equation and two unknowns. To address this problem and find a solution, we use a cost function J, which quantifies the discrepancies between the estimated values $\hat{y}$ and the corresponding measured value y_{real}. The cost function J is defined as the sum of squared errors, given by:

$$J\left(C_{\alpha,W}\right) = \frac{1}{2n}\sum_{i=1}^{n}(\hat{y}^{(i)} - y_{\text{real}}^{(i)})^2 \tag{14.19}$$

where n is the total number of data points acquired. By minimizing this cost function, we aim to optimize the parameters $C_{\alpha,W}$ and find the values that yield the closest match between the estimated and measured values.

To achieve the minimization of the cost function, we employ a gradient descent algorithm (Géron 2022), which involves iteratively adjusting the parameter values based on the gradient of the cost function J. To update these parameters, we compute the partial derivatives of with respect to $C_{\alpha,F}$; and $C_{\alpha,R}$:

$$\frac{\partial J}{\partial C_{\alpha,F}} = \frac{1}{n}\sum_{i=1}^{n}\left(\hat{y}^{(i)} - y_{\text{real}}^{(i)}\right)C_{\alpha,F} \tag{14.20}$$

$$\frac{\partial J}{\partial C_{\alpha,R}} = \frac{1}{n}\sum_{i=1}^{n}\left(\hat{y}^{(i)} - y_{\text{real}}^{(i)}\right)C_{\alpha,R} \tag{14.21}$$

At each iteration, we calculate the gradients, in Eqs. 14.20 and 14.21, using the current parameter values. Then, the parameter values are updated as follows:

$$C_{\alpha,F}(K+1) = C_{\alpha,F}(K) - \lambda\frac{\partial J}{\partial C_{\alpha,F}(K)} \tag{14.22}$$

$$C_{\alpha,R}(K+1) = C_{\alpha,R}(K) - \lambda\frac{\partial J}{\partial C_{\alpha,R}(K)} \tag{14.23}$$

Here, λ is the learning rate, which determines the step size in the parameter update. By iteratively performing these updates, we gradually move towards the optimal values of $C_{\alpha,F}$ and $C_{\alpha,R}$ that minimize the cost function J.

14.3 Performed Tests and Experimental Results

14.3.1 Mass and Center of Gravity

To measure the vehicle mass, four standard scales with a range of 0–10 kg and accuracy of ± 1 g were used. These scales were initially calibrated with an OHAUS precision scale model TP4000D with an accuracy of ± 0.01 g. These scales were placed under the vehicle tires and positioned on a flat surface. Five direct measurements were conducted to obtain the weight distribution between the scales and the associated uncertainty of each mean value of mass was calculated, with 95% of confidence (Figliola and Beasley 2021). In the experiments carried out, it is considered that the errors follow a normal distribution.

Using the measured mass values from each scale and applying them in Eq. 14.1, it was possible to determine the total mass of the vehicle and its uncertainty, by propagating the mass values uncertainties with the partial derivatives of Eq. 14.1 (Taylor 1997). Since the mass measurements were taken on a level surface, resulting in a uniform load distribution, these values can be applied in Eq. 14.2 to obtain a and b, representing the distances from the CG to the front and rear of the vehicle, respectively. Figure 14.8 shows the configuration used to obtain the mass and parameters a and b. The parameters c and d, which correspond to the position of the CG along the y axis, can be determined using the same experimental setup.

To determine the height h of the CG, eleven measurements were taken with the rear of the vehicle raised at an angle of $\Theta = 17\ ^{\circ}$. The measured results were used in Eq. 14.3, and then the average value of the height h was calculated. The obtained results for the total mass and CG position of the vehicle and their propagated uncertainties, calculated with the partial derivatives method (Figliola and Beasley 2021), are presented in Table 14.1.

The obtained values for the vehicle mass, a, b, c and d showed uncertainties values of less than 1%. The value of the height of the center of gravity had a reasonable

(a) Experiment on a level surface

(b) Inclined plane experiment

Fig. 14.8 Experiments for obtaining mass and CG

Table 14.1 Experimental results for the total mass and CG position with 95% of confidence

Dimension description	Variable	Values
Mass of the car	m	18.9032 ± 0.0416 kg
Distance from CG to the center of front wheels along the x axis	a	328.9 ± 1.2 mm
Distance from the center of rear wheels to the CG along the x axis	b	248.1 ± 1.3 mm
Distance from the middle of right wheels to the CG along the y axis	c	228.9 ± 0.8 mm
Distance from CG to the middle of left wheels along the y axis	d	Table. 227.1 ± 1.0 mm
Height of the CG	h	149.3 ± 7.3 mm

uncertainty of 5%, although higher than the other distances because of its Eq. 14.3, which involves more parameters. To determine the CG height of the vehicle we must consider multiple variables, such as the load distribution on each tire, the overall mass of the car, the angle of inclination observed during testing, and the previously calculated CG coordinate a. Therefore, the result of h has more associated errors than the other CG distances.

14.3.2 Moments of Inertia

As observed in Eqs. 14.11, 14.12, and 14.13, the moments of inertia around the x, y, and z axes are dependent on their respective oscillation periods. Therefore, four pendulum tests for each moment of inertia were conducted, and an XSENS inertial measurement unit (IMU) was used to generate the oscillation signals, with which the mean periods of oscillation were obtained for each experiment. The IMU was positioned near the vehicle CG to accurately measure the angular displacement. The experiment setups for the moments of inertia are shown in Fig. 14.9. Figure 14.10 shows the roll, pitch, and yaw angle signals obtained from the pendulum experiments described in Sect. 2.3.

(a) Single pendulum (b) Torsional pendulum

Fig. 14.9 Setup of pendulum experiments

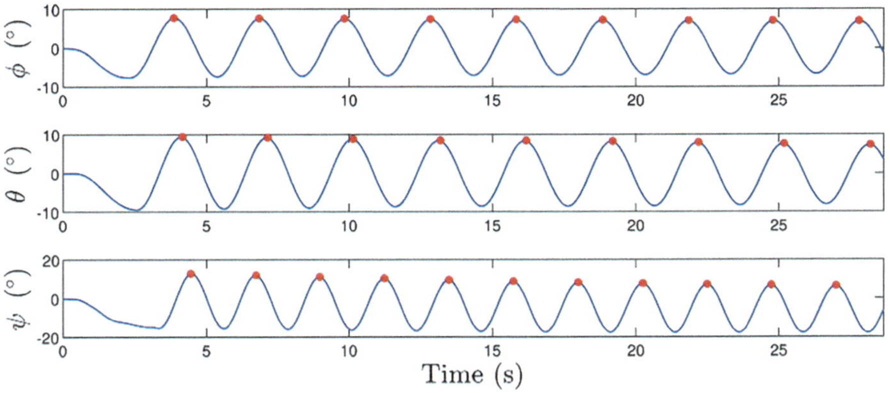

Fig. 14.10 Angular displacements of pendulum experiments

Observably, the amplitude of the pendulum oscillations decreases over time. However, the period of oscillation remains constant. Given that the time difference between successive peaks provides us with the oscillation period, the peaks of the oscillations were isolated and marked as red dots in Fig. 14.10. The first ten periods of time were averaged, and the frequencies obtained were then used in Eqs. 14.11, 14.12, and 14.13. For calculating the associated uncertainties of the time periods, where multiple experiments were done, the Maximum Likelihood Algorithm was used (Figliola and Beasley 2021). The resulting values and their associated uncertainties are presented in Table 14.2.

The measurements of I_z showed uncertainty less than 0.5% while the I_x and I_y moments of inertia uncertainties were much greater, this can be explained by the fact that the I_z Eq. 14.13 has less parameters, that is, fewer sources of error than the I_x and I_y equations, Eqs. 14.11 and 14.12 respectively, that contains the squared height

Table 14.2 Results for the I_x, I_y and I_z pendulum experiments, with 95% of confidence

Parameter description	Variable	Values
Average time period of the oscillation on the I_x experiment	$T_\emptyset$	2.9949 ± 0.0057 s
Principal moment of Inertia along the x axis	I_x	0.9659 ± 0.5430 kg m^2
Average time period of the oscillation on the I_y experiment	T_θ	3.0050 ± 0.0057 s
Principal moment of Inertia along the y axis	I_y	1.5913 ± 0.5409 kg m^2
Average time period of the oscillation on the I_z experiment	T_ψ	2.2492 ± 0.0012 s
Principal moment of inertia along the z axis	I_z	1.1713 ± 0.0045 kg m^2

of the CG in their equations, which by its own, has greater associated uncertainty than the other CG distances.

14.3.3 Cornering Stiffness

The cornering stiffness problem, as described in Sect. 2.4, exhibits an underdeterminate nature that necessitates persistent excitation for accurate results (Sierra et al. 2006). To achieve this, a varying steering signal was applied to the vehicle using a chirp function as input (Ribeiro et al. 2021). The amplitude of the signal was adjusted within the limits of the vehicle's steering system (up to 0.52 rad), and an example of the input signal is shown in Fig. 14.11.

Two classes of sensors are used to collect the data: the E2 incremental optical enconder, responsible for acquiring the values of the longitudinal velocity (u), and the Xsens MTI-G-710 IMU unit, which collects lateral and longitudinal acceleration data and the angular velocity r, which is directly measured by the IMU's gyroscopes.

The tests were performed on wet asphalt, the longitudinal speed was set to 1 m/s. Lateral velocity was obtained through an Extended Kalman Filter (EKF). Figure 14.12 shows the signal acquired during the experiment and the values estimated by the EKF. The lateral velocity obtained through the EKF is shown in the Fig. 14.13.

Solving the optimization problem presented in Sect. 2.3 with the data acquired, the optimum estimated parameters obtained are $C_{\alpha F} = 26.5588$ N/rad and $C_{\alpha,R} =$

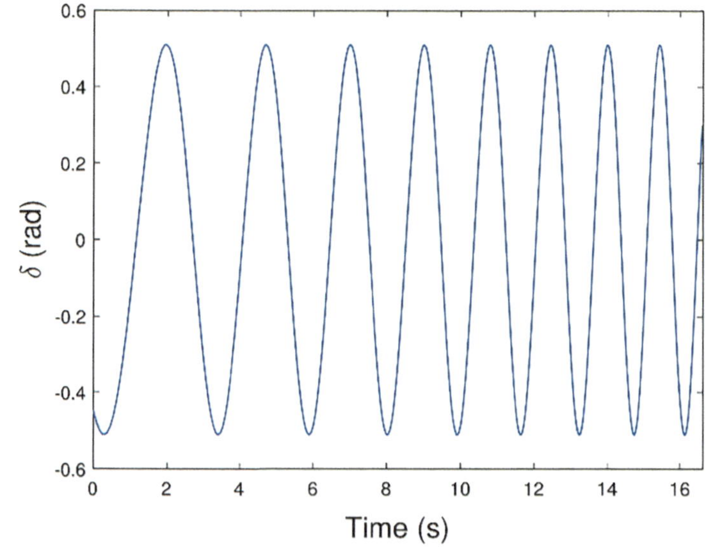

Fig. 14.11 Input signal

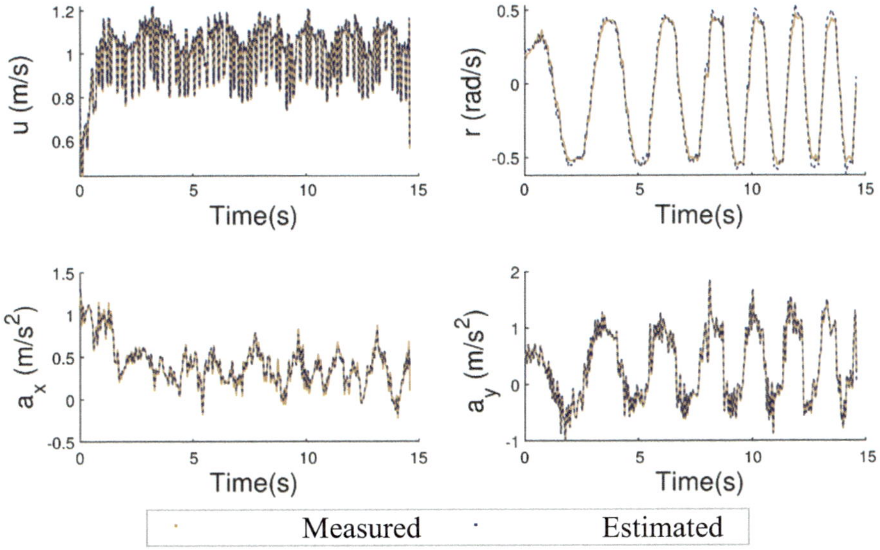

Fig. 14.12 Comparison of measured and EKF-estimated values for a longitudinal velocity of 1 m/s

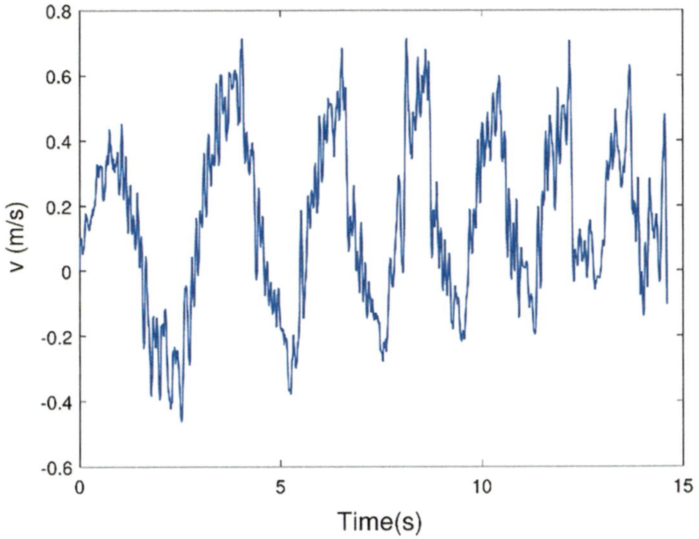

Fig. 14.13 Lateral velocity

34.2981 N/rad. The values obtained through the optimization technique demonstrate a good agreement with the result found in the EKF estimations. For the filter implementation, it was assumed that both rear and front tires had equal cornering stiffness values, and the obtained equivalent tire stiffness was approximately 30 N/rad. Finally, the root mean square error (RMSE) was calculated, Equation, resulting in an RMSE of 1.5929.

$$\mathrm{RMSE} = \sqrt{\sum_{i=1}^{n} \left(\hat{y}^i + y^i_{\mathrm{real}}\right)^2} \tag{14.24}$$

14.4 Conclusions

This work outlines methods employed on a scaled electric vehicle in order to identify fundamental mechanical parameters of vehicle dynamics. Experimental setups were proposed, using standard weighing scales and pendulums, to obtain accurate measurements of the vehicle mass, center of gravity (CG), and principal moments of inertia.

A 2D bicycle dynamic model was used to estimate the cornering stiffness of the front and rear vehicle wheels. These methods were able to identify key model parameters that can greatly impact controller design and simulator development. Furthermore, the proposed methods have the additional benefit of having a low-cost computational implementation, using regular sensors. The obtained results are consistent with prior knowledge of the system and align with factors such as vehicle structure and geometry.

The results of this paper corroborate and extend previous studies of the authors for larger vehicles (Cordeiro et al. 2014), showing that the proposed strategy is scalable for different vehicles. Therefore, the knowledge gained through scaled vehicles can be extrapolated for full-scale vehicles, indicating that design and optimization strategies can be first experimented with scaled vehicles, contributing for enhancing performance and efficiency in developing new commercial vehicle strategies.

In today's rapidly evolving engineering landscape, the need for practical solutions to advance vehicle dynamics is more pressing than ever. By employing those methods on scaled vehicles, this study not only contributes to the fundamental understanding of vehicle dynamics but also holds significant implications for real-world applications. The methodologies outlined herein offer a cost-effective approach to parameter identification, utilizing readily available equipment and computational techniques. Such accessibility is crucial in driving forward the integration of autonomous technologies into various industries, from agriculture to transportation. Moreover, the scalability of these methods, as demonstrated through corroboration with previous research on larger vehicles, underscores their potential for widespread adoption across diverse vehicle platforms. As such, this work not only enhances our understanding of vehicle dynamics but also serves as a catalyst for innovation and progress in engineering technology and industry.

Acknowledgements The authors thank the funding support from INCT-SAC Project through agencies Fapesp (2014/50851-0) and CNPq (465755/2014-3).

References

Bonacini L, Tronco ML, Higuti VAH, Velasquez AEB, Gasparino MV, Peres HEN, Oliveira RPd, Medeiros VS, Silva RPd, Becker M (2023) Selection of a navigation strategy according to agricultural scenarios and sensor data integrity. Agronomy 13(3)

Cordeiro RA (2013) Modelagem e Controle de Trajetória de um Veículo Robótico Terrestre de Exterior. Master's thesis

Cordeiro RA, Bueno SS, Azinheira JR, de Paiva EC, Pablo S, Vivan R, Azevedo H, Koyama MF (2014) Determinação experimental de parâmetros para a modelagem dinâmica de um veículo robótico terrestre. In XX Congresso Brasileiro de

Dunmore J (1961). Handbook on torsional vibrations. British Internal Combustion Engine Research Association. Cambridge University Press, London (1958), 664 pp. illustrated. 110s. Aeronaut J 65(601):71–71

Figliola RS, Beasley D (2021) Theory and design for mechanical measurements. John Wiley & Sons, s.l.

Géron A (2022) Hands-on machine learning with Scikit-Learn, Keras, and TensorFlow. O'Reilly Media Inc., s.l.

Gobbi M, Mastinu G, Previati G (2011) A method for measuring the inertia properties of rigid bodies. Mech Syst Signal Process 25(1):305–318

Goldfain B, Drews P, You C, Barulic M, Velev O, Tsiotras P, Rehg JM (2019) Autorally: an open platform for aggressive autonomous driving. IEEE Contr Syst Mag 39(1):26–55

Jazar RN (2008) Vehicle dynamics. Springer, s.l.

Lemos RA, Sobral G, Mirisola L, Martins R, Koyama M, de Paiva E (2017) Estratégia de navegaçao autônoma entre fileiras de cultivares baseada em lasers. In XIII Simpósio Brasileiro de Automaçao Inteligente, vol 1

Lutin JM (2018) Not if, but when: autonomous driving and the future of transit. J Public Transp 21(1):92–103

Nestorides EJ (ed) (1958) A handbook on torsional vibration. Cambridge University Press

Nogueira LADO, Koyama MF, Cordeiro RDA, Ribeiro AM, Bueno SS, De Paiva EC (2019) A miniaturized four-wheel robotic vehicle for autonomous driving research in off-road scenarios. In Congresso Brasileiro de Automática-CBA, vol 1

Polley M, Alleyne AG (2004) Dimensionless analysis of tire characteristics for vehicle dynamics studies. In: Proceedings of the 2004 American control conference, vol 4. IEEE, pp 3411–3416

Rajamani R (2011) Vehicle dynamics and control. Springer Science & Business Media

Ribeiro A, Koyama M, Moutinho A, de Paiva E, Fioravanti A (2021) A comprehensive experimental validation of a scaled car-like vehicle: lateral dynamics identification, stability analysis, and control application. Control Eng Pract 116:104924

Ribeiro AM, Fioravanti AR, Moutinho A, de Paiva EC (2022). Nonlinear state-feedback design for vehicle lateral control using sum-of-squares programming. Veh Syst Dyn 60(3):743–769

Schedlinski C, Link M (2001) A survey of current inertia parameter identification methods. Mech Syst Signal Process 15(1):189–211

Sierra C, Tseng E, Jain A, Peng H (2006) Cornering stiffness estimation based on vehicle lateral dynamics. Veh Syst Dyn 44(sup1):24–38

Snider JM (2009) Automatic steering methods for autonomous automobile path tracking. *R*obotics Institute, Pittsburgh, PA, Tech. Rep. CMU-RITR-09-08

Taylor J (1997) Introduction to error analysis, the study of uncertainties in physical measurements. s.n., s.l.

Uys P, Els P, Thoresson M, Voigt K, Combrinck W (2006) Experimental determination of moments of inertia for an off-road vehicle in a regular engineering laboratory. Int J Mech Eng Educ 34(4):291–314

Chapter 15
Dynamical Analysis of a Helical Gear Pair with Backlash Under Variable Load Torque Conditions

Laís Bitencourt Visnadi and Aline Souza de Paula

Abstract Wind turbine gearboxes are subjected to fluctuating torque conditions, which may generate vibration problems. This paper analyzes the mesh displacement of a helical gear pair, inspired by the challenges posed in simulating the high-speed stage of the wind turbine gearbox. The detailed mesh model, considering static transmission error, time-varying mesh stiffness, and backlash, associated with a 12-degree-of-freedom gear pair model better represents this system. The model also takes into account a load torque factor to prevent over-speed conditions. Results show a comparison between the system's response in the absence of backlash and when backlash is introduced. When there is no backlash, both the driving and driven gears exhibit quasi-periodic behavior. When there is a backlash, single-sided impacts occur along the permanent response of the gear system, a severe vibration condition that may cause damage to the system.

Keywords Wind turbine · Helical gear pair · Transmission error · Gear backlash · Gear dynamics

15.1 Introduction

The aim to reduce gas emissions that cause climate change has raised interest in clean energy production. In this context, research about wind turbines has been increased, regarding both efficiency and reliability. The gearbox is an essential component of the wind turbine since it increases the rotor speed to the required generator speed, and its failure is a significant concern in the design and operation of these systems (Ragheb and Ragheb 2010).

In most cases, the gearbox consists of one or two planetary gear stages and one or two parallel shaft stages (Teng et al. 2014). Geared systems present internal excitation generated by tooth contact. This contact is often modeled by time-varying stiffness

L. B. Visnadi (✉) · A. S. de Paula
Department of Mechanical Engineering, University of Brasília, Brasília, Brazil
e-mail: lais.visnadi@unb.br

P. Kurka and M. Pereira (eds.), *Advances in Structural Vibration*, Lecture Notes in Mechanical Engineering, https://doi.org/10.1007/978-3-031-71540-2_15

due to gear tooth geometry and the number of pairs in contact at the same time (Yang and Lin 1987). Additionally, models also take into account assembling and manufacturing variability as a static transmission error. Imperfect gear assembly is further considered by incorporating backlash into the model (Özgüven 1991). Various authors, such as Wei et al. (2015) and Wang et al. (2014), have addressed these nonlinearities through torsional models. As high loads on the shafts may cause gear center displacements and the gear mesh helix angle couples all the directions of motion, a 12-degree of freedom model, as proposed by Kahraman (1994), may better represent this system dynamics.

Fluctuations in wind patterns result in varying torque levels on the rotor. To achieve maximum efficiency and to guarantee the generator operates within its safe limits, avoiding any over-speed conditions, speed control is essential (Apata and Oyedokun 2020). Large torques and torque variations can cause contact loss on the gearbox, generating vibration problems (Zhao and Ji 2015).

The present work proposes an analysis of the behavior of a gear pair, subjected to torque variations, simulating the high-speed parallel stage. In the context of wind turbines, this shaft is connected to the electric generator. The mesh model considers all the characteristics mentioned and a 12 DOF lumped parameter model for the gear pair, including gyroscopic effects is used. The model encompasses backlash phenomena and incorporates a load torque consideration to mitigate the risk of over-speed conditions.

15.2 Methodology

A lumped parameter model of a helical gear pair was considered to simulate the system. Figure 15.1 shows gear 1 and gear 2 translational coordinates (y_1, x_1, z_1, y_2, x_2, and z_2), gear mesh representation (the line composed by $2b$, $e(t)$ and $k_m(t)$), helix angle (β), position angle (Ψ), driving torque ($T_1(t)$) and load torque ($T_2(t)$).

Gears' supports may not be perfectly rigid, so both gear centers can translate and also rotate in all directions (i.e., there are six more degrees of freedom in this gear pair, which are θ_{y1}, θ_{x1}, θ_{z1}, θ_{y2}, θ_{xg2}, and θ_{z2}) and then gyroscopic effects may occur. The system's equations of motion, considering gravity forces, are obtained by Lagrange's Equation (Kubur et al. 2004):

$$
\begin{aligned}
&m_1\ddot{y}_1 + k_m(t)h\big[p(t)\big]\cos(\beta)\sin(\Psi) + C_{s1}\dot{y}_1 + K_{s1}y_1 = 0,\\
&m_1\ddot{z}_1 + k_m(t)h\big[p(t)\big]\cos(\beta)\cos(\Psi) + C_{s1}\dot{z}_1 + K_{s1}z_1 = -m_1 g,\\
&m_1\ddot{x}_1 + k_m(t)h\big[p(t)\big]\cos(\beta)\cos(\Psi) + C_{s1}\dot{x}_1 + K_{s1}x_1 = 0,\\
&I_1\ddot{\theta}_{y1} + r_{b1}k_m(t)h\big[p(t)\big]\sin(\beta)\sin(\Psi) + \dot{\theta}_{x1}J_1\dot{\theta}_{z1} + C_{\theta s1}\dot{\theta}_{y1} + K_{\theta s1}\theta_{y1} = 0,\\
&I_1\ddot{\theta}_{z1} + r_{b1}k_m(t)h\big[p(t)\big]\sin(\beta)\cos(\Psi) - \dot{\theta}_{x1}J_1\dot{\theta}_{y1} + C_{\theta s1}\dot{\theta}_{z1} + K_{\theta s1}\theta_{z1} = 0,\\
&J_1\ddot{\theta}_{x1} + r_{b1}k_m(t)h\big[p(t)\big]\cos(\beta) = T_1(t),
\end{aligned}
$$

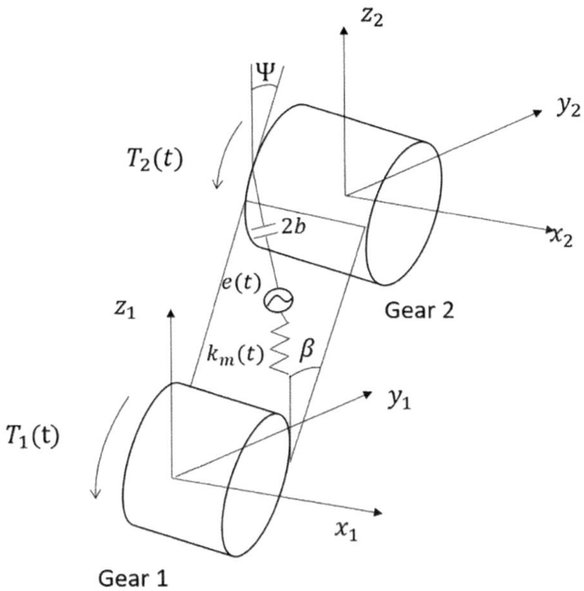

Fig. 15.1 Helical gear pair scheme. Adapted from Kubur et al. (2004)

$$
\begin{aligned}
&m_2\ddot{y}_2 - k_m(t)h\big[p(t)\big]\cos(\beta)\sin(\Psi) + C_{s2}\dot{y}_2 + K_{s2}y_2 = 0,\\
&m_2\ddot{z}_2 - k_m(t)h\big[p(t)\big]\cos(\beta)\cos(\Psi) + C_{s2}\dot{z}_2 + K_{s2}z_2 = -m_2g,\\
&m_2\ddot{x}_2 - k_m(t)h\big[p(t)\big]\cos(\beta)\cos(\Psi) + C_{s1}\dot{x}_2 + K_{s2}x_2 = 0,\\
&I_2\ddot{\theta}_{y2} + r_{b2}k_m(t)h\big[p(t)\big]\sin(\beta)\sin(\Psi) + \dot{\theta}_{x2}J_2\dot{\theta}_{z2} + C_{\theta s2}\dot{\theta}_{y2} + K_{\theta s2}\theta_{y2} = 0,\\
&I_2\ddot{\theta}_{z2} + r_{b2}k_m(t)h\big[p(t)\big]\sin(\beta)\cos(\Psi) - \dot{\theta}_{x2}J_2\dot{\theta}_{y2} + C_{\theta s2}\dot{\theta}_{z2} + K_{\theta s2}\theta_{z2} = 0,\\
&J_2\ddot{\theta}_{x2} + r_{b2}k_m(t)h\big[p(t)\big]\cos(\beta) = T_2(t),
\end{aligned}
\tag{15.1}
$$

where m indicates mass, C_s indicates support damping, K_s indicates support stiffness, g is the acceleration of gravity, I indicates moment of inertia, J indicates polar moment of inertia, $k_m(t)$ corresponds to mesh stiffness, r_b is the base radius, $h[p(t)]$ is the backlash function, and $p(t)$ is teeth deformation along the line of action (LOA). Subscripts 1 and 2 indicate gear 1 and gear 2 properties, respectively. Subscripts y, z, and x indicate the property direction. Since the gears' support flexibilities are considered in this paper, teeth deformation is given by:

$$
\begin{aligned}
p(t) = &\big[y_1\sin(\Psi) - y_2\sin(\Psi) + z_1\cos(\Psi)\\
&-z_2\cos(\Psi) + r_{b1}\theta_{x1} + r_{b2}\theta_{x2}\big]\cos(\beta)\\
&+\big[-x_1 + x_2 + r_{b1}\theta_{y1}\sin(\Psi) + r_{b2}\theta_{y2}\sin(\Psi)\\
&+r_{b1}\theta_{z1}\cos(\Psi) + r_{b2}\theta_{z2}\cos(\Psi)\big]\sin(\beta) - e(t),
\end{aligned}
\tag{15.2}
$$

where $e(t)$ is the static transmission error, considered here a composition of a mean value, e_m, and a floating value, e_r, which depends on the gear 1 angular position:

$$e(t) = e_m + e_r \sin(\theta_{x1}) \quad (15.3)$$

Since gear pair may be designed or assembled considering an initial gear tooth distance, called backlash, $2b$, a loss of teeth contact may occur. This condition is described by the backlash function, and is an important source of nonlinear internal excitation of the system Walha et al. (2006):

$$h[p(t)] = \begin{cases} p(t) - b, & \text{if } \ p(t) > b \\ 0, & \text{if } \ -b \le p(t) \le b \\ p(t) + b, & \text{if } \ p(t) < -b \end{cases} \quad (15.4)$$

Another internal excitation on geared systems is caused by the mesh stiffness, $k_m(t)$. This stiffness is time-varying due to gear tooth geometry and contact ratio (the number of pairs in mesh).

An analytical method to calculate the spur gear tooth stiffness is the potential energy method. This method considers that the tooth is a variable cross-section beam and the contact force is applied along the contact line at each time-instant Yang and Lin (1987). Then, tooth deformation potential energy yields stiffness calculation.

The total potential energy of a tooth subjected to an acting external force is the sum of its bending (b), axial compressive (a) (Yang and Lin 1987), and shear (s) potential energies (Tian 2004). Considering the contact of two surfaces, which is the case of a gear pair in mesh, there is also Hertzian energy, h_z (Yang and Lin 1987). The same idea can be applied to the helical gear tooth, considering it as a series of staggered spur gears (slices), taking into account the helix angle involved, as proposed by Wan et al. (2015). Helix angle also causes a variation in the contact line lengths, as shown in Fig. 15.2

Therefore, the stiffness of a helical tooth (k_{ht}) is:

Fig. 15.2 Helical gear tooth

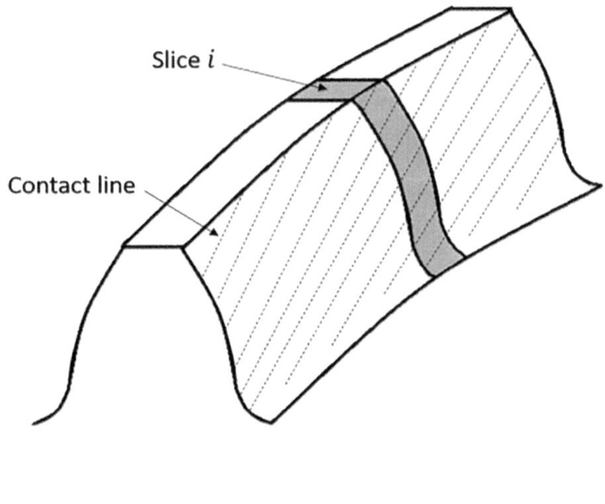

$$k_{\mathrm{ht}} = \left(\sum_{i=1}^{N} \frac{1}{k_{\mathrm{bi}}} \Delta y + \sum_{i=1}^{N} \frac{1}{k_{\mathrm{ai}}} \Delta y + \sum_{i=1}^{N} \frac{1}{k_{\mathrm{si}}} \Delta y \right)^{-1} \tag{15.5}$$

$$\Delta y = l_{\mathrm{ef}}/N, \tag{15.6}$$

where N is the number of spur gear teeth slices that were considered, k_b is the bending stiffness, k_a is the axial compressive stiffness, k_s is the shear stiffness and l_{ef} is the effective contact line length. In this paper, bending, axial, and shear stiffness of the spur gear tooth slice was calculated considering the improved method proposed by Ma et al. (2014). Details about contact length calculation are provided by Wan et al. (2015).

Tooth pair stiffness is calculated by a series association of pinion and gear stiffness and Hertzian stiffness. Total mesh stiffness, $k_m(t)$ is the sum of the stiffness of pairs in contact at the same time-instant.

15.3 Results

To simulate the gear pair that generates the high-speed motion for the wind turbine electric generator, a steel helical gear pair is considered (Teng et al. 2014). The helix angle is 15°, the gears' module is 7 mm and the face width is 100 mm. Mean (e_m) and floating (e_r) static transmission errors are 10 and 0.0001 μm, respectively. The driving gear (gear 1) has 54 teeth and an initial speed of 25 Hz, and the driven gear (gear 2) has 21 teeth, also based on Teng et al. (2014) data.

The wind turbine drive torque ($T_1(t)$) is considered as a sinusoidal function with a frequency matching the natural frequency of gear 1 torsional vibration mode (ω_n), in order to assess the system's response under critical conditions that could potentially lead to damage:

$$T_1(t) = T_m + T_f(1 + \sin(\omega_n t - \pi/2)), \tag{15.7}$$

where T_m is the medium torque on gear 1, T_f is the floating torque value and t is the time. In this work, $T_m = 50$ Nm, $T_f = 2.5$ Nm and $\omega_n = 1065$ [rad/s]. Additionally, an initial load torque of $T_2 = 19.44$ Nm (equilibrium condition at $t = 0$ s) is taken into account to prevent over-speed conditions. Moreover, it is established that this torque would experience a 10% increase if the driven gear rotating speed exceeds 1% of its initial value, which was set to the speed limit.

To solve the system of equations of motion, the Newmark method with middle point rule was used (Walha et al. 2006). Two conditions were simulated: without backlash ($2b = 0$) and with backlash, which was considered to be 2 μm. The time step was initially set at 1×10^{-5} s, but it was decreased by half during the transitions from contact to non-contact conditions and vice versa in the scenario that accounted for backlash. Results are shown and discussed in the following subsections.

15.3.1 Without Backlash

Figure 15.3 shows the driving torque (a), the load torque (b), the driven gear rotational speed (c), and the line of action (LOA) displacement (d), i.e., $h[p(t)]$, along the time interval considered, which in this case spanned 100 s.

Analyzing the LOA displacement (Fig. 15.3d), it is possible to notice a transient motion due to the simulation's initial conditions. Result shows that the driven gear rotational speed achieves its maximum at approximately 1 s and then the torque load is increased. After that, the driven gear speed oscillates around this maximum value, and the load torque presents variations to prevent the driven gear speed from exceeding the limit. As the main interest in this paper is the load torque variation effects, the subsequent analysis will focus on the results when the system stabilizes into a steady-state behavior.

Figure 15.4a shows five complete oscillations of the load torque ($T_2(t)$) in a steady-state regime. The same time interval is considered to show the driven gear rotational speed (θ_{x2}), as shown in Fig. 15.4b, and the LOA displacement ($h[p(t)]$), Fig. 15.4c. These figures show that in order to maintain the rotational speed of the driven gear within the specified limit, there are instances where the load torque is temporarily increased. When this occurs, the driven gear rotational speed decreases (in terms of absolute values), and then the load torque returns to its initial value. As a result, the driven gear rotational speed starts to increase again, leading to another increase in the load torque. As a consequence, the LOA displacement also shows an oscillating behavior.

To better understand the dynamical behavior of driving and driven gear, their horizontal (y), vertical (z), and rotation around x axis (θ_x) motions will be analyzed by their phase portrait and Poincaré map. The Poincaré map reference period is the

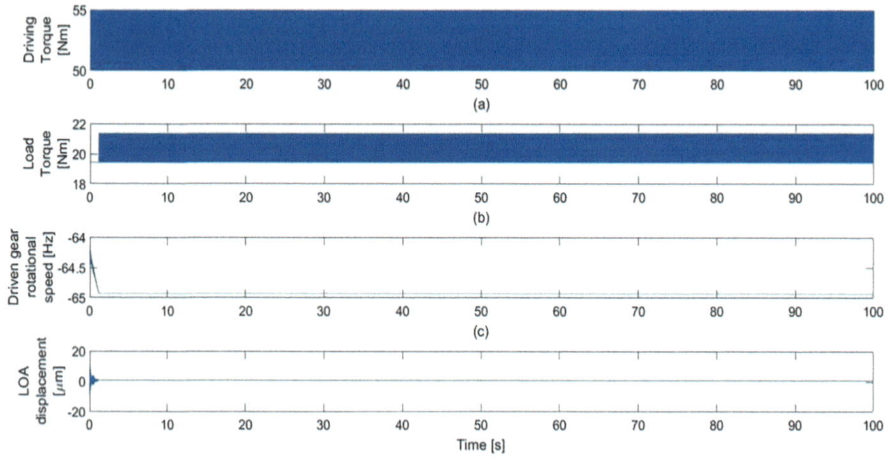

Fig. 15.3 Simulation results of **a** Driving torque; **b** Load torque; **c** Driven gear rotational speed; **d** LOA displacement for the case considering no backlash

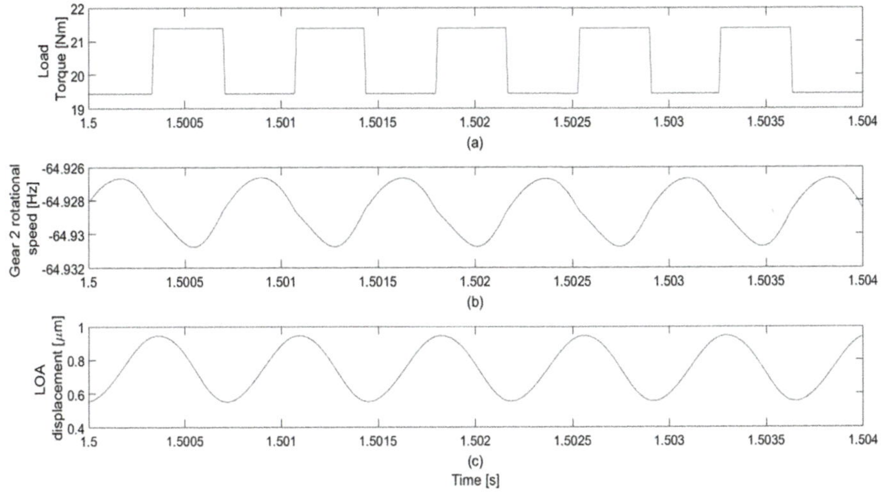

Fig. 15.4 Load torque oscillation (**a**), Driven gear rotational speed oscillation (**b**), and LOA displacement oscillation (**c**) for the case considering no backlash

same as the driving torque, $T_1(t)$, period. Axial motion (x_1 and x_2) and rotational motion around y and z axis (θ_{y1}, θ_{z1}, θ_{y2}, and θ_{z2}) are not analyzed in this paper, since simulation results showed that they can be neglected when compared to the motion in other directions.

Figure 15.5 displays the phase portrait and the Poincaré map of the driving gear's horizontal motion. The phase portrait shows displacement variations between approximately -1.08 and -0.88 μm and speed variations from -1.3×10^{-4} to 1.3×10^{-4} m/s. The light blue line shows the phase portrait of 0.1 s of motion (17 torque cycles). The Poincaré map forms a cloud of points around a closed curve, indicating a quasi-period motion.

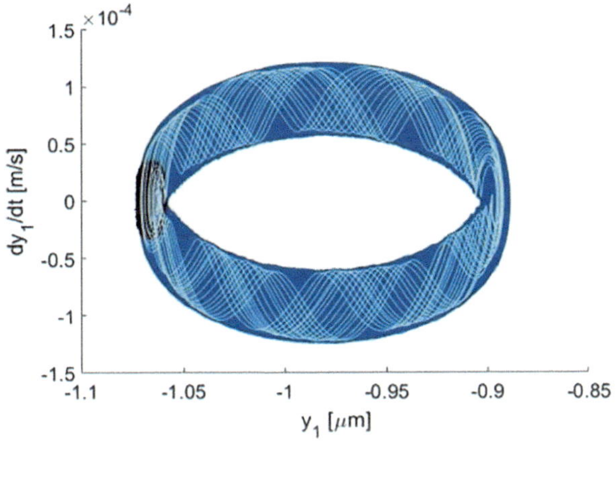

Fig. 15.5 Driving gear horizontal motion for the case considering no backlash. Dark blue: entire time interval phase portrait; Light blue: 0.1 s phase portrait; Black: Poincaré map

A similar result is observed in Fig. 15.6 for the driving gear vertical motion. Displacement variations are between approximately -11.6 and -11.1 μm and speed variations from -3.5×10^{-4} to 3.5×10^{-4} m/s. The phase portrait of the driving gear rotational motion (Fig. 15.7) is different from the translational ones: it is possible to see a relation between the angular displacement and the angular speed. The Poincaré map (Fig. 15.7), presents dots along the entire phase portrait, also indicating a quasi-periodic motion.

Driven gear phase portrait of horizontal (Fig. 15.8) and vertical motion (Fig. 15.9) present smaller displacements and velocities amplitudes when compared to the driving gear (Fig. 15.5 and 15.6, respectively). The phase portrait of the driven gear rotational motion (Fig. 15.10) displays a restrained trajectory and the Poincaré map presents dots along the entire phase portrait, also indicating a quasi-periodic motion.

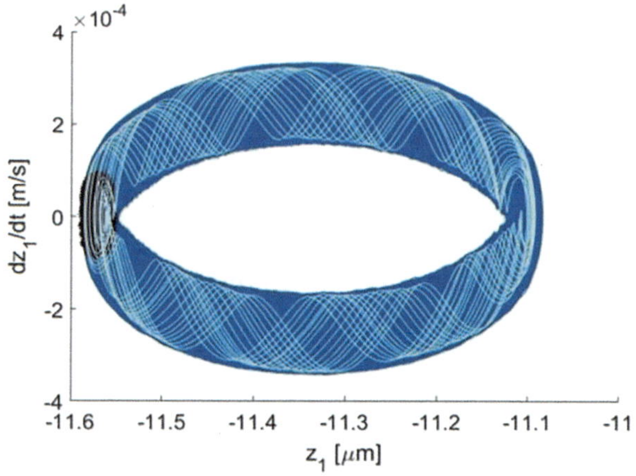

Fig. 15.6 Driving gear vertical motion for the case considering no backlash. Dark blue: entire time interval phase portrait; Light blue: 0.1 s phase portrait; Black: Poincaré map

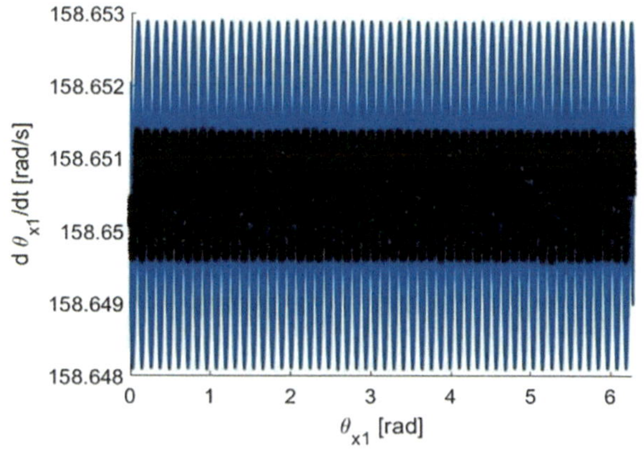

Fig. 15.7 Driving gear rotational motion for the case considering no backlash. Dark blue: entire time interval phase portrait; Black: Poincaré map

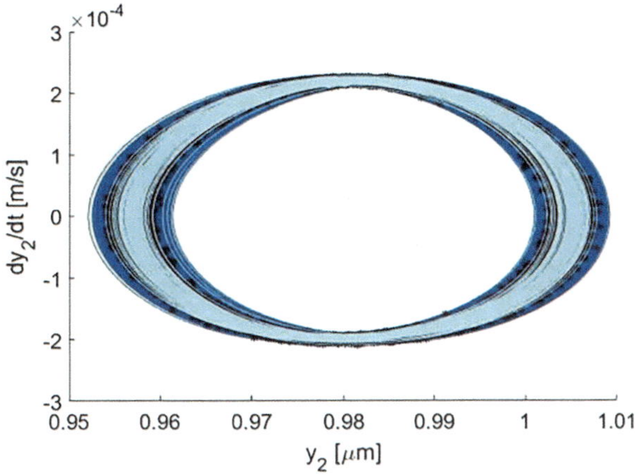

Fig. 15.8 Driven gear horizontal motion for the case considering no backlash. Dark blue: entire time interval phase portrait; Light blue: 0.1 s phase portrait; Black: Poincaré map

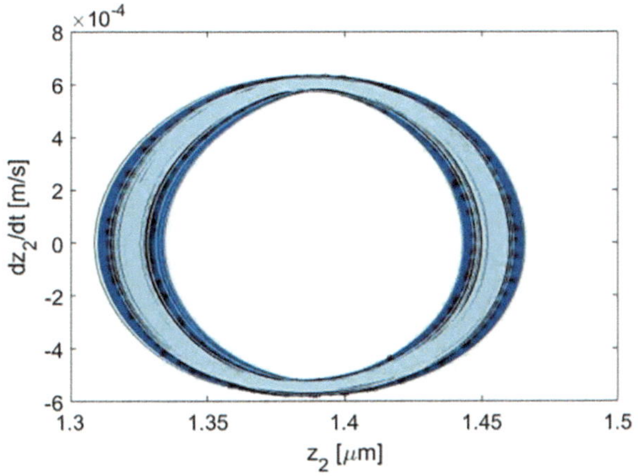

Fig. 15.9 Driven gear vertical motion for the case considering no backlash. Dark blue: entire time interval phase portrait; Light blue: 0.1 s phase portrait; Black: Poincaré map

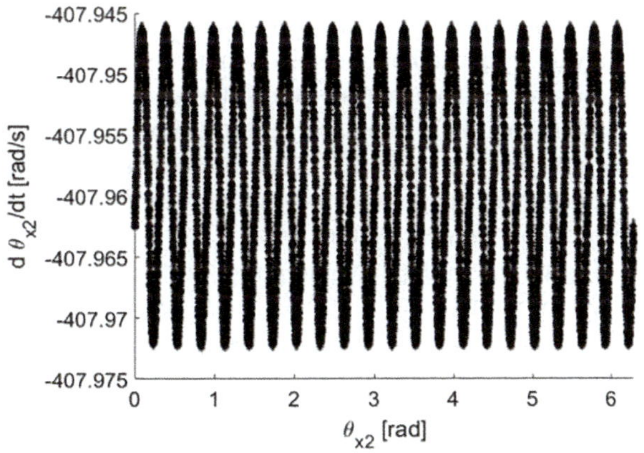

Fig. 15.10 Driven gear rotational motion for the case considering no backlash. Dark blue: entire time interval phase portrait; Black: Poincaré map

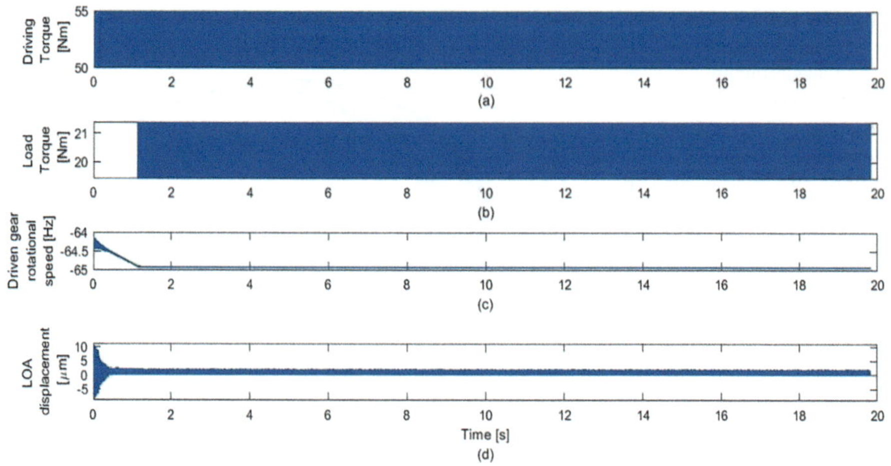

Fig. 15.11 Simulation results of (**a**) Driving torque; (**b**) Load torque; (**c**) Driven gear rotational speed; (**d**) LOA displacement for the case considering backlash

15.3.2 With Backlash

At this point, backlash is taken into consideration. Figure 15.11 shows the driving torque (a), the load torque (b), the driven gear rotational speed (c), and the line of action (LOA) displacement (d), i.e., $h[p(t)]$, along the time interval considered, which is this case spanned 19.8 s.

LOA displacement, shown in Fig. 15.11d, presents a transient motion within the first 0.4 s that occurs due to the simulation's initial conditions. The driven gear rotational speed achieves its maximum at approximately 1 s and then the torque load is increased. As a consequence of the load torque variation, both the load torque, Fig. 15.11b, and driven gear rotational speed, Fig. 15.11c, exhibit oscillatory behavior, similar to what is observed in the no backlash condition. The subsequent analysis is focused on the steady-state regime.

Figure 15.12 displays five oscillations of load torque (a), driven gear rotational speed (b) and LOA displacement (c), similarly to Fig. 15.4. In the presence of backlash, the difference is that LOA displacement exhibits not only regions of positive displacement but also periods with no contact ($h[p(t)] = 0$), indicating single-sided impacts on the gear system.

Figure 15.13 displays the phase portrait (dark blue line), 0.1 s system response phase portrait (light blue line), and the Poincaré map (black dots) of the driving gear's horizontal motion. The phase portrait shows displacement variations between approximately -1.1 and -0.88 μm and speed variations from -3×10^{-4} to 3×10^{-4} m/s. The velocity variation is more than twice as pronounced when compared to the case without backlash, and the phase space is entirely filled with trajectories. The Poincaré map corresponds to a circle in a limited region of the phase portrait, indicating quasi-periodic motion. A similar result is shown in Fig. 15.14, for the

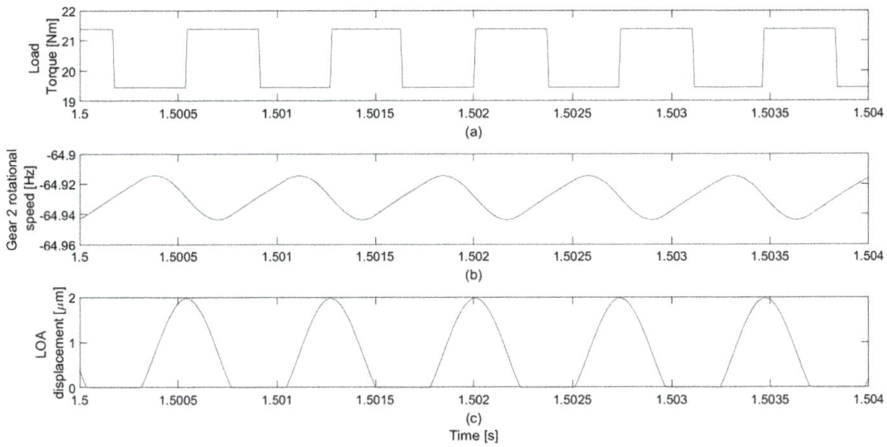

Fig. 15.12 Load torque oscillation (**a**), driven gear rotational speed oscillation (**b**), and LOA displacement oscillation (**c**)

driving gear vertical motion, with the phase portrait entirely filled with trajectories (dark blue line) and a circle restrained in a specific region in the Poincaré map. The phase portrait of the driving gear rotational motion and the Poincaré map, Fig. 15.15, are similar to the no-backlash condition, shown in Fig. 15.7.

Driven gear motion in this case is similar to the no-backlash condition. Horizontal and vertical motion phase portrait and Poincaré map (Figs. 15.16 and 15.17, respectively) exhibit an orbit, indicating quasi-periodic motion. The difference is that there are fewer variations, both in displacement and speed. Rotational motion (Fig. 15.18) also exhibits a restrained trajectory.

Fig. 15.13 Driving gear horizontal motion for the case considering backlash. Dark blue: entire time interval phase portrait; Light blue: 0.1 s phase portrait; Black: Poincaré map

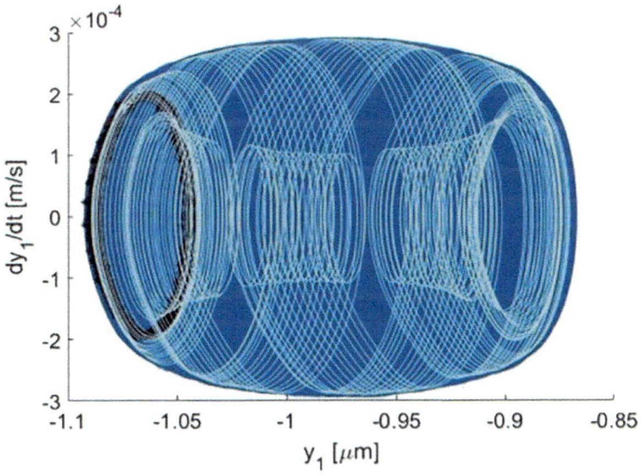

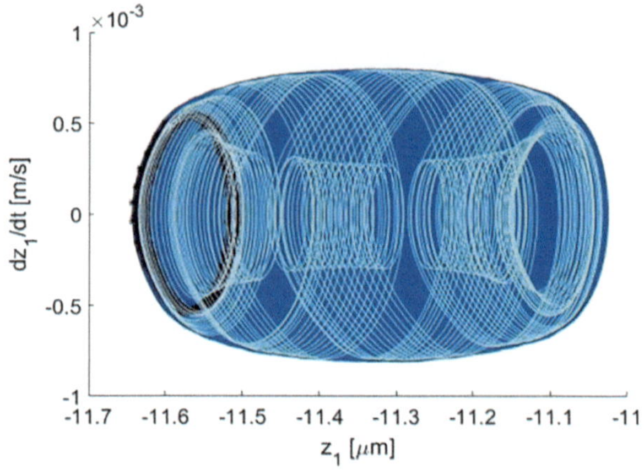

Fig. 15.14 Driving gear vertical motion for the case considering backlash. Dark blue: entire time interval phase portrait; Light blue: 0.1 s phase portrait; Black: Poincaré map

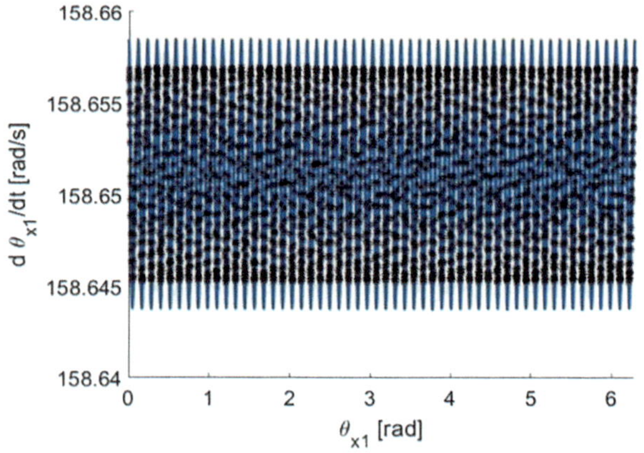

Fig. 15.15 Driving gear rotational motion for the case considering backlash. Dark blue: entire time interval phase portrait; Black: Poincaré map

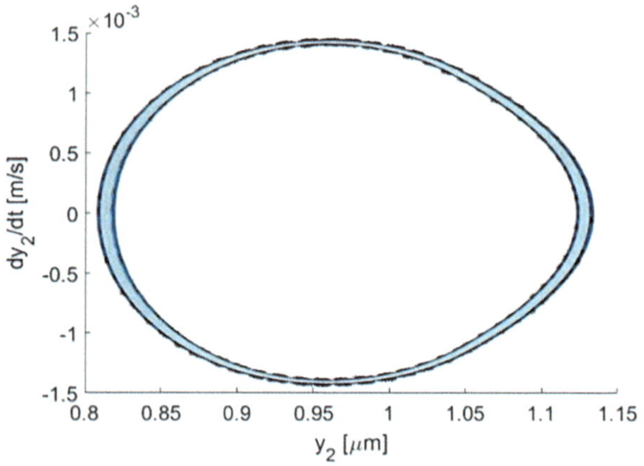

Fig. 15.16 Driven gear horizontal motion for the case considering backlash. Dark blue: entire time interval phase portrait; Light blue: 0.1 s phase portrait; Black: Poincaré map

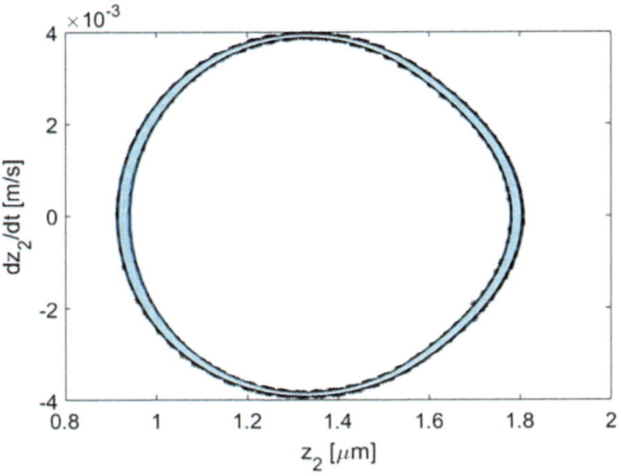

Fig. 15.17 Driven gear vertical motion for the case considering backlash. Dark blue: entire time interval phase portrait; Light blue: 0.1 s phase portrait; Black: Poincaré map

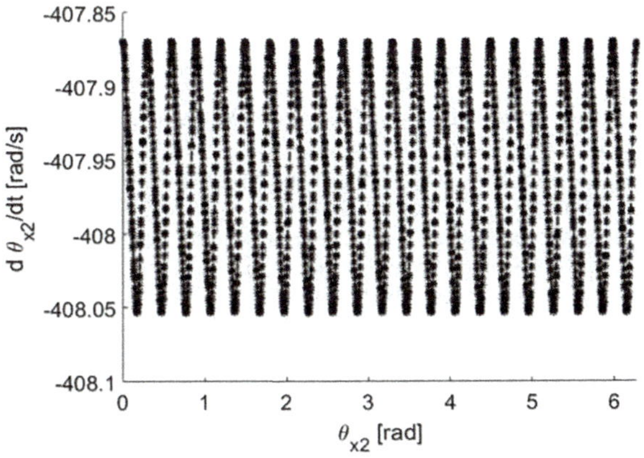

Fig. 15.18 Driven gear rotational motion for the case considering backlash. Dark blue: entire time interval phase portrait; Light blue: 0.1 s phase portrait; Black: Poincaré map

15.4 Conclusions

This paper introduces a lumped parameter model for a helical gear pair, which takes into account a detailed model of the gear mesh, flexibilities in the gear supports, and variable driving and load torques. The primary purpose of this model is to simulate the high-speed stage of a wind turbine gearbox. The gear mesh introduces internal non-linearity into the system, as it accounts for factors such as backlash, time-varying mesh stiffness, and static transmission error. Furthermore, the gear mesh also couples all directions of the gears' motion.

Two conditions were considered: with and without backlash. The presented results showed the time behavior of drive torque, load torque, and LOA displacement. Phase portraits and Poincaré maps of driving and driven gears were analyzed.

In the case without backlash, both driving and driven gears presented indications of quasi-periodic motion. When a backlash is considered, single-sided impacts occur during the entire period of steady-state response, and the driving gear translational motions exhibit a more severe vibration condition with a phase portrait entirely filled with trajectories. To achieve a more comprehensive and quantitative understanding of the system's behavior, it is advisable to employ additional analysis tools beyond the Poincaré map.

These preliminary results show that load torque variations, that may occur on wind turbines to avoid over-speed conditions, may cause damaging effects on the helical gear pair due to non-periodic motion. In the presence of a backlash, impacts and chaotic behavior may occur.

Acknowledgements The authors are grateful to the Brazilian National Council of Research (CNPq—Brazil), process 407978/2022-4, to CAPES and FAPDF, process 00193-00001139/2021-57, for their support.

References

Apata O, Oyedokun D (2020) An overview of control techniques for wind turbine systems. Sci Afr 10:e00566. https://doi.org/10.1016/j.sciaf.2020.e00566

Kahraman A (1994) Dynamic analysis of a multi-mesh helical gear train. J Mech Des Trans ASME 116(3):706–712. https://doi.org/10.1115/1.2919440

Kubur M, Kahraman A, Zini DM, Kienzle K (2004) Dynamic analysis of a multi-shaft helical gear transmission by finite elements: model and experiment. J Vibr Acoust Trans ASME 126(3):398–406. https://doi.org/10.1115/1.1760561

Ma H, Song R, Pang X, Wen B (2014) Time-varying mesh stiffness calculation of cracked spur gears. Eng Fail Anal 44:179–194. https://doi.org/10.1016/j.engfailanal.2014.05.018

Özgüven HN (1991) A non-linear mathematical model for dynamic analysis of spur gears including shaft and bearing dynamics. J Sound Vibr 145(2):239–260. https://doi.org/10.1016/0022460X(91)90590-G

Ragheb A, Ragheb M (2010) Wind turbine gearbox technologies. In: Proceedings of the 2010 1st international nuclear renewable energy conference (INREC), pp 1–8. https://doi.org/10.1109/INREC.2010.5462549

Teng W, Wang F, Zhang K, Liu Y, Ding X (2014) Pitting fault detection of a wind turbine gearbox using empirical mode decomposition. Strojniški Vestnik J Mech Eng 60(1):12–20. https://doi.org/10.5545/sv-jme.2013.1295

Tian X (2004) Dynamic simulation for system response of gearbox including localized gear faults. University of Alberta. https://books.google.com.br/books?id=X7PzSgAACAAJ

Walha L, Fakhfakh T, Haddar M (2006) Backlash effect on dynamic analysis of a two-stage spur gear system. J Fail Anal Prevent 6(3):60–68. https://doi.org/10.1361/154770206X107325

Wan Z, Cao H, Zi Y, He W, Chen Y (2015) Mesh stiffness calculation using an accumulated integral potential energy method and dynamic analysis of helical gears. Mech Mach Theory 92:447–463. https://doi.org/10.1016/j.mechmachtheory.2015.06.011

Wang L, Zuo S, Song YD, Zhou Z (2014) Variable torque control of offshore wind turbine on spar floating platform using advanced RBF neural network. Abstract Appl Anal 2014:3493. https://doi.org/10.1155/2014/903493

Wei S, Zhao J, Han Q, Chu F (2015) Dynamic response analysis on torsional vibrations of wind turbine geared transmission system with uncertainty. Renew Energy 78:60–67. https://doi.org/10.1016/j.renene.2014.12.062

Yang DC, Lin JY (1987) Hertzian damping, tooth friction and bending elasticity in gear impact dynamics. J Mech Des Trans ASME 109(2):189–196. https://doi.org/10.1115/1.3267437
Zhao M, Ji J (2015) Nonlinear torsional vibrations of a wind turbine gearbox. Appl Math Model 39(16):4928–4950. https://doi.org/10.1016/j.apm.2015.03.026

Chapter 16
Stability Assessment of the Resonant Linear Permanent Magnetic Generators Powered by Free Piston Engines

Vítor S. Medeiros, Solidônio R. Carvalho, and Marcelo B. Santos

Abstract A sustainable development is essential for the future of the planet. Therefore, the technology must satisfy current needs without compromising the ability of future generations to meet their own needs. Hence, this work assesses the stability conditions of a free piston engine powering a linear permanent magnetic generator. This kind of machine uses renewable fuels and has greater energy efficiency. The numerical model used to do it englobe three physical domains: electromagnetism, dynamics, and thermodynamics. The proposed machine works in its resonance frequency. Therefore, a sensitivity analysis was played out, interpreting how the three variables influence the engine cycle stability: the inlet pressure, air–fuel ratio and the associated mechanical spring stiffness. It was shown that when working in resonance the machine is more robust and less influenced by the thermodynamics conditions.

Keywords Free piston engine · Multi-domain simulation · Permanent magnetic linear generator · Resonance · Stability analysis

V. S. Medeiros
School of Mechanical Engineering, Federal University of Uberlandia, Uberlandia, Minas Gerais, Brazil
e-mail: vitormedeiros@ufu.br

S. R. Carvalho
Heat and Mass Transfer Laboratory, School of Mechanical Engineering, Federal University of Uberlandia, Uberlandia, Minas Gerais, Brazil
e-mail: solidonio@ufu.br

M. B. Santos (✉)
Mechanical System Laboratory, School of Mechanical Engineering, Federal University of Uberlandia, Uberlandia, Minas Gerais, Brazil
e-mail: marcelo.bragadossantos@ufu.br

P. Kurka and M. Pereira (eds.), *Advances in Structural Vibration*, Lecture Notes in Mechanical Engineering, https://doi.org/10.1007/978-3-031-71540-2_16

16.1 Introduction

A sustainable development is essential for the future of the planet. Therefore, the technology must satisfy current needs without compromising the ability of future generations to meet their own needs. In this way, fossil fuels in power generation are clearly an unsustainable source of energy. According to "Report of the World Commission on Environment and Development", edited in far 1987, the use of renewable fuels and greater energy efficiency are imperatives (Secretary-General 1987). The free piston engine was widely applied, in the mid-twentieth century until the 1960s, in hydraulic or pneumatic compression operations, including gases directed to generate electrical energy in industrial plants (Achten 1994; Mikalsen and Roskilly 2007). New power electronics, sensors, and actuators technologies bring greater interest in these types of engines, due to their thermomechanical efficiency, almost 56% of thermal efficiency for some projects and between 40 and 50% in the majority of other ones (Van Blarigan et al. 1998; Jia et al. 2016).

A crucial step in design and optimization of these machines is to calculate the electromagnetic field distribution in their interior. Several techniques exist in the literature to cover these demands. For design and optimization stage, commonly the subdomain method is chosen due to its small computation effort, even if that leads to less accurate results (Rodrigues et al. 2019a, b). Ergo, parametric analysis is a powerful tool that can lead to a wider comprehension of the system, and thus give us sufficient knowledge basis allowing us to trace the better strategical ways to obtain the best performance and design of a certain equipment. It can also provide information with which control strategy would be designed. A parametric analysis of a numerical model built for design and optimization purposes was developed for a free piston engine (FPE) used to power the linear permanent magnetic generator (LPMG). Using this analysis were observed how the electromechanical efficiency of the generator is affected by the variation of the electrical and mechanical parameters of the system (Rodrigues et al. 2019a, b). A LPMG powered by a FPE (FPE-LPMG) with a 10 kWe was optimized. A parametric model which includes dynamic, thermodynamic, and electrical principles was used to simulate the machine functioning dynamics. The thermodynamic model includes the scavenging process of the cylinder chamber. The results show a net efficiency of 49%, which is greater than the most efficient diesel generator produced by General Electric (Rigobello et al. 2021).

The main goal of this work is to explore the robustness of a free piston engine coupled to a specified linear generator, such that it can achieve maximum efficiency while driving an electrical generator of 10 kW. Based on the optimized parameters values obtained using a genetic algorithm, presented by Rigobello et al. (2021), this paper aims to assess the system robustness using a sensitivity analysis, in other words, evaluate the ability of the system to operate correctly in various thermodynamic conditions and air–fuel ratio (AFR) around their optimum values, mapping the failure regions due an unexpected situation.

The following sections are organized to show the physical model, including a brief explanation of the electrical machine subdomain model and the associated electrical circuit, the thermodynamical model used to estimate the pressure inside the cylinder chambers, and one degree of freedom model, as well as the domain coupling. The results are depicted and discussed in terms of the system's stability analysis when are varied the intake pressure, spring stiffness, and air/fuel ratio. Finally, are shown the conclusions.

16.2 Physical Model

The multi-domain model applied includes thermodynamic, mechanical vibration, and electromagnetism concepts (Rodrigues et al. 2019a, b). The integration between each domain is represented by the FPE-LPMG model diagram depicted in Fig. 16.1.

Both models, one which represents the electrical machine and other equivalent to the combustion chambers, receive the state variables, displacement, and speed of the 1 DOF system, returning the force to move the generator and the pressure inside each combustion chamber, respectively.

The engine is a dual-opposed cylinder and, its physical dynamic model is depicted in Fig. 16.2 and mathematic model of the system is defined by Eq. (16.1). The 1 DOF system is excited by the successive expansions of the burned gases in each one of the opposite combustion chambers, the additional elastic element is associated in series with the gas which is pressurizing in the cylinder opposed to one which has just burned and is expanding. Finally, the load imposed by the electrical generator over the motor is represented by its physical model or by an equivalent damping.

$$m\ddot{x}(t) + \underbrace{\left(c_{\text{loss}} + c_{\text{eq}}\right)}_{C}\dot{x}(t) + kx(t) = \underbrace{A_P(P_L(t) - P_R(t))}_{F_{\text{ex}}} \tag{16.1}$$

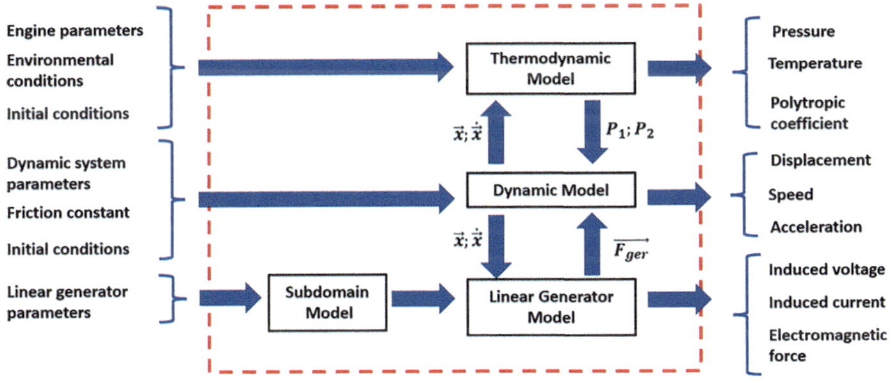

Fig. 16.1 Free piston electric generator model schematic diagram

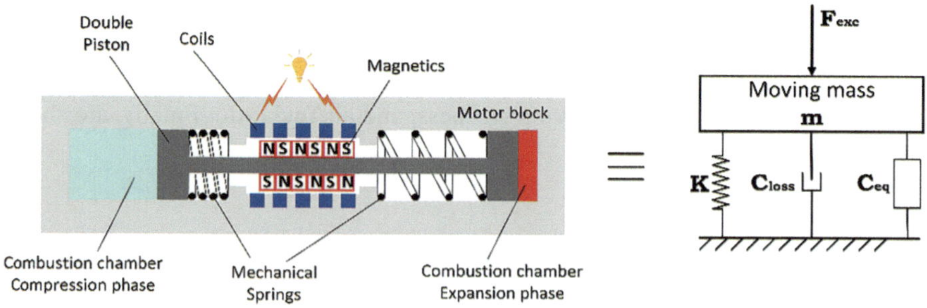

Fig. 16.2 Mass, spring, and damper dynamic system

In Eq. (16.1) m is the moving mass, k is the equivalent elastic constant, c_{loss} is the viscous damping which represents the losses due to viscous friction and c_{eq} is the equivalent damping representing the conversion of kinetical energy in electrical one by motion of the magnetics inside the coils, as depicted in Fig. 16.2. This simplification of the generator as a viscous damper is acceptable, since according to Lenz's Law, the generated voltage is proportional to the speed, consequently, for resistive circuits also the current. Ergo, there is a proportionality between the force to move the generator and x. Lorentz theorem also assures that magnetic force is proportional to the speed of magnetic field variation, as well as viscous damping (Rigobello et al. 2021).

The excitation force F_{ex} is due to the pressure inside the left cylinder $(P_L(t))$ and the right cylinder $(P_R(t))$, in that way, the opposed cylinders are producing a periodic excitation which fundamentally can be adjusted to be equal to the natural frequency of the vibratory system, as shown in Fig. 16.3. The last concept is responsible for improving the energy transfer from its thermochemical state to the mechanical one (kinetic plus potential energy), thus, improve improving the electrical energy generation.

The viscous damping c_{loss} is overestimated and the considered losses are around 2%, once which, 2% is the estimated value for friction losses in traditional ICE motors. Starting with thermodynamic model developed in Rodrigues et al. (2019a, b) the scavenging process, namely uniflow, was added to the model. Thus, the thermodynamic model used in this paper is a null dimensional model with homogeneous

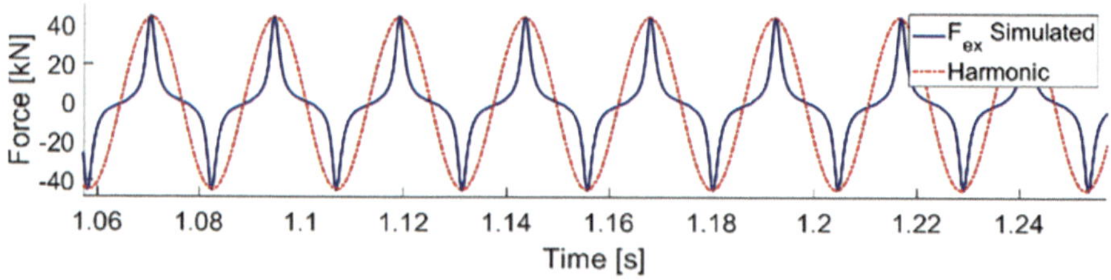

Fig. 16.3 Simulated excitation force and harmonic force

properties over the control volume. The heat release due to combustion is obtained using Wiebe function and the convective heat transfer to the cylinder walls has been considered imposing the wall temperature. The first thermodynamic law applied over an ideal gas was used to obtain the temperature, pressure, and polytropic coefficients. Finally, the inner cylinder mass balance has been divided into two stages: valves closed and intake/exhaust stages. In the first stage, the mass inside the control volume didn't change and in the second stage, the flows through the intake and exhaust valves had been considered in the control volume mass calculation by means of scavenging model (Rigobello et al. 2021). The dynamic system depicted in Fig. 16.2 and represented in Eq. (16.1) was solved using Runge–Kutta of the fourth-order method.

16.2.1 Thermodynamic Model

Since Uniflow system stands out for being the most efficient scavenging for two-stroke engines, presenting low air consumption, turbulence induction, and enabling asymmetric valve timing (Schafer and Basshuysen 2004; Blair 1996) it was chosen as scavenging process. Some physical parameters of the cylinder are defined, as shown in Fig. 16.4. The inlet position factor is defined as $\phi_{\text{in}} = L_{\text{in}}/S_{\text{cyl}}$, the exhaustion position factors are defined: $\phi_{\text{ex},1} = L_{\text{ex},1}/S_{\text{cyl}}$, for open positions and $\phi_{\text{ex},2} = L_{\text{ex},2}/S_{\text{cyl}}$, for close position. Where $L_{\text{ex},1}$ is the piston position during cylinder expansion equivalent to the exhaust valve opening position; and $L_{\text{ex},2}$ is the piston position during cylinder compression, equivalent to the exhaust valve closing position.

For the thermodynamic, it uses a null dimensional model, with homogeneous properties throughout the control volume (Rodrigues et al. 2019a, b). This contributes with a step forward in comparison with the one given by Jia et al. (2016), in which it was applied a unit step function to enable/disable the influence from the heat release, directly manipulating the internal pressure of each cylinder. In this work, the Wiebe function gives the rate of heat released by the combustion process, in the cylinder "i", as depicted by Eq. 16.2. Further, it is applied in the derived first law of thermodynamics.

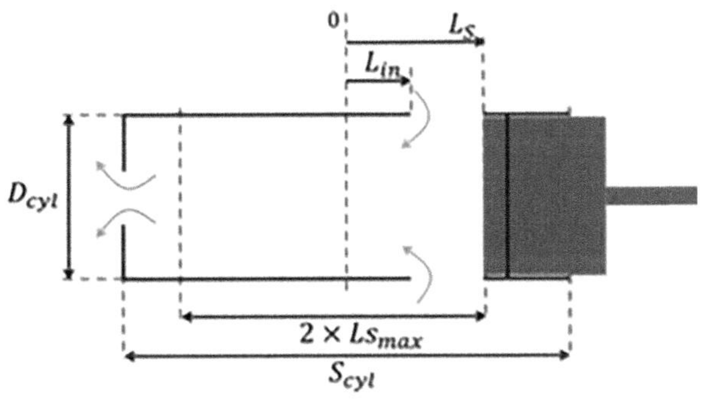

Fig. 16.4 Diagram of the cylinder and the main parameters

$$\frac{dQ_{c,i}}{dt} = a\frac{b+1}{C_d}\left(\frac{t-t_{s,i}}{C_d}\right)\exp\left[-a\left(\frac{(t-t_{s,i})}{2}\right)^{b-1}\right]Q_{\text{comb}} \tag{16.2}$$

The terms in the Eq. (16.2) are defined as: $Q_{c,i}$ is the effective heat produced by the combustion in the i-th chamber (i can be "left" or "right"), Q_{comb} is the heat released by the complete combustion of the mixture air/combustible, $C_d = 3.85$ ms is the duration of the combustion, $t_{s,i}$ is the instant of the ignition in the i-th chamber and is defined through the ignition position factor $\Phi_{ig} = 0.7049$, a and b are the Wiebe parameters and are settled to 5 and 2, respectively.

The heat transfer between the gas contained in the cylinder "i" and the walls of the cylinder $Q_{w,i}$ is modeled according to Eq. (16.3). Where the convection transfer coefficient $h_i(t)$ is estimated according to the expression given in Eq. (16.4) (Mikalsen and Roskilly 2008). The average temperature of the cylinder walls defined as $T_w =$ 75 °C.

$$\frac{dQ_{w,i}}{dt} = h_i(t)A_{\text{cyl}},\quad i(t)(T_i(t) - T_w) \tag{16.3}$$

$$h_i(t) = 130\, V_i^{-0.06}(t)\left(\frac{P_i(t)}{10^5}\right)T_i^{-0.4}(t)\left(v_p + 1.4\right)^{0.8} \tag{16.4}$$

The term V_i is the volume of gas inside the i cylinder, P_i is the pressure, T_i is the gas temperature and v_p is the mean speed of the piston. Based on the first law of thermodynamics and considering the mixture as a perfect gas, ideal gases law, the temperature Eq. (16.5), the pressure Eq. (16.6), and the polytropic coefficient γ Eq. (16.7) were calculated. Thermophysical properties were obtained from the JANAF table (Heywood 2018).

$$\frac{dT}{T} = \frac{dP}{P} + \frac{dV}{V} - \frac{dm}{m} \tag{16.5}$$

$$dP = \frac{1}{(1-\gamma)\frac{T}{R}\frac{dC_p}{dT} + 1}\left[\frac{\gamma - 1}{V}dQ - \frac{P(\gamma-1)}{V}\left(\frac{T}{R}\frac{dC_p}{dT} + \frac{\gamma}{\gamma-1}\right)dV - (\gamma-1)\frac{T^2}{V}\frac{dC_p}{dT}dm\right] \tag{16.6}$$

$$d\gamma = \frac{1-\gamma}{C_v}\frac{dC_p}{dT}dT \tag{16.7}$$

Finally, the inner cylinder mass flow is divided into two stages. The first stage is when the valves are closed, so there is no intake or exhaust process. In this case, it is assumed that $dm/dt = 0$. On the other hand, the second stage is regarding the intake and exhaust process, in which it presents variation in inner mass, because of the valve's flow. These equations had been modified by Rigobello (2020) which used the thermodynamics model (Rowland and Whitehouse 1979), but also the mass flow calculation (Ferguson and Kirkpatrick 2015), and also the valves geometry (Heywood 1988; Gordon 1999).

It simulated a free piston engine (FPE) to drive an electric generator of rated power of 10 kW with maximum efficiency. The FPE operation conditions were analyzed under dynamics constraints to maximize the efficiency; under some defined parameters, the exhaust pressure $P_{ex} = 135$ kPa; the inlet pressure $P_{in} = 1.5 \cdot 135$ kPa; the inlet temperature $T_{in} = 25\,°\mathrm{C}$; and the methane fuel main parameters: LHV $= 48500$ kJ/kg; $AFR = 17$, and $\eta_{comb} = 90\%$, the last one is The retained mass inside the cylinder (m_{mix}) is defined in Eq. (16.8) where R is the universal gas constant. The fuel mass (m_{fuel}) is defined in Eq. (16.9), in terms of m_{mix} and the air/fuel ratio ($\mathrm{AFR} = m_{air}/m_{fuel}$).

$$m_{mix} = \frac{p_{in}\left[\frac{\pi}{4}D_{cyl}^{2}\right]\frac{S_{cyl}}{2}\left[1 + \Phi_{ex,2}\right]}{RT_{in}} \tag{16.8}$$

$$m_{fuel} = \frac{m_{mix}}{\mathrm{AFR} + 1} \tag{16.9}$$

The heating generated by combustion in each cylinder has been written in Eq. (16.10), in terms of LHV and η_{comb}. Hence, given the last parameter to calculate the pressure inside each combustion chamber as function of the piston displacement and speed.

$$Q_{comb} = m_{fuel}\mathrm{LHV}\eta_{comb} \tag{16.10}$$

The optimum parameters for intake and exhaustion time are: $\Phi_{adm} = 0.64$, $\Phi_{ex,1} = 0.37$ and $\Phi_{ex,2} = 0.22$. These were taken as references for the best configuration of the free piston engine (Rigobello et al. 2021).

16.2.2 Linear Generator Electric Model

It is often necessary to estimate the rated power, efficiency, magnetic losses, and even magnetic vibrations of the electric machines to optimize their design. This estimation relies on the accurate computation of the machine characteristics, such as electromagnetic force, magnetic losses, and air-gap Maxwell forces. All these quantities can be computed if the magnetic field inside the machine is fully determined, meaning the space and time distribution for both radial and tangential components (Gysen et al. 2010).

For this purpose, the technique namely Subdomain Modeling (SDM) was selected between the different methods available in the scientific literature, due to the advantages it presents. The low computational effort combined with the accuracy of the results—which can be comparable with the solutions obtained via FEM for simple geometries—made SDM a suitable choice for this work. However, the limitations of SDM derive from the inherent assumptions that are made in its procedure, which are described further. SDM is a semi-analytical method that consists of dividing the problem into physical regions, called subdomains, in which Maxwell governing equations can be solved analytically. Then, Maxwell equations are written in each subdomain for scalar potential or vector potential, from which the flux density is derived. By expressing the boundary conditions of potential and magnetic field continuity at each interface between two subdomains, a linear system of analytical independent equations linking all unknown constants is obtained, Eq. (16.11) (Devillers et al. 2016).

$$[M]\{X\} = \{S\} \tag{16.11}$$

where the matrix $[M]$ is denominated as "topological matrix", and $\{S\}$ the "source vector". Obtaining the value of the constants $\{X\}$ enables us to compute the magnetic field's spatial distribution in each subdomain. The topological matrix and source vector depend on the topology of the electrical equipment and on the region division. There are numerous variants of tubular of Linear Permanent Magnetic Generator (LPMG), and for the purpose of this work, the selected topology for the LPMG is classified as radial magnetization, internal magnet topology, slotted armature, and three-phase winding. Figure 16.5 illustrates the mentioned topology.

The formulation proposed by Gysen et al. (2010) for the topological matrix and source vector is used in this paper, although the author proposed it for electrical machines used as actuators instead of generators. The division of the chosen topology and the applied boundary conditions used to achieve the mentioned formulation are illustrated in Fig. 16.6. The boundary condition is applied according to the material of the delimited regions. A Neumann boundary condition appears at the normal interface

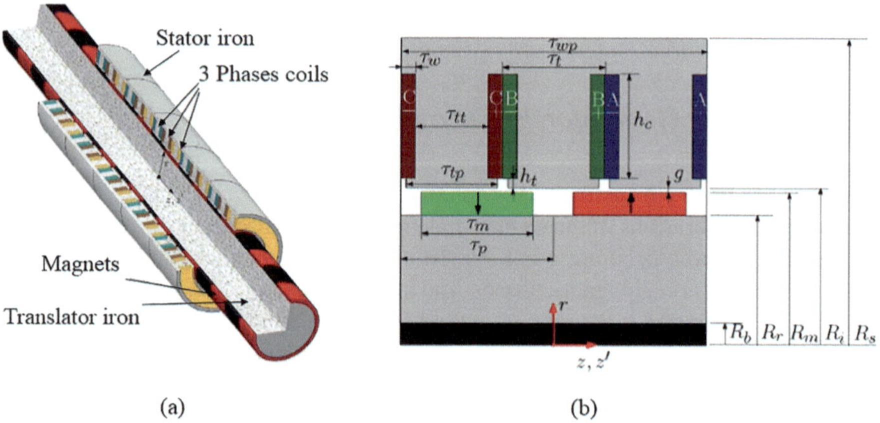

Fig. 16.5 Topology of the LPMG selected for this study (**a**); One active part of LPMG with the relevant dimensions (**b**). Adapted from Gysen et al. (2010)

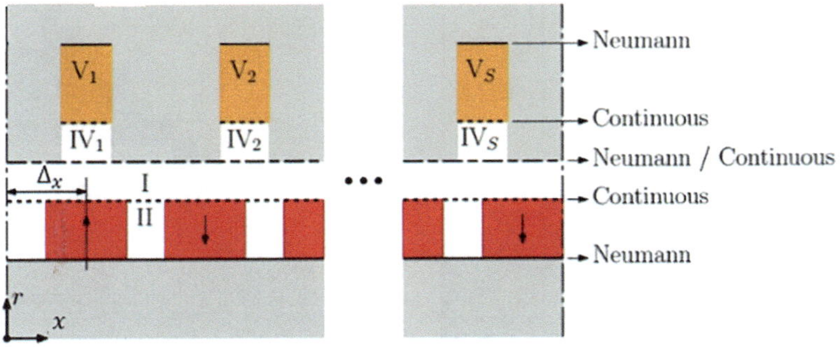

Fig. 16.6 Division in regions and boundary conditions for the chosen topology. Adapted from Gysen et al. (2010)

between an iron material and a soft-magnetic material. In this condition, the tangential magnetic field strength must be zero. For the continuous boundary condition, the normal component of the magnetic flux density needs to be continuous as well as the tangential component of the magnetic field strength at the delimited boundary. This type of condition is applied between two regions with finite permeability. A combination of Neumann and continuous boundary conditions occurs at an interface between regions which have unequal width and/or unequal offsets, as indicated in Fig. 16.6. The normal magnetic flux density component of every region should be equal to the normal magnetic field component of air-gap region on the boundary at $r = R_i$. Furthermore, the tangential magnetic field strength component H_z of the air-gap region must equal the tangential magnetic field strength component of every winding region, on respective boundary, and equal zero elsewhere.

It should be also highlighted that the solution for the magnetic field distribution, obtained via the SDM, in the foregoing machine topology is conditioned on the following assumptions:

1. The axial length of the machine is infinite.
2. The field distribution is axially symmetric and periodic in the z-direction.
3. The material properties are linear, homogeneous, and isotropic.
4. The permeability of the iron is infinity.
5. The effects due to the Eddy current are neglected.

To estimate the electromotive force $e_i(t)$ over each coil and consequently F_{ger}, necessary to move the translator of the linear generator is necessary to estimate the flux linkage over each coil as follows:

$$\Lambda_{i,\text{pm}}(x) = \frac{2\pi}{\tau_0(R_c - R_t)} \int_{R_t}^{R_c} \int_{0}^{\tau_0} A_\theta^{V,i}(r, \hat{x}, x) \cdot r \cdot \text{d}r \cdot \text{d}\hat{x} \tag{16.12}$$

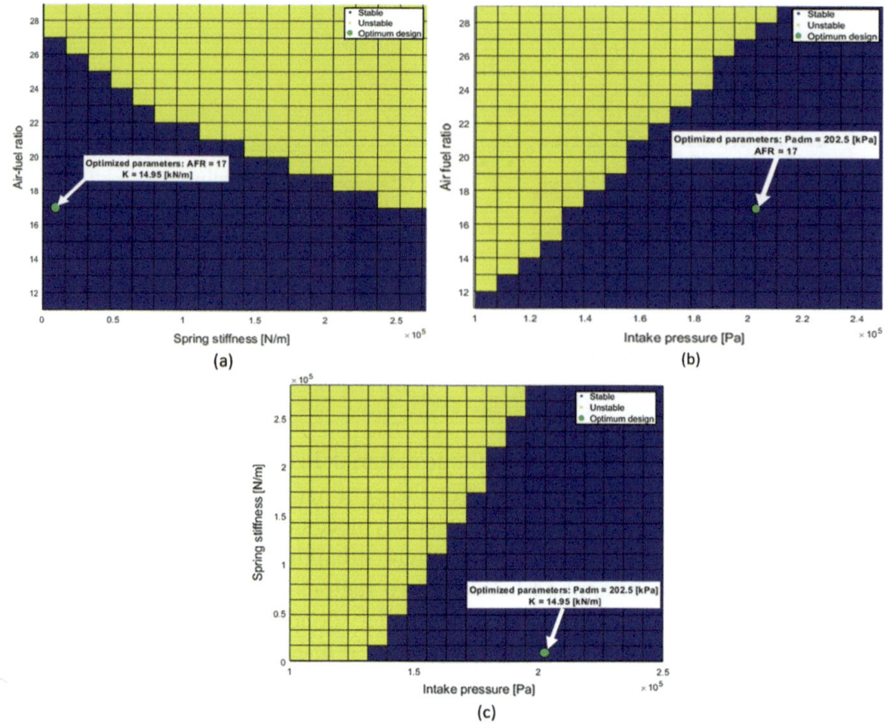

Fig. 16.7 Stability maps: **a** air–fuel ratio × spring stiffness (N/m); **b** Air–fuel ratio × intake pressure (Pa); **c** Intake pressure (Pa) × spring stiffness (N/m)

where $A_\theta^{V,i}(r, \hat{x}, x)$ is the magnetic potential vector at the V_i region, which is function of the constants $\{X\}$, determined by SDM technique. Must be observed that $\hat{x}$ is the coordinate relative to a fixed frame, and x is the relative displacement between the translator and the stator of the LPMG, as indicated in Fig. 16.6.

Hence, the electromotive voltage at the i-th coil can be calculated with Eq. (16.13).

$$e_i(t) = -N_c \frac{\mathrm{d}\Lambda_{i,\mathrm{pm}}(x)}{\mathrm{d}t} = -N_c \frac{\mathrm{d}\Lambda_{i,\mathrm{pm}}(x)}{\mathrm{d}x}\frac{\mathrm{d}x}{\mathrm{d}t} \tag{16.13}$$

where N_c is the number of turns of the coils.

In Eq. (16.14) it is depicted how to calculate the force necessary to move the translator of the linear generator as follows:

$$F_{\mathrm{ger}} = N_c \cdot N_{\mathrm{pa}}\{i_A i_B i_C\}\frac{\mathrm{d}}{\mathrm{d}x}\left\{\begin{array}{c}\Lambda_{A,\mathrm{pm}}(x)\\ \Lambda_{B,\mathrm{pm}}(x)\\ \Lambda_{C,\mathrm{pm}}(x)\end{array}\right\} \tag{16.14}$$

The term $\Lambda_{i,\text{pm}}(x)$ are the flux linkage at the i-th phase due to the magnets, N_{pa} is the number of active parts of the LPMG and i_i are the instantaneous current for the i-th phase.

To study the stability, in this work, the electrical circuit coupled with the LPMG is a three-phase resistive circuit as proposed by Rodrigues et al. (2019a, b). This is done since it is possible to approximate with a very good precision the F_{ger} by means of:

$$F_{\text{ger}} = c_{\text{eq}}\dot{x} \tag{16.15}$$

Combining Eqs. (16.14) and (16.15) is possible to define c_{eq} of the Eq. (16.1) and Fig. 16.2 as:

$$c_{\text{eq}} = N_c \cdot N_{\text{pa}}\{i_A i_B i_C\}\frac{\text{d}}{\text{d}x}\begin{Bmatrix} \Lambda_{A,\text{pm}}(x) \\ \Lambda_{B,\text{pm}}(x) \\ \Lambda_{C,\text{pm}}(x) \end{Bmatrix}\frac{1}{\dot{x}} \tag{16.16}$$

Therefore, solving the thermodynamic model Eqs. (16.5) and (16.16) is possible to determine all terms of the complete model depicted in Eq. (16.1).

16.3 Results and Discussion

To study the robustness of the proposed FPE-LPMG a sensitivity analysis method was carried out, to measure the model response to change in three input variables that could lead to an unstable condition, i.e., the Air–Fuel Ratio (AFR), the Intake Pressure and the associated mechanical spring stiffness (k). These parameters vary in the simulation from 11 up to 29, from 100 up to 250 (kPa), and from 0.78 up to 270.79 (kN/m), respectively. This analysis is useful because it improves the prediction of the model's stability. Additionally, the sensitivity analysis is employed to quantify the variation limits of each of the model's parameters around their optimum values, and their importance in the behavior of the system. The other parameters present in Eq. (16.1) and used in the simulation are shown in Table 16.1 (Rigobello et al. 2021).

Figure 16.7 depicts three stability maps where two of the chosen three variables are changed simultaneously while the third one remains in its optimal value.

The amount of energy produced during the combustion of air–fuel mixture is lower for low values of inlet pressure and for high values of AFR. So, it is possible to verify that the amount of energy generated during mixture combustion plays a great influence on the functioning stability. As higher energy is generated in the combustion, more stable is the machine. There are two extremes in AFR, lean mixtures and rich mixtures, and in these extremes, the energy free to produce work over the piston is low due to the small amount of fuel and due to the small amount of oxygen to burn the fuel, respectively. Blair (1996) adds that between the lean mixture misfire

Table 16.1 FPE-LPMG optimized parameters and main simulation inputs

Parameter	Value
Operation frequency (Hz)	41.5
Compression ratio	11.95
Half maximum piston stroke (mm)	32.71
System mass (kg)	10
Cylinder bore (mm)	106.83
Equivalent damping due to the electric generator (Ns/m)	272.36
Viscous damping (Ns/m)	51.07
Methane fuel lower heat value (kJ/kg)	48,500
Combustion efficiency (%)	10

and the stoichiometric ratio, there is an AFR for the maximum power release. At this point, which is around 0.86 of stoichiometric AFR, the excess fuel increases the combustion speed. Even though this is observed in real motors, here it happens because the Wiebe function, used to model the combustion process, depends on the total amount of fuel injected in the chamber, i.e., more fuel means higher rates of fuel heat release. Besides, optimum design gives robustness to the machine, once that is; this design admits a large variation in the studied parameters without occurrence of misfire.

It is depicted in Fig. 16.8 the pressure and temperature variation with piston displacement. It is clear the maximum cycle pressure is lowered as the AFR increases. The maximum pressure for AFR $= 28$ is too low to produce the compression of the gas inside the opposite chamber and of the associated spring, so that, it is possible to verify that this configuration is an unstable one as shown in Fig. 16.7a. Additionally, the area inside pressure $\times$ displacement curve decreases as AFR increases, which means less energy generated by combustion. The maximum temperature increases as AFR decreases; this phenomenon is also related to the application of the Wiebe function. The overlap between exhaust and intake process, an important characteristic of the Uniflow scavenging process, is also represented. This overlap means that the scavenging starts with the piston pushing out the burned gas, is augmented by the intake port open, which inserts fresh mixture in the chamber, and finishes with the piston movement.

Figure 16.9 depicts the piston displacements in three different conditions of AFR. It is possible to see that in the unstable condition (AFR $= 28$) the pistons reach a lower peak after each combustion. This happens due to the low amount of energy produced in these successive combustions and the energy decreases as the pressurization decreases until the piston does not pass anymore the ignition point. After this point, the pistons are deaccelerated more quickly. It is possible to verify that the displacement reaches a steady state for stable conditions. It should also be pointed out that the working frequency increases as the energy produced by the combustion

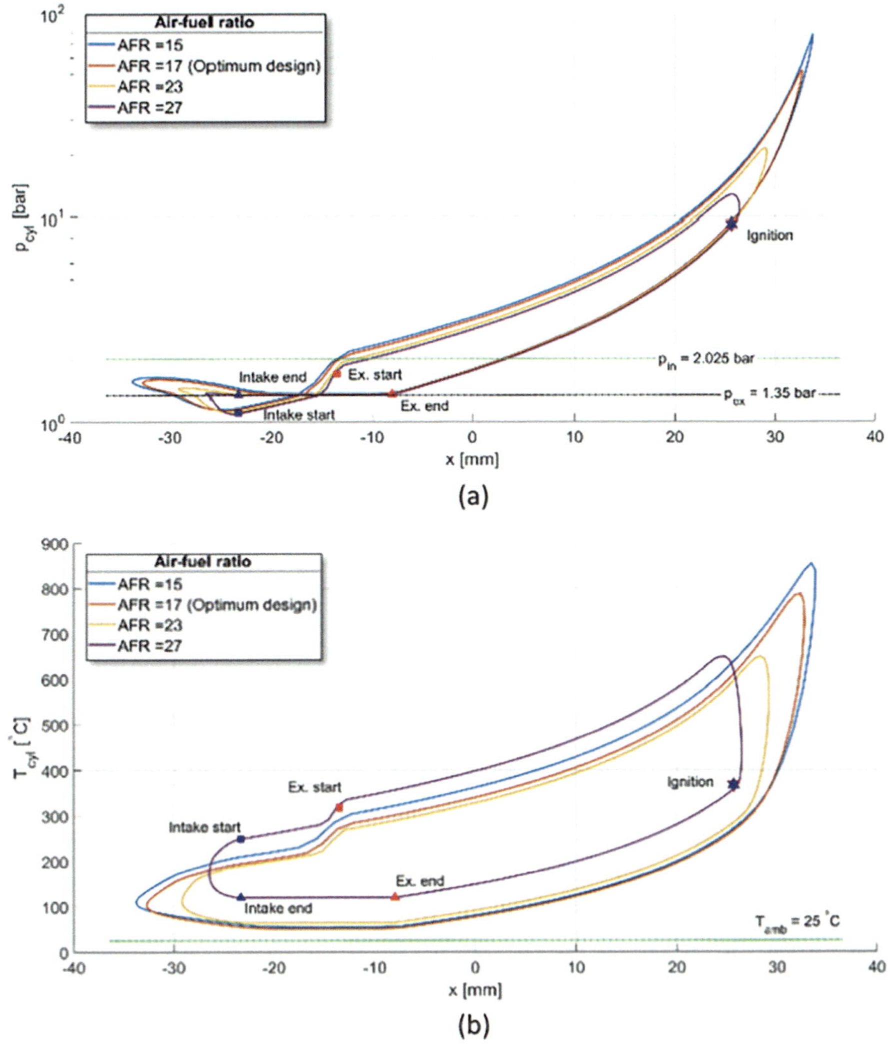

Fig. 16.8 Pressure (**a**) and temperature (**b**) as a function of piston displacement, considering the associated mechanical spring stiffness of 14.95 (kN/m)

increases, this is due to a speed enhancement which results in a short period to the piston runs over the stroke.

Figure 16.10 shows the simulated piston velocity as function of piston displacement for different air–fuel ratios Fig. 16.10a and spring stiffness Fig. 16.10b. It is possible to see that the operation cycle with AFR in its optimum value of 17 is more stable and reaches a steady state more rapidly, once that, the lines for each cycle do not diverge significantly. On the other hand, for the other configurations, it is observed a delay in the time taken to reach the steady state, represented by the

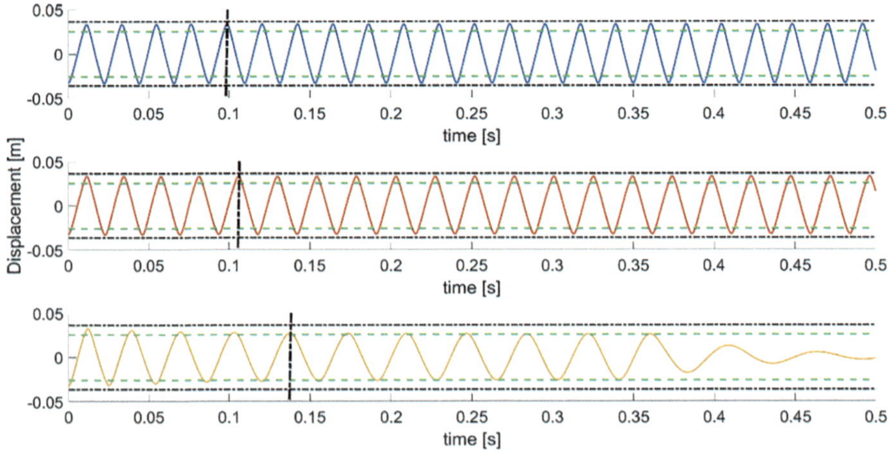

Fig. 16.9 Piston displacement as a function of time. Blue line (AFR = 15), Red line (AFR = 17), Yellow line (AFR = 28). Auxiliary lines: Dash green (Ignition point), Horizontal black dash-dot (Stroke limits), Vertical black dash-dot (Fifth period)

spread of the lines. Moreover, as observed in Figs. 16.9 and 16.10 also depicts the instability conditions (AFR = 29 and k = 300 kN/m) where the piston slows down until the operation stops.

Following, we perform an optimization, which is a technique to extract the maximum power from the generator even when the motor has variations in its working conditions. The results are shown in Fig. 16.11 for two input variables Air–Fuel Ratio (AFR) and Intake Pressure. Based on the results obtained, it is possible to verify that the FPE-LPMG produces more power when it is running a slightly rich mixture, and consequently a higher value for the optimal resistive coefficient ($C_{\text{opt}} = 196$ Ns/m) is observed. The lean mixture, when there's a higher concentration of air to fuel, enables the engine to be more efficient but less energy is available to be absorbed by the generator. Thus, a lower value for the optimal resistive coefficient is obtained ($C_{\text{opt}} = 115.15$ Ns/m), as shown in Fig. 16.11a.

Figure 16.11b depicts the impact of the intake pressure variation over the Copt value, and it is possible to see that the pressure is directly proportional to the absorbed power and the optimal resistive coefficient. The reason is that when the intake pressure is higher more air reaches the cylinders, as a result, more fuel can be burned, and the generator can absorb more power. The operation characteristics studied in this paper will be used as a reference for future optimization work, investigating the feasibility of different extremum-seeking (ES) control schemes to improve the conversion efficiency of a free piston engine used to power the linear permanent magnetic generator.

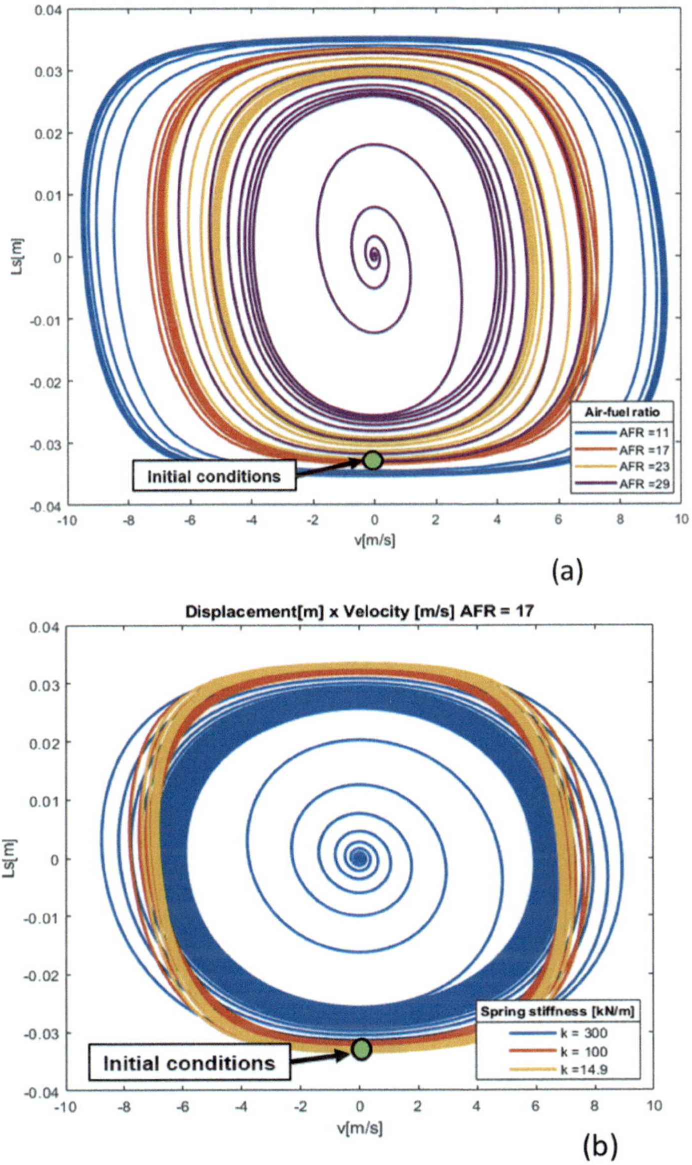

Fig. 16.10 Piston displacement as a function of piston velocity

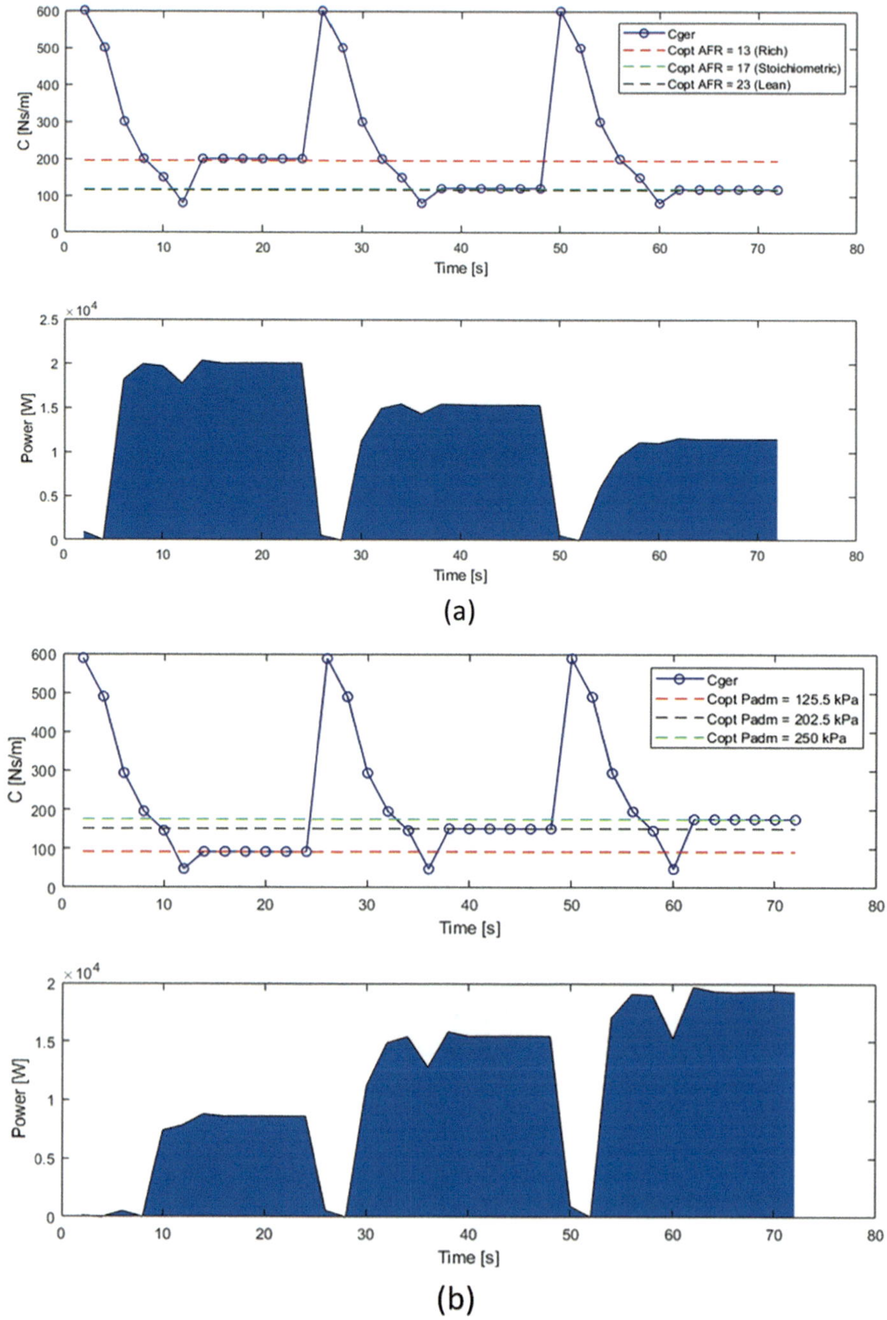

Fig. 16.11 Optimization of resistive coefficient C, for the FPE-LPMG subject to changes in input variables, **a** air–fuel ratio (AFR) and **b** intake pressure (Padm). The resistive power take-off coefficient Copt to optimize the system performance for varying external disturbances is indicated by dashed lines in the plots

16.4 Conclusion

This work presented an assessment of the FPE-LPMG working stability. The evaluation was based on the simulation results obtained through a sensitivity analysis, interpreting how the model input variables influence the engine cycle stability. Hence, the objective to evaluate the ability of the system to operate correctly in various thermodynamic conditions and provide information with which a control model would be designed has been reached.

The study shows the importance of energy produced during the combustion in the stability of motor generator working. It was demonstrated when it works in the resonance frequency, the FPE-LPMG has more robustness and tolerates more variations in the thermodynamic conditions. The numeric model proposed in previous works was useful to assess the stability and to represent the phenomena that were described in the literature for internal combustion motors when the intake pressure and AFR vary.

It highlighted the importance of setting the charge over the generator as the input conditions vary in the linear motor. This is the most important role played by the control strategy, i.e., the control strategy as demonstrated should be able to adjust in its optimal value the amount of energy generated by the machine assuring its working stability.

Acknowledgements The authors acknowledge the financial support given by FAPEMIG, CNPq, CAPES, and FINEP in terms of grants and support for this research.

References

Achten PAJ (1994) A review of free piston engine concepts. SAE International

Blair GP (1996) Design and simulation of two-stroke engines. Society of Automotive Engineers Inc., Warrendale

Devillers E, Besnerais JL, Lubin T, Hecquet M, Lecointe JP (2016) A review of subdomain modeling techniques in electrical machines: performances and applications. In: Proceedings of the 2016 XXII international conference on electrical machines (ICEM)

Ferguson CR, Kirkpatrick AT (2015) Internal combustion engines: applied thermosciences. Wiley

Gordon B (1999) Design and simulation of four-stroke engines. Design and simulation of four-stroke engines. SAE

Gysen BLJ, Meessen KJ, Paulides JJH, Lomonova EA (2010) General formulation of the electromagnetic field distribution in machines and devices using fourier analysis. IEEE Trans Magnet 46(1):39–52

Heywood JB (1988) Internal combustion engine fundamentals. Mcgraw-Hill, New York

Heywood JB (2018) Internal combustion engine fundamentals. McGraw-Hill Education, New York

Jia B, Smallbone A, Feng H, Tian G, Zuo Z, Roskilly AP (2016) A fast response free-piston engine generator numerical model for control applications. Appl Energy 162:321–329

Mikalsen R, Roskilly AP (2007) A review of free-piston engine history and applications. Appl Therm Eng 27(14–15):2339–2352

Mikalsen R, Roskilly AP (2008) The design and simulation of a two-stroke free-piston compression ignition engine for electrical power generation. Appl Therm Eng 28(5):589–600

Rigobello LS, Santos MB, Carvalho SR (2021) Analysis and design of a free piston engine generator with methane fuel. In: Proceedings of the 15th international conference on heat transfer, fluid mechanics and thermodynamics

Rigobello LS (2020) Projeto e dimensionamento de um motor de pistão livre para operação em associação com um gerador elétrico linear. Master, Federal University of Uberlandia

Rodrigues RV, Santos MB, Carvalho SR (2019a) A model of a linear permanent magnet altenator for design and optimization: part I. Modelling and validation. In: Proceedings of the 25th ABCM international congress of mechanical engineering (COBEM 2019). ABCM, Uberlandia

Rodrigues RV, Santos MB, Carvalho SR (2019b) A model of a linear permanent magnet altenator for design and optimization: part II. Parametric analysis. In: Proceedings of the 25th ABCM international congress of mechanical engineering (COBEM 2019). ABCM, Uberlândia

Rowland SB, Whitehouse ND (1979) Internal combustion engines. Pergamon

Schafer F, Basshuysen V (2004) Internal combustion engine handbook basics, components, systems, and perspectives. SAE International

Secretary-General U (1987) Report of the world commission on environment and development. New York

Van Blarigan P, Paradiso N, Goldsborough S (1998) Homogeneous charge compression ignition with a free piston: a new approach to ideal otto cycle performance. SAE International

Chapter 17
Magnetorheological Dampers in Landing Gears: Dynamic Analysis Focused on Shock Absorption and Vibration Comfort

Felipe Lima de Abreu, Murilo Cesar Filipus, Douglas D. Bueno, and Rodrigo B. Santos

Abstract This chapter investigates the use of Magnetorheological (MR) dampers in aircraft landing gear, aiming to improve safety, efficiency, and comfort during the landing. MR dampers can adjust the damping force in real time, adapting its behavior to specific needs during this critical phase of the flight. The study considers the Spencer dynamic model, which includes non-linear characteristics, such as that the hysteretic effect, and the decrease in the damping force if the velocity approaches zero. The work introduces force decomposition to the actuator into hydraulic, pneumatic, and active components. The landing gear system is modeled with two degrees of freedom, taking into account the vertical oscillations. The analysis is conducted by numerical simulations using the fourth-order Runge–Kutta integrator method. The results are evaluated in terms of shock absorption efficiency and vibration comfort. The results obtained demonstrate the potential of the technologies based on MR dampers to improve flight safety and efficiency. Furthermore, these technologies have potential industrial applications to make landings more comfortable for passengers.

Keywords MR damper · Landing gear · Dynamics analysis · Vibration comfort

F. L. de Abreu (✉) · M. C. Filipus · R. B. Santos
Federal University of Grande Dourados, Rodovia Dourados/Itahum, Km 12, Dourados, Mato Grosso do Sul 79.804-970, Brazil
e-mail: felipeabreu507@gmail.com

D. D. Bueno
São Paulo State University, Avenida Brasil nº 56—Centro, Ilha Solteira, São Paulo 15385-007, Brazil

P. Kurka and M. Pereira (eds.), *Advances in Structural Vibration*, Lecture Notes in Mechanical Engineering, https://doi.org/10.1007/978-3-031-71540-2_17

17.1 Introduction

In recent decades, innovative and promising technologies have emerged regarding the attenuation of vibrations using dampers, such that different applications involving Magnetorheological (MR) dampers. MR dampers are different from standard devices, and they are designed to take advantage of the MR fluid properties, whose characteristics can be controlled through a magnetic field. Discovered in the mid-1940s by Jacob Rabinow, these fluids are able to vary their resulting viscosity, presenting a semi-solid behavior in a magnetic field and an aspect similar to a Newtonian fluid in the absence of the applied magnetic field (Schwartz 2008). In the automotive and aeronautical industries, these dampers are gaining notoriety, mainly in car suspension applications (Masa'id et al. 2023) and in landing gears (Laporte 2021).

Landing gear is part of a typical aircraft that plays an extremely important role in aviation, providing a safe connection between the aircraft and the ground during landings, takeoffs, and even in ground maneuvers. This component supports the weight of the fuselage as well as contributes to attenuate the vibrations that occur during the landing. Typically, aircraft are equipped with passive oil-pneumatic landing gear that provides high-efficiency shock absorption (Currey 1988). However, it is difficult for passive landing gear to attend performance specifications in various landing conditions that are outside the designed conditions (Jo et al. 2021), Then, an alternative to these problems is found in semi-active dampers, which offer the advantages of passive and active characteristics (Wu et al. 2007).

The active dampers can adjust their damping force through control-based strategies (Dyke et al. 1999). They also can ensure stability by reverting to the passive mode if electric voltage is not applied to it. This behavior between passive and active from semi-active dampers offers more precise and efficient control of vehicle dynamic behavior, resulting in an improved stability and comfort while driving vehicles or aircraft. The performance and stability of the developed landing gears are evaluated through impact tests, in which they are validated by numerical simulations. Numerous impact test simulations are performed using a simplified dynamic model, to evaluate the behavior and investigate the supported loads, during the landing, known as touchdown. This simplified model consists of a 2 degrees of freedom (2DOF) system (Batterbee et al. 2007).

A dynamic model with high fidelity is commonly necessary to design controllers if MR dampers are employed in a particular application. Then, several models for this type of shock absorber have emerged over time, such as the Bingham model, Gamota and Filisko model, and the Bouc-Wen model (Spencer et al. 1997). Although each one of these models has some disadvantages, such as accurate hysteretic response representation, the change in force magnitude in relation to low speeds, and numerical instability in computational simulations, they provide interesting insight when investigating dynamics with dissipation devices. In this context, this chapter considers the dynamic analysis of the landing gear of a light aircraft, considering an MR damper instead of a conventional damper. The Spencer model is considered since it includes the non-linear behavior, which allows one to investigate the hysteretic effect, and exhibits the force-decreasing behavior if the involved velocity approaches zero. A landing gear model with 2 degrees of freedom is considered, which takes into account the vertical oscillations. The MR damper is applied between the first and second mass. The shock absorber force is rewritten and presented in terms of pneumatic, hydraulic, and active components (due to the acting magnetic field). The shock absorption efficiency is computed, as well as the Jerk parameter, which is used to evaluate the vibration comfort.

17.2 Spencer Model

Figure 17.1 shows a representation of an MR damper with a circular piston. The magnetic field is applied radially to the fluid, perpendicular to the direction of its displacement. The modeling of this system is carried out with the Spencer model, which is a phenomenological model that presents responses that realistically describe the experimental observation of the physical responses of the system. This type of modeling describes physical phenomena from empirical data. The model is useful to represent physical phenomena in a realistic way, it is able to perform consistent analysis, and it describes a good relation between measured data and system responses (Mcmullin 1968).

Spencer et al. (1997) observed the force response and noted that when no electric voltage is applied to the coil (no magnetic field), the damper response behaves as a purely viscous device, but if the voltage is applied, a Bingham plastic behavior parallels the behavior of a viscous fluid arises, and for a certain voltage level, the increase of this force is nearly linear. The model proposed by Spencer is a modified Bouc-Wen model, which manages to better represent the real response of this type of damper device. In the modified model, the classic Bouc-Wen model is placed in series with a linear damper to model the response of the device when its velocity is close to zero, and this is in parallel with a spring which is added to take into account

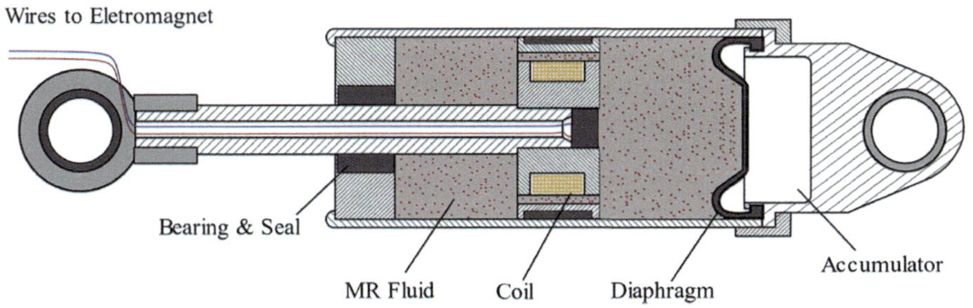

Fig. 17.1 MR damper (adapted from Dyke et al. (1999))

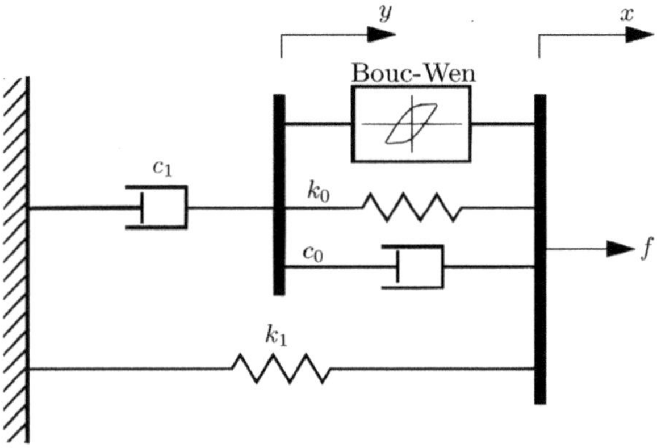

Fig. 17.2 Schematic representation of the Spencer MR fluid damper model, where x and y are respectively the displacements of the rightmost and leftmost bar, f is the force exerted by the damper, and c_1, c_0, k_1 and k_0 are model parameters

the accumulator from the device (Spencer et al. 1997). A schematic representation of the model is presented in Fig. 17.2.

17.2.1 Equation of MR Damper

The force exerted by the damper is given by Eq. 17.1 follows

$$f = \alpha z + c_0(\dot{x} - \dot{y}) + k_0(x - y) + k_1(x - x_0) \tag{17.1}$$

where z is an evolutionary variable (Wen 1976), and $\dot{y}$ is obtained from a sum of forces in the upper region of the schematic model, and it is given by,

$$\dot{z} = -\gamma(\dot{x} - \dot{y})|z||z|^{n-1} - \beta|\dot{x} - \dot{y}||z|^n + A(\dot{x} - \dot{y}) \tag{17.2}$$

$$\dot{y} = \frac{1}{(c_0 + c_1)}\{\alpha z + c_0\dot{x} + k_0(x - y)\} \tag{17.3}$$

the parameters change their values based on the applied electric voltage and consequently with the change in the magnetic field (Spencer et al. 1997), proposed a first-order filter that varies in terms of the electric voltage (v), such that,

$$\dot{u} = \eta(u - v) \tag{17.4}$$

in which the value of the parameters varies linearly with its response, as shown below by Eqs. 17.5 and 17.6.

$$\alpha = \alpha(u) = \alpha_a + \alpha_b u \tag{17.5}$$

$$c_1 = c_1(u) = c_{1a} + c_{1b}u, \quad c_0 = c_0(u) = c_{0a} + c_{0b}u \tag{17.6}$$

The stiffness term in the model due to the accumulator is represented by k_1 and x_0 represents an initial deflection in the spring, while the viscous damping observed at high speeds is represented by c_0. The c_1 component refers to a dashpot, which introduces a non-linear decrease in the force–velocity relationships observed at low speeds, and the parameter k_0 is used to control stiffness at high speeds. By manipulating the parameters γ, β, and A, it is possible to control the linearity of the unloading and the smoothness of the solidification of the fluid. These adjustments allow controlling of the gradual progression and the transition from the liquid to the solid phase, ensuring the desired behavior. At last, x describes the piston displacement.

Furthermore, the relationship between force and velocity does not show significant deviations when acceleration and velocity have opposite signs and the magnitude of the velocities is small. Then, the model is able to maintain a consistent response, even when changes in velocity direction or magnitude occur. The 14 model parameters, c_{0a}, c_{0b}, k_0, c_{1a}, c_{1b}, k_1, x_0, α_a, α_b, γ, μ, A, n, and η need to be adjusted from experimental data for different voltage values v. For the application of the model in the system of two degrees of freedom of the landing gear, there is only the need to change the displacement x where this is considered the relative motion between the two masses, i.e., $x = s$, where $s = x_1 - x_2$ and the velocity is given by $\dot{s} = \dot{x}_1 - \dot{x}_2$.

17.2.2 Simplified Model of Landing Gear

The modeling of the landing gear considers the instant in which the system touchdown occurs. In that instant, the horizontal energy is absorbed by the aircraft brakes, and the vertical energy is absorbed by the landing gear through the tire. Figure 17.3 presents a system of 2 degrees of freedom that represents a landing gear with an MR damper, which is inspired by the simplified model of 1/4 of the vehicle, where m_2 represents the tire mass, m_1 represents the mass of the fuselage supported by the actuator (Spencer et al. 1997), and f the actuator force given by the Spencer model Fig. 17.2. The tire reaction force is given by $F_{\text{tire}} = k_t \cdot x^b$ is the stiffness of the tire and b is an exponent for including nonlinearity in the force, and x_1 and x_2 are the masses displacements from m_1 and m_2, respectively.

The model presented above has its dynamics described by a system of 2nd-order equations given by,

$$m_1\ddot{x}_1 = -f + m_1 g \tag{17.6}$$

$$m_2\ddot{x}_2 = f - k_t x_2^b + m_2 g \tag{17.7}$$

The force of the shock can be rewritten in the form of Eq. 17.8 to present the force components in terms of the pneumatic force (F_p) and the MR fluid (F_{mr}) force.

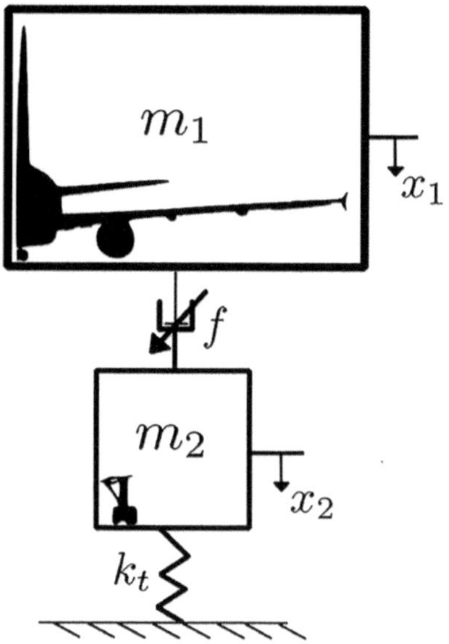

Fig. 17.3 Two degrees of freedom landing gear model

$$f = F_p + F_{\mathrm{mr}} \tag{17.8}$$

If the system is not influenced by a magnetic field (i.e., $v = 0\,V$) the forces acting on the system are the hydraulic force of the MR fluid (F_h) and the pneumatic forces due to the accumulator (F_p) such that.

$$F_p = k_1(x - x_0) \tag{17.9}$$

$$F_{\mathrm{mr}} = F_h, \quad \text{if} \;\; v = 0 \tag{17.10}$$

$$F_h = \alpha z + c_0(\dot{x} - \dot{y}) + k_0(x - y) \tag{17.11}$$

On the other hand, if a magnetic field is applied to the system ($v \neq 0\ V$), the MR fluid in contact with the field solidifies, presenting in parallel with the viscous damping, a damping similar to that provided by a Bingham fluid. In this model, the variation due to the magnetic field is projected into the variables c_0, c_1, and α, where they vary linearly with the voltage. The increase in hydraulic force relative to the fluid with no field applied is the active force (F_a), and it can be obtained by computing the difference between the forces $F_{\mathrm{mr\text{ - }on}}$(condition with magnetic field, $v \neq 0$) and $F_{\mathrm{mr\text{ - }off}}$ (condition without the magnetic field, $v = 0$), as presented in Eqs. 17.12 and 17.13.

$$F_{\mathrm{mr}} = F_h + F_a, \quad \text{if} \;\; v \neq 0 \tag{17.12}$$

$$F_a = F_{\text{mr - on}} - F_{\text{mr - off}} \tag{17.13}$$

The impact absorption efficiency is given by Eq. 17.14, where s is the relative displacement associated with the spring and F_s is the ground reaction force $(k_t \cdot x_2^b)$, which is also the force at the support.

$$E[\%] = \frac{\int_0^{s_{\max}} F_s \mathrm{d}s}{F_{s,\max} \cdot s_{\max}} \tag{17.14}$$

The device efficiency is computed based on the energy dissipated by the landing gear in relation to the total mechanical energy of the aircraft immediately before its contact with the ground (Jo et al. 2021). This parameter is essential to evaluate the performance of the MR fluid in comparison with other types of dampers, such as those purely hydraulic and pneumatic devices.

Another important performance indicator is the Jerk (j), which is the rate of acceleration over time, which can be determined by the following equation

$$j = \frac{\mathrm{d}}{\mathrm{d}t}\ddot{x}_1 \tag{17.15}$$

This parameter is computed in relation to the displacement of the mass m_1 (fuselage) and it provides interesting information to evaluate the level of comfort caused by the system acceleration over time.

17.2.3 System Parameters

The proposed methodology is investigated by considering the parameters of the landing gear previously considered by Jo et al. (2021). They are presented in Table 17.1. The MR damper device is described by the Spencer model, and the values of its parameters were adjusted in the model through algorithms of optimization, such as done by Spencer et al. (1997) and Wu et al. (2007). However, the authors sought from empirical methods to obtain values for the parameters so that the scale of the force generated by the shock absorber to be close to the results presented by Jo et al. (2021). The adjusted parameters are shown in Table 17.2.

Table 17.1 Landing gear parameter (Jo et al. 2021)

Parameters	Values
m_1	245 (kg)
m_2	15 (kg)
k_t	412 (kN/m)
b	1.13

Table 17.2 Parameters of Spencer's model

Parameters	Values	Parameters	Values
c_{0a}	4,737.6 (N.s/m)	α_a	0.056 (N/m)
c_{0b}	197.4 (N.s/V.m)	α_b	2,820.513 (N/V.m)
k_0	31,025.641 (N/m)	γ	363 (m^{-2})
c_{1a}	2,000.32 (N.s/m)	β	363 (m^{-2})
c_{1b}	987 (N.s/V.m)	A	301
k_1	19,743.59 (N/m)	n	2
x_0	0 (m)	η	190

17.3 Results and Discussion

Numerical simulations were carried out using an implemented Python-based algorithm. The integration of the system differential equations over time was performed using a fourth-order Runge–Kutta integrator from the SciPy library, where t taken into the range $0 < t \leq 2$ s, with a total of 65,536 points. Three different cases were evaluated considering different electric voltages, i.e., 0, 1.5, and 3 Volts, for applying passive control on and off. The analysis considers the time instant of the system touchdown, that is, at the exact moment that contact with the ground occurs, the initial conditions for the landing gear system were used equal to zero for the mass displacements and 3 m/s as sink speed, which is a commonly used standard as presented by Yoon et al. (2020). The damper initial conditions were taken equal to zeros.

Figure 17.4 presents the results for the case with three different electric voltages. Note that Fig. 17.4a, b show a reduction in the piston stroke and velocity, from the relative displacement and its velocity between the masses (s, $\dot{s}$) with increasing voltage and, consecutively, increasing the magnetic field. Figure 17.4c shows that an increase in the applied electric voltage, the ground reaction force also has growth, but this also causes greater damping of the oscillations in time, followed by the application of smaller forces in the next stages of compression and expansion. On the other hand, Fig. 17.4d shows the strut force by piston displacement. The force exerted by the structure has two distinct peak moments during compression (in $v = 0$ V).

The first peak is a result of hydraulic force (around 15 mm, $v = 0$ V), while the second peak is caused by the compressed air force (the region between $150 < s <$

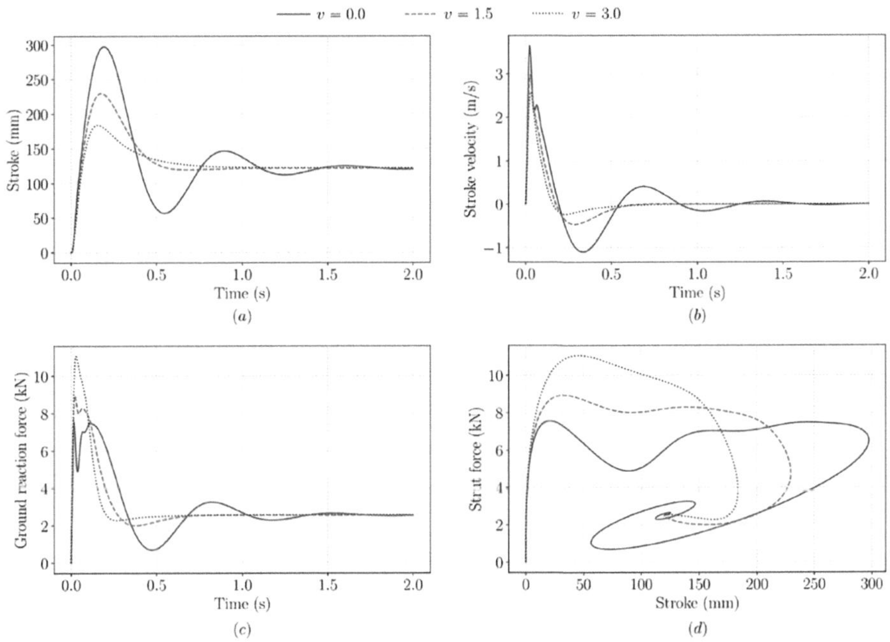

Fig. 17.4 Landing gear responses over time: **a** stroke, **b** stroke velocity, **c** ground reaction force, **d** strut force

290 mm, $v = 0$ V), which is better seen in the model present by Yoon et al. (2020), where the pneumatic force is given by an ideal gas. The efficiency is given by the area below the curve, where the distance between the peaks takes an important role, that is, the smaller the disparity between these two peaks implies to a more efficient system. Regarding the MR fluid, it can be seen that a higher system voltage, presents itself as more fluid solidified. The results show that the first peak is amplified due to the increase in electric voltage, consequently causing the graph to present a larger amplitude in the strut force direction. However, the pneumatic force is pulled to the left and the valley is suppressed, compelling the two peaks more parallel with each other, such that the area below the curve increases, and improves the landing efficiency.

Figures 17.5, 17.6 and 17.7 resent the force exerted by the damper considering different elements to compose it, i.e., the terms hydraulic, pneumatic, and magnetic with the three components to allow a visualization of the scale of their forces and the action of the tension in the total response of the system. The results show that if no voltage is applied Fig. 17.5, the system response to the model used responds similarly to that observed by alternative models as presented by Jo et al. (2021), Yoon et al. (2020). Figure 17.4a shows that the initial stage of the first compression is dominated by the hydraulic term due to the high velocity of the stroke, and at the end of the compression the pneumatic force acts due to the large displacement, the two peaks can be seen in Fig. 17.4b.

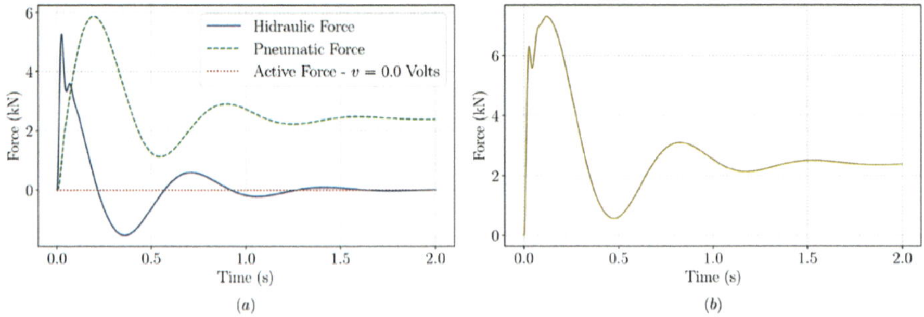

Fig. 17.5 Forces acting on the system: **a** force components, **b** actuator force ($v = 0$ v)

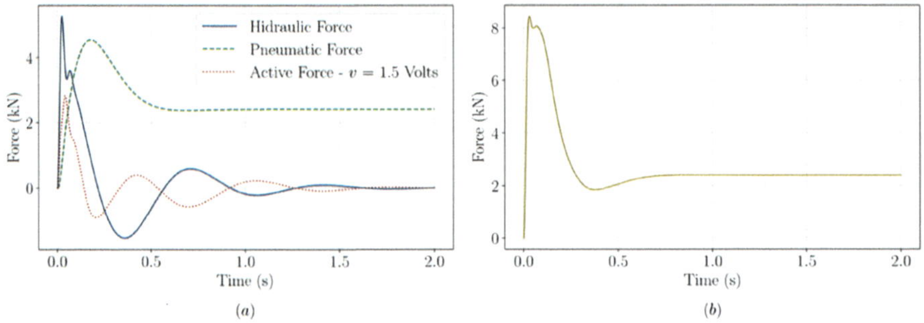

Fig. 17.6 Forces acting on the system: **a** force components, **b** actuator force ($v = 1.5$ V)

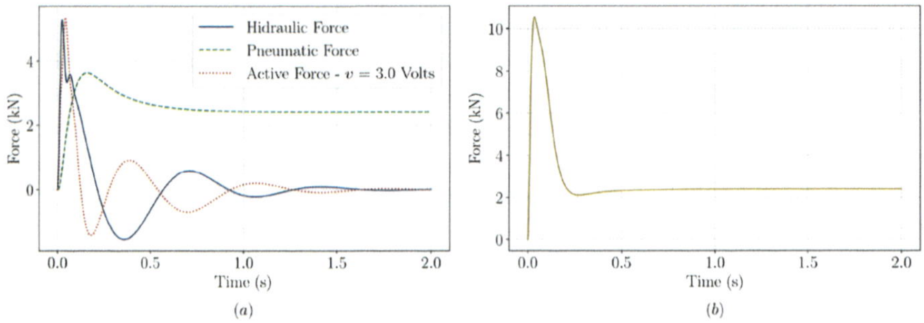

Fig. 17.7 Forces acting on the system: **a** force components, **b** actuator force ($v = 3.0$ V)

If an electric voltage is applied, Figs. 17.6 and 17.7, show a similar behavior, if an electric voltage is applied. However, this time decreasing the scale of force related to the pneumatic part, since the system has enough damping to reduce the displacement of the stroke in the first stage of compression due to the activation of the force provided by the action of the magnetic field, causing the system to oscillate with a small displacement and thus making the force produced by this term almost constant, the reduction of this force can be seen most prominently in Fig. 17.4d.

Figure 17.8a represents the displacement of the mass of the fuselage. It is possible to observe its movement at the time of landing, therefore, it is noted that the higher displacement level in the first impact that the aircraft experiences. Figure 17.8b presents the aircraft acceleration, and it becomes an important parameter in relation to the comfort of the passengers. So, it is possible to observe that in the first seconds, the acceleration is high, which suggests an important potential discomfort to the passengers. In addition, Fig. 17.8c shows its time variation rate, i.e., the system jerk, from which is possible to observe that the jerk decreases over time, thus the system comfort increases since the disturbances are minimized.

The drop in the peak at the first contact is due to the wheel with the ground, with the increase in voltage there is a more sudden drop, due to the MR fluid causing the system to stabilize faster, as the energy dissipation at the first contact is bigger than the other ones. On top of that, 8 days presents the efficiencies for each voltage applied to the system, in the case of 0 V the system efficiency is given only by the hydraulic and pneumatic forces. If the electric voltage is applied, the MR fluid starts to act, consequently increasing the efficiency of the system, presenting an efficiency of 69.5% for 0 V, 76.3% for 1.5 V, and 77.2% for 3 V.

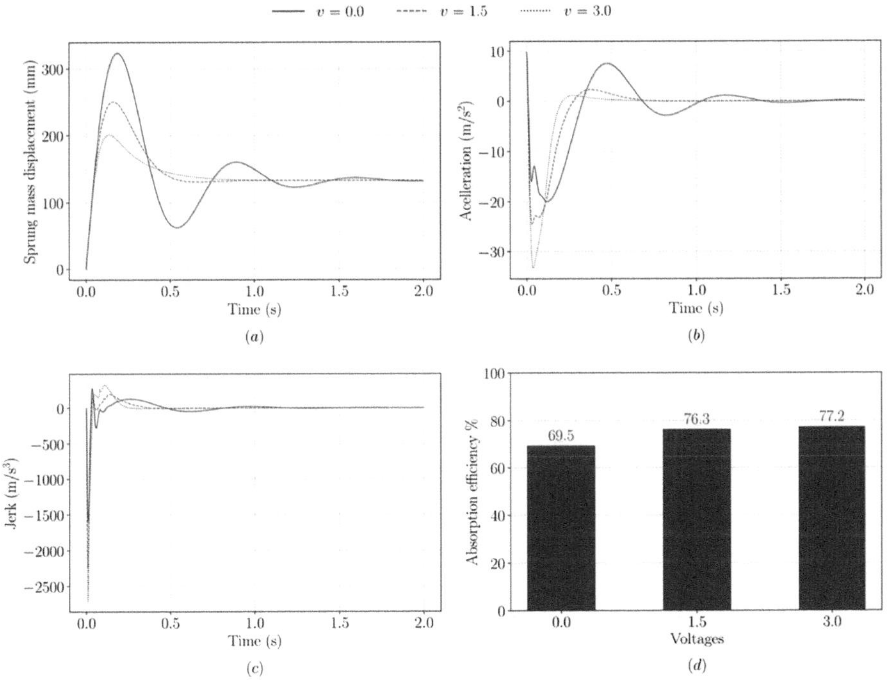

Fig. 17.8 Efficiency and comfort parameters: **a** sprung mass displacement, **b** sprung mass acceleration, **c** sprung mass jerk, **d** absorption efficiency

17.4 Conclusion

In this chapter, the Spencer MR damper model was used for modeling a magnetorheological damper device connected to a 2 DOF landing gear system. The model force components were evaluated to investigate the contributions of the pneumatic, hydraulic, and magnetic terms. Then, it was possible to understand the action of each one separately from each other. The Spencer model was used to adjust the geometric and physical parameters involved to obtain the MR force.

Numerical simulations were carried out focused on designing a passive controller. The response of the MR fluid in the landing gear system at the moment of landing due to an arbitrary voltage was obtained. In addition, three different electric voltages were used (0, 1.5, and 3 V). An increase in shock absorption was observed in comparison with the case of 0 V, such that 1.5 and 3 V of 9.8% and 11.2%, respectively. In relation to the jerk, this has a higher initial peak due to the deceleration in the initial instant provided by the active effect of the damper, but over time there is a decrease. Taking the area of the absolute value of the jerk over time, it is obtained that the jerk for the case of 1.5 and 3 V in relation to 0 V has a decrease of −19.5 and −1.6%, which is an interesting result since the minimization of jerk is favorable to increase comfort in terms of the vibrations. For future work, a parameter adjustment method to the Spencer model will be proposed and the use of a controller should be applied for a greater gain in efficiency and the minimization of jerk, where it can be observed that this last parameter of comfort, when having a higher voltage applied rises again, which can be improved using an active controller.

References

Batterbee D, Sims N, Stanway R, Wolejsza Z (2007) Magnetorheolocalanding gear: 1. A design methodology. Smart Mater Struct 16(6):2429
Currey NS (1988) Aircraft landing gear design: principles and practices. Aiaa
Dyke S, Spencer B, Sain M, Carlson J (1999) Modeling and control of magnetorheological dampers for seismic response reduction. Smart Mater Struct 5:565. https://doi.org/10.1088/0964-1726/5/5/006
Jo BH, Jang DS, Hwang JH, Choi YH (2021) Experimental validation for the performance of mr damper aircraft landing gear. Aerospace 8(9):272
Laporte DJ (2021) Redução de vibrações e supressão de shimmy de trens de pouso por designação de modos via pseudorealimentação
Masa'id A, Lenggana BW, Ubaidillah U, Susilo DD, Choi SB (2023) A review on vibration control strategies using magnetorheological materials actuators: application perspective. Actuators MDPI 12:113
Mcmullin E (1968) What do physical models tell us. Stud Log Found Math 52:385–396
Schwartz M (2008) Smart materials. CRC Press
Spencer D, SainM K et al (1997) Phenomenological model for magnetorheological dampers. J Eng Mech ASCE 123(3):230–238
Wen YK (1976) Method for random vibration of hysteretic systems. J Eng Mech Divis 102(2):249–263

Wu DS, Gu HB, Hui L (2007) Ga-based model predictive control of semi-active landing gear. Chin J Aeronaut 20(1):47–54

Yoon JY, Kang BH, Kim JH, Choi SB (2020) New control logic based on mechanical energy conservation for aircraft landing gear system with magnetorheological dampers. Smart Mater Struct 29(8):084003

Chapter 18
The Application of Magneto-Rheological Elastomer in a Base Insolation System for Seismic Mitigation of Highway Bridges

Amarildo Tabone Paschoalini, Fernanda Carolina de Almeida, Matheus Silva Proença, Henrique Edno Leoncini de Carvalho, Daniel Henrique de Sousa Obata, and Jeferson Camargo Fukushima

Abstract In recent years, the implementation of damping systems in civil structures has gained considerable attention, with emphasis on special building projects under dynamic loadings. In this scenario, due to their mechanical simplicity and controllable properties, Magneto-Rheological Elastomer (MRE) has stood out as providing an interesting alternative for vibration isolation. In these materials, the magnetic particles present in the elastomeric matrix are easily polarized in the presence of an external magnetic source, generating non-linear and reversible changes in the material, within a few milliseconds. In this way, the present work investigated numerically the efficiency of a certain damping system with MRE in isolating vibrations at the base of a bridge superstructure. The elastomer was simulated as a visco-elastic material of Kelvin–Voigt, and its stiffness and viscosity were regulated for five different scenarios. The unidimensional equivalent mechanical model was considered a single-degree-of-freedom (SDOF) system, and the ground motion generated by seismic excitations corresponded to shear excitations at the base. The base-isolated tests provided acceleration transmissibility under the different applied magnetic fields. It was observed that the viscoelastic support (VS) was able to shift the resonance frequency and the attenuation of transmissibility peak efficiently through field control. Moreover, with an adequate approach in the frequency domain, the random signal of a real earthquake was also inserted into the system for the evaluation of the isolator material. The findings demonstrated the good performance of the proposed MRE and its possibility of seismic vibration mitigation.

Keywords Magneto-rheological elastomers · Damping systems · Vibration isolation

---

A. T. Paschoalini (✉) · F. C. de Almeida · M. S. Proença · H. E. L. de Carvalho · D. H. de Sousa Obata · J. C. Fukushima
Department of Mechanical Engineering, Faculty of Engineering, UNESP, Ilha Solteira, Brazil
e-mail: amarildo.tabone@unesp.br

P. Kurka and M. Pereira (eds.), *Advances in Structural Vibration*, Lecture Notes in Mechanical Engineering, https://doi.org/10.1007/978-3-031-71540-2_18

18.1 Introduction

A Bridge's structure can be divided into the superstructure, composed of deck and beams, mesostructure (columns), and infrastructure. Between the superstructure and mesostructure are locating bearings. According to Niemierko (2016), bearings are intermediate structural elements that help in the bridge's durability, reliability, and safety behavior. These devices are responsible for linking elements such as beams and pillars. In recent decades, a great development of modern structural designs for bearings can be observed, Block et al. (2013 *apud* Niemierko 2016).

For this reason, smart materials have been integrated as bearings in the form of damping systems to optimize vibration isolation. Among these materials, Magneto-Rheological Elastomer (MRE) stands out, which is classified as a viscoelastic material (Jolly et al. 1996), whose mechanical and rheological properties can be modified through the application of an external magnetic field (Bastola and Hossain 2020). This classification of MRE as a viscoelastic material occurs because it presents characteristics between a viscous material and an elastic material when deformed. The elastic phase of the material is based on Hooke's law of isotropic elasticity and has no residual deformation, while, the viscous part is described by the Linear Law of Viscosity, defined by Newton. Thus, the graphic behavior of the MRE forms ellipsoidal curves showing slope, typical of elastic materials, and relaxation, typical of viscous materials.

Due to these properties, Tian and Nakano (2017) state that MREs have recently been used in a wide variety of applications, such as adapted vibration absorbers, dampers, sensors, and others. The damping occurs through the viscous flow of the rubber matrix and the inclusion of magnetic particles that allow additional, interfacial damping through the interaction of the particles (Khimi and Pickering 2015). At first, it is necessary to incorporate a magnetic field source to change the characteristics of the MRE. Using a controllable source, particularly an electromagnetic component, the supply of electric current to the coil generates a magnetic field, increasing the stiffness of the elastomer. These changes are non-linear, completely reversible, and occur in a few milliseconds (Ginder et al. 1999; Kukla et al. 2017).

First, to ensure the efficient insertion of this absorber, it is necessary to determine its internal forces and displacements, generally assuming that the structure is subjected to an arbitrary time-invariant load. One way to achieve this determination is to simplify the structure to perform a mechanical analysis, that is, to transform the structure into a system with n degrees of freedom (DOF) where the motion equations are obtained, thus enabling a prediction of the behavior along with some criteria such as natural frequency, stiffness, and damping. This is how this work intends to contribute to the studies of the present scenario.

This paper is organized as follows. Following this introduction, Sect. 18.2 describes the numerical modeling and procedures adopted for the simulation of the magneto-rheological elastomer. Exposure and discussion of the results are given in Sect. 18.3 and some conclusions are given in Sect. 18.4.

18.2 Materials and Methods

18.2.1 Structure Modeling

Due to the complexity of the real scale bridge system, this article uses only the structural form of a beam bridge to study the structure model with the vibration isolation system (Fig. 1a). Therefore, a system of concentrated masses with a single-degree-of-freedom (SDOF) was considered to model the superstructure and to analyze the similarity of the model with the best configuration of the MRE damper, as shown in Fig. 1b.

The modeling starts with the description of the structure, considering its elements in a simplified way and the relationships between them. This set of elements forms the spatial model which has information on the mass and resistance of the system. An MRE model and stiffness elements will be included in the system to describe the dampers. Finally, it will be assumed that the structure suffers displacement only in the longitudinal direction of the superstructure, in a shear way. The excitation force generated causes a pure shear stress τ in the MRE sample which deforms at a displacement x, shown in Fig. 18.2.

Consider a unidimensional system with single-degree-of-freedom of a viscoelastic material subjected to a harmonic force $F(t) = F_0 \cdot e^{j\omega t}$, as indicated in Fig. 18.1b. The equation of motion of the inertial mass, m, can be described as:

$$m\ddot{x} + k\left(1 + j\eta^*\right)x = F_0 e^{j\omega t} \tag{18.1}$$

where k is the material's stiffness, η^* its loss factor equivalent to $\eta^* = \eta/Gf$, where η is the viscosity, G is the shear modulus, f is the frequency, and $k(1 + j\eta^*) = k^*$ is the complex stiffness.

The complex stiffness formulation is derived from the stress–strain relationship for a linear viscoelastic material, treated by Kelvin–Voigt rheological model.

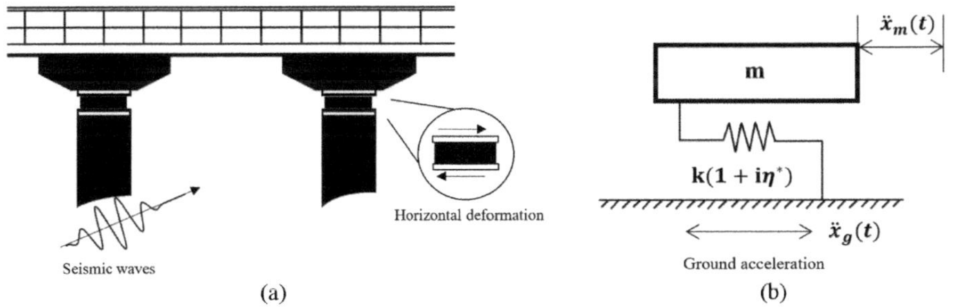

Fig. 18.1 **a** Illustrative image of the bridge structure, **b** mechanical system with viscoelastic damping

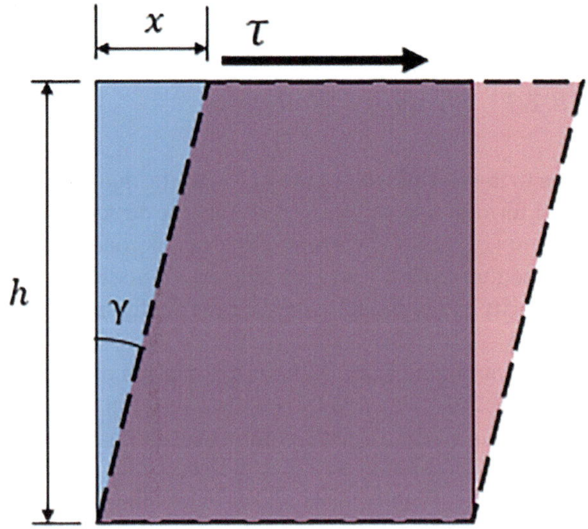

Fig. 18.2 Deformation due to shear

$$\tau = G\gamma + \eta\dot{\gamma} \tag{18.2}$$

Shear stress in its basic form is given by the relationship between the applied force and the area that is sheared. Moreover, the stress can be deduced by the constitutive relation of the material, i.e. the product of the shear modulus G and the angle γ generated by the displacement. For small angles, it is assumed that its tangent is equal to the angle itself ($\tan\gamma = x/h \approx \gamma$). Therefore, making such manipulations, the stiffness k of the material is given as:

$$G = \frac{\tau_{xy}}{\gamma_{xy}} = \frac{\frac{F}{A}}{\frac{x}{h}} = \frac{k \cdot h}{A}; \quad k = \frac{G \cdot A}{h} \tag{18.3}$$

where τ, F, and A are the shear stress, force, and area, respectively, G, x, d, h are the shear modulus, displacement, and sample length, respectively.

For the application of excitation, an earthquake signal was adopted to observe the behavior of the system when subjected to severe situations. The simulation was made with a real earthquake signal, which occurred in Iwate—Japan, in 2008. The signal was obtained from the PEER Ground Motion Database—NGA-West2 (2023) with a frequency within the range of 1–50 Hz. Generally, an earthquake is referred to by its size, for this reason, magnitude scales such as the Richter scale are used. However, for engineering applications, the best way to analyze a seismic event is by the variation of acceleration or displacement of the ground in time. From this signal it is possible to extract parameters that will serve as a basis for understanding the behavior of an earthquake, considering, for example, its response spectrum.

Table 18.1 Values for shear modulus, G (MPa)

Magnetic fields (T)	Frequency (Hz)										
	10	15	20	25	30	35	40	45	50	55	60
0.00	0.534	0.567	0.588	0.597	0.595	0.601	0.607	0.609	0.606	0.608	0.604
0.07	0.636	0.673	0.683	0.699	0.701	0.713	0.716	0.725	0.732	0.733	0.745
0.14	0.864	0.913	0.921	0.953	0.958	0.983	0.996	1.026	1.032	1.029	1.041
0.21	1.212	1.248	1.271	1.304	1.305	1.341	1.330	1.325	1.324	1.327	1.290
0.28	1.698	1.810	1.836	1.890	1.879	1.890	1.873	1.876	1.864	1.858	1.873
0.35	2.142	2.303	2.392	2.429	2.397	2.395	2.405	2.424	2.420	2.428	2.427

Table 18.2 Values for the viscosity, η

	Magnetic fields (T)					
	0.00	0.07	0.14	0.21	0.28	0.35
Viscosity (kPa.s)	1.8475	2.2081	3.1177	3.7668	4.9474	5.4649

18.2.2 MRE Modeling

The magneto-rheological elastomer's bearing used as a reference has a matrix consisting of white silicone rubber high flexibility Redeleaser and has a concentration of 33% carbonyl iron magnetic particles (Sigma > 97%) with area $A = 1200 \times 1200$ mm and thickness $e = 360$ mm—dimensions similar to those used by Block et al. (2013 *apud* Niemierko 2016) for common elastomeric bearings. An inertial mass on the material $m = 150$ kg was assumed, while the elastic and viscous properties of the MRE were obtained from a previous characterization carried out by Carvalho (2023), for shear stress of 0.49 Pa, a frequency range between 10 and 60 Hz and field magnetic between 0.00 and 0.35 T. Tables 18.1 and 18.2 show the properties used.

18.3 Results and Discussion

For each proposed arrangement, the absolute transmissibility was calculated (Eq. 18.4). The Motion Transmissibility Frequency Response Function is essential in vibration isolation studies since it defines the amount of motion transmitted from the base to the inertial mass, per unit of input motion in the support. Figures 18.3 and 18.4 show the results obtained.

$$\mathrm{Tr} = k\left(1 + i\eta^*\right)/ - m\omega^2 + k\left(1 + i\eta^*\right) \tag{18.4}$$

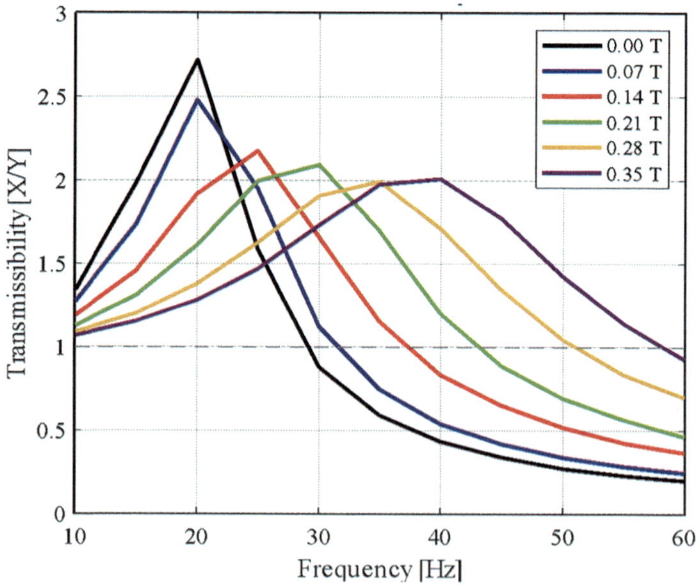

Fig. 18.3 Absolute transmissibility of displacement for each magnetic field

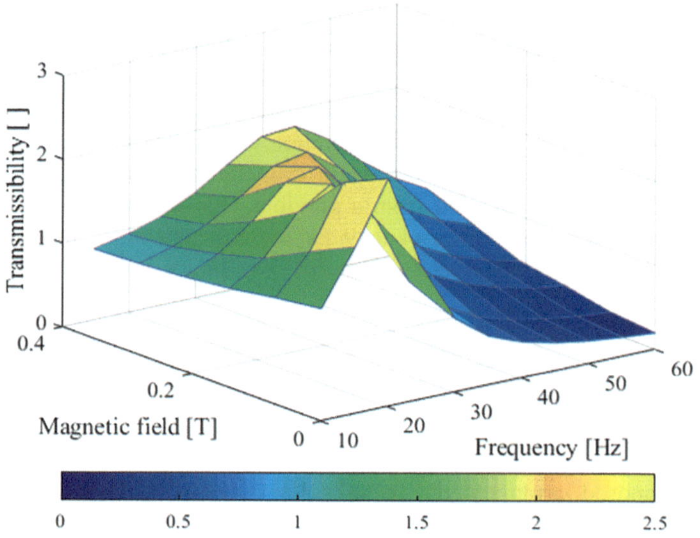

Fig. 18.4 Transmissibility surface

where k, η^*, m, and ω are stiffness, loss factor, mass, and angular frequency, respectively.

As can be seen, changing the magnetic field can control the amplitude and frequency of the material's resonance peak. From 0.00 to 0.35 T, the transmissibility was reduced by approximately 72%, and the resonance peak dropped from 20 to 40 Hz.

In addition to directly influencing the region controlled by damping by changing the viscosity of the material, the imposed magnetic field changes its stiffness, thus controlling the region of resonances.

Analyzing the system's Transfer Function, shown in Fig. 18.5—the ratio between the output displacement and the input force (Eq. 18.5)—it is also possible to show the control action of the magnetic field on the simulated MRE.

It is important to point out that the geometric parameters of the MRE are also of great importance in the design of the insulators, because, as shown in Eq. (18.3), they directly influence the material stiffness.

$$H = 1/-m\omega^2 + k\left(1 + i\eta^*\right) \quad (18.5)$$

Finally, to simulate a scenario close to reality, the random excitation of the collected real earthquake was inserted at the base of the insulator.

As depicted in Fig. 18.1, it was assumed that the earthquake generates pure shear in the MRE. Figure 18.6 shows the applied earthquake displacement and Fig. 18.7 presents the Fourier Transform of the signal.

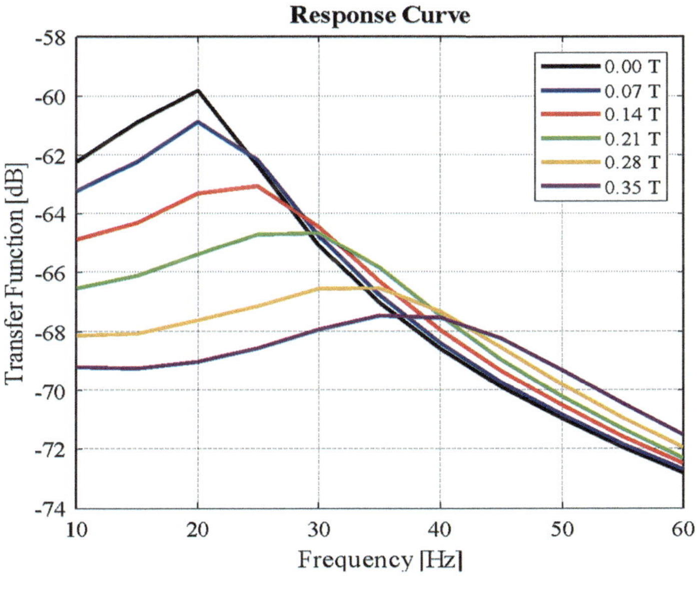

Fig. 18.5 System transfer function H

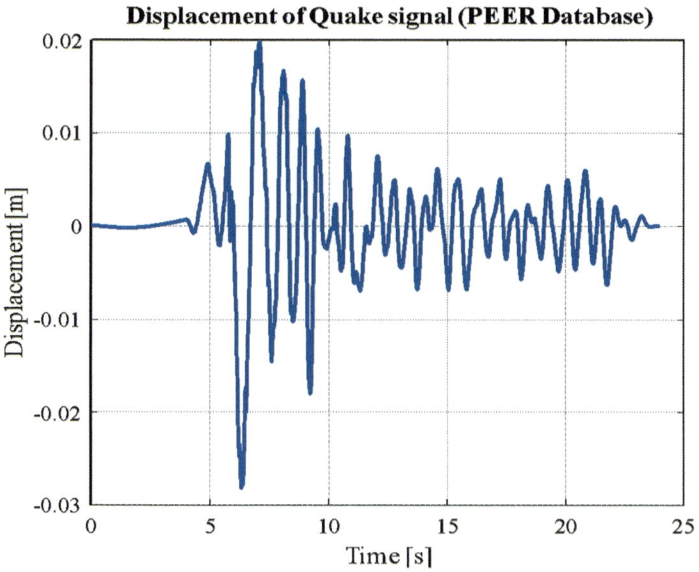

Fig. 18.6 Time signal of the earthquake displacement

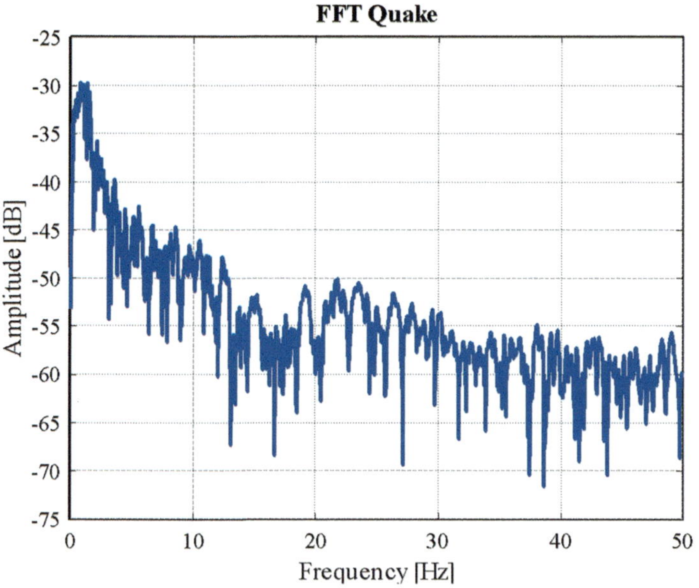

Fig. 18.7 Earthquake signal FFT

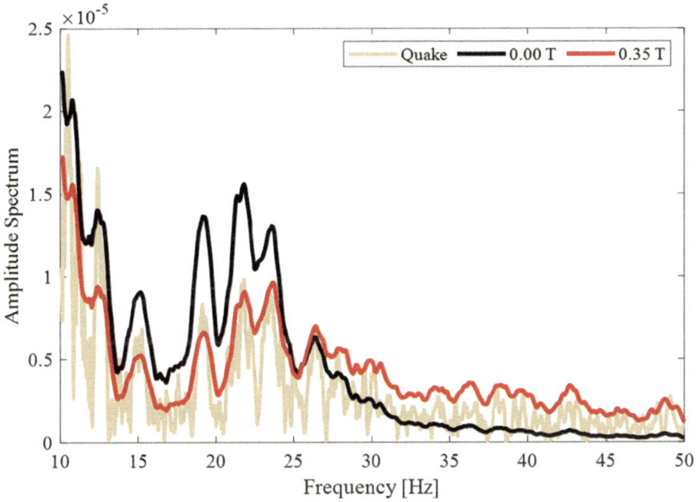

Fig. 18.8 System response to real earthquake excitation under different magnetic fields

Figure 18.8 shows the influence of the magnetic field in controlling the system response. As can be seen, under the larger field, the peaks of the earthquake signal were reduced in the range of up to approximately 20 Hz. In contrast, the signal gained more amplitude at frequencies above 27 Hz.

The existing peaks between 19 and 24 Hz in the array devoid of field action, are due to the resonance region of the material for the case, as observed in Figs. 18.3 and 18.4.

18.4 Conclusion

From the results obtained, it can be concluded that the MRE was able to change its mechanical properties when subjected to an applied magnetic field, which showed a good agreement of data about the literature.

Larger magnetic fields are expected to further control the system's resonance peaks. Future research will characterize the elastic and viscous properties of the material for a wider frequency range and under higher magnetic fields, to better investigate the application of the material as a bridge insulator.

However, the previous results obtained already indicate that the material is a good candidate for the development of a damping system for bridges. Its advantage over current systems is the effective control of the magnetic field over its stiffness and damping.

18.5 Responsibility Notice

The authors are the only ones responsible for the printed material included in this paper.

Acknowledgements This work was supported by the Coordination for the Improvement of Higher Education Personnel (CAPES).

References

Bastola AK, Hossain M (2020) A review on magneto-mechanical characterizations of magnetorheological elastomers. Compos B Eng 200:108348

Block T, Eggert H, Kauschke W (2013) Lager im Bauwesen. 3. Auflage. Wilhelm Ernst & Sohn, p 644. ISBN: 978-3-433-60323-9

Carvalho HEL (2023) Estudo do emprego de elastômero magneto-reológico para controle de vibrações em estruturas de engenharia civil. Tese (mestrado)—Universidade Estadual Paulista Júlio de Mesquita Filho, Faculdade de Engenharia de Ilha Solteira

Ginder JM et al (1999) Magnetorheological elastomers: properties and applications. Smart Struct Mater Smart Mater Technol 12:131–138. https://doi.org/10.1117/12.352787

Jolly MR, Carlson JD, Munoz BC (1996) A model of the behavior of magnetorheological materials. Smart Mater Struct 5(5):607–614

Khimi SR, Pickering K (2015) Comparison of dynamic properties of magnetorheological elastomers with existing antivibration rubbers. Compos B Eng Surv 83(1):175–183. https://doi.org/10.1016/j.compositesb.2015.08.033

Kukla M et al (2017) The determination of mechanical properties of magnetorheological elastomers (MREs). Proced Eng 177(1):324–330. https://doi.org/10.1016/j.proeng.2017.02.233

Niemierko A (2016) Modern bridge bearings and expansion joints for road bridges. Transp Res Proced 14:4040–4049. https://doi.org/10.1016/j.trpro.2016.05.501

Pacific Earthquake Engineering Research Center. PEER ground motion database. NGA West 2. https://ngawest2.berkeley.edu/site/documentation

Tian T, Nakano M (2017) Fabrication and characterization of anisotropic magnetorheological elastomer with 45° iron particle alignment at various silicone oil concentrations. J Intell Mater Syst Struct 29(2):151–159. https://doi.org/10.1177/1045389x17704071

Chapter 19
Dynamic Analyses (Modal and Force Vibration) in a Railway Vehicle Model with 25 Degrees of Freedom

Fernando Luiz Martinechen Beghetto

Abstract In this work, the modal and the forced vibration dynamic analyses are carried out in a railway vehicle model with 25 degrees of freedom. The objective of this work is to study the dynamic characteristics (modes and frequencies of vehicle vibration), as well as the analysis of forced vibration, allowing the simulation of resonance conditions and their critical speeds, important for the study of vehicle stability, safety, and passengers' comfort. The tridimensional model has a passenger car body, two bogies, and four wheelsets. Models of vertical, lateral, and rotational track irregularities are used. In mechanical contact between wheels and rails, the theories of Hertz and Kalker, and also Vermeulen and Johnson are used. The equations of motion of the coupled system are numerically integrated using the Newmark Method, in the MatLab software. It is possible to conclude that the railway track irregularities associated with critical traffic speeds produce vibrations in the vehicle. Due to the damping of the system, these vibrations are attenuated between the rigid bodies through the primary suspensions, between the wheels and bogies; and secondary suspensions, between the bogies and the passenger car body.

Keywords Railway vehicle · Track irregularities · Mechanical contact

19.1 Introduction

The railway transport in Brazil stands out for the number of passengers to be transported when compared to other transport types. In Rio de Janeiro's SuperVia, more than 1 million passengers are transported daily. In São Paulo's CPTM, more than 5.3 million passengers are transported daily. This modal is also used in other Brazilian cities: Fortaleza (CE), Natal (RN), João Pessoa (PB), Recife (PE), Maceió (AL), Salvador (BA), Brasília (DF), Belo Horizonte (MG) and Porto Alegre (RS).

F. L. M. Beghetto (✉)
UTFPR, Federal University of Technology—Paraná, Curitiba, Brazil
e-mail: flbeghetto@utfpr.edu.br

P. Kurka and M. Pereira (eds.), *Advances in Structural Vibration*, Lecture Notes in Mechanical Engineering, https://doi.org/10.1007/978-3-031-71540-2_19

The objective of this work is to carry out a modal dynamic analysis and forced vibration in a railway vehicle through computational models, considering track irregularities and mechanical contact between wheels and rails.

19.2 Railway Vehicle Model

Figure 19.1 illustrates the model with its respective 25 degrees of freedom.

The present model is conceived through the association of rigid bodies interconnected to suspension systems. It has four wheelsets interconnected to the two bogies through the primary suspensions. The bogies in turn are interconnected to the car body by the secondary suspensions. The equations of motion are obtained through dynamic equilibrium. This model is capable of representing 25 degrees of freedom of motion, namely: bouncing, rolling, and pitching of the car body, bouncing, rolling, and pitching of the bogies, and bouncing, lateral, rolling, and yawing of four wheelsets. These equations of motion can be written as matrix system of second-order differential equations below:

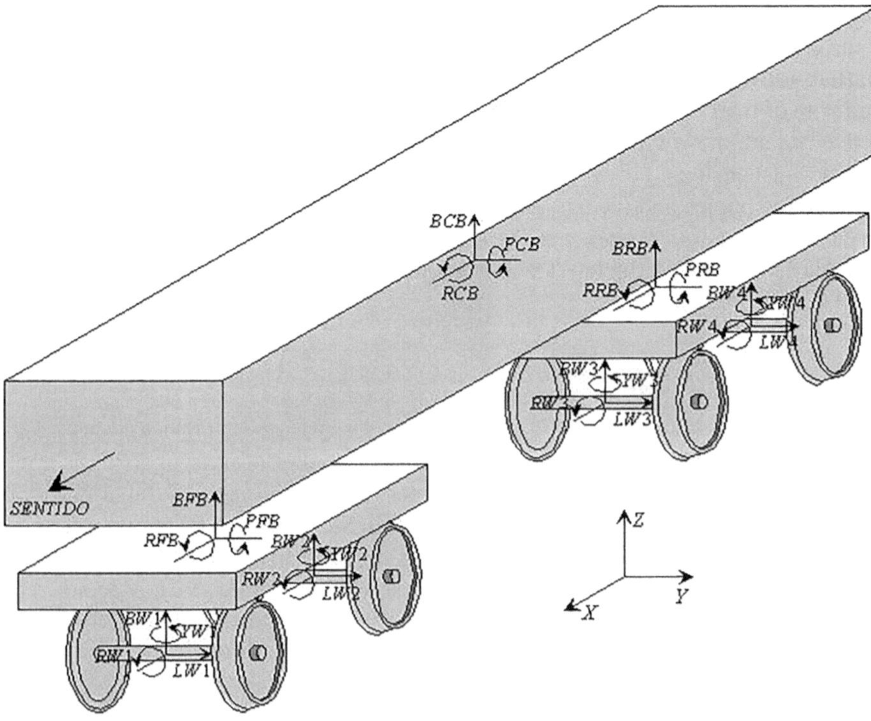

Fig. 19.1 Railway vehicle model

$$[M_V]\{\ddot{U}_V(t)\} + [C_V]\{\dot{U}_V(t)\} + [K_V]\{U_V(t)\} = \{P_V(t)\}. \qquad (19.1)$$

where $[M_V]$, $[C_V]$, and $[K_V]$ are the mass, damping, and stiffness matrices, respectively, and $\{\ddot{U}_V(t)\}$, $\{\dot{U}_V(t)\}$, $\{U_V(t)\}$, and $\{P_V(t)\}$ are the acceleration, velocity, displacement, and external forces vectors, respectively. The complete formulation and its respective developments are presented in detail in Beghetto (2011). The dynamic properties of the vehicle and the geometric distances are shown in Tables 19.1 and 19.2, respectively.

Table 19.1 Dynamic properties of the vehicle

Vehicle parts	Parameters	Values	Units
Car body	Mass	49,120	kg
	Mass moment of inertia (rolling)	64,050	kg m^2
	Mass moment of inertia (pitching)	2008,950	kg m^2
Front and rear bogies	Mass	11,000	kg
	Mass moment of inertia (rolling)	8750	kg m^2
	Mass moment of inertia (pitching)	20,800	kg m^2
Wheelsets	Mass	1500	Kg
	Mass moment of inertia (rolling)	989	kg m^2
	Mass moment of inertia (yawing)	989	kg m^2
Primary suspension	Spring coefficient	1637,275	N/m
	Damping coefficient	32,7455	N s/m
Secondary suspension	Spring coefficient	875,655	N/m
	Damping coefficient	17,5131	N s/m
Lateral links	Spring coefficient	100,000	N/m
	Damping coefficient	2000	N s/m
Longitudinal links	Spring coefficient	100,000	N/m
	Damping coefficient	2000	N s/m

Table 19.2 Geometric distances

Distances	Values	Units
Longitudinal distance of center of bogies	15	m
Longitudinal distance of center of wheelsets	2.6	m
Cross distance of center of wheelsets	2.5	m

19.3 Track Irregularities Models

The wheelsets' vibrations are produced by track irregularities through the mechanical contact between wheels and rails profiles. The wheelsets motions are absorbed by the primary suspensions, and transmitted to the bogies. These transmit the damped vibrations to the car body through secondary suspensions.

The length of the railtrack in this simulation is 300 m, and various track irregularities and model combinations are used to produce vibrations. The vertical model is based on sinusoidal harmonic functions, (Dahlberg 2006). The amplitudes used are 10 mm, and the lengths are varied to (1, 2, 2.5, 7.5, and 10 m), allowing to simulate the amplification of the dynamic responses of the vehicle.

For new profiles of wheels and rails, the variation of the gauge is assumed (9–15 mm), according to Brina (1979). The lateral model of track irregularities is also based on sinusoidal harmonic functions. The hunting motion of the wheelsets through the track is composed of a translation and rotation simultaneously, modeled by the cosinusoidal harmonic functions with 18.234 m of wavelength (Pombo et al. 2009).

19.4 Wheel-Rail Contact Models

According to Hertz's theory, when compressing two elastic bodies with curved surfaces, due to the mutual deformation of these bodies, the contact area generated will be an ellipse, (Timoshenko and Goodier 1980).

In Kalker (1990), contact with a certain amount of slippage between surfaces is considered, which is limited by Vermeulen and Johnson's model, (Iwnicki 2003; Pombo et al. 2009). In this model, the tangential contact force and the spin rotation moment are restricted to a cubic polynomial. Contemplating the geometric nonlinearities due to the profiles of the wheels and the rails, and the limits of adhesion in the contact between these bodies.

19.5 Modal Analyses of the Vehicle

Widely used for the characterization of dynamic systems, modal analysis determines the circular natural frequencies (eigenvalues) and their respective vibration modes (eigenvectors). One can find ways to solve the problem in question in Chopra (1995), and Bathe (1996).

Table 19.3 presents the circular natural frequencies (w_{n}), the damped circular natural frequencies (w_{d}), the cyclic natural frequencies (f_{n}), the damped cyclic natural frequencies (f_{d}), the periods of natural vibration (T_{n}), the periods of damped natural vibration (T_{d}), and the damping ratios (z_{v}) of the vehicle. The type of damping is classic.

Table 19.3 Frequencies, periods, and damping ratios

No	w_n (rad/s)	w_d (rad/s)	f_n (Hz)	f_d (Hz)	T_n (s)	T_d (s)	z_v
1st	8.165	8.138	1.300	1.295	0.770	0.772	0.082
2nd	13.923	13.787	2.216	2.194	0.451	0.456	0.139
3rd	14.84	14.676	2.362	2.336	0.423	0.428	0.148
4th	15.896	15.694	2.530	2.498	0.395	0.400	0.159
5th	17.776	17.493	2.829	2.784	0.353	0.359	0.178
6th	18.365	18.053	2.923	2.873	0.342	0.348	0.184
7th	52.107	44.474	8.293	7.078	0.121	0.141	0.521
8th	53.057	44.975	8.444	7.158	0.118	0.140	0.531
9th	71.926	49.97	11.447	7.953	0.087	0.126	0.719
10th	80.018	47.991	12.735	7.638	0.079	0.131	0.800

Here, only the first ten modes and frequencies are used, as the upper ones represent only rigid body motion. The natural, normalized modes of vibration are presented in the forced vibration analyses under the system's resonance conditions.

19.6 Forced Vibration Analyses of the Vehicle

The forced vibration of the vehicle will have as an external excitation source, track irregularities, associated with traffic speed. The equations of motion are integrated using the Newmark Method, with average acceleration (Bathe 1996). The system of linear equations is solved using the Gauss Elimination Method (Burden and Faires 2003).

Resonance will occur when the external excitation coincides with the damped circular natural frequencies (w_d). The vehicle will offer the resonant speeds (V_R) which are obtained through the damped circular natural frequencies (w_d) and the lengths of the irregularities (l_I), as follows:

$$\omega_d = \omega_I;\ \omega_I = 2 \cdot \pi \cdot f_I = \frac{2 \cdot \pi \cdot V_R}{l_I};\ V_R = \frac{\omega_d \cdot l_I}{2 \cdot \pi}. \tag{19.2}$$

Under the resonance condition, there will be an amplification of the dynamic, displacement (R_D), velocity (R_V), and acceleration (R_A) responses, which can be written as follows:

$$R_D = R_A = \frac{1}{2 \cdot \xi_v \cdot \sqrt{1 - \xi_v^2}};\ R_V = \frac{1}{2 \cdot \xi_v}. \tag{19.3}$$

With the values of the damping ratios listed in Table 19.3, the dynamic amplification factors for the displacement, velocity, and acceleration responses in the vehicle's vibration modes are determined. The critical speeds of the vehicle in resonance condition, and respective dynamic amplification factors are presented in Table 19.4. It is possible to observe that at low frequencies, greater amplification occurs since the damping ratio is lower in these situations.

In Fig. 19.2, it is observed that the 1st vibration mode is responsible for the lateral vibration of the wheelsets.

The transient dynamic response illustrates the harmonic oscillatory behavior, according to the irregularities model used. The behavior of the wheelsets is similar. The maximum displacement amplitudes are of the order of 3 mm, proving to be relatively low in this presented simulation.

In Fig. 19.3, it is observed that the 2nd mode of vibration is responsible for the vertical vibrations of the car body, the front and rear bogies, and the wheelsets.

The vibration behavior of the wheelsets is similar to each other, therefore, only the vibration of wheelset 1 was presented here. The same occurs between the front and rear bogies, therefore, only the vibration of the front bogie is presented here. It is noticed that the amplitudes of displacement of wheelset 1 are greater than those presented by the bogies, due to the presence of the primary suspensions. Those of the car body are smaller than those of the bogies, attenuated due to the secondary suspensions. Despite the resonant condition, the displacements are out of phase, due to the different geometry of the rigid bodies, under the irregularity model used. The displacement amplitudes presented are relatively small.

Table 19.4 Critical speeds, and dynamic amplification of the vehicle

Modes	1st	2nd	3rd	4th	5th	6th	7th	8th	9th	10th
V_R (km/h)	85	59	84	67	30	78	25	64	57	55
$R_D = R_A$	6.118	3.632	3.416	3.185	2.855	2.765	1.124	1.111	1.001	1.042
R_V	6.098	3.597	3.378	3.145	2.809	2.717	0.960	0.942	0.695	0.625

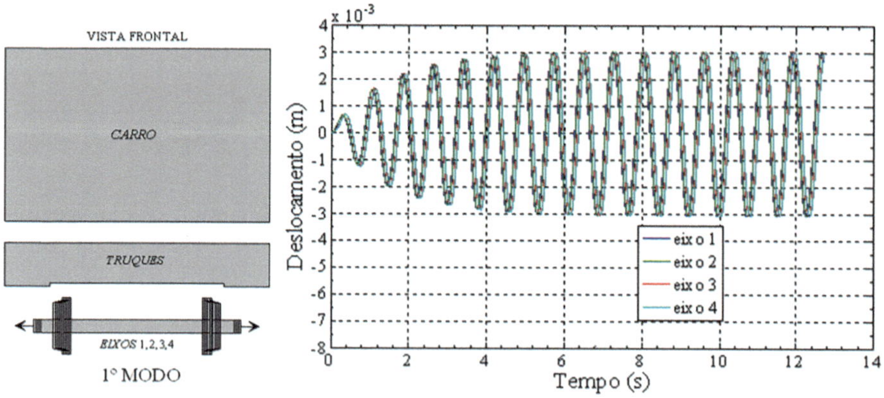

Fig. 19.2 Forced vibration in 1st mode

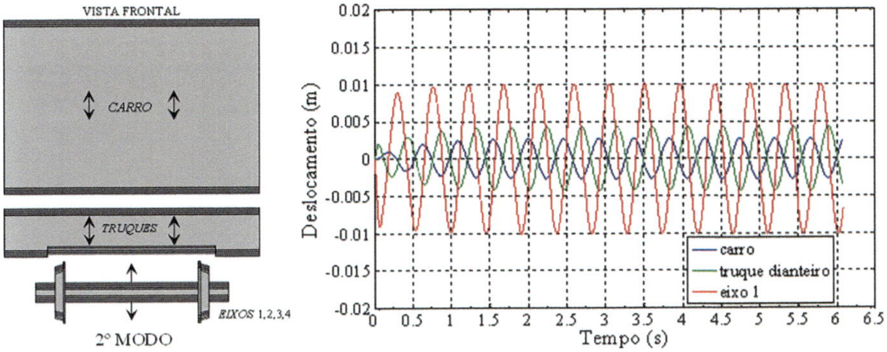

Fig. 19.3 Forced vibration in 2nd mode

In Fig. 19.4, it is observed that the 3rd mode of vibration is responsible for the vertical vibrations of the wheelsets, of the front and rear bogies, and the rotation movement (pitching) of the car body.

To induce the latter, without phase cancellation in the front and rear bogies, as well as in the wheelsets, a wavelength of 10 m and amplitude of 10 mm was used in the irregularities model.

Note an oscillatory movement according to the irregularity model used. The amplitudes of rotation (pitching) of the car body, as well as the produced angular accelerations, are relatively small, for the simulation carried out.

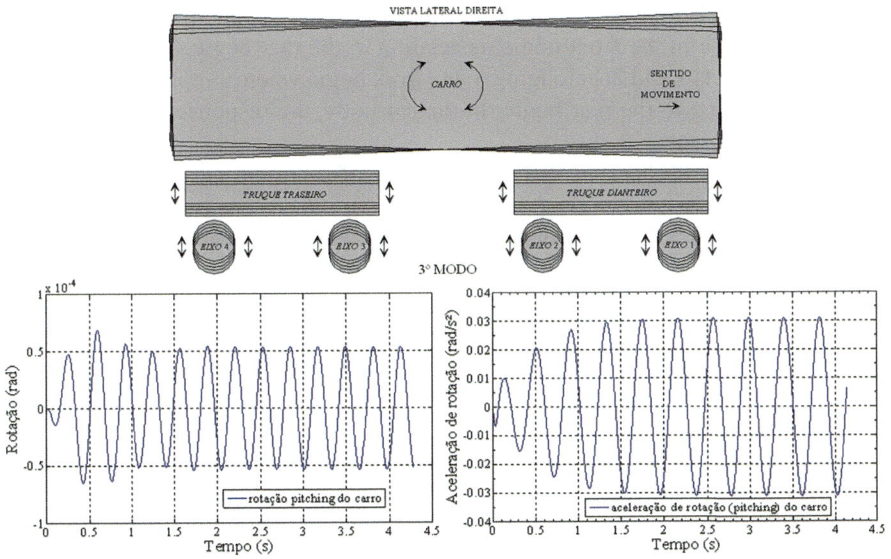

Fig. 19.4 Forced vibration in 3rd mode

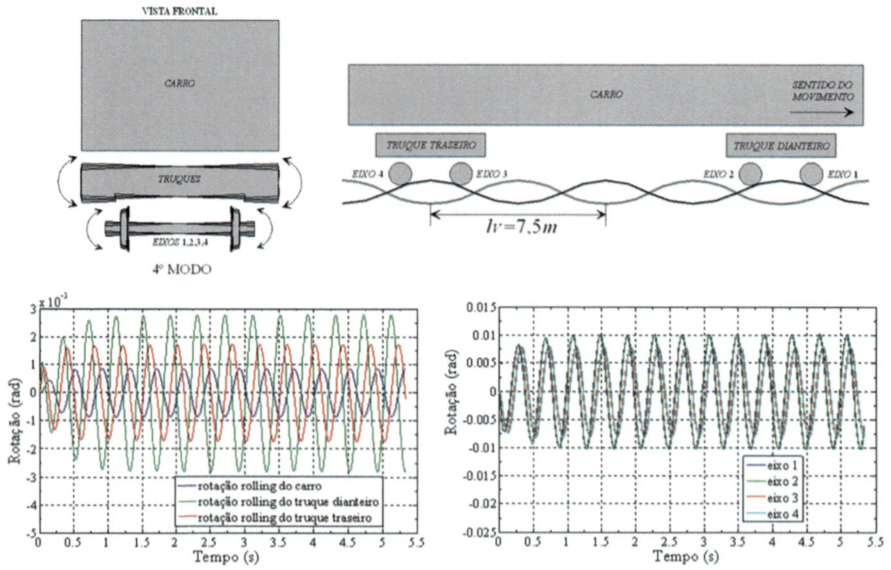

Fig. 19.5 Forced vibration in 4th mode

In Fig. 19.5, it is observed that the 4th mode of vibration is responsible for the rotational vibrations (rolling) of the front and rear bogies, and of the wheelsets. In this simulation, in the model of vertical irregularities used, the combination of lag between the amplitudes of the right and left rails was made, allowing, in this way, to analyze the behavior of this mode of vibration. There is an amplification in the dynamic response of the front bogie, in relation to the rear bogie. The same occurs between wheelsets 1 and 2, belonging to the front bogie when compared to wheelsets 3 and 4, belonging to the rear bogie. In the car body, the response is attenuated, in relation to the bogies.

In Fig. 19.6, it is observed that the 5th mode of vibration is responsible for the rotational vibrations (yawing) of the wheelsets. To induce this hunting behavior in the wheelsets, we used the calculation of the wavelength at which the phenomenon occurs, which was studied by Klingel and Boedecker, (Pombo et al. 2007), and results in 18.234 m, associated with the adopted rotational irregularities model. According to the studies by Mohan (2003), this phenomenon occurs in conditions of relatively low speed, in this case, for the excitation of the present mode, according to Table 19.4, there is the condition of critical speed (resonance) of 30 km/h. The amplitudes found are relatively small.

In Fig. 19.7, it is observed that the 6th mode of vibration is responsible for the vibrations of rotation (rolling) of the car body, the front and rear bogies, and the wheelsets. The vibration amplitudes of the front bogie are greater than those presented by the rear bogie, and they present a phase shift. Again the vibrations of the car body are less than those presented by the bogies. The angular acceleration is relatively small.

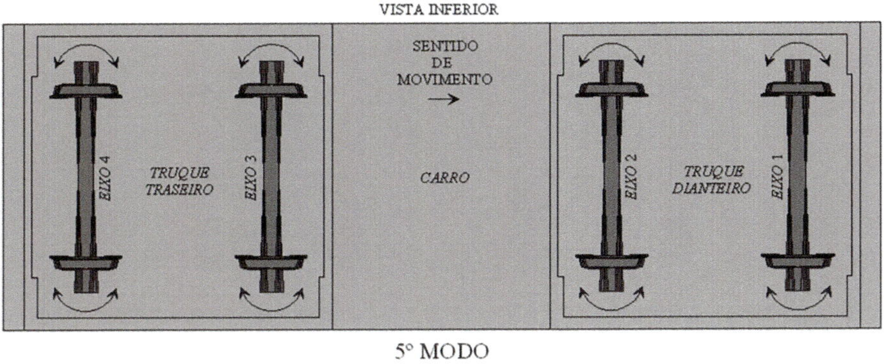

Fig. 19.6 Forced vibration in 5th mode

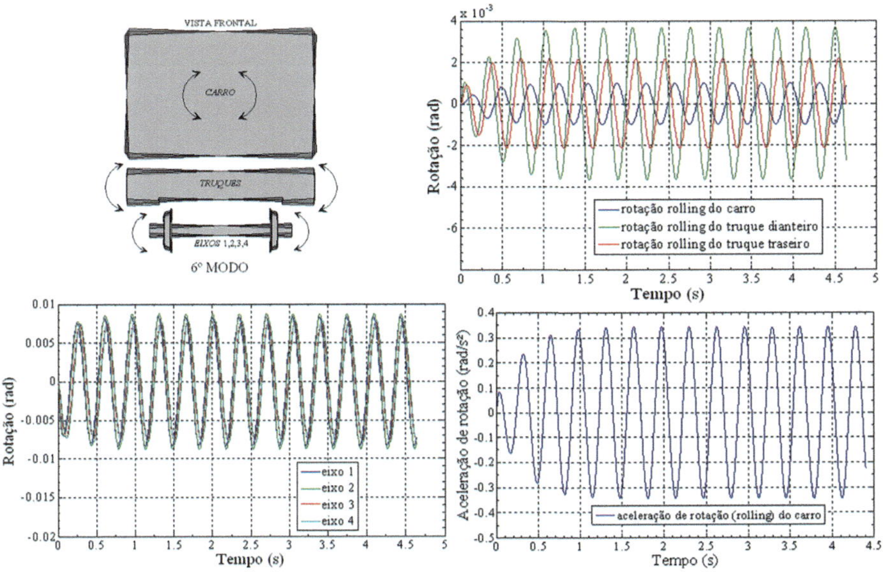

Fig. 19.7 Forced vibration in 6th mode

In Fig. 19.8, it is observed that the 7th mode of vibration.

This vibration mode is responsible for the vertical vibrations of the wheelsets, and rotations (pitching) of the front and rear bogies. In this simulation, amplitudes of 10 mm and lengths of 1 m were used in the model of vertical irregularities to induce the amplification of the rotation (pitching) of the front and rear bogies. There is a similar behavior between the wheelsets, presenting the same amplitudes, however lagged. The vertical displacements of the wheelsets are given in pairs, according to the front and rear bogies, respectively. Small displacement amplitudes are presented.

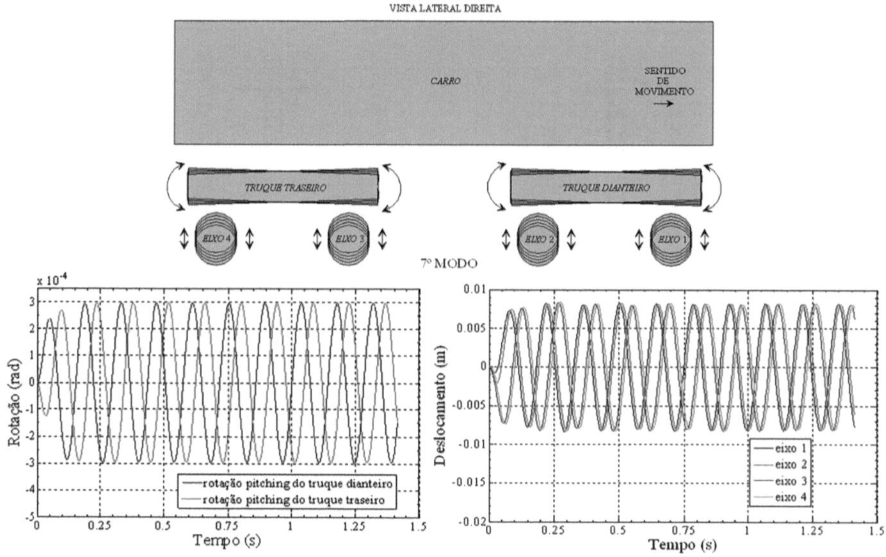

Fig. 19.8 Forced vibration in 7th mode

The relevance of choosing the wavelength of the vertical track irregularities model was noticed throughout the simulations, as it can lead to signal amplification or attenuation due to phase cancellation. It is noteworthy that the choice of wavelength will influence the critical (resonant) speed, according to Eq. (19.2) presented.

19.7 Conclusions

In this work, it was possible to analyze the dynamic behavior of a railway vehicle, in free vibration and in forced vibration produced by track irregularities associated with operating speeds.

In the free vibration analysis, the frequencies and respective modes of vibration of the system were obtained, which presented classical damping. The damping ratios showed an increase for higher natural frequencies. Vibration modes excite different vehicle degrees of freedom. The higher modes present only rigid body motion and were neglected in the study. In the forced vibration analysis, models of vertical, lateral, and rotational irregularities were used, combined with each other, allowing the simulation of the most unfavorable vibration conditions. The critical (resonant) speeds of the vehicle in unfavorable situations were determined.

In general, the dynamic responses presented are consistent with the track irregularities models used, presenting a harmonic and periodic behavior. The amplitudes of the dynamic responses presented in the proposed simulations were relatively small. The influence of the primary and secondary suspension systems was noticed, significantly reducing the dynamic responses of the bogies, and mainly of the car body.

Responsibility Notice The author is only responsible for the printed material included in this paper.

References

Bathe KJ (1996) Finite element procedures. Prentice-Hall, New Jersey

Beghetto FLM (2011) Modelagem tridimensional da interação dinâmica entre veículo e ponte ferroviária considerando contato roda-trilho, irregularidades da via e variação de velocidade. Tese (Doutorado), Pontifícia Universidade Católica do Paraná, Curitiba

Brina HL (1979) Estradas de Ferro. Volume 1. Via Permanente. Livros Técnicos e Científicos Editora

Burden RL, Faires JD (2003) Análise Numérica, Ed. Thomson

Chopra AK (1995) Dynamics of structures: theory and applications to earthquake engineering. Prentice-Hall, New York

Dahlberg T (2006) Handbook of railway vehicle dynamics, track issues. Taylor & Francis Group, LLC

Iwnicki S (2003) Simulation of wheel-rail contact forces. Fatigue Fract Eng Mater Struct 26:887–900

Kalker JJ (1990) Rolling contact phenomena: linear elasticity. Delft University of technology

Mohan A (2003) Nonlinear investigation of the use of controllable primary suspensions to improve hunting in railway vehicles. Master of Science in Mechanical Engineering Thesis, Faculty of Virginia Polytechnic Institute and State University, Blacksburg, Virginia, USA

Pombo J et al (2007) A new wheel-rail contact model for railway dynamics. IDMEC—Instituto Superior Técnico, Lisboa

Pombo J et al (2009) Wheel wear evolution and its influence on the dynamic behaviour of railway vehicles. In: Proceedings of the 7th EUROMECH solid mechanics conference, Lisbon

Timoshenko SP, Goodier JN (1980) Teoria da Elasticidade. 3º Edição. McGraw-Hill, Editora Guanabara Koogan

Chapter 20
Design of Passive Dynamic Absorbers to Attenuate Pathological Tremor of Human Upper Limb

Gabriel Guimarães de Souza Braga de Albuquerque and Marcela Rodrigues Machado

Abstract Limb tremors, prevalent in neurodegenerative diseases like Parkinson's disease, often lead to social discomfort and restrict daily activities. In addressing this challenge, using vibration-attenuating devices emerges as a promising technique, considering drug sensitivity, treatment invasiveness, and therapeutic efficacy. This study delves into optimizing the configuration of mechanical vibration absorbers to mitigate limb tremors in Parkinson's patients, exploring variations in absorber position and size. Controller efficacy is assessed by analyzing response magnitude and amplitude reduction across frequency and time domains. Findings underscore the significant tremor amplitude reduction achieved through vibration-absorbing devices, with parallel absorbers demonstrating the most effective configuration. Additionally, the study investigates the impact of employing more than two absorbers in parallel along the forearm and explores novel absorber designs tailored for limb tremor control. Analytical upper limb numerical models and passive controls are employed to estimate limb vibration characteristics and evaluate control efficiency across varying levels of tremor pathology excitation.

Keywords Parkinson's control · Rest tremor · Dynamic absorber

20.1 Introduction

The human body can exhibit tremors characterized by rhythmic oscillations of the limbs, leading to social discomfort and difficulty performing everyday tasks. While limb tremors can be easily observed in some cases, an accurate diagnosis requires careful investigation due to the variety of conditions that can manifest with this

G. Guimarães de Souza Braga de Albuquerque (✉) · M. Rodrigues Machado
Department of Mechanical Engineering, University of Brasília, Brasília 70910-900, Brazil
e-mail: gabriel.bsag@gmail.com

M. Rodrigues Machado
e-mail: marcelam@unb.br

P. Kurka and M. Pereira (eds.), *Advances in Structural Vibration*, Lecture Notes in Mechanical Engineering, https://doi.org/10.1007/978-3-031-71540-2_20

symptom, and treatment approaches may vary considerably from one condition to another, as discussed in Crawford III and Zimmerman (2018). Individuals with neurodegenerative diseases such as Parkinson's disease, white hand syndrome, essential tremor, and others (Chou 2004) increasingly require solutions for these conditions, which often present a range of unpleasant symptoms. Furthermore, certain medications can induce symptoms similar to those of Parkinsonism. The biomechanical dynamic of the limb system can be represented through the movement of masses (Gebai et al. 2018). In Hashemi et al. (2004), the authors introduced a two-degree-of-freedom (DOF) model of the upper limb with mass concentrated at the centroid and inertia. While this model described planar motion at the elbow and shoulder joints, it did not include wrist joint motion. The dynamic biomechanical model of an upper limb follows some literature that analyzes the upper limb based on its model characteristics (Adewusi et al. 2012; Gebai et al. 2016a, b, c; Hashemi et al. 2004).

Orthoses and wearable devices provide effective and noninvasive solutions (Mo and Priefer 2021) for managing tremors across different body parts, posing minimal risk to patients (Fromme et al. 2019). These devices employ passive, active, and semi-active vibrational control mechanisms. Passive controls dissipate energy without external power sources, responding to vibrations from the main system. Some researchers suggest using vibration absorbers to reduce tremors, such as a device with two tuned absorbers positioned on the arm's dorsal and ventral surfaces (Gebai et al. 2016a, b, c; Rudraraju and Nguyen 2018). Initially, addressing hand tremors involved the viscous beam orthosis (Kotovsky and Rosen 1998). Other proposed solutions include gloves embedded with metal disks for tremor reduction (Paulig et al. 2014) and particle dampers (Lu and Huang 2020). Additionally, solutions like the Vib-bracelet with a tuned mass damper target wrist pronation/supination tremors (Buki et al. 2017), while air dashpots integrated into orthoses address wrist and elbow tremors (Takanokura et al. 2011). Gyroscopic devices, including the GyroGlove with multiple gyroscopes on a fabric glove, aim to counteract tremors by resisting involuntary motions (De Panisse et al. 2018). Each gyroscope includes a rotatable disk capable of resisting involuntary movements. Recent metamaterial in designing a wearable device applied to the human limb was proposed in Machado and Moura (2023), also the technological efficiency was explored in Machado et al. (2020, 2022) and Moura et al. (2022, 2023).

This paper explores a device equipped with dual and triple passive dynamic absorber controllers to mitigate tremors induced by Parkinson's disease in the human upper limb. Leveraging detailed analytical modeling of the upper limb and passive control mechanisms, the study builds upon prior research in Gebai et al. (2016a, b, c) and Albuquerque et al. (2023). Through experimental evaluation across three stages of the disease, the device demonstrates notable effectiveness, yielding a substantial reduction in tremor amplitude. These findings promise to improve the quality of life for individuals with Parkinson's disease and underscore the potential of passive dynamic absorber technology in tremor management.

20.2 Human Upper Limb Mechanical Modeling

The analytical model of the upper limb utilized in this study is based on models proposed in Gebai et al. (2016a) and Hosseini et al. (2021). This model consists of a three-degree-of-freedom biodynamic representation, which includes flexion–extension planar motion across the shoulder, elbow, and wrist joints in the horizontal plane. Figure 20.1a illustrates the anatomical components of the human upper limb, while Fig. 20.1b depicts the schematic model with mechanical components. The model incorporates active input moments generated by the activation of shoulder, biceps, elbow, and wrist muscles at frequencies associated with resting tremors. The upper limb parameters, such as mass, inertia, length, mass distribution, density, and centroid position, are assumed to have similar values as provided in Contini (1972). The upper arm length is denoted as l_1, forearm length as l_2, and palm length as l_3, as illustrated in Fig. 20.1c. Table 20.1 contains the mass and inertia values for each segment. The total mass of the system under consideration is 3.77 kg. The centroids are labeled as a_1 (upper arm), a_2 (forearm), and a_3 (palm), with inertia denoted as I_1 for the upper arm, I_2 for the forearm, and I_3 for the palm.

Two configurations of dynamic vibration absorbers are tested: dual parallel absorber (DPA), consist of two single absorbers (SA1, SA2), positioned at a distance ab_i (where $i = 2, 3$) from point B (elbow), as depicted in Fig. 20.1. The triple parallel absorber (TPA) include an additional absorber on the upper arm at a distance ab_1

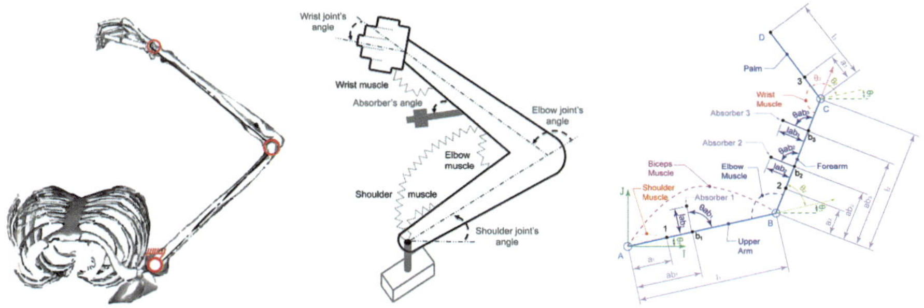

(a) (b) (c)

Fig. 20.1 Schematic of a human upper limb: **a** draft, **b** structural biomechanical model, and **c** dynamic biomechanical components with the triple parallel absorber [Adapted from Gebai et al. (2018)]

Table 20.1 Length and weight properties of each upper limb part (Gebai et al. 2016a)

Limb part	Length (cm)	Mass (kg)	Density (kg/m^3)	Centroid (m)	Moments of inertia (kgm^2/rd)
Upper arm	36.4	2.070	1088.0	0.427 l_1	0.0228
Forearm	29.9	1.160	1108.6	0.417 l_2	0.0082
Palm	20.3	0.540	1112.6	0.361 l_4	0.0012

Table 20.2 Joint muscles properties represent by stiffness and damping coefficient (Gebai et al. 2016a)

	Muscle Stiffness coefficient (Nm/rd)	Damping coefficient (Nms/rd)
Shoulder	170	0.002 $k1$
Elbow	150	0.002 $k2$
Biceps	40	0.002 $k3$
Wrist	10	0.001 $k4$

from the shoulder, resulting in an assembly of three single absorbers (SA1,SA2, SA3). The distance from the absorber's mass to its joint along the beam is denoted as l_{abi}, where $i = 1, 2, 3$. An additional mass of $m_\mathrm{d} = 0.0522$ kg is added to the absorbers at points b_i, where $i = 1, 2, 3$, to enhance the controller device (Gebai et al. 2016a). In both DPA and TPA configurations, the total device mass is 0.44751 kg. Table 20.2 lists the stiffness values assumed at the shoulder joint, indicated by k_1, elbow as k_2, biceps as k_3, and wrist as k_4.

20.3 Dynamic Model of the Upper Limb

The dynamic model of the upper limb, referred to as the principal system, consists of a three-degree-of-freedom (DOF) system. The motion begins from a state of stability when $\theta_1 = 0°$, $\theta_2 = 90°$, and $\theta_\mathrm{a} = 0°$, with zero angular velocities, as indicated in Fig. 20.1. The equations of motion follow those outlined in Gebai et al. (2016b). The kinematic equations utilize the Coriolis theorem to derive the three DOF systems concerning the three angular displacements relative to the global coordinate system, resulting in

$$v_1 = v_A + \Omega_{xyz/XYZ} \times r_{1/A} = a_1\dot{\theta}_1 j$$

$$v_B = v_A + \Omega_{xyz/XYZ} \times r_{B/A} = l_1\dot{\theta}_1 j$$

$$v_2 = v_B + \left(v_{2/A}\right) + \Omega_{xyz/XYZ} \times r_{2/B} = -a_2\left(\dot{\theta}_1 + \dot{\theta}_2\right)i + l_1\dot{\theta}_1 j$$

$$v_C = v_B + \left(v_{C/B}\right) + \Omega_{xyz/XYZ} \times r_{C/B} = -l_2\left(\dot{\theta}_1 + \dot{\theta}_2\right)i + l_1\dot{\theta}_1 j$$

$$v_3 = v_C + \left(v_{3/C}\right)_{x'y'z'} + \Omega_{x'y'z'/XYZ} \times r_{3/C} = -\left[(a_3 + l_2)\left(\dot{\theta}_1 + \dot{\theta}_2\right) + a_3\dot{\theta}_3\right]i + l_1\dot{\theta}_1 j$$

By employing the Lagrange equation and incorporating the Rayleigh dissipation function to account for damping effects, the equations of motion for the systems can be formulated as follows:

$$\frac{d}{dt}\left(\frac{\partial T}{\partial \dot{q}_i} - \frac{\partial U}{\partial \dot{q}_i}\right) - \frac{\partial T}{\partial q_i} + \frac{\partial U}{\partial q_i} = F_i + F_{ci},\ i = \{1, 2, 3\}. \tag{20.1}$$

The kinetic energy is represented as

$$T = \left[\frac{1}{2}I_1\dot{\theta}_1^2 + \frac{1}{2}m_1v_1^2\right] + \left[\frac{1}{2}I_2\left(\dot{\theta}_1 + \dot{\theta}_2\right)^2 + \frac{1}{2}m_1v_1^2\right] + \left[\frac{1}{2}I_3\dot{\theta}_3^2 + \frac{1}{2}m_4v_4^2\right] \tag{20.2}$$

where each segment (upper arm, forearm, and palm) was representing its portion in motion.

The potential energy is of the form

$$U = \frac{1}{2}k_1\theta_1^2 + \frac{1}{2}k_2\theta_2^2 + \frac{1}{2}k_3(\theta_1 + \theta_2)^2 + \frac{1}{2}k_4\theta_3^2 \tag{20.3}$$

and Rayleigh's dissipation function is represented by

$$C = \frac{1}{2}c_1\dot{\theta}_1^2 + \frac{1}{2}c_2\dot{\theta}_2^2 + \frac{1}{2}c_3\left(\dot{\theta}_1 + \dot{\theta}_2\right)^2 + \frac{1}{2}c_4\dot{\theta}_2^3. \tag{20.4}$$

The generalized equation of motion has the form

$$M\ddot{\theta} + C\dot{\theta} + K\theta = f, \tag{20.5}$$

where $\theta,\ \dot{\theta},\ \ddot{\theta}$ are the angular displacement, velocity, and acceleration vectors at the hand joints, respectively. For the uncontrolled hand system, $\theta_{1,\,2,\,3}$ corresponds to the flexion angle at the shoulder, elbow, and wrist joints, respectively. Hence, the matrix is expressed as

$$M = \begin{bmatrix} M_{11} & M_{12} & M_{13} \\ M_{21} & M_{22} & M_{23} \\ M_{31} & M_{32} & M_{33} \end{bmatrix},\ K = \begin{bmatrix} k_1 + k_3 & k_3 & 0 \\ k_3 & k_2 + k_3 & 0 \\ 0 & 0 & k_4 \end{bmatrix},\ C = \begin{bmatrix} c_1 + c_3 & c_3 & 0 \\ c_3 & c_2 + c_3 & 0 \\ 0 & 0 & c_4 \end{bmatrix}, \tag{20.6}$$

where

$$\begin{aligned} M_{11} &= \left(I_1 + m_1a_1^2\right) + \left(I_2 + m_2a_2^2\right) + m_2l_1^2 + I_3 \\ &\quad + m_3\left(l_1^2 + l_2^2 + a_3^2 + 2l_2a_3\right)M_{12} \\ &= \left(I_2 + m_2a_2^2\right) + m_4\left(l_2^2 + a_4^2 + 2l_2a_4\right) \end{aligned}$$

$$M_{13} = m_4\left(a_4^2 + l_2a_4\right)$$

$$M_{21} = M_{12}\ M_{22} = M_{12}\ M_{23} = M_{13}$$

$$M_{31} = M_{13}\ M_{23} = M_{13}\ M_{33} = I_4 + m_4 a_4^2$$

20.3.1 Excitation Force

The theory of the effects of Parkinson's disease on limb function examines how the illness influences muscle coordination and movement. It considers biological aspects of muscle structure and function, as well as the dynamics of the musculoskeletal system. The framework integrates knowledge related to how the nervous system adapts to utilize muscle and limb mechanics characteristics, along with assessing Parkinson's disease's influence on neural pathways and motor control. The dynamic numerical and experimental model of the forces of Parkinson's disease has been explored in Dong (2021). The force $f_{1,2,3}$ is assumed as the input moments due to shoulder, biceps, elbow, and wrist muscle activation, respectively. Those moments are driven at the first three resonance frequencies of the primary system as a triple harmonic function representing the resting tremor.

$$f^T = [f_1\ f_2\ f_3]^T$$

$$f_{1,2,3} = 1 \cos \cos(\omega_1 t) + 1 \cos \cos(\omega_2 t) + 1 \cos \cos(\omega_3 t) \tag{20.7}$$

for $\omega_1 = \omega_{n1}$, $\omega_2 = \omega_{n2}$ and $\omega_3 = \omega_{n3}$, where ω is the driving frequency of the muscles and ω_{ni}, $i = 1, 2, 3$ is the natural frequencies of the primary system.

20.3.2 Passive Control Systems

This paper explores two types of dynamic absorbers inspired by those proposed in Gebai et al. (2018) and shown in Fig. 20.2. These absorbers are affixed to the primary system to control undesired vibrations. The dual parallel and triple parallel absorber have a total mass of 0.44751 kg.

The dual parallel absorber (DPA), Fig. 20.2a, consists of two similar devices. The first device is tuned to the primary system's first natural frequency, while the second is tuned to the second natural frequency. On the other hand, the triple parallel absorber (TPA), Fig. 20.2b, incorporates three similar devices. The first two absorbers are tuned to the same frequency as the DPA, while the third device is tuned to the primary system's third natural frequency. The dual parallel absorber is attached to the forearm at distances of $ab_1 = 0.27$ m and $ab_2 = 0.29$ m from the elbow. In contrast,

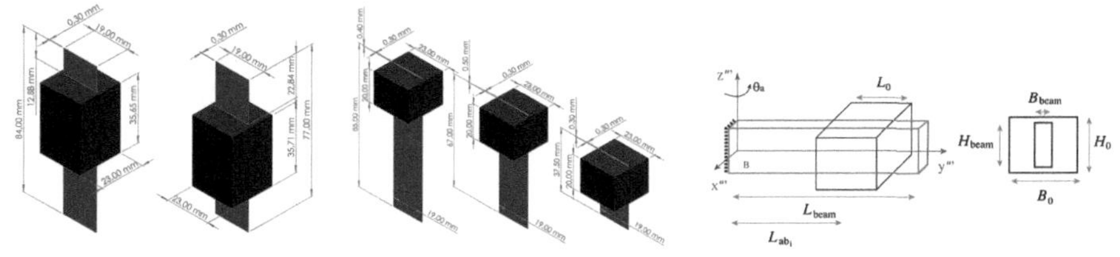

(a) (b) (c)

Fig. 20.2 Passive controllers modeled as a cantilevered beam with an attached mass: **a** DPA, **b** TPA, **c** parameter representation (Adapted from Gebai et al. (2016b))

the triple parallel absorber is attached to two different limbs: the first device is located on the upper arm ($ab_1 = 0.36$ m from the shoulder), while the other two devices are on the forearm, positioned at $ab_2 = 0.27$ m and $ab_3 = 0.29$ m from the elbow. These specific distances were chosen to maximize the absorption of undesired vibrations, especially when working close to the vibration source. In this case, the equation of motion for the primary system within the controls follows the same steps as the primary system. By using the DPA on the primary system, the matrices yield.

$$M = \begin{bmatrix} M_{11} & M_{12} & M_{13} & M_{14} & M_{15} \\ M_{21} & M_{22} & M_{23} & M_{24} & M_{25} \\ M_{31} & M_{32} & M_{33} & M_{34} & M_{35} \\ M_{41} & M_{42} & M_{43} & M_{44} & M_{45} \\ M_{51} & M_{52} & M_{53} & M_{54} & M_{55} \end{bmatrix},$$
$$K = \begin{bmatrix} k_1 + k_3 & k_3 & 0 & & 0 \\ k_3 & k_2 + k_3 & 0 & 0 & 0 \\ 0 & 0 & k_4 & 0 & 0 \\ 0 & 0 & 0 & k_{ab1} & 0 \\ 0 & 0 & 0 & 0 & k_{ab2} \end{bmatrix},$$
$$C = \begin{bmatrix} c_1 + c_3 & c_3 & 0 & & 0 \\ c_3 & c_2 + c_3 & 0 & 0 & 0 \\ 0 & 0 & c_4 & 0 & 0 \\ 0 & 0 & 0 & c_{ab1} & 0 \\ 0 & 0 & 0 & 0 & c_{ab2} \end{bmatrix}, \tag{20.8}$$

where

$$\begin{aligned} M_{11} = {} & \left(I_1 + m_1 a_1^2\right) + \left(I_2 + m_2 a_2^2\right) + m_2 l_1^2 + I_3 + m_3\left(l_1^2 + l_2^2 + a_3^2 + 2 l_2 a_3\right) \\ & + m_{b_2}\left(l_1^2 + a_{b2}^2\right) + m_{ab_1}\left(l_1^2 + a_{b1}^2 + l_{ab_1}^2 + 2 a_{b1} l_{ab_1}\right) \\ & + + m_{ab_2}\left(l_1^2 + a_{b2}^2 + l_{ab_2}^2 + 2 a_{b2} l_{ab_2}\right) \end{aligned}$$

$$M_{12} = \left(I_2 + m_2a_2^2\right) + I_3 + m_3\left(l_2^2 + a_3^2 + 2l_2a_3\right) + m_{b_1}a_{b1}^2 + m_{b_2}a_{b2}^2 + m_{ab_1}\left(l_{ab_1}^2 + a_{b1}^2 + 2l_{ab_1}a_{b1}^2\right) + m_{ab_2}\left(l_{ab_2}^2 + a_{b2}^2 + 2l_{ab_2}a_{b2}^2\right)$$

$$M_{13} = I_3 + m_3\left(a_3^2 + l_2a_3\right)$$

$$M_{14} = m_{ab_1}\left(l_{ab_1}^2 + l_{ab_1}a_{b1}\right)$$

$$M_{15} = m_{ab_2}\left(l_{ab_2}^2 + l_{ab_2}a_{b2}\right)$$

$$M_{21} = M_{12}, M_{22} = M_{12}, M_{23} = M_{13}, M_{24} = M_{14}, M_{25} = M_{15}, M_{31} = M_{13},$$

$$M_{32} = M_{23}, M_{33} = I_3 + m_3a_3^2, M_{34} = 0, M_{35} = 0, M_{41} = M_{14}, M_{42} = M_{24},$$

$$M_{43} = M_{34}, M_{44} = m_{ab_1}l_{ab_1}^2, M_{45} = 0, M_{51} = M_{15}, M_{52} = M_{25}, M_{53} = M_{34},$$

$$M_{54} = M_{45}, M_{55} = m_{ab_2}l_{ab_2}^2$$

20.3.3 Absorber Design

The physical design of the absorber consists of a beam attached to a mass, as shown in Fig. 20.2. The geometrical dimensions of each absorber are displayed in Table 20.3, where TPA and DPA are the dual and triple absorber designs, and SA1, SA2, and SA3 indicate each individual absorber (together they represent the entire absorber assembly). For validation of the theory and results, similar geometrical values presented in Gebai et al. (2018) are assumed in this study.

The position of the mass in each absorber was calculated following Dunkerley's semi-empirical formulation, which gives a lower band approximation (Jeffcott 1918) to satisfy the frequency tuning condition.

$$\frac{1}{\omega_{a_i}^2} \approx \frac{1}{\omega_{\text{beam}}^2} + \frac{1}{\omega_{m_0}^2}, i = \{1, 2, 3\} \tag{20.9}$$

where

$$\omega_{\text{beam}} = 3.5160\sqrt{\frac{E_{\text{beam}}I_{\text{beam}}}{m_{\text{beam}}L_{\text{beam}}^3}}, \quad \omega_{m_0} = \sqrt{\frac{6E_{\text{beam}}I_{\text{beam}}}{m_{a_i}a_i^2(3L_{\text{beam}} - a_i)}} \tag{20.10}$$

Table 20.3 Dimensions of the absorbers

Components	Absorber	L (cm)	H (cm)	B (cm)	Absorber	L (cm)	H (cm)	B (cm)
Beam	TPA-SA1	8.5	1.9	0.03	DPA-SA1	8.4	1.9	0.03
Attached mass		2	2.3	2.3		3.56	2.3	2.3
Beam	TPA-SA2	6.7	1.9	0.03	DPA-SA2	7.7	1.9	0.03
Attached mass		2	2.3	2.3		3.57	2.3	2.3
Beam	TPA-SA3	3.75	1.9	0.03				
Attached mass		2	2.3	2.3				

The parameters are $E_{\text{beam}} = 189.6$ GPa, $\rho_{\text{beam}} = 7800$ kg/m^3, and $\rho_{ma_i} = 8900$ kg/m^3. Hence, the DPA of the positions assumed by each absorber is $l_{ab1} = 0.0533$ m and $l_{ab2} = 0.0363$ m. The TPA of the positions assumed by each absorber is $l_{ab1} = 0.0746$ m, $l_{ab2} = 0.0565$ m, and $l_{ab3} = 0.0272$ m. The stiffness (k_{ai}) and damping (c_{ai}) coefficients are assumed to be proportional by a constant (Hashemi et al. 2004), such that $c_{ai} = 0.005\ k_{ai}$ Nms/rd, $i = \{1, 2, 3\}$, chosen to match the selected tuning frequency the system. The absorbers are designed to satisfy the loci frequency of A_1. The dynamic response is obtained using the transfer function $H(\omega) = \{(-\omega^2 M + K) + i\omega C\}^{-1})$ of the coupled system. The frequency domain response (θ) is obtained using the receptance transfer function (α) as follows:

$$\alpha_k = \left\{ \frac{A_{1_{ij}} + jB_{1_{ij}}}{A_2 + jB_2}, \ldots, \frac{A_{1_{nj}} + jB_{1_{nj}}}{A_2 + jB_2} \right\}^T = \left\{ \frac{\theta_{ij}}{F_j}, \ldots, \frac{\theta_{nj}}{F_j} \right\}^T \tag{20.11}$$

and the amplitude of angular displacement as

$$\theta_{ij} = \sum_{j=1}^{n} |\alpha_j| F_j, \tag{20.12}$$

where i, j are the i-th row and j-th column for the $n \times n$ transfer function of the n-DOF system. Controllers tuned to the wrist joint's response due to shoulder muscle activation must satisfy the following conditions: $A_{131} = 0$, $A_{132} = 0$, and $A_{133} = 0$. Hence, the stiffness of each absorber can be determined as shown in Table 20.4, in addition to the absorber's mass. Thereupon, one can simulate the respective absorbers from the parameters that satisfy the tuning conditions to evaluate their efficiency in tremor suppression.

The DVAs have suppressed the amplitude of the primary system, which can be estimated as the reduction percentage calculated by

Table 20.4 Stiffness of absorbers (Nm/rd)

	Absorber design				
Stiffness	DPA-SA1	DPA-SA2	TPA-SA1	TPA-SA2	TPA-SA3
Estimated values	0.2420	0.2454	0.2664	0.3332	0.4609

$$\text{Max.}(\%\text{Red.}) = \frac{\max.(\theta_{\text{uncontrolled}}) - \max.(\theta_{\text{controlled}})}{\max.(\theta_{\text{uncontrolled}})} \times 100 \qquad (20.13)$$

20.4 Results and Discussions

The properties of the upper limb are described in Sect. 20.3. The position of each absorber in the forearm influences the tremor attenuation. In this paper, in the DPA, the distance placed SA1 is $ab_1 = 0.27$ m and SA2 is $ab_3 = 0.29$ m from the elbow joint. On the other hand, in the TPA, the distance placed SA1 is $ab_1 = 0.36$ m from the shoulder joint, SA2 is $ab_2 = 0.27$ m and SA3 is $ab_3 = 0.29$ m from. Gebai attached most of the absorbers at the same position on the forearm, which are far 0.085 m from the wrist joint ($l_a = l_2 - 0.085$), decreasing the natural frequencies of the systems within absorbers. The natural frequency of each system mentioned in the previews section is calculated by solving eigenvalue problems. Table 20.5 gathers the frequencies for the hand and arm model without control and controlled by absorbers DPA and TPA. All frequency values follow the refereed paper except for the DPA's last frequency.

Graphs in Fig. 20.3 illustrate the system's behavior in frequency and time domains. A harmonic excitation, as expressed in Eq. 20.7, is applied at the shoulder joint, with each excitation frequency corresponding to the natural frequencies of the limb. The receptance and temporal angular displacements of the shoulder joint are depicted in Fig. 20.3a–b, those of the elbow joint in Fig. 20.3c–d, and those of the wrist joints in Fig. 20.3e–f. Additionally, the results display the dynamic responses of the shoulder, biceps, elbow, and wrist muscles. Figure 20.3a–f showcases the receptance and temporal responses observed at the upper limb joints for the system without control (blue dashed line) and with controllers using DPA (orange line) and TPA (green line).

Table 20.5 Natural frequencies of upper limb model

System	ω_{n1} (Hz)	ω_{n2} (Hz)	ω_{n3} (Hz)	ω_{n4} (Hz)	ω_{n5} (Hz)	ω_{n6} (Hz)
No control	3.5880	5.3010	14.3475			
DPA	3.2818	3.6244	4.8060	5.6709	14.2631	
TPA	3.4056	3.6027	5.0541	5.3666	11.7744	15.7094

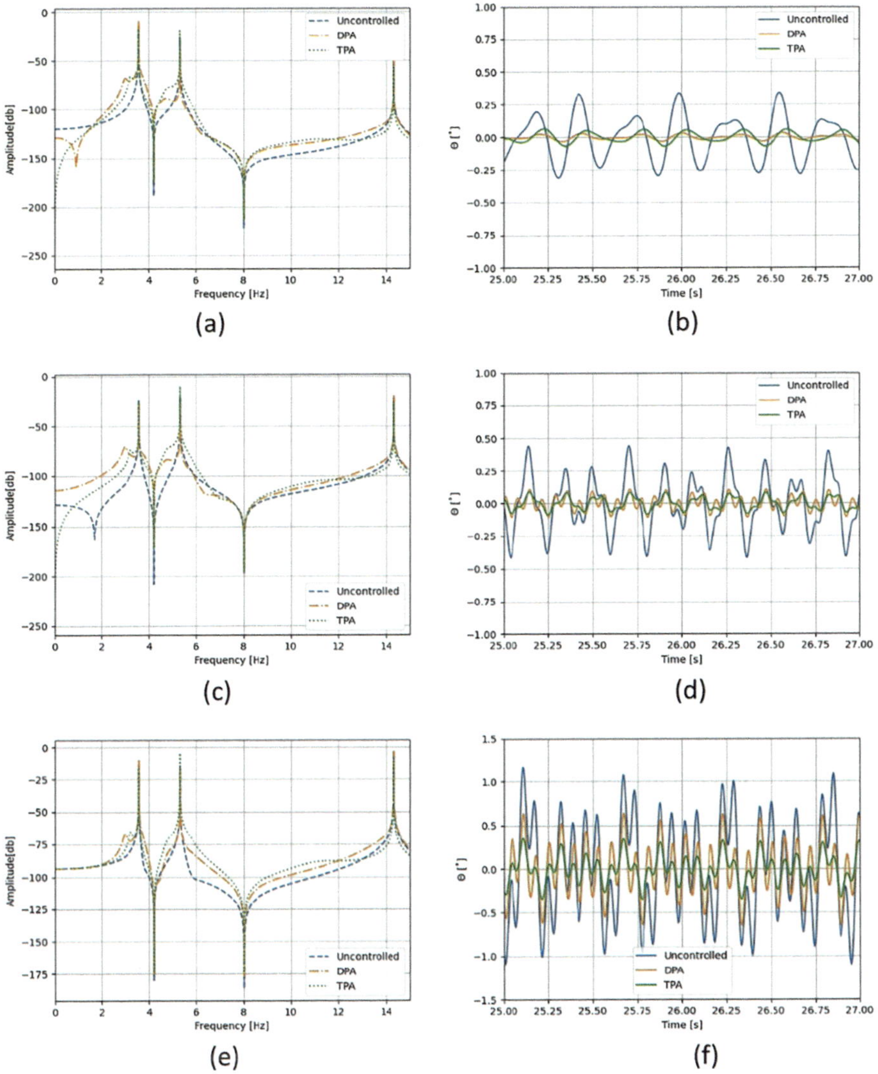

Fig. 20.3 Angular receptance and temporal responses obtained at **a–b** the shoulder joint, **c–d** the elbow joint, and **e–f** the wrist joint

The absorbers are designed with specific values for each driving frequency to mitigate tremors. All absorbers demonstrate a significant reduction in resonant frequency magnitudes. While DPAs induce resonant frequencies within certain bandwidths, they exhibit effective control at the tuned frequency, suggesting that the absorber can diminish tremor amplitudes around the tuning frequency. A similar analysis of these two joints reveals that the controllers provide approximate shoulder attenuation. Notably, the system controlled by the triple parallel absorber exhibits the

Table 20.6 Percentage reduction in tremor's amplitude

Absorbers	Shoulder joint (%)	Elbow joint (%)	Wrist joint (%)
DPA	90.14	75.30	50.05
TPA	79.82	79.64	72.85

highest reduction in tremor amplitude at the wrist joint compared to the dual parallel absorber.

The percentage of amplitude reduction at the shoulder, elbow, and wrist joints due to the dual and triple controllers is presented in Table 20.6. All those results were calculated using the particular solution, given that the homogeneous response ranged from approximately 0 to 15 s.

In the comparison between DPA and TPA, the dual parallel absorber demonstrates the greatest percentage reduction in shoulder tremors. Both absorbers show similar reduction percentages at the elbow joint, although TPA outperforms DPA.

However, TPA achieves the highest reduction at the wrist joint, reducing tremors by 72.85%, a difference of approximately 22.8% compared to DPA. DPA and TPA are efficient controllers, demonstrating significant reduction percentages across a wide frequency band. Consequently, compared to uncontrolled systems, the controlled system exhibits a considerable decrease in amplitude in both time and frequency domain responses at the joints. We conducted a position analysis further to refine our understanding of the optimal device coupling location. Using three different types of single absorbers, each targeting an identified natural frequency, we divided the arm into 30 equal sections from the shoulder to pinpoint ideal absorber placements. Consequently, we developed a heat map illustrating the results obtained at each position. Figure 20.4a–f demonstrate the system's consistent behavior even when employing different absorbers. However, it is crucial to note the system's response as the absorber approaches the wrist. Near the wrist is a splitting of the peak corresponding to the third natural frequency, as esvident in Fig. 20.4g-i. In positions immediately following the wrist, this split becomes more distinct. The closer the absorber is positioned to the wrist, the greater the vibration attenuation observed at the wrist joint. Upon analyzing the heat maps, it becomes evident that positioning the absorber closer to the wrist joint could significantly enhance attenuation.

20.5 Conclusion and Future Work

This paper aims to investigate and evaluate the optimal configuration of absorbers for mitigating tremors in Parkinson's patients. The responses of uncontrolled and controlled structures are compared in frequency and time domains. The efficacy of each controller is assessed based on response magnitude and the percentage of amplitude reduction. Positioning the absorber closer to the wrist joint along the forearm significantly decreases response amplitude. Palm vibration, identified as

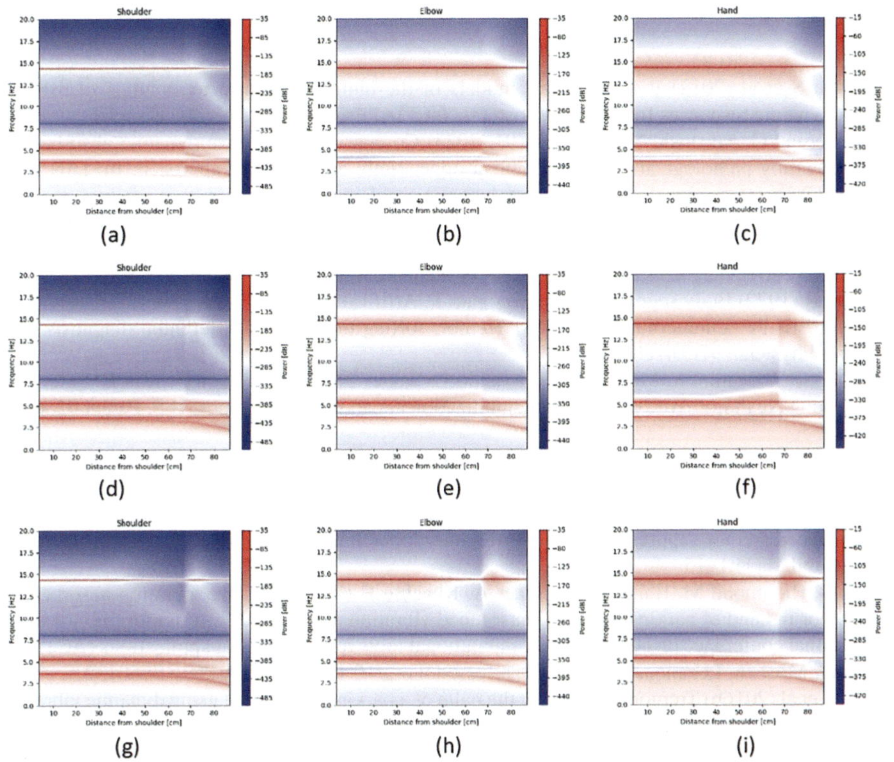

Fig. 20.4 Heatmap of the receptance response by the absorber positioning on **a** the upper arm, **b** the forearm, and **c** the hand, adjusted at the first natural frequency; **d** the upper arm, **e** the forearm, and **f** the hand, adjusted at the second natural frequency; and **g** the upper arm, **h** the forearm, and **i** the hand, adjusted at the third natural frequency

the symptom with the most substantial negative impact on patients' quality of life, underscores the importance of the triple parallel absorber (TPA) as the most effective controller discussed in this study. The TPA significantly reduces wrist joint amplitude and improves weight distribution. Furthermore, the potential benefits of employing more than two absorbers in parallel along the forearm are explored compared to the TPA, particularly in reducing angular motion at the wrist joint. This analysis can enhance our understanding of device efficacy and provide insights for designing novel absorbers tailored for controlling human limb tremors.

Acknowledgements Funded project by FAPDFno.435/2023 Process number 00193-00002143/2023-02, PUBLIC CALL 01/2023 BIO LEARNING/linked to notice 10/2023—FAPDF LEARNING PROGRAM Strategic Promotion Program in the macro area of the research line: BIO learning.

References

Adewusi S et al (2012) Biomechanical models of the human hand-arm to simulate distributed biodynamic responses for different postures. Int J Ind Ergon 42(2):249–260 Elsevier

Albuquerque G, Machado MR, Moura BB (2023) Comparison of passive dynamic absorbers in attenuating pathological tremor of human upper limb: an analytical approach. In: DINAME 2023 XIX international symposium on dynamic problems of mechanics

Buki E, Katz R, Zacksenhouse M, Schlesinger I (2017) Vib-bracelet: a passive absorber for attenuating forearm tremor. Med Biol Eng Comput 56:923–930

Chou KL (2004) Diagnosis and management of the patient with tremor. R I Med J 87(5):135

Contini R (1972) Body segment parameters, part ii. Artif Limbs 16(1):1–19

Crawford III PF, Zimmerman EE (2018) Tremor: sorting through the differential diagnosis. Am Fam Physician 97(3):180–186

De Panisse P, Ibrahim Y, Medeisis J, Tiarvando L, Vaklev NL, Ong JF, Gan B, Koh B, Soler XL, Choong Ngan Lou W (2018) Tremor stabilization apparatus and methods. U.S. Patent 0266820 A1

Dong J (2021) Real-time parkinsonian tremor signal identifier based on internal model principle. Electronic thesis and dissertation repository 8113. https://ir.lib.uwo.ca/etd/8113

Fromme NP, Camenzind M, Riener R, Rossi RM (2019) Need for mechanically and ergonomically enhanced tremor-suppression orthoses for the upper limb: a systematic review. J Neuroeng Rehabil 16(1):1–15

Gebai S, Hammoud M, Hallal A, Al Shaer A, Khachfe H (2016a) Passive vibration absorber for tremor in hand of parkinsonian patients: tuned vibration absorber for rest tremor. In: 3rd international conference on advances in computational tools for engineering applications (ACTEA), Zouk Mosbeh, Lebanon, pp 162–167. https://doi.org/10.1109/ACTEA.2016.7560132

Gebai S et al (2016b) Tremor reduction at the palm of a parkinson's patient using dynamic vibration absorber. Bioengineering 3(3):18. Multidisciplinary digital publishing institute

Gebai S, Hammoud M, Khachfe H (2016c) Using a dual vibration absorber to suppress rest hand tremor of elderly. In: The 5th international conference on global health challenges, GLOBAL HEALTH 2016, Venice, pp 9–13

Gebai S, Hammoud M, Hallal A, Al Shaer A (2018) Structural control and biomechanical tremor suppression: comparison between different types of passive absorber. J Vib Control 24(12):2576–2590

Hashemi SM, Golnaraghi MF, Patla AE (2004) Tuned vibration absorber for suppression of rest tremor in Parkinson's disease. Med Biol Eng Compu 42(1):61–70

Hosseini SM, Kalhori H, Al-Jumaily A (2021) Active vibration control in human forearm model using paired piezoelectric sensor and actuator. J Vib Control 27(19–20):2231–2242

Jeffcott H (1918) The periods of lateral vibration of loaded shafts—the rational derivation of dunkerley's empirical rule for determining whirling speeds. In: Proceedings of the royal society of London. series A, containing papers of a mathematical and physical character, vol 95, no. 666, pp 106–115

Kotovsky J, Rosen MJ (1998) A wearable tremor-suppression orthosis. J Rehabil Res Dev 35:373–387

Lu Z, Huang Z (2020) Analytical and experimental studies on particle damper used for tremor suppression. J Vib Control 1–11

Machado MR, Moura BB, Dey S, Mukhopadhyay T (2022) Bandgap manipulation of single and multi-frequency smart metastructures with random impedance disorder. Smart Mater Struct 31(10):105020

Machado MR, Fabro AT, Moura BB (2020) Spectral element for flexural waves control in smart material beams with single and multiple resonant impedance shunt circuit. 15. https://doi.org/10.1115/1.4047389

Machado MR, Moura BB (2023) Supressão e monitoramento de tremores do corpo humano a partir de metamaterial eletromecânico programável. Universidade de Brasília. BR n, Depositante, p 1020220039062
Mo J, Priefer R (2021) Medical devices for tremor suppression: current status and future directions. Biosens 11:99.https://doi.org/10.3390/bios11040099
Moura BB, Machado MR, Mukhopadhyay T, Dey S (2022) Dynamic and wave propagation analysis of periodic smart beams coupled with resonant shunt circuits: passive property modulation. Eur Phys J Spec Top 1:1–18
Moura BB, Machado MR, Dey S, Mukhopadhyay T (2023) Manipulating flexural waves to enhance the broadband vibration mitigation through inducing programmed disorder on smart rainbow metamaterials. Appl Math Model 12:650–671
Paulig J, Jabusch H-C, Groaybach M, Boullet L, Altenmüller E (2014) Sensory trick phenomenon improves motor control in pianists with dystonia: prognostic value of glove-effect. Front Psychol 5:1012
Rudraraju S, Nguyen T (2018) Wearable tremor reduction device (TRD) for human hands and arms. In: Proceedings of the 2018 design of medical devices conference, Minneapolis, MI, USA
Takanokura M, Sugahara R, Miyake N, Ishiguro K, Muto T, Sakamoto, K (2011) Upper-limb orthoses implemented with air dashpots for suppression of pathological tremor in daily activities. In: Proceedings of the 23rd congress of international society of biomechanics, Brussels, Belgium

Chapter 21
Comparing Analytical and Numerical Solutions for Fluid–Structure Problems with Free Surface Conditions: A Study on Tuned Liquid Dampers and Tuned Liquid Column Dampers

Pedro Falcomer Pontes Viégas, Marcus Vinicius Girão de Morais, Agnaldo Antônio Moreira Teodoro da Silva, and Raul Dario Durand Farfan

Abstract This paper presents a finite element analysis of sloshing in 3D water tanks using a pressure-based Eulerian approach. The fluid domain is discretized by isoperimetric elements (FLUID220) that exhibit a quadratic pressure behavior and are used for modeling the fluid domain. Three examples (a rectangular cavity, a square-tuned liquid column damper (TLCD), and a bidirectional-tuned liquid multicolumn damper (TMLCD)) were modeled with rigid contours and a free surface to study uncoupled liquid reservoirs. Free vibration and harmonic analysis were performed to determine dynamic parameters of the three cases. For the rectangular cavity, the numerical results are compared to analytical solutions and previous numerical and experimental studies. The numerical solutions present acceptable relative error (inferior to 0.2% relative error) when compared to analytical solutions. For the TLCD example, the numerical results were validated using experimental results and previous numerical studies, presenting acceptable errors (6% relative error). For the third case, the modal results were compared to the analytic solution and previous works returning an error of 2.75 and 3.14% relative to the experimental values and the harmonic analysis presented a reduction of 44.88% in the amplitude of displacement of the system, thus validating this approach of analysis in the development of this kind of device.

Keywords Sloshing · TLCD · Fluid–structure interaction · Acoustic finite element modeling · Structural vibration

P. F. P. Viégas · M. V. G. de Morais (✉) · R. D. D. Farfan
University of Brasilia, Faculdade do Gama, Gama 72444-240, Brazil
e-mail: mvmorais@unb.br

A. A. M. T. da Silva
University of Brasília, Campus Darcy Ribeiro, Asa Norte, Brasília, DF 70910-900, Brazil

P. Kurka and M. Pereira (eds.), *Advances in Structural Vibration*, Lecture Notes in Mechanical Engineering, https://doi.org/10.1007/978-3-031-71540-2_21

21.1 Introduction

Tuned liquid column damper (TLCD) is a kind of passive damper device used to control structural vibration in tall buildings, wind turbines, and ships. These passive devices are characterized as fluid–structure problems. Free surface problems can be modeled as TLD and TLCD to describe the behavior in top of tall and slender structures.

TLCDs were first proposed in tall buildings by Sakai et al. (1989). In the literature, this passive device is attractive for wind turbines due to its relatively low cost and good efficiency at low frequencies (Colwell and Basu 2009). Yalla and Kareem (2000) proposed an approximate optimized solution for the TLCD head loss gain for a white noise spectrum. Alkmin et al. (2018) optimized linearized TLCD parameters for a classic wind spectrum. Martins et al. (2019) experimentally showed the reduction of vibration of a TLCD in a reduced linear structure. There is also a small difference between theoretical frequencies and experimental results (Alkmim et al. 2018; Martins et al. 2019).

This study aims to compare analytical and numerical solutions of fluid–structure problems with free surface conditions, such as tuned liquid damper (TLD) and tuned liquid column damper (TLCD). The fluid domain is discretized by isoperimetric elements (FLUID220) that exhibit a quadratic pressure behavior and are used for modeling the fluid domain. Three problems were presented: (a) a rectangular water tank, (b) a U-shaped tube filled with a liquid, and (c) a multiple-column damper. For the rectangular cavity, the numerical results are compared to analytical solutions and previous numerical and experimental studies. The numerical solutions present acceptable relative error (inferior to 0.2% relative error) when compared to analytical solutions. For the TLCD example, the numerical results were validated using experimental results and previous numerical studies, presenting acceptable errors (6% relative error). For the third case, the results were compared to the analytic solution and previous works returning a relative error of 3.27%. The present numerical implementation presents a good agreement with the analytical solution and experimental results.

21.2 Wave Equation

For the discretization of the acoustic wave equation, the following assumptions are made:

i The fluid is compressible and homogeneous.
ii The fluid is inviscid and irrotational.
iii There is no mean flow of the fluid.
iv The acoustic pressure is defined as the pressure fluctuation about the mean.
v The change in density is small compared to the mean density.
vi The pressure and density can vary across elements.

In a compressible fluid, acoustic waves are described as pressure fluctuations, meaning they have infinitesimal amplitudes. According to assumptions (i), (iv), and (v), the mass density, pressure, and velocity are written as the sum of the mean values and a fluctuation, as shown by the equation.

$$\rho = \rho' + \rho_0 \left(\rho' \ll \rho_0\right) \tag{21.1}$$

$$p = p' + p_0 \left(p' \ll p_0\right) \tag{21.2}$$

$$u = u' + u_0 \left(u' \ll u_0\right), \tag{21.3}$$

where: ρ is the density, ρ_0 is the average density, and ρ' is the density fluctuation. As the flow is zero, the average flow $u = 0$ so the average velocity $u = u'$. For the description of the wave equation, a combination of Navier–Stokes, continuity, and state equations is used. These equations are presented below:

$$\frac{\mathrm{d}\rho}{\mathrm{d}t} + \nabla(\rho u) = 0 \tag{21.4}$$

$$\rho\left(\frac{\mathrm{d}\rho}{\mathrm{d}t} + u\nabla u\right) = -\nabla p + \mu\nabla^2 u + \frac{1}{3}\mu\nabla(\nabla \cdot u) + \rho g \tag{21.5}$$

$$\rho' = c^2\rho \tag{21.6}$$

where $c = \sqrt{k/\rho_o}$ is the velocity of propagation of sound in the fluid, K is the bulk modulus of the fluid, and μ is the viscosity coefficient of the fluid.

Applying the divergent properties in the continuity equation and applying the density and velocity equations in the continuity equation result in,

$$\frac{\mathrm{d}}{\mathrm{d}t}(\rho' + \rho_0) + \left(\rho' + \rho_0\right)\nabla \cdot u' + u'\nabla\left(\rho' + \rho_0\right) = 0 \tag{21.7}$$

The specific mass being independent of time in the first term and the third term of much lower order, the equation is rewritten in the form:

$$\frac{\mathrm{d}\rho'}{\mathrm{d}t} + \rho_0\nabla \cdot u' = 0 \tag{21.8}$$

Starting with the Navier–Stokes equation, the fluid is considered inviscid,

$$\rho\left(\frac{du}{dt} + u\nabla u\right) = -\nabla p + \rho g \tag{21.9}$$

As in this case there is no acoustic excitation, we have:

$$g\rho_0 = \nabla p_0 \tag{21.10}$$

Applying the pressure equation and the gradient properties results in the form:

$$\nabla p = \nabla p_0 + \nabla p' = g\rho_0 + \nabla p' \tag{21.11}$$

So, the continuity equation becomes:

$$\left(\frac{du'}{dt} + u\nabla u\right) = -\nabla p \tag{21.12}$$

where the nonlinear velocity term has negligible dimension and is rewritten as:

$$\rho\frac{du'}{dt} + \nabla p = 0 \tag{21.13}$$

Applying the divergence and deriving the continuity equation, we obtain the wave equation:

$$\frac{d^2p'}{dt^2} - \nabla^2 p' = 0 \tag{21.14}$$

or in linearized form:

$$\frac{1}{c^2}\frac{d^2p'}{dt^2} - \nabla^2 p' = 0 \tag{21.15}$$

considering the harmonic variation of the pressure

$$p' = \underline{p}pe^{-i\omega t} \tag{21.16}$$

where $\underline{p}$ stands for the pressure amplitude, f for pressure oscillation frequency, and $\omega = 2\pi f$ for angular frequency [rad/s]. By differentiating twice, the pressure equation, and substituting it into the prior wave equation, the Helmholtz equation is obtained.

$$\frac{\omega^2}{c^2}\underline{p} + \nabla^2\underline{p} = 0 \tag{21.17}$$

where velocity of sound $c = (K/\rho)^{0.5}$, in fluid medium, is function of density ρ and bulk modulus K.

21.2.1 Boundary Conditions

To carry out this study, the fluid domain (Ω_F) is limited by the following boundary conditions: free surface (Γ_{SL}), rigid wall (Γ_{PR}), and fluid–structure interface (Γ_{FE}), given by:

- Free surface (Γ_{SL})

$$\frac{dp}{d\vec{n}} = \frac{dp}{dz} = -\frac{1}{g}\frac{d^2p}{dt^2} \tag{21.18}$$

- Rigid wall (Γ_{PR})

$$\frac{dp}{d\vec{n}} = \nabla p \cdot \vec{n} = 0 \tag{21.19}$$

- Fluid–structure interface (Γ_{FE}), supposing $\ddot{u} = 0$

$$\frac{dp}{d\vec{n}} = \nabla p \cdot \vec{n} = -\rho_f \ddot{u} \tag{21.20}$$

where ρ_f, $\ddot{u}$, and g are the fluid density, the normal acceleration at the fluid–structure interface, and the acceleration due to gravity. Figure 21.1 illustrates the boundaries conditions described below.

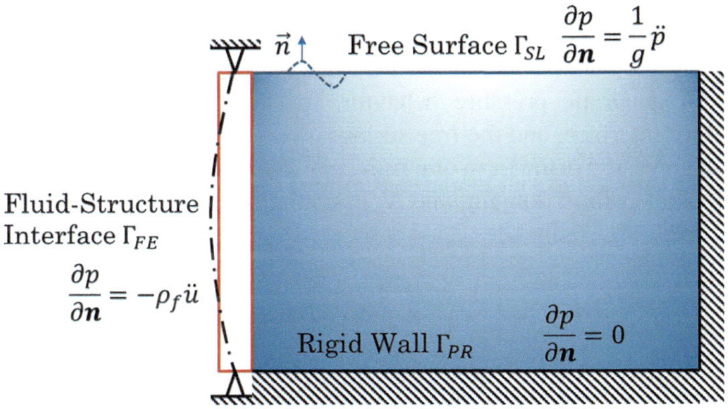

Fig. 21.1 Schematic representation of boundaries conditions of sloshing problems

21.3 Problem Discretization

The discretization of the wave equation, using the Galerkin method and an approximate function $\hat{p}$ that satisfies the boundary conditions, we have:

$$R(\hat{p}) = \nabla^2\hat{p} - \frac{1}{c^2}\ddot{\hat{p}} \neq 0 \tag{21.21}$$

By the method of weighted residues, the integral of the residue $R(\hat{p})$ with the approximation of $\hat{p}$ equals zero is expressed by:

$$\int_{\Omega_f} R(\hat{p})\hat{p}\mathrm{d}\Omega = \int_{\Omega_f} \left(\nabla^2\hat{p} - \frac{1}{c^2}\ddot{\hat{p}}\right)\hat{p}\mathrm{d}\Omega = 0 \tag{21.22}$$

Applying Green's theorem and the boundary conditions results in the weak form of the sloshing equation:

$$\int_{\Omega_f} \nabla\hat{p}\nabla\hat{p}\mathrm{d}\Omega + \oint_{\Gamma_{FE}} \rho(\ddot{u}\cdot\vec{n})\hat{p}\mathrm{d}\Gamma + \oint_{\Gamma_{SL}} \frac{1}{g}\ddot{\hat{p}}\hat{p}\mathrm{d}\Gamma + \int_{\Omega_f} \frac{1}{c^2}\ddot{\hat{p}}\hat{p}\mathrm{d}\Omega = 0 \tag{21.23}$$

21.4 Numerical Implementation

The weak form shown in Eq. 21.25 was discretized using isoparametric finite elements. The element adopted was FLUID220, a tridimensional 20-node solid element with quadratic pressure behavior, for modeling the fluid medium, the fluid–structure interface, and the free surface.

Assuming pressure in the form of $p \approx \hat{p} \simeq \sum N_i\hat{p}_i$, where $\hat{p}_i$ are the nodal pressure of finite element at fluid domain, and N_i is spatial interpolating shape function, the discretized form of,

$$K_f\hat{p} + \left\{\frac{1}{c^2}M_f + \frac{1}{g}M_{SL}\right\}\ddot{\hat{p}} = -\rho_f C_{FS}^{T} u_f \tag{21.24}$$

where $\hat{p}$ and $\hat{\ddot{u}}$ are the nodal acoustic pressure in fluid domain Ω_f and nodal solid acceleration in fluid–structure contour Γ_{FS}, respectively. The matrix K_f and M_f are the acoustic 'stiffness' matrix and acoustic 'mass' matrix, respectively, given by:

$$(K_f)_{ijk} = \int\int\int_{-1}^{+1} \left(\frac{\partial N_i}{\partial \xi}\frac{\partial N_j}{\partial \xi}\frac{\partial N_l}{\partial \xi} + \frac{\partial N_i}{\partial \eta}\frac{\partial N_j}{\partial \eta}\frac{\partial N_l}{\partial \eta} + \frac{\partial N_i}{\partial \zeta}\frac{\partial N_j}{\partial \zeta}\frac{\partial N_k}{\partial \zeta} \right) |J| d\xi d\eta d\zeta \tag{21.25}$$

$$(M_f)_{ijk} = \left(\frac{1}{c^2}\right) \int\int\int_{-1}^{+1} N_i N_j N_k |J| d\xi d\eta d\zeta \tag{21.26}$$

where J is the Jacobian matrix to transform global coordinate system (x, y, z) to natural coordinate system (ξ, η, ζ); $|J| = \det J$; the FLUID 220 shape function $N_i = \xi_i(2\xi_i - 1)\eta_i(2\eta_i - 1)\zeta_i(2\zeta_i - 1)$; and $i \in [1, 2, 3 \ldots, 20]$.

The matrix M_{SL} is the free surface boundary condition, and C_{FS}^{T} is the fluid–structure coupling matrix using isoparametric L^2 finite element, given by:

$$(M_{SL})_k = \left(\frac{1}{g}\right) \int_{-1}^{+1}\int_{-1}^{+1} N_k^T N_k |J| d\xi d\eta \tag{21.27}$$

$$\left(C_{FS}^{T}\right)_k = \rho_f \int_{-1}^{+1}\int_{-1}^{+1} N_k^T \cdot n N_k |J| d\xi d\zeta, \tag{21.28}$$

where n is the outwardly directed normal to the element surface along the interface. The numerical implementation was developed in ANSYS Workbench, harmonic acoustics module (ANSYS 2020R1)

21.5 Numerical Results

In this study, three cases are analyzed, a rectangular container (TLD), a U-shaped column damper (TLCD), and a multiple-column damper (TLMCD), which were analyzed in terms of the modal response and the forced harmonic response; these results were then compared, with the analytic solutions, the literature, and with simulations made in MATLAB and ANSYS APDL in previous works. All cases are evaluated with the same parameters, density $\rho = 1000$ kg/m^3, sound velocity $c = 1500$ m/s, gravity acceleration, $g = 9.81$ m/s^2, and the boundary conditions being composed of a rigid wall on all sides and free surface on the top faces.

21.5.1 Sloshing of Rectangular Container

For the first case, a rectangular container of dimensions $(L \times H) = (10 \times 10)\ [\mathrm{m}^2]$ in x and y (negligible z dimension), Table 21.1 resumes the natural frequencies simulated (for a converged mesh test) compared with previous studies and analytical data, done by expression:

$$\omega^2 = n\pi\left(\frac{g}{L}\right) \tanh \tanh\left[n\pi\left(\frac{H}{L}\right)\right] \tag{21.29}$$

where n is the number of sloshing natural frequency.

The results found show a good mesh convergence with the error relative lower than 0.2% as compared to analytical results. Figure 21.2 presents the first four sloshing modal shapes of rectangular container.

After the modal analysis, the harmonic analysis was performed with the mass source excitation with a value of 1 kg/mm^2 applied on the left side, with positive X direction. The dynamic response of the fluid domain was evaluated at a point on the extreme left of the free surface. Figure 21.3 shows the amplitude and phase of frequency response function of measured point for a circular excitation frequency ω varying between 1.5 and 5 rad/s.

Table 21.1 Convergence results for natural modes of vibration and frequencies of the rectangular container

Frequency modes	Present study	MATLAB*	APDL*	Analytic	Error P.S (%)	Error MATLAB Alkmim (2018) (%)	Error APDL Alkmim (2018) (%)
1	1.752	1.75	1.750	1.752	0.004	0.129	0.129
2	2.483	2.48	2.450	2.478	0.181	0.078	1.133
3	3.040	3.04	3.010	3.035	0.181	0.165	0.824
4	3.511	3.52	3.450	3.505	0.183	0.442	1.556

OBS: * da Silva and Morais (2018)

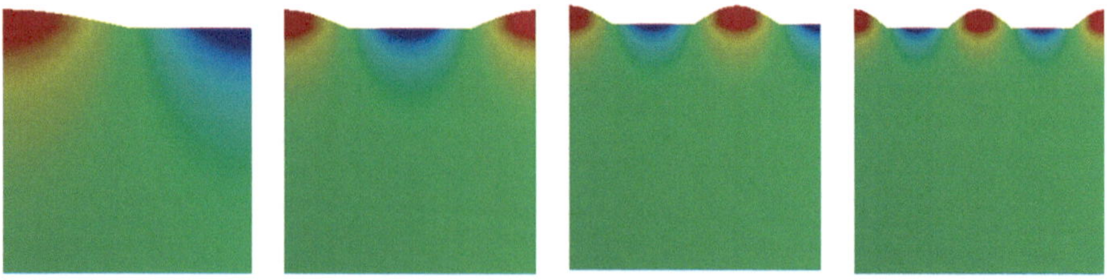

Fig. 21.2 First four modes of resonance of the rectangular tank TLD

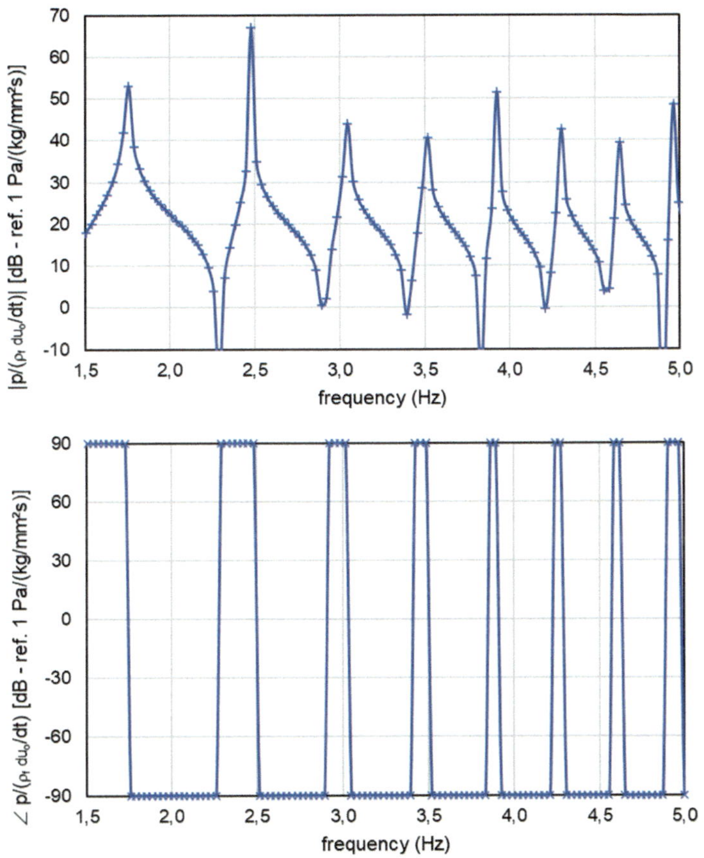

Fig. 21.3 Amplitude and phase of frequency response function on the extreme left point of the free surface of the container

21.5.2 Sloshing in TLCD

The analytical fundamental frequency of a TLCD is $\omega_1 = \sqrt{2g/L}$, where the total length $L = 2h_a + B$ of the TLCD. The others natural frequencies of TLCD $(n = 2, 3, 4, \ldots)$ are similar to a rectangular container, replacing n by $n - 1$, so:

$$\omega_n^2 = \pi(n-1)\left(\frac{g'}{L}\right) \tanh \tanh \left[\pi(n-1)\left(\frac{H'}{L'}\right)\right], n = 2, 3, 4, \ldots \quad (21.30)$$

where equivalent length $L' = 54$ mm. $H' = h_a + \left(\frac{54}{2}\right)$ [mm] becomes the total height of a column and L its length.

The first modal frequencies for different column heights obtained (Fig. 21.4) show reasonable concordance with the experimental results, 4.43% divergence, and the maximum divergences of 8.6% with the analytical results. The higher discrepancies are attributed to the mesh refinement level.

Table 21.2 presents all the values for the first natural frequency of TLCD as function of water level ha (mm) and the relative errors between when compared with the analytic and experimental values in comparison with previous studies.

Continuing the convergence analysis, Table 21.3 compares the converged firsts natural resonance frequencies founded in the present study for an uncoupled TLCD with $h_a = 105$ mm with previous studies (Alkmim et al. 2018; da Silva and Morais, 2018) and analytical solution.

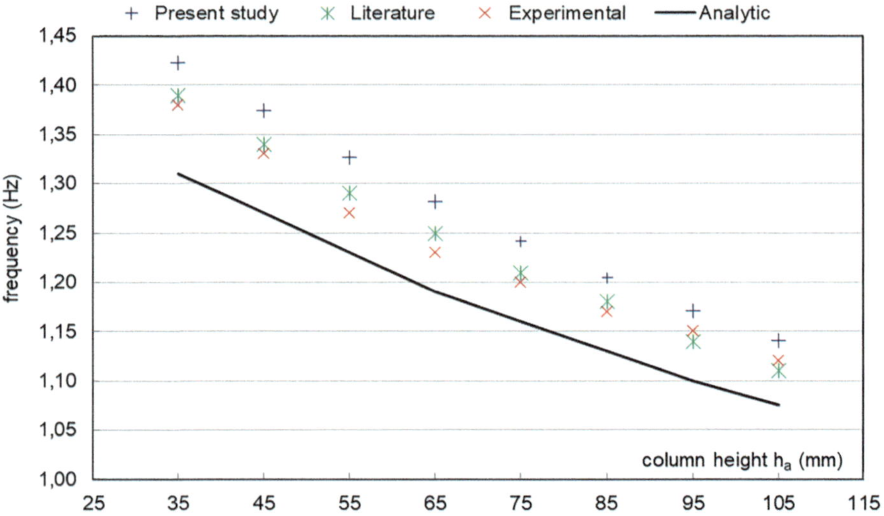

Fig. 21.4 First mode natural frequency in hertz, with respect to column height from different studies

Table 21.2 Convergence results for first natural frequency of the TLCD as function of water level ha (mm)

ha (mm)	Present study	MATLAB*	Analytic	Error (%) P.S	Experimental +	Error (%) P.S
35	1.42	1.39	1.31	8.60	1.38	3.09
45	1.37	1.34	1.27	8.19	1.33	3.31
55	1.33	1.29	1.23	7.82	1.27	4.43
65	1.28	1.25	1.19	7.73	1.23	4.23
75	1.24	1.21	1.16	7.05	1.20	3.48
85	1.21	1.18	1.13	6.65	1.17	3.01
95	1.17	1.14	1.10	6.47	1.15	1.84
105	1.14	1.11	1.08	6.07	1.12	1.80

OBS: *da Silva and Morais (2018), + Alkmin et al. (2018)

Table 21.3 Comparaison of results for firsts natural frequencies of the TLCD for ha = 105 mm related to previous studies

Frequency modes	Present study	MATLAB*	Analytic	Error (%) (P.S./ Ana.)	Error (%) (MATLAB/Ana.)
1	1.140	1.112	1.075	6.07	3.44
2	3.802	3.853	3.802	0.00	1.34
3	5.383	4.122	5.377	0.10	23.34
4	6.621	5.213	6.586	0.53	20.84

OBS: * da Silva and Morais (2018)

The results found for the natural frequencies in the X-direction showed good convergence when compared to the analytically calculated values, with the highest deviation shown being 6.07% in the first mode.

Figure 21.5 shows the first four modal shapes of TLCD with $h_a = 105$ mm. The harmonic analysis was performed with the same parameters as in the first case, mass source with a value of 1 kg/mm^2, applied to the left side, with a positive X-direction, with the application point of the frequency response function being the extreme point left of the free surface. Figure 21.6 illustrates the frequency response function of this point with the frequency varying between 0 and 7 Hz.

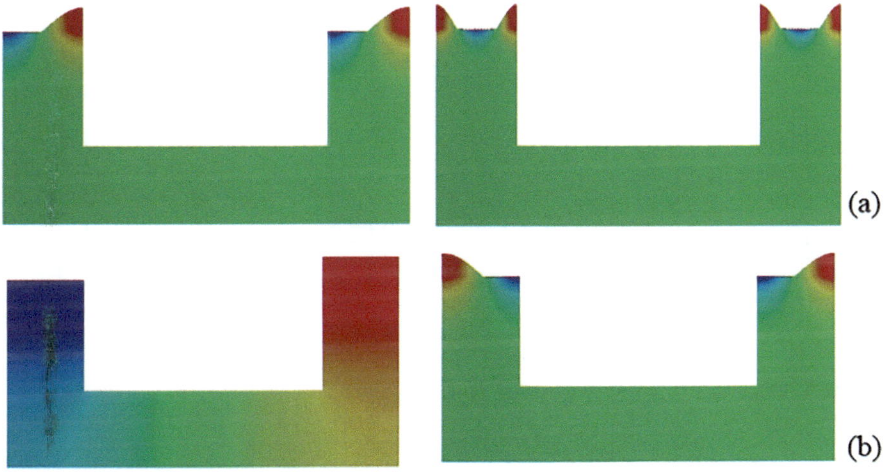

Fig. 21.5 First four modal shapes the U-shaped TLCD with $h_a = 105$ mm: **a** First and second modes, **b** third and fourth modes

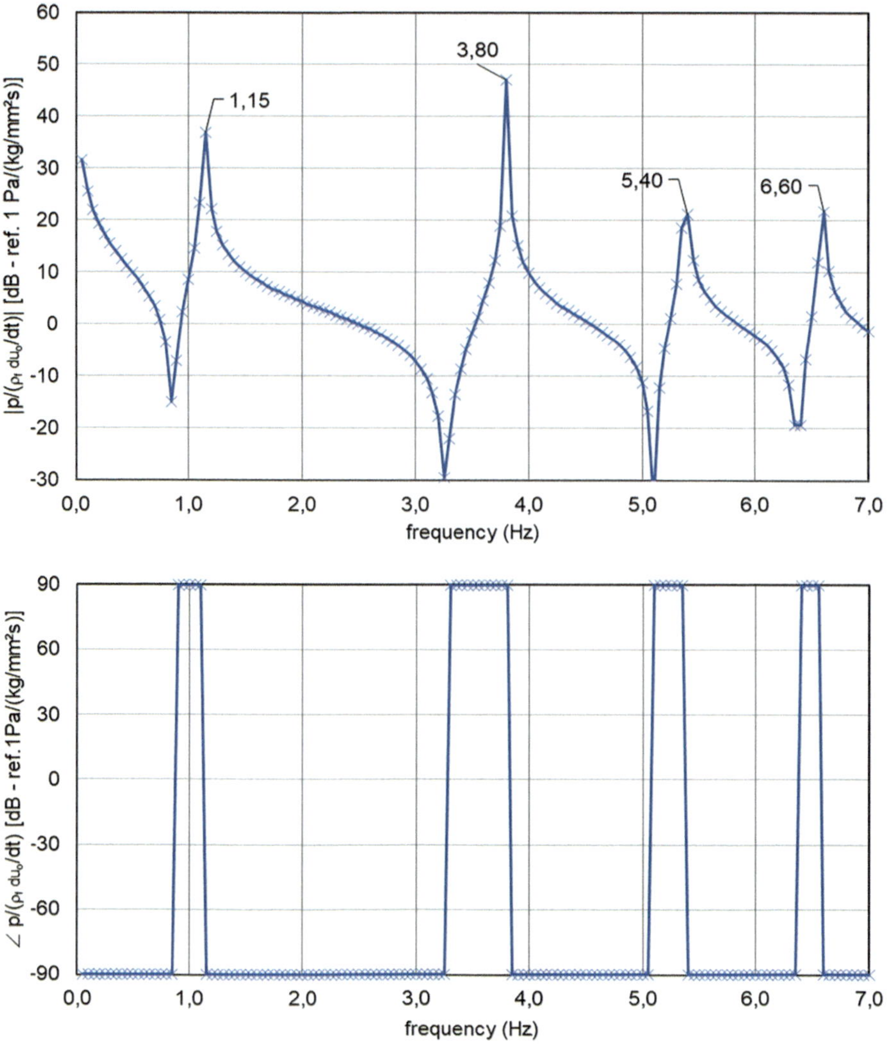

Fig. 21.6 Frequency response function on the left-most point of the free surface of the TLCD

21.5.3 Sloshing of TMLCD

In the third case, the selected tuned liquid multicolumn damper is a bidirectional TLCD presented in Rozas (2016), with the geometry configuration presented in Fig. 21.7.

As in a bidirectional study subject, the natural frequencies are now evaluated in both X and Y directions. The resultant modes and natural frequencies (Table 21.4)

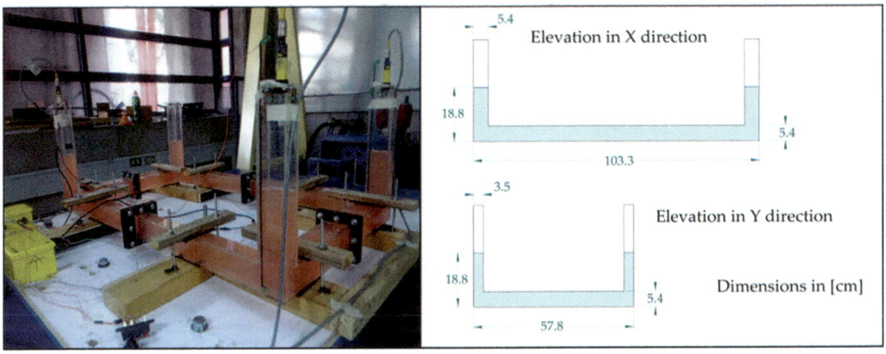

Fig. 21.7 Bidirectional TLCD main dimensions (Rozas 2016)

were then compared with the analytic solution and experimental data (Rozas et al. 2016).

The present results (Table 21.4) show reasonable concordance with both analytic and experimental results with divergences lower than 3.5% to the analytical solution and under 2.8% to the experimental results. Figure 21.8 shows the first two modal shapes of TMLCD in directions *X* and *Y*, respectively.

Table 21.4 Convergence results for natural modes of vibration and frequencies of the BTLCD

Direction	Analytic	Experimental	Present Study	Error analytic (%)	Error experimental (%)
X	0.618	0.617	0.632	2.26	2.48
Y	0.857	0.861	0.885	3.27	2.79

Fig. 21.8 Resonance modes of the BTLCD **a** First mode, in *X*-direction ($\omega_1 = 0.632\,\text{Hz}$), **b** second mode, in *Y*-direction ($\omega_2 = 0.885\,\text{Hz}$)

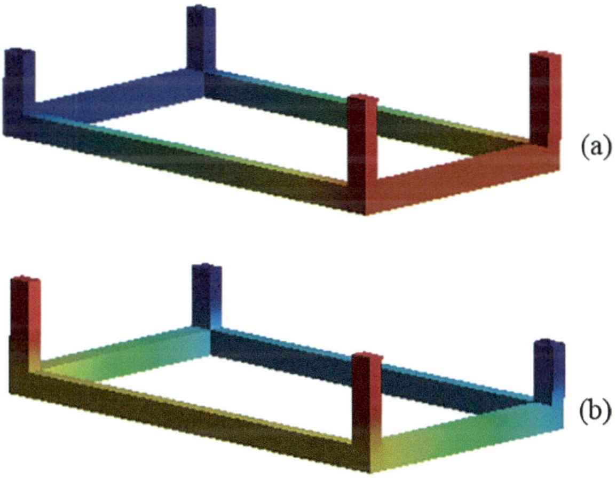

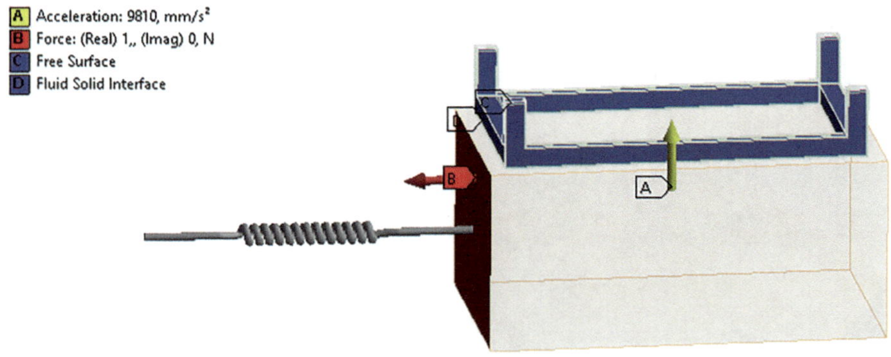

Fig. 21.9 Coupled system and boundary conditions

Coupling the acoustic body to the damper shell leads to some increase in the mass of the system and the addition of flexible walls, which influences the natural frequency of the set, from 0.632 and 0.885 Hz to 0.634 and 0.888 Hz leading to divergences with the experimental results of 2.75 and 3,14% in the X and Y directions, respectively.

Continuing the work, the harmonic analysis consists of coupling the bidirectional set to a spring-mass system of the same natural frequency ($K = 15.868$ N/m, mass $= 992.9$ kg) to evaluate the influence of the damper on the oscillation of this system in the X-direction.

Figure 21.9 shows the system coupled to the structure and its boundary conditions.

The harmonic analysis was performed discretizing the acoustic body with FLUID221 elements, which allow fluid–structure interactions, applied in the contact regions between the fluid and damper shell, the force applied to the spring mass system was 1 N applied on the left face of the mass, and the frequency response function was collected at the extreme right corner of the mass block.

The coupling effects of the device presented in Fig. 21.10 demonstrate accordance with the theory of mechanical vibrations showing a reduction of amplitude of 44.88% or 6.96 dB in the coupled system to the force of 1 N applied at the system, verifying the efficiency of this type of device.

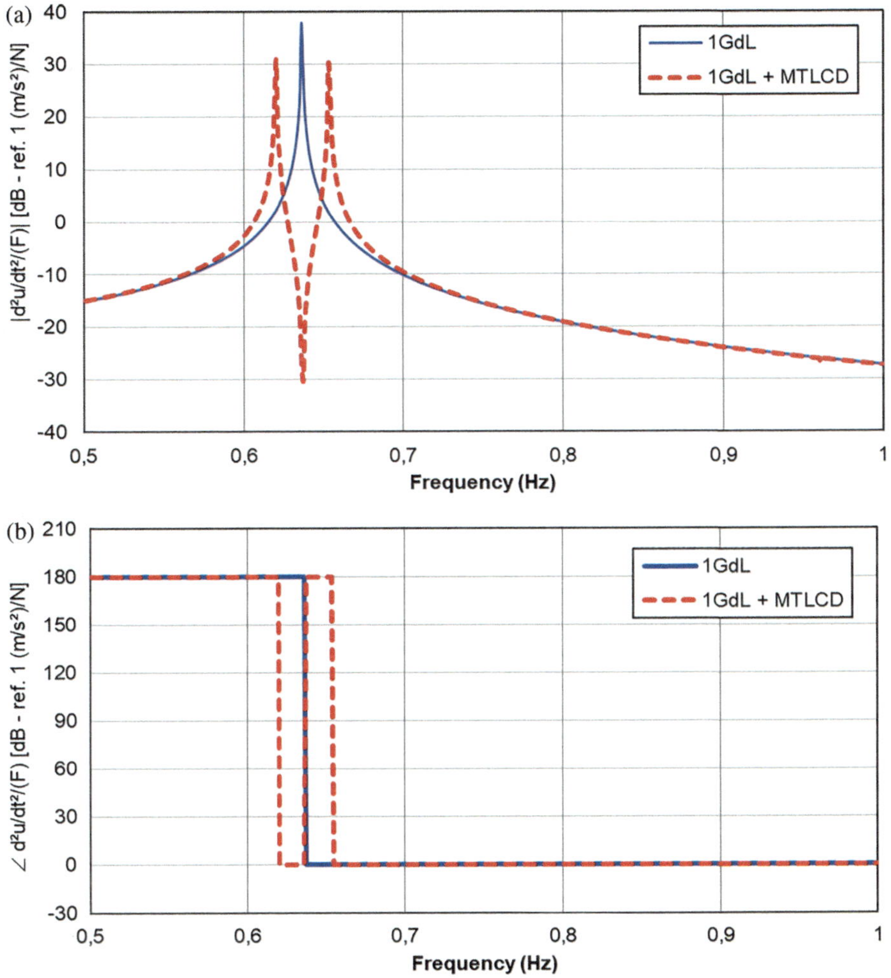

Fig. 21.10 Amplitude and phase of frequency response function of acceleration of mass-spring system for the following conditions: **a** Uncoupled 1GdL **b** coupled 1GdL + MTLCD

21.6 Conclusion

This study was comparing the ANSYS Workbench simulation of fluid–structure sloshing problems, such as rectangular container, TLCD, and TMLCD, with literature and analytical solutions. The fluid domain is discretized by isoparametric elements (FLUID220 and FLUID221).

Three problems were presented: (a) a rectangular water tank, (b) a U-shaped tube damper (TLCD), and (c) a multiple-column liquid damper (TMLCD). For the first example, the numerical solutions present acceptable relative error (inferior to 0.2% relative error) when compared to analytical solutions.

For the second case, the TLCD example, the numerical results were validated using experimental results and previous numerical studies, presenting acceptable errors (6% relative error).

For the last one, the results were compared to the analytic solution and previous works returning a relative error of 3.14%, relative to experimental results from Rozas, when coupled to the device's shell, and the coupled harmonic analysis presented satisfactory results, a reduction of 6.96 DB from the main to the residual peaks, proving the efficiency of this kind of device in dampening low frequencies.

The present case comparisons show reasonable agreement with literature and analytical solutions. Some numerical sources of errors must be explored par report to this present exploratory study. More tridimensional examples must be explored and coupled examples with nonlinear Navier–Stokes fluid equations must be explored and evaluated in fatigue response.

Acknowledgements The present study was financed by Brazilian National Council of Research and Development (Conselho Nacional de Pesquisa e Desenvolvimento—CNPq) with partnership with the University of Brasilia by Edital 2022/2023 PROIC/UnB-CNPq. The second author acknowledges CNPq PQ-2 fellowship support (process number 317383/2021-3) and FAP-DF (process number SEI 00193-00001807/2022-27) to carry out this work. The authors acknowledge to CNPq and UnB by resources that make possible the present work.

References

Alkmim MH, Fabro AT, de Morais MVG (2018) Optimization of a tuned liquid column damper subject to an arbitrary stochastic wind. J Braz Soc Mech Sci Eng 40(11):1–11. https://doi.org/10.1007/s40430-018-1471-3

Colwell S, Basu B (2009) Tuned liquid column dampers in offshore wind turbines for structural control. Eng Struct 31(2):358–368. https://doi.org/10.1016/j.engstruct.2008.09.001

da Silva AAMT, de Morais MVG (2018) FE sloshing modelling in bidimensional cavity using wave equation. MATEC Web of Conferences 211:07001. https://doi.org/10.1051/matecconf/201821107001

Martins JF, De Morais MVG, Avila SM (2019) Experimental study of equivalente 1DoF main structure coupled to TLCD: validation of optimum parameter obtained by parametric optimization. In: ABCM 25th ABCM international congress of mechanical engineering. UFU, Issue Unico, pp 1–4 https://doi.org/10.26678/abcm.cobem2019.cob2019-0812

Rozas L, Boroschek RL, Tamburrino A, Rojas M (2016) A bidirectional tuned liquid column damper for reducing the seismic response of buildings. Struct Control Health Monit 23(4):621–640. https://doi.org/10.1002/stc.1784

Sakai F, Takaeda S, Tamaki T (1989) Tuned liquid column damper-new type device for suppression of building vibrations. In: Proceedings of international conference on highrise buildings, pp 926–931

Yalla SK, Kareem A (2000) Optimum absorber parameters for tuned liquid column dampers. J Struct Eng 126(8):906–915. https://doi.org/10.1061/(ASCE)0733-9445(2000)126:8(906)

Chapter 22
Use of Metamaterials to Reduce Vibration of a Wind Tower Subjected to Arbitrary Stochastic Wind

Vinícius Gabriel Peixoto Borges, Nícolas da Silva Dias, Adriano Todorovic Fabro, and Aline Souza de Paula

Abstract To ensure that wind turbine towers remain economically viable, they must be able to operate for prolonged periods. Therefore, analyzing the dynamic behavior of the structure is crucial to extend its lifespan and optimize its efficiency. In this study, we investigate the attenuation of vibrations in a wind turbine tower using metamaterials. For this analysis, a discrete model of the tower is considered. The finite element method (FEM) is employed in the numerical simulation using a routine developed in Python. The wind effect is described by force signals obtained through a statistical method, using Weibull Probability Distribution as random distribution and Kaimal's power spectral density to simulate the speed wind fluctuations caused by turbulence. The action of the wind is represented by a concentrated force at the upper end of the tower, specifically concentrated at the nacelle. To reduce the tower vibration, identical metamaterials are placed in different positions along the tower and the performance is evaluated considering attenuation of the first and the fifth modes.

Keywords Wind turbine tower · Metamaterials · Dynamic response · Frequency domain · Discrete models

22.1 Introduction

One of the great debates of our time is certainly the pursuit of renewable energy sources, as they not only provide fewer environmental impacts and can be naturally replenished, but also contribute to reducing the world's dependence on non-renewable sources such as petroleum and coal. Since 2019, fossil fuel consumption accounted for over 81% of global energy production (IEA 2021). In this regard, wind turbines play a vital role in harnessing clean and renewable energy.

V. G. P. Borges (✉) · N. da Silva Dias · A. T. Fabro · A. S. de Paula
Department of Mechanical Engineering, University of Brasília, Brasília, Brazil
e-mail: viniborges.2000@gmail.com

P. Kurka and M. Pereira (eds.), *Advances in Structural Vibration*, Lecture Notes in Mechanical Engineering, https://doi.org/10.1007/978-3-031-71540-2_22

Onshore and offshore wind turbines are subjected to dynamic loads that may be induced by wind, wave, and seismic excitation, which potentially threaten their structural integrity and performance. To ensure the safe and efficient operation of wind towers, effective vibration control strategies are essential. In this context, different vibration control methods were proposed by various researchers (Jahani et al. 2022). Considering passive control strategies, the most studied devices are the tuned mass damper (TMD) (Sun and Jahangiri 2018; Yang et al. 2019; Colherinhas and Machado 2022), followed by the tuned liquid damper (Hemmati et al. 2019). As these devices would be positioned in the turbine nacelle, the practical implementation may be unfeasible due to space limitations.

One passive control strategy that has gained widespread application in recent times for reducing vibrations amplitude is the use of metamaterials. In summary, metamaterials are defined as materials or structures that exhibit some form of spatial periodicity, which are designed in such a way that the material possesses enhanced properties; in other words, in the study of metamaterials, the aim is to develop materials whose physical properties are not found in nature, like negative mass density and negative Poisson's ratio.

According to Hussein et al. (2014), the origin of the study of these structures' dates to Newton in his attempt to describe the propagation of sound in air (1686) and more recently with the work of Kock and Harvey (2005), who, based on the study of electromagnetic waves, developed structures capable of refracting and focusing sound waves using an array of small structures compared to the incident wavelength. Regarding periodic structures, Cremer and Leilich (1953) are among the first studies present in the literature. In their work, they sought to investigate harmonic flexural wave motion along a 1D periodic beam.

In general, it is possible to say that the development of metamaterials began in the field of electromagnetism and extended to various other engineering areas. Figure 22.1 illustrates a classification of metamaterials into four main categories: electromagnetic, optical, acoustic, and mechanical metamaterials. Presenting the distinct properties exhibited by each case.

The use of mechanical metamaterials to control vibrations has been vastly applied to beams (Gao et al. 2019; Sugino et al. 2016; Pai 2010). This concept of placing multiple resonators at periodic intervals along the structure may be interesting for wind turbine tower control. Compared to traditional TMDs, these resonators have a smaller mass and can be conveniently placed along the tower, possibly making them a more feasible option.

This paper presents an analysis of vibration reduction of a wind tower using metamaterials. To the force due to wind, a Weibull distribution was used as a random distribution to determine the average wind speeds, and the Kaimal power spectral density (PSD) was applied to simulate wind speed fluctuations caused by turbulence, following the study conducted by Alkmim (2017). The tower's structural representation involves a cantilever beam with an endpoint mass, and its discretized model is derived using the finite element method (FEM). The study evaluates the performance of the metamaterials for different positions of the resonators along the tower and considers attenuation of the first and the fifth modes.

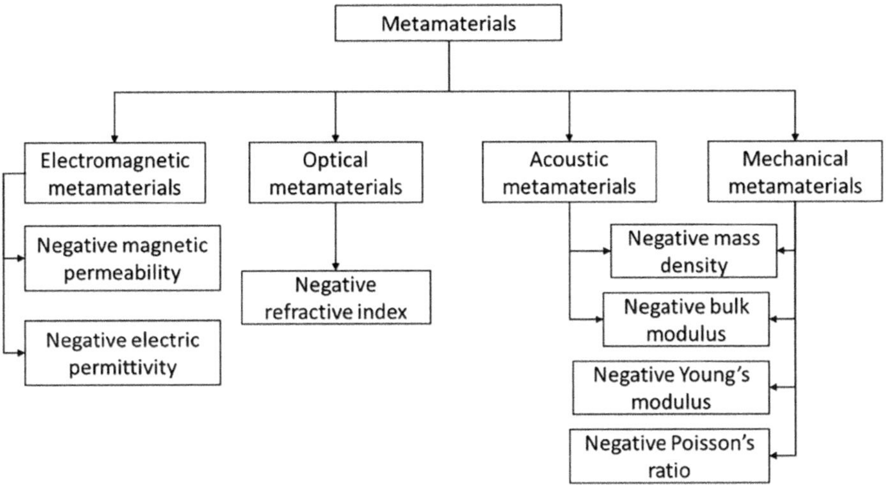

Fig. 22.1 Classification of metamaterials (Balaji et al. 2021)

22.2 Mathematical Model

This section is dedicated to presenting the mathematical model used to assess the dynamic behavior of the turbine, encompassing both the tower-nacelle assembly and the wind-induced forces.

22.2.1 Tower-Nacelle Assembly Model

The turbine's structural model, used to evaluate its dynamical behavior, is derived using the finite element method (FEM), where beam elements are considered. The simplified model of the tower and nacelle considers a cantilever uniform beam with endpoint mass at the top, as shown in Fig. 22.2.

According to Petyt (2010), the element matrices of mass and stiffness are given by:

$$[M_e] = \frac{\eta_e \rho A l_e}{420}\begin{bmatrix} 156 & 22l_e & 54 & -13l_e \\ 22l_e & 4l_e^2 & 13l_e & -3l_e^2 \\ 54 & 13l_e & 156 & -22l_e \\ -13l_e & -3l_e^2 & -22l_e & 4l_e \end{bmatrix}, \quad [k_e] = \frac{EI}{l_e^3}\begin{bmatrix} 12 & 6l_e & -12 & 6l_e \\ 6l_e & 4l_e^2 & -6l_e & 2l_e^2 \\ -12 & -6l_e & 12 & -6l_e \\ 6l_e & 2l_e^2 & -6l_e & 4l_e^2 \end{bmatrix} \tag{22.1}$$

where n_e is the number of elements, ρ is the density of the tower, A is the cross-sectional area of the tower, l_e is the length of the element, E is the elasticity modulus

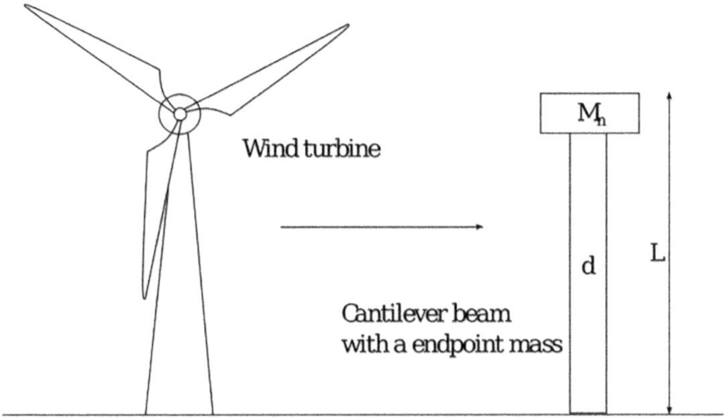

Fig. 22.2 Wind turbine simplified model

of the tower, and I is the cross-sectional moment of inertia of the tower. Global stiffness and inertia matrices are obtained from the element matrices presented in Eq. (22.1), incorporating the nacelle mass, M_{n}, and the linear resonators (modeled as mass-spring-damper oscillators). The equation of motion of the tower-nacelle metamaterials assembly is:

$$[M]\{\ddot{u}\} + [C]\{\dot{u}\} + [K]\{u\} = \{F\} \tag{22.2}$$

where $[M]$ is the global mass matrix, $[C]$ is the global damping matrix, $[K]$ is the global stiffness matrix, and $\{F\}$ is the global load vector.

22.2.2 *Wind Model*

To model the wind behavior, a random distribution function is employed to represent the probability distribution of wind speed. The Weibull distribution, widely recognized for this purpose, is adopted in the current study. The function is given by:

$$P(\underline{U}) = \frac{k}{c}\left(\frac{\underline{U}}{c}\right)^{k-1} \exp\left(\frac{\underline{U}}{c}\right) \tag{22.3}$$

Based on the data obtained by Ozawa (2017) through the least squares method for the city of Ventania (PR), it is used $k = 2.81$ and $c = 3.94$, where the former is the Weibull slop parameter and the latter is the Weibull mode parameter. The obtained Weibull distribution is shown in Fig. 22.3. Once the parameters of the Weibull distribution are known, it is possible to generate mean velocity values, $\underline{U}$, that follow this distribution using the Inverse Cumulative Distribution Function (ICDF) as follows:

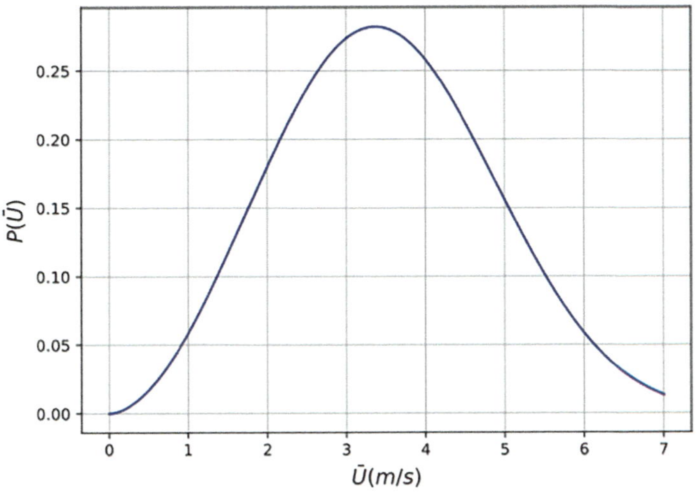

Fig. 22.3 Weibull distribution

$$\underline{U}(P) = -c[\ln(1 - P)]^{1/k} \tag{22.4}$$

The wind behavior data obtained by Ozawa (2017) were measured at a height of 10 m. Therefore, due to the impracticality of empirically measuring these values for all heights, Custódio (2007) proposed a mathematical model to determine wind velocities at any desired position:

$$\frac{v_1}{v_2} = \frac{\ln\left(\frac{h_1}{z_0}\right)}{\ln\left(\frac{h_2}{z_0}\right)} \tag{22.5}$$

where v is the velocity and h are the height at points 1 and 2, and z_0 is the roughness length at the location.

In the present study, the height of the wind tower is 44.075 m, and the roughness length used is 0.05, as described by Ferreira et al. (2010). Hence, utilizing Eq. (22.4) for a probability of 0.9, we obtain a wind velocity of Ozawa's tower. Substituting this value into Eq. (22.5), we derive a velocity of 6.78 m/s, as illustrated in Fig. 22.4.

Finally, it is possible to use the wind velocity to generate the wind force based on the Kaimal power spectral density (PSD), as described by Eq. 22.6:

$$S_{\text{xx}}(\omega) = \frac{4\sigma_u^2 L_{1u}/\underline{U}}{\left[1 + 6\omega\left(L_{1u}/\underline{U}\right)\right]^{5/3}}, \tag{22.6}$$

where $L_{1u} = 171.7$ is a scale parameter that involves the wind turbine, as specified by the International Electrotechnical Commission (Burton et al. 2011). The generated PSD is shown in Fig. 22.5.

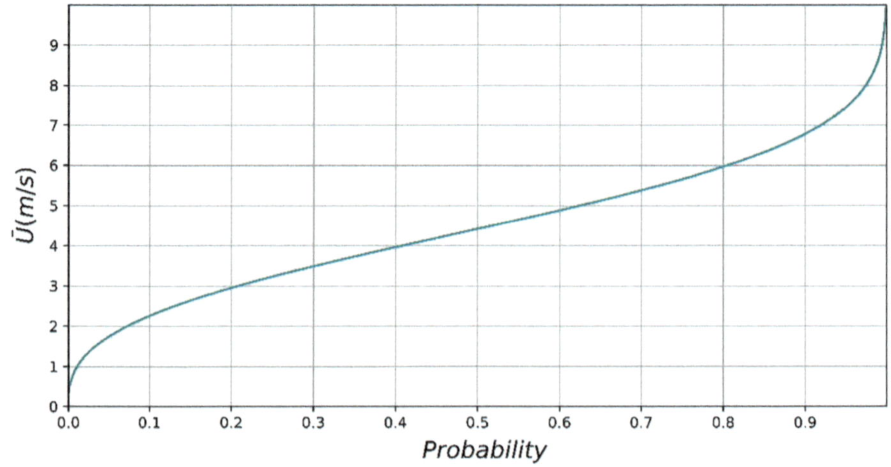

Fig. 22.4 Probability density for the studied tower

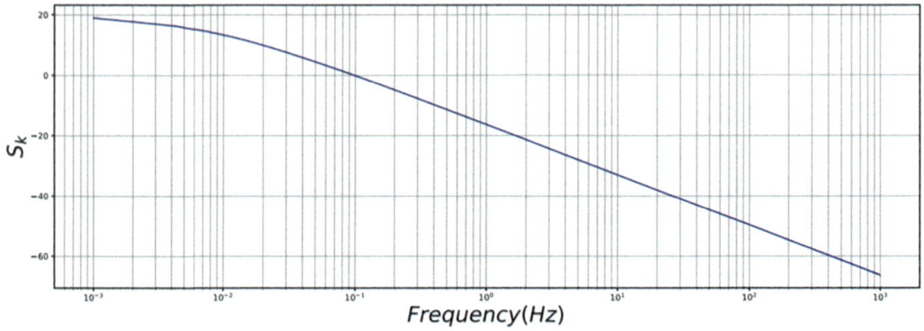

Fig. 22.5 Kaimal PSD

22.3 Results and Discussion

In this section, the dynamical response of the wind turbine tower under wind forces excitation is evaluated with and without control. The tower is discretized with 30 nodes, where each node has one rotational degree of freedom and one translational degree of freedom, resulting in 60×60 stiffness and inertia matrices. The parameters used in numerical simulations are presented in Table 22.1:

The damping matrix $[C]$ is defined proportionally to the stiffness matrix $[K]$. In this study, a negligible damping was considered as follows: $[C] = 10^{-12}[K]$.

As the wind model has a stochastic nature, the dynamical response of the system is evaluated by its spectral density function, S_{yy}, that is related with the spectral density function of the excitation, S_{xx}, as follows:

Table 22.1 Tower parameters

Parameter	Value
Elasticity modulus of the tower, E	2.1×10^{11} Pa
Tower diameter, d	2.715 m
Tower height, L	44.075 m
Nacelle's mass, M_n	95,000 kg
Density of the tower, ρ	7850 kg/m^3

$$S_{yy}(\omega) = |H(\omega)|^2 S_{xx}(\omega), \tag{22.7}$$

where $S_{xx}(\omega) = H(\omega)$ and the transfer function is

$$H(\omega) = \left[\omega^2 M + j\omega C + K\right]^{-1} \tag{22.8}$$

22.3.1 Response to Wind Excitation Without Control

The wind force is assumed to be concentrated at the nacelle, positioned at the free end of the tower, as depicted in Fig. 22.6a. The frequency response is presented in Fig. 22.6b. The response corresponds to the translational vibration of the top node, which is also the location where the excitation is applied. The first eight natural frequencies are observed, with the first one being very low at $\omega_1 = 0.463$ Hz.

22.3.2 Vibration Attenuation

To evaluate the performance of the metamaterials in reducing the vibration amplitudes, two frequencies are considered as target frequencies, one at a time, being the first and fifth natural frequencies of the turbine as follows:

- $\omega_{n1} = 0.463$ Hz
- $\omega_{n5} = 51.93$ Hz

Initially, only one dynamic vibration absorber (DVA), also called TMD, is considered, followed by the consideration of metamaterials. For both cases, a mass equal to 5% of the total mass of the tower is defined for the fifth mode, and 15% for the first mode.

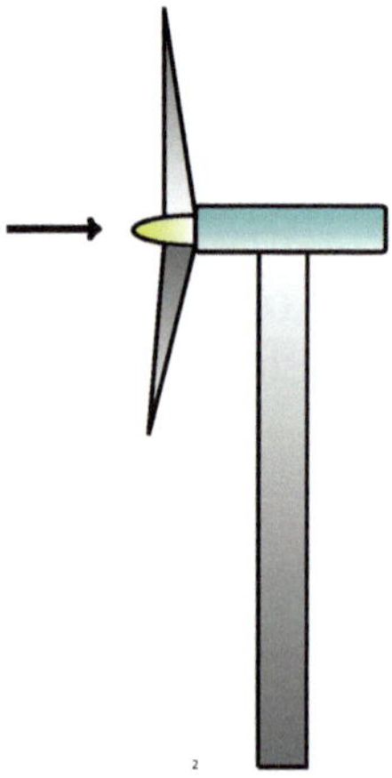

(a)

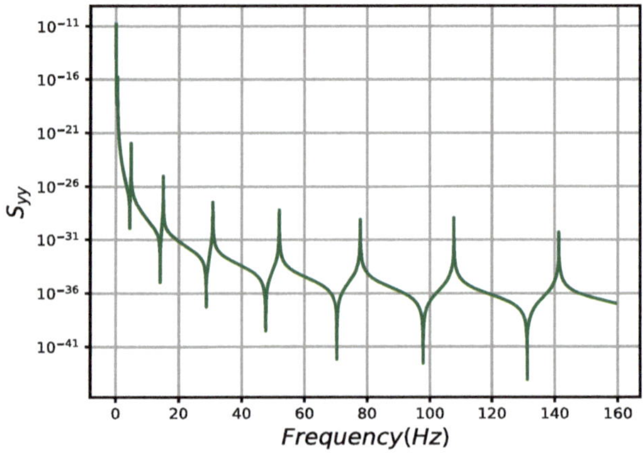

(b)

Fig. 22.6 **a** Schematic representation of the turbine with concentrated force action due to wind (Araújo 2021); **b** PSD response to the concentrated force applied at the top of the tower with an average wind velocity of 6.78 m/s

22.3.2.1 System Response with Dynamic Vibration Absorber

The dynamic vibration absorber (DVA) is modeled as a mass-spring oscillator. To target the first natural frequency, the DVA has a natural frequency of 0.463 Hz and a mass equivalent to 15% of the total tower mass, amounting to 14,250 kg. Conversely, when targeting the fifth mode, the DVA's natural frequency is set to 51.93 Hz, with a mass equal to 5% of the total tower mass, equivalent to 4750 kg. This higher mass used at the frequency of 0.493 Hz was employed to highlight the effects of the absorber, as these effects are less pronounced at lower frequencies.

The stiffness of the dynamic vibration absorber is then defined as follows:

$$K_{\mathrm{ab}} = \omega_{\mathrm{ab}}^2 M_{\mathrm{ab}}, \tag{22.9}$$

where ω_{ab} corresponds to each one of the target frequencies, i.e., first and fifth natural frequencies

Figure 22.7 shows the response of the top node without (in green) and with (in red) DVA, focusing solely on the region of the target frequency. Figure 22.7a illustrates the scenario for the first target frequency, while Fig. 22.7b corresponds to the fifth target frequency. Results show the classical performance of DVAs, exhibiting a reduction in vibration amplitudes within a narrow range around the target frequency.

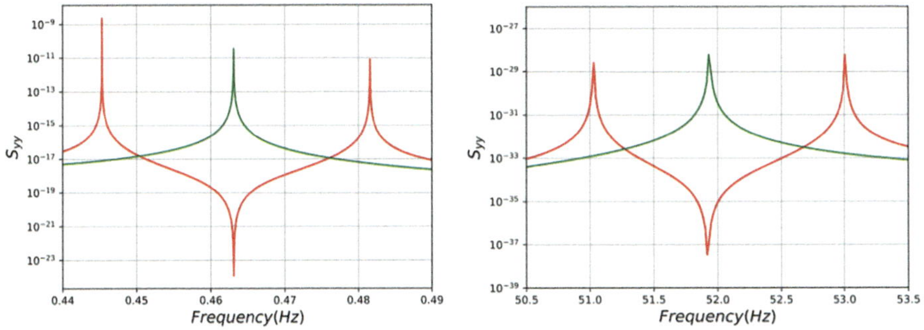

(a) (b)

Fig. 22.7 Response with (in red) and without (in green) the presence of the concentrated mass absorber: **a** First mode, and **b** fifth mode

22.3.2.2 System Response with Metamaterials

At this point, metamaterials are employed to reduce the vibrations of the tower-nacelle assembly. Three different configurations of the resonators are considered: at the base, distributed along the tower, and at the top. In the first case, the first resonator is positioned at the bottom of the structure, followed by the subsequent resonators in the subsequent nodes. The distributed case involves a periodic arrangement of the resonators along the tower. In the top case, the first resonator is placed at the top of the structure, followed by the subsequent resonators in the subsequent nodes. The distribution of the resonators is illustrated in Fig. 22.8.

The coupling of the resonators in the stiffness matrix can be exemplified through the schematic representation presented in Fig. 22.9, where in the first term, the matrix $[K_{\text{tower}}]_{NxN}$ represents the original matrix K, in addition to the matrices $[O]$ which are essentially matrices of 0. In the second term, the coupling matrix is represented, where $\left[K_{\text{res}}^{\text{diag}}\right]_{N_r \times N_r}$ represents the diagonal matrix formed by the stiffness terms of each resonator, $\left[K_{\text{res}}^{\text{coup}}\right]_{N \times N_r}$ is a matrix where each row contains the stiffness term of an absorber and the number of rows represents the quantity of resonators, and finally $[K_{\text{res}}]_{N \times N}$ is a matrix with the same dimensions as the original stiffness matrix and represents the sum of the stiffness elements of the resonators with the stiffness elements of the tower.

In all analyzed cases, the presented responses correspond to the top node, which corresponds to the nacelle vibration.

Initially, the first natural frequency of the structure is considered as the target frequency. Different numbers of resonators are considered, from 1 (which corresponds to an ADV) to 7. Figure 22.10 depicts the system response, in the region of the target frequency, considering the configuration with resonators positioned at the base. This arrangement exhibits the least effective performance of the metamaterials, which is expected due to the structure's smaller vibration amplitudes at the bottom. In the top and distributed cases, Figs. 22.11 and 22.12, respectively, it is evident that the

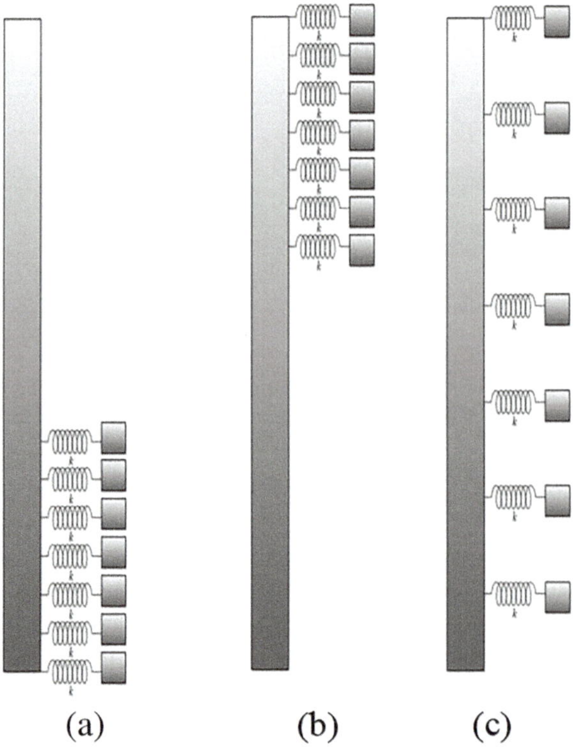

Fig. 22.8 Simplified model of resonator distribution: **a** At the base, **b** at the top, and **c** distributed along the tower

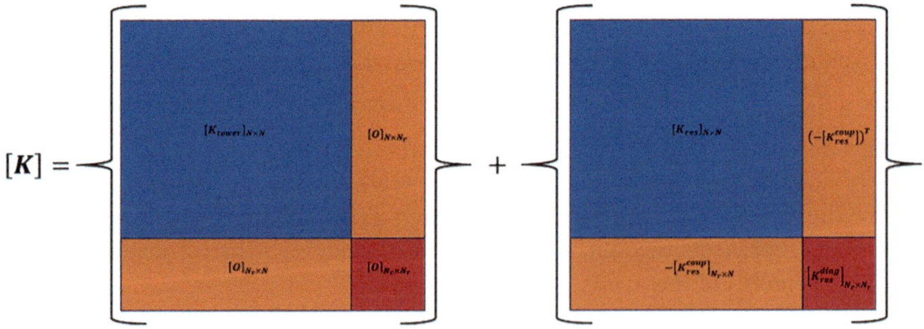

Fig. 22.9 Arrangement of the stiffness matrix in sparse representation

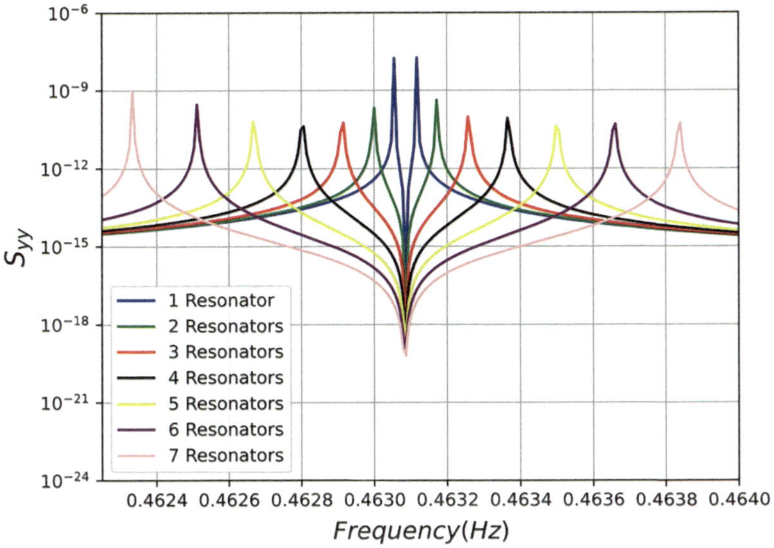

Fig. 22.10 PSD response of the structure when resonators are tuned to the first frequency and positioned at the base

configuration with only one resonator exhibits the best performance, and there is no significant difference observed when using 6 or 7 resonators. Further, increasing the number of resonators beyond 7 does not yield any noticeable difference compared to the 7-resonator case.

Now, the fifth natural frequency of the structure is the target one. The same analysis conducted for the first frequency is carried out again, as shown in Figs. 22.13, 22.14, and 22.15. Notably, for a higher frequency, the use of metamaterials proves to be more effective than the ADV. The results showcase the classical outcomes typically obtained in literature when metamaterials are employed for vibration reduction, with a broader attenuation band observed when the number of resonators is increased.

For a comprehensive comparison of the results, Figs. 22.16 and 22.17 illustrate the structural responses under different control scenarios: without control (in green), with ADV (in red), and with 7 resonators at the base (in purple), at the top (in blue), and distributed (in yellow). For the lower target frequency, as shown in Fig. 22.16, the ADV demonstrates superior performance in vibration attenuation. However, the case with metamaterials at the top proves to be an interesting alternative, as it exhibits a performance close to the ADV and may be more feasible to implement.

For the fifth frequency case, due to it being a higher frequency, the metamaterial demonstrates superior performance when compared to the case with ADV. In this scenario, the resonators at the base exhibited the best performance. The optimal positioning depends on the modal shape of the structure.

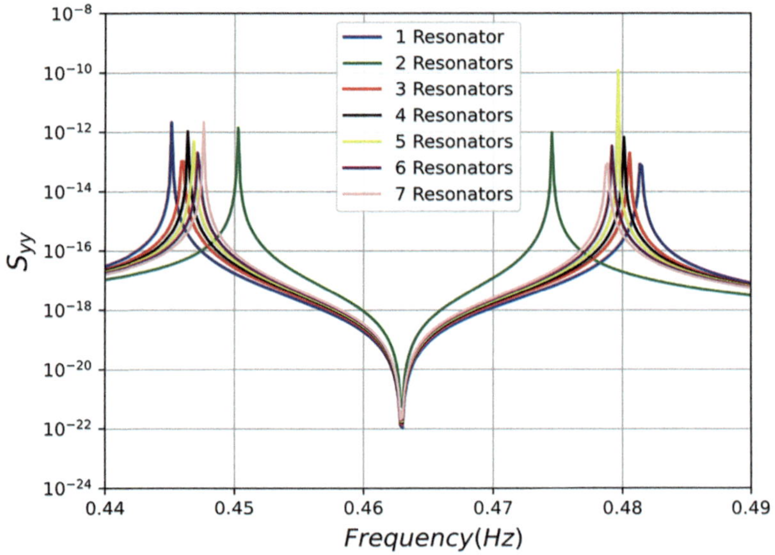

Fig. 22.11 PSD response of the structure when resonators are tuned to the first frequency and positioned at the top

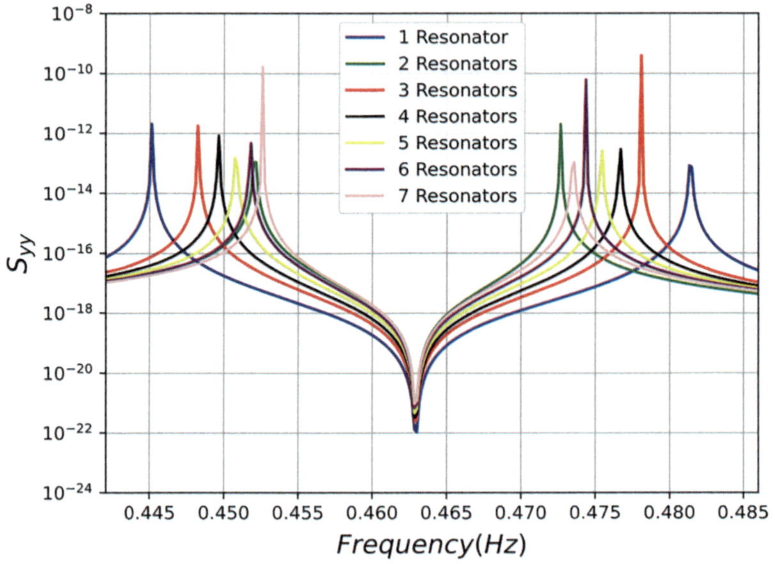

Fig. 22.12 PSD response of the structure when resonators are tuned to the first frequency and positioned distributed along the tower

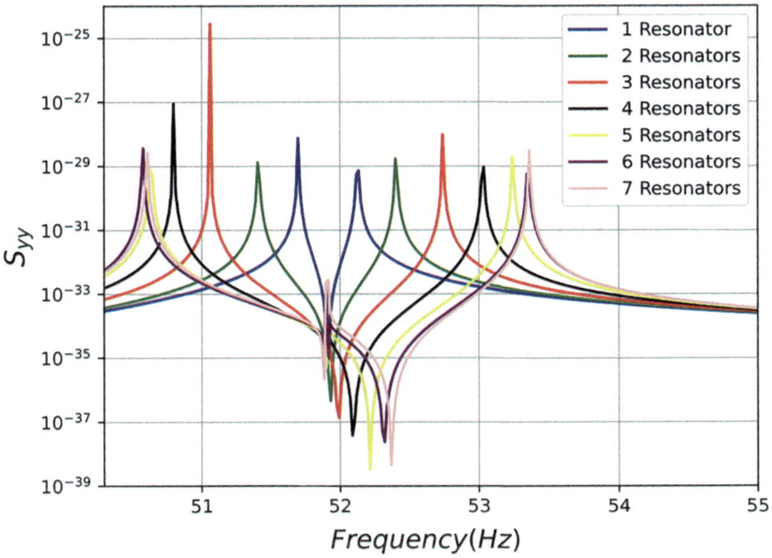

Fig. 22.13 PSD response of the structure when resonators are tuned to the fifth frequency and positioned at the base

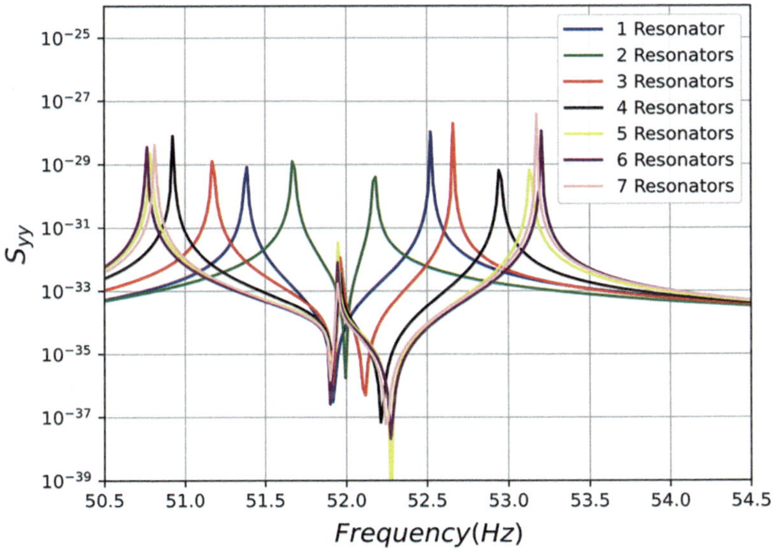

Fig. 22.14 PSD response of the structure when resonators are tuned to the fifth frequency and positioned at the top

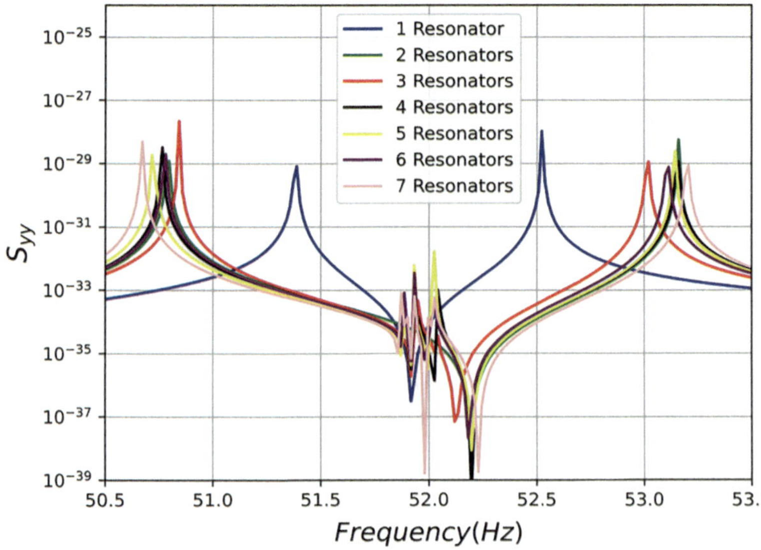

Fig. 22.15 PSD response of the structure when resonators are tuned to the fifth frequency and positioned distributed along the tower

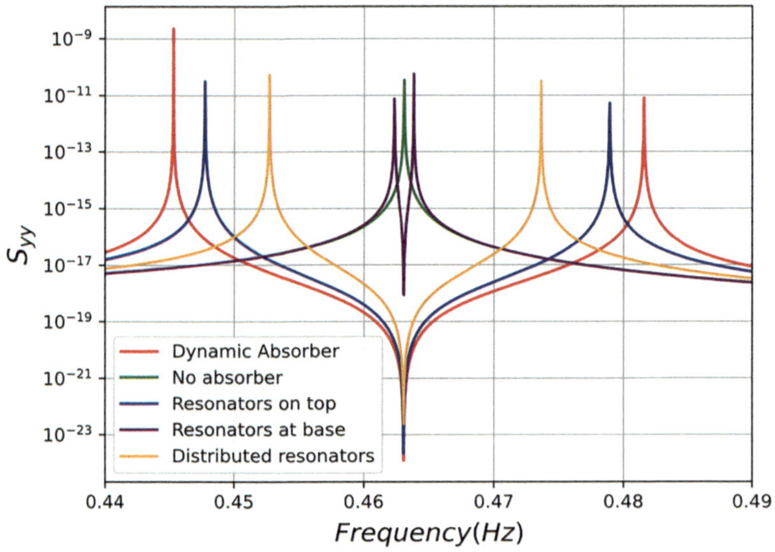

Fig. 22.16 Comparison between PDS responses without and with control when resonators are tuned to the first mode

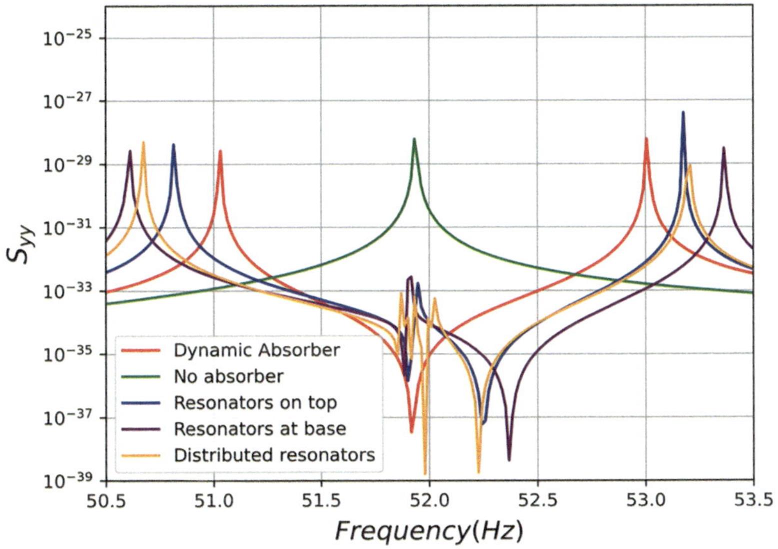

Fig. 22.17 Comparison between PDS responses without and with control when resonators are tuned to the fifth mode

22.4 Conclusion

Hence, this study aims to contribute further to the theory surrounding metamaterials, these intriguing and innovative materials. Moreover, it offers a possible solution for passive vibration control in wind towers.

Based on the above, when considering the metamaterials, three different configurations were considered concerning the resonators positioning: at the base, distributed along the tower, and at the top. Similar to the ADV case, when targeting the first natural frequency, the total of the resonators is equivalent to 15% of the total tower mass. When targeting the fifth mode, the total mass is equal to 5% of the total tower mass. These mass percentages are also employed to assess the comparative ADV performance.

For the fifth mode, a higher frequency, it is evident that the use of resonators yielded significantly better results than the DVA, especially when distributed along the base of the tower. However, at the lower frequency, the single resonator (ADV) was slightly more effective. Nevertheless, practical implementation of this DVA might be unfeasible due to its large mass, whereas in the second case, this mass is distributed among several resonators. Thus, for the first mode, despite the more modest outcome, the implementation of resonators may be an interesting alternative, as it exhibits a performance close to the ADV and may be more feasible to implement.

Therefore, this analysis provides important results for the optimization and selection of vibration attenuation systems in wind towers, considering different vibration modes and practical effectiveness in implementing control devices. Additionally, the study could be further explored for the case of low frequencies to achieve better results. This could be achieved by adding significant damping or exploring different resonator configurations.

Acknowledgements The authors are grateful to the Brazilian National Council of Research (CNPq—Brazil), process 407978/2022-4, to CAPES and FAPDF, process 00193-00001139/2021-57, for their support.

References

Alkmim MH (2017) Numerical and experimental analysis of a tuned liquid column damper in a wind turbine subject to stochastic load. Master's thesis, University of Brasília, Brasília

Araújo GB 2021 Investigação de vibração estrutural devido à ação do vento

Balaji P, Dalela S Jena D (2021) A review on application of mechanical metamaterials for vibration control

Burton T, Jenkins N, Sharpe D, Bossanyi E (2011) Wind energy handbook. John Wiley & Sons

Colherinhas GB, Machado MR (2022) Spectral model of offshore wind turbines and vibration control by pendulum tuned mass dampers. Int J Struct Stab Dyn 22:2250053. https://doi.org/10.1016/j.apor.2018.08.021. URL https://doi.org/10.1142/S0219455422500535

Cremer L, Leilich H (1953) Zur theorie der biegekettenleiter. Archiv der Elektrischen Ubertragung 7:261–270

Custódio Rd (2007) Energia eólica para produção de energia elétrica. Synergia Editora

Ferreira D, Ferreira C, Assis E (2010) "Classificação de rugosidade em tecido urbano. parte i: fundamentação teórica". In *CONGRESSO BRASILEIRO DE METEOROLOGIA*

Gao NS, Guo XY, Cheng BZ, Zhang YN, Wei ZY, Hou H (2019) Elastic wave modulation in hollow metamaterial beam with acoustic black hole. IEEE Access 7:124141–124146

Hemmati A, Oterkus E, Barltrop N (2019) Fragility reduction of offshore wind turbines using tuned liquid column dampers. Soil Dyn Earthq Eng 125:105705. ISSN 0267–7261. https://doi.org/10.1016/j.soildyn.2019.105705. URL https://www.sciencedirect.com/science/article/pii/S0267726118310170

Hussein MI, Leamy MJ, Ruzzene M (2014) Dynamics of phononic materials and structures: historical origins, recent progress, and future outlook. Appl Mech Rev 66(4):040802. ISSN 0003–6900. https://doi.org/10.1115/1.4026911. URL https://doi.org/10.1115/1.4026911

IEA (2021) "World Energy Balances: Overview" Disponível em: https://www.iea.org/reports/world-energy-balances-overview

Jahani K, Langlois RG, Afagh FF (2022) Structural dynamics of offshore wind turbines: a review. Ocean Engineering 251:111136. ISSN 0029–8018. https://doi.org/10.1016/j.oceaneng.2022.111136. URL https://www.sciencedirect.com/science/article/pii/S0029801822005467

Kock WE, Harvey FK (2005) Refracting sound waves. J Acoust Soc Am 21(5):471–481. ISSN 0001–4966. https://doi.org/10.1121/1.1906536. URL https://doi.org/10.1121/1.1906536

Mechanics of advanced materials and structures, vol 29. https://doi.org/10.1080/15376494.2021.1892244

Ozawa MT (2017) "Aplicação de parâmetros da distribuidção de weibull na análise do potencial energético de um microgerador eólico em cidades do estado do paraná"

Pai PF (2010) Metamaterial-based broadband elastic wave absorber. J Intell Mater Syst Struct 21(5):517–528

Petyt M (2010) Introduction to finite element vibration analysis. Cambridge University Press

Sugino C, Leadenham S, Ruzzene M, Erturk A (2016) On the mechanism of bandgap formation in locally resonant finite elastic metamaterials. J Appl Phys 120(13)

Sun C, Jahangiri V (2018) Bi-directional vibration control of offshore wind turbines using a 3d pendulum tuned mass damper. Mech Syst Signal Process 105:338–360. ISSN 0888–3270. https://doi.org/10.1016/j.ymssp.2017.12.011. URL https://www.sciencedirect.com/science/article/pii/S0888327017306465

Yang J, He E, Hu Y (2019) Dynamic modeling and vibration suppression for an offshore wind turbine with a tuned mass damper in floating platform. Appl Ocean Res 83:21–29. ISSN 0141–1187. https://doi.org/10.1016/j.apor.2018.08.021. URL https://www.sciencedirect.com/science/article/pii/S0141118718303079

GPSR Compliance

The European Union's (EU) General Product Safety Regulation (GPSR) is a set of rules that requires consumer products to be safe and our obligations to ensure this.

If you have any concerns about our products, you can contact us on ProductSafety@springernature.com

In case Publisher is established outside the EU, the EU authorized representative is:

Springer Nature Customer Service Center GmbH
Europaplatz 3
69115 Heidelberg, Germany

Batch number: 10755227

Printed by Printforce, the Netherlands